AF790105

Springer Series in Materials Science

Volume 184

Series Editors

Robert Hull, Charlottesville, VA, USA
Chennupati Jagadish, Canberra, ACT, Australia
Richard M. Osgood, New York, NY, USA
Jürgen Parisi, Oldenburg, Germany
Zhiming M. Wang, Fayetteville, AR, USA

For further volumes:
http://www.springer.com/series/856

The Springer Series in Materials Science covers the complete spectrum of materials physics, including fundamental principles, physical properties, materials theory and design. Recognizing the increasing importance of materials science in future device technologies, the book titles in this series reflect the state-of-the-art in understanding and controlling the structure and properties of all important classes of materials.

Alexey S. Kurlov · Aleksandr I. Gusev

Tungsten Carbides

Structure, Properties and Application in Hardmetals

Springer

Alexey S. Kurlov
Aleksandr I. Gusev
Institute of Solid State Chemistry
Ural Division of the Russian Academy of Sciences
Ekaterinburg
Russia

ISSN 0933-033X ISSN 2196-2812 (electronic)
ISBN 978-3-319-00523-2 ISBN 978-3-319-00524-9 (eBook)
DOI 10.1007/978-3-319-00524-9
Springer Cham Heidelberg New York Dordrecht London

Library of Congress Control Number: 2013942113

© Springer International Publishing Switzerland 2013
This work is subject to copyright. All rights are reserved by the Publisher, whether the whole or part of the material is concerned, specifically the rights of translation, reprinting, reuse of illustrations, recitation, broadcasting, reproduction on microfilms or in any other physical way, and transmission or information storage and retrieval, electronic adaptation, computer software, or by similar or dissimilar methodology now known or hereafter developed. Exempted from this legal reservation are brief excerpts in connection with reviews or scholarly analysis or material supplied specifically for the purpose of being entered and executed on a computer system, for exclusive use by the purchaser of the work. Duplication of this publication or parts thereof is permitted only under the provisions of the Copyright Law of the Publisher's location, in its current version, and permission for use must always be obtained from Springer. Permissions for use may be obtained through RightsLink at the Copyright Clearance Center. Violations are liable to prosecution under the respective Copyright Law.
The use of general descriptive names, registered names, trademarks, service marks, etc. in this publication does not imply, even in the absence of a specific statement, that such names are exempt from the relevant protective laws and regulations and therefore free for general use.
While the advice and information in this book are believed to be true and accurate at the date of publication, neither the authors nor the editors nor the publisher can accept any legal responsibility for any errors or omissions that may be made. The publisher makes no warranty, express or implied, with respect to the material contained herein.

Printed on acid-free paper

Springer is part of Springer Science+Business Media (www.springer.com)

Foreword

This monograph is entirely devoted only to tungsten carbides. At the same time it provides not only a deep theoretical insight into tungsten carbides and considers the physical essence of phenomena taking place in them, but also covers very important issues of tungsten carbide application for producing hardmetals.

The interest in this book is due in the first place to systematic description and critical analysis of the results of experimental and theoretical studies performed over the last 10–20 years reflecting the relationship between the structure, phase equilibria, and physicochemical properties of tungsten carbides.

Most of the original results for tungsten carbide described in the monograph were obtained after 1992 in the Laboratory of Refractory Compounds at the Institute of Solid State Chemistry of the Ural Division of the RAS (ISSC UD RAS), which in 2008 was renamed to the Laboratory of Nonstoichiometric Compounds. This is how the investigations into tungsten carbide began.

In 1993, the author of this foreword took part in the 13th International Plansee Seminar for the first time. Plansee Seminars deal with refractory compounds and their application in different industries. They are traditionally held each four years in a small Austrian town Reutte. Austria is the seminar site since it is in this country that the science of carbides and other refractory compounds emerged and firmed up.

In 1970–1980s, the research team of the Laboratory of Refractory Compounds at the ISSC UD was concerned with tungsten-free hardmetals based on titanium carbonitride. Their average grain size was several micrometers. My coauthors and I proposed nonstoichiometric carbides as an alternative to these and other traditional hardmetals. My report [1] at the Seminar on behalf of a group of scientists from Russia and Germany became later the basis of a large international project [2] resulting in a patent [3] for producing nonstoichiometric carbides with desired composition.

Among hundreds of reports presented at the Seminar, attention was drawn to the works of B. H. Kear, L. E. McCandlish, and etc. considering the methods for producing nanocrystalline powder compositions of tungsten carbide and cobalt, which are intended for the manufacture of nanostructured hardmetals. These works gave impetus to the beginning of examination of atomic defects in tungsten carbide [4] and to the production, by different methods, of nanocrystalline tungsten carbide powders and nanostructured hardmetals [5–8]. Earlier, these objects have not been studied at the Institute of Solid State Chemistry UD RAS.

In the present book, the authors involved in these studies have classified and generalized from a unified position the available experimental and theoretical data on the structure, physical and chemical properties of tungsten carbides, as well as on the application of nanocrystalline tungsten carbide for producing nanostructured hardmetals. Each chapter is concluded with a large list of references.

The book is meant first of all for researchers dealing with materials science, although for specialists in solid state physics, on the one hand, and for engineers engaged in the production of hardmetals, on the other hand, it will be of considerable interest. The book can be used as a teaching guide for students in higher training and for post-graduates specialized in chemistry, physics, and materials science.

A. A. Rempel
Corresponding Member of the RAS

References

1. Rempel AA, Gusev AI, Berger L-M. Nonstoichiometric transition metal carbides as alternative hard materials. In: Bildstein H, Eck R, editors. Proceedings of the 13th International Plansee Seminar. Reutte: Metallwerk Plansee; 1993. V.2, p. 671–684.
2. Berger L-M, Gusev AI, Rempel AA. Die Anwendung nichtstöchiometrischer Karbide des Titans, Niobiums und Tantals für die Herstellung von Hartmetallen mit verbesserten Eigenschaften. Projekt-Nr: 03M2103 (Programm Materialforschung des Bundesministeriums für Forschung und Technologie, Deutschland); 1993–1994.
3. Berger L-M, Hermann M, Gusev AI, Rempel AA. Verfahren zur Herstellung nichtstöchiometrischer Carbide definierter Zusammensetzung. Offenlegungsshrift DE 198 07 589 A 1. Int. $Cl.^6$: C 01 B 31/30 (C 04 B 35/36). Bundesrepublik Deutschland: Deutsches Patentamt: Anmeldetag 23.02.1998, Offenlegungstag 10.09.1998; p. 1–3.
4. Rempel AA, Würschum R, Schaefer H-E. Atomic defects in hexagonal tungsten carbide studied by positron annihilation. Phys Rev B. 2000;9 61:5945–5948.
5. Kurlov AS, Borisenko NI, Moldaver VA, Valeeva AA, Rempel AA. Certification of hard alloys produced from nanocrystalline tungsten carbide powder. Proceedings of the XVII mendeleev congress on general and applied chemistry (September 21–26, 2003, Kazan). Kazan: Russian Academy of Sciences; 2003. V.3, p. 358.
6. Kurlov AS, Leenaers A, Van den Berghe S, Van Houdt L, Scibetta M, Schröttner H, et al. Microstructure and strength of tungsten carbide WC-Co hard alloys sintered from nanopowders. Proceedings of the international conference on dislocation structure and mechanical properties of metals and alloys (April 10–14, 2008, Ekaterinburg). Ekaterinburg: Ural Division of the RAS; 2008. p. 137–138.
7. Kurlov AS, Rempel AA. Effect of WC nanoparticle size on the sintering temperature, density, and microhardness of WC-8 wt.% Co alloys. Inorg Mater. 2009;4 45:380–385.
8. Kurlov AS, Leenaers A, van den Berghe S, Scibetta M, Schröttner H, Rempel AA. Microstructure of nanocrystalline WC powders and WC-Co hard alloys. Rev Adv Mater Sci. 2011;2 27:165–172.

Preface

Since the discovery of tungsten carbide in 1893 by Henri Moissan and the first using of WC for industrial production of hardmetals at the beginning of twentieth century, basic research on the chemistry and physics of high-melting tungsten carbides and wear-resistant hardmetals has been carried out for three decades mainly in Austria, Germany, USA and USSR. In the 1950s, 1960s the interest in research of WC-based hardmetals, however, relaxed somewhat. Scientific and technological results of first stage of research and production of tungsten carbide and WC-based hardmetals were generalized in two monographs: Kieffer and Benesovsky [1] and Samsonov et al. [2]. Also, the structure and properties of tungsten carbides were partially and very small discussed in the following books on transition metal carbides: Goldschmidt [3]; Storms [4]; Toth [5]; Upadhyaya [6]. Noted that these books contain the limited data on tungsten carbides, and the part of these data is specified repeatedly.

But this situation abruptly changed at the beginning of the 1980s when the first indication for possible creation of fine-grained and nanocrystalline hardmetals have appeared. This aspect immensely stimulated research of phase equilibria in the "tungsten-carbon" system and of new methods for producing fine tungsten carbide powders and fine-grained hardmetals. Thousands original studies and articles appeared in the literature dealing with phase equilibria and phase diagrams of the tungsten-carbide system, with the crystal and electronic structure and properties of tungsten carbides in different structural states. In the last 30 years, the most active efforts were made in synthesis and application of the nanocrystalline WC powders for the production of nanostructured hardmetals.

This prompted to publish a volume devoted to tungsten carbides in the series "Materials Science". According to the purpose of this series, this volume must fill the gap between the scarce information about tungsten carbides in physicochemical textbooks and in monographs devoted to transition metal carbides and numerous review and original articles written for specialists. This monograph is the first work generalizing all up-to-date information about tungsten carbides.

The main subject of the monograph is comprehensive analysis of structure and properties of all stoichiometric and nonstoichiometric tungsten carbides including disordered and ordered phases. With this in mind, in Chap. 2 we considered in detail the phase equilibria and crystal structure of phases of the binary W–C and ternary W–Co–C systems, as well as the electronic structure of these phases.

Chapter 3 is devoted to discussing the phenomenon of ordering of tungsten carbides. Until now the literature data on the crystal structure of disordered and different ordered phases of W_2C_y carbide are limited and contradictory. In Chap. 3 the symmetry analysis of all possible superstructures of lower tungsten carbide W_2C_y is performed and the physically possible sequence of phase transformations in this carbide is established. The obtained theoretical results are compared with the experimental data on the crystal structure and temperature-concentration regions of existence of ordered and disordered phases of tungsten carbides. On the basis of this comparison, the phase boundaries are determined and phase diagram of the W-C system is refined.

According to the modern concepts, tungsten carbides in the nanocrystalline state, i.e. carbides, whose grains or particles are less than 80 or 100 nm, show the most promise. Therefore in Chap. 4 we described in detail the plasma-chemical and high-energy ball-milling methods for production of nanocrystalline tungsten carbide and consider the X-ray and electron microscopic methods for estimation of the average particle (grain) size. The model of mechanical grinding, which allows theoretical estimation of the ball-milling parameters necessary for production of powders with pre-assigned average particle size, is described for the first time. The application of this model is illustrated by the example of grinding of coarse-grained tungsten carbide powders.

Production of fine-grained WC–Co hardmetals has been examined repeatedly, but for a long time there were no available experimental data in the literature on the effect of particle size of tungsten carbide powders on the phase transformations taking place in hardmetals from the WC–Co system at different sintering temperature, on the peculiarities of interaction of nanocrystalline WC powders with cobalt, on thermal stability of phase and chemical composition, as well as on the particle size of nanocrystalline WC powders during their heating to 1400–1600 K in vacuum or in air. These problems as well as the peculiarities of sintering of WC–Co hardmetals with different cobalt content from nanocrystalline powders WC having different particle sizes, microstructure and mechanical properties of the sintered hardmetals are discussed in Chap. 5.

We hope that this monograph will be useful to specialists in solid state physics, to materials science engineers, as well as to students of physical, chemical and materials science faculties.

We wish to thank our nearest colleague Prof. A. A. Rempel for useful discussions, and the Executive Physics Editor of the publisher, Dr. Claus E. Ascheron, for an effective collaboration during the preparation of this monograph.

Ekaterinburg

Alexey S. Kurlov
Aleksandr I. Gusev

References

1. Kieffer R, Benesovsky FH. New York: Springer; 1965.
2. Samsonov GV, Vitryanuk VK, Chaplygin FI. Tungsten carbides. Kiev: Naukova Dumka; 1974.
3. Goldschmidt HJ. Interstitial alloys. New York: Plenum Press; 1967.
4. Storms EK. The refractory carbides. London: Academic Press; 1967.
5. Toth LE. Transition metal carbides and nitrides. London: Academic Press; 1971.
6. Upadhyaya GS. Nature and properties of refractory carbides. New York: Nova Science Publishers; 1996.

Contents

Notations

a, **b**, **c**	The fundamental translation vectors of crystal lattice
a, b, c	Lengths of basis vectors (lattice constants of an unit cell)
$\boldsymbol{b}_1$, $\boldsymbol{b}_2$, $\boldsymbol{b}_3$	Basis vectors of reciprocal lattice
d	Interplanar spacing
D, $\langle D \rangle$	Size or average size (diameter) of particle (grain)
E	Young's elastic modulus
E_F	Fermi energy
F	Free energy
FWHM	Full line width at half maximum
G	Shear modulus
h, k, l	Miller indices
$g(\theta)$	The Gauss function
H_k	Symmetry operation (element) of point group 6/*mmm*
H_V	Microhardness
k_B	Boltzmann constant
$\boldsymbol{k}$	Wave vector
$\{\boldsymbol{k}_s\}$	Wave vector star
$\boldsymbol{k}_s^{(j)}$	Ray of the star $\{\boldsymbol{k}_s\}$ (superstructural vector that belongs to the star $\{\boldsymbol{k}_s\}$)
K_c	Bulk modulus
K_{1c}	Fracture toughness (fracture strength)
K_{hkl}	Scherrer's constant
$l(\theta)$, $\ell(\theta)$	The Lorentz (Cauchy) function
m_0	Free electron mass
m^*	Effective electron mass
$n(\boldsymbol{r})$, $n(x_1, y_1, z_1)$	Distribution function of atoms on the crystal lattice sites
N_A	Avogadro's number
$N(E)$	Density of electronic states
$N(E_F)$	Density of electronic states on the Fermi level
$n(G)$	Order of the group G
P, $P_i^{(s)}$	Probability
$\boldsymbol{r}$	Radius vector that determines position of the crystal lattice site

$R^{(s)}$	The number of the ordering lattice sites in an s-type figure
s	Type of figure
$\{s\} = \{a, b, c\}$	Sequence of special figures
S_c	Configurational (combinatorial) entropy
S_{sp}	Specific surface
t	Time
T	Temperature
T_c	Superconducting critical temperature
T_C	Curie temperature
T_{trans}	Phase transformation (transition) temperature
x_I, y_I, z_I	Coordinates of the sites $\boldsymbol{r}$ of Ising lattice
y	Relative content of interstitial atoms X in nonstoichiometric compounds MX_y
$y^{(s)}$	Overestimate coefficient
z_j	Coordination number of the j-th coordination sphere
α_j	Short-range order parameter for j-th coordination sphere
β, $\beta(2\theta)$	Broadening of diffraction reflection
β_d	Deformation broadening (broadening caused by the lattice microstrains)
β_h	Inhomogeneity broadening (broadening due to the inhomogeneity of nonstoichiometric compounds or solid solutions)
β_s	Size broadening (broadening due to the small size of particles)
ε	Microstrain (microdeformation)
η, η_s	Long-range order parameter
θ	Bragg (diffraction, glancing) angle
λ	Radiation wavelength
$\lambda_i^{(s)}$	Multiplicity of i-configuration of the s-type figure
μ_B	Bohr magneton
ν	Poisson ratio
ρ	Density
σ_{bend}	Bending strength
τ	Positron lifetime
χ	Magnetic susceptibility
χ_d^-	The Landau diamagnetism of conduction electrons
χ_p^-	The Pauli paramagnetism of conduction electrons
$\chi_{VV\ orb}$	The Van Vleck orbital paramagnetism
$\varphi_s^{(j)}$	Phase shift of the concentration wave
ω	Angular speed
$\square$	Structural vacancy

Chapter 1
Introduction

Carbides of Group IV-VI transition *d*-metals belong to strongly nonstoichiometric interstitial compounds [1–5]. Transition metals of subgroups IVA and VA (Ti, Zr, Hf, V, Nb, Ta) form cubic carbides MC_y with carbon with the $B1$-type (NaCl) structure, while the metals of subgroups VA and VIA (V, Nb, Ta, Cr, Mo, W) form in addition lower hexagonal carbides M_2C_y with the $L'3$ (W_2C) structure (hereinafter in this book, the crystal lattice types of inorganic compounds are shown according to the handbook [6]).

A characteristic feature of the structures of these compounds is the presence of the face-centered cubic (fcc) or hexagonal (simple or closely-packed (hcp)) metallic lattice. Carbon atoms are located in the center of octahedral or trigonal interstitials of the metallic lattice. However in carbides, the metallic sublattice symmetry differs from that of transition metallic lattices. Thus the crystal structure of metals changes with the formation of carbides. Group IV metals (Ti, Zr, Hf), which have the hcp structure, form carbides with the fcc metallic sublattices. Metals with the body centred cubic (bcc) structure (V, Nb, Ta, Cr, Mo, W) form carbides with the cubic or hexagonal metallic sublattices. The changeover of the crystal structure of metals upon formation of carbides suggests strong metal—carbon interactions while direct interactions between carbon atoms are negligibly small [5].

In carbides, the carbon atoms that are located in the interstitial voids of the metallic sublattice form a nonmetallic sublattice. Unoccupied interstitials are called structural vacancies and designated as $\square$. Together with carbon atoms, they form an interstitial solution in the nonmetallic sublattice. The presence of structural vacancies is a characteristic feature of carbides that belong to the group of nonstoichiometric compounds. The concentration of vacancies and their disordered or ordered distribution in a crystal lattice have a strong effect on the properties of such carbides. The results of experimental and theoretical studies on how the distribution of atoms and vacancies affects the structure and properties of strongly nonstoichiometric compounds are generalized in monographs [1–5, 7, 8] and review article [9].

Carbides of Group IV-VI transition *d*-metals have the highest melting points and are the hardest among all the compounds known [3–5, 7–14]. Therefore, these

A. S. Kurlov and A. I. Gusev, *Tungsten Carbides*,
Springer Series in Materials Science 184, DOI: 10.1007/978-3-319-00524-9_1,

© Springer International Publishing Switzerland 2013

carbides are used in manufacturing structural and tool materials capable of operating at elevated temperatures, in aggressive media and under heavy loads. Carbides are used for deposition of protective and hard coatings, etc. Tungsten carbide WC is not the hardest and the most refractory compound, viz., its melting point (~3060 K) is low as compared with nonstoichiometric carbides TiC_y, ZrC_y, HfC_y, VC_y, NbC_y, TaC_y and its hardness (18–22 GPa at 300 K) is also low as compared with these carbides [3–5]. However, the hardness of WC is sufficiently stable and decreases relatively slightly with the increase in temperature from 300 to 1200–1300 K [15]. Moreover, as compared with carbides of other transition metals, tungsten carbide WC has a higher Young's modulus of elasticity $E \approx 700$ GPa which is twice as large as Young's modulus of other carbides. A thermal expansion coefficient of WC is equal to $\sim 5.5 \times 10^{-6}\ K^{-1}$ and is half as much as that of other transition metal carbides.

The mentioned properties of WC and especially their thermal stability during the heating up to 1000–1200 K determine the use of WC in the production of wear-resistant hardmetals, which form the major part of all tool materials. The direct use of WC carbide as cutting tool material is impossible because of its brittleness and sintering at very high temperatures.

Tungsten carbide was first synthesized by H. Moissan in 1893 [16]. Later, in 1897, Moissan managed to produce fused tungsten metal by the electrothermal way (earlier tungsten metal produced only in the form of a powder). The synthesis of high-melting carbides including WC from oxides, in which calcium carbide is used as the reactant, was described in work [17]. The industrial production of hardmetals based on WC started 20–25 years after discovery of tungsten carbide WC and goes on at present. The first period (1918–1923) of development of hardmetas on the basis of tungsten carbide was described in work [18] and the first patent for these hardmetals was received by K. Schrüter in the USA in 1925.

The practical interest in tungsten carbide determines the active fundamental investigations of its crystal and electronic structure and also of the physicochemical properties of both WC as such and related compounds and systems that contain it. In the past 20 years, the most active studies were associated with the synthesis and application of nanocrystalline tungsten carbide.

The studies of tungsten carbide [19] and WC-based hardmetals began at the Institute of Solid State Chemistry of the Ural Division of the RAS (ISSC UrD RAS) after 1992 on the initiative and with immediate participation of Prof. A. A. Rempel. He also took the lead in the research of nanocrystalline materials including nanocrystalline powders of tungsten carbide and nanostructured hardmetals WC-Co [20–23]. Earlier, tungsten carbide and nanocrystalline materials have not been studied at the Institute of Solid State Chemistry UD RAS.

The main aim of this book is to generalize the modern experimental and theoretical findings on tungsten carbide, including the results obtained at the ISSC UrD RAS under the direction of A. A. Rempel, and to promote the application of tungsten carbide for the production of nanostructured hardmetals.

References

1. Gusev AI, Rempel AA. Structural phase transitions in nonstoichiometric compounds. Moscow: Nauka; 1988. p. 308 (in Russian).
2. Gusev AI. Physical chemistry of nonstoichiometric refractory compounds. Moscow: Nauka; 1991. p. 286 (in Russian).
3. Gusev AI, Rempel AA. Nonstoichiometry, disorder and order in solids. Ekaterinburg: Ural Division of the Russian Academy of Sciences; 2001. p. 580 (in Russian).
4. Gusev AI, Rempel AA, Magerl AJ. Disorder and order in strongly nonstoichiometric compounds: transition metal carbides, nitrides and oxides. Berlin: Springer; 2001. p. 607.
5. Gusev AI. Nonstoichiometry, disorder, short-range and long-range order in solids. Moscow: Nauka-Fizmatlit; 2007. p. 856 (in Russian).
6. Pearson WB. The crystal chemistry and physics of metals and alloys. New York: Wiley - Interscience; 1972. p. 806.
7. Rempel AA. Effects of ordering in nonstoichiometric interstitial compounds. Ekaterinburg: Nauka; 1992. p. 232 (in Russian).
8. Lipatnikov VN, Gusev AI. Ordering of titanium and vanadium carbides. Ekaterinburg: Ural Division of the RAS; 2000. p. 265 (in Russian).
9. Rempel AA. Atomic and vacancy ordering in nonstoichiometric carbides. Uspekhi Fiz: Nauk. 1996;166(1):33–62 (in Russian). (Engl. Transl.: Phys Usp. 1996;39(1):31-56).
10. Goldschmidt HJ. Interstitial alloys. New York: Plenum Press; 1967. p. 820.
11. Kotel'nikov RB, Bashlykov SN, Galiakbarov ZG, Kashtanov AI. Fundamentals of refractory elements and compounds/handbook. Moscow: Metallurgiya; 1969. p. 376 (in Russian).
12. Storms EK. The refractory carbides. New York: Academic Press; 1967. p. 300.
13. Toth LE. Transition metal carbides and nitrides. New York: Academic Press; 1971. p. 280.
14. Holleck H. Binäre und Ternäre Carbid- und Nitridsysteme der Übergangsmetalle. Berlin: Gebrüder Borntraeger; 1984. p. 295.
15. Samsonov GV, Vitryanuk VK, Chaplygin FI. Tungsten carbides. Kiev: Naukova Dumka; 1974. p. 176 (in Russian).
16. Moissan H. Preparation au four électrique de quelques mètaux réfractaires: tungstène, molybdène, vanadium. C R. 1893;116:1225–7.
17. Moissan H. Nouvelle méthode de preparation des carbures par l'action du carbure de calcium sur les oxides. C R. 1897;125:839–44.
18. Schrüter K. The inception and development of hard metal carbides. Iron Age. 1934;133(2): 27–29.
19. Rempel AA, Würschum R, Schaefer HE. Atomic defects in hexagonal tungsten carbide studied by positron annihilation. Phys Rev B. 2000;61(9):5945–8.
20. Kurlov AS, Borisenko NI, Moldaver VA, Valeeva AA, Rempel AA. Certification of hard alloys produced from nanocrystalline tungsten carbide powder. Proceedings XVII mendeleev congress on general and applied chemistry, September 21–26, 2003, Kazan: Russian Academy of Sciences; 2003. vol. 3, p. 358.
21. Kurlov AS, Leenaers A, Van den Berghe S, Van Houdt L, Scibetta M, Schröttner H, Rempel AA. Microstructure and strength of tungsten carbide WC-Co hard alloys sintered from nanopowders. Proceedings international conference "Dislocation structure and mechanical properties of metals and alloys" April 10–14, 2008, Ekaterinburg: Ural Division of the RAS; 2008. pp. 137–8.
22. Kurlov AS, Rempel AA. Effect of WC nanoparticle size on the sintering temperature, density, and microhardness of WC-8 wt.% co alloys. Inorg Mater. 2009;45(4):380–5.
23. Kurlov AS, Leenaers A, van den Berghe S, Scibetta M, Schröttner H, Rempel AA. Microstructure of nanocrystalline WC powders and WC-co hard alloys. Rev Adv Mat Sci. 2011;27(2): 165–72.

Chapter 2
Phases and Equilibria in the W–C and W–Co–C Systems

Tungsten carbide phases which form in the W–C and W–Co–C systems belong to the group of nonstoichiometric interstitial compounds. Nonstoichiometric interstitial compounds include carbides, nitrides, and oxides of groups IV, V, and VI transition metals [1–5]. In accordance with modern ideas, a combined covalent-metallic-ionic type of chemical bonds is realized in nonstoichiometric interstitial compounds [1–6]. This is in line with such specific feature of nonstoichiometric compounds as the combination of main attributes of metals (a simple structure and high thermal and electric conductivity decreasing with temperature) and covalent compounds (high hardness and low plasticity) [3–12].

Hagg [13] proposed empirical rules of constructing crystal structures of nonstoichiometric interstitial compounds. In accordance with [13], nonstoichiometric interstitial compounds are formed if atomic radii of the metal R_M and nonmetal R_X meet the condition $0.41 < R_X/R_M < 0.59$. When this condition is fulfilled, nonmetal atoms are located at the largest interstitial sites of the metallic lattice, which are a little smaller than the interstitial nonmetal atoms. For tungsten carbides, an atomic radius ratio R_C/R_W is equal to 0.553 and agrees with the Hagg's rule. The change of symmetry and a slight expansion of the metallic lattice in carbide, nitride or oxide ensure stability of the structure.

For the first time tungsten carbide WC was synthesized by Moissan [14]. Mass production of hardmetals based on tungsten carbide started 20–25 years after discovery of WC carbide [15].

2.1 Phase Equilibria and the Crystal Structure of Phases in the W–C System

In addition to tungsten W and carbon C, the W–C system includes two phases, namely, W_2C and WC, each having several structural modifications, which are stable in certain temperature and concentration ranges (Fig. 2.1).

A. S. Kurlov and A. I. Gusev, *Tungsten Carbides*,
Springer Series in Materials Science 184, DOI: 10.1007/978-3-319-00524-9_2,
© Springer International Publishing Switzerland 2013

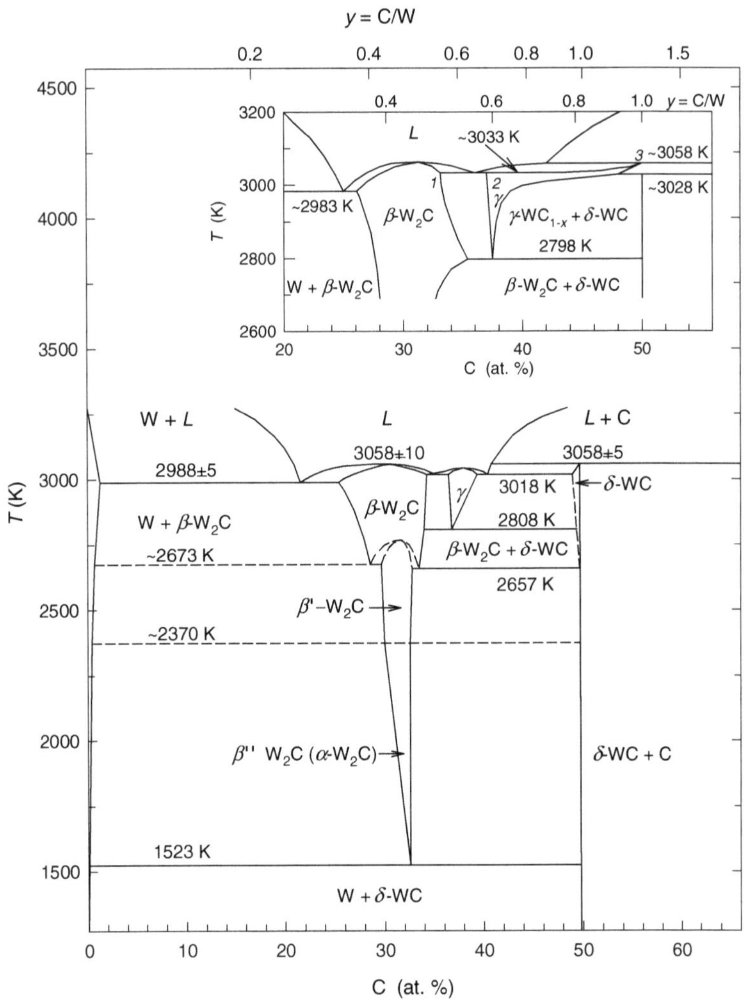

Fig. 2.1 Generalised phase diagram of the W–C system [16, 17] constructed according to data [18–21] with the partial account of results [22]. In contrast to study [21] where the temperature of eutectic $L \Leftrightarrow \gamma\text{-WC}_{1-x} + \delta - \text{WC}$ is determined equal to 2993 K, in studies [16, 17] temperature T_e = 3018 K is accepted for the same eutectic. The high-temperature part of the phase diagram of the W–C system, constructed in work [22] is shown on inset. Cubic (with $B1$-type structure) phase $\gamma\text{-WC}_{1-x} \equiv \gamma\text{-WC}_y$ is designated in figure as γ. Crystallographic characteristics of phases of the W–C system are given in Table 2.1

Two successive editions of a handbook on a phase diagrams of binary alloys [16, 17] show the generalized phase diagram of the W–C system, which is constructed based on earlier diagrams of this system [18–21] and takes into account in part the results of another publication [22]. A later handbook [23] represents a phase diagram from a paper by Rudy and Hoffman [21] while a Russian handbook [24] reproduces a phase diagram of the W–C system given in the handbook cited [16]. Relatively few authors dealt with phase equilibria in the W–C system; however, they used different designations for the discovered phases, which led to the confusion in the description of this system. Even in the handbooks [16, 17] the table that describes the phases of the W–C system confuses the designations for the disordered hexagonal and orthorhombic W_2C phases. To avoid confusion, in this section revealed mistake in the designations [17] is corrected and hereinafter corrected designations for the W–C system are used. In certain cases we also show the designations accepted in other sources.

Four modifications are described for lower tungsten carbide W_2C. Three of them are low-temperature β''-W_2C, intermediate β'-W_2C and high-temperature β-W_2C, which were also designated in works [12, 18, 21] as α-, β- and γ-W_2C, respectively. However, the structures of these modifications were proposed based on the limited experimental data. The structure of yet another modification designated as ε-W_2C [25] was revealed sufficiently reliably in neutron diffraction studies [25–29]. In all W_2C modifications, tungsten atoms form a hcp metallic sublattice, in which one-half of octahedral interstitials are occupied by carbon atoms. Depending on the distribution of carbon atoms C, the lower carbide W_2C can be disordered at a high temperature or ordered at a low temperature. The presence of different types of carbon atoms distribution causes the possibility of the formation of several structural modifications of W_2C. Higher carbide WC with the hexagonal structure is designated as δ-WC (or α-WC [22] or simply WC [12, 21]). Furthermore, the compositions between lower and higher tungsten carbides include the cubic phase γ-WC_{1-x} (designated also as β-WC [22] or α-WC_{1-x} [12, 21] or simply WC_{1-x}). For the first time, tungsten carbide with the cubic structure and the unit cell constant of 0.416 nm was synthesized in work [30] by its deposition from $W(CO)_6$ vapour. Authors [30] assumed that they have found a cubic modification of W_2C carbide. Later, cubic carbide with a unit cell constant of 0.425 nm was synthesized in a spark discharge between the electrodes made of hexagonal WC carbide, which were placed into a petroleum bath [31]. Authors [31] decided too that it is the lower W_2C tungsten carbide. Authors [32] who also detected the cubic modification of tungsten carbide identified it as yet another high-temperature modification of W_2C. At present, cubic carbide γ-WC_{1-x} is considered as a structural modification of higher WC carbide.

Now, we discuss the peculiarities of the crystal structure of phases in the W–C system.

Tungsten metal has the bcc structure (space group $Im\overline{3}m$ (O_h^9)) of its own structural type $A2$ (W-type). Such a structure is also typical of alkali metals, vanadium, niobium, tantalum, chromium and α-Fe. The crystal lattice basis contains two atoms in positions 2(a) with coordinates (0 0 0) and (1/2 1/2 1/2).

According to studies by Rudy et al. [18–21] and in line with the phase diagram of the W–C system (see Fig. 2.1), the most high-temperature phase β-W_2C has a hexagonal (space group $P6_3/mmc(D_{6h}^4)$) structure of the $L'3$-type (Fig. 2.2).

The existence of lower tungsten carbide W_2C with the $L'3$-type structure was directly proved in a neutron diffraction study [25] in which disordered W_2C phase was observed together with the ordered phase of this carbide. For high-temperature

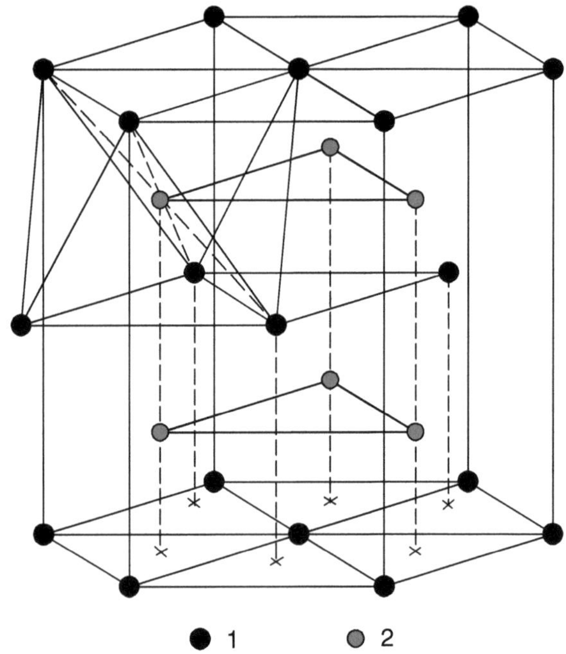

Fig. 2.2 Hexagonal structure of the $L'3$-type of lower disordered β-W_2C carbide [1, 2]. Tungsten atoms W • (*1*) form a hcp metallic sublattice the octahedral interstitials • (*2*) of which are randomly occupied by carbon atoms C with a probability of 1/2

completely disordered W_2C, the $L'3$-type structure is quite reasonable by analogy with other lower carbides and nitrides. Particularly, the $L'3$-type hexagonal structure is typical of most nonstoichiometric lower carbides and nitrides M_2X_y (M = V, Nb, Ta; X = C, N) with the exception of V_2N. In the unit cell of β-W_2C, two tungsten atoms W occupy the positions 2(*c*) with coordinates (1/3 2/3 1/4) and (2/3 1/3 3/4) and one carbon atom C is randomly distributed between positions 2(*a*) with coordinates (0 0 0) and (0 0 1/2). In other words, the hexagonal β-W_2C phase is disordered, and its carbon atoms randomly occupy one-half of all octahedral interstitials in the tungsten sublattice; hence, the probability of the occupation of any octahedral interstitial by C atom is equal to 1/2. Perpendicularly to the *c* axis of the β-W_2C carbide, the A and B layers formed by metal atoms alternate with the X layers formed by the nonmetallic lattice sites in the ...AXBXAXBX... sequence. The hexagonal β-W_2C phase has a homogeneity region from $WC_{0.34}$ to $WC_{0.52}$ and is stable in the temperature range from 2670 to 2720 K up to the melting temperature of 3000–3050 K. The unit cell constants of the β-W_2C phase with the composition $WC_{0.52}$ were recorded [9] to be $a = 0.2992$ and $c = 0.4722$ nm. According to data [19] the unit cell constants of the β-W_2C carbide are little bit more and are equal to $a = 0.3002$ and $c = 0.475 - 0.476$ nm. Note that the unit cell constants of all phases discussed were obtained for quenched samples at room temperature.

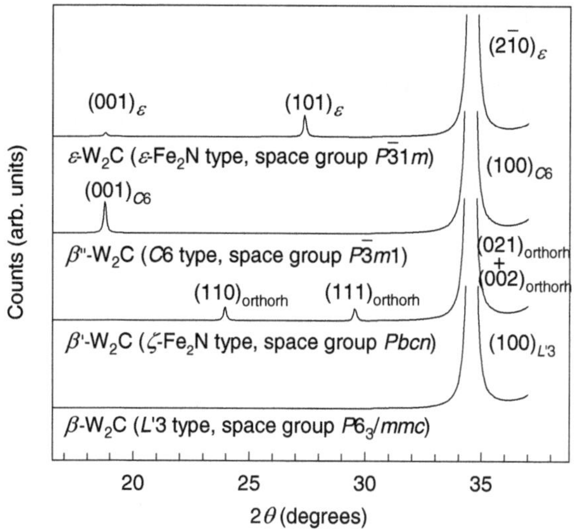

Fig. 2.3 XRD patterns of different β-W_2C, β'-W_2C, β''-W_2C and ε-W_2C modifications of lower W_2C tungsten carbide differ only in the region of small angles 2θ. Calculations of XRD patterns are performed for the perfect (without regard for atomic displacements) structures of lower tungsten carbide with $WC_{0.50}$ composition using Cu$K\alpha_{1,2}$ radiation

The modifications β-W_2C, β'-W_2C, β''-W_2C and ε-W_2C of the lower tungsten carbide W_2C are practically indiscriminate in powder X-ray diffraction (XRD) experiments. This is because they have similar hexagonal metallic sublattices, while the X-ray scattering amplitudes of atoms W is many times larger than that of atoms C. According to calculations, the change of the distribution of carbon atoms in the lattice of W_2C carbide influences X-ray diffraction patterns at small angles only (Fig. 2.3).

Intensities of the reflections $(110)_{orthorh}$ and $(111)_{orthorh}$, which are characteristic of the orthorhombic carbide β'-W_2C, the reflection $(001)_{C6}$ of the carbide β''-W_2C with the *C*6 structure, and the reflection $(101)_\varepsilon$ typical of the carbide ε-W_2C account for less than 0.5 % of the intensity of the strongest reflection $(101)_{L'3}$ (or reflections $(100)_{C6}$, $(002)_{orthorh}$, and $(2–10)_\varepsilon$ of the corresponding structures). In other words, these reflections are well at the background level and far below the experimental intensity error in a real X-ray diffraction study.

The modification β'-W_2C of lower tungsten carbide has the rhombic symmetry, although its metallic sublattice retains a hexagonal closely packed structure similar to the structure of high-temperature carbide β-W_2C. The structure of orthorhombic lower tungsten carbide β'-W_2C is not yet known with certainty, but can be assigned to the PbO_2 type with the space group *Pbcn* (D_{2h}^{14}). For this structure, the atomic positions can be specified only in general, namely, the carbon atoms C occupy positions 4(*c*) with coordinates (0 *y* 1/4) and the tungsten atoms W are in positions 8(*d*) with coordinates (*x y z*). Based on the comparison of experimental and theoretical positions of the reflections in the XRD pattern, it was concluded [18] that lower carbide β'-W_2C (β-W_2C) has an orthorhombic structure similar to the ζ-Fe_2N-type structure of ordered carbide Mo_2C with the same space group *Pbcn* [33]. As was noted above, the hexagonal disordered, trigonal and orthorhombic modifications of lower carbide W_2C with the same unit cell constants of the basic disordered phase are undistinguishable in powder XRD experiment from the position of their X-ray reflections at $2\theta > 30°$. In the range of small angles $2\theta < 30°$, in which the XRD patterns are slightly different (see Fig. 2.3), no reflections were observed [18]. No calculation of XRD pattern with allowance for possible intensities of diffraction reflections was carried out in study [18]. Thus, the literature contains no experimental structural data which would substantiate the detection of the orthorhombic lower tungsten carbide β'-W_2C. Moreover, the authors of the study [18] incorrectly determined the coordinates of tungsten atoms in the suggested structure of β'-W_2C carbide (the positions 8(*d*) occupied by tungsten atoms were claimed to have the coordinates (3/8 ~ 3/8 ~ 1/2)).

According to [18], the unit cell of orthorhombic β'-W_2C carbide has the constants $a = 0.4728$, $b = 0.6009$, and $c = 0.5193$ nm. Our analysis of the distribution of atoms in orthorhombic unit cell, which is a derivative of the hexagonal unit cell with $L'3$-type structure, and the calculation of XRD pattern for β'-W_2C carbide with ζ-Fe_2N-type structure have shown that tungsten atoms occupy 8(*d*) positions with

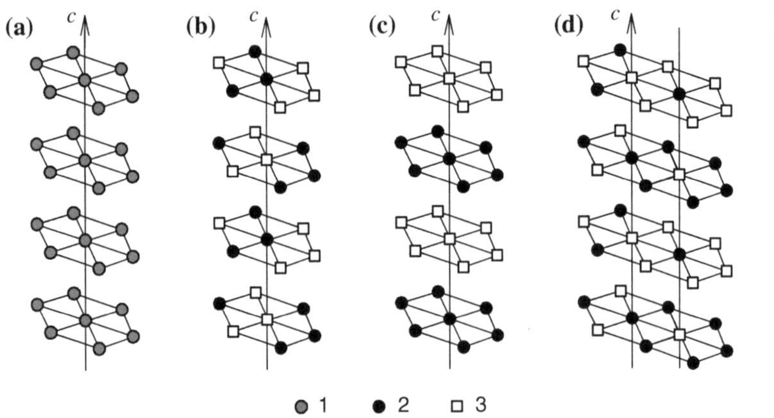

Fig. 2.4 Distribution of carbon atoms in lower carbides β-W_2C (**a**), β'-W_2C (**b**), β''-W_2C (**c**) and ε-W_2C (**d**), which are ideal as regards both their compositions and structures [2]. Planes formed by the nonmetallic sublattice sites are located normally to the *c* axis of the original disordered structure of the $L'3$-type structure; tungsten atoms W are not shown. **a** The $L'3$-type structure of β-W_2C carbide with the random distribution of carbon atoms C and vacancies □; **b** the Mo_2C (ζ-Fe_2N) type structure of β'-W_2C carbide with the ordered distribution of carbon atoms C and vacancies □ in each plane of the nonmetallic sublattice; **c** the *C*6-type structure of β''-W_2C (α-W_2C) carbide in which the layer of carbon atoms regularly alternate with the layer of structural vacancies; **d** the ε-Fe_2N-type structure of ε-W_2C carbide with alternating planes of the nonmetallic sublattice, in which each vacancy is surrounded by six carbon atoms or each carbon atom is surrounded by six vacancies. *1* the positions of nonmetal interstitial atoms •, which are randomly occupied by carbon atoms with a probability of 1/2; *2* carbon atoms C •; *3* vacancies □

coordinates (1/4 ~ 1/8 ~ 1/12) and carbon atoms are in 4(*c*) positions with coordinates (0 ~3/8 1/4). As compared with the high-temperature hexagonal phase β-W_2C (Fig. 2.4a), the β'-W_2C phase with the ζ-Fe_2N-type structure (Fig. 2.4b) is ordered phase with respect to the non-metallic sublattice. It is assumed that orthorhombic carbide β'-W_2C has the homogeneity region from $WC_{0.34}$ to $WC_{0.49}$ and exists in an intermediate temperature interval from ~ 2370 K to 2670–2750 K (see Fig. 2.1). The later studies failed to detect the orthorhombic (space group *Pbcn*) phase β'-W_2C.

The low-temperature phase β''-W_2C (α-W_2C) is ordered, but the temperature range of its existence and its crystal structure are not yet reliably determined. An electron diffraction study [34] dealt with lower tungsten carbide, which was synthesized by carburization of a thin metal tungsten film in the carbon monoxide atmosphere at ~1400 K for 5 min. The electron diffraction pattern of this carbide revealed 74 reflections, which pointed to the ordering of carbon atoms. Authors [34] have assumed that the lower tungsten carbide β''-W_2C phase has the *C*6 (anti-CdI_2)-type structure with the space group No. 164, $P\bar{3}m1(D^3_{3d})$ whereas other possible types of ordering were ignored. In the rhombohedral (trigonal) structure of the *C*6-type, the layer of carbon atoms regularly alternate with the layer of structural vacancies, being oriented normally to the *c* axis. The layers formed solely by carbon atoms C are located between the pairs of layers AB formed by tungsten atoms W. The layers of vacant interstitials, i.e., structural vacancies, are located between the pairs of metal layers BA (Fig. 2.4c). Note that according to data [34], tungsten atoms occupy virtually ideal positions in the lattice (z = 0.25), although the distance along the *c* axis between the metal layers separated by the layer of vacancies should be slightly shorter than the distance between the metal layers separated by a carbon-atom layer.

Thus, the perfect structure of β''-W_2C (α-W_2C) carbide contains two tungsten atoms W in positions 2(*d*) with coordinates (1/3 2/3 1/4) and (2/3 1/3 3/4), a carbon atom C in position 1(*a*) with coordinates (0 0 0) and a vacant site in position 1(*b*) with coordinates (0 0 1/2). A comparison of the considered structure with that of the high-temperature disordered phase β-W_2C clearly shows that, as a result of ordering, the positions 2(*a*) one-half of which are randomly occupied by carbon atoms are split into positions 1(*a*) and 1(*b*), where the former are occupied by carbon atoms and the latter remain vacant. At temperature ~2300 K, the homogeneity region of β''-W_2C (α-W_2C) phase extends from $WC_{0.34}$ to $WC_{0.48}$; a decrease in temperature is accompanied by narrowing of this homogeneity region and, finally, its disappearance at 1523 K and at a composition of $WC_{0.48}$ (see Fig. 2.1). According to [19], the β''-W_2C phase with the compositions $WC_{0.41}$ and $WC_{0.50}$ have the unit cell constants $a = 0.2985$, $c = 0.4717$ nm and $a = 0.3001$, $c = 0.4728$ nm, respectively. If the temperature drops below 1523 K, lower carbide β''-W_2C decomposes to give W and higher carbide δ-WC (WC). Note that later publications provided no information on the rhombohedral (trigonal) (space group $P\bar{3}m1$ (D^3_{3d})) modification β''-W_2C of lower tungsten carbide with the C6-type structure. It cannot be ruled out that the phases β'-W_2C and β''-W_2C described in the literature are in fact the same phase with different degrees of ordering of carbon atoms and vacancies.

Insofar as the X-ray diffraction technique fails to accurately determine the structure of possible modifications of lower tungsten carbide W_2C, the neutron diffraction analysis was applied for investigation W_2C carbide [25–29]. Carbide W_2C was synthesized by sintering of a pressed powder mixture of tungsten and carbon black (W:C = 2 : 1) at high temperature in vacuum for 3 h; then, the samples were quenched from temperatures of 2670, 2570, 2470 and 2370 K by quick immersion into a bath filled with liquid tin [25]. According to XRD data, at temperature 293 K all the samples synthesized had a hexagonal metallic sublattice with the lattice constants $a = 0.2992$ and $c = 0.4721$ nm. The neutron diffraction patterns of W_2C carbide were recorded with the use of a monochromatic neutron beam ($\lambda = 0.1131$ nm). Along with structural reflections, the neutron diffraction patterns contained weak superstructural reflections which indicated the ordering of carbon atoms and vacancies with the formation of a trigonal ε-W_2C phase which pertains to the ε-Fe_2N structural type. According to [25], the unit cell constants of the ε-W_2C phase are equal to $a = a_{L'3}\sqrt{3} = 0.5184$ and $c = c_{L'3} = 0.4721$ nm.

In study [26] tungsten carbide samples with the composition $\sim WC_{0.43-0.45}$ were synthesized by sintering of pressed mixtures of tungsten and carbon black powders at 1920 K for 30–50 h followed by cooling to room temperature during 10 min. Neutron diffraction analysis of these samples has shown that carbon atoms are ordered according to the ε-Fe_2N structure type (i.e., the same type as that reported in Ref. [25]). According to data [26], the ε-W_2C phase has a trigonal structure and belongs to space group No. 162, $P\overline{3}1m$ (D^1_{3d}). The unit cell constants of ε-W_2C carbide phase are as follows: $a = 0.5185$ and $c = 0.4723$ nm. In the ideal trigonal structure of this carbide, all tungsten atoms W occupy 6(k) positions with coordinates (1/3 0 1/4) and carbon atoms C occupy 1(a) positions with coordinates (0 0 0) and 2(d) positions with coordinates (1/3 2/3 1/2), while positions 1(b) (0 0 1/2) and positions 2(c) (1/3 2/3 0) remain vacant. Figure 2.4d illustrates the distribution of carbon atoms and vacancies in the nonmetallic sublattice of ideal trigonal (space group $P\overline{3}1m$)ε-W_2C carbide. However, the best agreement between the experimental and theoretical neutron diffraction patterns was achieved [26] when the population of the positions in the nonmetallic sublattice by C atoms was varied, and a small number of carbon atoms occupied positions 2(c).

Neutron diffraction analysis ($\lambda = 0.1225$ nm) was also used [27] for studying the structure of $WC_{0.45}$ carbide, which was produced by the arc fusion of pellets pressed from a mixture of tungsten and carbon powders followed by quenching to room temperature at a rate of $150\ K \cdot s^{-1}$. According to data [27], the studied carbide had the same trigonal (space group $P\overline{3}1m$) crystal structure as in studies [25, 26]; the unit cell constants of carbide studied are $a = 0.5181$ and $c = 0.4722$ nm. The best agreement with the experimentally found intensities of reflections (as in Ref. [26]) was achieved with allowance made for the different population of nonmetallic sublattice sites with carbon atoms. With regard to their population, the sample composition was $WC_{0.42}$. The trigonal (space group $P\overline{3}1m$) structure of the lower carbide ε-W_2C with unit cell constants $a = 0.5190$ and $c = 0.4724$ nm was also revealed in studies [28, 29]. According to [28, 29], a small number of carbon atoms occupies positions 1(b) and 2(c), which are completely vacant in an ideal ε-Fe_2N-type structure.

Thus, neutron diffraction studies [25–29] of lower tungsten carbide samples, which were quenched from the arc fusion temperature [27], from the temperatures of 2670–2370 K [25] and from temperature 1920 K [26] revealed only a trigonal (space group No. 162, $P\overline{3}1m$)) ordered phase with the ε-Fe_2N -type structure. In all the cases, the ordering degree did not reach its maximum value. The orthorhombic (space group *Pbcn*) β'-W_2C phase with the Mo_2C-type structure and the rhombohedral (trigonal) (space group No. 164, $P\overline{3}m1$) β''-W_2C phase with the *C*6-type structure were not found by the neutron diffraction analysis.

The temperatures of consecutive phase transformations β-W_2C $\Leftrightarrow$ β'-W_2C and β'-W_2C $\Leftrightarrow$ β''-W_2C and eutectoid decomposition β''-W_2C $\Leftrightarrow$ W + δ-WC (1530 K), namely, ~2750, ~2410 K, and 1530 K, respectively, were determined [18] using specially designed instruments [35]. The high-temperature differential thermal analyzer [35] involved a highly sensitive optical (photodiodes and micropyrometer) system for measuring the temperature and the temperature difference ΔT between the working and reference samples. This system allowed one to perform measurements up to 3900 K. With the use of such instruments, the temperatures of phase transformations in hafnium, zirconium and titanium and also in lower carbides of tantalum, molybdenum, and tungsten, i.e., Ta_2C, Mo_2C and W_2C, were determined [35]. The high-temperature differential thermal analyzer [35] allowed measuring the absolute difference of temperatures ΔT between the reference sample and the sample under study, which depends on the initial experimental conditions; hence, the thermal effect can have the opposite sign as compared with the measurements carried out on a standard DTA set-up.

Figure 2.5a shows a DTA curve [18] which confirms the presence of consecutive transformations of lower tungsten carbide. The authors only noted that all the effects observed at temperature decrease are exothermic. The small magnitude of observed thermal effects (Fig. 2.5a) could be caused by the incomplete transformation due to the rapid rate of cooling. With allowance for the results of neutron diffraction studies [25–29] it is more likely that the effects *1* and *3*, which are observed in the DTA curve at ~2750 and 1530 K, correspond to the transformation β-W_2C $\Leftrightarrow$ ε-W_2C and eutectoid decomposition ε-W_2C $\Leftrightarrow$ W + δ-WC.

The γ-WC_{1-x} phase has a cubic (space group $Fm\overline{3}m$ (O_h^5)) structure of the *B*1 (NaCl) type (Fig. 2.6), which is typical of all nonstoichiometric carbides and nitrides MX_y of Group IV and V transition metals. The researchers who discovered [30–32] this phase considered it as new high-temperature phase of lower carbide W_2C. However, based on the theoretical possibility that carbon atoms can occupy all octahedral interstitials in the fcc sublattice of tungsten, it seems more reasonable to consider this phase as a structural modification of higher tungsten carbide. Cubic tungsten carbide was also produced [32] as a result of the electric-spark treatment of tungsten foil, i.e., by generating electric discharge between the foil and a graphite electrode. The existence of cubic WC carbide was confirmed in work [22]. A phase γ-WC_{1-x} (β-WC) with the *B*1-type cubic structure was revealed [36] in the products of the plasma-chemical synthesis of tungsten carbide. The synthesis was carried out in the argon arc plasma from the tungsten and carbon powders or from a tungsten powder and methane CH_4. The synthesis with methane gave a product containing from 74 to 96 % cubic WC_{1-x} phase with particles 2–16 nm in size. No data on the

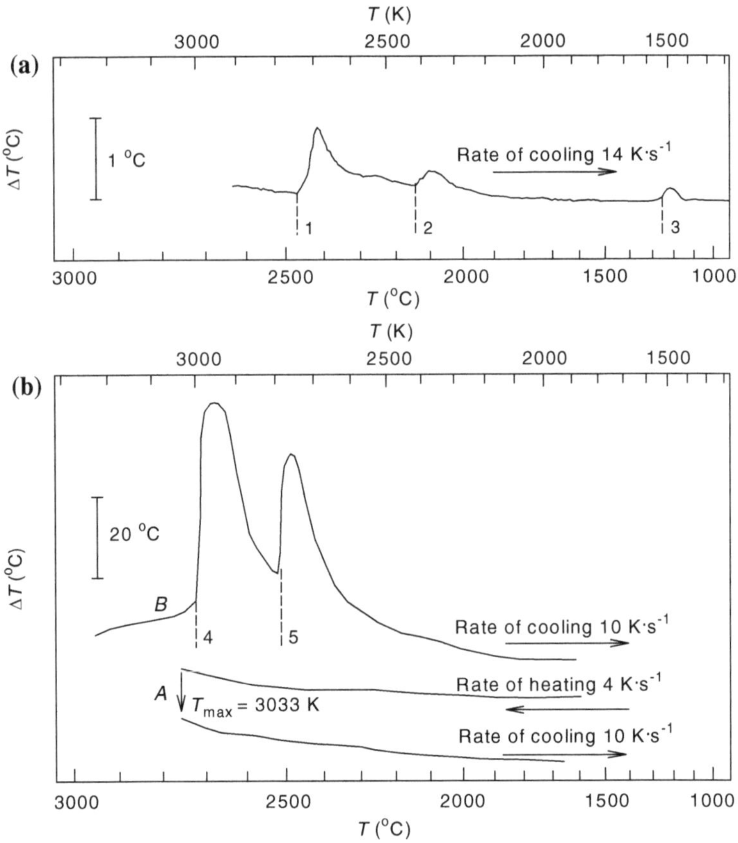

Fig. 2.5 DTA analysis of phase equilibria in the W–C system [18, 21]. **a** DTA curve of tungsten carbide $WC_{0.48-0.49}$ containing 32.7 at.% of carbon: according to data [18], the effects *1*, *2*, and *3*, which are observed in the DTA curve at ~2750, ~2410 and 1530 K, correspond to the transformation β-$W_2C \Leftrightarrow \beta'$-$W_2C$, β'-$W_2C \Leftrightarrow \beta''$-$W_2C$ and eutectoid decomposition β''-$W_2C \Leftrightarrow W + \delta$-WC, respectively; cooling rate was equal to $14\,K \cdot s^{-1}$. With allowance for the neutron diffraction data [25–29] it is more likely that the effects *1* and *3*, which are observed at ~2750 and 1530 K, correspond to the transformation β-$W_2C \Leftrightarrow \varepsilon$-$W_2C$ and eutectoid decomposition ε-$W_2C \Leftrightarrow W + \delta$-WC. **b** DTA curves of tungsten carbide containing 51.0 at.% of carbon [21]. Melting is not reached at heating the sample along curve *A* up to 3030 K. Curve *B* is measured after brief exposure of the sample at ~3270 K and following cooling of the sample from its melt; according to [21], exothermic heat effects *4* and *5* which are observed at temperatures of ~2990 and ~2790–2810 K, correspond to the transformation $L \Leftrightarrow \gamma$-$WC_{1-x} + \delta$-WC and the eutectoid decomposition γ-$WC_{1-x} \Leftrightarrow \beta$-$W_2C + \delta$-WC, respectively. The arrows indicate the direction of temperature change

chemical composition of the synthesized cubic phase γ-WC_{1-x} were presented in study [36].

Superconducting transition temperatures T_c of cubic carbides MoC and WC and carbide solid solutions of the systems NbC-MoC, NbC-WC, NbC-TaC, TaC-MoC,

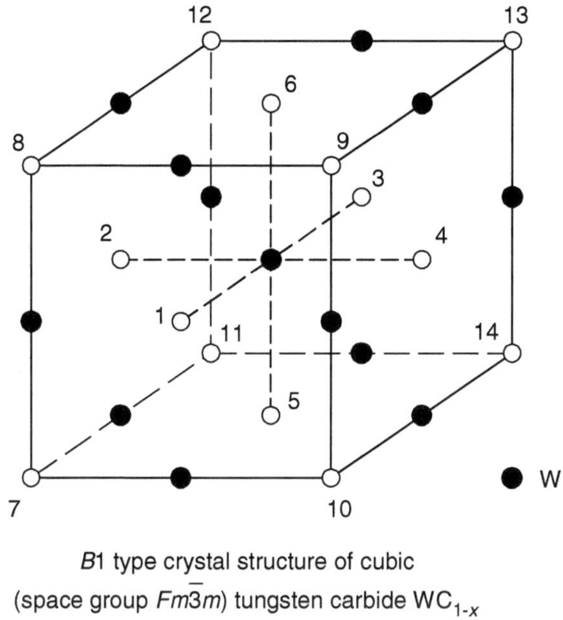

Fig. 2.6 The $B1$-type (NaCl) structure [1, 2] of cubic carbide γ-WC_{1-x} and nearest environment of W atom in the unit cell center. Nonmetallic sublattice sites are numbered. The first coordination sphere of a tungsten atom W is formed by six sites (from 1 to 6) of the nonmetallic sublattice, which are located in the centers of unit cell faces. The third coordination sphere is formed by eight sites (from 7 to 14) of the same sublattice, which are located in the vertices of cubic unit cell. The second coordination sphere of the central tungsten atom is formed by twelve metal atoms located in the middle of edges

TaC-WC, MoC-WC with the $B1$ structure were determined in works [37, 38]. Cubic tungsten carbide was synthesized in these works by rapid quenching from the molten state. According to data [37], the T_c value is sufficiently high, namely, 10 K for stoichiometric cubic carbide $WC_{1.0}$ and 9 K for nonstoichiometric cubic carbide $WC_{0.85}$. Yet another sample, which contained cubic and lower hexagonal tungsten carbides and a large amount of free (uncombined) carbon, has $T_c = 8.1$ K [37]. If as a first approximation to assume that the lattice constant and T_c of cubic WC_{1-x} tungsten carbide depend linearly on the carbon content, then $T_c = 8.1$ K corresponds to cubic carbide $WC_{0.71-0.72}$ with the unit cell constant $a_{B1} = 0.424$ nm. Note that the superconducting transition temperatures of hexagonal carbides β-W_2C and δ-WC are equal to 3.6 and 0.3 K, respectively [37]. This is substantially lower than those of the cubic phase γ-WC_{1-x}.

The electric-spark treatment of the molten samples of tungsten carbide and some WC-based hardmetals is accompanied by an appearance of cubic carbide γ-WC_{1-x} on a surface of the samples [39]. A cumulative explosion of tungsten, graphite and ammonium nitrate powders allowed to produce the samples that contained up to

50–70 vol.% of γ-WC_{1-x} phase [40]. The explosion conditions were equivalent to the temperature of steady-state sintering which equals to 2770–3270 K. Note that in studies [39, 40] cubic carbide γ-WC_{1-x} was designated as β-WC_{1-x}.

According to the phase diagram of the W-C system (see Fig. 2.1), the single-phase cubic carbide γ-WC_{1-x} exists in the composition range of $WC_{0.58} - WC_{0.65}$ at temperatures above 2790–2810 K. Moreover, γ-WC_{1-x} was found in the two-phase regions (β-$W_2C+\gamma$-WC_{1-x}) and (γ-$WC_{1-x}+\delta$-WC). According to [22], the position of the lower boundary of the homogeneity region of cubic carbide γ-WC_{1-x} is independent of the temperature and corresponds to the $WC_{0.59}$ composition. The upper boundary shifts to higher carbon contents with the increase in the temperature and can reach the stoichiometric composition $WC_{1.00}$ see Fig. 2.1, inset). In particular, at 3000–3040 K, cubic carbide γ-WC_{1-x} has a homogeneity region $WC_{0.59}-WC_{0.92}$ which is much wider than reported in works [19–21]. At temperature ~2800–2810 K eutectoid decomposition of cubic carbide γ-WC_{1-x} to β-W_2C and δ-WC (WC) takes place.

Figure 2.5b shows DTA curves of cubic tungsten carbide containing 51.0 at.% carbon [21]. As is seen, heating of cubic carbide along curves *A* up to ~3030 K without any exposure at the maximum temperature failed to melt the sample and subsequent cooling from 3030 to 1940 K revealed no heat effects. However, curve *B* measured after the sample was cooled upon its brief exposure at temperature ~3270 K, i.e., from its melt, demonstrates peaks corresponding to strong exothermic heat effects *4* and *5* at temperatures of ~2990 and ~2790–2810 K, respectively. From the viewpoint of the authors [21], these effects were caused by the following reversible transformations: the eutectics formation $L \Leftrightarrow \gamma$-$WC_{1-x}+\delta$-WC, and the eutectoid decomposition γ-$WC_{1-x} \Leftrightarrow \beta$-$W_2C+\delta$-WC. According to the conditions under which curves *A* and *B* were recorded, cubic carbide γ-WC_{1-x} is formed relatively easily from the melt, but the direct solid-phase transformation δ-WC $\Leftrightarrow \gamma$-WC_{1-x} is either hindered (for it to happen, the temperature should change slowly near the transformation temperature T_{trans}) or the temperature range of existence of cubic carbide near the stoichiometric composition WC is very narrow.

The phase equilibria in the W–C system were studied [22] in the temperature interval from 2770 to 3100 K (see Fig. 2.1, inset) with the use of the XRD method, chemical and metallographic analyses and also of the DTA method. To determine the phase composition, the samples heated to 2870 K (or higher temperature) were quenched by cooling to 1800 K during 20 s and lower to 1300 K during 40 s. Samples with the compositionbI between W_2C and WC were quenched from $T \geq 2800$ K by quick immersion into liquid tin with the temperature of 600–650 K. In study [22], high-temperature DTA was carried out with the use of specially designed instruments. The temperature interval used in study [22] was narrower than that used by authors of investigations [18–21]. This is why those phase transitions of lower carbide W_2C, which are associated with its ordering, could not be detected at temperatures below 2750 K.

By and large, the results of independent studies [18–21] and [22] on the determination of the temperature-concentration position of phase boundaries in the W–C system are almost identical. For example, the deviations of the found phase-transformation

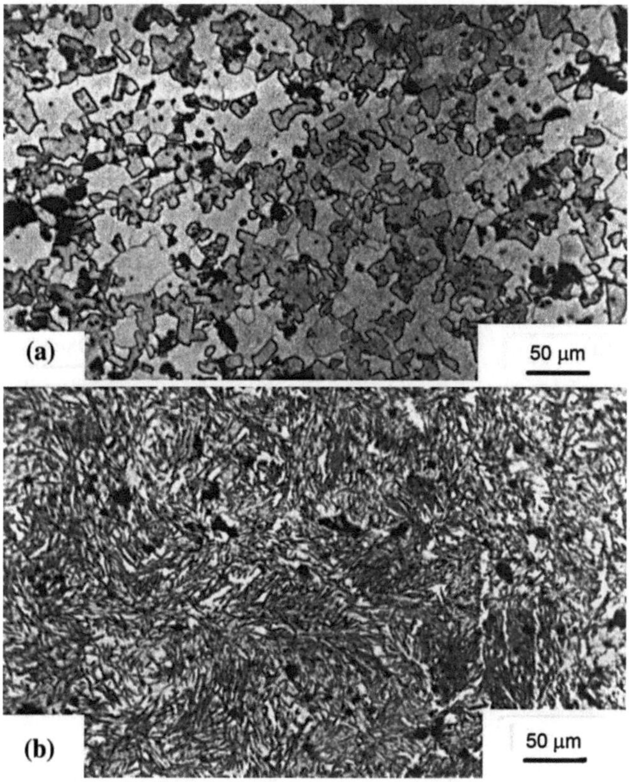

Fig. 2.7 Microstructures of carbide samples with 60 at.% of tungsten W and 40 at.% of carbon C quenched from temperatures 2778 and 2808 K [22]. **a** The sample, quenched from temperature 2778 K, contains hexagonal phases β-W_2C (light grains) and δ-WC (dark grains). **b** The sample, quenched from temperature 2808 K contains β-W_2C and δ-WC phases too, but has a pearlite-like microstructure which has appeared as a result of the eutectoid decomposition of quickly cooled cubic γ-WC_{1-x} carbide formed at temperature $T > 2798$ K

temperatures do not exceed 10 K, i.e., lie within the experimental error range. However, the region of existence of high-temperature cubic monocarbide γ-WC_{1-x} in the phase diagram was studied in more detail by author of study [22]. Particularly, a thorough metallographic and XRD study of samples quenched from the temperatures above and below 2798 K since the DTA method has shown that a reversible phase transformation takes place at 2798 K in samples containing from 35 at.% to 50 at.% carbon. Figure 2.7 illustrates the changes in the microstructures of samples containing 40 at.% carbon, which were quenched from 2778 and 2808 K. A sample quenched from 2778 K contained two hexagonal phases β-W_2C and δ-WC (Fig. 2.7a).

A sample quenched from 2808 K also contained two phases β-W_2C and δ-WC but had a different pearlite-like microstructure (Fig. 2.7b), which appeared as a result of the eutectoid decomposition of cubic carbide γ-WC_{1-x} formed at $T > 2798$ K.

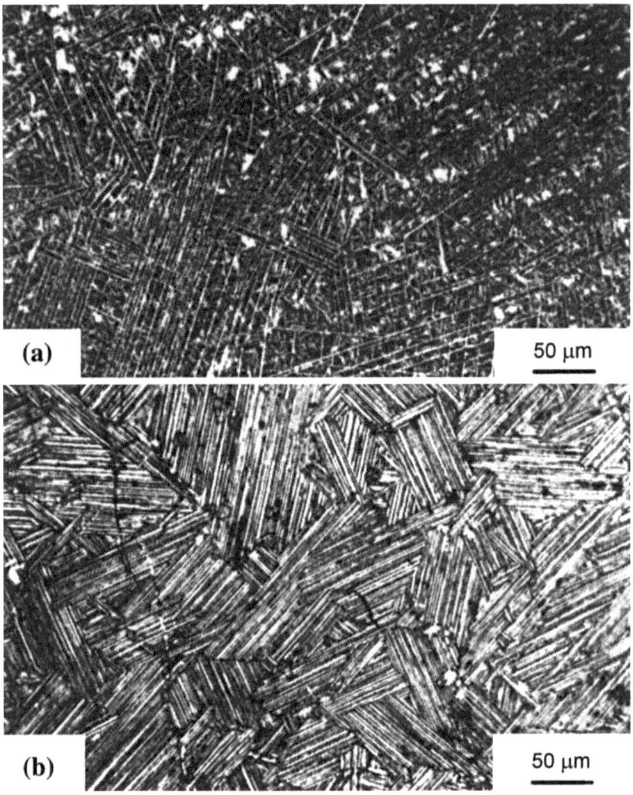

Fig. 2.8 Microstructures of samples with cubic carbide γ-WC_{1-x} carbide obtained by quenching from the melt [22]. **a** A sample with 45 at.% of C quenched from 3043 K contains inclusions of hexagonal monocarbide δ-WC in the matrix of cubic γ-WC_{1-x} carbide; **b** eutectic (β-W_2C + γ-WC_{1-x}) microstructure of a sample with 36 at.% of carbon quenched from 3053 K. The both sample surface is etched by a mixture of 10 % NaOH and 30 % $K_3Fe(CN)_6$ taken in the ratio 1 : 1

Cubic carbide γ-WC_{1-x} can be directly visualized in the microstructure of samples quenched from higher temperature 2988–3023 K in the solid state, and also in the microstructures of samples quenched from the melt from temperature of 3043–3063 K (Fig. 2.8).

It follows from result [22] that cubic γ-WC_{1-x} carbide has a wide homogeneity region, namely, from $WC_{0.59}$ to $WC_{0.98-1.00}$ at temperature 3030–3055 K (see Fig. 2.1, inset).

High temperature of the transition into the superconducting state, namely, 10 K [37, 38], confirms that the composition that corresponds to the upper boundary of the homogeneity region of γ-WC_{1-x} carbide is stoichiometric, i.e., $WC_{1.0}$. It is known [3, 4, 41] that, among the cubic carbides of Group IV–VI transition metals, stoichiometric carbides $NbC_{1.0}$ and $TaC_{1.0}$ have the highest T_c equals 10–12 K. Upon deviation of composition of these carbides from stoichiometry to $NbC_{0.9}$ or $TaC_{0.9}$,

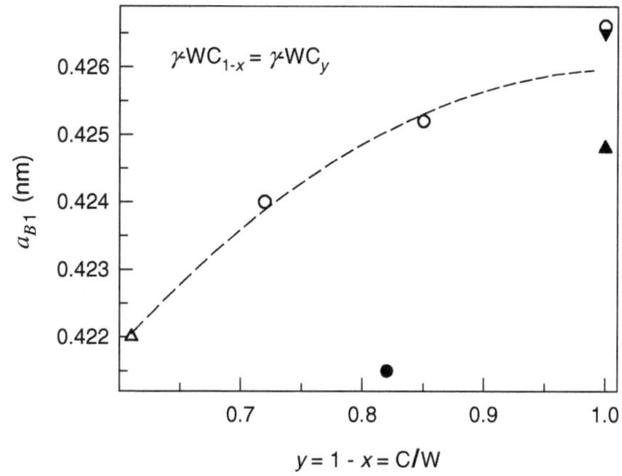

Fig. 2.9 Changes in the unit cell constant a_{B1} of cubic γ-WC$_{1-x}$ $\equiv$ γ-WC$_y$ ($y \equiv 1 - x$) carbide with $B1$-type structure in its homogeneity region [43]: (△) [19, 21], (○) [37, 38], (•) [22], (▲) [39], (▼) [42]. Approximation of data [19, 21, 37–39, 42] by function $a_{B1}(y) = 0.4015 + 0.0481y - 0.0236y^2$ (nm) is shown by a *dotted line*

their T_c quickly drops to ~4 K. Carbides NbC$_{0.8}$ and TaC$_{0.8}$ have the $T_c \leq 1$ K. According to Willens and Buehler [37], a decrease in the carbon content in cubic tungsten carbide causes a relatively smaller decrease in T_c.

The unit cell constant of γ-WC$_{1-x}$ carbide is $a_{B1} \approx 0.427$ nm [32] and cubic carbide WC$_{0.82}$ has a lattice constant $a = 0.4215$ nm [22]. Cubic tungsten carbide with an unknown composition was found to have the unit cell constant of 0.4229 nm [36], while almost stoichiometric cubic carbide have the unit cell constant $a = 0.4248$ nm [39]. The unit cell constants of cubic carbides WC$_{1.0}$, WC$_{0.85}$ and WC$_{0.72}$, were shown [37] to be 0.4266, 0.4252 and 0.4240 nm, respectively. For cubic carbide WC$_{0.61}$, $a = 0.4220$ nm was found [19, 21] and the extrapolation of this value to defect-free cubic carbide gave lattice constant $a = 0.4265$ nm [42]. The unit cell constant of cubic carbide γ-WC$_{1-x}$ of the unknown composition was found to vary from 0.4222 to 0.4263 nm [40]. Such a small unit cell constant of cubic carbide γ-WC$_{1-x}$ could be caused by oxygen impurity because the sample was obtained in a cumulative explosion in ammonium nitrate which contains oxygen. As is seen from Fig. 2.9, the best agreement is observed for the unit cell constant values found for cubic carbide γ-WC$_{1-x}$ $\equiv$ γ-WC$_y$ ($y \equiv 1 - x$) in studies [19, 21, 37–39, 42] (Fig. 2.9). Analysis performed in study [43] is shown that these data are adequately described by a quadratic function $a_{B1}(y) = 0.4015 + 0.0481y - 0.0236y^2$ nm and unambiguously point to the increase in the unit cell constant of the γ-WC$_{1-x}$ phase with the increase of carbon content.

Higher tungsten carbide δ-WC (WC) has a negligibly small homogeneity region or does not have it at all. Higher tungsten carbide δ-WC have hexagonal (space group $P\overline{6}m2$ (D^1_{3h})) crystal lattice with own structural type B_h (the WC-type) and with the unit cell constants $a = 0.2906$ and $c = 0.2837$ nm [7, 9, 10, 19]. In this structure,

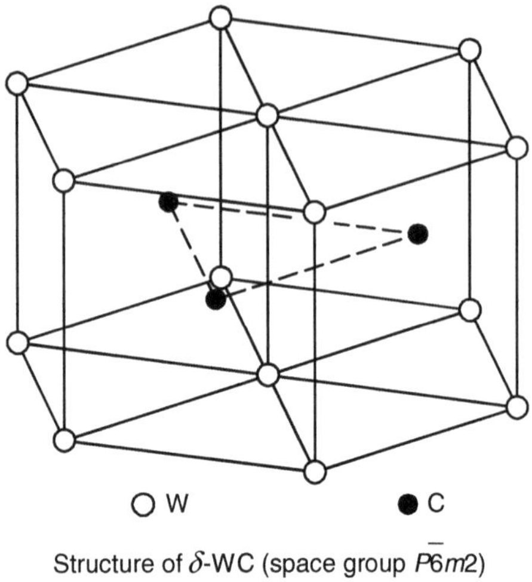

Fig. 2.10 Hexagonal structure of higher δ-WC (WC) tungsten carbide: (•)C, (○)W

tungsten atoms W occupy positions 1(*a*) with coordinates (000) and carbon atoms are located in positions 1(*d*) with coordinates (1/32/31/2) [44]. Tungsten and carbon atoms in higher δ-WC (WC) carbide form simple hexagonal sublattices, in which C atoms are located in the centers of trigonal prismatic interstitials of the tungsten sublattice (Fig. 2.10). The results of simulations of the crystal and electronic structures of hexagonal WC carbide by a pseudopotential local-orbital method employing the local-density approximation confirmed the stability of this compound [45]. The estimated unit cell constants $a = 0.288$ and $c = 0.281$ nm [45] well agree with experimental results.

A positron lifetime studies [46, 47] are the last experimental confirmation for the stoichiometry of hexagonal monocarbide δ-WC. Samples of powdery and sintered WC and also sample of a WC–Co hardmetal were studied. The positron lifetime spectra of WC carbide [46] revealed a short component with lifetime 124 ± 10 ps and more long component with lifetime 175 ± 20 ps. The former component $\tau^{\text{free}} = 124 \pm 10$ ps corresponds to the annihilation of positrons from the free (delocalized) state, the latter component $\tau_{\Box W} = 175 \pm 20$ ps corresponds to the capture and annihilation of positrons in the metal vacancies the concentration of which is very low. No components that would correspond to the positron annihilation in the carbon-sublattice vacancies are found. Under the irradiation of δ-WC carbide with electrons having energy of 1 MeV, the positron lifetime spectrum contained only one component with a lifetime of ~136 ps. It was shown [46] that this component corresponds to the capture and annihilation of positrons in the carbon vacancies, which have appeared under the irradiation. Indeed, the kinetic energy, which is transferred by the incident 1-MeV electrons to C and W atoms in the collisions, is higher than the displacement threshold of the carbon atom but lower than the displacement threshold of the tungsten atom. Hence, only carbon atoms are displaced in tungsten carbide and the vacancies are induced only in the carbon sublattice of WC carbide. Due to the higher transferred energy, the irradiation of δ-WC carbide with electrons with the energy of 2.5 MeV leads to the displacement of both carbon and tungsten atoms and the appearance of vacancies in both sublattices [47]. A comparison of the experimental results [46, 47] with those obtained in the calculation [48] of the positron lifetime by the method of linear muffin-tin orbitals in the atomic sphere approximation (LMTO-ASA) demonstrated good agreement of calculations with experimental data for $\tau_{\square W}$ (positron lifetime in a metal vacancy) and a noticeable deviation in the positron lifetime $\tau_{\square C}$ in a carbon vacancy. Hexagonal tungsten carbide WC was modeled by carbide with a cubic $B1$-type lattice [48]. On the whole, the results [46, 47] confirmed that under normal conditions higher hexagonal tungsten carbide δ-WC is stoichiometric compound and has no carbon vacancies in carbon sublattice.

Thus, among tungsten carbides, only higher hexagonal carbide δ-WC (WC) has virtually no homogeneity region whereas lower carbides W_2C and cubic γ-WC_{1-x} have relatively wide homogeneity regions.

The data on the crystal structures of the phases in the W–C system are given in Table 2.1.

Thus, among tungsten carbides, only higher hexagonal carbide δ-WC (WC) has virtually no homogeneity region whereas lower carbides W_2C and cubic γ-WC_{1-x} have relatively wide homogeneity regions. It should be borne in mind that the conclusions made in those studies in which the structures of nonstoichiometric tungsten carbides were determined only from the XRD data are merely the assumptions. Particularly, this concerns the lower W_2C carbide because ordering of atoms and vacancies can occur in the nonmetallic sublattice of this carbide. The structures of nonstoichiometric tungsten carbides found by the neutron diffraction analysis are more confident and reliable.

The main difference between the phase diagrams of the W–C system proposed by Rudy et al. [19–21], on the one hand, and Sara [22], on the other hand, concerns the position and width of the homogeneity region of cubic carbide γ-WC_{1-x}. The authors of the studies [19–21] had the limited experimental results, which were not sufficient to determine the position of cubic carbide in the phase diagram and used the analogy with the Mo–C system studied earlier [49, 50]. In the Mo–C system, they detected the β-MoC_{1-x} (Mo_3C_2 or $MoC_{0.67}$) phase with a narrow homogeneity region and assumed that the WC_{1-x} phase in the W–C system has the same stoichiometric composition W_3C_2($WC_{0.67}$) and the same narrow homogeneity region. It is no mere chance that the article by Rudy et al. [19] with the first version of the W–C phase diagram, presented on Fig. 2.1, is entitled "W–C system: supplemental information on the Mo–C system."

At the same time, it follows from the study [22] that the homogeneity region of cubic γ-WC_{1-x} tungsten carbide is much wider than that given in studies [19–21] and reaches the $WC_{1.0}$ composition. This agrees with the results of later studies [32, 36–39]. However, for a cubic carbide of the stoichiometric composition $WC_{1.0}$, the phase diagram (see Fig. 2.1, inset) [22] contains no temperature region of its existence. The region of the existence of the γ-WC_{1-x} phase has such boundaries at (1-x) $\rightarrow$ 1 that the temperature decrease merely from 3058 to 3028 K causes the following consecutive transformations in the $WC_{1.0}$ carbide: first, a peritectic

Table 2.1 Crystal structure of phases[a] which existence was supposed in the W–C system at $T > 1300$ K according to [17] (see Fig. 2.1)

Phase: Designations as in Fig. 2.1 [17]	Phase: Designations in other studies	Existence region (at.% C)	Lattice constants (nm)	Space group	Pearson symbol	Lattice type and symmetry
W	W	0–1.0	$a = 0.3165$	$Im\bar{3}m\,(O_h^9)$	$cI2$	$A2$ (W type), bcc
β-W_2C	γ-W_2C [18, 21]	~25.5–34.0	$a = 0.3002$, $c = 0.475 - 0.476$ [19]	$P6_3/mmc\ (D_{6h}^4)$	$hP3$	$L'3$ (W_2C type), hexagonal
β'-W_2C	β-W_2C [18, 21]	~29.5–33.0	$a = 0.4728$, $b = 0.6009$, $c = 0.5193$ [18]	$Pbcn\ (D_{2h}^{14})$	-	PbO_2 or Mo_2C (ζ-Fe_2N) type, orthorhombic
β''-W_2C	α-W_2C [18, 21]	~29.5–32.5	$a = 0.2985$, $c = 0.4717$ ($WC_{0.41}$) [19], $a = 0.3001$, $c = 0.4728$ ($WC_{0.50}$) [19]	$P\bar{3}m1\ (D_{3d}^3)$	$hP3$	$C6$ (anti-CdI_2 type), trigonal
[b]γ-WC_{1-x}	α-WC_{1-x}[21], β-WC [22, 36]	~ 37.0–39.5 [21], ~ 37.1–50.0 [22]	$a = 0.4266$ ($WC_{1.0}$), $a = 0.4252$ ($WC_{0.85}$) [37]	$Fm\bar{3}m\ (O_h^5)$	$cF\,8$	$B1$ (NaCl type), cubic
δ-WC	WC [21], α-WC [22]	~50	$a = 0.2906$, $c = 0.2837$ [19]	$P\bar{6}m2\ (D_{3h}^1)$	$hP2$	B_h (WC type), hexagonal
C	C	100	$a = 0.142$, $c = 0.339$	$P6_3/mmc\ (D_{6h}^4)$	$hP4$	$A9$ (C, graphite), hexagonal

[a] In the study [17], the phase diagram of the W–C system takes into account the structure data [18–22, 34], some of which were not confirmed later; trigonal (space group) phase ε-W_2C phase the existence of which was reliably confirmed [25–29] is not shown in the phase diagram of the W–C system in the study [17]

[b] γ-$WC_{1-x} \equiv \gamma$-$WC_y (y \equiv 1 - x)$

transformation and then a peritectoid one. This is highly improbable. Moreover, the line of phase equilibrium *1-2-3* (see Fig. 2.1, inset) is shown incorrectly in the cited study [22], namely, from a horizontal that corresponds to the nonvariant eutectic reaction $L \Leftrightarrow \beta\text{-}W_2C + \gamma\text{-}WC_{1-x}$ this line gradually transforms into a monovariant curve, which is physically impossible and incorrect from the viewpoint of geometrical thermodynamics. The phase equilibrium shown could be possible if point *2* was a singular point, i.e., a kink point. This in turn, would have led to the narrowing of the temperature interval of the existence of a two-phase region $L + \gamma - WC_{1-x}$ and to the shift of its lower boundary to higher temperatures, which contradicts the results of DTA and metallographic analysis [22]. If a cubic tungsten carbide with the stoichiometric composition $WC_{1.0}$ has at least a narrow temperature range of stability and the two consecutive phase transitions are somewhat shifted in the composition one with respect to another, then the noted inaccuracies and contradictions could be eliminated and the total body of experimental results could be taken into account.

The phase diagram of the W–C system with allowance made for the presence of the temperature interval of existence of cubic carbide $WC_{1.0}$ and also for experimental results [18–22, 32, 36–39] was constructed [43]. However, the phase diagram of the W–C system was constructed in study [43] without allowing for the neutron diffraction data [25–29]. Therefore later in review [51], the phase diagram of the W–C system published in study [43] was supplemented by some corrections in the positions of phase boundaries of region of existence of the lower tungsten carbide W_2C, namely, the neutron diffraction results [25–29] was taken into account and the $\varepsilon\text{-}W_2C$ phase was shown instead of phases $\beta'\text{-}W_2C$ and $\beta''\text{-}W_2C$, the existence of which is not confirmed experimentally. These corrections are reflected in Fig. 2.11 [51].

It is seen from the phase diagram of the W–C system (Fig. 2.11) that in the temperature range of 3058–2993 K, in the vicinity of tungsten carbide with $WC_{1.0}$ composition, a decrease in temperature leads first to the peritectic formation of cubic tungsten carbide from the melt $L + C \Leftrightarrow \gamma\text{-}WC_{1-x}$, and then to the transformation of cubic monocarbide into the hexagonal modification $\gamma\text{-}WC_{1-x} \Leftrightarrow \delta\text{-}WC$. The temperature of the eutectoid decomposition $\gamma\text{-}WC_{1-x} \Leftrightarrow \delta\text{-}WC + C$, equal to 2993 K (see Fig. 2.11) was found from the results of DTA [21] for a tungsten carbide sample with 51 at.% carbon. This value was assigned by authors of study [21] to the temperature of the eutectic $L \Leftrightarrow \beta - W_2C + \gamma\text{-}WC_{1-x}$. However, even according to the phase diagram version proposed in study [21] carbide with such composition falls beyond a region where this phase equilibrium is permitted. Therefore, this temperature cannot be the temperature of the eutectic. Although the temperature interval of the equilibrium existence of $\gamma\text{-}WC_{1-x}$ is small (Fig. 2.11), the cubic carbide of the stoichiometric composition $WC_{1.0}$ can at least be quenched from the melt. This is agrees with the experimental results [37, 38].

The special points in the phase diagram of the W–C system (Fig. 2.11) at temperatures above 1300 K are described in Table 2.2.

Table 2.2 Special points in the phase diagram of the W–C system (see Fig. 2.11) at a temperature above 1300 K [51]

Reaction	Composition of phases involved in the reaction (at.% of C)			T (K)	Reaction type
$L \Leftrightarrow W$	0	0	-	3755 ± 5	Melting
$L \Leftrightarrow \beta\text{-}W_2C$	~ 30.6	~ 30.6	-	3058 ± 10	Congruent melting
$L + C \Leftrightarrow {}^a\gamma\text{-}WC_{1-x}$	~ 42.0	100.0	50.0	3058 ± 5	Peritectics
$L \Leftrightarrow W + \beta\text{-}W_2C$	~ 23.5	~ 1.2	~ 25.5	2988 ± 5	Eutectics
$L \Leftrightarrow \beta\text{-}W_2C + {}^a\gamma\text{-}WC_{1-x}$	~ 37.0	~ 34.3	~ 37.8	3028 ± 5	Eutectics
${}^a\gamma\text{-}WC_{1-x} \Leftrightarrow \delta\text{-}WC$	~ 49.3	~ 49.3	-	3008 ± 5	Polymorphic transformation
${}^a\gamma\text{-}WC_{1-x} \Leftrightarrow \delta\text{-}WC + C$	50.0	~ 49.8	100.0	2993 ± 5	Eutectoid decomposition
${}^a\gamma\text{-}WC_{1-x} \Leftrightarrow \beta\text{-}W_2C + \delta\text{-}WC$	~ 38.2	~ 34.0	~ 49.5	2798 ± 5	Eutectoid decomposition
$\beta\text{-}W_2C \Leftrightarrow \varepsilon\text{-}W_2C$	~ 31.6	~ 31.6	-	2728 ± 10	Disorder-order transformation
$\beta\text{-}W_2C \Leftrightarrow W + \varepsilon\text{-}W_2C$	~ 28.6	~ 0.7	~ 29.7	2673 ± 10	Eutectoid decomposition
$\beta\text{-}W_2C \Leftrightarrow \varepsilon\text{-}W_2C + \delta\text{-}WC$	~ 33.5	~ 32.8	~ 49.8	2657 ± 10	Eutectoid decomposition
$\varepsilon\text{-}W_2C \Leftrightarrow W + \delta\text{-}WC$	~ 32.6	0	50.0	1523 ± 5	Eutectoid decomposition

[a] $\gamma - WC_{1-x} \equiv \gamma - WC_y \ (y \equiv 1 - x)$

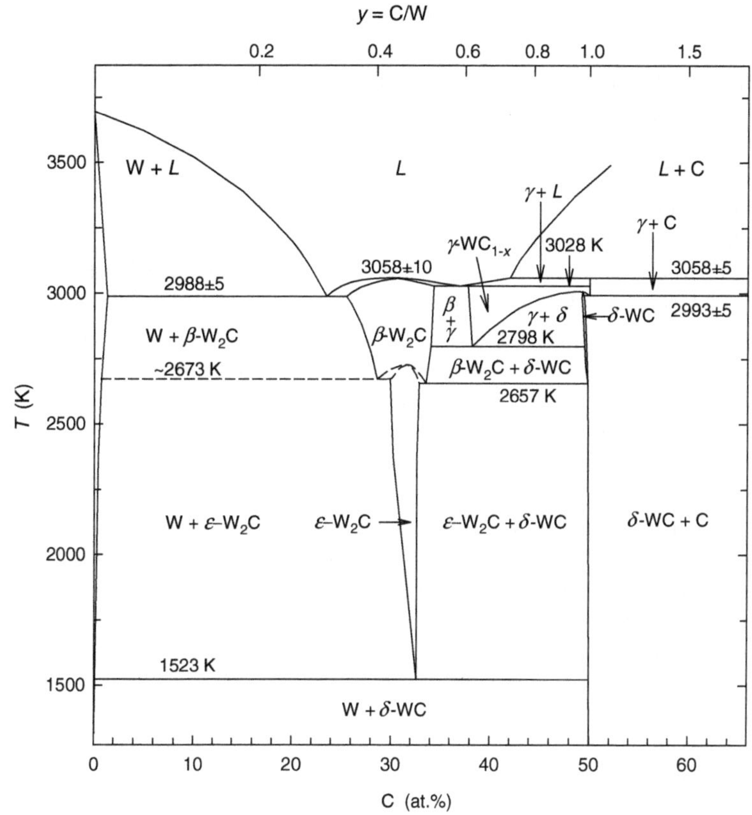

Fig. 2.11 Phase diagram of the W–C system [51]. The trigonal (space group $P\bar{3}1m$)) ε-W_2C phase which was observed in the neutron diffraction studies [25–29] is shown instead of phases β'-W_2C and β''-W_2C, the existence of which is not confirmed experimentally. The special points of the phase diagram are described in Table 2.2

2.2 Phase Equilibria in the W–Co–C System

The wide application of high-strength metal-cutting hardmetals formed in the pseudobinary system WC–Co brought about the investigations of the phase equilibria in the ternary system W–Co–C.

Information on the phases in the W–Co–C system which was obtained before 1952 is generalized by authors of study [52] in which new diffraction, metallographic and DTA investigations of this system were carried out. For the W–Co–C system, the liquidus surface projection, the total isothermal cross-section at temperature of 1673 K, and also the pseudobinary *c*–*T* diagrams along the cross-sections C–$Co_{43}W_{57}$ and WC–Co were constructed, in which the stable and metastable phase equilibria were specified [52]. In the pseudobinary cross-section WC_y–Co (with 16 wt.% Co), the effect of the carbon content on the phase equilibria and structure of alloys was studied in work [53]. The results of the metallographic and XRD studies of the M_6C-type cubic η-carbides, which are formed in the W–Co–C system, were discussed [54]. Particularly, η-carbide Co_2W_4C with a cubic unit cell constant, which varied from 1.107 to 1.125 nm depending on the carbon content, was revealed. Besides, two total isothermal cross-sections of the W-Co–C system were constructed at temperatures 1273 and 1673 K [54]. Crystal structure of binary and ternary phases formed in the W–Co–C system was described in study [55].The stability of η-carbides of the M_6C-type in the W–Co–C system was also considered by other authors [56] who showed additional isothermal cross-sections in this system at 1373 and 1573 K. The experimental investigation of phase equilibria in the W–Co–C system at 1423 K which was supplemented by thermodynamic calculations was performed in study [57].

According to the results of numerous studies, the system W–Co–C forms several ternary compounds, viz., cubic (space group $Fd\overline{3}m$ (O_h^7)) η-carbides Co_2W_4C [54] and Co_4W_2C [58] of the M_6C-type and also two hexagonal (space group $P6_3/mmc$ (D_{6h}^4)) carbides (CoW_3C with the structure of $Co_3W_9C_4$-type [59] and $Co_6W_{20}C_7$ [60]. Note that the $Co_3W_9C_4$ structure has two types of metallic sublattice interstitials, in which carbon atoms are located as follows: in the unit cell, two carbon atoms of eight are located in the centre of trigonal prisms formed by six tungsten atoms, while the remaining six C atoms are in the centers of octahedra formed of five tungsten and one cobalt atoms. A comprehensive study of these carbides showed that all of them can exhibit a noticeable carbon deficiency, i.e., they are nonstoichiometric compounds. Hence, their compositions should be written with allowance made for the possible nonstoichiometry, namely, $Co_2W_4C_y$, $Co_4W_2C_y$ and CoW_3C_y. Carbides $Co_2W_4C_y$ and $Co_4W_2C_y$ are often united into a single phase $Co_3W_3C_y$. It is evident that here we have, in fact, a pseudobinary solid solution between $Co_2W_4C_y$ and $Co_4W_2C_y$. For the case of deficient carbon content, this variable-composition phase corresponds to the $\sim Co_6W_6C_y$ carbide in the ternary phase diagram.

The results of the aforementioned studies are taken into account in a comprehensive survey [61] which is devoted to the thermodynamic properties and phase equilibria in the W–Co–C system. Moreover, the author of this survey carried out calculations to refine the phase equilibria by means of the CALPHAD method [62] using the approximations and models described elsewhere [63]. Figure 2.12 shows a series of isothermal cross-sections of the ternary system W–Co–C at 2073, 1773, 1698, 1673, 1548 and 1443 K, which were constructed based on these calculations and are shown in a handbook [64]. All isothermal cross-sections contain carbide $\sim Co_6W_6C$ and a solid solution $Co_3W_3C_y$ (in the survey cited [61], they appear as $M_{12}C$ and M_6C, respectively). In the isothermal cross-section at 1273 K, the solid solution $Co_3W_3C_y$ is absent but a η-carbide Co_2W_4C appears [61]. As is seen from Fig. 2.12, the ternary compounds are formed if the carbon atom content in an alloy is lower than the tungsten atom content. Note that the isothermal cross-section of the W–Co–C system at 1273 K ignores the existence of the hexagonal intermetallic phase Co_3W, which stabilizes at $T \leq 1350$ K and decomposes via the peritectoid reaction at temperature of $\sim$1360 K.

Section of the ternary W–Co–C system in the WC–Co cross section [61] that is relevant for hardmetals is shown in Fig. 2.13.This section contains two- and three-

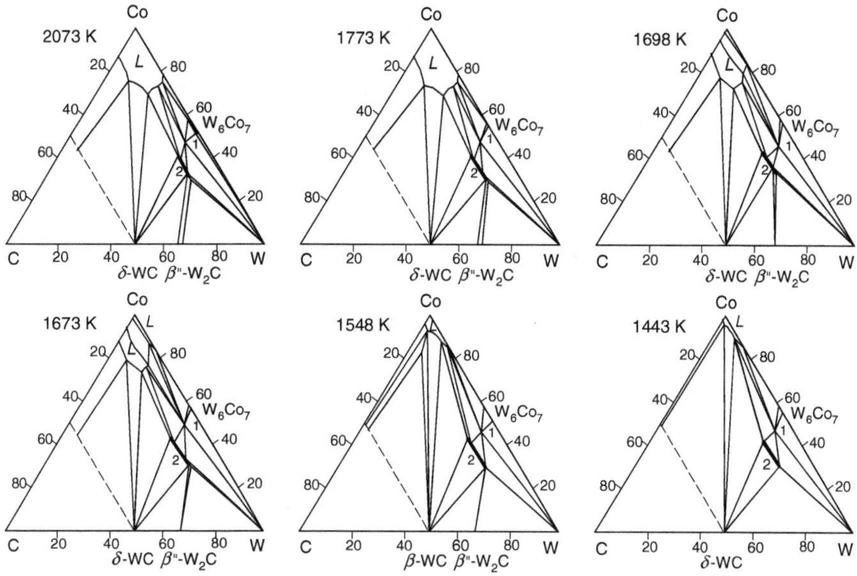

Fig. 2.12 Isothermal cross-sections of the ternary W–Co–C system at 2073, 1773, 1698, 1673, 1548, and 1443 K [64]. The concentrations are shown in at.%, the triple point position corresponds to the results of thermodynamic calculations [61]. (*1*) $Co_6W_6C_y$, (*2*) $Co_3W_3C_y$

Fig. 2.13 Section of the ternary W–Co–C system cut along the WC–Co cross section [61]

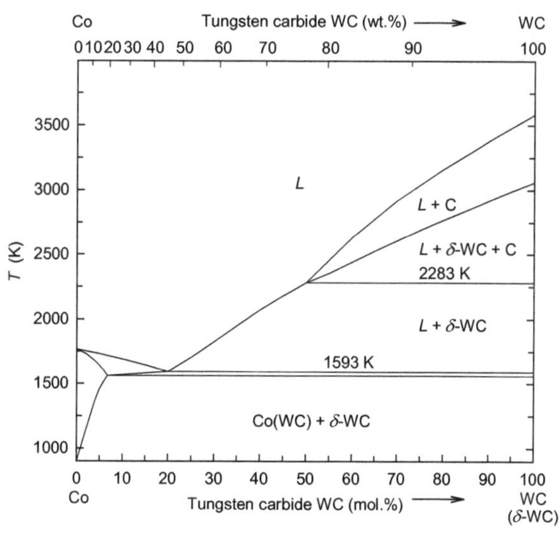

phase regions. The liquid phase appears in the WC–Co system at 1560–1593 K. According to [61], the maximum solubility of WC in solid cobalt does not exceed 6.7 mol.% (or 19.3 wt.%) and the solubility does not exceed 3 mol.% ($\sim$ 9 wt.%) at decreasing temperature up to 1270 K. According to the earlier data (see Ref. [7], pp. 268–274), the maximum solubility of WC carbide in cobalt reaches 22 wt.%.

A solid solution Co(WC) of tungsten carbide in cobalt is formed as a result of the diffusion dissolution and appears when the sintering temperature of the WC–Co system reaches 1420–1570 K. A part of cobalt which does not belong to the Co(WC) solid solution passes into the liquid phase that wets the carbide grains.

2.3 Electronic Structure of Tungsten Carbides

The band structures of hexagonal and cubic carbides W_2C and WC were calculated in studies [45, 65–69]. The electronic structure of hexagonal WC carbide was calculated [65] by a non-self-consistent method of augmented plane waves (APW). A more correct investigation of the band structure in the bulk and on the surface of hexagonal WC carbide was performed by a self-consistent linear method of augmented plane waves (LAPW) [66].

Figure 2.14a shows the calculated curve of the density of electronic states (DOS) for hexagonal WC carbide. The Fermi level is located near the curve minimum, and the density of electronic states on this level, $N(E_F)$, is relatively low, viz., approximately $\sim$0.34 eV^{-1}($\sim$4.6 Ry^{-1}) per unit cell. The Fermi surface of hexagonal WC carbide has small hole and electron 'pockets' in the vicinity of K and L points of the Brillouin zone [66].

The electronic structure of hexagonal WC was calculated [45] by the pseudopotential method of local orbitals. Figure 2.14b shows the resulting curves of the density of electronic states (it is the density of spin states $N_{sp}(E)$ that was calculated; to make the comparison easier, the figure shows the density of electronic states $N(E) = 2N_{sp}(E)$ in place of the density of spin states). It is evident that the calculated density-of-states curves of hexagonal WC [45, 66] are in adequate agreement one with another. Both in calculation [45] and in calculation [66] the Fermi level is located in a distinct density-of-states minimum and the quantity $N(E_F)$ is $\sim$0.3 eV^{-1} per unit cell. Presumably [45, 66], it is the small density of electronic states on the Fermi level that determines the high stability of hexagonal WC carbide. According to the authors [45], the low superconducting transition temperature of this carbide (about 1.3 K [70]) is also associated with the low quantity of density of electronic states on the Fermi level, $N(E_F)$.

The linear method of muffin-tin orbitals in the atomic sphere approximation (LMTO-ASA) allowed the densities of states to be calculated [67] not only for higher hexagonal tungsten carbide WC, but also for cubic (with $B1$-type structure) monocarbide WC. As was shown above (see Sect. 2.1), to date, nonstoichiometric cubic carbide WC_y with $y \leq 0.82$ is experimentally detected, although according to [22] the upper boundary of the homogeneity region of cubic tungsten carbide approaches $WC_{1.0}$. The calculated curves of the density of states *N*(*E*) are shown in Fig. 2.15.

The calculated curves of the density of electronic states $N(E)$for hexagonal WC carbide which are shown in Figs. 2.14 [45, 66] and 2.15a [67] differ, first of all, in the Fermi level position. According to calculations [45, 66], the Fermi level lies at a pronounced minimum in the DOS curve, between valence-binding and high-energy metal bands and the density-of-states on the Fermi level is small and is not more than $0.35\ eV^{-1}$ ($4.6\ Ry^{-1}$) per unit cell. According to calculations [67], the Fermi level of hexagonal WC carbide is located in a band formed largely by the 5*d* states of tungsten, and $N(E_F) = 0.78\ yA^{-1}$ ($10.6\ Ry^{-1}$) per unit cell, i.e., almost twice greater than it follows from other calculations [45, 66]. Note that in their later publication [68] the authors of a study [67] showed a different density-of-states curve for hexagonal WC carbide, which is closer to those calculated in the studies [45, 66].

The curve of the density of electronic states for cubic WC (Fig. 2.15b) is on the whole similar to the corresponding curves of cubic carbides of other transition metals MC (M = Ti, Zr, Nb, Ta) and contains a low-energy 2*s*(C) band with small contributions of *s*-, *p*- and *d*-states of tungsten, a ground-state valence-binding band formed due to the hybridization of 2*p*(C) and 5*d*(W) states and a high-energy band largely formed by the 5*d*(W) states with an admixture of 6*s*(W) states. As compared with the *N*(*E*)curve of TaC carbide (a typical nonstoichiometric carbide, the closest to

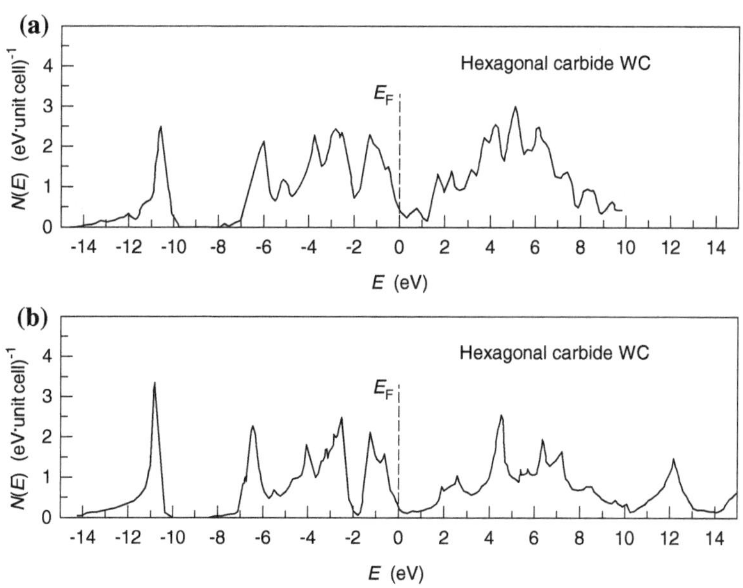

Fig. 2.14 Curves of the total density of electronic states *N*(*E*) for hexagonal (space group) tungsten monocarbide WC. **a** LAPW calculations [66], **b** calculations by the preudopotential method of local orbitals [45]

WC), the 2s(C) and 2p(C)-5d(W) bands of the cubic tungsten carbide are somewhat broadened due to the smaller WC unit cell constant.

The band structure of W_2C carbide is intermediate between the band structure of bcc tungsten metal and hexagonal WC carbide. The main difference in the DOS curves of lower hexagonal W_2C carbide (Fig. 2.15c) and hexagonal and cubic monocarbides WC (see Fig. 2.15a and b) is that W_2C has narrower bands. Hence, the hybridized valence-binding and metal 5d(W) bands have the density of states much higher as compared with monocarbides WC. The density of electronic states at the Fermi level, $N(E_F)$, is also higher for W_2C carbide as compared with hexagonal and cubic WC monocarbides. The $N(E_F)$ values calculated in study [67] for hexagonal and cubic WC monocarbides and also for lower hexagonal carbide W_2C are equal to 0.78, 0.86 and 2.17 eV^{-1}(10.6, 11.7 and 29.5 Ry^{-1}), respectively.

Based on the results of energy band structure calculations, the unit cell constants, the bulk (compression) modulus B (also conventionally designated as K) and the Debye temperature θ_D were calculated for these carbides [67]. According to [67], the bulk modulus of hexagonal WC carbide is 655 GPa (6.55 Mbar) and substantially exceeds $B = 490$ GPa of lower carbide W_2C. For hexagonal WC carbide, the same authors showed a different calculated value of bulk modulus equal to 699 GPa [68].

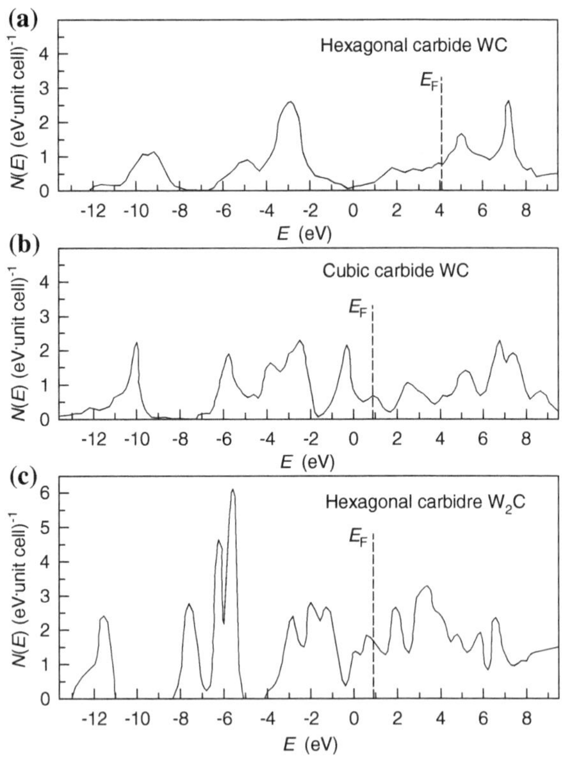

Fig. 2.15 Curves of density of electronic states $N(E)$ of tungsten carbides in the LMTO-ASA approximation [67]. **a** Hexagonal (space group $P\overline{3}m1$) monocarbide WC, **b** cubic (space group, $B1$-type structure) monocarbide WC, **c** lower hexagonal (space group $P6_3/mmc$) carbide W_2C

These bulk modulus values are 60–70 % higher than the value 413 GPa (4.13 Mbar) calculated for the same carbide in study [45]. The lowest bulk modulus equal to 375 GPa, was obtained in calculation [67] for cubic WC carbide; the unit cell constant a_{B1} was calculated to be 7.98 a.u. (0.422 nm). According to [5, 67, 69], the small bulk modulus of cubic tungsten carbide is associated with its metastable state, because the energy of valence electrons in cubic WC carbide is higher than that in hexagonal WC carbide by ~12.2 eV (~0.9 Ry).

As noted above, the measurements of the positron lifetime allowed the vacancies to be revealed in the carbon and metallic sublattices of irradiated hexagonal WC carbide [46]. Taking into account this experimental result, an attempt was undertaken [71] to elucidate how the metal and carbon vacancies affect the WC band structure. The calculations were performed using a first-principles self-consistent nonempirical full-potential linear muffin-tin orbital (FP-LMTO) method. The simulations of stoichiometric and defective carbides involved the use of a supercell formed by eight sites of tungsten sublattice and eight sites of carbon sublattice. These sites are occupied by W and C atoms according to the compositions W_8C_8(WC), W_8C_7 ($WC_{0.875}$) and W_7C_8 ($W_{0.875}C$) of the supercells. The calculated total densities of electronic states for tungsten carbides are shown in Fig. 2.16. According to earlier calculations [45, 66–69], the valence band of the hexagonal stoichiometric monocarbide WC consists of two subbands separated by a forbidden gap. The lower subband *A* is largely formed by states 2*s*(C), while the high-energy subband *B* is formed by hybridized states 2*p*(C)-5*d*(W). The Fermi level ($E_F = 25.3$ eV, $N(E_F) = 0.23$ eV^{-1} per unit cell) is located near the minimum of the DOS curve (Fig. 2.16a). The introduction of a carbon vacancy leads to the appearance of a vacancy peak *C* a little higher than the Fermi level, increase of $N(E_F)$ up to 0.61 eV^{-1} per unit cell and to formation of finer structure of the band *B* (Fig. 2.16b). The presence of a vacancy in the tungsten sublattice also results in both the appearance in the Fermi level vicinity of a small vacancy peak and the increase in $N(E_F)$ up to 0.69 eV^{-1} per unit cell (Fig. 2.16c). The introduction of a carbon or a tungsten vacancy reduces the Fermi energy E_F to 25.0 and 22.0 eV, respectively.

The calculated in study [71] distribution of electronic density in the spheres of the tungsten and carbon vacancies and the charges of these vacancies ($q_{\square C} = 0.67$ *e*, $q_{\square W} = 0.51$ *e*) are in qualitative agreement with the positron lifetime in these vacancies [46]: the charge of an carbon vacancy is higher and the positron lifetime in it is shorter as compared with the vacancy in the tungsten sublattice ($\tau_{\square C} = 136$ps and $\tau_{\square W} = 175$ ps, respectively).

Systematic investigation of the band structure and relative stability of four supposed polymorphous modifications of lower tungsten carbide W_2C (see Sect. 2.1) was carried out by the authors [72]. The band structure of lower W_2C carbides was calculated using the full-potential linearized augmented plane wave (FLAPW) method within the generalized gradient approximation (GGA) of the exchange–correlation potential. Distribution of tungsten and carbon atoms in the crystal lattices of different W_2C modifications, which was used in calculations [72], is described in Sect. 2.1. The optimized structural parameters for all the W_2C modifications according to

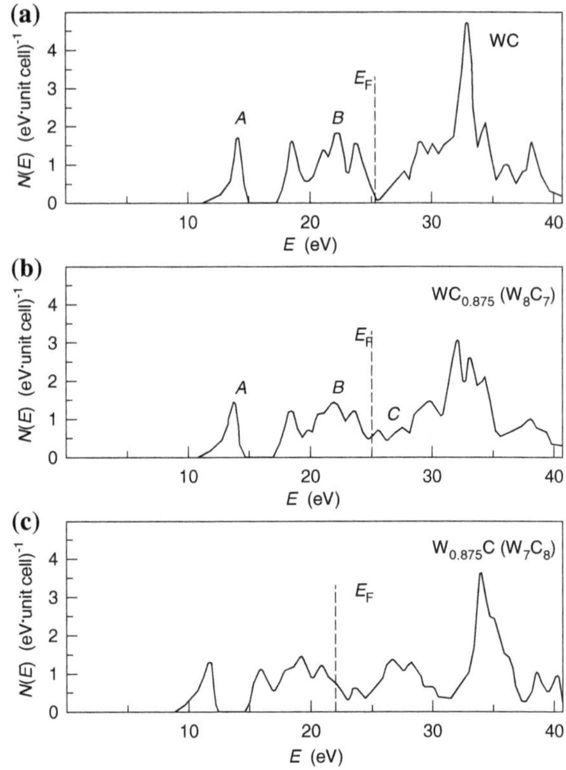

Fig. 2.16 The effect of vacancies on the total density of electronic states $N(E)$ of hexagonal tungsten carbide according to FP-LMTO calculations [71]. **a** Stoichiometric carbide WC, **b** carbide $WC_{0.875}$ (W_8C_7) with carbon vacancies, **c** carbide $W_{0.875}C$ (W_7C_8) with vacancies in the tungsten sublattice

the FLAPW–GGA calculations [72] are in reasonable agreement with the available experimental data (the discrepancy does not exceed 2.5 %).

The relative stability of different W_2C modifications was estimated by their formation energy E_{form}. Energies of formation, E_{form}, of lower W_2C carbides were calculated from the relationship

$$E_{form\ W2C} = E_{tot\ W2C} - 2E_{tot\ W} - E_{tot\ C(gr)} \tag{2.1}$$

which corresponds to the reaction 2W + C(gr) → W_2C for the formation of lower W_2C carbide from body-centered cubic tungsten metal and carbon in the form of graphite C(gr). The quantities $E_{tot\ W2C}$, $E_{tot\ W}$, and $E_{tot\ C(gr)}$ in (2.1) are the total energies of the lower tungsten carbide, tungsten metal, and graphite, respectively, according to the self-consistent FLAPW-GGA calculations. In this approach, negative energies E_{form} suggest that the formation of the lower tungsten carbide from a mechanical mixture of the initial reactants is energetically favorable. By contrast, at $E_{form} > 0$, the corresponding W_2C modification should be attributed to unstable systems. Besides, the energies of formation of hexagonal δ-WC and cubic γ-WC_{1-x} (with $x = 0$) monocarbides were calculated in work [72].

According to calculations [72], the ε-W_2C and β'-W_2C (β-W_2C) are the most stable modifications of lower tungsten carbide with close energies of formation -0.04 and -0.02 eV per formula unit, respectively. The β''-W_2C (α-W_2C) modification with *C*6-type crystal structure has the positive energy of formation $E_{form} = +\ 0.19$ eV per formula unit and is unstable. The disordered hexagonal (space group $P6_3/mmc$) lower tungsten carbide β-W_2C (γ-W_2C) has E_{form} ~0 and, hence, is metastable at low temperatures. Actually, according to the data [43, 51], the disordered carbide β-W_2C is in a thermodynamic equilibrium at temperatures from ~2300 K to the melting temperature and, at temperatures below 2000–2300 K, undergoes sequential disorder–order and order–order transformations with the formation of the ordered ε-W_2C and β'-W_2C phases. Thus, the most energetically favorable modifications of the lower carbide W_2C correspond to the variants of the ordering of carbon atoms and vacancies in trigonal (space group) phase ε-W_2C and orthorhombic (space group *Pbcn*) phase β'-W_2C. The rhombohedral β''-W_2C (α-W_2C) modification is characterized by a maximum instability and has a crystal structure in which the layer of carbon atoms regularly alternate with the layer of structural vacancies, being oriented perpendicularly to the *c* axis.

As follows from the data [72], the δ-WC modification with the energy of formation $E_{form} = -0.339$ eV per formula unit is the most stable tungsten carbide, whereas the cubic modification γ-$WC_{1.0}$ of the monocarbide has a positive energy formation of +0.562 eV per formula unit and is most unstable. The results obtained allow us to arrange the polymorphous modifications of the higher and lower tungsten carbides in order of decreasing the stability as follows: δ-WC > ε-W_2C > β'-W_2C > β-W_2C > β''-W_2C > γ-$WC_{1.0}$. The first three carbides can be considered to be stable ($E_{form} <$ 0), disordered β-W_2C carbide can be treated as metastable ($E_{form} \sim 0$), whereas β''-W_2C and γ-$WC_{1.0}$ modifications appear to be unstable ($E_{form} > 0$).

According to [72], the specific feature of the electron energy spectra of all W_2C modifications (Fig. 2.17) is a group of low-energy bands that are located below the Fermi level E_F and formed predominantly by the C 2*s* states. These C 2*s* quasi-core bands are separated from the group of occupied bands that are adjacent to the Fermi level and composed predominantly of 2*p*(C) and 5*d*(W) states. The upper occupied and lower vacant bands are predominantly formed by the 5*d*(W) states with insignificant contributions of the 2*p*(C) states. The Fermi level is located in the range of the 5*d*(W)-like bands that corresponds to metallic conductivity of all modifications of lower W_2C tungsten carbide. The maximum density of states on the Fermi level, $N(E_F)$, is observed for the most stable trigonal (space group $P\overline{3}1m$) ε-W_2C modification.

According to the results of calculations [45, 65–69, 71, 72], the band structure of studied tungsten carbides is basically the same and contains three main bands, namely, the low-energy 2*s*(C) band, the ground-state hybridized valence-binding band 2*p*(C)-5*d*(W) and the high-energy band formed largely by the 5*d* states of tungsten.

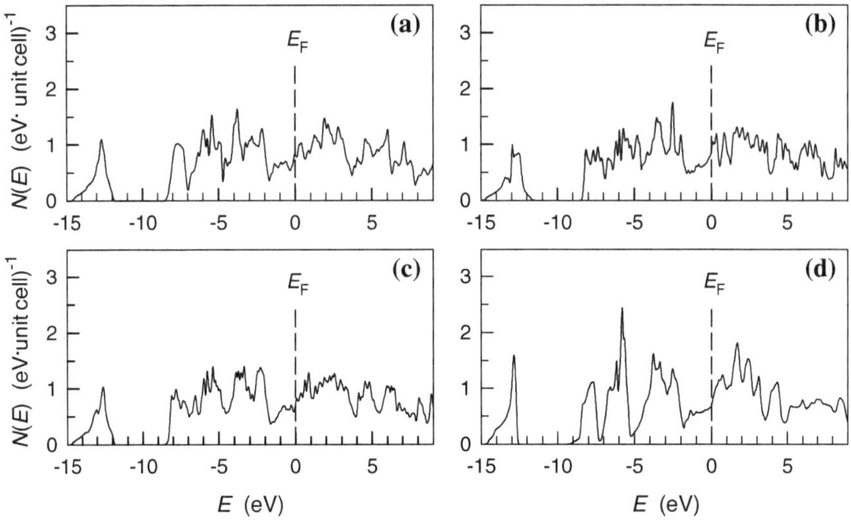

Fig. 2.17 The total densities of electronic states N(*E*) of polymorphous modifications of lower W_2C tungsten carbide [72]. **a** Disordered hexagonal (space group *P*6$_3$/*mmc*) carbide β-W_2C, **b** trigonal (space group) carbide ε-W_2C, **c** orthorhombic (space group *Pbcn*) carbide β'-W_2C, **d** trigonal (space group) carbide β''-W_2C with *C*6-type crystal structure

2.4 Methods of Preparation of Tungsten Carbide and WC–Co Hardmetals with Different Grain Sizes

Polycrystalline metal carbides are usually synthesized by the powder metallurgy methods, which are described in detail in monographs [3, 4]. Different methods of preparation of carbides are also described in works [73–75]. The most popular methods are those that involve the direct synthesis from a metal and carbon or from the compounds containing the metal (for example, metal oxides or chlorides) and carbon (e.g., a hydrocarbon or CO).

A conventional method of the synthesis of hexagonal carbide WC involves heating of tungstic acid H_2WO_4 or ammonium tungstate $(NH_4)_2WO_4$ in a mixture of hydrogen H_2 and methane CH_4 at 1020–1270 K. When tungstic acid is used, the metal is completely reduced even before the carburization. The carbothermal reduction of WO_3 by carbon in the inert or reductive atmosphere proceeds in a similar way. Tungsten carbide synthesized by carbothermal reduction usually contains traces of oxygen and oxycarbides. Therefore, synthesis of a pure tungsten carbide requires that the carbothermal reduction product was annealed in addition in vacuum at a temperature not lower than 1800 K.

The purest hexagonal WC carbide is synthesized by the solid-phase vacuum sintering of tungsten metal and carbon (carbon black) at a temperature not lower than 1400 K. Hexagonal carbide W_2C is synthesized by the solid-phase vacuum sintering of tungsten W and carbon C powders at 850–900 K. However, the solid-state carburization at this low temperature proceeds only in the presence of hydrogen or a gaseous hydrocarbon; moreover, the final product contains a relatively high amount of free (uncombined) carbon. Synthesis of a single-phase WC powder with the grain size of 5–6 μm from the tungsten metal powder in a mixture of hydrogen with methane CH_4 is described [76]. At temperature 2300–2500 K and a gas mixture with 2 vol.% methane, WC with the free carbon content less than 0.1 wt.% could be obtained. High temperature (up to ~2400 K) is required for synthesis of homogeneous tungsten carbide if the solid-state sintering is carried out in vacuum, although the carburization starts at 1400 K.

Tungsten carbides can also be synthesized using such methods as deposition from the gas phase in the reaction of $W(CO)_6$ or tungsten chloride with a mixture of a hydrocarbon and hydrogen, thermal decomposition of various precursors [77], self-propagating high-temperature synthesis and mechanochemical synthesis [78]. For example [79], a mixture of carbon black and tungsten powder with the grain size of up to 10 μm was ground in an eccentric-type mill. During the grinding, the average load in the ball impact zone was 10^8–10^9 Pa and the quantity of mechanical energy absorbed by the powder during the grinding reached 30 $kJ{\cdot}g^{-1}$. As a result of such treatment, lower hexagonal W_2C tungsten carbide was obtained. Mechanochemical synthesis of carbides from tungsten and carbon elements in a planetary ball mill with the use of an inert copper matrix was described [80]; according to the authors estimates, the average size of W_2C and WC particles formed did not exceed 0.5 μm.

We have already mentioned that hexagonal WC carbide is not the hardest and most refractory among carbides. Its melting point ~3060 K is lower as compared with melting points of TiC, ZrC, HfC, NbC, and TaC carbides. According to different sources, at 300 K the hardness of WC carbide is equal to 18–22 GPa and is less than the hardness of nonstoichiometric TiC_y, ZrC_y, HfC_y, VC_y, NbC_y, and TaC_y carbides at the same temperature [3, 4]. However, the hardness of WC is sufficiently stable. For example, the microhardness H_V of hexagonal WC carbide decreased from ~18 to ~12 GPa at heating from room temperature to 1300 K [81] whereas H_V of other carbides decreased under the same conditions from their maximum values to 3–8 GPa (Fig. 2.18).

According to [82], the temperature increase from 300 to 1100 K is accompanied by a more pronounced decrease (approximately three-fold) in the WC hardness; however, the hardness of other carbides undergoes a still sharper decrease (approximately five-fold) in the same temperature range. The sharp decrease in the WC hardness with the increase in the temperature follows from the experimental results [83] which substantially differ from those given by other authors [81, 82]. The same study [83] also shows the hardness value for vanadium monocarbide at ~900–1500 K, which is lower as compared with that found in other literature sources. Apparently, the value of WC hardness measured in the study cited [83] is severely underestimated. Moreover, as compared with other carbides, WC carbide has Young modulus of elasticity (~700 GPa [84]) higher by a factor of 1.5–2.0 and a thermal expansion coefficient (5–$6 \times 10^{-6} K^{-1}$ [84]) lower by a factor of 1.5.

The mentioned properties of WC carbide and their thermal stability during the heating up to 900–1000 K determined its application as the basis of wear-resistant

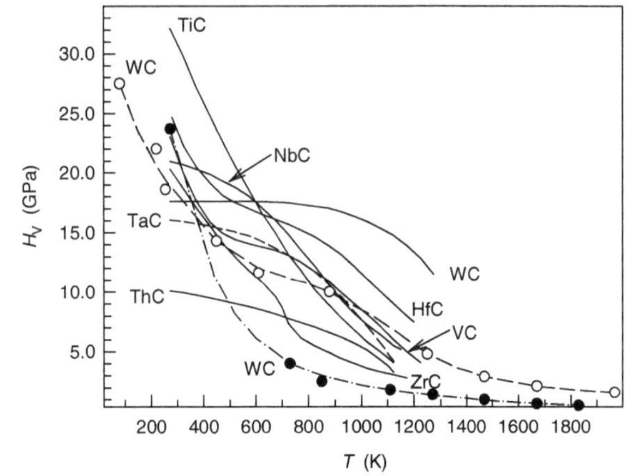

Fig. 2.18 The effect of temperature on the Vickers hardness of hexagonal WC tungsten carbide and carbides of other transition metals [81]. For comparison, the known $H_V(T)$ dependences of monocarbide WC are shown: (○) [82] and (●) [83]. The value of WC hardness measured in the study [83] is severely underestimated

hardmetals that are used in metal-working and drilling operations. The direct use of WC carbide for these purposes is impossible because of its brittleness and ability to sintering only at very high temperatures. Hence, for a given carbide to be applied in practice, it should be imparted certain viscosity without loss in hardness. This is achieved by using a metal binder together with WC that provides the plasticity of the composition material and, at the same time, has sufficiently high strength, melting point and impact elasticity. The choice of a metal for the binder is restricted by the following conditions: this metal does not form carbides and this metal in molten state should adequately wet the carbide which is a base of hardmetal. Moreover, a partial solubility of WC carbide in the binding metal is required. This provides strong adhesion between carbide and binding metal. In the case of tungsten carbide WC, best of all these requirements are met by cobalt Co. That is why the first hardmetals that have found wide use in the industry were synthesized in the WC–Co system.

The brief analysis of the data published from 1918 to 1923 on the development of hardmetals based on tungsten carbide in which such metals as Fe, Ni and Co were used as the binder was given in review [15]. The author of this review, K. Schrüter, in 1925 patented the WC–M (M=Co, Ni, Fe) hardmetals in the USA with the priority from 1923. Later, similar hardmetals were developed in Sweden, Austria and other countries. In the Soviet Union, hardmetals based on tungsten carbide were manufactured under a trademark "Pobedit".

2.4.1 Hardmetals WC–Co

According to the phase diagram of the ternary W–Co–C system (see Fig. 2.12), at the low carbon content, the ternary compounds of the η-carbide type or the solid solution $Co_3W_3C_y$ based on η-carbides Co_4W_2C and Co_2W_4C can be formed. Further decrease of carbon content in the $Co_3W_3C_y$ solid solution leads to the appearance of carbide $\sim Co_6W_6C_y$ on the phase diagram of the ternary system. These ternary compounds and solid solutions exhibit enhanced brittleness. Therefore the WC–Co cross-section region where no ternary compounds can be formed is used actually for the synthesis of hardmetals. In commercial WC–Co hardmetals, the cobalt content varies from 2 to 18 wt.% depending on their application.

The most important traditional operation in the technology of hardmetals is the sintering of green samples pressed from a powder mixture with an addition of a plasticizer [85, 86]. Hardmetals are sintered at a temperature 1600–1800 K. During the sintering of a WC–Co hardmetal, a certain amount of liquid phase, namely, molten cobalt, is formed. The wetting angle of WC with cobalt is zero, i.e., molten cobalt completely wets and partly dissolves the tungsten carbide grains with their subsequent crystallization under cooling. The sintering of a hardmetal includes the following stages which take place at increasing temperature: (1) at 470–670 K, the plasticizer is removed; (2) at 670–970 K, the admixtures of tungsten and cobalt oxides are reduced; (3) at 1050–1400 K, individual carbide grains are welded together in their contact points; (4) at 1400–1570 K, a solid solution of tungsten carbide in cobalt is formed, the green sample shrinks and all cobalt passes to the luquid phase; (5) at 1550–1670 K, due to the migration of carbide grains in the liquid phase, the densification of sample comes to end and the sample acquires its final compact shape. Further increase of temperature up to 1770–1800 K is accompanied by additional dissolution of the WC grains in cobalt, on the one hand, and the growth of carbide grains due to the recrystallization via the liquid phase, on the other hand; simultaneously, complete degassing of the hardmetal occurs. After the exposure at the maximum sintering temperature, hardmetal is cooled. During the cooling, the dissolved tungsten carbide deposits from the liquid phase onto the surface of carbide grains, which results in their growth.

In the ideal case, the sintered WC–Co hardmetal is two-phase and contains only WC grains and the metal binder between them. At a deficit or an excess of carbon, the alloy becomes three-phase and contains additionally either $Co_3W_3C_y$ η-phase (at carbon deficit) or graphite inclusions (at carbon excess). The graphite inclusions as well as the presence of very coarse (several times larger than the average size) WC grains substantially reduce the hardmetals service life [87].

After sintering, the average size of carbide phase grains increases noticeably as compared with the particle size in the original WC powder. Fine-grained hardmetals exhibit better performance and the use of fine-grained mixtures for the sintering allows producing the higher quality hardmetals. All other conditions being equal, the grain size in a hardmetal decreases with the decrease of the cobalt content due to the decreasing the liquid phase content and the deceleration of the grain growth by recrystallization via the liquid phase. The grain size also decreases at a lowered content of carbon, which, however, leads to the appearance of the brittle η-phase. Deceleration of the growth of carbide-phase grains in WC–Co hardmetals is achieved by their doping with other carbides. The use of grain growth inhibitors is widely practiced for sintering fine-grained WC–Co hardmetals in the industry. Commonly used inhibitors include vanadium, tantalum, and chromium carbides. Their relative effectiveness can be ranked in the order VC > Cr_3C_2 > NbC > TaC, although consideration of such hardmetal properties as corrosion resistance and hot hardness may affect the choice of grain growth inhibitors. In the last years, the first steps for using the rare earth elements and their compounds as a grain growth inhibitors are made.

The most efficient dopant in this respect is vanadium carbide. Introduction of not more than 0.3 wt.% of vanadium carbide helps to the production of fine-grained hardmetals with the average grain size of the carbide phase about $\sim 1\ \mu m$. To obtain fine-grained hardmetals, the prolonged (up to 48 h) grinding of WC and Co powders in ethanol is used at the stage of mixture preparation [88].

Over the past 20 years, the methods of synthesis of various substances and materials in the nanocrystalline state are actively developed. The size of particles (grain, crystallites) in nanocrystalline substances does not exceed 40 nm [89–92]. Substances with particles measuring from 40 to 100 nm are called submicrocrystalline [91–93]. Different methods of the synthesis of nanocrystalline powders and compact nanomaterials, the peculiarities of their microstructures, the methods of certification of nanomaterials, the effect of the particle (grain) size on the physical properties of substances are described in detail in monographs [90–92]. The nanoscale effects which are observed in the structure and properties of metals, compounds and materials were reviewed [94, 95]. Application of carbides in nanocrystalline state is very perspective.

2.4.2 Nanostructured Hardmetals

The most promising way of the production of nanostructured hardmetals with the reduced sintering temperature and the enhanced (due to the superfine-grained structure) hardness, strength, fracture toughness (crack-resistance) is to use nano- and submicrocrystalline carbides as the components of these alloys [96, 97]. The decrease in the WC particle size even in the micrometer scale is accompanied by a noticeable increase in the wear resistance of WC–Co hardmetals as well as the increase in their hardness and the compression and bending strength (see Table 2.3 [98] and Fig. 2.19 [99, 100]). The only exclusion known is the decrease in the bending strength of WC–6 wt.% Co hardmetals with the decrease in the grain size [98] (see Table 2.3).

The results of other studies demonstrate the only improvement of the mechanical properties with the transition from coarse-grained to fine-grained WC–Co hardmetals. According to [96], upon sintering of WC-15 wt.% Co nanopowders with the average particle size of 970, 490 and 30 nm at temperature 1690 K the resulting hardmetals with the average grain size of 1350, 950 and 620 nm exhibited hardness of 13.8, 14.7 and 17.0 GPa, respectively. A hardmetal WC-10 wt.% Co with the carbide phase of superfine structure (the WC grain size is smaller than $0.5\ \mu m$) exhibits microhardness of 19.3 GPa and a bending strength of 3890 MPa, whereas the same characteristics of a similar alloy with the grain size of $0.5–1.0\ \mu m$) are smaller and equal to 17.8 GPa and 3780 MPa [101]. Hardmetals WC–6 wt.% Co manufactured by 'Krupp Widia GmbH', which contain carbide-phase grains measuring 1.4–1.8, 0.8–0.9 and 0.5–0.6

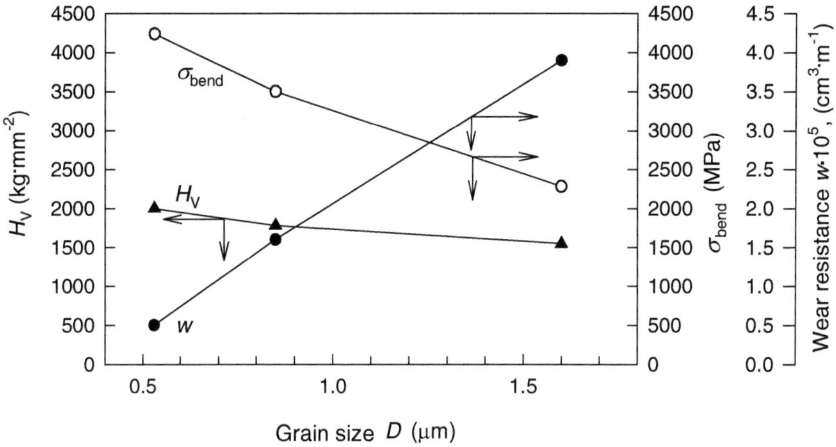

Fig. 2.19 Bending strength σ_{bend}, wear resistance w and hardness H_V as a function of the grain size D of the carbide phase of WC-6 wt.% Co hardmetals. It is constructed in [100] on a base of data [99]

μm), exhibit the bending strength of 2000, 3000 and 3800 MPa, the hardness of 15.8, 18.0 and 20.4 GPa, the abrasive wear resistance of 1.26 2.08 and 2.78 %, respectively. A similar improvement in the mechanical properties with the decrease in the carbide-phase grain size was observed for hardmetals with large (up to 15 wt.%) cobalt content. The strength characteristics of conventional (with micrometer-scale grains) hardmetals based on WC carbide are discussed in a monograph![87].

Table 2.3 The effect of the WC grain size on certain properties of WC–Co hardmetals [98]

Properties	Composition of hardmetals (wt.%)					
	97WC-3Co	94WC-6Co	94WC-6Co	94WC-6Co	90WC-10Co	90WC-10Co
Grain size (μm)	1.0–2.0	0.5–1.0	1.0–2.0	2.0–5.0	0.5–1.0	2.0–5.0
Rockwell hardness (HRA)	92.5–93.2	92.5–93.1	91.7–92.2	90.5–91.5	90.7–91.3	87.4–88.2
Density (g·cm^{-3})	15.3	15.0	15.0	15.0	14.6	14.5
Bending strength (MPa)	1590	1790	2000	2210	3100	2760
Compression strength (MPa)	5860	5930	5450	5170	5170	4000
Young modulus of elasticity (GPa)	641	614	648	641	620	552
Relative wear resistance	100	100	58	25	22	7

Let us consider the main methods for production of tungsten carbide and WC–Co hardmetals in nanocrystalline or submicrocrystalline states.

Plasma-chemical synthesis is among the most widely used chemical methods for producing fine-grained powders of nitrides and carbides [102, 103]. The main conditions for the production of fine-grained powders by this method are as follows: the reaction should proceed far from its equilibrium, the new-phase nucleation rate should be high whereas the nuclei growth rate should be small. Under real conditions of the plasma-chemical synthesis, nanoparticles are produced upon fast cooling of the plasma flow in which the gas phase is condensed. This reduces the sizes of the particles formed and suppresses their coalescence upon collisions. The plasma-chemical synthesis of carbides makes use of the low-temperature (3000–8000 K) hydrocarbon and argon plasma of the arc, glowing arc, high-frequency and microwave discharges. Elements and their halides and other compounds are used as the starting substances in plasma-chemical synthesis [89]. The characteristics of the powders synthesized depend on the starting materials, the technology of synthesis and the type of reactor. The powder particles obtained in plasma-chemical synthesis represent single crystals, have regular shapes and the sizes from 10 to 100–200 nm (and higher).

The plasma-chemical synthesis is characterized by high rates of formation and condensation of compounds and sufficiently high production capacity. The major disadvantages of this method are as follows: the wide size distribution of particles and, as a consequence, the presence of sufficiently coarse (up to 1–5 μm) particles, which means that the selectivity of the process is low, and also high content of admixtures in the synthesized powder. To date, fine-grained powders of titanium, niobium, tantalum and tungsten carbides have been synthesized by the plasma-chemical method [89, 103–106]. For instance, the synthesis of a submicrocrystalline WC powder with a specific surface area of 15 $m^2 \cdot g^{-1}$ and a particle size not exceeding 300 nm, which was conducted in an argon-hydrogen plasma from methane CH_4 and tungsten hexafluoride WF_6, was reported [106]. Note that WC carbide has a density $\rho = 15.67$ $g \cdot cm^{-3}$ therefore average particle size (diameter) $D = 6/\rho S_{sp}$ in the WC powder with a specific surface $S_{sp} = 15\ m^2 \cdot g^{-1}$ is equal to ~30 nm. This particle size is 10 times smaller than a particle size reported in study [106].

Synthesis of fine-grained tungsten carbide powders with the use of an arc plasmatron is described in [107]. The plasma-chemical synthesis produced a material of the W–C–O system with a definite carbon-to-oxygen ratio, while the synthesis of the single-phase carbide as such was carried out in a pipe furnace by the reduction of this material with hydrogen and CO. The specific surface area of synthesized WC carbide varied from 5–6 to 50 $m^2 \cdot g^{-1}$. To produce the WC–Co furnace mixture, a cobalt powder with a specific surface area of 8–10 $m^2 \cdot g^{-1}$, which was synthesized by plasma-chemical reduction of the oxide, was used. Sintering of the furnace mixture was carried out by hot pressing at temperature 1370–1670 K and at compacting pressure 10–50 MPa. The WC grain size in the samples sintered in the solid phase was ~100 nm.

A design of an electro-arc plasmatron for the plasma-chemical synthesis of nanorystalline powders of refractory compounds including tungsten carbide with the particle sizes from 10 to 100 nm was reported [108]. Hydrogen plasma with the temperature of 3000–4000 K is used in this plasmatron.

Low-temperature gas-phase synthesis of WC nanopowders is described [109, 110]. A two-step process, was originally proposed [109] which included reduction of tungsten W nanopowder from tungstic acid salts, i.e., the reductive decomposition of the precursor, followed by carburization of tungsten in a CO-containing gas mixture. The synthesis produced a WC powder with the average grain size of 8 nm when conducted at 900–950 K, and metastable lower W_2C carbide with particle measuring 13–19 nm when conducted at 600–800 K. A more straightforward one-step synthesis of the nanocrystalline WC carbide powder was described [110]. An H_2/CO gas mixture (2 : 1 or 1 : 1 molar ratios) was passed through ammonium tungstate $(NH_4)_2WO_4$ or tungsten oxide WO_3 powder at 800–1000 K; this directly led to the formation of nanocrystalline WC carbide with particle size <10 nm. One important parameter of the single-step synthesis is the heating rate, which should provide the kinetic balance between the reduction and carburization reactions. According to [110], the heating rate of 2 K·min^{-1} is close to the optimum and heating even up to 1000 K resulted merely in a minor coarsening of WC particles as compared with heating to 800 K.

Mechanochemical synthesis in an eccentric-type mill produced hexagonal W_2C carbide with the average particle size of 10 nm from a powder mixture of tungsten with the average particle size of up to 10 mm and carbon [79].

To preserve the positive effect achieved due to the use of nanocrystalline powders, the hardmetals discussed should retain the small sizes of WC particles during their sintering. In studies [111–113], authors used the X-ray diffraction and magnetic susceptibility measurements to follow the changes in the thermal stability of a WC nanopowder that was synthesized by the plasma-chemical method from tungsten oxide WO_3 and propane C_3H_8 at a temperature of about 3300–3800 K in hydrogen plasma, followed by annealing in an argon atmosphere at a temperature of 1100–1300 K. The average particle size in the nanopowder was 50–55 nm. According to studies [111–113], the particles in this nanocrystalline powder retained the stable size when annealed in the temperature range from 300 to 1200 K, while the relaxation of microstrains began at ~600 K and ended at ~900 K.

The fine-grained structure of WC–Co hardmetals synthesized from WC powders is determined not only by the small size of the furnace mixture particles, but also by the way their sintering proceeds. Upon formation of the liquid phase, first, the smallest particles and, then, medium size particles pass to the melt, while more coarse particles survive. Insofar as most nanoparticles are of the medium size, the number of crystallization centers in the hardmetal will be sufficiently high. In the recrystallization, tungsten carbide deposits on the surface of non-dissolved WC particles. Due to the excessive surface energy of small particles, they acquire a thicker layer of deposited tungsten carbide as compared with coarse particles. As a result, the size of carbide-phase particles levels off. In an ordinary (conventional) hardmetal, the number of crystallization centers is substantially smaller than in a hardmetal synthesized from the WC nanopowder. Hence, this hardmetal would have the more fine-grained and uniform structure as compared with an ordinary hardmetal. However, it should be borne in mind that the growth of grains in nanocrystalline WC–Co composite can occur at temperature ~1070 K [114], i.e., below the eutectic temperatureof the W–Co–C system, which is equal to ~1600 K. The nanopowders WC-15 wt.% Co with the average particle size of 30 nm were studied in [114], the liquid phase appeared at ~1420 K during their sintering. This means that, during the liquid-phase sintering of WC–Co nanopowders, the initial growth of carbide-phase grains occurs in the range of solid-phase sintering temperatures, whereas the subsequent liquid-phase sintering only leads to the coarsening of WC grains [115].

Also note that the most compact (non-porous) hardmetals can be obtained provided the furnace mixture is not monodispersed, because obviously the total volume of a sample cannot be filled without voids with unisized particle. More compact packing is achieved due to the presence of different-sized particles in the furnace mixture, where the finer particles occupy the voids between the coarser particles. Analysis of the three-dimensional packing of spherical particles has shown [116] that a very high degree of the free space filling (~0.9) can be reached for their bimodal size distribution where, along with medium-size particles D_{aver} that have the maximum share, the fraction of particles with a diameter $\sim 0.1 D_{aver}$ is fairly high. In other words, the size-distribution curve of the particles should possess two maxima, the first of which (abundance) corresponds to the particles with the medium size D_{aver}, while the second corresponds to the particles with the size $\sim 0.1 D_{aver}$.

Analysis of the literature data has shown that nanocrystalline or submicrocrystalline tungsten carbide can be synthesized by different methods. However, for the production of nanostructured WC–Co hardmetals, recommended better that not only tungsten carbide, but the whole powder mixture of WC and Co was in the nanocrystalline state.

Such a mixture of WC and Co can be obtained by thermolysis of various precursors [77]. Particularly, heating can induce the following decomposition reactions resulting in the formation of WC and Co metal:

$$(NH_4)_8[H_2Co_2W_{11}O_{40}]R \rightarrow WC + Co, \tag{2.2}$$

$$[CoEn]_3WO_4 \rightarrow WC + Co, \tag{2.3}$$

where R and En are alkyl and ethylenediamine, respectively.

Modern technologies of the production of nanocomposite WC–Co mixtures include several consecutive operations, namely, deposition from colloid solutions, cryogenic sputtering and drying, carbothemal reduction. Thus an efficient method for the production of nanocrystalline mixtures of tungsten carbide and cobalt for manufacturing hardmetals is based on the co-deposition of WC and Co precursors from colloid solutions [117, 118]. According to this method, the colloid solutions of tungsten and cobalt salts (e.g., $(NH_4)_6(H_2W_{12}O_{40})\cdot 4H_2O$, $CoCl_2$, $Co(NO_3)_2$, $Co(CH_3COO)_2$) are dried by sputtering, then the obtained powder is subjected to the low-temperature carbothermal reduction in the suspension layer and thus retains its high-dispersion state. This technology was patented [119] as the "Spray Conversion Process (SCP)". Insofar as the synthesis starts in the solution, the components W and

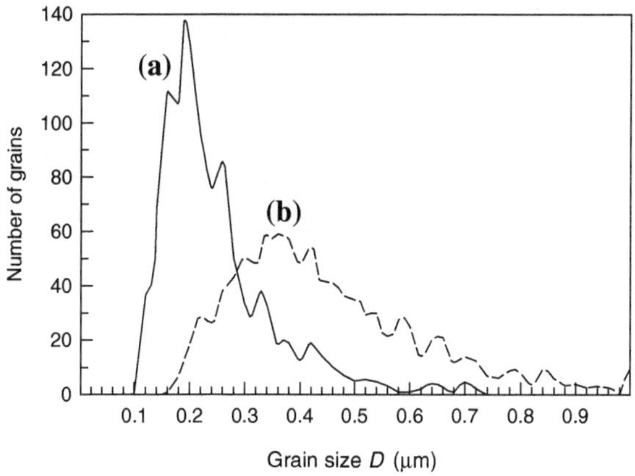

Fig. 2.20 Size distribution of grains in the samples of WC-6 wt.% Co hardmetal [121]. **a** Alloy sintered from a nanocomposite mixture of WC, Co and 0.83 wt.% of VC, which was obtained by the co-deposition from colloid solution followed by drying and carbothermal reduction; the sample contained 1000 grains. **b** Conventional fine-grained WC-6 wt.% Co hardmetal with 0.32 wt.% of VC; the studied sample contained 948 grains

Co are mixed on the molecular level. After carburization of produced WC–Co mixture, obtained carbide powder is amorphous. The growth of grains and suppression of the WC solubility in cobalt can be inhibited by adding, up to 1 wt.% of non-stoichiometric vanadium carbide to the mixture. The hardmetal obtained from this nanocrystalline mixture demonstrates the optimum combination of high hardness and high strength. For example, its Vichers hardness reached 21.5 GPa, whereas the maximum hardness of a hardmetal with the same composition, which was obtained from a conventional WC powder, did not exceed 19.5 GPa [117–120]. The increase in the hardness and the decrease in WC grain size correlate with the increase in the vanadium-carbide content in the alloy. Figure 2.20 illustrates the size distribution of grains in the samples of the WC-6 wt.% Co hardmetal obtained from a nanocomposite mixture prepared by the deposition method described above and containing 0.83 wt.% of vanadium carbide, and in a sintered sample of a conventional alloy WC-6 wt.% Co containing 0.32 wt.% of VC. As is seen, the average size of WC grains in the hardmetal synthesized from the amorphous nanocomposite powder is 250 nm, whereas a conventional hardmetal has the grains almost twice larger (450 nm).

It was shown in studies [121, 122] that each nanocomposite WC–Co particle with a size of ~75 mm consists of several millions of nanocrystalline WC grains measuring <50 nm, which are distributed throughout the cobalt matrix (Fig. 2.21).

Sintering of a nanocomposite tungsten carbide mixture with 6.8 wt.% of Co and 1 wt.% of vanadium carbide VC produced the hardmetals which contained 60 % of the WC grains measuring <250 nm and 20 % measuring <170 nm. A structure with still finer grains was obtained for a hardmetal which, in addition to tungsten carbide, contained 9.4 wt.% of Co, 0.8 wt.% of Cr_3C_2 and 0.4 wt.% of VC. Upon sintering at temperature 1670 K, this hardmetal contained 60 % of the tungsten carbide grains with a size smaller than 140 nm and 20 % of the WC grains with a size less 80 nm.

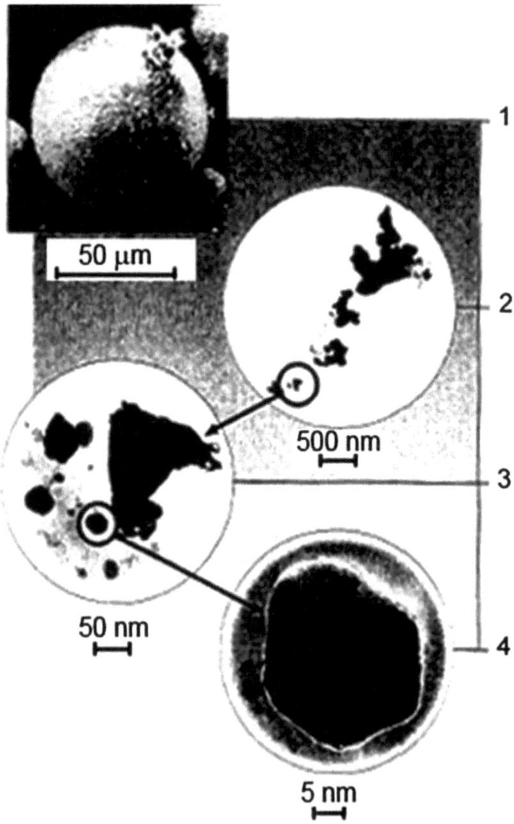

Fig. 2.21 Distribution of nanoscaled WC grains in coarse particles of the amorphous nanocomposite mixture WC-6 wt.% Co with 0.83 wt.% of VC, which was obtained by the deposition from a colloid solution followed by drying and carbothermal reduction [121, 122]. (*1*) A particle of a composite powder WC–Co measuring 75 μm, (*2*) particle fragments after grinding of the powder, (*3*) a polycrystalline cluster formed by WC grains in a cobalt matrix, (*4*) a WC grain measuring 30 nm

A comparison of the nanostructured hardmetal with a conventional polycrystalline hardmetal, having equal hardness, have shown [122] that the crack resistance of the nanostructured hardmetal is higher by a factor of 1.2–1.4 as compared with conventional coarse-grained hardmetal.

The described nanocrystalline WC–Co composite powder with the carbide grain size of 20–40 nm is manufactured on the commercial scale in the USA as NANOCARB™. For the deposition of protective coatings by the thermal sputtering method, the nanocomposite multimodal (containing several fractions of carbide grains, viz., from 8 to 50 nm) WC–Co powder NANOMYTE™ is used.

Synthesis of a one-phase nanocrystalline WC powder by grinding of tungsten and carbon in a ball mill at room temperature for 2.95×10^5 s is described in study [123]. The increase of the grinding time to 4.32×10^5 s afforded a nanopowder with the average grain size <5 nm. Then different amounts of cobalt were added to the WC nanopowder, and grinding was continued for 2.6×10^5 s. The resulting nanocomposite mixture WC–Co contained different amounts of cobalt. The particles of this mixture had the average size less than 5 nm and represented WC carbide covered with a cobalt layer. After pressing and hot sintering of nanocomposite mixture, the hardmetal with carbide phase grains with the average size of ~100 nm was obtained. The hardness and the elasticity modulus of the hardmetals, which contained from 5 wt.% to 12 wt.% of Co, ranged within 20–12 GPa and 740–580 GPa, respectively.

It was already mentioned that the initial and most active growth of carbide phase grains occurs in the sintering of WC–Co nanopowders in the temperature range of solid-state sintering. To decelerate the growth of grains in this temperature region, it was proposed [124] to synthesize the nanostructured hardmetals WC–Co in an electric arc discharge. In this case, the stage of compacting of the nanopowder mixture is combined with its fast sintering and subsequent cooling of the sintered sample. The heating and cooling rates in this method are sufficiently high to avoid the sintering in the solid state and prevent the grain growth during the cooling. Thus a hardmetal was obtained [124] from a nanopowder mixture of tungsten carbide with 10 wt.% of Co prepared by the SCP technology (see Refs. [117–119]) and had, on average, 50-nm coherent-scattering regions (according to XRD data) or the average particle size of 200 nm (according to the data from scanning electron microscopy). The compacting pressure reached 300 MPa. When a current pulse was passed for several microseconds through a sample with the mass of 2.5 g, the energy about 3 $kJ{\cdot}g^{-1}$ was liberated, which led to sintering. As a result, a hardmetal was obtained that had a density equal to 95 % of its theoretical value and contained only WC and a fcc solid solution of tungsten carbide in cobalt. The average grain size in the carbide phase of this hardmetal was 120 nm according to data from electron microscopy; the measured modulus of elasticity was 550 GPa [124].

According to [125], a hardmetal WC–Co synthesized from a tungsten carbide powder with particle size of 30–50 nm had a more uniform fine-grained structure and exhibited higher hardness and strength as compared with conventional hardmetal of the same composition. It was mentioned [125] that the introduction of 3–5 wt.% of tungsten carbide nanopowder to the standard furnace mixture containing coarse (1–4 μm) WC grains levels off the deviations in the hardness and strength of the resulting WC–Co hardmetal, i.e., stabilizes the hard-alloy properties.

A virtually ideal version of mechanochemial synthesis, which is coupled with the synthesis of a nanocomposite mixture, was described [126]. A mixture of coarse (~75 mm) tungsten, graphite and cobalt powders was ground in an argon-filled ball mill for 100 h; the resulting nanocomposite WC–Co mixture contained cobalt and tungsten-carbide grains with the average diameter of 11–12 nm. In the hardmetal, which was obtained by cold pressing of the mixture followed by its sintering at 1310 K, most WC grain had a size less than 200 nm, i.e., were several times smaller as compared with conventional alloys of the same composition. The sintered samples of hardmetal demonstrated the hardness of ~18 GPa and the relative density equal to 80 % of the theoretical value. The possibility of sintering of a nanocomposite WC–Co powder mixture at lower temperatures as compared with coarse-grained mixtures is determined by the lower melting point of nanocrystalline cobalt as compared with coarse-grained cobalt.

Bearing in mind that even a small amount of nonstoichiometric vanadium carbide inhibits the growth of the carbide phase grains during the sintering, it was proposed in study [127] to synthesize V_8C_7 + WC powder mixtures as the first stage of the

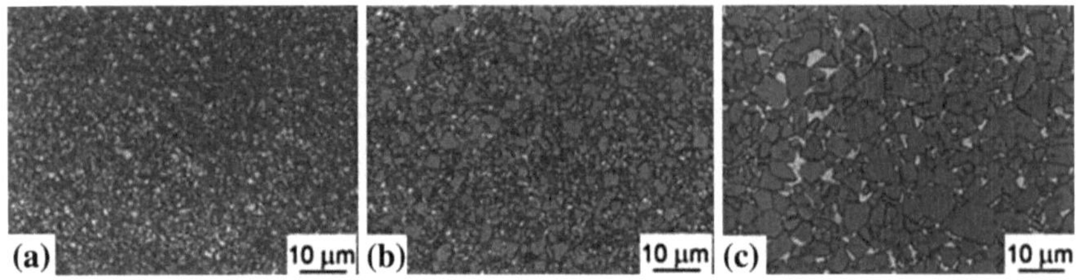

Fig. 2.22 A typical microstructure of WC-6 wt.% Co hardmetals with the average size *D* of carbide phase grains of 0.6 μm (**a**), 1.4 μm (**b**), and 3.6 μm (**c**) [130]

synthesis of WC–VC–Co hardmetals. Joint grinding of WC and V_8C_7 powders with an addition of 0.5 wt.% of Co for 48 h followed by their 3-h exposure at temperature 1470 K to form their mixed (V, W)C carbide produced a mixture with the average particle size of 1.4 μm. Joint grinding of vanadium oxide, tungsten carbide and carbon was also carried out, which was followed by the carburization of the resulting mixture at ∼1470 K in the H_2/Ar atmosphere. In this case, the obtained powder had the particles with the average size of 1.8 μm.

The synthetic methods, the structure and mechanical properties of fine-grain hardmetals based on tungsten carbides were discussed briefly in a review [128]. Studies devoted to the microstructure, hardness and fracture toughness of WC–Co hardmetals [129] have shown that the hardness increases with a decrease in the thickness of the metal binder layer. It was also noted that the hardness measurements carried out at low loads or using a nanoindenter overestimate its value by ∼2 GPa. The effects of the microstructure and the grain size of alloys WC–6 wt.% Co on their hardness in the temperature range from 80 to 1170 K was studied [130]. The examined hardmetals had the microstructure typical of conventional WC–Co hardmetals (Fig. 2.22).

Figure 2.23 illustrates the changes in the hardness H_V of hardmetals with the average grain size of 0.6, 1.4 and 3.6 μm as a function of the temperature. It is evident that the hardmetals with a fine-grained structure are harder throughout this temperature range.

Recently, nanopowders of the WC–Co system came into use for synthesizing nanocomposites that contain a diamond phase [131, 132]. A nanopowder WC–15 wt.% Co with WC grains measuring 100 nm on average was pressed into samples of 8-mm diameters under a pressure of 100 and 300 MPa [131]. The samples were first heated to 1170 K in hydrogen atmosphere to remove the oxide film and then sintered in vacuum at 1270–1470 K with 30-min exposure at the maximum sintering temperature. By varying the sintering time at the lowest temperature used, one could control the size and the arrangement of pores in the sample. Then, the samples were kept for 3 h at temperature 1170 K in a flow of $H_2 + CH_4$ gas mixture to infiltrate the gas into the pores, after which they were simultaneously subjected to impact of high temperatures and pressures in order to reduce methane to carbon and transform carbon (graphite) into the diamond phase. The resulting superhard nanocompos-

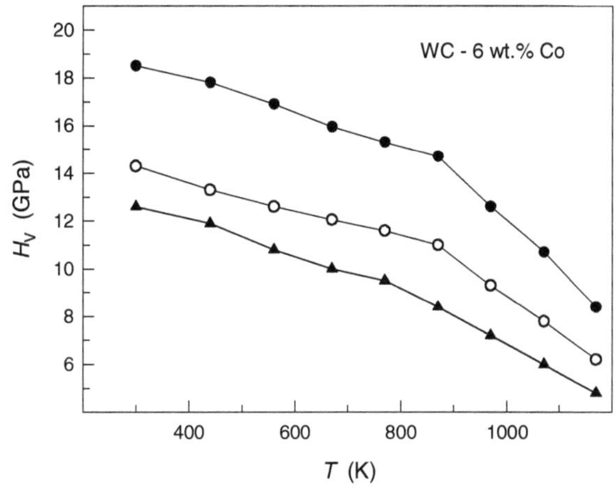

Fig. 2.23 The effect of temperature of the Vickers hardness, H_V, of WC–6 wt.% Co hardmetals with the average size, D, of carbide phase grains of 0.6 μm (•),1.4 μm (∘), and 3.6 μm (▲) [130]. The measurements were performed under a load of 6 kg

ite material contained ~45 vol.% of the diamond phase [131]. The diamond-phase content in the synthesized nanocomposite seems to be significantly overrated.

A somewhat different method for the synthesis of WC–Co-diamond nanocomposites was proposed [132]. The initial mixture represented a powder mixture of WC with 15 wt.% of Co; the average size of WC particles was 100 nm. The samples compacted under a pressure of 200 MPa were heated to 870 K in hydrogen atmosphere to remove the oxide film and then sintered in vacuum at temperature 1370 K for 30 min. The prepared porous samples with the 70 % relative density were subjected to electrochemical etching in an aqueous $NaC_4H_5O_6$ solution. After the etching, the samples were washed with water, dried and then saturated with a gas mixture H_2 + CH_4, which contained 15 % of methane. The infiltration of the gas into pores was carried out at 1170 K for 4 h. Then hydrostatic pressure of 8 GPa was applied at temperature 1770 K to decompose methane in a catalytic reaction $CH_4 \rightarrow C + 2H_2$, deposit carbon on the pore walls and transform carbon from the graphite into the diamond phase and also to densify the samples. The hardness of samples near their edges reached 32 ± 3 GPa, which was five times higher than in the bulk.

The results of sintering nanocrystalline WC–Co powders was discussed in review [133]. It was shown that controlling the rapid grain growth during the early stage of sintering is the key problem to the production of bulk nanostructured WC–Co hardmetals. According to [133], use of grain growth inhibitors and pressure during sintering helps to control grain growth in hardmetals. Use of fast sintering techniques allows to to reduce duration of sintering and to decrease grain growth. Indeed, compared to conventional sintering, such fast sintering techniques as SPS, HFIHS etc. are characterized by shorter densification time at lower temperature. SPS, for example, by application of electrical fields, can generate rapid and simultaneous heating within samples due to the pulse electrical current through the graphite die and/or samples. Table 2.4, which was represented in review [131], summarizes the results of various sintering techniques with respect to densification and grain sizes of WC–Co hardmetals.

Table 2.4 Results of various sintering techniques with respect to densification and grain size [133]

Consolidation process	Process temperature and hold time	Heating rate ($K \cdot min^{-1}$)	Pressure (MPa)	Initial powder and grain size	Grain growth inhibitor	Relative density (%)	Sintered grain size (nm)
Hot pressing	1570K, 20 min	[a]N/A	30	WC-10Co-0.6VC	0.6VC	100	169
Hot pressing	1970K, 43 Ks	N/A	1500	WC-14Co, 7 nm	No	100	95
Hot pressing	1570K, 90 min	N/A	25	WC-11Co, 80 nm	No	100	780
[b]SPS	1370K, 5 min	N/A	25	WC-11Co, 80 nm	No	100	780
[c]HIP	1270K, 1 h	N/A	150	WC-10Co	No	100	~400
HIP	1270K, 1 h	N/A	150	WC-10Co-0.4VC	0.4VC	100	~200
SPS	1420K, 10 min	N/A	60	WC-10Co	No	98	230
SPS	1370K, 10 min	150	60	WC-12Co, 40–250 nm	No	99.89	800
SPS	1370K, 10 min	150	60	WC-12Co, 40–250 nm	1VC	95.94	470
SPS	1510K, 2 min	100	60	WC-10Co, 50 nm	No	N/A	280
SPS	1510K, 2 min	100	60	WC-10Co, 50 nm	0.9VC	N/A	170
SPS	1420K, 5 min	100	10 kN	WC-8Co, 33 nm	No	N/A	200
SPS	1470 K, 5 min	N/A	30	WC-10Co-0.8VC, 11 nm	0.8VC	95.1	<100 nm
SPS	1270K, 10 min	100	100	WC-10Co, 100 nm	No	N/A	~350
[d]HFIHS	1470K, 55 s	1400	60	WC-10Co, 100 nm	No	99.4	323
[e]PPS	1370K, 5 min	1000–1200	60	WC-12Co, 60 nm	No	98	50
[f]ROC	1270K, 10 s	N/A	830	WC-8Co, Nanocarb™	No	14.6	150
[g]UPRC	1470K, 15 min	200	1000	WC-10Co, 50 nm	No	99.4	97

[a] not available
[b] Spark plasma sintering
[c] Hot isostatic pressing
[d] High frequency induction-heated sintering
[e] Pulse plasma sintering
[f] Rapid omni compaction
[g] Ultrahigh pressure rapid hot consolidation

Thus, nowadays, there are different methods for the synthesis of nano- or submicrocrystalline tungsten carbides, WC–Co composite nanopowders and nanostructured WC–Co hardmetals. The available data suggest (see Table 2.3, Fig. 2.19) that the decrease in the size of carbide grains in WC–Co hardmetals is accompanied by noticeable increase in the abrasive wear resistance (by a factor of 2–3). Indeed, when drilling the printed-circuit boards, the drills made of ultrafine-grained hardmetals prove to be 2–3 times more stable as compared with those made of standard (conventional) hardmetals [128]. Therefore, it is reasonable to use the hardmetals with ultradispersed and nanocrystalline structures, which combine the enhanced wear resistance, strength and fracture toughness, in those operations that require high wear resistance. However, it deserves mentioning that for conventional hardmetals the same and even greater effect of the increase in the wear resistance can be achieved by the deposition of a wear-resistant coating [134]. This should be taken into account when assessing the economic profits associated with the use of nanostructured hardmetals.

* * *

During more than eighty years hexagonal tungsten monocarbide WC is the most extensively used among all carbides of Group IV–VI transition metals. Originally, in the W–C system, only two hexagonal carbides W_2C_y (β-W_2C) and WC (δ-WC) were known, which despite their similar symmetry differ significantly, namely, lower carbide W_2C_y has a noticeable homogeneity region $WC_{0.34} - WC_{0.52}$, while hexagonal monocarbide δ-WC can have only the stoichiometric composition and contains no vacancies in the carbon sublattice. The nonstoichiometry of disordered lower W_2C_y carbide, which is in equilibrium at temperature from 2700 K up to the melting point, is a prerequisite for its ordering with respect to both atoms and vacancies at lower temperatures. According to the neutron diffraction data, with the decrease in the temperature, the following disorder-order transformation β-$W_2C \Leftrightarrow \varepsilon$-$W_2C$ takes place in β-W_2C carbide. The earlier concepts on the two consecutive transformations, viz., the disorder-order transformation β-$W_2C \Leftrightarrow \beta'$-$W_2C$ with the formation of the orthorhombic β'-W_2C phase of the PbO_2 type and the order-order transformation β'-$W_2C \Leftrightarrow \beta''$-$W_2C$ with the formation of the trigonal β''-W_2C phase of the C6 type, were not confirmed by the neutron diffraction analysis of the structure of lower W_2C tungsten carbide. One peculiarity of lower tungsten carbide is that at temperature 1523 K it undergoes the eutectoid decomposition to give tungsten and δ-WC and thus does not exist at lower temperatures. However, there are experimental data which argue against the existence of the eutectoid decomposition of the lower tungsten carbide on W and WC at 1523 K. These experimental data will be discussed in detail in Sect. 3.2.

In the 1960s, a new nonstoichiometric compound with structural vacancies in the carbon sublattice, namely, cubic γ-WC_{1-x} carbide was discovered in the W–C system. Under equilibrium conditions, the region of the existence of this carbide is very wide with respect to composition (from $WC_{0.58-0.59}$ to $WC_{1.00}$) and very narrow with respect to temperature (from ~3060 to ~2990–3020 K). However, due to quenching from high temperature, this carbide can be preserved in the metastable state at

300 K. The temperature of the equilibrium transformation γ-$WC_{1-x} \Leftrightarrow \delta$-WC is 3000–3010 K; however, the temperature of the nonequilibrium transformation of the quenched cubic γ-WC_{1-x} carbide to give hexagonal δ-WC carbide, which occurs on heating, is still unknown.

Studies of the W–C system are also important from the viewpoint of practice, because tungsten carbide is used in the production of hardmetals applied in the metal working and drilling. The traditional basis for such hardmetals is hexagonal stoichiometric carbide WC(δ-WC). To date, no attempts were undertaken to use cubic carbide γ-$WC_{1.0}$ of the stoichiometric composition or non-stoichiometric cubic carbide γ-WC_{1-x} for these purposes. The cubic structure of tungsten carbide can apparently be stabilized by the addition of other cubic carbides, for example, VC_y. The use of cubic tungsten carbide with VC_y additions as the basis for hardmetals can also provide the thermal stability of the carbide phase and the fine-grained microstructure of the hardmetal.

Yet another direction in the application of tungsten carbide, which was actively developed in the past 25–30 years, concerns the nanostructured hardmetals. Nanostructured WC–Co hardmetals have been the subject of interests and focus of research programs around the world for the past three decades owing to the expectations that the mechanical properties of the hardmetals may improve significantly when grain sizes reduce to nanometer scale. At present, WC–Co nanocomposites for producing hardmetals can be prepared by such methods as co-deposition of tungsten and cobalt from colloid solutions, thermal decomposition of precursors, plasmachemical reduction, mechanochemical synthesis, etc. Although the particle sizes in nanocomposites can be very low (10–40 nm), obtaining true nanostructured WC–Co hardmetals (with average WC grain size <100 nm) remains a great problem due to the difficulties of controlling grain growth during sintering. Nonetheless, the microstructure of newly developed fine-grained hardmetals contains carbide-phase grains with the average size of 100–250 nm. This provides the noticeable increase in their hardness, wear resistance and other operational properties. The introduction of small amounts of vanadium carbide favors the retention of small size of grains in nanostructured hardmetals. However, the problem of the conservation of the small particles during sintering of a hardmetal is not yet solved completely. Note that evaluation of the mechanical properties of nanocrystalline WC–Co hardmetals is also difficult because there is little published data that are obtained on hardmetal samples with truly nanosized grains of carbide phase.

In the systems W–C and WC–Co–C, the nanostructured carbides or nanocomposites of the WC–Co and WC–Co-diamond types were obtained. Their synthesis became possible due to the use of new and, as a rule, combined synthetic methods and non-traditional technologies.

On the whole, the studies of tungsten carbides and the related phase equilibria in both the W–C system and more complex systems containing tungsten carbides contribute to the development of fundamental concepts on the nature of refractory compounds of transition metals, nonstoichiometry in these compounds and serves as the basis for further wide application of refractory compounds including tungsten carbide.

References

1. Gusev AI, Rempel AA. Structural phase transitions in nonstoichiometric compounds. Moscow: Nauka; 1988. p. 308 (in Russian).
2. Gusev AI. Physical chemistry of nonstoichiometric refractory compounds. Moscow: Nauka; 1991. p. 286 (in Russian).
3. Gusev AI., Rempel AA. Nonstoichiometry, disorder and order in solids. Ekaterinburg: Ural Division of the Russian Academy of Sciences; 2001. p. 580 (in Russian).
4. Gusev AI, Rempel AA, Magerl AJ. Disorder and order in strongly nonstoichiometric compounds: transition metal carbides, nitrides and oxides. Berlin: Springer; 2001. p. 607.
5. Gusev AI. Nonstoichiometry, disorder, short-range and long-range order in solids. Moscow: Nauka-Fizmatlit; 2007. p. 856 (in Russian).
6. Gubanov VA, Ivanovsky AL, Zhukov VP. Electronic structure of refractory carbides and nitrides. Cambridge: Cambridge University Press; 1994. p. 256.
7. Goldschmidt HJ. Interstitial alloys. New York: Plenum Press; 1967.
8. Kotel'nikov RB, Bashlykov SN, Galiakbarov ZG, Kashtanov AI. Fundamentals of refractory elements and compounds/handbook. Moscow: Metallurgiya; 1969. p. 376 (in Russian).
9. Storms EK. The refractory carbides. New York: Academic Press; 1967.
10. Toth LE. Transition metal carbides and nitrides. New York: Academic Press; 1971.
11. Andrievski RA, Lanin AG, Rymashevskii GA. Strength of high-melting compounds. Moscow: Metallurgiya; 1974. p. 232 (in Russian).
12. Holleck H. Binäre und Ternäre Carbid- und Nitridsysteme der Übergangsmetalle. Berlin–Stuttgart: Gebrüder Borntraeger; 1984. p. 295SS.
13. Hägg G. Gezetsmässigkeiten im Kristallbau bei Hydriden, Boriden, Karbiden und Nitriden der Übergangselemente. Zeitschr Phys Chem 1931;12(1):33–56.
14. Moissan H. Preparation au four électrique de quelques mètaux réfractaires: tungstène, molybdène, vanadium. Compt Rend. 1893;116:1225–27.
15. Schrüter K. The inception and development of hard metal carbides. The Iron Age. 1934;133(2):27–9.
16. C-W (carbon–tungsten). In: Massalski TB, Murray JL, Bennet LH, Baker H, Kasprzak L, editors. Binary alloy phase diagrams. Vol 1. Metals Park (Ohio): ASM Internship Publications; 1987. p. 599–600.
17. C-W (carbon–tungsten). In: Massalski TB, Subramanian PR, Okamoto H, Kasprzak L, editors. Binary alloy phase diagrams. 2nd ed., Vol 1. Metals Park (Ohio): ASM Internship Publications; 1990. p. 895–896.
18. Rudy E, Windisch S. Evidence to zeta Fe_2N-type sublattice order in W_2C at intermediate temperatures. J Am Ceram Soc. 1967;50(5):272–3.
19. Rudy E, Windisch S, Hoffman JR In: Ternary phase equilibria in transition metal-boron-carbon-silicon systems (Part I. Related Binary Systems. Vol VI). USAF Technical Document Report AFML-TR-65-2. Wright-Patterson Air Force Base (Ohio): Metals and Ceramics Division, Air Force Materials Laboratory; 1966. p. 1–50.
20. Rudy E. Compendium of phase diagram data. Ternary phase equilibria in transition metal-boron-carbon-silicon systems (Part V). Final Technical Report AFML TR-65-2. Wright-Patterson Air Force Base (Ohio): Metals and Ceramics Division, Air Force Materials Laboratory; 1969. p. 735.
21. Rudy E, Hoffman JR Phasengleichgewichte im Bereich der kubischen Karbidphase im System Wolfram-Kohlenstoff. Planseeber. Pulvermet. 1967;15(3):174–8.
22. Sara RV. Phase equilibrium in the system tungsten–carbon. J Am Ceram Soc. 1965;48(5):251–7.
23. C-W. In: McHale AE, editor. Phase equilibria diagrams. Phase diagrams for ceramists. Vol X. Westerville (Ohio): American Ceramic Society Publications; 1994. p. 272–273.
24. C-W (carbon–tungsten). In: Lyakishev NP, editor. Phase diagrams of the binary metal systems/handbook. Vol 1. Moscow: Mashinostroenie; 1996. p. 778–779 (in Russian).

25. Yvon K, Nowotny H, Benesovsky F. Zur Kristallstruktur von W_2C. Monatsh. Chemie. 1968;99(2):726–9.
26. Harsta A, Rundqvist S, Thomas JO. A neutron powder diffraction study of W_2C. Acta Chem Scand A. 1978;32A(9):891–2.
27. Lönnberg B, Lundström T, Tellgren R. A neutron powder diffraction study of Ta_2C and W_2C. J Less-Common Met. 1986;120(2):239–45.
28. Epicier T, Dubois J, Esnouf C, Fantozzi G. Identification de la phase β-W_2C, type ε-Fe_2N dans l'hémicarbure de tungsten. Compt Rend Acad Sci Paris Ser II. 1983;297(3):215–18.
29. Epicier T, Dubois J, Esnouf C, Fantozzi G, Convert P. Neutron powder diffraction studies of transition metal hemicarbides M_2C_{1-x}. II. In situ high temperature study of W_2C_{1-x} and Mo_2C_{1-x}. Acta Metall. 1988;36(8):1903–21.
30. Lander JJ, Germer LH. Plating molybdenum, tungsten, and chromium by thermal decomposition of their carbonyls. Trans AIME. 1948;175:661–92.
31. Lautz G, Schneider D. Über die Supraleitung in den Wolframkarbiden W_2C und WC. Zeitschr. Naturforsch. A. 1961;16A(12):1368–72.
32. Goldschmidt HJ, Brand JA. The tungsten-rich region of the system tungsten–carbon. J Less-Common Met. 1963;5(2):181–94.
33. Parthe E, Sadagopan V. The structure of dimolybdenum carbide by neutron diffraction technique. Acta Crystallogr. 1963;16(3):202–05.
34. Butorina LN, Pinsker ZG. Electron diffraction study of W_2C. Kristallografiya. 1960;5(4):585–588 (in Russian) (English Translation: Sov Phys Crystallogr. 1960;5(4):560–62).
35. Heetderks HD, Rudy E, Eckert T. Differential thermal analysis apparatus for high temperatures. High temperature phase reactions in refractory carbide systems. Planseeber. Pulvermet. 1965;13(2):104–25.
36. Ronsheim P, Toth LE, Mazza A, Pfender E, Mitrofanov B. Direct current arc-plasma synthesis of tungsten carbides. J Mater Sci. 1981;16(10):2665–74.
37. Willens RH, Buehler E. The superconductivity of the monocarbides of tungsten and and molybdenum. Appl Phys Lett. 1965;7(1):25–6.
38. Willens RH, Buehler E, Matthias BT. Superconductivity of the transition-metal carbides. Phys Rev. 1967;159(2):327–30.
39. Krainer E, Robitsch J. Röntgenographischer Nachweis des kubischen Wolframkarbides in funkenerosiv bearbeiteten Hartmetallen und in reinen Wolframschmelkarbiden. Planseeber. Pulvermet. 1967;15(1):46–56.
40. Gromilov SA, Kinelovskii SA. X-ray investigation of tungsten carbides synthesized by cumulative explosion. Zh Struct Khimii. 2003;44(3):486–93 (in Russian) (English Translation. J Struct Chem. 2003;44(3):434–40).
41. Vonsovskii SV, Izyumov YA, Kurmaev EZ. Superconductivity of transition metals, their alloys and compounds. Moscow: Nauka; 1977. p. 384 (in Russian).
42. Krainer E, Robitsch J. Zur Frage des kubischen Wolframkarbids. Planseeber. Pulvermet. 1967;15(3):179–80.
43. Kurlov AS, Gusev AI. Tungsten carbides and W-C phase diagram. Neorgan. Materialy. 2006; 42(2):156–163. (in Russian) (English Translation. Inorg Mater. 2006;42(2):121–7.
44. Butorina LN. An electron diffraction study of the tungsten carbide WC. Kristallografiya. 1960;5(2): 233–7 (in Russian) (English Translation: Sov Phys Crystallogr. 1960;5(2):216–20).
45. Liu AY, Wentzcovitch RM, Cohen ML. Structural and electronic properties of WC. Phys Rev B. 1988;38(14):9483–8.
46. Rempel AA, Würschum R, Schaefer H-E. Atomic defects in hexagonal tungsten carbide studied by positron annihilation. Phys Rev B. 2000;61(9):5945–8.
47. Rempel AA., Schaefer H-E, Forster M, Girka AI. Atomic defects in transition metal carbides and SiC studied by positron annihilation. In: Barron AR, Fischman GS, Fury MA, Hepp AF, editors. Proceedings of the materials research society symposium. Vol 327 (Covalent ceramics II: non-oxides). Symposium held November 29–December 2, 1993, Boston, Massachusetts, USA. Pittsburgh (Pennsylvania): Materials Research Society; 1994. p. 301–4.

48. Puska MJ, Šob M, Brauer G, Korhonen T. Positron annihilation characteristics in perfect and imperfect transition metal carbides and nitrides. J Physique IV. 1995;5(1):C1-135–42.
49. Rudy E, Windisch S, Chang YA. Mo- C system. In: Ternary phase equilibria in transition metal-boron-carbon-silicon systems (Part I. Related Binary Systems. Vol I). USAF Technical Document Report AFML-TR-65-2. Wright-Patterson Air Force Base (Ohio): Metals and Ceramics Division, Air Force Materials Laboratory; 1965. p. 1–164.
50. Rudy E, Windisch S, Stosick AJ, Hoffman JR. The constitution of binary molybdenum—carbon alloys. Trans AIME. 1967;239(8):1247–67.
51. Kurlov AS, Gusev AI. Phase equilibria in the W-C system and tungsten carbides. Uspekhi Khimii. 2006;75(7):687–708 (in Russian) (English Translation: Russ Chem Rev. 2006;75(7):617–36).
52. Rautala P, Norton JT. Tungsten-cobalt-carbon system. Trans AIME. 1952;194(4):1045–50.
53. Gurland J. A study of the effect of carbon content on the structure and properties of sintered WC- Co alloys. Trans AIME. 1954;200(3):285–90.
54. Pollock CB, Stadelmaier HH. The eta carbides in the Fe-W-C and Co-W-C systems. Metall Trans. 1970;1(4):767–70.
55. Ramhath V, Jayaraman N. Quantitative phase analysis by X-ray diffraction in Co-W-C system. J Mater Sci Lett. 1987;6(12):1414–8.
56. Ettmayer P, Suchentrunk R. Über die thermische Stabilität der Eta-Carbide. Monatsch Chemie. 1970;101(4):1098–1103.
57. Johansson T, Uhrenius B. Phase equilibria, isothermal reactions, and a thermodynamic study in the Co-W-C system at 1150 ° C. Metal Sci. 1978;12(1):83–94.
58. Adelsköld V, Sundelin A, Westgren A. Carbide in kohlenstoffhaltigen Legierungen von Wolfram und Molybdän mit Chrom, Mangan, Eisen, Kobalt und Nickel. Zeitschr Anorg Allgem Chemie. 1933;212(4):401–9.
59. Schönberg N. The structure of the $Co_3W_9C_4$ phase. Acta Metall. 1954;2(6):837–40.
60. Harsta A, Johansson T, Rundqvist N, Thomas JO. A neutron powder diffraction study of the κ-phase in the Co-W-C system. Acta Chem Scand A. 1977;31A(4):260–4.
61. Guillermet AF. Thermodynamic properties of the Co-W-C system. Metall Trans AIME A. 1989;20A(5):935–56.
62. Kaufman L, Bernstein H. Computer calculation of phase diagrams. New York: Academic Press; 1970. p. 344.
63. Sundman B, Ansara I, Hillert M, Inden G, Lukas H-L, Kumar KCH. Contributions to the thermodynamic modelling of solution. Zeitschr Metallkunde. 2001;92(6):526–32.
64. Villars P, Prince A, Okamoto H, Handbook of ternary alloy phase diagrams. Vol 5. Metals Park (Ohio): ASM Publication; 1995. p. 6585–605.
65. Alekseev ES, Arkhipov PG, Popova SV. Band structure of hexagonal tungsten carbide. Phys Status Solidi (b). 1982;110(2):K151–4.
66. Mattheiss LF, Hamann DR. Bulk and surface electronic structure of hexagonal WC. Phys Rev B. 1984;30(4):1731–8.
67. Zhukov VP, Gubanov VA. Energy band structure and thermo-mechanical properties of tungsten and tungsten carbides as studied by the LMTO-ASA method. Solid State Commun. 1985;56(1):51–5.
68. Zhukov VP, Gubanov VA, Jarlborg T. Study of energy-band structures, some thermo-mechanical properties and chemical bonding for a number of refractory metal carbides by the LMTO-ASA method. J Phys Chem Solids. 1985;46(10):1111–6.
69. Zhukov VP, Gubanov VA. Study of the energy-band structure and chemical-bond in ZrC, NbC, and WC by the LMTO-ASA method. Izv AN SSSR Neorgan Materialy. 1986;22(10):1665–71 (in Russian) (English Translation: Inorg Mater. 1986;22(10):1458–63).
70. Matthias BT, Hulm JK. A search for new superconducting compounds. Phys Rev. 1952;87(5):799–806.
71. Medvedeva NI, Ivanovskii AL. Effect of metal and carbon vacancies on the band structure of hexagonal tungsten carbide. Fiz Tverd Tela. 2001;43(3):452–5 (in Russian) (English Translation. Phys Solid State. 2001;43(3):469–72).

72. Suetin DV, Shein IR, Kurlov AS, Gusev AI, Ivanovskii AL. Band structure and properties of polymorphic modifications of lower tungsten carbide W_2C. Fiz Tverd Tela. 2008;50(8):1366–72 (in Russian) (English Translation. Phys Solid State. 2008;50(8):1420–6).
73. Weimer AW, editor. Carbide, nitride and boride materials synthesis and processing. London: Chapman and Hall; 1997. p. 671.
74. Gogotsi YG, Andrievski RA, editors. Materials science of carbides, nitrides and borides. Dordrecht: Kluwer Academic Publishers; 1999. p. 360.
75. Berger L-M, Hermann M, Gusev AI, Rempel AA. Verfahren zur Herstellung nichtstöchiometrischer Carbide definierter Zusammensetzung Offenlegungsshrift DE 198 07 589 A 1. Int. Cl.[6]: C 01 B 31/30 (C 04 B 35/36). Bundesrepublik Deutschland: Deutsches Patentamt (Anmeldetag 23.02.1998, Offenlegungstag 10.09.1998.), 1998. p. 3.
76. Bondarenko VP, Pavlotskaya EG. High-temperature synthesis of tungsten carbide in a methane-hydrogen gas medium. Poroshk Metallurgiya. 1995;34(9–10):21–6 (in Russian) (English Translation: Powder Metall Met Ceram. 1995;34(9–10). p. 508–12).
77. Kuznetsov NT. Precursors for carbide, nitride and boride synthesis. In: Gogotsi YG, Andrievski RA, editors. Materials science of carbides, nitrides and borides. Dordrecht: Kluwer Academic Publishers; 1999. p. 223–46.
78. Avvakumov EG, editor. Mechanical alloying in inorganic chemistry. Novosibirsk: Nauka; 1991. p. 305 (in Russian).
79. Davydkin VY, Trusov LI, Butyagin PY, Moskvin VV, Kolbanev IV, Novikov VI, Plotkin SS. Structure of refractory carbides synthesized by mechanochemical method. In: Avvakumov EG, editor. Mechanical alloying in inorganic chemistry.Novosibirsk: Nauka; 1991:183–5 (in Russian).
80. Baikalova YV, Lomovsky OI. Solid state synthesis of tungsten carbide in an inert copper matrix. J Alloys Comp. 2000;297(1). p. 87–91.
81. Westbrook JH, Stover ER. Carbides for high-temperature materials. In: Campbell IE, Sherwood EM, editors. High-temperature materials and technology. New York: Wiley;1967:312–48.
82. Kovalchenko MS, Dzhemelinskii VV, Skuratovskii VN, Tkachenko YG, Yurchhenko DZ, Alekseev VI. Microhardness of some carbides at various temperatures. Poroshk Metallurgiya. 1971;10(8):87–91 (in Russian) (English Translation: Powder Metall Met Ceram. 1971;10(8):665–8).
83. Atkins AG, Tabor D. Hardness and deformation properties of solids at very high temperatures. Proc Roy Soc Lond. 1966;292(1431):441–59.
84. Samsonov GV, Vitryanuk VK, Chaplygin FI. Tungsten carbides. Kiev: Naukova Dumka; 1974. p. 176 (in Russian).
85. Kieffer R, Benesovsky F. Hartmetalle. Wien-New York: Springer; 1965. p. S541.
86. Romanova NI, Chekulaev PG, Dusev VI, Livshits TA, Kurdov MN. Cermet hard alloys. Moscow: Metallurgiya; 1970:144–74 (in Russian).
87. Loshak MG. Strength and operating life of hard alloys. Kiev: Naukova Dumla; 1984. p. 328 (in Russian).
88. Emelyanova TA, Kobitskaya NB, Gorbacheva TB. Kinetics of wet grinding of submicron tungsten-carbide powder. Metally. 1992;(1):75–80 (in Russian) (English Translation. Russ Metall. 1992;(1):64–8).
89. Gusev AI. Nanocrystalline materials: preparation and properties. Ekaterinburg: Ural Division of the Russian Academy of Science; 1998. p. 200 (in Russian).
90. Gusev AI, Rempel AA. Nanocrystalline materials. Moscow: Nauka-Fizmatlit; 2000. p. 224 (in Russian).
91. Gusev AI, Rempel AA. Nanocrysnalline materials. Cambridge: Cambridge Internship Science Publications; 2004. p. 351.
92. Gusev AI. Nanomaterials, nanostructures, and nanotechnologies. 2nd ed. Moscow: Nauka-Fizmatlit; 2007. p. 416 (in Russian).
93. Valiev RZ, Aleksandrov IV. Nanostructured materials obtained by severe plastic deformation. Moscow. Logos; 2000. p. 272 (in Russian).

94. Gusev AI. Effects of the nanocrystalline state in solids. Uspekhi Fiz Nauk. 1998;168(1):55–83 (in Russian) (English Translation: Phys Uspekhi. 1998;41(1):49–76).
95. Gusev AI. Nanocrystalline materials: synthesis and properties. In: Schwarz JA, Contescu C, Putyera K, editors. Dekker encyclopedia of nanoscience and nanotechnology, in 5 volumes. Vol 3. New York: Marcel Dekker Inc.; 2004:2289–304.
96. Berger S, Porat R, Rosen R. Nanocrystalline materials: a study of WC-based hard metals. Prog Mater Sci. 1997;42(1–4):311–20.
97. Gusev AI, Kurlov AS. Hardmetals today and tomorrow. Metally Evrazii. 2005;(2):42–45 (in Russian).
98. Scussel HJ. Friction and wear of cemented carbides. In: Metals handbook (ASM) 10th edition in 23 volumes, 1990–1997/friction, lubrication and wear technology (ASM Handbook, Vol 18). Vol 18. Metals Park (Ohio): ASM Internship Publications; 1992:795–6.
99. Daub HW, Dreyer K, Happe A, Holzhauer H, Kassel D. Leistungspotentiale von Feinst- und Ultrafeinstkorn-Hartmetallen und Herstellung/Pulvermetallurgie in Wissenschaft und Praxis. Bd.11. Ed. H. Kolaska. Oberursel: DGM Informationsgesellschaft Verlag, 1995. p. 285–306.
100. Vaβen R., Stöver D. Processing and properties of nanophase ceramics. J Mater Process Technol. 1999;92–93:77–84.
101. Schubert WD, Bock A, Lux B. General aspects and limits of conventional ultrafine WC powder manufacture and hard metal production. In: Bildstein H, Eck R, editors. Proceedings of the 13th Internship on Plansee Seminar. Reutte: Metallwerk Plansee, 1993;4:283–305.
102. Miller TN. Plasmochemical synthesis and properties of powders of high-melting compounds. Izv AN SSSR Neorgan Materialy. 1979;15(4):557–62 (in Russian) (English Translation. Inorg Mater. 1979;15(2):476–80).
103. Kosolapova TY, Makarenko GN, Zyatkevich DP. Plasmochemical synthesis of high-melting compounds. Zh Vsesouzn Khimichesk Obschestva. 1979;24(3):228–33 (in Russian).
104. Kear BH, Strutt PR. Chemical processing and applications for nanostructured materials. Nanostruct. Mater. 1995;6(1–4):227–36.
105. Blinkov IV, Ivanov AV, Orekhov IE. Synthesis of ultradispersed carbide powders in pulsed plasma. Fizika i Khimiya Obrabotki Mater. 1992;(2):73–6 (in Russian).
106. Shishkovskii VI, Kopytin YD, Gubaidullin NE. Preparation of ultradisperse powders of refractory nitrides and carbides in plasma of high-frequency discharge. In: Ultradisperse powders, materials and nanostructures. Production, properties, application/transactional interreg conference, December 17–19, 1996, Krasnoyarsk: Krasnoyarsk State Technical University; 1996. P. 56 (in Russian).
107. Blagoveshchenskii YV, Danilkin EA, Egorikhina TP, Terekhov VI. Nanocrystalline composite WC-Co. In: Physics and chemistry of ultradisperse systems/Trans. 4th all-Russian conference, Obninsk/Russia, June 29–July 3, 1998. Moscow: Moscow Engineering Physics Institute; 1998. p. 274 (in Russian).
108. Lebedev AV, Blagoveshchenskii YV, Moldaver VA. Creation of plasma-chemical equipment for manufacture of ultrafine powders of refractory metals and compounds. In: Physics and chemistry of ultradisperse systems/trans. 5th all-Russian conference, Yekaterinburg/Russia, October 9–13, 2000, Moscow: Moscow Engineering Physics Institute; 2000. p. 145–7 (in Russian).
109. Gao L, Kear BH. Low temperature carburization of high surface area tungsten powders. Nanostruct Mater. 1995;5(5):555–69.
110. Gao L, Kear BH. Synthesis of nanophase WC powder by a displacement reaction process. Nanostruct Mater. 1997;9(3):205–8.
111. Kurlov AS, Nazarova SZ, Rempel AA. Production, structure and properties of nanocrystalline tungsten carbide. In: Ustinov VV, Noskova NI, editors. Nanotechnology and physics of functional nanocrystalline materials. Vol 1. Ekaterinburg: Ural Division of the Russsian Acadamey of Science; 2005, p. 59–67 (in Russian).
112. Kurlov AS, Nazarova SZ, Gusev AI. Magnetic susceptibility and thermal stability of nanocrystalline tungsten carbide. Doklady Akad Nauk. 2005;405(2):218–23 (in Russian) (English Translation. Dokl Phys Chem. 2005;405(1):229–34).

113. Kurlov AS, Nazarova SZ, Gusev AI. Magnetic susceptibility of tungsten carbide: relaxation and impurity effects. Pis'ma v ZhETF. 2005;82(8):567–72 (in Russian) (English Translation. JETP Lett. 2005;82(8):509–14).
114. Arató P, Bartha L, Porat R, Berger S, Rosen A. Solid or liquid phase sintering of nanocrystalline WC/Co hardmetals. Nanostruct Mater. 1998;10(2):245–55.
115. Sommer M, Schubert W-D, Zobetz E, Warbichler E. On the formation of very large WC crystals during sintering of ultrafine WC-Co alloys. Int J Refract Met Hard Mater. 2002;20(1):41–50.
116. Anishchik SV, Medvedev NN. Three-dimensional Apollonian packing as a model for dense granular systems. Phys Rev Lett. 1995;75(23):4314–17.
117. McCandlish LE, Kear BH, Kim BK. Processing and properties of nanostructured WC-Co. Nanostruct Mater. 1992;1(1):119–24.
118. Wu L, Lin J, Kim BK, Kear BH, McCandlish LE. Grain growth inhibition of nanostructured WC-Co alloys. In: Bildstein H, Eck R, editors. Proceedings of the 13th internship plansee seminar. Vol 3. Reutte: Metallwerk Plansee; 1993. p. 667–75.
119. McCandlish LE, Kear BH, Bhatia SJ. Spray conversion process for the production of nanophase composite powders. 1994. U.S. patent No 5352269.
120. Fang Z, Eason JW. Study of nanostructured WC-Co composites. In: Bildstein H, Eck R, editors. Proceedings of the 13th internship plansee seminar. Vol 3. Reutte: Metallwerk Plansee, 1993; p. 625–38.
121. Seegopaul P, McCandlish LE. Nanostructured WC-Co powders: review of application, processing and characterization properties. Adv Powder Metall Particular Mater. 1995;3:13-3–13-15.
122. Seegopaul P, McCandlish LE, Shinneman FM. Production capability and powder processing methods for nanostructured WC-Co powder. Int J Refract Met Hard Mater. 1997;15(1–3):133–8.
123. El-Eskandarany MS, Mahday AA, Ahmed HA., Amer AH. Synthesis and characterizations of ball-milled nanocrystalline WC and nanocomposite WC-Co powders and subsequent consolidations. J Alloys Compd. 2000;312(1–2):315–25.
124. Wu XY, Zhang W, Wang W, Yang F, Min JY, Wang BQ, Guo JD. Ultrafine WC-10Co cemented carbides fabricated by electric-discharge compaction. J Mater Res. 2004;19(8):2240–4.
125. Borisenko NI, Moldaver VA, Lebedev AV, Kobzarev NV. Experience of using of nano-sized powders in technology of hardmetals. In: Physics and chemistry of ultradisperse systems/Transactional 6th all-Russian conference. Moscow: Moscow Engineering Physics Institute; 2003. p. 505–9 (in Russian).
126. Xueming MA, Gang JI. Nanostructured WC-Co alloy prepared by mechanical alloying. J Alloys Compd. 1996;245:L30–2.
127. Osborne C, Cornish L. On the preparation of fine V_8C_7-WC and V_4C_3-WC powders. Int J Refract Met Hard Mater. 1997;15(1–3):163–8.
128. Klyachko LI, Fal'kovskii VA, Khokhlov AM. Hard alloys based on tungsten carbide with fine structure. Moscow: Ruda i, Metally; 1999. p. 48 (in Russian).
129. Jia K, Fischer TE, Gallois B. Microstructure, hardness and toughness of nanostructured and conventional WC-Co composite. Nanostruct Mater. 1998;10(5):875–91.
130. Milman YV, Chugunova S, Goncharuck V. Low and high temperature hardness of WC-6 wt.% Co alloys. Int J Refract Met Hard Mater. 1997;15(1–3):97–101.
131. Sadangi RK, Voronov OA, Kear BH. WC-Co-diamond nanocomposites. Nanostruct Mater. 1999;12(5–8):1031–4.
132. Jain M, Sadangi RK, Cannon WR, Kear BH. Processing of functional graded WC/Co/diamond nanocomposites. Scripta Mater. 2001;44(8–9):2099–103.
133. Fang ZZ, Wang X, Ryu T, Hwang KS, Sohn HY. Synthesis, sintering, and mechanical properties of nanocrystalline cemented tungsten carbide—a review. Int J Refract Met Hard Mater. 2009;27(2):288–99.
134. Tret'yakov VI. Fundamentals of physical metallurgy and manufacturing technology of sintered hard alloys. Moscow: Metallurgiya; 1976. p. 528 (in Russian).

Chapter 3
Ordering of Tungsten Carbides

Before discussing the phenomenon of ordering of tungsten carbides, let us consider main concepts associated with a disorder in solids or a disorder in crystals.

The most widespread type of the disorder in solids is a substitutional disorder. This type of disorder is realized in substitutional solid solutions A_yB_{1-y} whose lattice sites are occupied by atoms A or B. In the model of ideal solutions, which disregards the interatomic interaction, substitution of atoms of one species by atoms of another species does not lead to distortion of the crystal lattice. This model describes an ideal substitutional disorder when probabilities of sites being filled with atoms of particular species coincide with their concentration.

The concept of a crystal as a three-dimensional periodic and fully ordered assembly of particles represents the physical model of an ideal solid. Real crystals often contain various types of defects. Examinations of solids, performed on different materials by different methods, showed that many properties of crystal are caused by such deviations of their real structure from perfect structure as the presence of defects and nonstoichiometry. A significant evidence of nonstoichiometry is the disagreement between the chemical composition of a compound and the concentration of lattice sites occupied by the compound components. Nonstoichiometry is possible only in two- and multicomponent substances. Concept of a nonstoichiometry is associated with a notion about structural vacancies □ [1, 2].

From crystallographical point of view the presence of structural vacancies is due to the mismatch between the chemical composition, i.e. the relative number of atoms of different species, and the relative number of sites in different crystal sublattices occupied by these atoms. From physical or chemical point of view the reason of structural vacancy formation is following. If the chemical composition do not corresponds to the crystal structure and atoms from one species sublattice cannot occupy the sites of other species sublattice, i.e. formation of antisite defects is energetically not possible, and then structural vacancies are formed in a crystal.

A specific feature of structural vacancies is that their concentration is directly related with the chemical composition of nonstoichiometric compounds and it can

A. S. Kurlov and A. I. Gusev, *Tungsten Carbides*,
Springer Series in Materials Science 184, DOI: 10.1007/978-3-319-00524-9_3,

© Springer International Publishing Switzerland 2013

reach up to the range of a dozen of atomic percent. In nonstoichiometric compounds structural vacancies represent analogs of atoms, i.e. are quasiparticles.

Formally, vacant sites of the crystal lattice behave themselves as atoms occupying sites in the same lattice. For this reason structural vacancies □ are viewed as some analog of atoms rather than "holes" in the crystal lattice. The deviation from stoichiometry and the resulting interval of homogeneity can be considered as a substitutional solid solution whose components are atoms and vacancies. Thus, the atoms X and vacancies □ form a substitutional solid solution in the nonmetal sublattice of nonstoichiometric compounds $MX_y(MX_y\square_{1-y})$.

The phenomenon of nonstoichiometry is intimately connected with ordering and disordering processes. Moreover, the existence of nonstoichiometry is a prerequisite of a disorder or order in the distribution of atoms and vacancies in the structure of nonstoichiometric compounds [1, 2].

Disorder in solids is disturbance of crystal regularity. However in substitutional solid solutions, the crystal lattice preserves its symmetry even in the presence of disorder. Indeed, the occupation probability of any site $\boldsymbol{r}$ by atom of a given species is quite defined magnitude which is equal to the concentration of these atoms in substitutional solid solutions. Thus in a case of substitutional solid solutions, there is a certain lattice of probabilities that possesses all the symmetry elements of a crystal.

The number of positions in the nonmetallic sublattice of nonstoichiometric compounds MX_y is larger than that of interstitial atoms X. Under certain conditions, interstitial atoms will be distributed randomly over available positions and will occupy sites of the nonmetallic sublattice with an equal probability. For example, in the nonstoichiometric compound $MX_y\square_{1-y}$, the probability of occupation of a site of the nonmetallic sublattice with an atom X is equal to y and the probability that this site is not occupied with an atom, i.e., is vacant, is $(1 - y)$. This means that all sites of the nonmetallic sublattice are crystallographically equivalent in a disordered nonstoichiometric compound. As the temperature decreases, a redistribution of atoms and vacancies over the lattice sites can occur; as a result, some lattice sites become predominantly occupied with atoms X, whereas other sites are vacant. Thus, two opposite tendencies can be distinguished in the structure of nonstoichiometric compounds: the tendencies to ordering and to disordering. An ordered vacancy distribution is most probable at low temperatures; a disordered distribution exists at high temperatures, when the entropy contribution to the free energy of a nonstoichiometric compound is sufficiently large.

The completely ordered and completely disordered distributions are the limit states of a nonstoichiometric compound.

Most nonstoichiometric compounds with broad homogeneity intervals have a high-symmetry structures of the $B1$ and $L'3$ types.

The ordered state of nonstoichiometric compounds with a high concentration of structural vacancies is the thermodynamically stable equilibrium one at a temperature below the disorder–order transition temperature T_{trans}. This state is normally achieved thanks to long-term annealing or other heat treatment procedures providing diffusion redistribution of atoms over the lattice sites. On the contrary, the disordered state is realized easily enough thanks to quenching: for example, when compounds produced by high-temperature sintering are cooled from the synthesis temperature to the room temperature ~300 K at a rate of 25 K min^{-1} or higher.

Atomic ordering of strongly nonstoichiometric compounds like carbides, nitrides and oxides of transition metals is a very widespread, but poorly understood phenomenon. The study of atomic ordering in relation to nonstoichiometry is part of such general problem of solid state physics as disorder-order phase transformations in defect solids.

The MC_y and M_2C_y carbides of transition Group IV–VI *d* metals are strongly nonstoichiometric interstitial compounds [1, 2] and are interesting as convenient model objects for the study of the relation between nonstoichiometry and ordering. The study of carbides opened up fresh opportunities for the modification of the structure and properties of strongly nonstoichiometric compounds via ordering. In carbide, carbon atoms occupy the octahedral interstices of the metallic sublattice. Vacant interstices are called structural vacancies □. Vacancies and carbon atoms in the nonmetallic sublattice form a substitutional solid solution. This substitutional solid solution may be disordered or ordered.

The ordering of the higher cubic MC_y carbides has been studied in detail in [1–5], whereas the ordering of the lower carbides M_2C_y $(0.7 < y \le 1.0)$ with a hexagonal structure is still not clearly understood. Among the hexagonal carbides, the lower tungsten carbide W_2C is the least known.

The lower tungsten carbide W_2C_y belongs to the group of strongly nonstoichiometric interstitial compounds [1, 2] and has a homogeneity interval from $WC_{0.34}$ to $WC_{0.52}$ at a temperature about 3000 K [6]. As the temperature decreases, the homogeneity region narrows. Four modifications of the lower tungsten carbide W_2C_y are known. In all these modifications, tungsten atoms form a hexagonal close-packed (hcp) metallic sublattice in which from 34 to 52 % of the octahedral interstices can be occupied by C atoms. Depending on the distribution of C atoms and □ vacancies, the lower carbide W_2C_y can be disordered at a high temperature or ordered at a low temperature.

The high-temperature basic phase β-W_2C has a hexagonal (space group $P6_3/mmc(D^4_{6h})$) structure of the $L'3$-type (see Fig. 2.2) with a disordered arrangement of carbon atoms C and structural vacancies □ in the nonmetallic sublattice. A unit cell of the hexagonal phase β-W_2C is shown in Fig. 3.1. This phase is stable over the temperature interval starting from 2300–2400 or 2670–2720 K and ending with the melting temperature of 3000–3050 K. Direct experimental evidence to the existence of the lower tungsten carbide with the $L'3$-type structure is adduced by a neutron diffraction study [7], in which the disordered carbide β-W_2C was observed simultaneously with an ordered phase of this carbide.

Three ordered phases of the carbide W_2C are mentioned in the literature: orthorhombic β'-W_2C (space group No. 60, *Pbcn* (D^{14}_{2h})) with the ζ-Fe_2N (Mo_2C) structure; rhombohedral β''-W_2C (space group No. 164, $P\overline{3}m1$ (D^3_{3d})) with the *C*6 (anti-CdI_2) structure; and trigonal ε-W_2C (space group No. 162, $P\overline{3}1m$ (D^1_{3d})) with the ε-Fe_2N structure. In some studies the orthorhombic carbide β'-W_2C is labeled ζ-W_2C and the rhombohedral carbide β''-W_2C is labeled α-W_2C. Some researchers

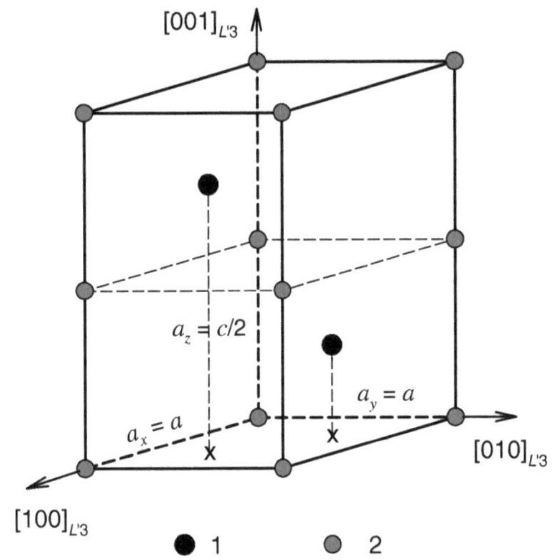

Fig. 3.1 Unit cell of the basic disordered hexagonal (space group $P6_3/mmc$) phase β-W_2 C of the lower tungsten carbide with the structure of the $L'3$-type (lattice constants of the unit cell are equal to a in the directions $[100]_{L'3}$ and $[010]_{L'3}$, and c in the direction $[001]_{L'3}$). The primitive (concerning the nonmetallic sublattice) cell is shown with a *dashed line*. Lattice constants a_x and a_y of the primitive cell coincide with the corresponding lattice constants of the unit cell; the lattice constant a_z of the primitive cell in the direction $[001]_{L'3}$ is twice as small as the lattice constant c of the unit cell, i.e. $a_z = c/2$. *1* Tungsten atoms W •, *2* nonmetallic sublattice sites • statistically (with the probability of 1/2) occupied by carbon atoms C

determined the structure of the ordered phases of the lower tungsten carbide by the X-ray diffraction (XRD) method using powdered samples, while others used the neutron diffraction study. In this connection, before we proceed to discussion of the literature data, let us dwell briefly on capabilities of these methods as applied to the tungsten carbide W_2C.

As is noted in Sect. 1.1 (see also Fig. 1.3), the modifications β-W_2C, β'-W_2C (ζ-W_2C), β''-W_2C and ε-W_2C of the lower tungsten carbide W_2C are practically indiscriminate in powder X-ray diffraction experiments. This is because they have similar hexagonal metal sublattices, while the X-ray scattering amplitudes of atoms W is many times larger than that of atoms C. In this connection, results of some investigations, in which the structure of the ordered phases of the lower tungsten carbide was determined only by the powder X-ray diffraction method, are not quite reliable. However, the presence of atomic displacements in the ordered phases can lead to specific splitting of some diffraction reflections. Then the symmetry of a particular phase can be assessed qualitatively.

A more informative method for analysis of ordering in the nonstoichiometric tungsten carbide is neutron diffraction because the intensities of neutron scattering by tungsten and carbon nuclei are comparable. In neutron diffraction experiments the redistribution of carbon atoms in the carbide W_2C is followed by a considerable change of the intensity of diffraction reflections and appearance of measurable superstructure reflections specific of each ordered phase. Preliminary calculations of neutron diffraction patterns of the disordered hexagonal lower carbide β-W_2C ($WC_{0.50}$) and perfect ordered phases β'-W_2C (ζ-W_2C), β''-W_2C and ε-W_2C having the orthorhombic, the rhombohedral and the trigonal symmetry demonstrated that superstructure reflections characteristic of these phases can be observed over the interval of small angles $2\theta < 33°$ (Fig. 2.3).

Taking into account the aforementioned capabilities of the X-ray and neutron diffraction methods, let us consider literature data on ordering in the lower tungsten carbide.

The researchers [8–10] claimed the existence of the orthorhombic phase β'-W_2C when they analyzed the XRD powder pattern of a W_2C sample quenched from 2630 K and reviewed DTA data. According to [8], the phase β'-W_2C is stable at temperatures of ~2370–2750 K. The structure of the orthorhombic (space group

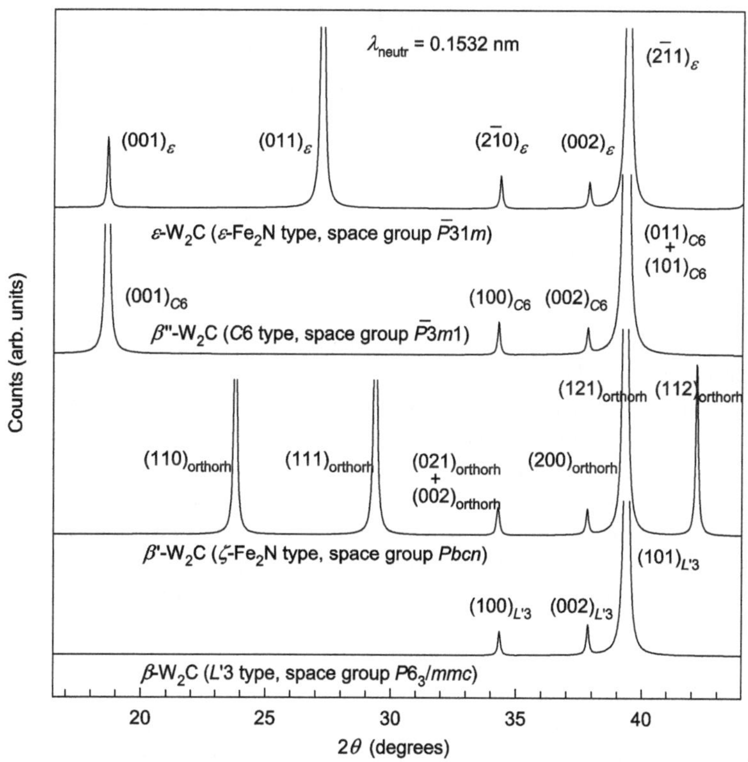

Fig. 3.2 Calculated neutron diffraction patterns ($\lambda = 0.1532$ nm) of the disordered hexagonal ($L'3$-type, space group $P6_3/mmc$) carbide β-W_2C ($WC_{0.50}$) and perfect ordered phases β'-W_2C (ζ-W_2C), β''-W_2C and ε-W_2C with the orthorhombic (ζ-Fe_2N type, space group $Pbcn$), rhombohedral ($C6$-type, space group $P\bar{3}m1$) and trigonal (ε-Fe_2N type, space group $P\bar{3}1m$) symmetries

Pbcn) phase β'-W_2C was proposed [8] by analogy with the ordered molybdenum carbide Mo_2C [11]. The same assumption on the crystal structure of the orthorhombic phase β'-W_2C was done by authors [12] which studied phase equilibria in the Cr–W–C system. However, in powder XRD experiments the change of the carbon atom distribution in the nonmetallic sublattice of the lower tungsten carbide has little effect on the XRD patterns at $2\theta < 30°$ only (see Fig. 2.3), whereas in study [8] the first reflection was observed at $2\theta = 34.55°$. Thus, the study [8] did not provide any experimental structural data confirming the existence of the orthorhombic tungsten carbide β'-W_2C at temperatures of 2370–2750 K.

Authors [13, 14] prepared W_2 C ($WC_{0.50}$) samples by hot pressing of a mixture of tungsten and carbon powders and its subsequent smelting in an electric-arc furnace. Then the W_2C ($WC_{0.50}$) samples were annealed in a vacuum at 1470, 1370, 1270, 1170 and 1070 K for 600 h. Some diffraction reflections of the basic hexagonal lattice were split in XRD patterns of samples annealed at $T \leq 1270$ K. Calculations [13] showed that the observed XRD pattern corresponds to the ordered orthorhombic phase β'-W_2C having the structure of the ζ-Fe_2 N type. The ordered orthorhombic (space group *Pbcn*) phase β'-W_2C was still present in W_2C samples [14] annealed at 870, 1070 and 1270 K for 3000, 1500 and 750 h respectively. Samples of W_2C ($WC_{0.50}$), which were annealed at 1070–1270 K for 600 h, were studied [15] in addition by the neutron diffraction method. The analysis revealed the presence of the ordered orthorhombic (space group *Pbcn*) phase β'-W_2C, in which ~85 % carbon atoms were ordered. However, the authors [15] gave incorrect coordinates (1/4 1/4 1/2) of positions 8(*d*) occupied by atoms W in the β'-W_2C structure. Previously, other inaccurate coordinates (3/8 ~ 3/8 ~ 1/2) for W atoms, occupying 8(*d*) positions in the suggested orthorhombic (space group *Pbcn*) β'-W_2C phase, were specified by authors of study [8]. Really, atoms in the orthorhombic unit cell have the following coordinates: Tungsten atoms W occupy positions 8(*d*) with the coordinates (1/4 ~ 1/8 ~ 1/12), while carbon atoms C are in positions 4(*c*) with the coordinates (0 ~3/8 1/4).

The researchers [16] detected the orthorhombic phase β'-W_2C in samples obtained by carburization of tungsten wires in xylene at ~2300–2800 K and subsequent annealing of wires carburized in a vacuum at 1170 K for 250 h. Later [17] the orthorhombic phase β'-W_2C was observed by the XRD method. Samples of W_2C were prepared by arc-melting of W and WC mixtures with subsequent annealing at 2970 K for 1 h, lowering of the temperature to 1120 K during 418 h, and final annealing at 1120 K for 146 h. Single crystals of W_2C having the size of 0.2–0.4 mm were taken from the annealed W_2C samples. These single crystals were used [17] for the X-ray determination of the ordered phase symmetry.

According to neutron diffraction data [18, 19], the orthorhombic phase β'-W_2C with the ζ-Fe_2N structure was observed in W_2C samples only after annealing at $T < 1300$ K. Considering the results [13, 19], the orthorhombic (space group *Pbcn*) modification β'-W_2C (ζ-W_2C) exists at temperatures below 1300 K.

The authors [8, 9] supposed the presence of the ordered rhombohedral phase β''-W_2C (α-W_2C) without adducing any structural evidence, while making reference to the paper [20]. This paper is dedicated to electron diffraction analysis of the lower carbide W_2C prepared by carburization of a thin film of metallic tungsten in the carbon oxide CO at ~1400 K for 5 min. With limited data and without looking at other models, the authors [20] conjectured that the lower carbide β''-W_2C has a rhombohedral (space group No. 164, $P\overline{3}m1$ (D^3_{3d})) structure of the *C*6-type (anti-CdI_2). According to [8], the phase β''-W_2C exists at temperatures of 2300 to 1500 K. Without adducing any structural evidence, the authors [16] report observation of the phase β''-W_2C (α-W_2C) in samples of the lower tungsten carbide prepared at temperatures of 2420–2520 K. The existence of the rhombohedral phase β''-W_2C was not confirmed afterwards.

The trigonal (space group $P\overline{3}1m$) phase ε-W_2C was detected in W_2C samples synthesized by solid-phase sintering of tungsten and carbon mixtures at temperatures of 2370–2670 K [7], and 1920 K [21] and 2070 K [22] and those produced by arc-melting [23] of tungsten and carbon mixtures. In Refs. [18, 19] and [24] the samples of the lower tungsten carbide were produced by annealing of carbon-saturated tungsten plates at 2370–2470 K. Then the synthesized W_2C samples were ground and the powders were annealed at 1470 K for 100 h. Neutron diffraction patterns of the annealed powders of the lower tungsten carbide were taken at room temperature of 300 K and over the temperature interval of 1520–2070 K. The neutron and the electron diffraction analysis revealed [23] the trigonal carbide ε-W_2C with the structure of the ε-Fe_2N type in the annealed samples. Detailed neutron diffraction measurements [19] showed that the ordered trigonal (space group $P\overline{3}1m$) carbide ε-W_2C is the main phase and the hexagonal tungsten monocarbide WC is the impurity phase in the annealed powders. According to [19], the β-$W_2C \leftrightarrow \varepsilon$-$W_2C$ disorder–order transformation is the second-order (second-kind) transition, while the transition temperature changes between 1920 and 2100 K depending on the concentration of carbon in the lower tungsten carbide. As to the ordered orthorhombic (space group *Pbcn*) carbide β'-W_2C, it follows from neutron diffraction patterns [19] that a small amount of this carbide is formed in W_2C samples annealed at 1190–1300 K or lower. Annealing of the W_2C powder at 1170 K for 550 h led to formation of the stable orthorhombic phase β'-W_2C.

The dependence of the unit cell lattice constants on the composition of the trigonal (space group $P\overline{3}1m$) carbide ε-W_2C [7, 18, 19, 21, 23] is given in Table 3.1.

Samples of W_2C prepared by arc-melting of W and WC mixtures with subsequent annealing at 1720–2120 K for 1.5 h and quenching to room temperature were studied [25, 26] by the XRD method. The structure was determined in W_2C single crystals 0.2–0.5 mm in size taken from quenched samples. The authors [25, 26] found that at a temperature of ~2120 K the disordered lower carbide β-W_2C transforms to the trigonal ordered phase ε-W_2C. According to [26], the trigonal phase ε-W_2C is thermodynamically stable at temperatures of 2120–1520 K.

In addition to the assumption of the sequential phase transformations "disordered hexagonal phase β-W_2C → ordered orthorhombic phase β'-W_2C → ordered rhombohedral phase β''-W_2C", which take place in the lower tungsten carbide with decreasing temperature, it was conjectured [8, 9] that at $T \leq 1530$ K the carbide β''-W_2C decomposes to tungsten W and the higher carbide WC. To confirm their conjecture, the authors [8, 9] referred to the study [27], in which the lower carbide

Table 3.1 Dependence of the lattice constants of a unit cell on the composition of trigonal carbide ε-W_2C (WC_y)

ε-W_2C (WC_y) carbide composition obtained from		Lattice constants of a unit cell (nm)		References
Chemical analysis	Occupancies of nonmetal atomic positions of a unit cell according to XRD data	*a*	*c*	
$WC_{0.415}$	–	0.5182	0.4720	[18, 19]
$WC_{0.43}$	–	0.51833	0.47240	[21]
$WC_{0.445}$	$WC_{0.423}$	0.51852	0.47232	[21]
$WC_{0.45}$	$WC_{0.42}$	0.51809	0.47216	[21]
$WC_{<0.5}$	–	0.5184	0.4721	[7]
$WC_{0.460}$	$\sim WC_{0.50}$	0.5190	0.4724	[18, 19]
$WC_{0.468}$	–	0.5194	0.4721	[18, 19]

W_2C could not be synthesized by sintering of a W and C mixture at a temperature of 1320 K for ~100 h.

The possible solid-phase decomposition of the lower tungsten carbide is also suggested by results reported in Ref. [26]: Two-phase (W + W_2C) and (W_2C + WC) samples, which were prepared by arc-melting, were annealed at 1170 K for 850 h. The higher carbide WC was formed after low-temperature annealing of (W + W_2C) samples and tungsten W appeared after annealing of (W_2C + WC) samples. Therefore, the authors [26] assumed that at a temperature of about 1520 K the ordered trigonal phase ε-W_2C decomposes by an eutectoid reaction to tungsten W and the higher tungsten carbide WC, with the intermediate decomposition product being the metastable ordered orthorhombic (space group *Pbcn*) phase β'-W_2C having the ζ-Fe_2N structure. The same authors supposed the solid-phase decomposition of W_2C in their earlier study [17]. Notice that the assumption [17, 26] of the eutectoid decomposition of the lower tungsten carbide contradicts results of the neutron diffraction [19]: There was no evidence that W_2C decomposed to W and WC when a single-phase sample of W_2C ($WC_{0.50}$) was annealed at 1170 K for a long time.

Although new experimental data on phase transformations in the lower tungsten carbide W_2C have appeared in the literature, handbooks on phase diagrams give 40-year-old results for the W–C system that are not quite reliable. For example, the handbook [28] contains a generalized phase diagram of the W–C system, which was constructed using earlier phase diagrams [8–10] of this system and considering in part results of the study [29]. The phase diagram of the W–C system in the handbook [30] is adopted from [9, 10].

Thus, the literature data on ordering of the lower tungsten carbide W_2C and the temperature range of W_2C stability are contradictory and ambiguous. In this connection, the theoretical symmetry analysis of possible ordered phases of W_2C carbide was performed in the studies [22, 31–33]. For detection of the possible ordered phases, the crystal structure of the lower tungsten carbide, which was synthesized in different conditions over a wide temperature interval of ~3600 to 1370 K, was examined [22, 31–33] by the neutron and X-ray diffraction methods. The authors [32, 33] proposed a renovated phase diagram of the W–C system considering literature data and obtained experimental and theoretical results. Let us discuss the results of studies [22, 31–33] in more detail.

3.1 Atomic and Vacancy Ordering of Lower W_2C Tungsten Carbide

Disorder–order or order–order transformations, which take place with decreasing temperature, represent transitions from the state with a large free energy to the state with a smaller free energy. The state of solids during atomic or atomic-vacancy ordering can be characterized by the Landau thermodynamic potential, which in this case is the functional of probabilities of finding some species of atoms at lattice sites, coordinates of sites and the temperature. In turn, the probabilities are the functions of the long-range order parameters η. The Landau potential has several minimums, which correspond to high-symmetry disordered and low-symmetry ordered phases, and depends on the crystal lattice symmetry. As the temperature decreases, the transition from the disordered phase to any of the ordered phases or from one ordered phase to another takes place only if the symmetry is reduced. The symmetry analysis provides the quantitative evaluation of the symmetry reduction after formation of some superstructure and allows determining the physically reasonable sequence of its formation.

The symmetry analysis of possible ordered phases of the lower tungsten carbide W_2C, which was carried out in the studies [32, 33], is described below.

3.1.1 Symmetry Analysis of Possible Ordered Phases in the Lower Tungsten Carbide

Let us perform the symmetry analysis of ordered phases of the lower tungsten carbide W_2C, i.e. establish corresponding channels of the disorder–order transition, calculate the distribution functions of carbon atom, and determine the change of the symmetry upon transition from one phase to another. The procedure of calculation of the superstructure vectors of the reciprocal lattice, the disorder-order phase transition channel and the distribution function is described in detail elsewhere [1, 2].

The high-temperature phase β-W_2C has the hexagonal structure of the $L'3$-type with a disordered arrangement of carbon atoms C and structural vacancies $\square$ at sites 2(a) having the coordinates (0 0 0) and (0 0 1/2). The volume of the unit cell of β-W_2C carbide is $V = (\sqrt{3}/2)a^2c$. In the disordered lower carbide β-W_2C with the $L'3$ structure the Ising lattice, in which atomic-vacancy ordering can take place, is the hexagonal nonmetallic sublattice. In the carbide β-W_2C the lattice constants

$a_x = a_y = a$, while the lattice constant a_z of the translational primitive (concerning the nonmetallic sublattice) cell in the direction [0 0 1] is twice as small as the lattice constant c of the unit cell, i.e. $a_z = c/2$ (see Fig. 3.1). Therefore, structural vectors of the reciprocal lattice of the basic hexagonal lattice are $\boldsymbol{b}_1 = (1,\ 0,\ 0)$ and $\boldsymbol{b}_2 = (0,\ 1,\ 0)$ in units $4\,\pi/(a\sqrt{3})$, while $\boldsymbol{b}_3 = (0,\ 0,\ 1)$ in units $4\pi/c$.

The assumed modification β'-W_2C with the structure of the ζ-Fe_2N type has the orthorhombic (space group *Pbcn*) symmetry, although the metal sublattice remains to be a hexagonal closely packed one similarly to the lattice of the high-temperature carbide β-W_2C. In this structure tungsten atoms W are at sites 8(d) with the coordinates ($\sim$ 1/4 $\sim$ 1/8 $\sim$1/12), while atoms C occupy positions 4(c) with the coordinates (0 $\sim$3/8 1/4). Positions 4(c) with the coordinates (0 $\sim$7/8 1/4) are vacant. Lattice constants of the perfect orthorhombic phase β'-W_2C are $a_{\mathrm{orthorh}} = c_{L'3}$, $b_{\mathrm{orthorh}} = 2a_{L'3}$ and $c_{\mathrm{orthorh}} = (\sqrt{3})a_{L'3}$, while the volume of the unit cell of this phase is $V_{\mathrm{orthorh}} = (2\sqrt{3})a^2c$. The phase β'-W_2C (Fig. 3.3) is ordered by the nonmetallic sublattice as compared to the high-temperature hexagonal phase β-W_2C (see Fig. 3.1).

Calculations show that one nonequivalent superstructure vector $-(1/2\,1/2\,-1/2)$, which corresponds to the ray $\boldsymbol{k}_{14}^{(3)} = -(\boldsymbol{b}_1 + \boldsymbol{b}_2 - \boldsymbol{b}_3)/2$ of the three-ray star $\{\boldsymbol{k}_{14}\}$, is present in the first Brillouin zone of the disordered nonmetallic hexagonal sublattice when the orthorhombic phase is formed (here and henceforth the stars $\{\boldsymbol{k}_s\}$ of the wave vectors of the hexagonal lattice are numbered and described according to [34, 35]). Thus, the β-W_2C (space group $P6_3/mmc$) $\rightarrow$ β'-W_2C (space group *Pbcn*) disorder–order transition channel includes one superstructure vector $\boldsymbol{k}_{14}^{(3)}$.

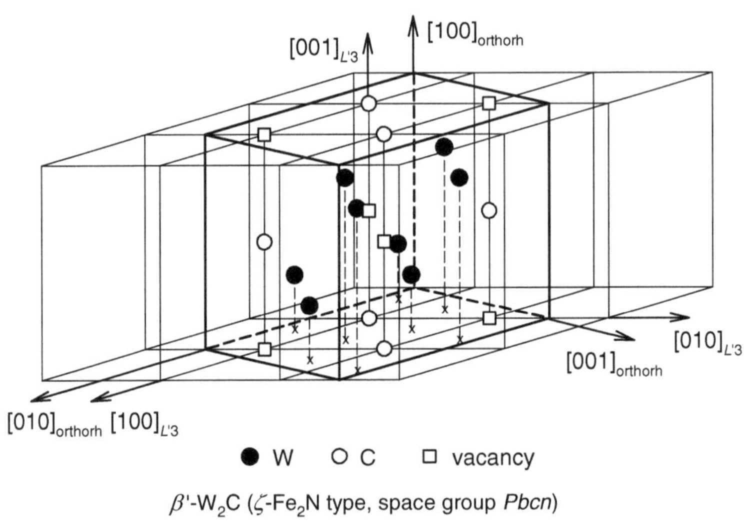

Fig. 3.3 Position of the unit cell of the ordered orthorhombic (space group *Pbcn*) phase β'-W_2C (ζ-W_2C) of the ζ-Fe_2N type in the basic lattice with the $L'3$ structure. (•) tungsten atoms W, (∘) carbon atoms C, and (□) vacant sites of the nonmetallic sublattice

The structure of ordered phases is conveniently described by the distribution function $n(\boldsymbol{r})$, which is the probability that an atom of the given species is located at a site $\boldsymbol{r}$ of the ordering sublattice. The deviation of the probability $n(\boldsymbol{r})$ from its value for a disordered (statistical) distribution can be presented as a superposition of several plane concentration waves [36]. The wave vectors of these waves are superstructure vectors forming the disorder–order transition channel [1, 2]. In the method of static concentration waves [36] the distribution function $n(\boldsymbol{r})$ has the form (0, 0, 1)

$$n(\boldsymbol{r}) = y + \frac{1}{2}\sum_{s}\sum_{j\in s}\eta_s\gamma_s\left[\exp(i\varphi_s^{(j)})\exp(i\boldsymbol{k}_s^{(j)}\boldsymbol{r}) + \exp(-i\varphi_s^{(j)})\exp(-i\boldsymbol{k}_s^{(j)}\boldsymbol{r})\right], \tag{3.1}$$

where y is the fraction of sites occupied by atoms of the given species in the ordering sublattice; η_s is the long-range order parameter corresponding to the star $\{\boldsymbol{k}_s\}$; $\boldsymbol{k}_s^{(j)}$ is the superstructure vector of the star $\{\boldsymbol{k}_s\}$ generating the concentration wave; $\eta_s\gamma_s$ and $\varphi_s^{(j)}$ are the amplitude and the phase shift of the concentration wave respectively.

Considering (3.1), the distribution function of carbon atoms in the orthorhombic (space group *Pbcn*) phase β'-W_2C depends on one long-range order parameter η_{14}, which corresponds to the star $\{\boldsymbol{k}_{14}\}$. In accordance with calculations, this function has the form

$$n(x_I, y_I, z_I) = y + (\eta_{14}/2)\cos\pi(x_I + y_I - 2z_I), \tag{3.2}$$

where $y \le 0.5$ is the relative concentration of carbon in the carbide β'-$W_2C(WC_y)$; x_I, y_I and z_I are coordinates of sites $\boldsymbol{r}$ in the nonmetallic sublattice of the basic disordered hexagonal phase. The distribution function (3.2) takes two values (Table 3.2) at sites relating to different positions in the nonmetallic sublattice of the orthorhombic β'-W_2C superstructure. Values of the distribution function at sites relating to different positions in the nonmetallic sublattice of any ordered phase are the occupancy of these sites by carbon atoms or, what is the same, the probability that carbon atoms are present at these sites.

In the case of disorder–order or order–order transformations, the change of the point (rotational) symmetry N_p upon transition from the phase 1 to the phase 2 equals the ratio of the order of point symmetry groups of the phases 1 and 2, i.e. $N_p = n_p(G_1)/n_p(G_2)$. The order $n_p(G)$ of a point symmetry group G equals the number of symmetry elements in this group. The change of the translational symmetry, N_t, is equal to the ratio between the volumes V_2 and V_1 of the unit cells of the phases 2 and 1 or the ratio of the number of crystallographic sites m_2 and m_1 in the unit cells of these phases, i.e. $N_t = V_2/V_1 \equiv m_2/m_1$. The total change of the symmetry N upon transition from the phase 1 to the phase 2 equals the product of the change of the point symmetry N_p by the change of the translational symmetry N_t, i.e. $N = N_pN_t = n_p(G_1)V_2/n_p(G_2)/V_1 \equiv [n_p(G_1)/V_1]/[n_p(G_2)/V_2]$.

The point symmetry group $mmm(D_{2h})$ of the orthorhombic carbide β'-W_2C includes 8 symmetry elements H_1, H_4, H_7, H_9, H_{13}, H_{16}, H_{19} and H_{21}, while the point group $6/mmm(D_{6h})$ of the basic (parent) disordered phase β-W_2C has 24 elements $H_1 - H_{24}$ [1, 2, 34, 35]. Therefore, the point (rotational) symmetry

Table 3.2 Orthorhombic (ζ-Fe_2N-type, space group No. 60, *Pbcn* (D_{2h}^{14})) ordered phase β'-W_2C (ζ-W_2C), $Z = 4$: $\boldsymbol{a}_{\rm orthorh} = \langle 001\rangle_{L'3}$, $\boldsymbol{b}_{\rm orthorh} = 2\langle 100\rangle_{L'3}$, $\boldsymbol{c}_{\rm orthorh} = \langle 120\rangle_{L'3}$

Atom	Position and multiplicity	Atomic coordinates in the basic disordered structure			Atomic coordinates in the ideal ordered structure			Values for the distribution function $n(x_{\rm I}, y_{\rm I}, z_{\rm I})$
		$x_{\rm I} = x/a_{L'3}$	$y_{\rm I} = y/b_{L'3}$	$z_{\rm I} = z/c_{L'3}$	$x/a_{\rm orthorh}$	$y/b_{\rm orthorh}$	$z/c_{\rm orthorh}$	
C1 (vacancy)	4(*c*)	1	0	0	0	7/8	1/4	$n_1 = y - \eta_{14}/2$
C2	4(*c*)	0	0	0	0	3/8	1/4	$n_2 = y + \eta_{14}/2$
W	8(*d*)	−2/3	−1/3	1/4	1/4	1/8	1/12	

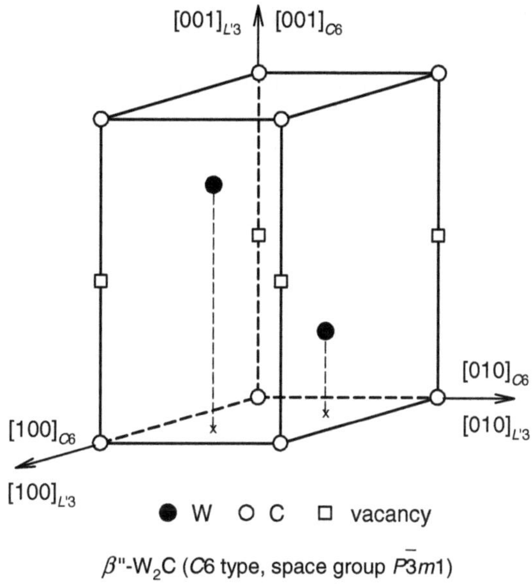

Fig. 3.4 Position of the unit cell of the ordered rhombohedral (space group $P\overline{3}m1$) phase β''-W_2C of the $C6$-type in the basic lattice with the $L'3$ structure. (•) tungsten atoms W, (∘) carbon atoms C, and (□) vacant sites of the nonmetallic sublattice

reduction N_p equals 3. When going from the disordered carbide β-W_2C to the orthorhombic carbide β'-W_2C, the unit cell volume increases by a factor of 4 and, hence, the translational symmetry reduction N_t is equal to 4. Therefore, the total symmetry reduction is $N = N_pN_t = 12$ upon the β-W_2C (space group $P6_3/mmc$) → β'-W_2C (space group $Pbcn$) transition.

Translation vectors of the unit cell of the rhombohedral (space group $P\overline{3}m1$) phase β''-W_2C with the structure of the $C6$-type coincide with translation vectors of the unit cell of the disordered hexagonal phase (Fig. 3.4) and are equal to $\boldsymbol{a}_{C6} = [100]_{L'3}$, $\boldsymbol{b}_{C6} = [010]_{L'3}$ and $\boldsymbol{c}_{C6} = [001]_{L'3}$.

Therefore, the volume of the unit cell of the rhombohedral phase $V_{C6} = (\sqrt{3}/2)a^2c$ equals the volume of the unit cell of the disordered carbide β-W_2C. In a ideal case, two atoms W occupy positions 2(d) with the coordinates (1/3 2/3 z) and (2/3 1/3 $-z$), where $z = 0.25$, the atom C is in the position 1(a) with the coordinates (0 0 0), and the vacant site is in the position 1(b) with the coordinates (0 0 1/2) in the phase β''-W_2C. The comparison with the disordered carbide β-W_2C (see Fig. 3.1) clearly shows that the positions 2(a), which are statistically half-filled with atoms C, split to the positions 1(a) and 1(b) (one position being occupied by carbon atoms, while the other being vacant) as a result of rhombohedral ordering.

According to calculations, the phase β''-W_2C is formed by the disorder–order transition channel including the ray $\boldsymbol{k}_{17}^{(1)} = \boldsymbol{b}_3/2$ of the star $\{\boldsymbol{k}_{17}\}$. Therefore, the carbon atom distribution function at sites $\boldsymbol{r}$ in the nonmetallic sublattice of the ordered rhombohedral phase β''-$W_2C(WC_y)$ with any degree of order has the form

$$n(x_I, y_I, z_I) = y + (\eta_{17}/2)\cos 2\pi z_I \quad (3.3)$$

Table 3.3 Rhombohedral ($C6$-type (anti-CdI_2)), space group No. 164, $P\overline{3}m1(D^3_{3d})$ ordered phase β''-W_2C, $Z = 1 : \boldsymbol{a}_{C6} = \langle 100\rangle_{L'3}, \boldsymbol{b}_{C6} = \langle 010\rangle_{L'3}, \boldsymbol{c}_{C6} = \langle 001\rangle_{L'3}$

Atom	Position and multiplicity	Atomic coordinates in the ideal ordered structure			Values for the distribution function $n(x_I, y_I, z_I)$
		$x/a_{C6} \equiv x/a_{L'3}$	$y/b_{C6} \equiv y/b_{L'3}$	$z/c_{C6} \equiv z/c_{L'3}$	
C1 (vacancy)	$1(b)$	0	0	1/2	$n_1 = y - \eta_{17}/2$
C2	$1(a)$	0	0	0	$n_2 = y + \eta_{17}/2$
W	$2(d)$	1/3	2/3	1/4	

and depends on one long-range order parameter η_{17}. Values of the distribution function (3.3) at sites in the nonmetallic sublattice of the rhombohedral superstructure are given in Table 3.3.

The point symmetry group $\overline{3}m(D_{3d})$ of the rhombohedral carbide β''-W_2C includes 12 symmetry elements H_1, H_3, H_5, H_8, H_{10}, H_{12}, H_{13}, H_{15}, H_{17}, H_{20}, H_{22} and H_{24} out of 24 elements of the hexagonal group 6/*mmm* [1, 2, 34, 35]. Therefore, the rotational symmetry reduction is $N_p = 2$. Since the unit cell volume does not change in going from the disordered to the ordered state, the translational symmetry reduction is 1, while the total symmetry reduction upon the β-W_2C (space group $P6_3/mmc$) $\rightarrow$ β''-W_2C (space group $P\overline{3}m1$) transformation equals 2.

The trigonal (space group $P\overline{3}1m$) phase ε-W_2C has the unit cell (Fig. 3.5) with translation vectors $\boldsymbol{a}_\varepsilon = [1\ -10]_{L'3}$, $\boldsymbol{b}_\varepsilon = [120]_{L'3}$ and $\boldsymbol{c}_\varepsilon = [001]_{L'3}$ and the volume $V_\varepsilon = (3\sqrt{3}/2)a^2c$.

In the perfect trigonal superstructure, atoms W occupy positions 6(k) with the coordinates (1/3 0 1/4), atoms C are at sites 1(a) with the coordinates (0 0 0) and 2(d) with the coordinates (1/3 2/3 1/2), whereas site 1(b) with the coordinates (0 0 1/2) and sites (2(c) with the coordinates (1/3 2/3 0) are vacant.

Calculations of the superstructure vectors of the reciprocal lattice of the trigonal phase ε-W_2C and their translation show that three nonequivalent superstructure vectors are present in the first Brillouin zone of the disordered nonmetallic hexagonal sublattice. The first vector corresponds to the ray $\boldsymbol{k}_{17}^{(1)} = \mathrm{b}_3/2$ of the Lifschitz one-ray star $\{\boldsymbol{k}_{17}\}$, while the other two vectors correspond to the rays $\boldsymbol{k}_{15}^{(1)} = (\boldsymbol{b}_1+\boldsymbol{b}_2)/3+\boldsymbol{b}_3/2$ and $\boldsymbol{k}_{15}^{(2)} = -\boldsymbol{k}_{15}^{(1)}$ of the Lifschitz two-ray star $\{\boldsymbol{k}_{15}\}$. Thus, the disorder–order phase transition channel, which is connected with formation of the ordered trigonal phase ε-W_2C, includes three superstructure vectors $\boldsymbol{k}_{15}^{(1)}, \boldsymbol{k}_{15}^{(2)}$ and $\boldsymbol{k}_{17}^{(1)}$. The presence of Lifschitz star rays alone in the transition channel means that formation of the trigonal superstructure satisfies the Landau criterion for second-order (second-kind) phase transitions. This is in agreement with results of the studies [19, 22], according to the β-$W_2C \leftrightarrow \varepsilon$-$W_2C$ disorder-order transformation is a second-order phase transition. However, the fulfillment of the Landau criterion does not exclude the possibility that the β-$W_2C \leftrightarrow \varepsilon$-$W_2C$ transformation is realized by the mechanism of the first-order phase transition.

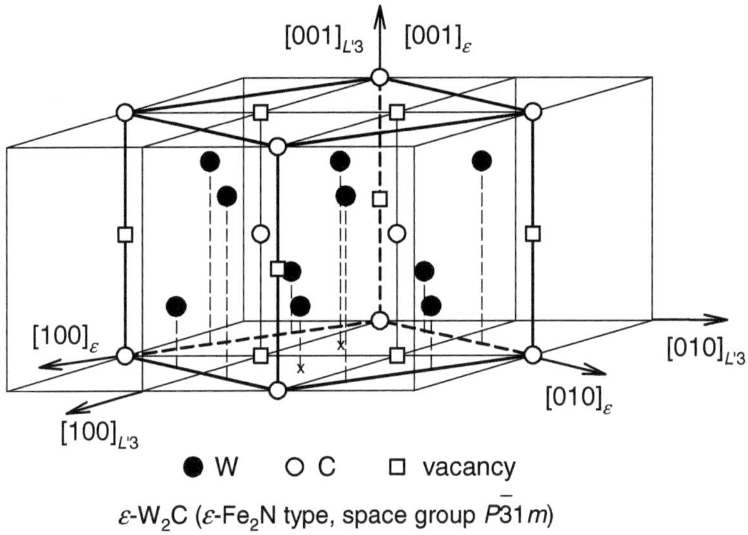

Fig. 3.5 Position of the unit cell of the ordered trigonal (space group $P\bar{3}1m$) phase ε-W_2C of the ε-Fe_2N type in the basic lattice with the $L'3$ structure. (•) tungsten atoms W, (○) carbon atoms C, and (□) vacant sites of the nonmetallic sublattice

The carbon atom distribution function for the ordered trigonal phase ε-$W_2C_{y'}$ (WC_y, $y \le 0.5$) has the form

$$n(x_I, y_I, z_I) = y - (\eta_{17}/6)\cos 2\pi z_I + (2\eta_{15}/3)\cos[(2\pi/3)(x_I + y_I + 3z_I)]. \quad (3.4)$$

This function depends on two long-range order parameters η_{17} and η_{15} corresponding to the stars $\{\boldsymbol{k}_{17}\}$ and $\{\boldsymbol{k}_{15}\}$. It takes four values (Table 3.4) at sites in the nonmetallic sublattice of the trigonal phase ε-W_2C.

The ray $\boldsymbol{k}_{17}^{(1)}$ of the star $\{\boldsymbol{k}_{17}\}$ and the long-range order parameter η_{17} determine the alternation of nonmetal atomic planes $(00z)_\varepsilon$, which are perpendicular to the c-axis and differ in the occupancy by carbon atoms C: the planes $(00z)_\varepsilon$ with the occupancy 1/3 (at $z = 0$) and 2/3 (at $z = 1/2$) alternate in the perfect ordered phase ε-W_2C (Fig. 3.5). The rays $\boldsymbol{k}_{15}^{(1)}$ and $\boldsymbol{k}_{15}^{(2)}$ of the star $\{\boldsymbol{k}_{15}\}$ and the long-range order parameter η_{15}determine the location of carbon atoms C and vacancies □ in nonmetal planes $(00z)_\varepsilon$.

The rotational symmetry reduction N_p is 2 and the unit cell volume increases by a factor of 3 upon the β-W_2C (space group $P6_3/mmc$) → ε-W_2C (space group $P\bar{3}1m$) transition. Taking this into account, when going from the disordered carbide β-W_2C to the trigonal carbide ε-W_2C, the total symmetry reduction $N = N_p N_t$ is equal to 6.

The change of the symmetry during formation of possible orthorhombic and rhombohedral phases is such that the sequence of "disordered hexagonal phase

Table 3.4 Perfect trigonal (ε-Fe_2N-type, space group No. 162, $P\overline{3}1m(D_{3d}^1)$) ordered phase ε-W_2C, $Z = 3$: $\boldsymbol{a}_\varepsilon = \langle 1-10 \rangle_{L'3}$, $\boldsymbol{b}_\varepsilon = \langle 120 \rangle_{L'3}$, $\boldsymbol{c}_\varepsilon = \langle 001 \rangle_{L'3}$

Atom	Position and multiplicity	Atomic coordinates in the basic disordered structure			Atomic coordinates in the ideal ordered structure			Values for the distribution function $n(x_I, y_I, z_I)$
		$x_I = x/a_{L'3}$	$y_I = y/b_{L'3}$	$z_I = z/c_{L'3}$	x/a_ε	y/b_ε	z/c_ε	
C1 (vacancy)	1(*b*)	0	0	1/2	0	0	1/2	$n_1 = y + \eta_{17}/6 - 2\eta_{15}/3$
C2 (vacancy)	2(*c*)	1	1	0	1/3	2/3	0	$n_2 = y - \eta_{17}/6 - \eta_{15}/3$
C3	1(*a*)	0	0	0	0	0	0	$n_3 = y - \eta_{17}/6 + 2\eta_{15}/3$
C4	2(*d*)	1	1	1/2	1/3	2/3	1/2	$n_4 = y + \eta_{17}/6 + \eta_{15}/3$
W	6(*k*)	1/3	−1/3	1/4	1/3	0	1/4	

β-W_2C $\rightarrow$ ordered orthorhombic phase β'-W_2C $\rightarrow$ ordered rhombohedral phase β''-W_2C" transformations, which is proposed in Refs. [8] and [10], is physically impossible, because the symmetry increases rather than decreases during the β'-W_2C $\rightarrow$ β''-W_2C transition. If all the three superstructures are formed at different stages of ordering of W_2C carbide, the symmetry analysis suggests the only possible sequence of phase transformations, which take place with decreasing temperature and do not contradict the symmetry variation: Disordered hexagonal (space group $P6_3/mmc$) phase β-W_2C $\rightarrow$ ordered rhombohedral (space group $P\overline{3}m1$) phase β''-W_2C $\rightarrow$ ordered trigonal (space group $P\overline{3}1m$) phase ε-W_2C $\rightarrow$ ordered orthorhombic (space group *Pbcn*) phase β'-W_2C. In this case, the symmetry is reduced twice on transition from the disordered hexagonal to the rhombohedral carbide, then thrice on transition from the rhombohedral to the trigonal carbide, and finally twice on transition from the trigonal to the orthorhombic carbide. If some ordered phase is not detected in experiment, the transformation sequence is still physically correct even without this phase.

Considering the above analysis, let us discuss results of the experimental study into the structure of the lower tungsten carbide W_2C.

3.1.2 Real Structure of Ordered Phase in the Lower Tungsten Carbide

Samples of the lower tungsten carbide W_2C were prepared by solid-phase sintering of W+C pressed powder mixtures in a vacuum of 0.0013 Pa (10^{-5} mm Hg) under three different temperature routes: (I) sintering at a temperature of 2070 K for 10 h; (II) annealing of the samples, which were sintered from tungsten and carbon at 2070 K, in a vacuum at $T = 1370$ K for 35 h; (III) sintering of W+C powder mixtures in a vacuum at 1370 K for 50 h. A tungsten powder with particles 3–5 μm in size and purity 99.95 % and carbon black with purity 99.999 % were the starting materials for solid-phase vacuum sintering. The cylindrical samples 15 mm in diameter and 15–20 mm in length were fabricated by cold-pressing of W+C powder mixtures at a pressure of 0.7 GPa and then were placed in a graphite crucible for sintering. The W_2C powder was also produced by plasma chemical synthesis from the oxide WO_3 and propane C_3H_8 in hydrogen plasma at a temperature of ~3800 K. The reaction products condensed and were quenched near the reactor walls from a temperature of 2300–2400 K. The composition of the tungsten carbide samples obtained was certified by chemical and spectral methods. The metal impurity content of the starting powders and of the sintered samples was determined by inductively coupled plasma-mass spectrometry on a Spectromass 2000. The total impurity content of the samples as determined from the results of chemical and spectral analysis did not exceed 0.25 wt.%.

The phase composition of the samples and the lattice constants of the phases were determined by the XRD method in a DRON-UM1 diffractometer using $CuK\alpha_{1,2}$

radiation at the angles 2θ between 10° and 140° in steps $\Delta 2\theta = 0.03°$ with the scan time of 2 s at each point.

The structure of the ordered phases of the lower carbide W_2C was studied by the neutron diffraction method using *D*7a ($\lambda = 0.1532$ nm) and *D*2 ($\lambda = 0.1805$ nm) diffractometers installed in the horizontal channel of the research water-water atomic reactor IVV-2M (Zarechny, Russia). The diffractometer resolution $\Delta d/d$ was equal to 0.3 %. The primary beam was monochromated by reflection of neutrons from two monochromators: neutrons were reflected in sequence from the (002) face of pyrolytic graphite and the (333) face of a Ge single crystal in the case of the *D*7a diffractometer; in the case of the *D*2 diffractometer neutron beam was reflected from the (111) face of a germanium single crystal and the (002) face of pyrolytic graphite. Neutron diffraction patterns were recorded at room temperature in a step-scanning mode with steps of $\Delta(2\theta) = 0.1°$ within the region 2θ from 5–10° to 115–125°. The neutron diffraction analysis of the plasma chemical W_2C powder was performed using the exposure time of 260 and 70 s at each point with neutrons having the wavelength of 0.1532 and 0.1805 nm respectively. Powder samples were kept in vanadium containers. Neutron diffraction measurements in compact sintered samples of W_2C were made with neutrons of wavelength 0.1532 nm at the exposure time of 150 s per point. Sintered bulk samples of W_2C were fixed were installed on a vertical aluminium holder. The GSAS program package [37] was used for phase structure refinements from powder data. Background was fitted by the Chebyshev 6th order polynomial. The reflection shape was modeled as a pseudo-Voigt function.

The average size of coherent scattering regions $\langle D \rangle$ was estimated from diffraction reflection broadening by the Warren method [38] from the formula

$$\langle D \rangle = \frac{K_{hkl} \cdot \lambda}{\cos\theta \cdot \beta(2\theta)}, \tag{3.5}$$

where λ is the radiation wavelength, θ is the Bragg angle, $K_{hkl} \approx 1$ is the Scherrer constant whose value depends on the crystallite (domain) shape and reflection indices (*hkl*), and $\beta(2\theta) = [(\mathrm{FWHM_{exp}})^2 - (\mathrm{FWHM_R})^2]^{1/2}$ is the broadening of diffraction reflection ($\mathrm{FWHM_{exp}}$ being the full width of the reflection at half maximum (height) and $\mathrm{FWHM_R}$ the instrumental angular resolution function of the diffractometer). The instrumental angular resolution function $\mathrm{FWHM_R}(2\theta) = (u\tan^2\theta + v\tan\theta + w)^{1/2}$ of the DRON-UM1 X-ray diffractometer was determined in a special diffraction experiment with the cubic lanthanum hexaboride LaB_6 (NIST Standard Reference Powder 660a) having the lattice constant $a = 0.41569162$ nm. This angular resolution function has the following parameters: $u = 0.0041$, $v = -0.0021$ and $w = 0.0093$. Instrumental resolution functions of the *D*7a and *D*2 neutron diffractometers were determined using a standard sample of the aluminum oxide Al_2O_3 (corundum) with the lattice constants $a = 0.4789$ nm and $c = 1.2991$ nm.

The microstructure of compact polished samples of the lower tungsten carbide was analyzed in reflected polarized white light under a Leica DM 2500M optical polarization microscope at a magnification of 100 to 1000 times. The microhardness

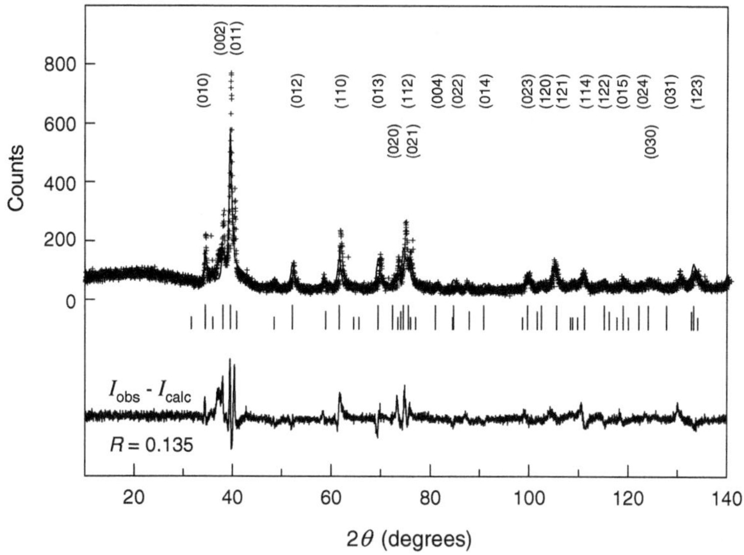

Fig. 3.6 Experimental (*crosses*) and calculated (*solid line*) XRD patterns of unannealed nanocrystalline powder. According to minimization of the XRD pattern, this nanocrystalline sample contains ~75 wt.% W_2C (space group $P6_3/mmc$), ~15 wt.% W and ~10 wt.% WC. According to the results of chemical analysis, this nanopowder also contains ~3.5 wt.% free (uncombined) amorphous carbon. The long, medium and short ticks correspond to reflections of W_2C, W and WC phases. Indices of diffraction reflections of disordered lower tungsten carbide W_2C are pointed. The bottom curve shows the difference ($I_{obs} - I_{calc}$) between the observed and calculated XRD patterns. Experimental XRD pattern is recorded in Cu$K\alpha_{1,2}$ radiation

H_V of disordered and ordered samples of the carbide W_2C was measured on polished sections by the Vickers method.

The XRD pattern of powder, produced by plasma-chemical synthesis, contains diffraction reflections of three phases: The lower W_2C and the higher WC tungsten carbide, and tungsten W (Fig. 3.6). All the reflections are very broad. Broadening is due to a small size of particles of the synthesized powder. The average size of coherent scattering regions was 55 ± 10 nm as estimated from diffraction reflection broadening. Minimization of the XRD pattern showed that the nanocrystalline sample contains ~75 wt.% W_2C (space group $P6_3/mmc$), ~15 wt.% W and ~10 wt.% WC. Under these conditions the reliability factor $R_I(R_{Bragg})$ is 0.135. Taking into account the presence of ordered phases (ε-W_2C, β'-W_2C or β''-W_2C) did not improve convergence because of large broadening of the diffraction reflections. The unit cell lattice constants for the disordered carbide W_2C are equal to $a_{L'3} = 0.3027$ and $c_{L'3} = 0.4766$ nm. The unit cell lattice constant a in the metallic tungsten is 0.3158 nm. The lattice constants of the unit cell of tungsten monocarbide WC are $a = 0.2896$ and $c = 0.2865$ nm.

The chemical analysis showed that the total content of carbon in the nanopowder is ~6.2 wt.% including ~3.6 wt.% free (uncombined) amorphous carbon. According to results of the XRD and the chemical analysis, the nanocrystalline powder contains ~72.5 wt.% W_2C, ~14.5 wt.% W, ~9.5 wt.% WC and ~3.5 wt.% C.

The nanocrystalline powder was studied by neutron diffraction using neutrons of wavelengths 0.1532 and 0.1805 nm. Measurements with neutrons having the large wavelength were made for a more detailed analysis of the small-angle scattering region, which could contain reflections characteristic of some ordered phase. Neutron diffraction patterns of the nanocrystalline powder are given in Fig. 3.7.

Neutron diffraction patterns of the nanocrystalline powder under radiation with $\lambda = 0.1532$ were taken over the interval of interplanar spacings $0.9 \geq d \geq 0.09$ nm. Measurements under long-wave radiation with $\lambda = 0.1805$ nm covered a wider interval of $1.7 \geq d \geq 0.1$ nm. At $2\theta < 32°$ both neutron diffraction patterns clearly exhibit the superstructure reflection $(011)_\varepsilon$ with $d \approx 0.3304$ nm, which is characteristic of the ordered carbide ε-W_2C (Fig. 3.7). Both neutron diffraction patterns contain the same set of diffraction reflections of four phases: the disordered lower hexagonal (space group $P6_3/mmc$) tungsten carbide W_2C; the ordered trigonal (space group $P\overline{3}1m$) carbide ε-W_2C; metallic tungsten W with the bcc (space group $Im\overline{3}m$) structure; and the hexagonal (space group $P\overline{6}m2$) monocarbide WC. Reflections of the ordered orthorhombic (space group $Pbcn$) and rhombohedral (space group $P\overline{3}m1$) carbides β'-W_2C and β''-W_2C are not detected.

A neutron diffraction pattern, which was collected with a large signal accumulation under radiation with $\lambda = 0.1532$ nm, was used for a more detailed analysis of the structure of the observed phases. According to calculations, the content of the detected crystal phases in the nanocrystalline powder is ~67.5 wt.% W_2C, ~9.8 wt.% ε-W_2C, ~14.5 wt.% W, and ~8.2 wt.% WC. Taking into account the presence of free carbon in the nanopowder, the phase content is ~65.0 wt.% W_2C, ~9.5 wt.% ε-W_2C, ~14.0 wt.% W, ~8.0 wt.% WC, and ~3.5 wt.% C. These values are in agreement with the data obtained by minimization of the XRD pattern of the same powder. A comparatively small content of the ordered carbide ε-W_2C and a large (~65.0 wt.%) content of the disordered carbide W_2C in the nanocrystalline powder are explained by quenching of the nanopowder from a high temperature of ~2400 K. The multiple phases of the nanopowder result from the fact that the plasma chemical synthesis reaction is realized far from equilibrium conditions.

The neutron diffraction pattern was used for the structure refinement by the GSAS program [37]. It was established that the unit cell lattice constants in the tungsten monocarbide WC are equal to $a = 0.2893(8)$ and $c = 0.2818(9)$ nm, while the unit cell lattice constant in metallic tungsten is 0.3155(2) nm. As for the lower tungsten carbide W_2C, $a_{L'3} = 0.29947(6)$ and $c_{L'3} = 0.4728(1)$ nm for the disordered phase. These lattice constants correspond to carbide ~$WC_{0.46}$. The lattice constants and the volume of the unit cell in the ordered phase ε-W_2C are equal to $a_\varepsilon = 0.5152(6)$ nm, $c_\varepsilon = 0.4681(2)$ nm, and $V_\varepsilon = 107.568 \times 10^{-3}$ nm^3. Thus, the lattice constants a_ε and c_ε are smaller than the theoretical values of these constants $a_{\varepsilon\ \mathrm{theor}} = a_{L'3}\sqrt{3}$ and $c_{\varepsilon\ \mathrm{theor}} = c_{L'3}$ in the perfect ordered phase. Minimization of the neutron diffraction pattern taking into account the ordered phase ε-W_2C with a fixed occupancy of sites equal to their occupancy in the perfect trigonal superstructure provides satisfactory agreement of the calculated and experimental spectra: the factor $R_I = 0.052$.

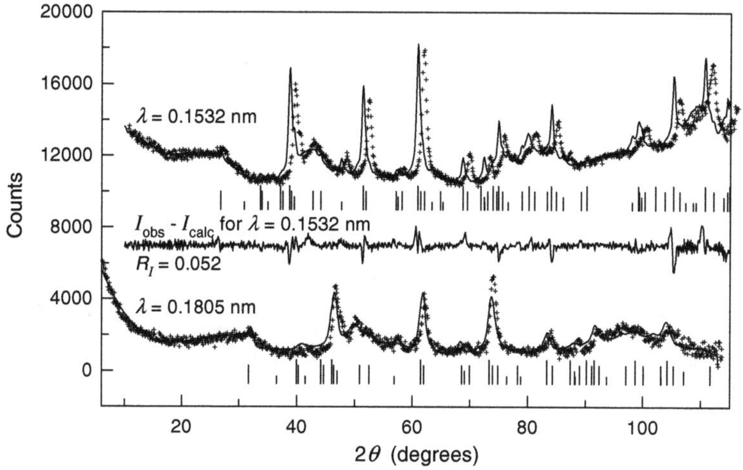

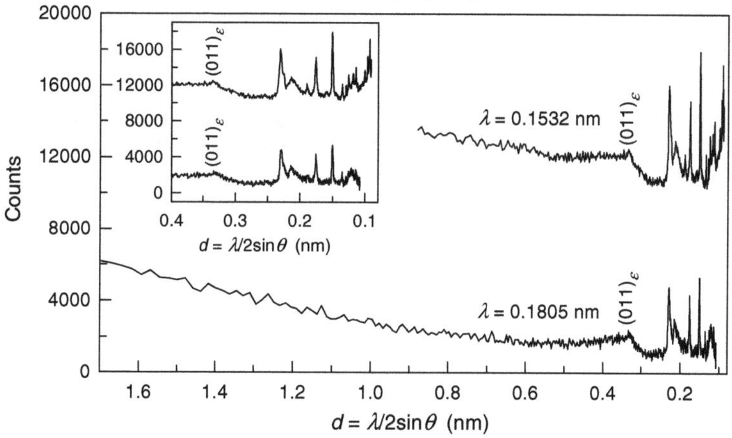

Fig. 3.7 Experimental (*crosses*) and calculated (*solid line*) neutron diffraction patterns ($\lambda = 0.1532$ and 0.1805 nm) of the nanocrystalline powder containing ~65.0 wt.% disordered lower tungsten carbide W_2C, ~9.5 wt.% ordered trigonal (space group $P\overline{3}1m$) carbide ε-W_2C, ~14.0 wt.% W, ~8.0 wt.% WC and ~3.5 wt.% free (uncombined) amorphous carbon C. Difference neutron diffraction pattern is shown for neutrons of wavelength 0.1532 nm only. For comparison, the experimental neutron diffraction patterns are shown as a function of the interplanar spacing $d = \lambda/(2\sin\theta)$. The inset presents the neutron diffraction patterns in the interval $0.5 > d > 0.08$ nm. At $2\theta < 32°$ both neutron diffraction patterns clearly exhibit the superstructure reflection $(011)_\varepsilon$ with $d \approx 0.3304$ nm, which is characteristic of the ordered trigonal carbide ε-W_2C. The longest ticks denote diffraction reflections of the disordered phase W_2C; long, medium and short ticks correspond to reflections of ε-W_2C, W and WC, respectively

Varying of the occupancy of sites in the trigonal superstructure did not lead to considerable improvement of the agreement between experimental and calculated neutron diffraction patterns because of multiple phases of the sample, large broadening of diffraction reflections and a small content of the ordered carbide ε-W_2C. The reli-

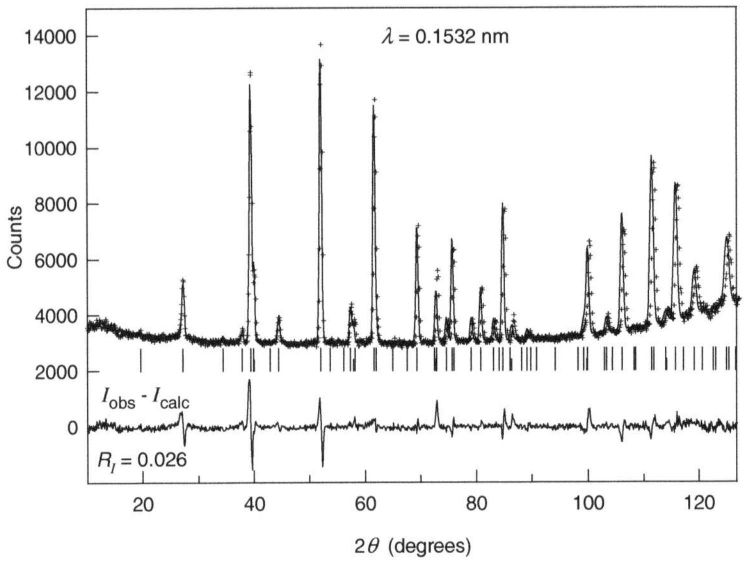

Fig. 3.8 Experimental (*crosses*), calculated (*solid line*) and difference neutron diffraction pattern ($\lambda = 0.1532$ nm) of the lower carbide W_2C synthesized by sintering of W + C powdered mixture at 2070 K for 10 h. The *long* and *short ticks* correspond to reflections of the ordered phase ε-W_2C and tungsten W. The *bottom curve* shows the difference ($I_{obs} - I_{calc}$) between the observed and calculated neutron diffraction patterns

ability factor R_I equals 0.083 for the neutron diffraction pattern ($\lambda = 0.1805$ nm) collected with a small signal accumulation.

To determine the temperature range of existence of the lower tungsten carbide W_2C and the ordered phase ε-W_2C, samples of the lower tungsten carbide were synthesized additionally by solid-phase sintering of W and C powders at temperatures of ~2070 and 1370 K. The neutron diffraction pattern of the lower carbide W_2C, which was synthesized at 2070 K for 10 h, is shown in Fig. 3.8.

The neutron diffraction pattern (Fig. 3.8) contains a set of diffraction reflections, which is characteristic of the ordered trigonal (space group $P\overline{3}1m$) phase ε-W_2C. Also, reflections of cubic (space group $Im\overline{3}m$) tungsten with the lattice constant $a = 0.31651(1)$ nm are observed. The best agreement between the experimental and theoretical spectra ($R_I = 0.0258$, $R_p = 0.0353$, $\omega R_p = 0.0448$) was achieved by taking into account different occupancies of sites by atoms C in the nonmetallic sublattice (Table 3.5).

The occupancy of the sites 1(*a*) and 2(*d*) proved to be less than 1, whereas a small number of carbon atoms occupied the sites 1(*b*) and 2(*c*), which are fully vacant in the perfect ordered phase. The occupancies of sites in the nonmetallic sublattice differ from the ideal values of 1 and 0. It is explained both by the deviation of the composition of the ordered carbide from the stoichiometric composition W_2C, and

Table 3.5 Real structure of trigonal (ε-Fe_2N-type, space group No. 162, $P\bar{3}m1(D^1_{3d})$) ordered carbide ε-$W_2C_y(Z = 3)$ before and after annealing

Atom	Position and multiplicity	[b]Carbide ε-W_2C_y obtained by sintering for 10 h at a temperature of 2070 K: $W_2C_{0.953}$, $a_\varepsilon = b_\varepsilon = 0.51813(3)$ nm, $c_\varepsilon = 0.47283(8)$ nm, $V_\varepsilon = 109.932 \times 10^{-3}$ nm^3					[b]The same carbide ε-W_2C_y after annealing for 35 h at 1370 K: $W_2C_{1.00}$, $a_\varepsilon = b_\varepsilon = 0.51880(2)$ nm, $c_\varepsilon = 0.47273(5)$ nm, $V_\varepsilon = 110.192 \times 10^{-3}$ nm^3				
		Atomic coordinates in the ordered structure			B_{iso} (pm^2)	Occupancy	Atomic coordinates in the ordered structure			B_{iso} (pm^2)	Occupancy
		x/a_ε	y/b_ε	z/c_ε			x/a_ε	y/b_ε	z/c_ε		
C1 ([a]□)	1(*b*)	0	0	1/2	380	0.0175	0	0	1/2	0	0
C2 ([a]□)	2(*c*)	1/3	2/3	0	1330	0.2656	1/3	2/3	0	1120	0.2316
C3	1(*a*)	0	0	0	1150	0.7983	0	0	0	1280	0.7253
C4	2(*d*)	1/3	2/3	1/2	1030	0.7565	1/3	2/3	1/2	910	0.9093
W	6(*k*)	0.328(1)	0	0.253(5)	980	1	0.331(7)	0	0.252(4)	840	1

[a] □ is vacant site (vacancy) of carbon sublattice

[b] The final values of reliability factors R_I, R_p and ωR_p are given in text (see Sect. 3.1.2)

by the fact that the degree of the long-range order in the carbide is less than the maximum one.

Indeed, the relative concentration of carbon in the ordered trigonal phase equals $y = (n_1 + 2n_2 + n_3 + 2n_4)/6$, where n_1, n_2, n_3 and n_4 denote occupancies of nonmetallic sublattice sites located in positions $1(b)$, $2(c)$, $1(a)$ and $2(d)$ respectively (see Tables 3.4 and 3.5). Considering the occupancy of these positions (Table 3.5), the ordered trigonal phase ε-W_2C under study has the composition $W_2C_{0.953}$ ($WC_{0.48}$). Values of the long-range order parameters η_{17} and η_{15} can be determined from the occupancies considering values taken by the distribution function that describes the trigonal superstructure ε-W_2C:

$$\eta_{15} = (n_3 + n_4 - n_1 - n_2)/2, \eta_{17} = n_1 + 2n_4 - 2n_2 - n_3. \tag{3.6}$$

According to the estimates, the long-range order parameters in the ordered trigonal carbide ε-$W_2C_{0.953}$ ($WC_{0.48}$), which was synthesized at ~2070 K, are $\eta_{15} = 0.636$ and $\eta_{17} = 0.201$. The lattice constants and the volume of the unit cell in this carbide ε-W_2C are equal to $a_\varepsilon = b_\varepsilon = 0.51813$ nm, $c_\varepsilon = 0.47284$ nm and $V_\varepsilon = 109.932 \times 10^{-3}$ nm^3, i.e. are larger than the lattice constants a_ε and c_ε and the volume of the unit cell of the same ordered phase in the nanocrystalline powder.

The W_2C sample synthesized at 2070 K for 10 h was annealed additionally at a temperature of 1370 K for 35 h. The X-ray (Fig. 3.9) and neutron (Fig. 3.10) diffraction patterns of the annealed sample exhibit the same set of diffraction reflections of only two phases ε-W_2C and W as in the neutron diffraction pattern of the unannealed sample (see Fig. 3.8).

The annealed and synthesized samples have the same phase composition within the calculation error and contain ~85 ± 2 wt.% ε-W_2C and ~15 ± 2 wt.% W. Refinement of the structure of the ordered phase in the annealed sample showed that the best agreement of the experimental and theoretical neutron diffraction patterns ($R_I = 0.0252$, $R_p = 0.0299$, $\omega R_p = 0.0390$) is achieved when the carbon atom occupancy of vacancy positions $1(b)$ and $2(c)$ decreases and the occupancy of positions $2(d)$ increases considerably (Table 3.5). Annealing changes the ordered phase composition from ε-$W_2C_{0.953}$ ($WC_{0.48}$) to ε-$W_2C_{1.00}$ ($WC_{0.50}$). The long-range order parameters of the trigonal carbide ε-$W_2C_{1.00}$ increase to $\eta_{15} = 0.702$ and $\eta_{17} = 0.630$ after annealing. The lattice constants and the volume of the unit cell of the annealed carbide ε-W_2C equal $a_\varepsilon = b_\varepsilon = 0.51880(2)$ nm, $c_\varepsilon = 0.47273(5)$ nm and $V_\varepsilon = 110.192 \times 10^{-3}$ nm^3, i.e. are larger than those of the same ordered phase before annealing. The lattice constants increase because the carbon concentration of the ordered trigonal carbide changes from $WC_{0.48}$ to $WC_{0.50}$.

The full width at half maximum (FWHM) of the diffraction reflections in X-ray and neutron diffraction patterns of the sintered W_2C samples is not larger than the instrumental width of the corresponding reflections. The absence of broadening means that the size of grains of the ordered phase ε-W_2C in the sintered samples is larger than 200–300 nm [39, 40]. We failed in observing the boundaries of domains of the ordered trigonal phase ε-W_2C by optical polarization microscopy therefore the domain size was supposed to be less than 1 μm. It may be assumed therefore that the

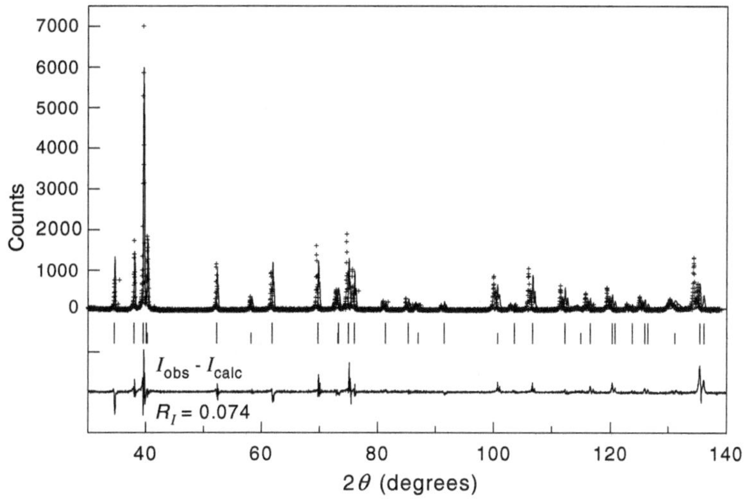

Fig. 3.9 Experimental (*crosses*), calculated (*solid line*) and difference XRD patterns of the lower carbide W_2C after additional annealing at 1370 K for 35 h. The *long* and *short ticks* correspond to reflections of the ordered phase ε-W_2C and tungsten W. The *bottom curve* shows the difference ($I_{obs} - I_{calc}$) between the observed and calculated XRD patterns. Experimental XRD pattern is recorded in $CuK\alpha_{1,2}$ radiation

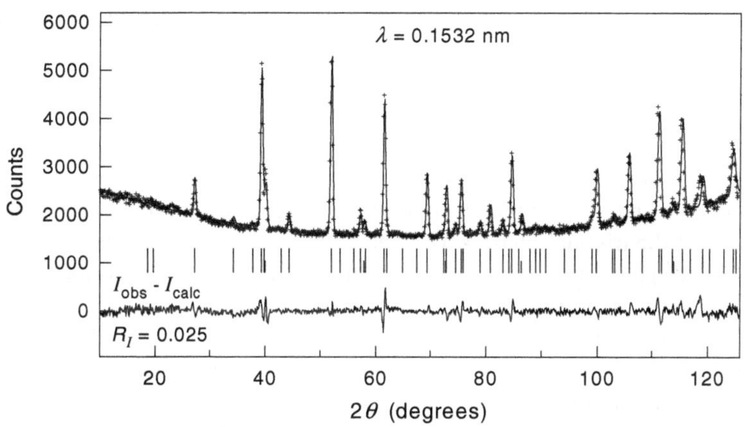

Fig. 3.10 Experimental (*crosses*), calculated (*solid line*) and difference neutron diffraction pattern ($\lambda = 0.1532$ nm) of the lower carbide W_2C after additional annealing at 1370 K for 35 h. The *long* and *short ticks* correspond to reflections of the ordered phase ε-W_2C and tungsten W. The *bottom curve* shows the difference ($I_{obs} - I_{calc}$) between the observed and calculated neutron diffraction patterns

size of domains of the ε-W_2C phase in the sintered samples is $0.3 < \langle D \rangle < 1.0\,\mu m$. According to microscopy observations, the size of carbide grains in the sintered polycrystalline samples of W_2C is 5–25 μm (Fig. 3.11).

Fig. 3.11 Microstructure of the W_2C ($WC_{0.50}$) sample synthesized by sintering of a W + C powdered mixture at 2070 K for 10 h and subsequent annealing at 1370 K for 35 h. The size of carbide grains is 5–25 μm

The grains have an irregular prismatic shape characteristic of crystals of the hexagonal or the trigonal system. The examination of the microstructure of the sintered samples in reflected polarized white light showed that all carbide grains have an interference color, i.e. are anisotropic. When nonstoichiometric carbides are ordered by the mechanism of the first-order (first-kind) phase transition (e.g., ordering in titanium TiC_y and vanadium VC_y carbides [41, 42]), the optical microscopy easily reveals interfaces between the disordered and ordered phases even if the size of domains of the ordered phase is small (100 nm or smaller). According to diffraction data, the disordered phase is absent in the sintered W_2C samples before and after annealing. As for the ordered phase, the degree of the long-range order in this phase increases after annealing. The presence of the ordered carbide phase alone is in agreement with the relevant microscopy observations, which did not reveal interfaces between the disordered and ordered phases. In turn, the absence of interfaces means that the formation of the trigonal phase ε-W_2C is realized most likely as the second-order (second-kind) phase transition. The symmetry analysis also suggests the possibility of the second order for the β-W_2C (space group $P6_3/mmc$) $\rightarrow$ ε-W_2C (space group $P\overline{3}1m$) disorder–order transition.

The measured microhardness of the annealed W_2C sample, which contained the ordered trigonal carbide ε-W_2C, was $H_V = 15.3 \pm 0.3$ GPa. According to the literature data, the microhardness of the lower tungsten carbide in an unknown structural state is 14.5 to 18.1 GPa.

As noted above, nonmetal atomic planes $(00z)_\varepsilon$ with $z = 0$ and $z = 1/2$, which differ in their carbon atom occupancies, alternate at the right angle to the direction $[001]_\varepsilon$ in the ordered trigonal carbide (see Fig. 3.5). The probabilities that a carbon atom is located in these planes are $n_{z=0} = (2n_2 + n_3)/3$ and $n_{z=1/2} = (n_1 + 2n_4)/3$. These probabilities are equal to 1/2 in the disordered basic hexagonal phase β-W_2C. In the perfect ordered phase ε-W_2C probabilities are $n_{z=0} = 1/3$ and $n_{z=1/2} = 2/3$. In turn, they are $n_{z=0} = 0.443$ and $n_{z=1/2} = 0.510$ in the carbide ε-$W_2C_{0.953}$ synthesized at ~2070 K. When this carbide was annealed additionally at a temperature of ~1370 K for 35 h, the long-range order parameters η_{15} and η_{17} increased and the probabilities $n_{z=0}$ and $n_{z=1/2}$ reached 0.396 and 0.606, which are close to the values of $n_{z=0}$ and $n_{z=1/2}$ in the perfect ordered phase ε-W_2C. The comparison of the values of $n_{z=0}$ and $n_{z=1/2}$ shows that carbon atoms are redistributed between adjacent nonmetal atomic planes $(00z)_\varepsilon$ with $z = 0$ and $z = 1/2$ during trigonal ordering in the lower tungsten carbide.

An analysis of the distribution functions calculated allows us to establish the relationship between the maximum value of the long-range order parameter and the composition of the nonstoichiometric carbide $W_2C_{y'} \equiv WC_y$ ($y \leq 0.5$). This relationship has the form

$$\eta^{\max}(y) = \begin{cases} 2(1-y), & \text{if } y \geq 0.5 \\ 2y, & \text{if } y < 0.5 \end{cases}. \tag{3.7}$$

The dependence of the maximum value of any long-range order parameter on the composition of the carbide WC_y, which is ordered after the W_2C type, is defined by Eq. 3.7) and the minimum value of the long-range order parameters is zero. Therefore, any order parameter η describing the above superstructures meets the condition

$$0 \leq \eta_s \leq m^*, \tag{3.8}$$

where $m^* = 2(1-y)$ if $y \geq 0.5$ and $m^* = 2y$ if $y < 0.5$.

The condition (3.8) uniquely defines one-dimensional intervals of admissible values of the long-range order parameters for superstructures, which are described by one parameter η_s. The ordered trigonal phase ε-W_2C is described by the distribution function 3.4), which depends on two long-range order parameters η_{17} and η_{15}. In the case of superstructures described by several long-range order parameters, one should be careful that the distribution function values fall always within 0 and 1. Taking into account values of the distribution function (Table 3.4) and the aforementioned constraints, the interval of admissible values of long-range order parameters $\eta_{17}(y)$ and $\eta_{15}(y)$ for the ordered trigonal (space group $P\overline{3}1m$) phase ε-W_2C is defined by the conditions

$$\left.\begin{array}{l} -m^* \leq -\eta_{17}(y) + 4\eta_{15}(y) \leq 3m^* \\ 0 \leq \eta_{17}(y) \leq m^* \end{array}\right\}. \tag{3.9}$$

The interval of admissible values of the order parameters $\eta_{17}(y)$ and $\eta_{15}(y)$ for the ordered trigonal phase ε-W_2C is shown in Fig. 3.12.

The distribution function (3.4) describing the ordered phase ε-W_2C is a superposition of two static concentration waves related to the long-range order parameters η_{17} and η_{15}. Figure 3.13 presents the superposition of these concentration waves calculated for the plane $(1\ -1\ 1)_{L'3}$ taking some admissible values of the parameters η_{17} and η_{15}.

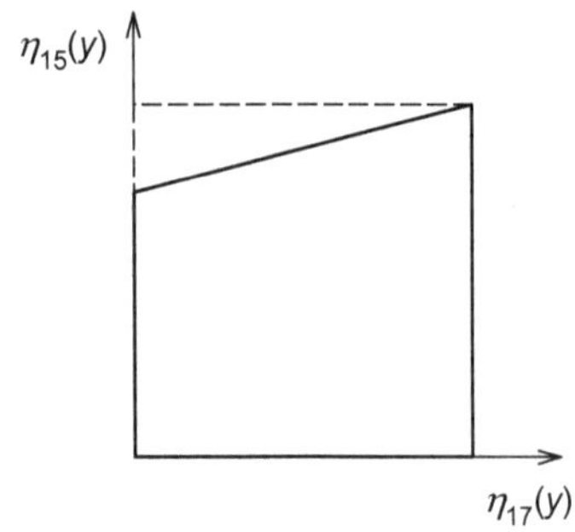

Fig. 3.12 Two-dimensional interval of admissible values of the long-range order parameters η_{17} and η_{15} for the ordered trigonal (space group $P\bar{3}1m$) tungsten carbide ε-W_2C

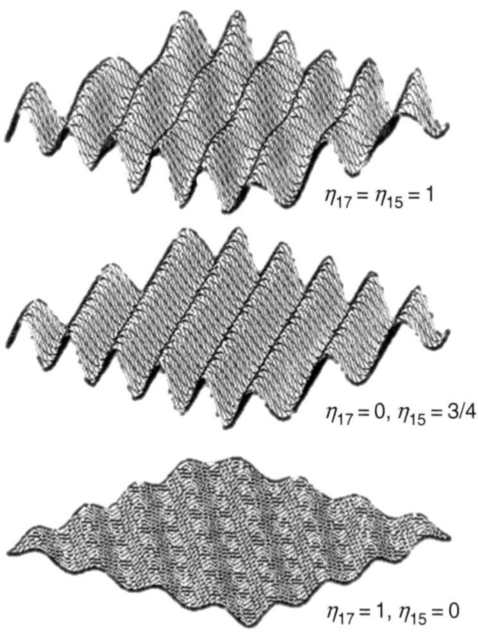

Fig. 3.13 Concentration waves in the nonmetal plane $(1\ -1\ 1)_{L'3}$ of the ordered trigonal (space group $P\bar{3}1m$) carbide ε-W_2C at different values of the long-range order parameters η_{17} and η_{15}

3.1.3 Low-Temperature Limit of the Existence Region of W_2C Carbide

The X-ray and neutron diffraction analyses of the phase composition and the structure showed that the W_2C sample synthesized at ~2070 K and the same sample, which was annealed additionally at a temperature of 1370 K for 35 h, contain only the ordered trigonal carbide ε-W_2C and tungsten W (see Figs. 3.8, 3.9, and 3.10). The concentrations of these phases in the sample before and after annealing are nearly equal, but the degree of order in the trigonal carbide ε-W_2C increases after annealing. The authors [10, 26, 27] think W_2C decomposes to W and WC at temperatures of 1530 K and lower. According to [26], the ordered orthorhombic (space group

Pbcn) phase β'-W_2C is formed as an intermediate product of decomposition. If decomposition occurs, annealing of the W_2C sample at 1370 K (150 K below the temperature of the supposed solid-phase decomposition) for 35 h should lead to formation of the carbide WC. However, the X-ray and neutron diffraction analyses did not reveal any traces of WC in the sample. Diffraction reflections of the ordered orthorhombic (space group *Pbcn*) phase β'-W_2C were not observed also. The absence of reflections of the orthorhombic (space group *Pbcn*) phase β'-W_2C in the neutron diffraction pattern is in agreement with results [14, 15, 19]. According to these data, the orthorhombic phase β'-W_2C with the ζ-Fe_2N structure is observed in W_2C samples only after annealing at $T < 1300$ K. Thus, annealing at 1370 K does not lead to formation of the ordered orthorhombic phase and decomposition of the lower tungsten carbide to WC and W.

Samples of $WC_{0.35}$, $WC_{0.50}$ and $WC_{0.70}$ were synthesized by sintering of a W+C mixture in a vacuum at 1370 K. The sintering time was 30 h. Then the sintered samples were ground, compacted again, and sintered at the same temperature of 1370 K for 20 h. XRD patterns of all the samples sintered for 30 h contain reflections of the lower tungsten carbide W_2C along with those of tungsten W and the higher carbide WC (Fig. 3.14).

The concentration of W_2C is the maximum in the $WC_{0.50}$ sample since its composition corresponds to the stoichiometric lower carbide. As the sintering time increased to 50 h, the concentrations of W and W_2C decreased and increased respectively. The observed formation of W_2C in the samples sintered from tungsten and carbon at 1370 K and the growth of the W_2C concentration in these samples with the increase of sintering time contradicts the assumption [10, 26, 27] of the eutectoid decomposition of the lower carbide W_2C to W and WC at $T \leq 1530$ K. Tungsten is present in all the samples because they were synthesized at a low temperature and synthesis was not complete even after 50 h.

For comparison, samples of $WC_{0.35}$, $WC_{0.50}$ and $WC_{0.70}$ were synthesized from tungsten and carbon powder mixtures at 1870 K for 5 h. Synthesis was almost complete at this temperature and, therefore, just two phases were present in the samples: ~43 wt.% W and ~57 wt.% W_2C in $WC_{0.35}$; ~9 wt.% W and ~91 wt.% W_2C in $WC_{0.50}$; ~86 wt.% W_2C and ~14 wt.% WC in $WC_{0.70}$.

Thus, the experiment demonstrated that the lower tungsten carbide W_2C is thermodynamically stable from the melting point of ~3050 K to 1370 K and does not decompose over this temperature interval. At temperatures of ~2300 to 1370 K the only ordered phase of the lower tungsten carbide W_2C is the trigonal (space group $P\bar{3}1m$) phase ε-W_2C. Traces of the ordered orthorhombic (space group *Pbcn*) phase β'-W_2C were not detected even after long-time annealing at 1370 K. The transition from the ordered trigonal phase ε-W_2C to the ordered orthorhombic (space group *Pbcn*) phase β'-W_2C is probable to occur at a temperature below 1370 K. According to [14, 15, 19], the orthorhombic (space group *Pbcn*) phase β'-W_2C with the ζ-Fe_2N structure indeed is observed in samples of the lower tungsten carbide at a temperature below 1300 K.

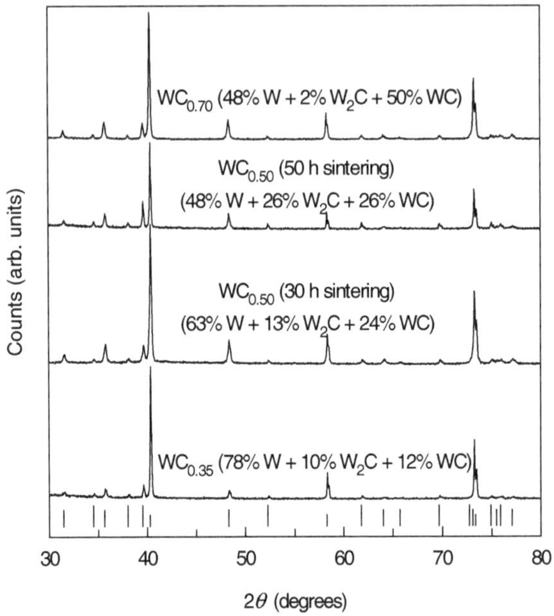

Fig. 3.14 XRD patterns of $WC_{0.35}$, $WC_{0.50}$ and $WC_{0.70}$ samples synthesized from a W+C mixture by sintering at 1370 K for 30 h and the $WC_{0.50}$ sample prepared by sintering at the same temperature for 50 h. The lower tungsten carbide W_2C is present in all the samples. The content of the W_2C phase is the maximum in $WC_{0.50}$ sample, which is the closest to the stoichiometric composition of the lower tungsten carbide. The W_2C content grows as the sintering time increases to 50 h. The phase composition is given in wt.%. *Long, middle* and *short ticks* correspond to reflections of W_2C, WC and W phases, respectively. All XRD patterns are recorded in $CuK\alpha_{1,2}$ radiation

3.1.4 Refinement of Phase Diagram of the W–C System

Refinement of the phase diagram of the W-C system is related essentially to the existence range of the carbide W_2C and its ordered phases. It follows from the above X-ray and neutron diffraction analyses that the ordered trigonal (space group $P\overline{3}1m$) phase ε-W_2C exists in samples synthesized over a wide temperature interval of ~2300–1370 K. No traces of other ordered phases were detected. The concentration of the ordered phase ε-W_2C was much smaller (nearly 7 times) than the concentration of the disordered phase β-W_2C in samples prepared by plasma chemical synthesis at a high temperature. The phase ε-W_2C appears in these samples quenched from 2300–2400 K because the reaction of synthesis occurs in strongly non-equilibrium conditions. The obtained results are in agreement with the literature data [7, 19, 21–25, 31–33], according to which the only ordered trigonal phase ε-W_2C is detected in samples of the lower tungsten carbide prepared by different methods at temperatures of ~1900–2300 K. Thus, trigonal (space group $P\overline{3}1m$) phase ε-W_2C is formed at a temperature of ~1900–2300 K most likely by the mechanism of the second-order (second-kind) phase transition.

Long-time low-temperature (1370 K) annealing of samples containing the ordered phase ε-W_2C did not change their phase composition. It means that the ε-W_2C superstructure is stable at this temperature. When WC_y samples ($y = 0.35, 0.50, 0.70$) were synthesized at 1870 K and at a lower temperature of 1370 K by sintering of W+C powder mixtures, all of them contained the lower tungsten carbide W_2C. The content of W_2C phase depends on the sample composition and is the maximum in the $WC_{0.50}$ sample. Direct synthesis of the carbide W_2C from tungsten and carbon at 1370 K and 1870 K means that the lower carbide is in thermodynamic equilibrium at these temperatures. Taking into account the data on low-temperature annealing of W_2C it can be stated that the lower tungsten carbide is thermodynamically stable at ≥1370 K and does not undergo solid-phase decomposition at ~1530 K although such decomposition was supposed in [8–10, 26, 27].

Taking into account experimental results [22, 31–33] related to the crystal structure and the temperature range of existence of the lower tungsten carbide, which is described in Sects. 3.1.1 to 3.1.3, and considering the literature data [7, 19, 21–24], the positions of phase boundaries of the lower carbide W_2C are corrected as shown in Fig. 3.15. Over the temperature interval of ~2300–1370 K the phases β'-W_2C and β''-W_2C, not found experimentally, are replaced by the observed ordered phase ε-W_2C. This ordered phase does not undergo eutectoid decomposition at ~1530 K and is thermodynamically stable at least to a temperature of ~1370 K.

The authors [43] analyzed a large number of literature data on the cubic tungsten carbide γ-WC_{1-x} and showed that this carbide exists as a one-phase compound over a narrow temperature interval of 3058–2798 K. The cubic tungsten carbide has a wide homogeneity interval from $WC_{0.61}$ to $WC_{1.00}$ at a temperature of ~3025 K, but a homogeneity interval quickly narrows with decreasing temperature and only the cubic carbide $WC_{0.62}$ exists at ~2800 K. At a temperature of ~2798 K the cubic tungsten carbide γ-$WC_{0.62}$ undergoes eutectoid decomposition to the lower hexagonal W_2C and the higher hexagonal WC tungsten carbide. The phase boundaries of the cubic tungsten carbide γ-WC_{1-x} are shown in Fig. 3.15 as they were determined by authors [6, 43].

Special points of the phase diagram of the W-C system (Fig. 3.15) are given in Table 3.6

On the whole, experimental and theoretical studies [32, 33] of the structure of the lower carbide W_2C demonstrated that this carbide is thermodynamically stable from the melting point of ~3050 K to a temperature of 1370 K and does not decompose over this temperature interval. At temperatures of ~2300 to 1370 K the only ordered phase of the lower tungsten carbide W_2C is the trigonal (space group $P\overline{3}1m$) phase ε-W_2C. Trigonal phase ε-W_2C is formed at a temperature of ~1900–2300 K most likely by the mechanism of the second-order phase transition.

Thus, the experiments and theory show that the following sequence of phase transformations "disordered hexagonal (space group $P6_3/mmc$) phase β-W_2C → ordered trigonal (space group $P\overline{3}1m$) phase ε-W_2C" is realized. Probably, at temperatures below 1370 K the trigonal phase ε-W_2C transforms to the orthorhombic phase

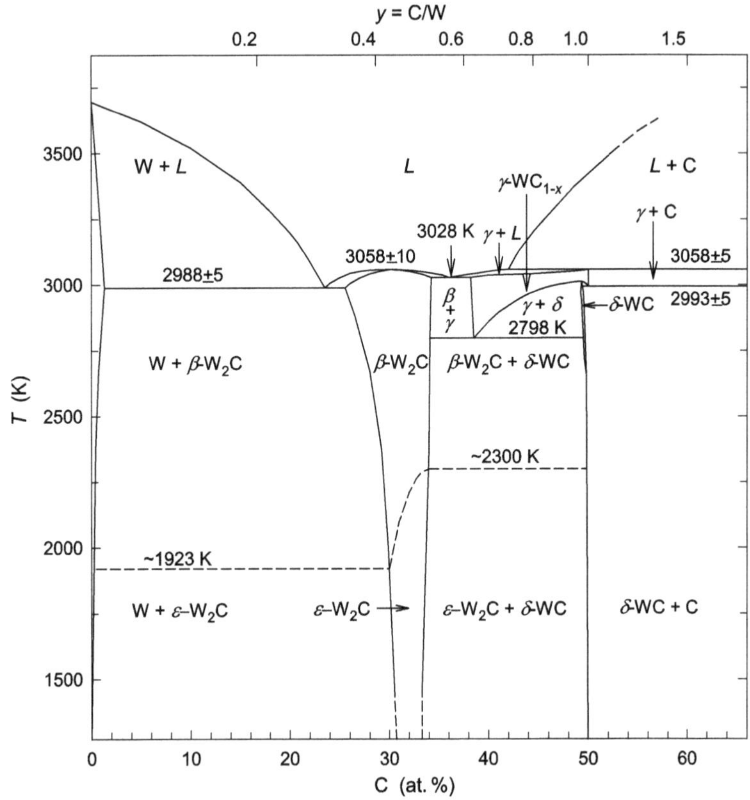

Fig. 3.15 Phase diagram of the W-C system. Instead of the phases β'-W_2C and β''-W_2C, not found experimentally, ordered trigonal phase ε-W_2C experimentally detected in studies [7, 19, 21–25, 31–33] is shown. Trigonal phase ε-W_2C is thermodynamically stable in the temperature interval of ~2300–1300 K. Phase boundaries of the cubic tungsten carbide γ-$WC_{1-x} \equiv \gamma$-WC_y are shown as reported in Refs. [6] and [43]. Special points of the diagram are given in Table 3.6

β'-W_2C. This assumption does not contradict the theoretically possible sequence of phase transformations and is confirmed by results obtained in studies [14, 15, 19]. According to these results, the orthorhombic (space group *Pbcn*) phase β'-W_2C with the ζ-Fe_2N structure was observed in samples of the lower tungsten carbide annealed at temperatures below 1300 K. For the addition and specification of these literature data, *in situ* high temperature neutron diffraction measurements are required.

The eutectoid decomposition of lower W_2C carbide to tungsten W and the higher WC carbide at temperature 1370 K and higher was not observed [32, 33]. However, vacuum annealing of nanocrystalline WC powders at temperature 1470 K, which was accompanied by partial loss of carbon, caused shift of carbide composition in region of existence of W_2C and appearance of tungsten W along with WC and W_2C [44, 45] (see Sect. 4.3.3). The appearance of tungsten metal might be caused by the eutectoid decomposition $W_2C \rightarrow W + WC$ which occurs at $T \leq 1530$ K [8, 10, 26,

Table 3.6 Special points in the phase diagram of the W–C system (see Fig. 3.15) at a temperature above 1300 K

Reaction	Composition of phases involved in the reaction (at.% of C)			T (K)	Reaction type
$L \Leftrightarrow W$	0	0	–	3755 ± 5	Melting
$L \Leftrightarrow \beta\text{-}W_2C$	~30.2	~30.2	–	3058 ± 10	Congruent melting
$L + C \Leftrightarrow {}^*\gamma\text{-}WC_{1-x}$	~42.0	100.0	50.0	3058 ± 5	Peritectics
$L \Leftrightarrow W + \beta\text{-}W_2C$	~23.5	~1.2	~25.5	2988 ± 5	Eutectics
$L \Leftrightarrow \beta\text{-}W_2C + {}^*\gamma\text{-}WC_{1-x}$	~36.0	~34.2	~38.2	3028 ± 5	Eutectics
${}^*\gamma\text{-}WC_{1-x} \Leftrightarrow \delta\text{-}WC$	~49.3	~49.3	–	3012 ± 5	Polymorphic transformation
${}^*\gamma\text{-}WC_{1-x} \Leftrightarrow \delta\text{-}WC + C$	50.0	~49.8	100.0	2993 ± 5	Eutectoid decomposition
${}^*\gamma\text{-}WC_{1-x} \Leftrightarrow \beta\text{-}W_2C + \delta\text{-}WC$	~38.5	~34.0	~49.5	2798 ± 5	Eutectoid decomposition
$\beta\text{-}W_2C \Leftrightarrow \varepsilon\text{-}W_2C$	~30.0–34.0	~30.0–34.0	–	~1920–2300	Disorder–order transformation

${}^*\gamma\text{-}WC_{1-x} \equiv \gamma\text{-}WC_y$ $(y \equiv 1 - x)$

27]. Thus, the problem of a presence or absence of eutectoid decomposition needs further investigation.

3.2 Ordering of Nonstoichiometric Hexagonal Compounds M_2X

The ordering of nonmetallic interstitial atoms and structural vacancies in lower carbides $M_2C_{y'}$ and nitrides $M_2N_{y'}$ $(0.7 < y' \leq 1.0)$ with a hexagonal structure of the $L'3$ type has been poorly investigated. Only in the last years, the sequence of phase transformations in one of these compounds, i.e., tungsten carbide $W_2C_{y'}$, was revealed using the Landau phenomenological theory of phase transitions and symmetry analysis [32, 33]. For other lower carbides and nitrides $M_2X_{y'}$, there are several fragmentary experimental investigations with contradictory and mutually excluding results that were performed more than thirty years ago and indicated the possibility of forming superstructures of the $M_{2t}X_{2t-1}$ $(t = 1)$ type with different symmetries. Previously, one of the variants of the statistical theory of ordering, namely, the order-parameter functional (OPF) method, was proposed to describe structural disorder–order phase transformations in strongly nonstoichiometric interstitial compounds [46]. This method was developed and successfully used for studying the atomic-vacancy or atomic ordering in cubic monocarbides MC_y $(MC_y\square_{1-y})$, mononitrides MN_y $(MN_y\square_{1-y})$, and substitutional solid solutions A_yB_{1-y} [1, 2, 5, 47]. The physical ground of the order-parameter functional method is the mean field approximation. Because of its formalism, the OPF method belongs to the group of cluster methods but favourably differs from these in the possibility of thoroughly accounting for the symmetry of a crystal with any degree of long-range order. Up to now, this method has not been applied to lower carbides and nitrides $M_2X_{y'}$ and, in general, to compounds with a hexagonal structure.

According to [48], in the OPF method, calculation of the disorder–order phase transition consists of the following independent successive problems: determination of the disorder–order phase transition channel, calculation of the distribution function, choice of the basis cluster figure and determining all sequence of special figures, representing the cluster probabilities by long-range order parameters, and calculation of the cluster energy. The successful solution of these problems allows the direct computation of phase transition under study. The solution of first two problems for the lower tungsten carbide $W_2C_{y'}$ is described in Sect. 3.1.1. Technique of calculations of phase equilibria and phase boundaries in the systems with ordering of nonstoichiometric compounds was considered in detail in monographs, reviews, and articles [1, 2, 5, 47, 49–52]. The choice of the basis cluster figure and determination of the sequence of special figures will be considered below.

3.2.1 Sequence of Special Figures

With the aim of analyzing the thermodynamics of a system of many interacting particles with the use of the OPF method, the basis cluster figure and an ensemble of figures uniquely describing the atomic distribution in the crystal should be determined from the crystal lattice geometry of the compound under consideration. In study [53], this problem is solved for nonstoichiometric compounds $M_2X_{y'}$ with a basis hexagonal structure of the $L'3$-type.

Let us consider crystal structure of disordered nonstoichiometric compound M_2X with a basis hexagonal structure of the $L'3$-type.

The unit cell of an M_2X compound with an $L'3$ hexagonal (space group $P6_3/mmc$) structure is shown in Fig. 3.16a. In disordered nonstoichiometric compounds M_2X with the $L'3$ structure, metal atoms *M* form a metallic hcp sublattice, whereas nonmetal atoms X statistically (with a probability of 1/2) occupy octahedral interstitial positions of the metal sublattice (crystallographic positions 2(*a*) with the coordinates (0 0 0) and (0 0 1/2)) and form a nonmetallic simple hexagonal sublattice (Fig. 3.16b). Layers A and B consisting of metal atoms M and layers of sites of the nonmetallic sublattice X in the general sequence AXBXAXBX... alternate in the direction of the *c* axis. In the M_2X compounds with the $L'3$ structure and the unit cell parameters a_h and c_h, the octahedral interstitial positions of the metal sublattice are distorted: the

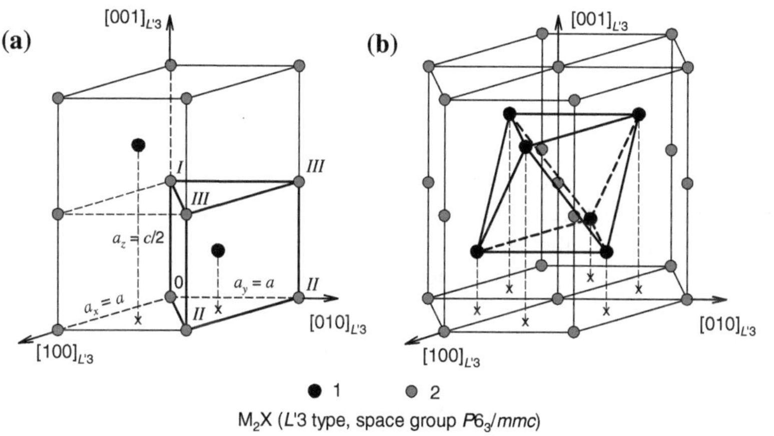

Fig. 3.16 Crystal structure of a disordered nonstoichiometric hexagonal (space group *P63/mmc*) compound M_2X: **a** Unit cell of a disordered compound M_2X with the $L'3$-type structure; **b** general view of the crystal lattice with the $L'3$-type hexagonal structure. *1* metal atoms M •, *2* nonmetallic sublattice sites •, statistically occupied (with a probability of 1/2) by nonmetal atoms X. *Thick lines* indicate the basis cluster in the form of a trigonal prism consisting of six sites of the nonmetallic sublattice with the metal atom M at the center and the octahedral interstice formed in the hcp metal sublattice by six metal atoms M with the site of the nonmetallic sublattice at the center. (*I*), (*II*), and (*III*) are the nonmetallic sublattice sites located in the first, second and third coordination spheres, respectively, with respect to the site 0

length of six edges of the octahedron is equal to the unit cell parameter a_h, and the length of the other six edges is $(a_{\mathrm{h}}^2/3 + c_{\mathrm{h}}^2/4)^{1/2}$.

In nonstoichiometric compounds M_2X with this structure, the first coordination sphere of any metal atom is a trigonal prism formed by six sites of the nonmetallic sublattice (Fig. 3.16a). Each of these sites can be vacant or occupied by the interstitial atom. The distance from the metal atom to any one of the six nearest sites of the nonmetallic sublattice is $(a_{\mathrm{h}}^2/3 + c_{\mathrm{h}}^2/16)^{1/2}$. If the atomic displacements are disregarded, six sites of the nonmetallic sublattice form a regular trigonal prism MX_6 with the metal atom at the center (Fig. 3.16a). The length of the side edges of the prism is $c_h/2$, and the length of the edges of the upper and lower bases is equal to the lattice spacing a_h. The choice of the cluster in the form of the regular trigonal prism as the basis figure allows us to take into account the metal-nonmetal (M-X) interatomic interactions most important for nonstoichiometric compounds M_2X and many-particle correlations in the first coordination sphere of the metal atom. Moreover, the choice of the trigonal prism as the basis cluster figure makes it possible to include all crystal lattice sites without exception, because each lattice site belongs to at least one trigonal prism. It should be noted that, in compounds M_2X with the $L'3$ structure, the MX_6 trigonal prisms are separated from each other by trigonal holes.

The requirements imposed on the basis cluster figure in the OPF method are identical to those in the cluster variation (CV) method [54–58]. In particular, the basis figure should be "closed", i.e., it should not contain separately bound sites [56]. In order to verify the fulfillment of this condition, it is necessary to find the sequence of overlapping figures. In the lattice with the $L'3$ structure, the first generation of overlapping figures is the edges of the trigonal prism that couple the two nearest sites of the nonmetallic sublattice, i.e., paired bonds. The trigonal prism contains the paired bonds with lengths $c_h/2$ (side edges) and a_h (base edges). The sites of the nonmetallic sublattice serve as the overlapping figures of paired bonds. Therefore, in the case under consideration, the sequence of special figures $\{s\}$ involves four figures: the nonmetallic sublattice site (a-type figure), the bond between the neighboring sites located in the nonmetallic sublattice at the distance $c_h/2$ (b_1-type figure), the bond between the neighboring sites located in the nonmetallic sublattice at the distance a_h (b_2-type figure), and the trigonal prism consisting of six sites of the nonmetal sublattice with the metal atom at the center (c-type figure). The trigonal prism satisfies the definition of the "closed" figure [56], because any one of its six sites is a vertex and belongs simultaneously to three edge bonds (one figure of the b_1-type and two figures of the b_2-type), i.e., to more than one special figure b. The basis cluster figures in the form of the MX_6 trigonal prisms with the metal atom at the center do not overlap over the triangular bases or side rectangular faces, because they are separated by holes.

The relation between the special figures of the sequence $\{s\}$ is determined with the use of the matrix $|u^{(m)(s)}|$, in which the elements indicate the number of m-type figures contained in the s-type figure: for example, we have element $|u^{(a)(c)}| = 6$, because the figure c (trigonal prism) contains six figures a (sites). In order to avoid overestimation of lattice sites due to the overlapping figures (for example, each site of the nonmetallic sublattice simultaneously enters into six adjacent trigonal prisms

MX_6 and eight bonds, namely, two bonds b_1 and six bonds b_2), it is necessary to determine the overestimation coefficients $y^{(s)}$. These coefficients can be found by solving the linear system constructed from the equations

$$\sum_{s=a}^{q} u^{(m)(s)} y^{(s)} = x^{(m)} \quad (m = a, b, c, \ldots, q), \tag{3.10}$$

where $x^{(m)}$ is the number that indicates how many times the number of m-type figures is larger than the number of sites N in the ordering Ising lattice. The coefficient $x^{(m)}$ is equal to the number of m-type figures simultaneously containing any arbitrarily chosen site of the lattice under consideration that is divided by the number of sites in the m-type figure: for example, $x^{(b_2)} = 6/2 = 3$ (because any site of the nonmetallic sublattice belongs to six bonds b_2 containing two sites each) and $x^{(c)} = 6/6 = 1$ (because any site of the nonmetallic sublattice in the $L'3$ structure belongs to six c-type figures, i.e., trigonal prisms with the metal atom at the center, each of which involves six sites of the nonmetal sublattice).

For the lattice under consideration with the $L'3$ structure and the chosen sequence of the special figures $\{s\}$, including the basis cluster figure c in the form of the trigonal prism, the system of linear equations has the form

$$\left.\begin{array}{l} u^{(a)(a)}y^{(a)} + u^{(a)(b_1)}y^{(b_1)} + u^{(a)(b_2)}y^{(b_2)} + u^{(a)(c)}y^{(c)} = x^{(a)} \\ u^{(b_1)(a)}y^{(a)} + u^{(b_1)(b_1)}y^{(b_1)} + u^{(b_1)(b_2)}y^{(b_2)} + u^{(b_1)(c)}y^{(c)} = x^{(b_1)} \\ u^{(b_2)(a)}y^{(a)} + u^{(b_2)(b_1)}y^{(b_1)} + u^{(b_2)(b_2)}y^{(b_2)} + u^{(b_2)(c)}y^{(c)} = x^{(b_2)} \\ u^{(c)(a)}y^{(a)} + u^{(c)(b_1)}y^{(b_1)} + u^{(c)(b_2)}y^{(b_2)} + u^{(c)(c)}y^{(c)} = x^{(c)} \end{array}\right\}. \tag{3.11}$$

The elements $|u^{(m)(s)}|$ of the matrix that relate the special figures of the sequence $\{s\}$, the coefficients $x^{(m)}$, and the calculated overestimation coefficients $y^{(s)}$ for the $L'3$-type lattice are listed in Table 3.7.

The sites of the same figure can be occupied by atoms and vacant. Therefore, to each figure in the chosen sequence $\{s\}$ there correspond several variants of the mutual arrangement of occupied and vacant sites, i.e., different i-configurations with

Table 3.7 The matrix elements $|u^{(m)(s)}|$, the values of $x^{(m)}$ and overestimate coefficients $y^{(s)}$ for a $L'3$-type crystal lattice in the case of basis cluster in the form of a trigonal prism with the metal atom at the center

Figure of m-type	Figure of s-type				$x^{(m)}$
	a	b_1	b_2	c	
a	1	2	2	6	1
b_1	0	1	0	3	1
b_2	0	0	1	6	3
C	0	0	0	1	1
$y^{(s)}$	5	−2	−3	1	

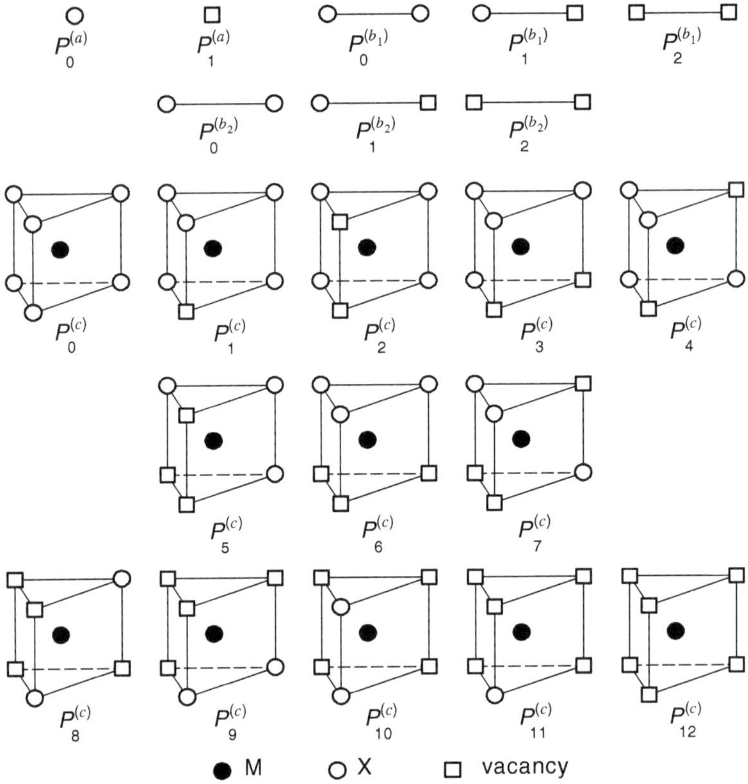

Fig. 3.17 Nonequivalent configurations of the figures a(site), b_1 and b_2 (paired bonds), and c (trigonal prism) of the sequence $\{s\}$ necessary to describe nonstoichiometric interstitial compounds $M_2X_{y'}$ $(M_2X_{y'}\square_{1-y'})$ with the basis hexagonal structure $L'3$: (•) metal atoms M, (○) interstitial atoms X, and (□) vacant sites of the nonmetallic sublattice

the probabilities $P_i^{(s)}$. For the figure a (site), two nonequivalent configurations are possible: the site occupied by the X atom (with the configuration probability $P_0^{(a)}$) and the vacant site □ (with the configuration probability $P_1^{(a)}$). The probabilities $P_0^{(b_1)}$, $P_1^{(b_1)}$, $P_2^{(b_1)}$ and $P_0^{(b_2)}$, $P_1^{(b_2)}$, $P_2^{(b_2)}$ correspond to the figures b_1 and b_2 (paired bonds with different lengths) with the configurations X-X, X-□ and □-□. The basis cluster figure in the form of the trigonal prism can exhibit thirteen nonequivalent configurations: complete cluster (all six sites are occupied by X atoms), cluster with one vacancy, and so on up to the completely defect cluster (all six sites are vacant) with the probabilities from $P_0^{(c)}$ to $P_{12}^{(c)}$. All nonequivalent configurations of the chosen sequence $\{s\} = \{a, b, c\}$ are shown in Fig. 3.17.

Apart from the nonequivalent configurations, there are equivalent configurations coinciding with each other under the action of symmetry operations of the point group of the complete figure s. The site (occupied or vacant) has no equivalent

Table 3.8 The multiplicities $\lambda_i^{(s)}$ of i-configurations of figures of the sequence $\{s\}$ (see Fig. 3.17), used for describing a lattice with the $L'3$ structure

Type of figure s	i												
	0	1	2	3	4	5	6	7	8	9	10	11	12
a	1	1	–	–	–	–	–	–	–	–	–	–	–
b_1	1	2	1	–	–	–	–	–	–	–	–	–	–
b_2	1	2	1	–	–	–	–	–	–	–	–	–	–
c	1	6	3	6	6	12	2	6	6	6	3	6	1

configurations. The paired bonds in which both sites are occupied by identical atoms or vacant have no equivalent configurations. The bond in which one site is occupied and the second site is vacant has two equivalent configurations X–□ and □–X. The complete basis cluster $P_0^{(c)}$ in the form of the regular trigonal prism has symmetry point group $\overline{6}2m$ (D_{3h}), which includes 12 symmetry operations (elements) H_1, H_3, H_5, H_8, H_{10}, H_{12}, H_{14}, H_{16}, H_{18}, H_{19}, H_{21}, and H_{23} (the point groups and their symmetry elements H_i are described in [1, 2, 35]). The multiplicity $\lambda_i^{(s)}$ (the number of equivalent configurations of some i-configuration) is calculated from the equation

$$\lambda_i^{(s)} = n(G_0^{(s)})/n(G_i^{(s)}), \tag{3.12}$$

where $G_0^{(s)}$ and $G_i^{(s)}$ are the groups of the complete figure s and the figure s with the i-configuration, respectively, and $n(G)$ is the order of the group G that is equal to the number of symmetry elements in this group. For all nonequivalent configurations of the basis cluster in the form of the trigonal prism and overlapping figures (Fig. 3.17) describing the lattice with the $L'3$ structure, the corresponding multiplicities $\lambda_i^{(s)}$ are presented in Table 3.8. For the i-configurations with identical number m of occupied sites, we have $\sum_{i\in m}\lambda_i^{(s)} = C^m_{R^{(s)}}$, where $R^{(s)}$ is the number of sites of the figure $s(R^{(a)} = 1, R^{(b)} = 2, R^{(c)} = 6)$.

With due regard for the definition of the multiplicities $\lambda_i^{(s)}$ and the overestimation coefficients $y^{(s)}$, it is possible to write the normalization condition that should be satisfied in the description of the $MX_{y'/2}$ ($M_2X_{y'}$) crystal. The probability normalization condition has the form

$$\sum_s y^{(s)} \sum_i \lambda_i^{(s)} P_i^{(s)} = 1. \tag{3.13}$$

For compounds $MX_{y'/2} \equiv MX_{y'/2}\square_{1-y'/2}$, it is possible to write the equation that relates the probabilities $P_i^{(s)}$ of all configurations of the figures in the sequence $\{s\}$ to the composition of the $MX_{y'/2}$ compound, i.e., to the quantity $y'/2$, which is the relative content of occupied sites in the nonmetallic sublattice. In actual fact, this equation is the normalization condition for the composition

$$\sum_s y^{(s)} \sum_i l_i^{(s)} \lambda_i^{(s)} P_i^{(s)} = y'/2, \tag{3.14}$$

where $l_i^{(s)}$ is the fraction of the nonmetallic sublattice sites occupied in the by X atoms in the figure s with the i-configuration.

The normalization conditions (3.13) and (3.14) are written for all figures of the sequence $\{s\}$, i.e., in the very general form. However, they can also be written for figures of one type; in this case, the summation over s is removed in formulas (3.13) and (3.14).

In the order-parameter functional method, the probabilities $P_i^{(s)}$ of the i-configuration of any figures are represented through the long-range order parameters with the use of the distribution function $n(\boldsymbol{r})$ of atoms over sites $\boldsymbol{r}$ in the ordering lattice. The distribution function (3.1) represents a probability of finding an atom of a particular species in any site $\boldsymbol{r}$ and expressed directly through the long-range order parameters. In the case of ordering of lower carbides $M_2C_{y'}$ and nitrides $M_2N_{y'}$ with the $L'3$ hexagonal structure, the composition in formula (3.1) and subsequent formulas (3.15)–(3.18), (3.21)–(3.30) is $y = y'/2 \leq 0.5$.

In [1, 2, 46, 47], it was demonstrated that, for a crystal in an equilibrium state, all values of any distribution function $n(\boldsymbol{r})$ degenerate into two values. This is always valid for ordered structures described by one long-range order parameter. For superstructures described by more than one long-range order parameter, the condition for the degeneracy of the distribution function $n(\boldsymbol{r})$ is the equality of the long-range order parameters $\eta_{s1} = \eta_{s2} = \cdots = \eta_{s\,m} = \eta \cdot 0$ Therefore, under equilibrium conditions, any distribution function $n(\boldsymbol{r})$ over the entire set of sites of the ordered lattice of the $M_{2t}X_{2t-1}$ ($t = 1, 1.5, 2, 3, 4$) superstructures formed in the MX_y and $M_2X_{y'}$ compounds takes on two values

$$\left.\begin{aligned} n_1^{(d)} &= y - \eta(2t-1)/2t, \\ n_2^{(d)} &= y + \eta/2t. \end{aligned}\right\}. \tag{3.15}$$

The value of $n_1^{(d)}$ is the probability of finding the interstitial atom at the site of the vacancy sublattice, and $n_2^{(d)}$ is the probability of finding interstitial atoms at the site of the sublattice of interstitial atoms of the ordered structure $M_{2t}X_{2t-1}$.

The free energy reaches a minimum when the long-range order parameters are equal to each other [1, 2, 46, 47]. In this case, all probabilities $P_i^{(s)}$ depend only on two values (3.15) of the distribution function. In particular, according to [2], the probability $P_0^{(2)(s)}$ of the complete pair with allowance made for the values of $n_1^{(d)}$ and $n_2^{(d)}$ can be analytically represented as

$$P_0^{(2)} \equiv P_0^{(2)(s)} = y^2 - a^{(s)}\eta^2/4t^2, \tag{3.16}$$

where the coefficient $a^{(s)}$ depends on the type and structure of the ordered phase, as well as on the size and shape of the basis cluster [1, 47, 49, 59].

Similarly, the probabilities of the paired bonds X-□ and □-□ are written in the form

$$P_1^{(2)} \equiv P_1^{(2)(s)} = y(1-y) + a^{(s)}\eta^2/4t^2 \tag{3.17}$$

and

$$P_2^{(2)} \equiv P_2^{(2)(s)} = (1-y)^2 - a^{(s)}\eta^2/4t^2. \tag{3.18}$$

The long-range order parameter reaches the maximum value of $\eta = 1$ for the ideal completely ordered phase of the stoichiometric composition $M_{2t}X_{2t-1} \equiv MX_{y_{st}}$. Under these conditions, the coefficient $a(s)$ can be determined from relationships (3.16) and (3.18); that is,

$$a^{(s)} = 4t^2\left[y_{st}^2 - P_0^{(2)}(\eta = 1)\right]/\eta^2 \tag{3.19}$$

or

$$a^{(s)} = 4t^2\left[(1-y_{st})^2 - P_2^{(2)}(\eta = 1)\right]/\eta^2. \tag{3.20}$$

It can be seen from expressions (3.16)–(3.18) that, for the known quantities y and t, i.e., the composition of the ordered nonstoichiometric compound $M_2X_{y'}$ and the type of the formed superstructure $M_{2t}X_{2t-1}$, the coefficient $a^{(s)}$ determines the probabilities $P_i^{(2)(s)}$.

In the OPF method [1, 2, 46, 47], the free energy of the crystal with any degree of long-range order η is written through the probabilities $P_i^{(s)}$ as follows:

$$F(y,\eta,T) = N\left\{\sum_s y^{(s)} \sum_{i\in s} [\varepsilon_i^{(s)}(T) - TS_i^{(s)}(T)]\lambda_i^{(s)} P_i^{(s)} - TS_c(y,\eta)\right\}, \tag{3.21}$$

where $\varepsilon_i^{(s)}(T)$ and $S_i^{(s)}(T)$ are the temperature-dependent energies and vibrational entropies of different i-configurations of figures s, respectively; $P_i^{(s)}$ are the probabilities of the i-configurations of figures s expressed through the values of the distribution function $n(\boldsymbol{r})$ of atoms that are located over sites $\boldsymbol{r}$ of the nonmetallic sublattice by ordered manner; and $S_c(y,\eta)$ is the configurational (combinatorial) entropy. In the OPF-method, configurational entropy S_c is expressed through distribution function $n(\boldsymbol{r})$ as

$$S_c(y,\eta) = -k_B \sum_{\boldsymbol{r}} \{n(\boldsymbol{r}) \ln n(\boldsymbol{r}) + [1 - n(\boldsymbol{r})] \ln[1 - n(\boldsymbol{r})]\} \tag{3.22}$$

instead of its representation in the form

$$S(P_i^{(n)}) = -k_B N \sum_{s=a}^{n} y^{(s)} \sum_{i=s} \lambda_i^{(s)} P_i^{(s)} \ln P_i^{(s)}, \tag{3.23}$$

which used in the cluster variation method. In expression (3.22), the summation is performed over all sites $\boldsymbol{r}$ of the ordering Ising lattice (in discussed case, the nonmetallic sublattice is the Ising lattice). It is clear, when the cluster probabilities and the configurational (combinatorial) entropy are represented through the distribution function $n(\boldsymbol{r})$, the number of independent variables used for minimizing the free energy $F(y, \eta, T) = E(y, \eta, T) - TS_c(y, \eta)$, coincides with the number of long-range order parameters and is much less than the number of cluster probabilities $P_i^{(s)}$.

In [1, 3, 47], it was demonstrated that, when the enthalpy $H(y, 0, T)$ of the disordered crystal MX_y or $M_2X_{y'}$ is expanded into a power series in the composition y

$$H(y, 0, T) = N \sum_{n=0}^{R^{(s)}} y^n H_n(T), \tag{3.24}$$

the energies $\varepsilon_i^{(s)}$ and the entropies $S_i^{(s)}$ of clusters can be derived analytically. With due regard for this circumstance and the limitation of expansion (3.24) of the enthalpy for real nonstoichiometric compounds by the second-order terms, the free energy (3.21) of compounds MX_y or $M_2X_{y'}$ with any degree of long-range order η is described by the relationship

$$F(y, \eta, T) = F_0(T) + yF_1(T) + P_0^{(2)(s)} F_2(T) - TS_c(y, \eta), \tag{3.25}$$

where $P_0^{(2)(s)}$ is the probability of complete (containing no vacancies) paired bond, which is the subfigure of the basis cluster s (in our case, $s \equiv c$); $F_0(T)$, $F_1(T)$, and $F_2(T)$ are the parameters of the expansion of the free energy (without regard for the configurational entropy term $TS_c(y, \eta)$) of the disordered nonstoichiometric compound MX_y or $M_2X_{y'}$ ($y = y'/2$) into the power series in the composition y. It should be noted that, for nonstoichiometric compounds MX_y and $M_2X_{y'}$, the energy parameter $F_2(T)$ is always positive [1, 2]: $F_2(T) > 0$. The configurational entropy $S_c(y, \eta)$ depends only on two degenerated values of the distribution function

$$\begin{aligned} S_c(y, \eta) = &-(k_B/2t)\{n_1^{(d)} \ln n_1^{(d)} + (1 - n_1^{(d)}) \ln(1 - n_1^{(d)}) \\ &+ (2t - 1)[n_2^{(d)} \ln n_2^{(d)} + (1 - n_2^{(d)}) \ln(1 - n_2^{(d)})]\}. \end{aligned} \tag{3.26}$$

In order to calculate the free energy of the crystal from relationship (3.21), the probabilities $P_i^{(s)}$ of all configurations of the basis figure and the overlapping figures should be determined through the distribution function $n(\boldsymbol{r})$. The transformation of the free energy into formula (3.25) reduces the amount of calculations, because, in this case, only the probabilities $P_0^{(2)(s)}$ of the complete paired bonds of the basis cluster should be expressed through the distribution function.

Expression (3.16) describing the probability $P_0^{(2)(s)}$ of the complete pair contains the coefficient $a^{(s)}$, which depends on the type and structure of the ordered phase, as well as on the size and shape of the basis cluster. Actually, depending on the size, the basis cluster can involve sites belonging to several coordination spheres (from the first to k-th sphere). In the hexagonal lattice with the $L'3$ structure under consideration, the basis cluster in the form of the trigonal prism contains sites of three coordination spheres in the nonmetallic sublattice: with respect to the site 0 (Fig. 3.16a), the first coordination sphere involves one site *I* located at the distance $r_1 = c_h/2$ from the site 0, the second coordination sphere involves two nearest sites *II* of the base that are located at the distance $r_2 = a_h$ from the site 0, and the third coordination sphere involves two sites *III* located at the distance $r_3 = (a_\mathrm{h}^2 + c_\mathrm{h}^2/4)^{1/2}$ from the site 0. The coordination numbers z_j of the first, second, and third coordination spheres with the radii r_1, r_2, and r_3 are equal to 2, 6, and 12, respectively. The pair interatomic interactions in the nonmetallic sublattice of the nonstoichiometric compounds under investigation occur only in the first and second coordination spheres. Therefore, only these coordination spheres should be taken into account in the calculation of the coefficient $a^{(s)}$.

At the disorder–order transformation temperature T_trans, the free energies of the disordered and ordered phases are equal to each other and, hence, $\Delta F = F(y, \eta, T) - F(y, 0, T) = 0$. At the same time, for the equilibrium value of η, the free energy $F(y, \eta, T)$ is minimum and, hence, $\partial F(y, \eta, T)/\partial \eta = 0$. Therefore, the equilibrium conditions for the disorder–order transformation with allowance for the probability $P_0^{(2)(s)}$ in the form of expression (3.16) are determined by the equations

$$\Delta F(\eta) = -a^{(s)}\eta^2 F_2(T)/4t^2 - T[S_c(y, \eta) - S_c(y, 0)] = 0, \tag{3.27}$$

$$\partial F(y, \eta, T)/\partial \eta = -(a^{(s)}/2t^2)\eta F_2(T) - T\partial S_c(y, \eta)/\partial \eta = 0. \tag{3.28}$$

It follows from relationships (3.27) and (3.28) that, at the known quantities y and t, i.e., the composition of nonstoichiometric compound $M_2X_{y'}$ and the type of the formed superstructure $M_{2t}X_{2t-1}$, the coefficient $a^{(s)}$ determines the equilibrium conditions for the disorder–order transformation.

The solution to Eqs. (3.27) and (3.28) are the equilibrium value of the long-range order parameter at the disorder–order transformation point, η_trans, and the disorder–order transformation temperature T_trans. The equation for the long-range order parameter η_trans has the form

$$(\eta_\mathrm{trans}/2)[\partial S_\mathrm{c}(y, \eta)/\partial \eta]_{\eta=\eta_\mathrm{trans}} - S_\mathrm{c}(y, \eta_\mathrm{trans}) + S_\mathrm{c}(y, 0) = 0. \tag{3.29}$$

By determining the long-range order parameter η_{trans} from Eq. (3.29) and substituting it into relationship (3.28), we find the disorder–order transformation temperature T_trans. If the disorder–order transformation is a first-order (first-kind) phase transition, the temperature T_trans is defined by the expression

$$T_\mathrm{trans} = -\frac{a^{(s)}}{2t^2} F_2(T_\mathrm{trans})\eta_\mathrm{trans} \left[\frac{\partial\, S_\mathrm{c}(y, \eta)}{\partial \eta}\right]^{-1}_{\eta=\eta_\mathrm{trans}}. \tag{3.30}$$

Since the derivative $\partial S_c(y,\eta)/\partial\eta$ is always less than zero, it is clear from expressions (3.28) and (3.30) that, at $F_2(T) > 0$, the ordering is possible only at $a^{(s)} > 0$. At the negative coefficient $a^{(s)}$, the ordering is impossible because nonstoichiometric compounds with $F_2(T) < 0$ do not exist [1, 2, 47] .

For second-order disorder–order transitions when the long-range order parameter continuously varies from $\eta = 0$ to η_{equil} at $T \leq T_{\text{trans}}$, formula (3.30) can be rewritten in the form

$$T_{\text{trans}} = -\frac{a^{(s)}}{2t^2} F_2(T_{\text{trans}}) \lim_{\eta\to 0}\left[\frac{\eta}{\partial S_c(y,\eta)/\partial\eta}\right]. \qquad (3.31)$$

After necessary transformations, we obtain the following relationship for the transition temperature:

$$T_{\text{trans}} = \frac{2a^{(s)}}{(2t-1)} y(1-y)\frac{F_2(T_{\text{trans}})}{k_B}, \qquad (3.32)$$

which is similar to the formula for temperature of the disorder–order transformation of the second-order (second-kind) in the mean-field approximation. In the case of ordering of lower hexagonal carbides $M_2C_{y'}$ and nitrides $M_2N_{y'}$, the composition in formulas (3.30)–(3.32) is $y = y'/2 \leq 0.5$.

It should be noted that the conclusion regarding the impossibility of forming superstructures with $a^{(s)} < 0$ corresponds only to the case where the enthalpy of formation of compounds MX_y or $M_2X_{y'}$ is a quadratic function of the composition y and the free energy $F(y,\eta,T)$ for the compound with an arbitrary degree of long-range order η is written in the form of relationship (3.25). If the expansion of the formation enthalpy in a power series in y contains terms of higher orders than the second-order terms, the equilibrium conditions for the disorder–order transformation will be different and the restrictions imposed on the possibility of formation of ordered phases with $a^{(s)} < 0$ can be less rigid.

3.2.2 Probabilities of Pair Interactions in the $M_{2t}X_{2t-1}$ Superstructures of Nonstoichiometric Compounds MX_y

The analytical determination of the probabilities $P_i^{(s)}$, including the probability $P_2^{(2)} \equiv P_2^{(2)(s)}$, is one of the main operations of the OPF method. In this respect, a simple analytical expression for the calculations of the coefficient $a^{(s)}$ and the probabilities $P_0^{(2)}$, $P_1^{(2)}$ and $P_2^{(2)}$ of existence of the pair interactions X-X, X-□, and □-□ was obtained in study [60].

It can be seen from expressions (3.16)–(3.18) that, for the known quantities y and t, i.e., the composition of the nonstoichiometric compound MX_y and the type of the formed superstructure $M_{2t}X_{2t-1}$, the coefficient $a^{(s)}$ determines the probabilities $P_i^{(2)(s)}$ (multiplicities $\lambda_0^{(2)} = 1$, $\lambda_1^{(2)} = 2$ and $\lambda_2^{(2)} = 1$).

Let us determine the coefficient $a^{(s)}$ in the analytical form. The formula for the probability of the complete pair as a subfigure of the basis cluster can be derived using the concept of the superstructural short-range order, which was introduced in studies [61, 62].

The superstructural short-range order is a part of the short-range order that is directly associated with the presence of the long-range order. In the absence of correlations in the arrangement of atoms, the probability of finding the interstitial atom *X* (or the A atom) in the *j*th coordination sphere constructed around the X atom of the MX_y compound (or the A atom of the A_yB_{1-y} solid solution) ordered according to the $M_{2t}X_{2t-1}$ (or $A_{2t-1}B$) type is determined only by the superstructural short-range order

$$P_{X(X)\,j} = P_{A(A)\,j} = y + \alpha_j(y,\eta)(1-y). \tag{3.33}$$

In the lattice, we choose the basis cluster *s* that includes $R^{(s)}$ sites of the ordering lattice and satisfies the requirements imposed on the basis cluster figure in the cluster variation method [54, 56] and the order-parameter functional method [1, 2, 46, 47, 63]. In the cluster *s*, the sites of the ordered lattice and the atoms that occupy these sites are located at different distances from each other, i.e., in different coordination spheres from the first sphere to the *j*th sphere. Let $z_j^{(s)}$ be the number of sites of the basis cluster *s* in the *j*th coordination sphere of an arbitrary site. The probability $P_{\mathrm{X(X)}}^{\mathrm{aver}}$ average within the cluster *s* can be determined by the sequential averaging over the coordination spheres *j* of cluster *s*; that is,

$$P_{\mathrm{X(X)}}^{\mathrm{aver}} = \Big(\sum_j z_j^{(s)} P_{\mathrm{X(X)j}}\Big) \Big/ \sum_j z_j^{(s)}. \tag{3.34}$$

Substituting the probability $P_{X(X)}$ from expression (3.33) into relationship (3.34) gives

$$P_{\mathrm{X(X)}}^{\mathrm{aver}} = y + (1-y)\sum_j z_j^{(s)} \alpha_j(y,\eta) \Big/ \sum_j z_j^{(s)}. \tag{3.35}$$

Now, we determine the probability $P_0^{(2)(s)}$ of existence of the complete paired bond X-X within the cluster *s*. Since the neighboring clusters overlap with each other over paired bonds, the paired bond in the *j*th coordination sphere can belong to several clusters. In order to avoid the overestimation of the number of paired bonds, we introduce the overestimation coefficient $f_j^{(2)(s)}$ equal to unity divided by the number of clusters containing the given bond. Any site of the basis cluster form $z_j^{(s)} f_j^{(2)(s)}$ paired bonds with sites of the *j*th coordination sphere and $\sum_j z_j^{(s)} f_j^{(2)(s)}$ paired bonds with sites of all coordination spheres of the basis cluster from the first sphere to the *j*th coordination sphere. For the MX_y compound, the probability that the interstitial atom X occupies the site that is the center of the *j*th coordination sphere is *y*. Therefore, the probability $P_0^{(2)(s)}$ is equal to $y P_{\mathrm{X(X)}}^{\mathrm{aver}} f_j^{(2)(s)}$ or with allowance made for expression (3.35) is given by the formula

$$P_0^{(2)(s)} = y^2 + y(1-y)\frac{\sum_j z_j^{(s)} \alpha_j(y,\eta) f_j^{(2)(s)}}{\sum_j z_j^{(s)} f_j^{(2)(s)}}. \tag{3.36}$$

According to [61], the parameter of the superstructural short-range order $\alpha_j(y,\eta)$ in the *j*th coordination sphere of the $M_{2t}X_{2t-1}$ ($A_{2t-1}B$) superstructure is described by the formula

$$\alpha_j(y,\eta) = -\frac{1}{y(1-y)}\frac{\eta^2}{4t^2}\left[1 - 2t + \frac{2t\sum_f g_{(1f)} m_{2(1f)}^{(j)}}{\sum_f g_{(1f)}}\right], \tag{3.37}$$

where $m_{2(1f)}^{(j)}$ is the relative number of occupied sites in the *j*th coordination sphere in which the center is a vacant site that can be located at one or several nonequivalent crystallographic positions (1*f*) with the multiplicity $g_{(1f)}$. At $y = y_{st} \equiv (2t-1)/2t$ and $\eta = \eta_{max} \equiv 1$, from relationship (3.37), we find

$$m_{21}^{(j)} \equiv m_{2(1f)}^{(j)} = y_{\rm st}[1 - \alpha_j(y_{\rm st}, \eta_{\rm max})]. \tag{3.38}$$

With allowance made for expression (3.37) for $\alpha_j(y,\eta)$, Eq. (3.36) for probability of complete pair can be written as

$$P_0^{(2)(s)} = y^2 - (\eta^2/4t^2)\left\{1 - 2t + \frac{2t\sum_f g_{(1f)}\sum_j z_j^{(s)} m_{2(1f)}^{(j)} f_j^{(2)(s)}}{\sum_f g_{(1f)}\sum_j z_j^{(s)} f_j^{(2)(s)}}\right\}. \tag{3.39}$$

In formula (3.39), the summation over *f* is associated with the fact that, in the most general case, vacant sites can be located at different crystallographic positions (1*f*). Each this position can have a different radial environment by sites of the nonmetallic sublattice and, in this case, will be characterized by different sets of coefficients $m_{2(1f)}^{(j)}$.

A comparison of relationships (3.16) and (3.39) demonstrates that the coefficient $a^{(s)}$ is represented in the form

$$a^{(s)} = 1 - 2t + \frac{2t\sum_f g_{(1f)}\sum_j z_j^{(s)} m_{2(1f)}^{(j)} f_j^{(2)(s)}}{\sum_f g_{(1f)}\sum_j z_j^{(s)} f_j^{(2)(s)}}. \tag{3.40}$$

In the cases where the vacant site can be located at only one crystallographic position (this is the most frequent case), the averaging over the crystallographic positions *f* is eliminated and formula (3.40) is simplified:

$$a^{(s)} = 1 - 2t + \frac{2t\sum_j z_j^{(s)} m_{2(1)}^{(j)} f_j^{(2)(s)}}{\sum_j z_j^{(s)} f_j^{(2)(s)}}. \tag{3.41}$$

Expressions (3.40) and (3.41) allow us to calculate the coefficients $a^{(s)}$ and, correspondingly, the probabilities $P_i^{(2)(s)}$ for any superstructures. The quantities $m_{21}^{(j)} \equiv m_{2(1f)}^{(j)}$ that are necessary for the calculations and characterize the short-range order in superstructures with cubic basic lattices can be found with relationship (3.38) by using the numerical parameters of the superstructural short-range order, $\alpha_j(y_{st}, \eta_{max})$, [1, 2, 47] in the coordination spheres from the first sphere to the *j*th sphere in different superstructures $M_{2t}X_{2t-1}$.

The nonstoichiometric compounds MX_y with cubic (space group $Fm\overline{3}m$) and with hexagonal (space group P_3/mmc) basic structures have the highest symmetry: the point group $m\overline{3}m(O_h)$ of the cubic disordered compound MX_y contains all 48 elements $h_1 - h_{48}$ of the cubic symmetry group, and the point group $6/mmm\,(D_{6h})$ of the disordered compound $M_2X_{y'}$ includes all 24 elements H_1–H_{24} of the hexagonal symmetry group [34, 35]. Any superstructure that is formed with a reduction of symmetry can be arise on a basis of these high-symmetry disordered nonstoichiometric compounds.

The ordering in nonstoichiometric cubic (space group $Fm\overline{3}m$) compounds MX_y with the *B*1 basic structure is described by the OPF method with the use of the cluster in the form of an octahedron, which consist of six sites of the nonmetallic sublattice ($R^{(s)} = 6$) with the metal atom *M* at the center (Fig. 3.18). In all known superstructures of the $M_{2t}X_{2t-1}$ type that are formed in the cubic basic lattice of the *B*1 type and have several types of crystallographic positions of vacant sites, the radial environment of vacant sites by sites of the nonmetallic sublattice is identical

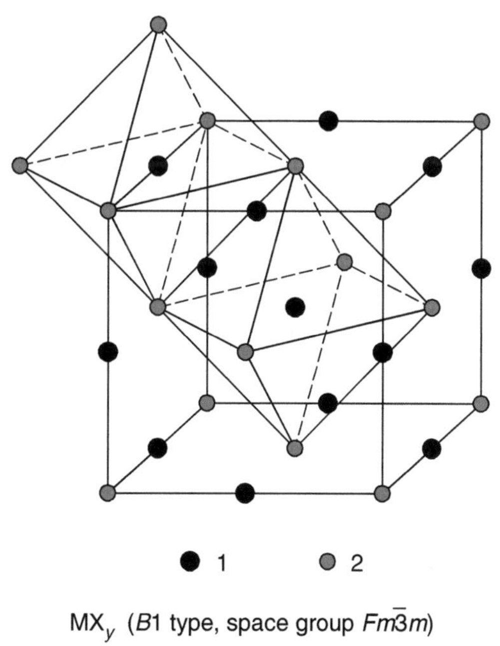

Fig. 3.18 Unit cell of a disordered nonstoichiometric compound MX_y with the *B*1 cubic structure (space group $Fm\overline{3}m$): *1* metal atoms *M* •, *2* nonmetallic sublattice sites • statistically occupied (with a probability of *y*) by nonmetal atoms X. Adjacent octahedral clusters consisting of six sites of the nonmetallic sublattice with the metal atom M at the center are shown

[1, 2, 47]: the *j*th (for example, seventh) coordination spheres constructed around the vacant sites located at different crystallographic positions contain identical number of occupied sites. Consequently, for these superstructures, the summation and averaging over the crystallographic positions f in expression (3.40) are eliminated and the coefficient $a^{(s)}$ can be calculated from formula (3.41).

The ordering in lower hexagonal (space group $P6_3/mmc$) nonstoichiometric $M_2X_{y'}$ (or $MX_{y'/2}\square_{1-y'/2}$, where $0.35 < y'/2 \leq 0.50$) compounds with the $L'3$ structure is described using the basis cluster in the form of a regular trigonal prism MX_6, which is formed by six sites of the nonmetallic sublattice ($R^{(s)} = 6$) and contains the metal atom M at the center (see Fig. 3.16a). In superstructures of the $M_{2t}X_{2t-1}$ type that are formed in the hexagonal basic lattice of the $L'3$ type and have several types of crystallographic positions of vacant sites, the coefficient $a(s)$ is calculated from formula (3.40).

The choice of the clusters in the form of the trigonal prism and octahedron when describing the ordering in the hexagonal and face-centered cubic basic lattices are justified in detail in [53, 64], respectively. The choice of the clusters in the form of the octahedron and regular trigonal prism as the basic figures allows us to account for the most important interatomic interactions metal-nonmetal M-X and many-particle correlations in the first coordination sphere of the metal atom in the MX_y and M_2X_y nonstoichiometric compounds. Moreover, the choice of the octahedron and trigonal prism as the basis cluster figures makes it possible to take into account all (without exception) crystal lattice sites, because each lattice site belongs, at least, to one octahedron (for lattices with the $B1$ structure) or one trigonal prism (for lattice with the $L'3$ structure). With the use of the cluster in the form of the octahedron or trigonal prism, the summation in formulas (3.40) or (3.41) is performed over the first and second coordination spheres of the nonmetal sublattice. The characteristics $R^{(s)}, z_1^{(s)}, f_1^{(2)(s)}, z_2^{(s)}$, and $f_2^{(2)(s)}$ of the basis clusters used for describing the ordering in structures of the $B1$ and $L'3$ types are listed in Table 3.9.

The probabilities $P_i^{(2)(s)}$ should be calculated taking into account that, for any $M_{2t}X_{2t-1}$ superstructures, the minimum value of the long-range order parameter η is equal to zero and the maximum value of the long-range order parameter depends on the composition of the MX_y nonstoichiometric compound:

$$\eta^{\max}(y) = \begin{cases} 2t(1-y), & \text{if } y \geq (2t-1)/2t, \\ 2ty/(2t-1), & \text{if } y < (2t-1)/2t. \end{cases} \tag{3.42}$$

Table 3.9 Characteristics of basis clusters for the $B1$-type and $L'3$-type structures

Basic structure type	Basis cluster	$R^{(s)}$	$z_1^{(s)}$	$f_1^{(2)(s)}$	$z_2^{(s)}$	$f_2^{(2)(s)}$
Cubic $B1$	Octahedron	6	4	1/2	1	1
Hexagonal $L'3$	Trigonal prism	6	1	1/3	2	1/2

This considerably limits the range of variation in the probabilities $P_i^{(2)(s)}$ as functions of the composition y of the ordering compound MX_y, the long-range order parameter η and the type of the formed superstructure $M_{2t}X_{2t-1}$.

If the superstructure $M_{2t}X_{2t-1}$ with identical coefficient t have identical coefficients $a^{(s)}$ then it follows from relationships (3.42) that the dependences of the maximum value of the long-range order parameter $\eta_{max}(y)$ on the composition of the ordering compound MX_y for these superstructures are also identical. Therefore, it is clear from expressions (3.16)–(3.18) that the such superstructures are characterized by identical dependences of the probabilities $\lambda_0^{(2)} P_0^{(2)}$, or $\lambda_1^{(2)} P_1^{(2)}$, or $\lambda_2^{(2)} P_2^{(2)}$ of the pair interaction X-X, or X-□, or □-□ on the composition y and the long-range order parameter η.

The coefficients $a^{(s)}$ are calculated by Eqs. (3.40) and (3.41) for all discussed superstructures. The quantities $m_{21}^{(1)}$ and $m_{21}^{(2)}$ that are necessary for these calculations are given in Table 3.10.

The calculated values of the coefficients $a^{(s)}$ for superstructures of the $M_{2t}X_{2t-1}$ type formed in the $M_2X_{y'}$ and MX_y nonstoichiometric compounds with the hexag-

Table 3.10 Structural characteristics of $M_{2t}X_{2t-1}$ ordered phases of $M_2X_{y'}$ and MX_y nonstoichiometric compounds with hexagonal ($L'3$ type) and cubic ($B1$ type) basic structures

Ordered phase type	Space group	Translation vectors[a] *a*	*b*	*c*	N[b]	t	$g_{(1f)}$	$m_{21}^{(1)}$	$m_{21}^{(2)}$	$a^{(s)}$
Hexagonal basic structure of the $L'3$ (W_2C) type										
M_2X	$P\bar{3}m1$	$\langle 100\rangle_{L'3}$	$\langle 010\rangle_{L'3}$	$\langle 001\rangle_{L'3}$	1	1	–	1	0	−0.5
M_2X	$Pbcn$	$\langle 001\rangle_{L'3}$	$2\langle 100\rangle_{L'3}$	$\langle 120\rangle_{L'3}$	4	1	–	1	2/3	0.5
M_2X	$P\bar{3}1m$	$\langle 1\bar{1}0\rangle_{L'3}$	$\langle 120\rangle_{L'3}$	$\langle 001\rangle_{L'3}$	3	1	$g_{(11)} = 1$	1	1	0.5
Cubic basic structure of the $B1$ (NaCl) type										
M_2X	$R\bar{3}m$	$\frac{1}{2}\langle 1\bar{2}1\rangle_{B1}$	$\frac{1}{2}\langle 2\bar{1}1\rangle_{B1}$	$\frac{1}{2}\langle 1\bar{1}2\rangle_{B1}$	1	1	–	1/2	1	1/3
M_2X	$Fd\bar{3}m$	$\langle 200\rangle_{B1}$	$\langle 020\rangle_{B1}$	$\langle 002\rangle_{B1}$	16	1	–	1/2	1	1/3
M_2X	$I4_1/amd$	$\langle 100\rangle_{B1}$	$\langle 010\rangle_{B1}$	$\langle 002\rangle_{B1}$	4	1	–	2/3	1/3	1/9
M_2X	$P4/mmm$	$\frac{1}{2}\langle 1\bar{1}0\rangle_{B1}$	$\frac{1}{2}\langle 110\rangle_{B1}$	$\langle 001\rangle_{B1}$	1	1	–	2/3	0	−1/9
M_3X_2	$Immm$	$\frac{1}{2}\langle 1\bar{1}0\rangle_{B1}$	$\frac{3}{2}\langle 110\rangle_{B1}$	$\langle 001\rangle_{B1}$	2	1.5	–	5/6	2/3	1/3
M_3X_2	$C2$	$\frac{1}{2}\langle 112\rangle_{B1}$	$\langle 11\bar{1}\rangle_{B1}$	$\frac{3}{2}\langle 1\bar{1}0\rangle_{B1}$	6	1.5	–	5/6	2/3	1/3
M_3X_2	$C222_1$	$\langle 1\bar{1}0\rangle_{B1}$	$\langle 330\rangle_{B1}$	$\langle 002\rangle_{B1}$	16	1.5	–	3/4	1	0.5
M_3X_2	$P\bar{3}m1$	$\frac{1}{2}\langle 10\bar{1}\rangle_{B1}$	$\frac{1}{2}\langle 011\rangle_{B1}$	$\langle 1\bar{1}1\rangle_{B1}$	1	1.5	–	1/2	1	0
M_4X_3	$Pm\bar{3}m$	$\langle 100\rangle_{B1}$	$\langle 010\rangle_{B1}$	$\langle 001\rangle_{B1}$	1	2	–	1	0	−1/3
M_4X_3	$I4/mmm$	$\langle 100\rangle_{B1}$	$\langle 010\rangle_{B1}$	$\langle 002\rangle_{B1}$	2	2	–	1	1/3	1/9
M_6X_5	$C2/m$	$\frac{1}{2}\langle 1\bar{1}\bar{2}\rangle_{B1}$	$\frac{3}{2}\langle 110\rangle_{B1}$	$\frac{1}{2}\langle 1\bar{1}2\rangle_{B1}$	2	3	–	1	1	1.0
M_6X_5	$P3_1$	$\frac{1}{2}\langle 21\bar{1}\rangle$	$\frac{1}{2}\langle \bar{1}12\rangle_{B1}$	$2\langle 1\bar{1}1\rangle_{B1}$	3	3	–	1	1	1.0
M_6X_5	$C2$	$\frac{1}{2}\langle 1\bar{1}\bar{2}\rangle_{B1}$	$\frac{3}{2}\langle 110\rangle_{B1}$	$\langle 1\bar{1}2\rangle_{B1}$	4	3	–	1	1	1.0
M_8X_7	$Fm\bar{3}m$	$\langle 200\rangle_{B1}$	$\langle 020\rangle_{B1}$	$\langle 002\rangle_{B1}$	4	4	–	1	1	1.0
M_8X_7	$P4_332$	$\langle 200\rangle_{B1}$	$\langle 020\rangle_{B1}$	$\langle 002\rangle_{B1}$	4	4	–	1	1	1.0

[a]Translation vectors of the unit cell of the $M_{2t}X_{2t-1}$ ordered phase in the hexagonal ($L'3$ type) or or cubic ($B1$ type) basic lattice

[b]N is the number of $M_{2t}X_{2t-1}$ formula units in the unit cell of the ordered phase

onal ($L'3$ type) and cubic ($B1$ type) basic structures, respectively, are presented in Table 3.10.

References

1. Gusev AI, Rempel AA, Magerl AJ. Disorder and order in strongly nonstoichiometric compounds: transition metal carbides, nitrides and oxides. Berlin: Springer; 2001. 607 pp.
2. Gusev AI. Nonstoichiometry, disorder, short-range and long-range order in solids. Moscow: Nauka-Fizmatlit; 2007. 856 pp. (in Russian).
3. De Novion CH, Beuneu B, Priem T, Lorenzelli N, Finel A. Defect structures and order-disorder transformations in transition metal carbides and nitrides. In: Freer R, editor. The physics and chemistry of carbides, nitrides and borides. Netherlands: Kluwer; 1990. p. 329–55.
4. Gusev AI, Rempel AA. Superstructures of non-stoichiometric interstitial compounds and the distribution functions of interstitial atoms. Phys Stat Sol. (a) 1993;135(1):15–58.
5. Gusev AI. Order-disorder transformations and phase equilibria in strongly nonstoichiometric compounds. Uspekhi Fiz Nauk. 2000;170(1):3–40 (in Russian) (Engl. Transl.: Physics - Uspekhi. 2000;43(1):1–37).
6. Kurlov AS, Gusev AI. Phase equilibria in the W-C system and tungsten carbides. Uspekhi Khimii. 2006;75(7):687–708 (in Russian) (Engl. Transl.: Russian Chem Rev. 2006;75(7):617–636).
7. Yvon K, Nowotny H, Benesovsky F. Zur Kristallstruktur von W_2C. Monatsh Chemie. 1968;99(2):726–729.
8. Rudy E, Windisch S. Evidence to zeta Fe_2N-type sublattice order in W_2C at intermediate temperatures. J Am Ceram Soc. 1967;50(5):272–3.
9. Rudy E, Hoffman JR. Phasengleichgewichte im Bereich der kubischen Karbidphase im System Wolfram-Kohlenstoff. Planseeber Pulvermet. 1967;15(3):174–8.
10. Rudy E. Compendium of phase diagram data. Ternary phase equilibria in transition metal-boron-carbon-silicon systems (Part V). Final Tech. Report AFML TR-65-2. Wright-Patterson Air Force Base (Ohio, USA): Metals and Ceramics Division, Air Force Materials Laboratory; 1969. 735 pp.
11. Parthe E, Sadagopan V. The structure of dimolybdenum carbide by neutron diffraction technique. Acta Crystallogr. 1963;16(3):202–5.
12. Stecher P, Benesovsky F, Nowotny H. Chromium-tungsten-carbon system. Planseeber Pulvermet. 1964;12(2):89–95.
13. Telegus VS, Gladyshevskii EI, Kripyakevich PI. Rhombic modifications of the compounds W_2C and Mo_2C. Kristallografiya 1967;12(5–6):936–9 (in Russian) (Engl. Transl.: Sov Phys Crystallogr. 1967;12(6–6):813–816).
14. Telegus VS, Kuz'ma YuB, Marco MA. Phase equilibria in the systems molybdenum-manganese-carbon and tungsten-manganese-carbon. Poroshk Metallurgiya. 1971;10(11):56–63. (in Russian) (Engl. Transl.: Powder Metallurg Metal Cer. 1971;10(11):898–903).
15. Nozik YuZ, Lipin YuF, Kuvaldin BV. A neutron diffraction study of the orthorhombic modification of W_2C. Izv Akad Nauk Latv SSR, Ser Fiz-Tekh. 1968;(6):30–33 (in Russian).
16. Morton N, James BW, Wostenholm GH, Hepburn DCB. Superconductivity and the structure of W_2C. J Less-Common Metals 1972;29(4):423–426.
17. Kublii VZ, Velikanova TYa, Gnitetskii OA, Makhovitskaya SI. Structural parameters of the low-temperature metastable form of the carbide W_2C. Poroshk Metallurgiya. 2000;39(3–4):46–53 (in Russian) (Engl. Transl.: Powder Metallurg Metal Cer. 2000;39(3–4):151–6).
18. Dubois J, Epicier T, Esnouf C, Fantozzi G, Convert P. Neutron powder diffraction studies of transition metal hemicarbides M_2C_{1-x}. I. Motivation for a study on W_2C and Mo_2C experimental background for an in situ investigation at elevated temperature. Acta Metallurg. 1988;36(8):1891–1901.

19. Epicier T, Dubois J, Esnouf C, Fantozzi G, Convert P. Neutron powder diffraction studies of transition metal hemicarbides M_2C_{1-x}. II. In situ high temperature study of W_2C_{1-x} and Mo_2C_{1-x}. Acta Metallurg. 1988;36(8):1903–21.
20. Butorina LN, Pinsker ZG. Electron diffraction study of W_2C. Kristallografiya 1960;5(4):585–8 (in Russian) (Engl. Transl.: Sov Phys Crystallogr. 1960;5(4):560–62).
21. Harsta A, Rundqvist S, Thomas JO. A neutron powder diffraction study of W_2C. Acta Chem Scand A. 1978;32A(9):891-92.
22. Gusev AI, Kurlov AS. Ordering of the lowest tungsten carbide W_2C. Pis'ma v ZhETF. 2007;85(1):40–45 (in Russian) (Engl. Transl.. JETP Lett. 2007;85(1):34-39).
23. Lönnberg B, Lundström T, Tellgren R. A neutron powder diffraction study of Ta_2C and W_2C. J Less-Common Metals. 1986;120(2):239–45.
24. Epicier T, Dubois J, Esnouf C, Fantozzi G. Identification de la phase β-W_2C, type $\varepsilon - Fe_2N$ dans l'hémicarbure de tungsten. Compt Rend Acad Sci Paris Ser II. 1983;297(3):215–8.
25. Kublii VZ, Velikanova TYa, Khaenko BV. Single-crystal x-ray study of the W_2C carbide with structure of the ε-Fe_2N type. Metallofizika i Noveish. Tekhnologii (Metal Phys Adv Technol) 1999;21(9):26–8.
26. Kublii VZ, Velikanova TYa. Ordering in the carbide W_2C and phase equilibria in the tungsten-carbon system in the region of its existence. Poroshk Metallurgiya. 2004;43(11–12):101–16 (in Russian) (Engl. Transl.: Powder Metallurg Metal Cer. 2004;43(11–12):630–44).
27. Gleiser M, Chipman J. Free energy of formation of tungsten carbide WC. Trans Metallurg Soc AIME. 1962;224(6):1278–79.
28. C - W (carbon - tungsten). In: Massalski TB, Subramanian PR, Okamoto H, Kasprzak L, editors. Binary alloy phase diagrams. 2nd ed, vol. 1. Metals Park: ASM; 1990. p. 895–6.
29. Sara RV. Phase equilibrium in the system tungsten-carbon. J Am Ceram Soc. 1965;48(5):251–7.
30. C-W. In: McHale AE, editor. Phase equilibria diagrams. Phase diagrams for ceramists, vol. X. Westerville: American Ceramic Society; 1994. p. 272–3.
31. Kurlov AS, Gusev AI. Phase transitions in the lowest tungsten carbide W_2C. Doklady Akad Nauk 2007;417(5):616–23 (in Russian) (Engl. Transl.. Doklady Phys. 2007;52(12):656–62).
32. Kurlov AS, Gusev AI. Atomic-vacancy ordering of lower tungsten carbide W_2C. Zh Eksp Teor Fiz. 2007;132(4):812–26 (in Russian) (Engl/ Transl.: J Exp Theor Phys. 2007;105(4):710–21).
33. Kurlov AS, Gusev AI. Neutron and x-ray diffraction study and symmetry analysis of phase transformations in lower tungsten carbide W_2C. Phys Rev B. 2007;76(17). Paper 174115. p. 174115-01–174115-16.
34. Kovalev OV. Irreducible representations of the space groups. New York: Gordon & Breach; 1965. p. 210.
35. Kovalev OV. Representations of the crystallographic space groups: irreducible representations, induced representations and corepresentation. Yverdon: Gordon & Breach; 1993. p. 390.
36. Khachaturian AG. Theory of structural transformations in solids. New York: John Wiley; 1983. p. 574.
37. Larson AC, von Dreele RB. General structure analysis system (GSAS). Los Alamos National Laboratory Report LAUR 86–748. Los Alamos, 2004.
38. Warren BE, Averbach BL, Roberts BW. Atomic size effect in the X-ray scattering by alloys. J Appl Phys. 1951;22(12):1493–6.
39. Gusev AI, Rempel AA. Nanocrystalline materials. Cambridge: Cambridge International Science Publishers; 2004. p. 149.
40. Gusev AI. Nanomaterials, nanostructures, and nanotechnologies. 2nd ed. Moscow: Nauka - Fizmatlit; 2007. p. 186 (in Russian).
41. Zueva LV, Lipatnikov VN, Gusev AI. Ordering effects on the microstructure and microhardness of nonstoichiometric titanium carbide TiC_y. Neorgan Materialy. 2000;36(7):836–40 (in Russian) (Engl. Transl. Inorganic Mater. 2000;36(7):695–8.
42. Lipatnikov VN, Gusev AI. Annealing-induced ordering of bulk nonstoichiometric vanadium carbide. Neorgan Materialy. 2006;42(1):16–21 (in Russian) (Engl. Transl. Inorganic Mater. 2006;42(1):14–8).

43. Kurlov AS, Gusev AI. Tungsten carbides and W - C phase diagram. Neorgan Materialy. 2006;42(2):156–63 (in Russian) (Engl. Transl. Inorganic Mater. 2006;42(2):121–7).
44. Kurlov AS, Gusev AI. Pecualirities of vacuum annealing of nanocrystalline WC powders. Int J Refr Met Hard Mater. 2012;32(5):51–60.
45. Kurlov AS, Gusev AI. Vacuum annealing of nanocrystalline WC powders. Neorgan Materialy. 2012;48(7):781–91 (in Russian) (Engl. Transl.: Inorg Mater. 2012;48(7):680–90).
46. Gusev AI, Rempel AA. Structural phase transitions in nonstoichiometric compounds. Moscow: Nauka, 1988. p.308 (in Russian).
47. Gusev AI, Rempel AA. Nonstoichiometry, disorder and order in solids. Ekaterinburg: Ural Division of the Russian Academy of Sciences; 2001. p. 580 (in Russian).
48. Gusev AI. Physical chemistry of nonstoichiometric refractory compounds. Moscow: Nauka; 1991. p. 286 (in Russian).
49. Gusev AI. Phase diagrams for ordering systems in order-parameter functional method. Fiz Tverd Tela 1990;32(9):2752–60 (in Russian) (Engl. Transl.: Sov Phys Solid State. 1990;32(9):1595–9).
50. Gusev AI, Rempel AA. Calculation of phase diagrams of interstitial compounds. J Phys Chem Solids. 1994;55(3):299–304.
51. Gusev AI, Rempel AA. Phase diagrams of metal-carbon and metal-nitrogen systems and ordering in strongly nonstoichiometric carbides and nitrides. Phys Stat Sol (a) 1997;163(2):273–304.
52. Gusev AI, Rempel AA. Atomic ordering and phase equilibria in strongly nonstoichiometric carbides and nitrides. In: Gogotsi YG, Andrievski RA, editors. Materials science of carbides, nitrides and borides. Dordrecht: Kluwer; 1999. p. 47–64.
53. Kurlov AS, Gusev AI. Ordering of nonstoichiometric hexagonal compounds M_2X: a sequence of special figures. Fiz Tverd Tela. 2009;51(10):1933–9 (in Russian) (Engl. Transl.: Phys Solid State. 2009;51(10):2051–7).
54. Kikuchi R. Theory of cooperative phenomena. Phys Rev. 1951;81(6):988–1003.
55. Kurata M, Kikuchi R, Watari T. Theory of cooperative phenomena. Detailed discussions of the cluster variation method. J Chem Phys. 1953;21(3):434–48.
56. Hijmans J, de Boer J. Approximation method for order-disorder problem. Physica. 1955;21(6):471–516.
57. Golosov NS, Ushakov AV. Model of alloy in statistical theory of ordering. Fiz Tverd Tela. 1976;18(5):1262–8 (in Russian).
58. Sanchez JM, de Fontaine D. The f.c.c Izing model in the cluster variation approximation. Phys Rev B. 1978;17(12):2926–36.
59. Gusev AI, Rempel AA. Thermodynamic model of atomic ordering. Phase diagrams of ordered systems. Zh Fiz Khimii. 1991;65(3):625–33 (in Russian) (Engl. Transl.: Russ J Phys Chem. 1991;65(3):330–34).
60. Kurlov AS, Gusev AI. Determination of the probability of existence of pair interactions in the formation of $M_{2t}X_{2t-1}$ superstructures in MX_y nonstoichiometric compounds. Fiz Tverd Tela. 2010;52(2):342–50 (in Russian) (Engl. Transl.: Phys Solid State. 2010;52(2):370–76).
61. Rempel AA, Gusev AI. Short-range order in ordered alloys and interstitial phases. Fiz Tverd Tela. 1990;32(1):16–24 (in Russian) (Engl. Transl.: Sov Phys Solid State. 1990;.32(1):8–13).
62. Rempel AA, Gusev AI. Short-range order in superstructures. Phys Stat Sol. (b). 1990;160(2):389–402.
63. Gusev AI. Disorder and long-range order in nonstoichiometric interstitial compounds: transition metal carbides, nitrides and oxydes. Phys Stat Sol. (b). 1991;163(1):17–54.
64. Gusev AI, Rempel AA. Calculating the energy parameters for CV and OPF methods. Phys Stat Sol. (b). 1987;140(2):335–46.

Chapter 4
Nanocrystalline Tungsten Carbide

The production of ultrafine powders of different substances and nanostructured materials for different areas of technology has been discussed in literature for many years now. In the last decade, the interest in this subject has greatly increased because it was found that a decrease in the size of particles (grains, crystallites) below some threshold value may result in a large change of the properties [1–15]. These effects appear when the average size of particles or grains does not exceed 100 nm, and are most evident when the particle or grain size is smaller than 10 nm. When examining the properties of superfine materials, it is necessary to take into account not only their structure and composition but also dispersion.

Polycrystalline superfine materials with an average grain size of 300 to 40 nm are referred to as submicrocrystalline materials, and those with an average grain size of less than 40 nm as nanocrystalline or nanostructured materials. The conditional classification of materials on the basis of the size D of particles (grains) is shown in Fig. 4.1.

4.1 X-ray Diffraction Method of Determination of the Small Particle Size and Microstrains

The synthesis of small particles, clusters, and nanocrystals has gained increased interest after the discovery of the fact that a decrease in crystallite size below some threshold results in both a significant alteration of crystal structure and the manifestation of quantum properties. When particle size is less than 10 nm, quantum effects are the most pronounced [13].

Therefore during recent twenty or twenty five years, nanocrystalline materials have been extensively investigated and found a wide application. In this respect, there arises a need for their characterization according to the particle size. The size of small particles can be determined using electron microscopy, sedimentation, photon correlation spectroscopy, gas adsorption, magnetic measurements, and other

A. S. Kurlov and A. I. Gusev, *Tungsten Carbides*,
Springer Series in Materials Science 184, DOI: 10.1007/978-3-319-00524-9_4,
© Springer International Publishing Switzerland 2013

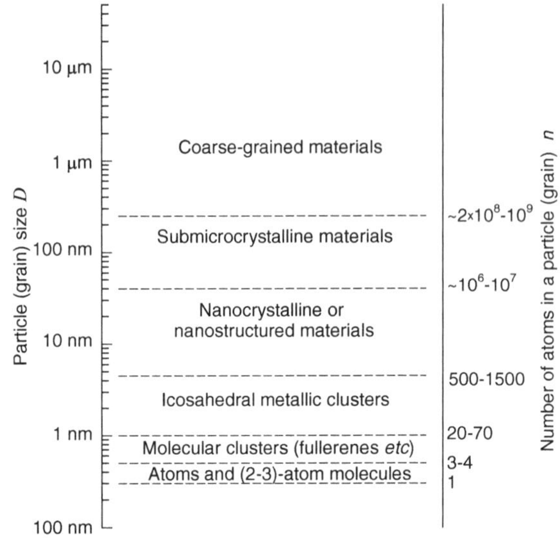

Fig. 4.1 Classification of substances and materials on the basis of the particle (grain) size D

methods. One of the most accessible and universally employed methods for determining the particle size and microstrains in bulk and powdered nanocrystalline substances consists in analyzing the broadening of X-ray diffraction reflections, i.e., X-ray diffraction (XRD) method.

The point is that any defects bring about displacements of atoms from the lattice sites. This results in a change in the intensity and/or the shape of diffraction reflections in the diffraction experiment. Krivoglaz [16] obtained the known relationship for the intensity of Bragg reflections from defect crystals, which makes it possible to conventionally separate all defects into two groups. Defects of the first group only decrease the intensity of diffraction reflections, whereas defects of the second group are responsible for the broadening of reflections. The defects of the second group are associated primarily with the small size D of grains (crystallites, particles) and microstrains ε of the crystal lattice. The broadening β_s due to the small size D of particles (grains, crystallites) and stacking faults is proportional to $\sec\theta$, where θ is the glancing angle in the diffraction experiment. The broadening β_d caused by the lattice microstrains ε and randomly distributed dislocations is proportional to $\tan\theta$.

One more type of defects of the second group, such as an inhomogeneity, is less known. Inhomogeneity causes a variation of the compound composition (i.e., the non-uniform composition of a compound) over the volume of the specimen. The broadening of diffraction reflections due to the inhomogeneity can be observed for compounds and materials with atom or atom–vacancy substitution, i.e., in the nonstoichiometric compounds MX_y ($MX_y\square_{1-y}$ where $\square$ is a structural vacancy) and solid solutions (alloys) A_yB_{1-y}. In the presence of an inhomogeneity, the composition of the $MX_y\square_{1-y}$ nonstoichiometric compound or the A_yB_{1-y} solid solution varies in the range $y \pm \Delta y$, where the quantity Δy is the degree of inhomogeneity (deviation from the homogeneity). An expression according to which the broadening β_h associated with the degree of inhomogeneity Δy is proportional to $(\sin^2\theta)/\cos\theta$ was derived in [17, 18]. The degree of inhomogeneity as a factor responsible for the broadening of diffraction reflections is described in detail in the monographs [14, 15, 19–21].

Different angular dependences of the size, strain, and inhomogeneity broadenings make it possible to separate these types of broadenings and, hence, to determine the particle size, microstrains, and the degree of inhomogeneity.

When defects are examined using the broadening of diffraction reflections, the width of reflections from a substance in the nanocrystalline state (i.e., the state with small-sized (less than 150 nm) grains or particles) in some cases is compared with the width of reflections from the same substance with large-sized grains (particles). This determination of the broadening and subsequent estimations of the average particle size, microstrains, and the degree of inhomogeneity can lead to a very large error (up to several hundred percent). The point is that the broadening should be determined with respect to the diffraction reflections from an infinitely large crystal. This means that the measured width of the diffraction reflections should be compared with the instrumental width, i.e., with the width of the diffractometer resolution function, which should be preliminarily obtained in a special diffraction experiment. Moreover, the diffraction reflection width can be accurately determined only with the theoretical reconstruction of the shape of the experimental reflections.

The separation of the contributions from the small size of coherent scattering regions (particles, grains), the microstrains, and the degree of inhomogeneity to the broadening of diffraction reflections requires the use of different algorithms.

According to [14, 15, 22, 23], the sequence of the performance of the diffracion experiment on the determination of the average size of coherent scattering regions (particle sizes), microstrains, and the degree of inhomogeneity from the broadening of diffraction reflections involves the following stages:

(1) the experimental measurement of the XRD pattern of a reference compound, numerical analysis of the profile of diffraction reflections and the determination of the reflection width, the construction of the experimental dependence of the reflection width on the diffraction angle 2θ, and the calculation of the diffractometer resolution function approximating this dependence;
(2) the measurement of the XRD pattern of the compound under investigation, numerical analysis of the profile of diffraction reflections, and the determination of the reflection width;
(3) the determination of the dependence of the reflection width on the diffraction angle for the compound under investigation;
(4) the separation of the contributions made to the broadening by the small size of particles, strain induced lattice distortions, and the degree of inhomogeneity of the compound under investigation; and
(5) the determination of the average size of coherent scattering regions (grains, particles, crystals), microstrains, and the degree of inhomogeneity.

An X-ray diffraction method of determination of the size of small particles, microstrains and degree of inhomogeneity is described in detail in works [14, 15, 22–24] using carbides and sulphides as the model objects.

4.1.1 The Pseudo-Voight Function

The diffraction reflection width can be described by the Gauss function

$$g(\theta) = A\exp[-(\theta - \theta_0)^2/(2\theta_G^2)] \tag{4.1}$$

or by the Lorentz function (this function is named the Cauchy distribution in mathematics)

$$\ell(\theta) = A\left[1 + \frac{(\theta - \theta_0)^2}{\theta_L^2}\right]^{-1}. \tag{4.2}$$

However, the diffraction reflection profile can be most accurately described with a linear superposition of these functions, which is known as the pseudo-Voigt function [25, 26]

$$V(\theta) = cA\left[1 + \frac{(\theta - \theta_0)^2}{\theta_L^2}\right]^{-1} + (1 - c)A\exp\left[-\frac{(\theta - \theta_0)^2}{2\theta_G^2}\right], \tag{4.3}$$

where c is the relative contribution of the Lorentz function to the pseudo-Voigt function, i.e. to the total intensity of diffraction reflection; θ_0 is the position of the maximum of the function, and A is the normalizing factor.

If an X-ray diffraction pattern is recorded using $K\alpha_{1,2}$-radiation with two wavelengths λ_1 and λ_2, each reflection in the XRD pattern is a doublet. Therefore, each diffraction reflection (*hkl*) is described by two pseudo-Voigt functions. According to the Wulff-Bragg law $2d_{hkl}\sin\theta_{hkl} = n\lambda$, the positions θ_1 and θ_2 of the maxima of the pseudo-Voigt functions are related through the interplanar spacing d_{hkl}, corresponding to the given reflection, by the expression

$$\theta_2 = 2\arcsin\left[\frac{\lambda_2}{\lambda_1}\sin(\theta_1/2)\right]. \tag{4.4}$$

In formula (4.4), $2\theta_1$ and $2\theta_2$ are the angles corresponding to the maxima of the doublet lines of the (*hkl*) reflection. The calculations are performed taking into account that the intensity of the $K\alpha_2$-line in the doublet is equal to 0.497 of the intensity of the $K\alpha_1$-line.

The description of diffraction reflections with the use of the model function (4.3) enables one to determine more accurately their position, the integral intensity, and the full width at half-maximum (FWHM). By knowing the full width at half-maximum, it is possible to calculate the broadening of the diffraction reflection. In turn, the average size of coherent scattering regions, the microstrains, and the degree of inhomogeneity can be determined from the angular dependence of the broadening of the diffraction reflections. The positions of the reflections make it possible to determine the unit cell lattice constants for a chosen structural model, to calculate the diffraction reflection intensities for this model, and to assess mathematically the validity of the structural model by comparing the calculated and experimental intensities.

Let us determine the full widths at half-maximum for the Gauss and Lorentz functions.

The Gauss function $g(\theta)$ at the maximum is equal to A. The Gauss function takes the value equal to half the maximum value, i.e., $g(\theta) = A/2$, at $\theta^{(1)} = \theta_0 + \theta_{\mathrm{G}}\sqrt{2\ln 2}$ and $\theta^{(2)} = \theta_0 - \theta_{\mathrm{G}}\sqrt{2\ln 2}$. Correspondingly, we find $\mathrm{FWHM}_G = \theta^{(1)} - \theta^{(2)} = 2\theta_{\mathrm{G}}\sqrt{2\ln 2} \approx 2.355\theta_G$ for the Gauss function (Fig. 4.2a).

The full width at half-maximum for the Lorentz function $\ell(\theta)$ is determined in the same manner as for the Gauss function. The maximum value of the Lorentz function $\ell(\theta)$ is equal A. Let the Lorentz function $\ell(\theta)$ be equal to half the maximum value, i.e., $\ell(\theta) = A/2$. In this case from relationship (4.2), we obtain $\theta^{(1)} = \theta + \theta_{\mathrm{L}}$ and $\theta^{(2)} = \theta - \theta_{\mathrm{L}}$. Since the full width at half-maximum is equal to $\theta^{(1)} - \theta^{(2)} = 2\theta_{\mathrm{L}}$, we have $\mathrm{FWHM}_L = 2\theta_{\mathrm{L}}$ for the Lorentz function (Fig. 4.2b).

The pseudo-Voight function (4.3) is the superposition of the Lorentz and Gauss functions. As the zero approximation, the full width at half-maximum of the pseudo-Voight function may be approximated as

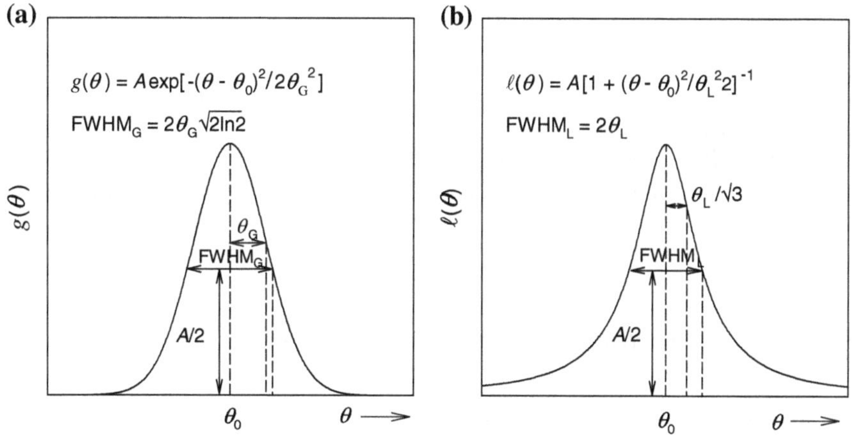

Fig. 4.2 Gauss $g(\theta)$ and Lorentz (Cauchy) $\ell(\theta)$ distributions, used for model description of the shape of the diffraction lines: **a** relationship between the parameter θ_{G} of the Gauss function and the full width of the function $g(\theta)$ at half of its maximum, $\mathrm{FWHM}_{\mathrm{G}}$; **b** for the Lorentz function parameter $\theta_L = \mathrm{FWHM}_{\mathrm{L}}/2$, and the second momentum of the Lorentz $\ell(\theta)$ function is equal to $\theta_L/\sqrt{3}$

$$\begin{aligned}\mathrm{FWHM_V} &= c\mathrm{FWHM}_L + (1-c)\mathrm{FWHM}_G = 2c\theta_\mathrm{L} + 2\sqrt{2\ln 2}(1-c)\theta_\mathrm{G} \\ &\equiv 2c\theta_\mathrm{L} + 2.355(1-c)\theta_\mathrm{G}.\end{aligned} \tag{4.5}$$

Under the assumption that the parameters θ_L and θ_G for the Lorentz and Gauss functions in the pseudo-Voigt function are identical $\theta_\mathrm{L} = \theta_\mathrm{G} = b$, the exact value of full width at half-maximum FWHM_V for the pseudo-Voigt function is determined from the solution to (4.3) at $V(\theta) = A2$. In this case, relationship (4.3) takes the form

$$c\left[1+\left(\frac{\theta-\theta_0}{b}\right)^2\right]^{-1} + (1-c)\exp\left[-\frac{1}{2}\cdot\left(\frac{\theta-\theta_0}{b}\right)^2\right] = \frac{1}{2}. \tag{4.6}$$

Equation (4.6) is nonlinear and has no exact solution. The graphical solution to this equation has the form

$$\mathrm{FWHM_V} = b(2.355 - 0.276c - 0.079c^2). \tag{4.7}$$

Taking into account Eq. (4.5), a good approximation for the full width at half-maximum for the pseudo-Voigt function is

$$\mathrm{FWHM_V} = c\cdot\mathrm{FWHM_L} + (1-c)\cdot\mathrm{FWHM_G} = 2b[c + (1-c)\sqrt{2\ln 2}], \tag{4.8}$$

where $\mathrm{FWHM_L}$ and $\mathrm{FWHM_G}$ are the full width at half-maximum for the Lorentz and the Gauss functions, respectively, and c is the relative contribution of the Lorentz function to the pseudo-Voigt function.

For the calculations from relationships (4.7) and (4.8), the maximum difference between the full widths at half-maximum $\mathrm{FWHM_V}$ for the pseudo-Voigt function is equal to 0.9 % at $c = 0.5$.

In order to determine the reflection broadening, the experimental reflection width should be compared with the instrumental function of the angular resolution of the diffractometer.

4.1.2 The Diffractometer Resolution Function

The instrumental function of the angular resolution of the diffractometer is determined in a special experiment by measuring the full widths at half-maximum FWHM of diffraction reflections over a wide range of 2θ angles for a compound containing no defects that can lead to an additional broadening. In other words, a well annealed and completely homogeneous substance with particles (grains) more than 1 μm in size should be used as a reference compound. In this case, only the instrumental broadening of diffraction reflections will be observed. Lanthanum hexaboride LaB_6 (NIST Standard Reference Powder 660a) with the cubic lattice

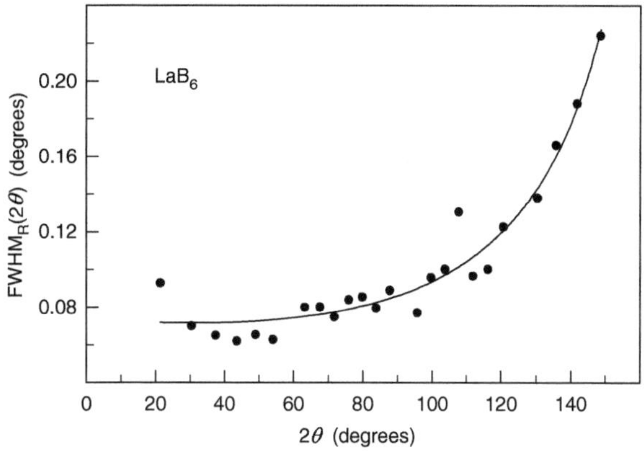

Fig. 4.3 Instrumental function of the angular resolution of the Siemens D-500 diffractometer according to the experimental data obtained for lanthanum hexaboride LaB_6

constant $a = 415.69162 \pm 0.00097$ pm can be used for determining the diffractometer resolution function. The X-ray diffraction patterns of the standard reference compound LaB_6 were recorded on a Siemens D-500, DRON-UM1 and Shimadzu XRD-7000 diffractometers using $CuK\alpha_{1,\,2}$ radiation. Each reflection was described by the pseudo-Voigt function (4.3), and the full width at half-maximum $FWHM_R$ was obtained from formula (4.7). Then, the angular dependence of the full width at half-maximum $FWHM_R(2\theta)$ was constructed (Fig. 4.3).

The instrumental function of angular resolution was analytically described by the relationship

$$FWHM_R(2\theta) = \sqrt{u \cdot \tan^2\theta + v \cdot \tan\theta + w}, \tag{4.9}$$

which was proposed by Cagliotti et al. [27] and introduced into a wide scientific use by Rietveld [28]. The approximation of the experimental dependences $FWHM_R(2\theta)$ by relationship (4.9) allowed finding the parameters u, v, and w of the instrumental functions of the angular resolution (Table 4.1) for different diffractometers.

Table 4.1 Parameters of the instrumental angular resolution functions

Diffractometer	Parameters of the $FWHM_R(2\theta)$ function		
	u	v	w
Siemens D-500	0.0043	−0.0024	0.0055
DRON-UM1	0.0041	−0.0021	0.0093
Shimadzu XRD-7000	0.00618	−0.00457	0.00778

4.1.3 Determination of the Reflection Broadening, Average Particle Size, and Microstrains

In order to determine the broadening of diffraction reflections, the X-ray diffraction pattern of the compound under investigation is measured, each reflection is described by the pseudo-Voigt function (4.3), and the full width at half-maximum $\mathrm{FWHM}_{\mathrm{exp}}$ is determined from formula (4.5), (4.7) or (4.8).

For the first time, the idea of determination of the size of small particles by an X-ray method was stated by P. Scherrer. In 1918 he proposed the formula connecting the width of diffraction reflection on half of its height with the size of a particle of investigated substance [29]. Later, in 1925, the article [30] appeared in which the similar relationship between the reflection width and the size of a particle was derived by soviet scientist N.Y. Selyakov. Therefore in Russian literature this formula sometimes (very seldom) is named as Selyakov-Scherrer formula. Nevertheless the priority belongs to P. Scherrer undoubtedly.

Using the Wulff-Bragg condition it may easily be shown that $\Delta d_{hkl}/d_{hkl} = -[\mathrm{cotan}(\theta_{hkl})]\Delta\theta$; on the other hand, $\Delta d_{hkl}/d_{hkl} = \Delta\lambda/\lambda$, and therefore $\Delta\lambda/\lambda = -[\mathrm{cotan}(\theta)]\Delta\theta$. Whence it follows that the broadening $\beta_{hkl}(2\theta) = 2(\Delta\lambda/\lambda)\tan\theta_{hkl}$. We shall assume that the least reflection width is equal to the spectral width $\Delta\lambda/\lambda \approx 10^{-3}$ of the reflection. Then from the expression obtained it is easy to find the largest size of the particles (grains, crystallites), which cause a measurable reflection broadening. We shall assume that the smallest reflection width is equal to the spectral width $\Delta\lambda/\lambda \approx 10^{-3}$ of the reflection. In this case, the broadening $\beta_{hkl} = 2 \times 10^{-3} \tan\theta_{hkl}$ and $\langle D\rangle = \lambda/(\beta_{hkl}\cos\theta_{hkl}) = 10^3\lambda/[(2\tan\theta_{hkl}) \times (\cos\theta_{hkl})] = 10^3\lambda/(2\sin\theta_{hkl}) = 10^3 d_{hkl} \approx 200$ nm. Thus, the diffraction method allows to determine the size only of such particles which have size less than 200 nm.

In a general case, when the particles of the substance have an arbitrary shape, the average particle size $\langle D\rangle \approx V^{1/3}$ (V is volume of the particle) can be determined by the Debye-Scherrer formula [14, 15, 29]:

$$\langle D\rangle = \frac{K_{hkl} \cdot \lambda}{\cos\theta \cdot \mathrm{FWHM}_{\mathrm{exp}}(2\theta)} \equiv \frac{K_{hkl} \cdot \lambda}{2\cos\theta \cdot \mathrm{FWHM}_{\mathrm{exp}}(\theta)}, \tag{4.10}$$

where λ is the radiation wavelength, and $K_{hkl} \approx 1$ is the Scherrer's constant whose value depends on the shape of the particle (crystallite, domain) and on diffraction reflection indices (*hkl*).

According to [31], in a cubic crystal lattice, the crystallites have the size of the same order in three perpendicular directions. In this case, Scherrer's constant K_{hkl} for reflections with different crystallographic Miller indices (*hkl*) of a cubic crystal lattice may be calculated from the equation:

$$K_{hkl} = \frac{6\,|h|^3}{(h^2+k^2+l^2)^{1/2}(6h^2 - 2\,|hk| + |kl| - 2\,|hl|)}. \tag{4.11}$$

Table 4.2 Scherrer's constant K_{hkl} for the diffraction reflections (*hkl*) from crystals with fcc lattice

(*hkl*)	(111)	(200)	(220)	(311)	(222)	(400)
K_{hkl}	1.1547	1.0000	1.0606	1.1359	1.1547	1.0000
(*hkl*)	(331)	(420)	(422)	(333)	(511)	(440)
K_{hkl}	1.1262	1.0733	1.1527	1.1547	1.1018	1.0607
				For (333) and (511) reflections average value of K_{hkl} is 1.2113		

Values of K_{hkl} for reflections (*hkl*) of fcc lattice are presented in Table 4.2.

In a real experiment, the reflection is broadened as a result of a finite resolution of a diffractometer and cannot be less than the instrumental reflection width. In other words, in formula (4.10) one should use the reflection broadening β with respect to the instrumental width $\mathrm{FWHM_R}$, rather than the experimental width $\mathrm{FWHM_{exp}}(2\theta)$ of reflection

$$\beta(2\theta) = [(\mathrm{FWHM_{exp}})^2 - (\mathrm{FWHM_R})^2]^{1/2}. \quad (4.12)$$

The formula (4.12) should be used, if physical broadening β and the instrumental width $\mathrm{FWHM_R}$ are described by the Gauss or the pseudo-Voight functions with large contribution of the Gauss functions. In practice just such description of the experimental broadening and the instrumental resolution function of a diffractometer is the most widespread. If physical broadening and the instrumental width are described by the Lorentz (Cauchy) functions, then the physical broadening is equal to a difference of full width of experimental reflection and the instrumental width of this reflection, i.e. $\beta = \mathrm{FWHM_{exp}} - \mathrm{FWHM_R}$.

According to [16], the broadening of diffraction reflections due to the small size of particles (size broadening) is represented in the form

$$\beta_s(2\theta) = \frac{K_{hkl}\lambda}{\langle D\rangle\cos\theta} \approx \frac{\lambda}{\langle D\rangle\cos\theta} \quad [\mathrm{rad}] \quad (4.13)$$

and the broadening caused by the strain-induced distortions of the crystal lattice (strain broadening) can be written in the following form:

$$\beta_d(2\theta) = 4\varepsilon\ \mathrm{tg}\,\theta \quad [\mathrm{rad}]. \quad (4.14)$$

With allowance for relationship (4.13), the average particle size D in the diffraction experiment is determined using the Warren method [32]

$$\langle D\rangle = \frac{K_{hkl}\lambda}{\cos\theta\cdot\beta_s(2\theta)} \equiv \frac{K_{hkl}\lambda}{2\cos\theta\cdot\beta_s(\theta)}. \quad (4.15)$$

Note that $\beta(2\theta) \equiv 2\beta(\theta)$.

For nanostructured materials, the broadening as a rule is caused by both the small size of particles and the microstrains. In order to separate these contributions to the experimental reflection broadening, the extrapolation Williamson-Hall method [33] is used. It is based on the construction of the dependence of the reduced broadening $\beta^*(2\theta) = [\beta(2\theta)\cos\theta]/\lambda$ of the (*hkl*) diffraction reflection on the scattering vector $s = (2\sin\theta)/\lambda$. The foundation of the Williamson-Hall method is the assumption that size and strain broadenings are described by the Lorentz (Cauchy) functions. As a result, total reflection broadening is a sum of size and strain broadenings, i.e. $\beta(2\theta) = \beta_s(2\theta) + \beta_d(2\theta) = \lambda/[\langle D\rangle\cos\theta] + 4\varepsilon\tan\theta$. In this case the reduced total reflection broadening is $\beta^*(2\theta) = [\beta(2\theta)\cos\theta]/\lambda = 1/\langle D\rangle + 2\varepsilon \times [2\sin\theta]/\lambda = 1/\langle D\rangle + 2\varepsilon = 1/\langle D\rangle + s \times \tan\varphi$ with $2\varepsilon \equiv \tan\varphi$ and is a linear function of the scattering vector s.

The average particle size $\langle D\rangle$ is determined by extrapolating the dependence of the reduced broadening $\beta^*(2\theta) = 1/\langle D\rangle + 2\varepsilon s$ on the scattering vector s to the value $s = 0$; that is,

$$\langle D\rangle = \frac{1}{\beta^*(2\theta)|_{s=0}}. \tag{4.16}$$

The average value of the microstrains ε is obtained with the use of the expression

$$\varepsilon = [(\tan\varphi)/2] \cdot 100\,\% \tag{4.17}$$

from the slope angle φ of the straight line, which approximates the dependence $\beta^*(s)$. In the physical meaning, the quantity $\varepsilon = \Delta d/d_0$ is the microdeformation of the lattice and it characterizes the uniform strain averaged over the crystal volume, i.e., the relative variation in the interplanar spacing in comparison with the perfect crystal. Note that d_0 and Δd are the interplanar spacing in a perfect crystal and the mean variation in the distance between the (*hkl*) planes in the volume V of the crystal, respectively.

It is clear that the Williamson-Hall method should be used if reflection broadening β is described by the Lorentz function or the pseudo-Voight function with large contribution c of the Lorentz function (at $c \geq 0.7$).

As an example, the numerical analysis of the diffraction reflection profile is performed and the reflection width is determined for a nanocrystalline powder of cubic (space group $Fm\overline{3}m$) lead sulphide PbS. The lead sulphide nanopowder n-PbS was synthesized by the chemical deposition from aqueous solution. Figure 4.4 depicts the XRD pattern of the synthesized nanocrystalline sulphide n-PbS in the range of 2θ angles from 18° to 100°. For comparison, the XRD pattern of the lanthanum hexaboride LaB_6 used as the reference specimen is also shown in Fig. 4.4. It can be clearly seen from this figure that the diffraction reflections of nanocrystalline n-PbS sulphide are strongly broadened.

As an example, Fig. 4.5 shows the $(111)_{B1}$ and $(200)_{B1}$ doublet diffraction reflections of the nanocrystalline cubic lead sulphide PbS and their approximation by two pseudo-Voigt functions. The parameters of the pseudo-Voigt functions are calculated by the least-squares method.

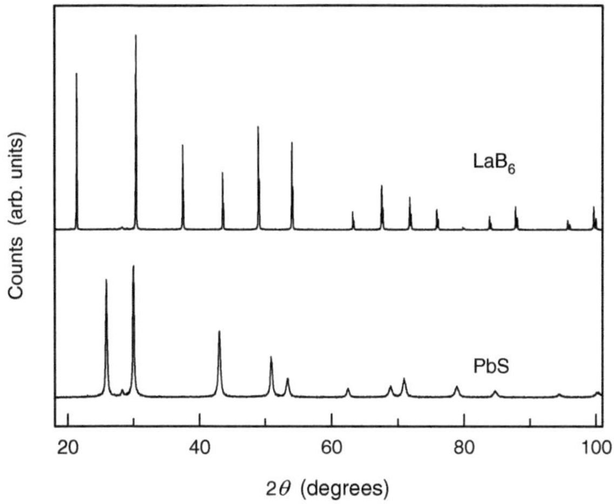

Fig. 4.4 XRD pattern of the nanocrystalline powder of cubic (space group $Fm\overline{3}m$) lead sulphide n-PbS. XRD pattern of the lanthanum hexaboride LaB_6 used as the reference sample for determining the diffraction reflection broadening is shown for comparison. Diffraction reflections of the n-PbS nanopowders are considerably broadened. XRD measurements are performed using Cu$K\alpha_{1,2}$ radiation

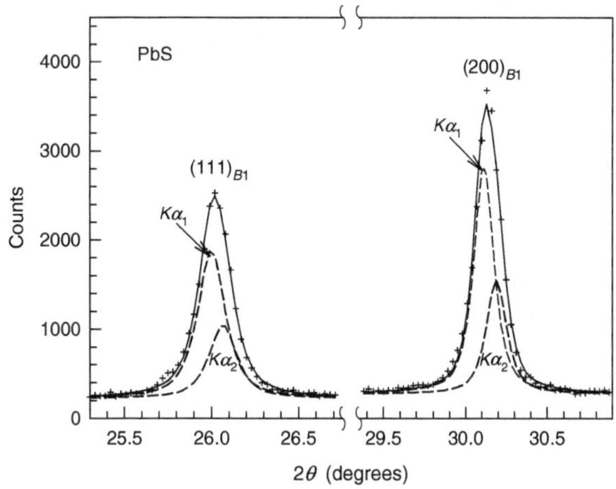

Fig. 4.5 Experimental profiles (*crosses*) of the $(111)_{B1}$ and $(200)_{B1}$ doublet diffraction reflections of the nanocrystalline lead sulphide PbS and their computational description by two pseudo-Voigt functions. XRD measurements were performed using Cu$K\alpha_{1,2}$ radiation

The dependence of the reduced broadening $\beta^*(2\theta)$ of the reflections on the scattering vector s for the n-PbS nanopowder is plotted in Fig. 4.6.

If the reflection broadening was caused only by small particle size, the experimental points would be scattered around horizontal lines: The higher the line is, the smaller the particles are. However, Fig. 4.6 shows that the broadening increases with

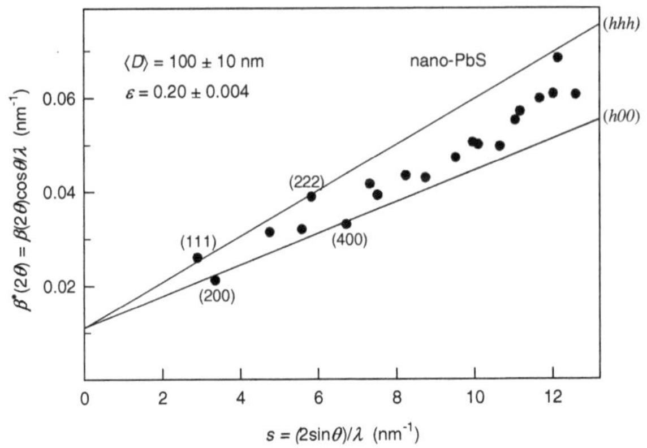

Fig. 4.6 Estimation of the particle size D and microstrains ε from the dependence of the reduced broadening $\beta^*(2\theta) = [\beta(2\theta)\cos\theta]/\lambda$ of the diffraction reflections on the scattering vector $s = (2\sin\theta)/\lambda$ for the nanocrystalline lead sulphide n-PbS: $\langle D \rangle = 100 \pm 10$ nm, $\varepsilon = 0.20 \pm 0.004$ %

scattering vector length. This means that in addition to small size, there are microstrains in the particles. It is seen from Fig. 4.6 that experimental points of dependence $\beta^*(s)$ are scattered in the sector formed by two lines. The top line passes through reflections (hhh), and the bottom line passes through reflections (h00). In this connection for correct determination of the particle size and microstrains it is necessary to take into account Scherrer's constant (anisotropy coefficient) K_{hkl} and to calculate the reduced reflection broadening by the Debye-Scherrer formula $\beta^*(2\theta) = [\beta(2\theta)\cos\theta]/K_{hkl}\lambda$. Coefficient K_{hkl} for crystals with cubic symmetry is calculated by the formula (4.11) (see also Table 4.2).

According to the calculations, the average size of individual particles (coherent scattering regions) in the synthesized n-PbS nanopowder is equal to 100 ± 10 nm and the microstrains amount to $\varepsilon = 0.20 \pm 0.004$ %.

4.1.4 Degree of Inhomogeneity

For substitutional solid solutions $A_{1-y}B_y$ and nonstoichiometric interstitial compounds MX_y, the broadening can also be caused by the inhomogeneity (i.e., a nonuniform composition of a compound over this volume) [14, 17–21].

In [18], it was demonstrated that the broadening β_h of a diffraction reflection due to the inhomogeneity Δy of nonstoichiometric compounds MX_y or substitutional solid solutions $A_{1-y}B_y$ with a cubic basic structure $B1$, whose composition varies in the range $y \pm \Delta y$, is proportional to $(\sin^2\theta)/\cos\theta$; that is,

$$\beta_h(2\theta) = \frac{8\sqrt{2\ln 2}\cdot \left|a'_{B1}(y)\right|_{y=y_0}}{\lambda\sqrt{h^2+k^2+l^2}} \frac{\sin^2\theta}{\cos\theta}\Delta y \quad [\text{rad}], \tag{4.18}$$

where $a'_{B1}(y) \equiv da(y)/dy$ is the derivative of the concentration dependence of the lattice constant for the nonstoichiometric compound or solid solution with respect to the composition y; $K_\mathrm{f} = \sqrt{h^2 + k^2 + l^2}$ is a quadratic form for the (*hkl*) reflection of a compound with a cubic $B1$-type structure. It should be noted that the broadening β_h coincides with the full width at half-maximum $\mathrm{FWHM_h}$ if it is caused only by the inhomogeneity.

When the broadenings associated with the small size of particles and the microstrains are absent, the degree of inhomogeneity Δy of the cubic nonstoichiometric compound or solid solution can be defined as follows:

$$\Delta y = \frac{\lambda\sqrt{h^2 + k^2 + l^2}}{8\sqrt{2\ln 2}\cdot \left|a'_{B1}(y)\right|_{y=y_0}} \frac{\cos\theta}{\sin^2\theta}\beta_\mathrm{h}(2\theta). \tag{4.19}$$

For the broadening β_h caused by the inhomogeneity of nonstoichiometric compounds MX_y or solid solutions $A_{1-y}B_y$ with an arbitrary structure, relationship (4.18) takes the form

$$\beta_\mathrm{h}(2\theta) = \frac{8\sqrt{2\ln 2}\cdot |da(y)/dy|_{y=y_0}}{\lambda K_\mathrm{f}} \frac{\sin^2\theta}{\cos\theta}\Delta y \quad [\mathrm{rad}], \tag{4.20}$$

where K_f is the quadratic form, which corresponds to the crystal system of the compound under investigation and relates one of the unit cell parameters (a, b, or c) to the interplanar spacings d_{hkl}.

In the absence of the size and strain broadenings, the degree of inhomogeneity Δy of the nonstoichiometric compounds or solid solutions with an arbitrary crystal structure is determined by the expression

$$\Delta y = \frac{\lambda K_\mathrm{f}}{8\sqrt{2\ln 2}\cdot |da(y)/dy|_{y=y_0}} \frac{\cos\theta}{\sin^2\theta}\beta_\mathrm{h}(2\theta). \tag{4.21}$$

The quantitative parameter of the degree of inhomogeneity Δy permits one to control and optimize the conditions of synthesis of nonstoichiometric compounds and substitutional solid solutions (alloys) with a high degree of homogeneity. The degree of inhomogeneity is a useful parameter for the characterization of disordered and ordered nonstoichiometric compounds of the carbide and nitride types, because it provides a means for differentiating the inhomogeneous and two-phase substances.

According to [22, 23], in the presence of the broadening associated with the small size of particles (crystallites), microstrains, and the degree of inhomogeneity, the total broadening can be determined as

$$\beta(2\theta) = \left[(\mathrm{FWHM_{exp}})^2 - (\mathrm{FWHM_R})^2\right]^{1/2} = \left[\beta_\mathrm{s}^2(2\theta) + \beta_\mathrm{d}^2(2\theta) + \beta_\mathrm{h}^2(2\theta)\right]^{1/2}. \tag{4.22}$$

The three contributions to the broadening of diffraction reflections are separated as follows. In the case of cubic crystals, the dependences of the experimental broadening $\beta(2\theta)$ are numerically constructed as a function of such arguments as sec θ, tanθ, and $(\sin^2\theta)/(\sqrt{h^2+k^2+l^2}\cos\theta)$, which correspond to the size, strain, and inhomogeneity broadenings. For crystalline substances of other crystal systems, the dependences of the experimental broadening $\beta(2\theta)$ are constructed as a function of the arguments sec θ, tan θ, and $(\sin{}^2\theta)/(K_f\cos\theta)$. Then, each of the three dependences is approximated by a linear function and the root-mean-square deviation (variance) of the experimental values $\beta(2\theta)$ from the approximating linear function is determined. If any found dependences are similar to a linear dependence, the total broadening $\beta(2\theta)$ involves the corresponding contributions. A large deviation of the experimental values $\beta(2\theta)$ from a linear behavior indicates the absence of the contribution described by this function. This technique for separating different contributions to the experimentally observed broadening of diffraction reflections was successfully tested in works [34, 35] for carbide solid solutions of the ZrC-NbC system.

The method proposed for estimating the average particle (grain) size, the microstrains, and the degree of inhomogeneity from the broadening of diffraction reflections can be applied to both dispersed and compacted materials. It should be noted that the accuracy of this method primarily depends on the quality of the experimental XRD patterns of the compounds under investigation, the accuracy in the determination of the instrumental resolution function, and the quality of mathematical processing. In contrast to electron microscopy, which is a local visual technique for estimating the particle sizes, X-ray diffraction is an integrated method, which makes it possible to determine the particle size averaged over the entire volume. Moreover, the XRD method allows to determine quantitatively not only the particle sizes but also the microstrains and atomic displacements. Furthermore, the XRD method is more accessible because X-ray structure analysis is widely used and the necessary equipment is available.

4.2 Production of Nanocrystalline Powders of Tungsten Carbide

The problem of production of ultrafine powders of metals, alloys and compounds for different areas of technology has been discussed in literature for many years now. In this section, attention will be given mainly to the production of nanocrystalline tungsten carbide WC powders by means of a plasma-chemical synthesis and high-energy ball milling. Particle size and microstructure of nanocrystalline and submicrocrystalline powders of tungsten carbide WC will be discussed also.

4.2.1 Plasma-Chemical WC Powder

Nanocrystalline powder WC-plasm of hexagonal (space group $P\overline{6}m2$) tungsten carbide was synthesized by the plasma-chemical method from tungsten oxide WO_3 and propane C_3H_8 at a temperature of about 3300–3800 K in a low-temperature hydrogen plasma. Synthesized nanopowder had non-uniform phase composition and contained three crystal phases (W_2C, WC and W) and about 1 wt.% of free (uncombined) carbon. Therefore synthesized nanopowder was subjected to additional homogenization in an argon atmosphere at a temperature of 1100 K. The total content of carbon in homogeneous WC-plasm nanopowder was equal to ~6.14 wt.%. The design of the electroarc plasmatron for plasma-chemical synthesis of tungsten carbide nanopowder is presented in Ref. [36]. In order to relieve stresses, the synthesized single-phase nanopowder was additionally annealed at a residual pressure of 0.13 Pa (10^{-3} mm Hg) and a temperature up to 1000 K.

Figure 4.7 depicts the XRD patterns of the synthesized WC-plasm nanocrystalline powder, the same WC-plasm nanopowder after his homogenization in argon atmosphere at temperature 1100 K and homogeneous carbide powder after annealing at 1000 K for stress relief. As is seen from Fig. 4.7, synthesized WC-plasm nanopowder is heterogeneous and contains 58 wt.% W_2C, 29 wt.% WC and 12 wt.% W. Both WC-plasm nanopowders which were obtained after homogenizing annealing in argon and after annealing for stress relief contain only hexagonal (space group $P\overline{6}m2$) carbide WC with the unit cell constants $a = 0.29060$ and $c = 0.28375$ nm. For comparison, XRD pattern of the lanthanum hexaboride LaB_6 used as the reference specimen is also shown in Fig. 4.7. It can be clearly seen from this figure that all diffraction reflections of both annealed WC-plasm nanopowders are strongly broadened in comparison with reflections of lanthanum hexaboride LaB_6.

As an example, Fig. 4.8 shows the (300) doublet diffraction reflection of the nanocrystalline powder WC-plasm of hexagonal tungsten carbide and its approximation by two pseudo-Voigt functions. The parameters of the pseudo-Voigt functions are calculated by the least-squares method.

The dependences of the reduced broadening $\beta^*(2\theta)$ of the reflections on the scattering vector s for the plasma-chemical WC-plasm nanopowder before and after annealing at temperature 1000 K are plotted in Fig. 4.9.

If the reflection broadening was caused only by small particle size, the experimental points would be scattered around horizontal line: The higher the line is, the smaller the particle size is. However, Fig. 4.9 shows that the broadening increases with scattering vector length. This means that in addition to small size, there are microstrains $\varepsilon = \Delta d/d$ in small particles. Lattice microdeformations ε induce strain broadening $\beta_d(2\theta) = 4\varepsilon \tan\theta$ of diffraction reflections. Thus, experimental broadening $\beta(2\theta)$ is a superposition of broadenings β_s and β_d, i.e. $\beta = \sqrt{\beta_s^2 + \beta_d^2}$ [14]. Small particle size and microstrains contributions in broadening of a physical profile of diffraction reflections were determined by the Williamson-Hall method [33, 37, 38] which is described in detail in Sect. 4.1.3.

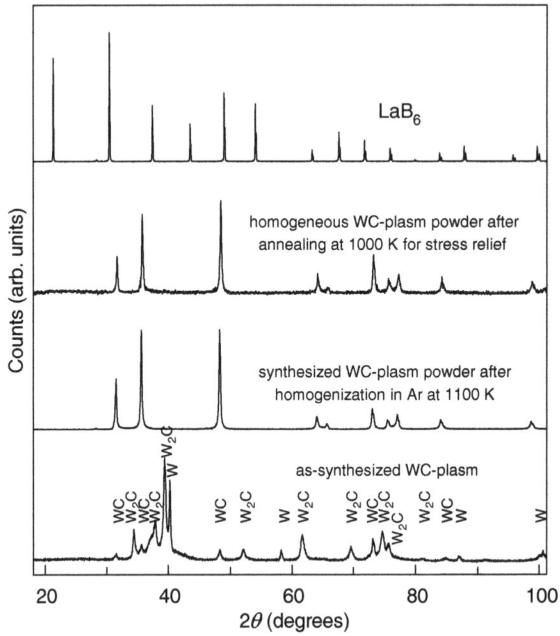

Fig. 4.7 XRD patterns of the nanocrystalline WC-plasm powder immediately after the plasma-chemical synthesis, after his homogenization in argon at temperature 1100 K and after annealing at 1000 K for stress relief. XRD pattern of the lanthanum hexaboride LaB_6 used as the reference sample for determining the diffraction reflection broadening is shown for comparison. Synthesized WC-plasm nanopowder contains three crystal phases W_2C, WC and W. Annealed WC-plasm powders contain only hexagonal (space group $P\overline{6}m2$) tungsten carbide WC. Diffraction reflections of the WC-plasm nanopowders are considerably broadened. XRD measurements are performed using $CuK\alpha_{1,2}$ radiation

It is seen from Fig. 4.9 that after annealing the average size of nanopowder particles remains virtually unchanged, whereas the microstrains decrease considerably. According to the calculations, the average size of individual particles (coherent scattering regions) in the synthesized WC-plasm nanopowder is equal to 55 ± 10 nm and the microstrains amount to $\varepsilon = 0.06 \pm 0.005$ %. After annealing, the microstrains decrease to 0.015 ± 0.002 % and the average particle size remains constant to within the experimental error.

The particle size distribution in the tungsten carbide nanopowder WC-plasm was determined also using the laser diffraction method on a HORIBA partica LA-950 V2 Analyzer. According to the laser diffraction analysis the smallest particles in the WC-plasm nanopowder are ~50 nm in size, half of all the particles are less than 350 nm in size, and average size is equal to 510 nm. This means that the nanopowder particles are agglomerated.

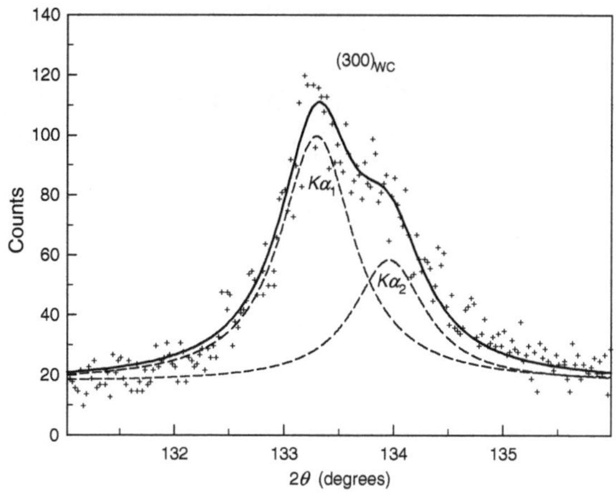

Fig. 4.8 Experimental profile (*crosses*) of the (300) doublet diffraction reflection of the nanocrystalline powder WC-plasm and its computational description by two pseudo-Voigt functions. XRD measurements were performed using Cu$K\alpha_{1,2}$ radiation

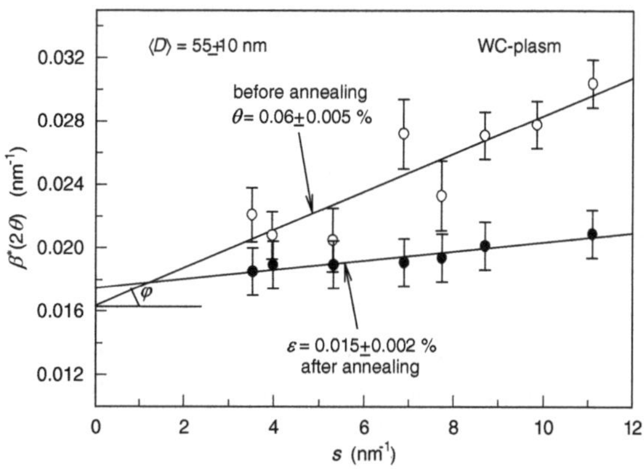

Fig. 4.9 Estimation of the particle size $\langle D\rangle$ and microstrains ε from the dependence of the reduced broadening $\beta^*(2\theta) = [\beta(2\theta)\cos\theta]/\lambda$ of the diffraction reflections on the scattering vector $s = (2\sin)\theta/\lambda$ for the nanocrystalline tungsten carbide n-WC: (○) before annealing ($\langle D\rangle = 55 \pm 10$ nm, $\varepsilon = 0.06 \pm 0.005$ %) and (•) after annealing at temperature of 1000 K ($\langle D\rangle = 55 \pm 10$ nm, $\varepsilon = 0.015 \pm 0.002$ %)

According to result of scanning electron microscopy, the WC-plasm nanopowder produced by plasma-chemical synthesis has a microstructure (Fig. 4.10) with well-observed large (up to 10 μm) loose agglomerates composed of smaller particles.

The discrepancy in averaged particle sizes observed by XRD, SEM, and laser diffraction is evidence for a high degree of agglomeration of the WC-plasm nanopowder. Thus, the coarse particles, up to several micrometers in size, observed by SEM are agglomerates of small particles with tens of nanometers in size.

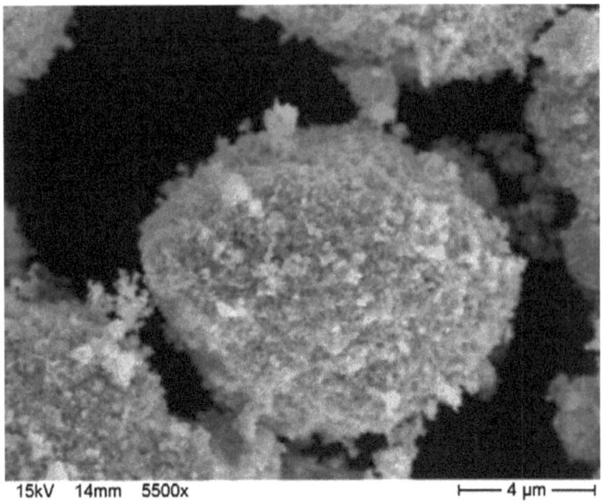

Fig. 4.10 Scanning electron microscopy of annealed nanocrystalline WC-plasm tungsten carbide powder, synthesized by plasma-chemical method. Tungsten carbide nanoparticles with a size about 100 nm form well-observed large loose agglomerates in the size up to 10 μm. SEM image is obtained in JEOL 6310 electron microscope (SCK-CEN, NMS Institute, Mol, Belgium)

On the whole, the estimation of average particle size by XRD method for WC-plasm nanopowder is in qualitative agreement with the relevant results of the electron microscope examination and the laser diffraction.

4.2.2 Production of Nanocrystalline Powders by High-Energy Ball Milling

In the recent decade, the preparation of substances and materials in the nanocrystalline state is one of the actively advancing directions in materials science. This is due to a considerable modification of properties of solids in the nanocrystalline state [14, 15, 39]. For example, powders having different properties are required for production of ceramic products used in different applications. Size and shape of the particles, the value of the specific surface and chemical composition are the important characteristics of powders for their following application. The fracture of solids as a result of long-term mechanical compressive and bending strains is reduced to its gradual fragmentation to a powder-like state. The powder particle size can fall in the micro- or nanometer range depending on the applied energy. Mechanical fragmentation of powders turned out to be one of the effective and productive methods for producing various substances and materials in a nanocrystalline state and is widely used in modern physical metallurgy [14, 15, 39, 40].

Ball milling is a simple and efficient method for making various nanocrystalline powders [14, 41–48] with particles of up to 20 nm in size intended for superconducting ceramic materials, materials with magnetic anisotropy, superhard metal-cutting materials, nanocomposites, and superplastic ceramics. High-energy planetary, vibratory, whirlwind and others mills and also attritors of different types are used for grinding.

The strength and the mechanical failure of solids are the objects in a number of original and generalizing studies (see, e.g., [49–53]) focusing on the description and the analysis of the theory of strength and the failure mechanism (the interatomic bond breakage kinetics during plastic deformation, the formation and the growth of cracks, stress relaxation, etc.). As to ball milling of powders and its final result, i.e. the particle size, the studies of this kind are still performed at the empirical level since models relating the size of particles in prepared nanocrystalline powders to the milling energy are absent. Ball milling is still studied on an empirical level in spite of the presence of several milling models [54–56]. According to [54], deformation during mechanical milling of powders is localized in shear bands with a high dislocation density. The dislocation model [55] is based on the assumption that any material (substance) is characterized by a certain minimum particle size D_{min} achievable upon milling [57, 58]. According to [55, 57], size D_{min} is determined by equilibrium between the appearance of a dislocation structure in milling and its relaxation at a certain reached level of stresses due to dislocation annihilation and recombination into low-angle boundaries separating individual grains. The model [55] also takes into account the data [57, 58] that the size of metallic particles that form upon milling under the same conditions is inversely proportional to melting temperature T_{melt} and shear and bulk moduli G and K_c, respectively. However, the model [55] only describes experimental results and postulates the dependence of D_{min} on some properties of a material (hardness, shear modulus, activation energy of vacancy migration) and does not establish a relationship between the binding energy or the milling time and the particle size reached upon milling. Based on X-ray diffraction data, the authors [56] described the size distribution of powder particles and estimated the degree of refining and did not try to propose any predicting model. At first the dependence of the size of nanocrystalline powder particles on the milling parameters was briefly considered in [59]. Milling model [59] has been developed with taking into account physical properties of milled substance in following works [60–62]. Later authors [63, 64] have improved milling model [59–62] and have established relationship between the particle size of the ball-milled powders, from the one hand, and the milling energy, the particle size and the mass of the initial powder, from the other hand.

In this section, a model for mechanical milling of powders to a small particle size is described in detail and the applicability of this model for estimating the powder particle size as a function of the applied energy, initial powder mass M, and the initial particle size is considered.

The Milling Model

Let us establish the relationship between the milling energy E_{mill} and the average size of the particles formed during milling. While grinding the decrease in the particle size is accompanied by the initiation of microstrains in the particles.

Below we consider the initial powder with particles having the average linear size D_{in}. The volume of a particle in the initial powder and its surface area are equal to

$V_{\rm in} = f_{\rm v} D_{\rm in}^3$ and $S_{\rm in} = f_{\rm s} D_{\rm in}^2$, respectively, where $f_{\rm v}$ and $f_{\rm s}$ are the form factors of the volume and the surface area (i.e. the proportionality factors controlled by the body shape). For spherical particles of the diameter $D_{\rm in}$ the form factors are $f_{\rm v} = \pi/6$ and $f_{\rm s} = \pi$, whence $f_{\rm s}/f_{\rm v} = 6$; for cubic particles with the edge length $D_{\rm in}$ the form factors $f_{\rm v} = 1$ and $f_{\rm s} = 6$, whence $f_{\rm s}/f_{\rm v} = 6$ too; for particles shaped as thin plates $f_{\rm s}/f_{\rm v} \approx 2$. If in the first approximation particles of the initial and ball-milled powders have the same shape (or the particle shape distribution in the initial and ball-milled powders is the same), the ratio between the form factors of the volume and the surface area $f_{\rm s}/f_{\rm v}$ is some constant value.

Let the initial powder have the density d and the mass M. Then the number of particles in the initial powder is $M/dV_{\rm in} = M/(df_{\rm v}D_{\rm in}^3)$. According to [51], when a powder is being milled, the energy is consumed for the rupture of interatomic bonds in crystals and formation of the additional surface by cleavage of crystal particles. Therefore, the milling energy $E_{\rm mill}$ can be written in the form

$$E_{\rm mill} = \frac{M}{df_{\rm v}D_{\rm in}^3}(E_{\rm rupt} + E_{\rm surf}), \tag{4.23}$$

where $E_{\rm rupt}$ is the energy consumed for the rupture of interatomic bonds in one particle of the initial powder and $E_{\rm surf}$ is the energy consumed for formation of the additional surface upon the cleavage of one particle of the initial powder.

Grinding of one particle of the initial powder provides n smaller particles with the average linear dimension D; the volume and the surface area of one particle are $f_{\rm v}D^3$ and $f_{\rm s}D^2$. Since $V_{\rm in} = nf_{\rm v}D^3$, then $n = D_{\rm in}^3/D^3$. In this case, the surface area of all the particles formed by grinding equals $S = nf_{\rm s}D^2 = f_{\rm s}D_{\rm in}^3/D$, while the surface area increment is

$$\Delta S = S - S_{\rm in} = f_{\rm s}D_{\rm in}^2(D_{\rm in} - D)/D \tag{4.24}$$

A crystal fails along slip planes. Let the slip plane area per unit cell of the crystal be $s_{\rm f}$; then, the number of planes along which cleavage proceeds is $\Delta S/s_{\rm f}$. If q interatomic bonds with the binding energy u pass through every slip plane per unit cell, the energy consumed for the rupture of bonds during grinding of one particle of the initial powder is

$$E_{\rm rupt} = quf_{\rm s}D_{\rm in}^2(D_{\rm in} - D)/s_{\rm f}D. \tag{4.25}$$

The increment in the surface energy related to an increase in the particle surface area by ΔS is $\Delta E_{\rm s} = \gamma\Delta S$, where γ is the specific (per unit interface area) excess energy caused by a random network of edge dislocations. According to [51, 65], energy $E_{\rm surf}$ consumed for creating the additional surface upon the cleavage of crystalline particles is several hundred times higher than the change $\Delta E_{\rm s}$ of the surface energy; i.e., $E_{\rm surf} = \beta\Delta E_s = \beta\gamma\Delta S$, where β is the coefficient of proportionality. As a result, we have

$$E_{\text{surf}} = \beta E_{\text{s}} = \beta\gamma\Delta S = \beta\gamma f_{\text{s}} D_{\text{in}}^2 (D_{\text{in}} - D)/D. \tag{4.26}$$

In milling, as during other methods of mechanical deformation, grain or particle boundaries (i.e., interfaces) are nonequilibrium. The authors [66] proposed a model for nonequilibrium boundaries that takes into account the presence of chaotic ensembles of introduced grain-boundary dislocations in them. Therefore, they were able to obtain an expression for the root-mean-square strain and to estimate the excess energy of an interface directly related to the appearance of elastic stress fields. In particular, they derived an expression for specific excess energy γ of the interface created by a chaotic network of edge dislocations with a Burgers vector $\boldsymbol{b} = (\pm b, 0, 0)$,

$$\gamma = \frac{Gb^2 \rho \ln(D_{\text{in}}/2b)}{4\pi(1-\nu)}, \tag{4.27}$$

where G and ν are the shear modulus and the Poisson ratio of the material, respectively, $b = |\boldsymbol{b}|$ is the Burgers vector magnitude, and $\rho \approx \rho_{\text{V}} D/3$ and ρ_{V} are the linear and volume density of dislocation, respectively.

As follows from [67], the volume density of dislocations which are randomly distributed in the grain bulk equals the geometric mean of the density of dislocations, $\rho_D = 3/D^2$, connected with the grain size D, and the density of dislocations, $\rho_{\text{s}} = C\varepsilon^2/b^2$, connected with the microstrains ε, i.e.

$$\rho_{\text{V}} = (\rho_D \rho_{\text{s}})^{1/2}. \tag{4.28}$$

With allowance for (4.28), the linear density of dislocations ρ is

$$\rho \approx \rho_V D/3 = [(3/D^2)(C\varepsilon^2/b^2)]^{1/2} D/3 = \sqrt{3C}\varepsilon/3b, \tag{4.29}$$

where C is a constant for a given material and takes value from 2 to 25 [67].

Writing the specific excess energy γ (4.27) with allowance for (4.29) and substituting obtained energy γ into (4.26), gives the energy E_{surf} consumed for the formation of an additional surface upon the cleavage of one crystalline particle [59–62]:

$$E_{\text{surf}} = \beta\gamma f_{\text{s}} D_{\text{in}}^2 (D_{\text{in}} - D)/D = \frac{\sqrt{3C}\beta f_{\text{s}}}{12\pi} \frac{GbD_{\text{in}}^2 (D_{\text{in}} - D)\ln(D_{\text{in}}/2b)}{(1-\nu)} \frac{\varepsilon}{D}. \tag{4.30}$$

Substitution of (4.25) and (4.30) into (4.23), we obtain a formula relating the milling energy E_{mill} to the average size of particles D after milling:

$$E_{\text{mill}} = \frac{Mf_{\text{s}}}{df_{\text{v}}} \frac{(D_{\text{in}} - D)}{DD_{\text{in}}} \left[\frac{12\pi(1-\nu)qu + \sqrt{3C}\beta Gbs_{\text{f}}\varepsilon \ln(D_{\text{in}}/2b)}{12\pi(1-\nu)s_{\text{f}}} \right]. \tag{4.31}$$

For a particular powdered material, the values of d, f_{s}, f_{v}, q, u, C, β, G, ν, b, and s_{f} are fixed, while the milling energy E_{mill}, the particle size and the microstrains are a functions of the milling time t and powder mass (charge) M. Therefore, in works [63, 64] formula (4.31) has been converted to the form

$$E_{\mathrm{mill}}(t) = M[D_{\mathrm{in}} - D(t, M)][A + B\varepsilon(t, M)\ln(D_{\mathrm{in}}/2b)]/[D(t, M)D_{\mathrm{in}}], \quad (4.32)$$

from which it follows

$$D(t, M) = \frac{M[A + B\varepsilon(t, M)\ln(D_{\mathrm{in}}/2b)]}{E_{\mathrm{mill}}(t) + M[A + B\varepsilon(t, M)\ln(D_{\mathrm{in}}/2b)]/D_{\mathrm{in}}} \quad (4.33)$$

where $A = \frac{f_s}{f_v}\frac{qu}{s_f d}$ and $B = \frac{f_s}{f_v}\frac{\beta Gb\sqrt{3C}}{12\pi(1-\nu)d}$ are certain constants characterizing a given substance. It is easily seen that the formula (4.33) satisfies the edge condition $D(0, M) = D_{\mathrm{in}}$, since at the initial moment of time $t = 0$ the milling energy $E_{\mathrm{mill}}(0) = 0$ and the microstrains $\varepsilon(0, M) = 0$. It is seen from (4.33) that microstrains ε retard powder fragmentation. Formula (4.33) is the basic expression of the milling model, which defines the average size of powder particles as a function of the applied milling energy, $E_{\mathrm{mill}}(t) \sim kt$. Coefficient k in milling energy $E_{\mathrm{mill}}(t)$ has different forms depending on the mill design (planetary, ball, and vibration mills, disintegrators) and the mechanics of milling body motion.

At a given initial particle size D_{in} and a fixed milling time t, the dependence of milled particle size D on initial powder mass (charge) M is described by function

$$D(t = const, M) = D_{\mathrm{in}}M/(KD_{\mathrm{in}} + M), \quad (4.34)$$

where $K = \frac{E(t=const)}{A+B[\ln(D_{\mathrm{in}}/2b)]\varepsilon(t=const, M)}$ is a constant for given time t. As follows from (4.34) the smaller the charged powder mass at a given milling time is, the smaller the particle size D in the milled powder is.

The value of the microstrains $\varepsilon = \Delta l/l \equiv \Delta d/d$ characterize the uniform deformation averaged over the crystal volume, i.e., relative change Δd in interplanar spacing d as compared to its change in a perfect crystal. According to Hooke's law, we have $\Delta l/l = \sigma/E$ in the general case. Fracture begins at critical stress σ_{max}, which is equal to the substance strength for a given type of deformation; therefore, we have $\varepsilon_{\mathrm{max}} = \sigma_{\mathrm{max}}/E$, where E is the corresponding elastic modulus. Microstrains ε change from zero at $t = 0$ to some limiting value $\varepsilon_{\mathrm{max}}$, above which the crystal lattice of the ball-milled material changes its symmetry and fails. At a fixed milling time the larger the milled substance mass is, the smaller the microstrains are. Therefore, the dependence of microstrains ε on milling time t and substance mass M can be described by the empirical function $\varepsilon(t, M) = \varepsilon_{\mathrm{max}}[t/(t + \tau)][M/(M + p)] \equiv (\sigma_{\mathrm{max}}/E)[t/(t + \tau)][M/(M + p)]$, where τ and p are the normalizing parameters.

Now we find the theoretical dependence of average particle size D on milling time t and milled substance mass M during grinding in a PM-200 Retsch planetary ball mill.

Mechanics of Motion for the Grinding Balls

There are two grinding bowls of the "Comfort" type in a PM-200 Retsch planetary ball mill. Each bowl has the holding capacity (volume) of 50 ml. The bowls are placed symmetrically and oppositely to each other at the periphery of the bearing disk. Grinding bowls of the "Comfort" type are intended especially for operation in extreme conditions (long-term grinding at a maximum rotation speed and under high mechanical loadings) and had internal lining with hard alloy WC-6 wt. % Co based on tungsten carbide with a cobalt binder. The grinding balls of diameter 3 mm were made of the same hard alloy. This allowed using a PM-200 Retsch planetary ball mill for grinding of superhard materials.

By design of the mill, the bearing disk and the grinding bowls rotated in opposite directions.

We shall consider the general form of the basic equations of motion for the grinding balls in a bowl during the milling process [61] for analytical determination of the energy $E_{\mathrm{mill}}(t)$ consumed in the mill for grinding of the powder. The scheme of motion of both bearing disk and grinding bowl is shown in Fig. 4.11a. As the bowls rotate, a centrifugal force acts on the powder and the grinding balls, therefore balls are in contact with the powder particles on the inner wall of the bowl.

Let R_c be the radius of the circle described by the bowl axis, r be the internal radius of the bowl, ω be the angular speed of rotation of the bearing disk, $\varphi = 2\pi\omega t$

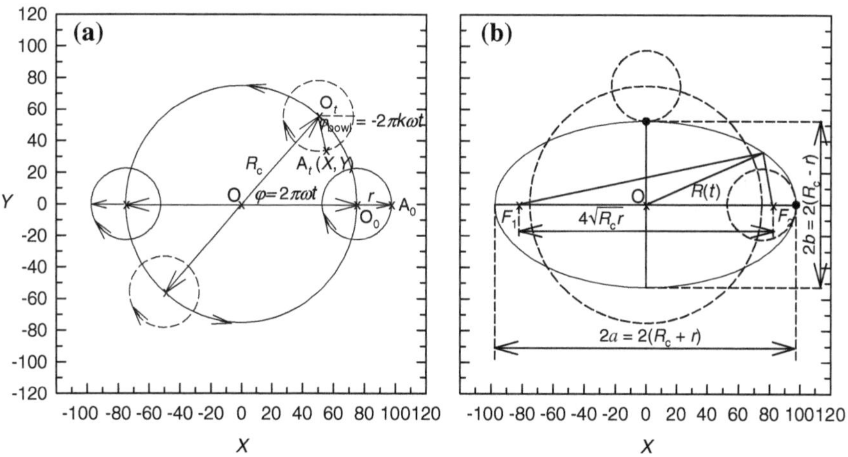

Fig. 4.11 The scheme of the motion of both bearing disk and grinding bowl in a Retsch PM-200 mill (**a**) and the elliptic trajectory of the motion of grinding balls (**b**) at $k = 1$. The bearing disk and the grinding bowl are rotated in opposite directions. $R_c = 0.075$ m is the radius of a circle around which the grinding bowl axis moves; $r = 0.0225$ m is the internal radius of the bowl; ω is the angular speed of rotation of the bearing disk; k is the ratio of the bowl angular speed to the bearing disk angular speed; $\varphi = 2\pi\omega t$ and $\varphi_{\mathrm{bowl}} = -2\pi k\omega t$ are the angles through which the bearing disk and the grinding bowl rotate over time t, respectively; F_1 and F_2 are the foci of elliptic trajectory

be the turn angle of the bearing disk of the mill in the time t, $\omega_c = -k\omega$ be the angular speed of rotation of the grinding bowl, and k be the coefficient showing how many times the grinding bowl angular speed is higher than the angular speed of the bearing disk. As the grinding balls experience a centrifugal force induced by the grinding bowl rotation, each ball touches the internal wall of the bowl. The coordinates of the bowl axis change with rotation of the bearing disk, as do the coordinates of a ball relative to the bowl rotation axis during the milling process. Let a certain ball be at the point A_0 in the bowl at the initial time $t_0 = 0$. Then, at an arbitrary moment of time t, the position of the bowl axis O_t is defined by the coordinates

$$\left.\begin{aligned} x_0 &= R_c \cos(2\pi\omega t) \\ y_0 &= R_c \sin(2\pi\omega t) \end{aligned}\right\}. \tag{4.35}$$

Over the same time t, as a result of the bowl rotation relative to its axis in the opposite direction with the angular velocity -$k\omega$, the coordinates of the ball relative to the bowl rotation axis will have changed and will then be

$$\left.\begin{aligned} x &= r\cos(-2\pi\omega t) = r\cos(2\pi\omega t) \\ y &= r\sin(-2\pi\omega t) = -r\sin(2\pi\omega t) \end{aligned}\right\}. \tag{4.36}$$

With allowance for (4.35) and (4.36), first ball, which at the initial moment of time was in position A_0, over the time t, will have taken position A_t with the coordinates

$$\left.\begin{aligned} X &= x_0 + x = R_c\cos(2\pi\omega t) + r\cos(2\pi\omega t) \\ Y &= y_0 + y = R_c\sin(2\pi\omega t) - r\sin(2\pi\omega t) \end{aligned}\right\}. \tag{4.37}$$

The distance from the center of the trajectory (the pivot point of the bearing disk) to any arbitrary point of the trajectory, which is occupied by a ball at a moment of time t, equals

$$R(t) = [X^2 + Y^2]^{1/2} = \{R_c^2 + r^2 + 2R_c r\cos[2\pi(k+1)\omega t]\}^{1/2}. \tag{4.38}$$

At the moment the mill starts revolving the balls touch the internal wall of the bowl and are displaced from aforesaid ball on an arbitrary angle α. It is clearly that the path of the ball, which is displaced from the position of the first ball on the angle α, will be turned to the path of the first ball on the same angle α. The coordinates of such a ball can be written as

$$\left.\begin{aligned} X &= R_c\cos(2\pi\omega t) + r\cos(2\pi\omega t - \alpha) \\ Y &= R_c\sin(2\pi\omega t) - r\sin(2\pi\omega t - \alpha) \end{aligned}\right\}. \tag{4.39}$$

As a result, the combined motion of all the balls in the approximation of their independent motion represents the multitude of similar trajectories, which are turned to each other.

The design of the PM-200 Retsch mill provides angular speeds of rotation of the bearing disk and the grinding bowls that are equal in magnitude, i.e. the coefficient $k = 1$ for this mill. When $k = 1$, the path of a grinding ball is an ellipse with the axes $2(R_c + r)$ and $2(R_c - r)$ (Fig. 4.11b). For the elliptic trajectory of the ball, the current radius $R(t) \equiv R(\varphi)$, i.e. the distance between any arbitrary point of the trajectory and the center is

$$R(t) = [X^2 + Y^2]^{1/2} = \{R_c^2 + r^2 + 2R_c r \cos[4\pi\omega t]\}^{1/2}$$
$$\equiv [R_c^2 + r^2 + 2R_c r \cos(2\varphi)]^{1/2}. \tag{4.40}$$

The ball acceleration $a(t)$ is a periodic variable, which depends on the turn angle φ or, what is the same, the value of the current radius $R(t)$:

$$a(t) = [(d^2X/dt^2)^2 + (d^2Y/dt^2)^2]^{1/2}$$
$$= 4\pi^2\omega^2\{R_c^2 + r^2 + 2R_c r \cos[4\pi\omega t]\}^{1/2} \equiv 4\pi^2\omega^2 R(t). \tag{4.41}$$

Taking into account the acceleration $a(t)$, a grinding ball of the mass m, which follows an elliptical path with the current radius $R(t)$, acts on a powder particle or the mill wall with the force

$$f(t) = ma(t) = 4\pi^2 m\omega^2 R(t). \tag{4.42}$$

At the ball mass $m = 0.00025$ kg and the maximum angular speed of rotation equal to 10 revolutions per second (rps) for the PM-200 Retsch mill the maximum, medium and minimum values of the force $f(t)$ are 0.179, 0.089 and 0.028 N respectively (Fig. 4.12). When a ball hits a particle of the initial powder, which has the linear average size D_{in}, the particle experiences a pressure $P \sim f(t)/D_{in}^2$. Then the medium and maximum pressures P acting on a particle with the linear size $D_{in} = 10\,\mu$m during the milling process are ~0.9 and ~1.8 GPa. These impact pressures are sufficient for crushing of hard particles. As the powder is milled, the number of particles in the zone of the impact point contact with a grinding ball and the surface area of the contact zone increase, leading to the decrease in the pressure P and the gradual cessation of the grinding action.

As the bearing disk performs one revolution, a grinding ball travels a distance equal to the ellipse perimeter $p = 4(R_c + r)\boldsymbol{E}(e)$, where $\boldsymbol{E}(e)$ is the general elliptical integral of the 2nd kind for the ellipse with the eccentricity $e = 2\sqrt{R_c r}/(R_c + r)$. Expanding $\boldsymbol{E}(e)$ in a series gives

$$p = 2\pi(R_c + r)\left[1 - \left(\frac{1}{2}\right)^2 e^2 - \left(\frac{1\cdot 3}{2\cdot 4}\right)^2 \frac{e^4}{3} - \left(\frac{1\cdot 3\cdot 5}{2\cdot 4\cdot 6}\right)^2 \frac{e^6}{5} - \dots\right]. \tag{4.43}$$

Limiting ourselves to the first expansion terms in (4.43), we obtain

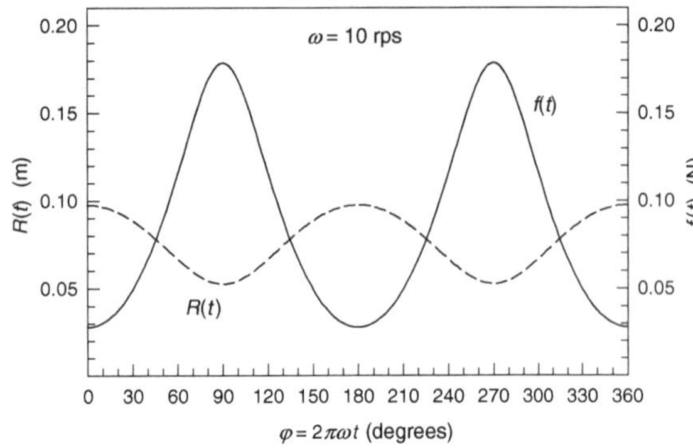

Fig. 4.12 The dependences of the current radius $R(t)$ of the trajectory and force $f(t) \equiv f(\varphi)$ from turn angle $\varphi = 2\pi\omega t$ at the angular rotation speed $\omega = 10$ rps for the PM-200 Retsch mill with $R_c = 0.075$ m, $r = 0.0225$ m and $k = 1$

$$p \approx 2\pi R_c \frac{64 - 3(r/R_c)^4}{64 - 16(r/R_c)^2}. \tag{4.44}$$

The full path traversed by a grinding ball in the time t is $p\omega t$. If the kinetic energy $E(t)$ of all N_b balls is taken equal to the sum of kinetic energies of each ball, then

$$E(t) = N_b f(t) p\omega t = 8\pi^3 N_b m\omega^3 R_c \frac{64 - 3(r/R_c)^4}{64 - 16(r/R_c)^2} R(t)t, \tag{4.45}$$

where ω is the angular speed of rotation in rps and t is the milling time. Substituting the average value of $R(t) \approx R_{av} = (R_c^2 + r^2)^{1/2}$ for the current radius $R(t)$ of the trajectory in (4.45) we obtain

$$E(t) = 8\pi^3 N_b m\omega^3 (R_c^2 + r^2)^{1/2} R_c \frac{64 - 3(r/R_c)^4}{64 - 16(r/R_c)^2} t. \tag{4.46}$$

In the course of crushing and milling most of the energy is consumed for elastic deformation of the milling system, i.e. for the interaction of the grinding bodies with the walls of the grinding chamber, while less than 3–5 % of the total kinetic energy is consumed for milling of the powder [54, 68]. Therefore, the milling energy E_{mill} can be defined as $E_{mill} = a_k E(t)$ [59–62], where $a_k \ll 1$ is the coefficient showing the portion of the energy consumed for the ball-milling of the powder. It follows from (4.46) that, given the same milling conditions, the energy consumed for milling of the powder is proportional to the cube of the angular rotation speed, ω^3, and the milling time t, i.e.

$$E_{mill} = \kappa\omega^3 t, \tag{4.47}$$

where $\kappa = 8\pi^3 a_k N_b m (R_c^2 + r^2)^{1/2} R_c \frac{64-3(r/R_c)^4}{64-16(r/R_c)^2}$ is a constant parameter typical of this mill with the dimensions R_c and r, the definite number N_b of grinding balls, and the mass m of each ball (ω is the angular speed of mill rotation in rps, R_c is the radius of the circle along which the bowl axis moves, r is the inside bowl radius). In

Fig. 4.13 Scanning electron microscope image of the initial coarse-grained powder of tungsten carbide WC with the average particle size $D_{in} \approx 6\,\mu m$

the case of the PM-200 Retsch planetary ball mill $R_c = 0.075$ m and $r = 0.0225$ m and the total mass of the grinding balls is $N_b m = 0.1$ kg; if we take $a_k = 0.01$, then the coefficient $\kappa \approx 0.0015\,kg{\cdot}m^2$.

Let us construct the theoretical dependence of average particle size D on milling time t and milled substance mass M using a coarse-grained tungsten carbide WC powder with average grain size $D_{in} \approx 6\,\mu m$ (Fig. 4.13) as an example.

Hexagonal (space group $P\bar{6}m2$) tungsten carbide WC with the unit cell constants $a = 0.29060$ and $c = 0.28375$ nm has the density $d = 15.8\,g{\cdot}cm^{-3}$, the bulk modulus $K_c = 630$ GPa, the shear modulus $G = 274$ GPa, ultimate compression strength $\sigma_c \approx 2.7$ GPa, and the Poisson ratio $\nu = 0.31$ [69, 70]. As fracture begins at critical stress σ_{max}, which is equal to the substance strength for a given type of deformation, so $\varepsilon_{max} = \sigma_{max}/E$, where E is the corresponding elastic modulus. Therefore, the limiting value of microstrains for WC fall in the range $0.0045 < \varepsilon_{max} < 0.01$. An electron-microscopic study of dislocation formation during the plastic deformation of WC [71] shows that its main slip system is $\{10-10\}\langle 0001\rangle$, i.e., deformation on the $\{10-10\}$ plane in the $\langle 0001\rangle$ direction. With allowance for lattice constants a and c of the WC unit cell, the area s_f of this plane of sliding is equal to $ac \approx 0.0825\,nm^2$ for hexagonal tungsten carbide. Moreover, slip in ⟨11-23⟩ and ⟨2-1-10⟩ directions in the same $\{10-10\}$ plane can also take place. The Burgers vectors corresponding to slip in these directions are $\boldsymbol{b} = \langle 0001\rangle$, (1/3)⟨11-23⟩ and (1/3) ⟨2-1-10⟩ and their absolute values are $b \approx 0.26$–0.29 nm for hexagonal WC [71].

The quantity qu can be evaluated from the atomization energy E_{at} of tungsten carbide WC. A unit cell of tungsten carbide includes one WC formula unit and 6 paired bonds W-C with lengths $(a^2/3 + c^2/4)^{1/2}$. Therefore energy of one bond is $q = E_{at}/6N_A$, where N_A is the Avogadro number. Within an unit cell, four paired bonds W-C pass through a plane {1 0 -1 0}, i.e. $u = 4$. Taking this into account, $qu = 4E_{at}/6N_A = 2E_{at}/3N_A$. The atomization energy of hexagonal tungsten carbide WC, which is determined from thermodynamic data [69, 72–74], is $E_{at} = 1600 \pm 50\,\text{kJ} \cdot \text{mol}^{-1}$, whence $qu = 1.77 \times 10^{-18}$ J. With allowance for the values s_f, qu, G, b, ν, d and $f_s/f_v = 6$, $C = 18$ and $\beta = 100$, constants $A = \frac{f_s}{f_v}\frac{qu}{s_f d}$ and $B = \frac{f_s}{f_v}\frac{\beta G b \sqrt{3C}}{12\pi(1-\nu)d}$ in formula (4.33) are equal $A = 0.008$ and $B = 0.85\,\text{J}\cdot\text{m}\cdot\text{kg}^{-1}$ for hexagonal tungsten carbide WC.

According to [64], with allowance for $E_{mill}(t) = \kappa\omega^3 t$ (4.47) and the dependence of microstrains ε on milling time t and powder mass M, we can rewrite the formula (4.33) in the form

$$D(t, M) = \frac{M\{A + B[\ln(D_{in}/2b)]\varepsilon_{max}[t/(t+\tau)][M/(M+p]\}}{\kappa\omega^3 t + M\{A + B[\ln(D_{in}/2b)]\varepsilon_{max}[t/(t+\tau)][M/(M+p)]\}/D_{in}}. \tag{4.48}$$

Given the same angular rotation speed ω the relationship (4.48) takes the form

$$D(t, M) = \frac{M[a_D + b_D\varepsilon(t, M)]}{t + M[a_D + b_D\varepsilon(t, M)]/D_{in}}, \tag{4.49}$$

where $a_D = \frac{A}{\kappa\omega^3} = \frac{f_s}{f_v}\frac{qu}{s_f d\kappa\omega^3}$, $b_D = \frac{B\ln(D_{in}/2b)}{\kappa\omega^3} = \frac{f_s\sqrt{3C}}{f_v}\frac{\beta G b\ln(D_{in}/2b)}{12\pi(1-\nu)d\kappa\omega^3}$ and $\varepsilon(t, M) = \varepsilon_{max}[t/(t+\tau)][M/(M+p)]$. At $\kappa \approx 0.0015\,\text{kg}\cdot\text{m}^2$, angular speed of rotation $\omega = 8.33$ rps, and initial particle size $D_{in} = 6\cdot10^{-6}$ m, we obtain $a_D = 0.00933$ m·s/kg and $b_D = 9.0899$ m·s/kg from the theoretical estimation of coefficients A and B.

The theoretical three-dimensional dependence $D_{theor}(t, M)$ which is calculated by the formula (4.48) or (4.49) using theoretical values of A and B (or a_D and b_D) at $\varepsilon_{max} = 0.008$, angular speed of rotation $\omega = 8.33\ rps$, and particle size $D_{in} = 6000$ nm of initial tungsten carbide powder is shown in Fig. 4.14.

It is seen that particle size D after milling decreases with increasing milling time t, decreasing powder mass M or particle size D_{in} in the initial powder. The particle size decreases rapidly at the initial stage, when milling time t does not exceed 10000 s. Upon the further increase in the milling time and fixed mass M, the $D(t,M)$ dependence tends asymptotically to a certain limiting value.

Comparison of Milling Model with Experimental Results

For comparison of the proposed model with experiment, the initial coarse-grained (with average particle size $D_{in} \approx 6\,\mu\text{m}$) powder of tungsten carbide WC was ground down to the nanocrystalline state with the particle size down to 10 nm.

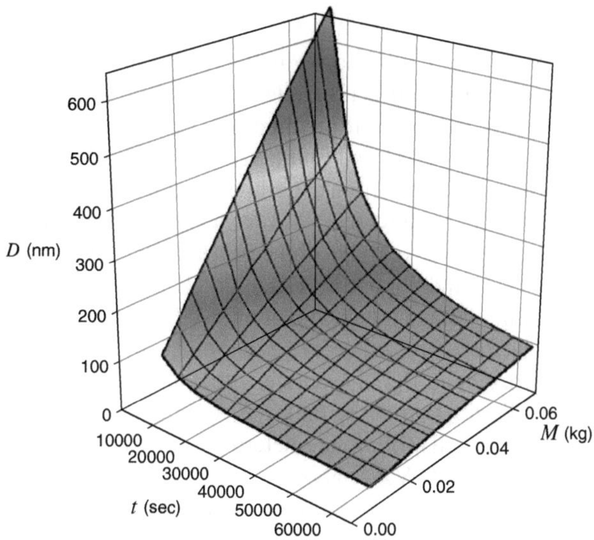

Fig. 4.14 Theoretical three-dimensional dependence $D(t, M)$ of the powder particle size D on the milling time t and mass M of initial coarse-grained WC powder. Dependence $D(t, M)$ is calculated by the formula (4.48) using theoretical values of A and B at $\varepsilon_{max} = 0.0078$, angular speed of rotation $\omega = 8.33$ rps, and particle size $D_{in} = 6000$ nm

Milling was performed in an automatic mode at angular rotation speed $\omega =$ 8.33 rps. The rotation direction changed every 15 min, and there were 5 s stops between changes in the rotation direction. The ball charge, i.e. the total mass of the grinding balls was ~100 g, and the number of grinding balls was N_b ~450.

The milling time was from 10 min to 15 h. Isopropanol $CH_3CH(OH)CH_3$ was added as milling liquid in amount of 5 ml per 10 g of the initial material. In comparison with other possible liquids (butanol, isobutanol, hexane, etc) isopropanol has the higher self-ignition temperature 456 °C and does not react with ring seals of grinding bowls. The powder was dried after milling. Charge mass M, i.e., the powder mass taken for ball milling, was 10, 20, 25, and 33.3 g in various experiments.

While grinding the decrease in the particle size is accompanied by the initiation of microstrains in the particles. The only experimental method which allows to determine the size D of the small particles (more specifically, the size of coherent scattering domains) and the value of the microstrains ε is an X-ray diffraction (XRD) method. Therefore the average particle size $\langle D \rangle$ and the value of the microstrains ε in ball-milled tungsten carbide powder were determined by XRD method from broadening of diffraction reflections.

X-ray diffraction patterns were recorded on a Shimadzu XRD-7000 and a DRON-UM1 diffractometers by the Bragg-Brentano method using Cu $K\alpha_{1,2}$ radiation at the angles 2θ of 10° to 140° with the scan step $\Delta(2\theta) = 0.03°$ and the exposure time of 2 s at each point. The diffraction reflections were described by the pseudo-Voigt function (4.6) with identical parameters $\theta_L = \theta_G = b$.

The diffraction reflection broadening $\beta(2\theta)$ was defined by formula (4.12) as $\beta(2\theta) = \sqrt{(\mathrm{FWHM_{exp}})^2 - (\mathrm{FWHM_R})^2}$, where $\mathrm{FWHM_{exp}}$ is the full width at half-maximum of an experimental diffraction reflection and $\mathrm{FWHM_R}$ is the instrumental angular resolution function of the diffractometer. The angular resolution functions $\mathrm{FWHM_R}(2\theta)$ of the DRON-UM1 and Shimadzu XRD-7000 X-ray diffractometers were determined in a special diffraction experiments with the cubic lanthanum hexaboride LaB_6 (NIST Standard Reference Powder 660a) and were represented in the form (4.12). The parameters u, v, and w of the instrumental angular resolution functions for the DRON-UM1 and Shimadzu XRD-7000 X-ray diffractometers are given in Table 4.1. The using of standard specimen allowed to determine the lattice constants of the investigated compound accurate within 5 significant digits and the diffraction reflection broadening $\beta(2\theta)$ accurate within 4 significant digits.

The particle size distribution in the tungsten carbide nanopowder n-WC, which was made by high-energy ball milling, was determined also using the laser diffraction method on a HORIBA LA-950V2 Analyzer.

Specific features of the microstructure of micro- and nanocrystalline powders and compact sintered specimens, and the size of particles, agglomerates and grains (crystallites) were determined by high-resolution scanning electron microscopes of the JEOL 6310 (SCK-CEN, Mol, Belgium) and BS-301 Tesla types.

Scanning electron microscopy and the laser diffraction method were used in our study only as the auxiliary methods for confirmation of results of an XRD method.

According to the energy dispersive X-ray spectroscopy (EDS/EDX), the initial coarse-grained WC powder contains 93.4 wt.% of tungsten W and 6.6 wt.% of carbon C, oxygen has not been found out. The ground powder WC-mill contains 91.2 wt.% W, 5.9 wt.% C and 2.9 wt.% of oxygen O (Fig. 4.15).

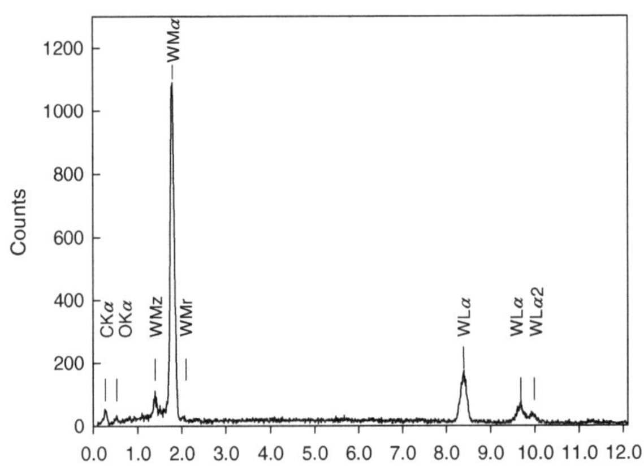

Fig. 4.15 EDX analysis of nanocrystalline WC-mill powder prepared by high-energy ball milling of 10 g of the initial coarse-grained WC powder during 10 h

Oxygen in ground powder WC-mill is adsorbed on a surface of carbide nanoparticles. Data of the chemical analysis of the ground powder WC-mill are close to EDX results: the content of the total carbon amounts to 6.12 wt.%, free (uncombined) carbon is absent. The identical content of tungsten and carbon in initial coarse-grained and ground nanocrystalline powders means that the composition of a powder has not changed as a result of the ball-milling and is close to the stoichiometric tungsten carbide WC.

Figure 4.16 presents X-ray diffraction patterns of the initial coarse-grained WC powder and nanocrystalline powders WC-mill of tungsten carbide made by milling for 5, 10 and 15 h. All the powders only contain hexagonal carbide WC, but the diffraction reflections become much wider after milling (Fig. 4.16). The quantitative analysis of broadening of the reflections obtained for the nanocrystalline powders showed that it was due to both the small size of the particles and the microstrains. Values of the particle size and strain broadening were separated and the size $\langle D \rangle$ of the coherent scattering domains and the value of the microstrains ε were determined by the Williamson-Hall method [14, 15, 33, 37, 75] using the dependence of the reduced broadening $\beta^*(2\theta) = [\beta(2\theta)\cos\theta]/\lambda$ of the reflections (hkl) on the scattering vector $s = (2\sin\theta)/\lambda$ (see Sect. 4.1.3). The value of the microstrains ε characterized the relative change of the interplanar spacing. The value of the microstrains in relative units was found from the slope φ of the straight line approximating the dependence of β^* on the scattering vector s by the formula $\varepsilon = \{[\beta^*(2\theta)]/2s\} \equiv [(tg\varphi)/2]$ (4.17). Note that the quantity of microstrains has been wrongly determined as $\varepsilon = \{[\beta^*(2\theta)]/4s\} \equiv [(tg\varphi)/4]$ in works [59–62] and it is underestimated in 2 times. According to performed estimate, for 20 g of the WC-mill powder ground for 10 h the mean size $\langle D \rangle$ of the coherent scattering regions and the value of the microstrains ε are 37 ± 5 nm and 0.0075 ± 0.0002 (or 0.75 ± 0.02 %), respectively (Fig. 4.17).

Figure 4.18 shows the dependence of the reduced broadening $\beta^*(2\theta)$ of the diffraction reflections on the scattering vector s for nanocrystalline WC-mill powders ground for 5, 10 and 15 h. According to the estimates, the average size $\langle D \rangle$ of the coherent scattering domains in the n-WC powders ground for 5, 10 and 15 h is 32 ± 10, 20 ± 10 and 11 ± 5 nm, respectively. The value of the microstrains ε in the powders ground for 5, 10 and 15 h is as large as 0.0066 ± 0.0003, 0.0070 ± 0.0003 and 0.0060 ± 0.0004 (or 0.66 ± 0.03, 0.70 ± 0.03 and 0.60 ± 0.04 %).

The particle size distribution in the ball-milled powder of tungsten carbide WC-mill was determined also by the laser diffraction method on a HORIBA LA-950V2 Analyzer. It is seen (Fig. 4.19) that according to the laser diffraction analysis the smallest particles in the n-WC nanopowder are ~80 nm in size, half of all the particles are less than 170 nm in size, and 95 % of all the particles has the size of not over 500 nm. This means that the nanopowder particles are agglomerated.

Observations in a scanning electron microscope (Fig. 4.20) also pointed to easy agglomeration of the ball-milled powder: the agglomerates are 100–400 nm in size.

On the whole, the average size of the particles, which was estimated from broadening of the X-ray diffraction reflections obtained for the tungsten carbide nanopowder, is in qualitative agreement with the relevant results of the electron microscope examination and the laser diffraction. The particle size determined from the XRD

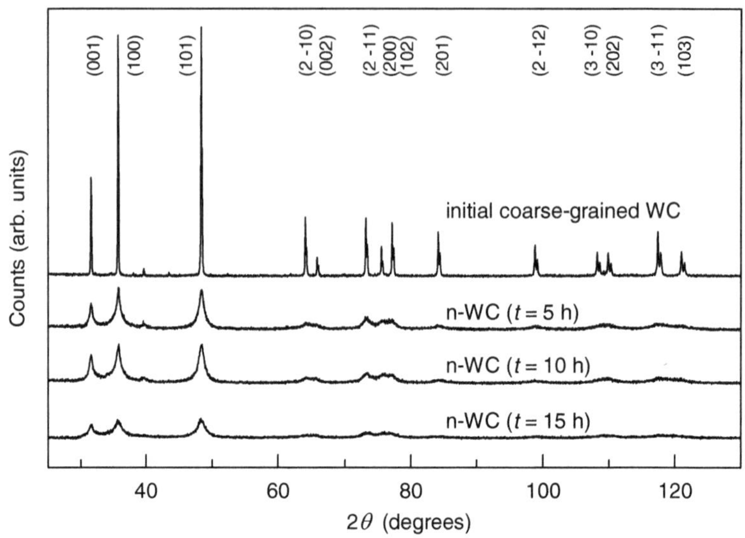

Fig. 4.16 X-ray diffraction patterns of the initial coarse-grained ($D_{in} \approx 6\,\mu m$) WC powder and nanocrystalline WC-mill powders prepared by high-energy ball-milling of the initial WC powder in a PM-200 Retsch planetary ball mill during 5, 10 and 15 h. All the powders only contain hexagonal (space group $P\overline{6}m2$, $a = 0.29060$ and $c = 0.28375$ nm) tungsten carbide WC with particles of different sizes. Diffraction reflections of the nanocrystalline WC-mill powders are broadened considerably. XRD measurements were performed using Cu$K\alpha_{1,2}$ radiation

Fig. 4.17 Determination of mean particle size $\langle D \rangle$ and microstrains ε in nanocrystalline WC-mill powder prepared by ball milling of 20 g of the initial coarse-grained WC powder during 10 h: $\langle D \rangle = 37 \pm 5$ nm, $\varepsilon = 0.75 \pm 0.02\,\%$

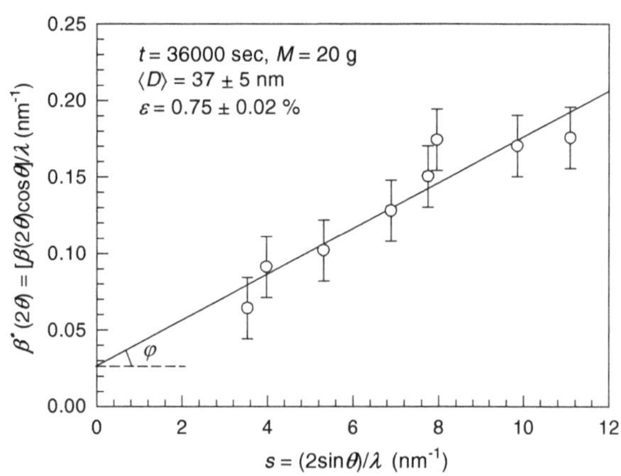

analysis is smaller because the size of agglomerates rather than separate particles is estimated by the microscope and the laser method. It should be noted also that the X-ray diffraction method is a volume technique and, hence, the size of particles is volume-averaged in this method. Furthermore, the value of microstrains, along with the particle size, is determined quantitatively by the XRD method.

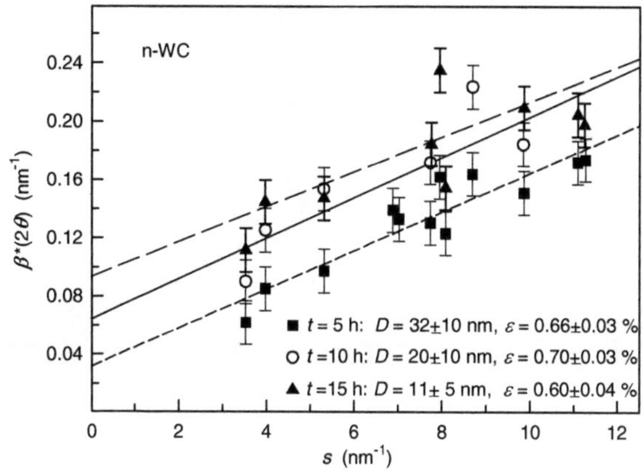

Fig. 4.18 Average size of particles D and microstrains ε in nanocrystalline powders WC-mill prepared by ball milling of 10 g of the initial coarse-grained WC pow-der during 5, 10, and 15 h

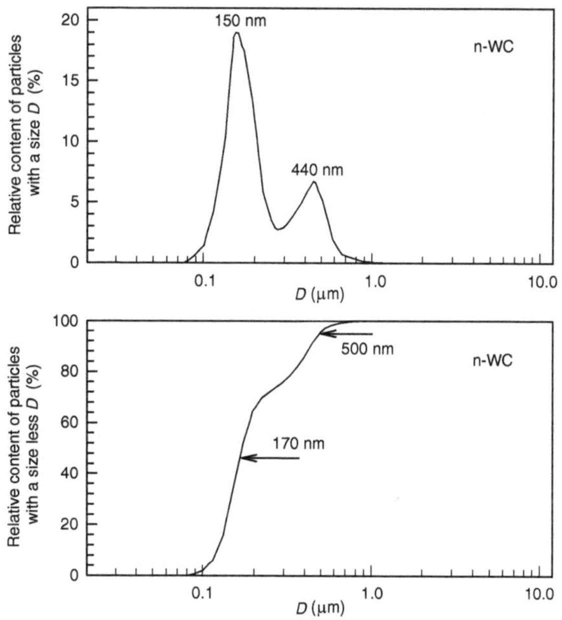

Fig. 4.19 Particle size distribution in the nanocrystalline powder WC-mill of tungsten carbide prepared by ball milling during 10 h. The distribution is determined by the laser diffraction method on a HORIBA LA-950V2 Analyzer

Experimental dependences of the average size D of the WC particles and the microstrains ε on the milling time t and the powder mass M [64] are shown in Fig. 4.21. It is seen that, at the experimental milling parameters, the particle size decreased quickly and the microstrains were building up during the first 100–150 min

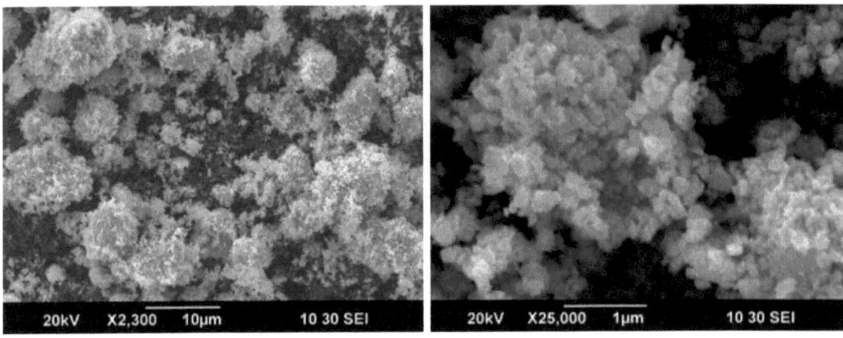

Fig. 4.20 Scanning electron microscope image of a nanocrystalline powder WC-mill prepared by ball milling of 10 g of the initial coarse-grained WC powder during 10 h in a PM-200 Retsch planetary ball mill. The tungsten carbide nanoparticles are agglomerated and the agglomerates are 100 to 300–400 nm in size

of the milling process. As the milling time increased further, the dependences $D(t)$ and $\varepsilon(t)$ asymptotically approached some limiting values. A decrease in initial powder charge M at a constant milling time is accompanied by a decrease in the particle size and an increase in the microstrains.

It is seen from Fig. 4.21 that the experimental dependence $D_{\text{exp}}(t, M = const)$ is approximated well by the function (4.49) with the parameters $a_D = 0.00199$ m·s/kg and $b_D = 7.6514$ m·s/kg (ε measured in relative units). The agreement between the theoretically estimated values of parameters a_D and b_D and the experimental values is satisfactory. The microstrains ε (in relative units) are described by the empirical dependence $\varepsilon(t, M) = \varepsilon_{\max}[t/(t + \tau)][M/(M + p)]$, where $\varepsilon_{\max} = 0.007842$, $\tau = 7335$ s and $p = -0.00164$ kg.

The dependence $D_{\text{theor}}(t, M = const)$ (Fig. 4.21, the dashed line), which was calculated by the formula (4.49) using theoretical values of a_D and b_D, is slightly above its experimental counterpart. This observed discrepancy between the experimental and calculated results is explained primarily by the rough estimation of qu value and the empirical coefficients a_k, f_s/f_v, C and β.

The proposed model of ball milling of powders allows substituting the empirical selection of milling conditions for the theoretical determination of milling parameters.

This substitution is possible taking into account from the one hand, the physical characteristics of the initial powder, and from the other hand, the value of the milling energy. The derived analytical expression allows estimating the value of milling energy of planetary ball mill. It is shown, that the milling energy is proportional to a cube of angular speed of rotation and milling duration. An analytical expression has been deduced describing the size of nanocrystalline powder particles as a function of the milling time and powder mass. For a coarse-grained tungsten carbide WC powder the good consent between theoretical and experimental dependences of the particle size on mass of initial powder and milling duration is shown. According

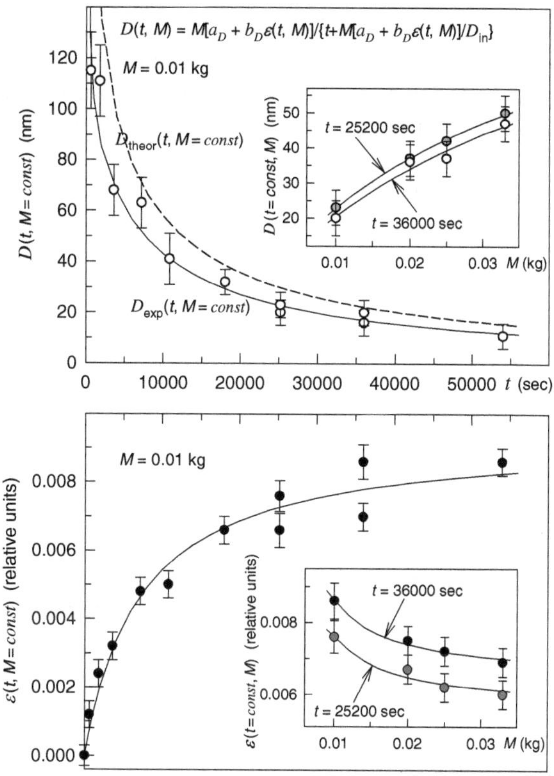

Fig. 4.21 Dependences of (o) average particle size D and (•) microstrains ε on milling time t and mass M of an initial coarse-grained WC pow-der in an PM-200 Retsch planetary ball mill (the ball mass is 100 g, the powder mass is 10 g, the angular rotation speed is 8.33 rps). The experimental data of particle size D_{exp} are approximated by (4.49) with parameters $a_D = 0.00199$ and $b_D = 7.6514$ m·s/kg (ε is measured in arbitrary units). The change in the microstrains ε is described by the empirical relationship $\varepsilon(t, M) = \varepsilon_{max}[t/(t + \tau)][M/(M + p)]$ with $\varepsilon_{max} = 0.007842$, $p = -0.00164$ kg, and $\tau = 7335$ s. Calculated $D_{theor}(t)$ dependence is shown as a *dashed line*

to the offered model after milling within 15 h the average size of particles of a nanocrystalline powder WC-mill is 15 nm, while experimentally determined average size $\langle D \rangle$ of particles is 11 ± 5 nm. The model proposed is applicable to single-phase solids. The presence of many phases in powder mixtures should be considered by special manner.

4.3 Properties of Nanocrystalline Powders of Tungsten Carbide WC

The difference in the properties of nanocrystalline and coarse-grained powders is associated with different size of particles and extremely developed interfaces of the nanopowders. The interfaces of nanopowders can include up to 50 % of all atoms, fraction of atoms which are on a surface of particles is negligibly small in coarse-grained powders. At present, many researchers of nanocrystalline powders assume that their physical and chemical properties are determined primarily by the large specific surface and large area of the interfaces [12–15, 76–79].

The properties of nanocrystalline powders are very sensitive to the impurities, especially to the impurities of oxygen and ferromagnetic metals. The effect of impurities on the properties of nanocrystalline substances can be stronger than that of small grain sizes.The unusually large area of the surface results in high chemical activity of nanocrystalline powders. For example, many extraordinary results obtained in early studies on nanocrystalline powders and explained by size effects had appeared to be a consequence of the contamination of the objects under study by oxygen, hydrogen, ferromagnetic, and other impurities.

4.3.1 Magnetic Susceptibility and Thermal Stability of Particle Size of Nanocrystalline Tungsten Carbide WC

The advantages due to the small size of particles (grains, crystallites) of nanostructured materials can be retained and realized if the small particle sizes are thermally stable. However, studies of many nanocrystalline metals and alloys [80–86] showed that the particle (grain) sizes increase noticeably even at room temperature 300 K, not to mention higher temperatures. In oxide materials (e.g., MgO [87]), the grain sizes remain almost unchanged upon annealing up to 800–900 K. This indicates their increased thermal stability in comparison with metals. In general, studying the thermal stability of particle sizes in nanocrystalline substances and nanostructured materials is of interest in itself.

The literature contains no data on the magnetic properties and thermal stability of nanocrystalline tungsten carbide. There is only one study [88] on the magnetic susceptibility χ of coarse-grained carbide WC which was preliminarily washed in hydrochloric acid HCl to remove ferromagnetic impurities. The susceptibility of WC carbide was measured in a magnetic field of 3.6 kGs at 293, 513, and 723 K. Measured susceptibility values are equal to 0.07×10^{-6}, 0.06×10^{-6} and $0.02 \times 10^{-6}\,\mathrm{cm^3 \cdot g^{-1}}$, respectively. Judging from these data, tungsten carbide is a weak paramagnet. Detected decrease in the susceptibility of WC with an increase in temperature [88] indicates that WC contains a ferromagnetic impurity (most likely iron).

Later in our works [89–91], the magnetic susceptibility of nanocrystalline tungsten carbide WC was studied in situ to determine the temperature range of thermal stability of particle sizes and to detect possible effects of the nanocrystalline state at temperatures of 300–1250 K.

To detect the direct or indirect influence of the small particle size on the magnetic susceptibility of tungsten carbide and analyze the thermal stability of the nanoparticle size, a nanocrystalline WC-plasm powder was studied. Synthesis conditions of WC-plasm nanopowder are described in Sect. 4.2.1.

To exclude any effects caused by the small particle size, the magnetic susceptibility of coarse-grained powder of hexagonal tungsten carbide WC was measured additionally. This WC powder was washed in hydrochloric acid HCl and did not contain ferromagnetic impurities. The size of the particles of the coarse-grained WC powder was determined by the Fischer method and was equals to 4–9 μm.

The powder composition before and after measurements was determined by chemical and spectral methods. In both WC-plasm and coarse-grained WC carbides, the contents of tungsten and carbon are equal to 93.80 ± 0.05 and 6.12 ± 0.01 wt.%, respectively. The oxygen and nitrogen impurity contents are 0.14 and 0.02 wt.% in WC-plasm and 0.1 and less than 0.02 wt.% in a coarse-grained WC powder, respectively. According to spectral analysis, WC-plasm carbide contains 0.0008, 0.002, and less than 0.0001 wt.% of Fe, Co, and Ni, respectively. In coarse-grained carbide WC, the contents of these elements are less than 0.00001 wt.% each. The content of each of the ferromagnetic rare-earth metals Gd, Dy, Tb, Ho, Er, and Tm in coarse-grained WC and nanocrystalline WC-plasm carbides did not exceed 0.000005 wt.%.

Milling of a coarse-grained WC powder in a PM-200 Retsch planetary ball mill too allows producing tungsten carbide nanopowder with the average size of particles ~20–40 nm. Grinding bowls of a ball mill had internal lining with hard alloy WC-Co. Therefore milled WC powder contains a ferromagnetic impurity of cobalt Co and it cannot be used for high-precision measurements of a magnetic susceptibility of nanocrystalline tungsten carbide.

The structure and phase composition of tungsten carbide powders were studied by an X-ray diffraction method on a DRON-UM1 diffractometer using Cu$K\alpha_{1,2}$ radiation at the angles 2θ of 10 to 140° with the scan step $\Delta(2\theta) = 0.03°$ and the exposure time of 2 s at each point. To accurately determine the position, width, and intensity of each reflection, the diffraction reflections were described by the pseudo-Voigt function. The nanopowder particle size was determined from broadening of the diffraction reflections (see Sect. 4.1.3) and by means of scanning electron microscopy.

The broadening $\beta(2\theta)$ is quantitatively determined by the formula (4.12) by comparing the experimental width of the diffraction reflections, $\mathrm{FWHM_{exp}}$, of WC-plasm carbide with the instrumental resolution function $\mathrm{FWHM_R}$ of the DRON-UM1 diffractometer. The parameters u, v, and w of the resolution function written in the form (4.9) are given in Table 4.1.

The magnetic susceptibility χ was measured by the Faraday method on a Domenicalli pendulum magnetic balance that has a sensitivity of no worse than $0.5 \times 10^{-8}\,\mathrm{cm^3{\cdot}g^{-1}}$ and is intended for studying weak paramagnetic and diamagnetic materials. The measurements were carried out in a vacuum of 10^{-3} Pa at a temperature of 300–1250 K in magnetic fields of 7.2, 8.0, and 8.8 kGs.

The magnetic susceptibility χ of nanocrystalline WC-plasm powder was measured in situ in the following order: specimen heating from 300 K to annealing temperature T; annealing the specimen at this temperature for 1 h and measurement of the susceptibility $\chi(T)$ at the end of the annealing; cooling from the annealing temperature T to 300 K and measurement of the susceptibility $\chi(300,T)$ at 300 K; heating to the next annealing temperature, and so on (Fig. 4.22). The average temperature step was 50 K. The total duration of the susceptibility measurement cycle was 60 h.

The measurements made directly at the annealing temperatures relate to the temperature dependence of the susceptibility and are denoted by $\chi(T)$. The susceptibility measurements carried out after specimen annealing and subsequent cooling to 300 K

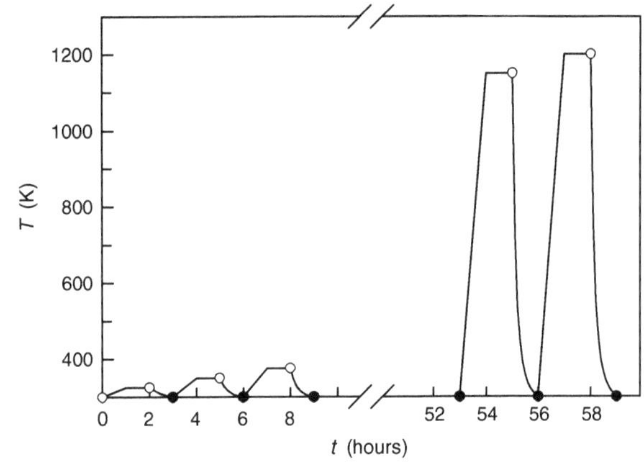

Fig. 4.22 Sequence of magnetic susceptibility measurements in annealing experiments: (o) measurements in situ (at annealing temperature *T*), (•) measurements after cooling from annealing temperature to 300 K

refer to the annealing-temperature dependence and are denoted as $\chi(300, T)$. This susceptibility measurement method was proposed in [83] for studying structural features of nanocrystalline weak magnetic substances and applied in works [92–95] to study the microstructure and magnetic properties of submicrocrystalline copper Cu, palladium Pd, and titanium Ti.

Particle Size and Magnetic Susceptibility of WC Carbide

The XRD patterns of coarse-grained and nanocrystalline (both initial and annealed during susceptibility measurements) tungsten carbide powders are shown in Fig. 4.23. Coarse-grained and initial nanocrystalline tungsten carbide powders contain only hexagonal (space group $P\overline{6}m2$) carbide WC with unit cell constants $a = 0.29060$ nm and $c = 0.28375$ nm. Nanopowder WC-plasm, annealed during susceptibility measurements at temperature 1100 K, contains 79 wt.% of tungsten carbide WC and 21 wt.% of tungsten W. The diffraction reflections of the initial and annealed nanocrystalline WC-plasm powders (Fig. 4.23) are broadened.

When analyzing the diffraction reflections, the size and strain broadenings [14] were taking into account. Two types of broadening were separated and the size $\langle D \rangle$ of the coherent scattering region and the microstrains ε were determined by the Williamson-Hall method [33, 37] which is described in Sect. 4.1.3.

A quantitative analysis of the diffraction reflection broadening for the WC-plasm nanopowders is described in Sect. 4.2.1. Calculation showed that the average size of coherent scattering regions is 55 ± 10 nm in WC-plasm powders and does not change during annealing when measuring χ in the temperature range of 300–1000 K. However, the annealing in the range of 300–1000 K causes relaxation of the microstrain ε which reduces from 0.06 ± 0.005 to 0.015 ± 0.002 % (see Fig. 4.9). Subsequent measurement of a susceptibility χ of WC-plasm nanopowder in the temperature range from 1000 to 1250 K causes increasing the average size of coherent scattering

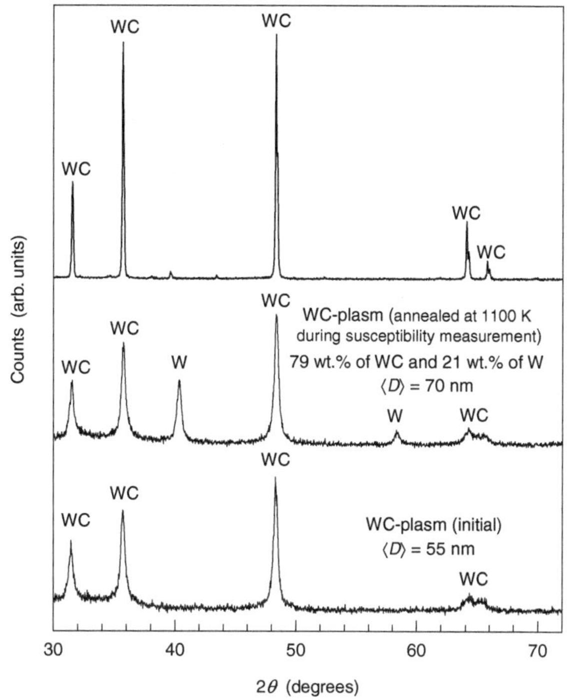

Fig. 4.23 XRD patterns of initial nanocrystalline WC-plasm powder and the same powder after annealing at temperature 1100 K during susceptibility measurements. XRD pattern of coarse-grained tungsten carbide WC is shown for comparison. Coarse-grained and initial nanocrystalline tungsten carbide powders contain only hexagonal (space group $P\bar{6}m2$) carbide WC. Nanopowder WC-plasm, annealed during susceptibility measurements at temperature 1100 K, contains 79 wt.% of tungsten carbide WC and 21 wt.% of tungsten W. Diffraction reflections of the WC-plasm nanopowders are considerably broadened. XRD measurements are performed using $\mathrm{Cu}K\alpha_{1,2}$ radiation

regions to 70–80 nm while the microstrains remain unchanged. No diffraction reflection broadening is observed in coarse-grained carbide WC. According to the scanning electron microscopy data, the average size of particles in WC-plasm nanopowder before annealing is ~60 nm. Thus, the nanoparticle size as estimated by the XRD method is in agreement with the microscopy data.

Measurements of the magnetic susceptibility χ of coarse-grained WC powder showed (Fig. 4.24, curve *1*) that tungsten carbide WC is a weak paramagnet with a susceptibility $\chi(300) \approx 0.058 \times 10^{-6}\,\mathrm{cm^3{\cdot}g^{-1}}$ ($11.4 \times 10^{-6}\,\mathrm{cm^3{\cdot}mol^{-1}}$).

The susceptibility $\chi(T)$ of this WC carbide in the range 300–1250 K has no specific features and is described by the function $\chi(T) = \chi(0) + BT^2$, typical of Pauli paramagnetism, where $\chi(0) \approx 0.0565 \times 10^{-6}\,\mathrm{cm^3{\cdot}g^{-1}}$ and $B \approx 1.30 \times 10^{-14}\,\mathrm{cm^3{\cdot}g^{-1}{\cdot}K^{-2}}$.

According to [20, 21, 96, 97], the magnetic susceptibility of tungsten carbide WC which is an electronic conductor can be written as a sum of few contributions:

$$\chi(T) = \chi_\mathrm{p}^-(T) + \chi_\mathrm{d}^- + \chi_\mathrm{VVorb} + \chi_\mathrm{p\,W}^+ + \chi_\mathrm{d\,W}^+ + \chi_\mathrm{p\,C}^+ + \chi_\mathrm{d\,C}^+, \tag{4.50}$$

where χ_p^- is the Pauli paramagnetism of conduction electrons; $\chi_\mathrm{d}^- = -(m_0/m^*)^2 \chi_\mathrm{p}^-(0)/3$ is the Landau diamagnetism (m_0 is the mass of a free electron and m^* is the effective mass), which is proportional to the Pauli paramagnetism $\chi_\mathrm{p}^-(0)$ at

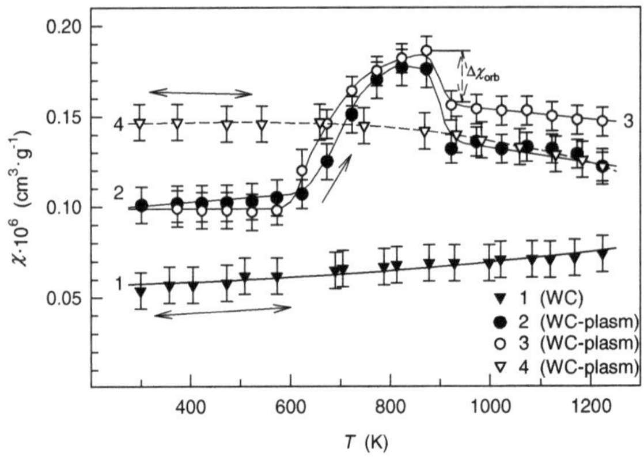

Fig. 4.24 Temperature dependences of the magnetic susceptibility χ of WC carbide in a magnetic field $H = 8.8$ kOe: (▼) susceptibility $\chi_{WC}(T)$ of pure coarse-grained carbide WC; (•) susceptibility $\chi_{\text{WC-plasm}}(T)$ of nanocrystalline carbide WC-plasm; (○) annealing-temperature dependence $\chi(300, T)$ for nanocrystalline carbide WC-plasm; (∇) susceptibility $\chi^{(\text{ann})}_{\text{WC-plasm}}(T)$ of carbide WC-plasm, annealed during susceptibility measurement. Value $\Delta\chi_{\text{orb}} \approx 0.03 \times 10^{-6}\ \text{cm}^3\cdot\text{g}^{-1}$ is the Van Vleck orbital paramagnetic contribution to the susceptibility of nanocrystalline carbide WC-plasm. The directions of temperature changes during χ measurements are indicated by arrows

0 K; χ_{p}^{+} and χ_{d}^{+} denote para- and diamagnetisms of ion cores ($\chi_{\text{p W}}^{+} \equiv 0$ and $\chi_{\text{p C}}^{+} \equiv 0$; $\chi_{\text{d W}}^{+} = -13 \times 10^{-6}$ and $\chi_{\text{d C}}^{+} = -0.15 \cdot 10^{-6}\ \text{cm}^3\cdot\text{mol}^{-1}$ for W^{6+} and C^{4+} [98], respectively); $\chi_{\text{VV orb}} = \left[\left|s|\hat{\text{M}}|0\right|\right]^2/(\varepsilon_s - \varepsilon_0)$ is the Van Vleck orbital paramagnetism; ε_s and ε_0 are the energies of the excited (s) and ground (0) states of the atom, respectively; $\left|s|\hat{\text{M}}|0\right|$ is the modulus of the off-diagonal matrix element of the magnetic moment operator (physically, the quantity $\left[\left|s|\hat{\text{M}}|0\right|\right]^2$ is proportional to the probability of a quantum transition from the ground state to the excited state caused by system perturbations).

Among all the terms of Eq. (4.50), only the Pauli paramagnetism χ_{p}^{-} depends on temperature [97]:

$$\chi_{\text{p}}^{-}(T) = N_{\text{A}}\mu_{\text{eff}}^2 N(E_{\text{F}}) + \frac{\pi^2 k_{\text{B}}^2}{6} N_{\text{A}}\mu_{\text{eff}}^2 \left\{ N''(E_{\text{F}}) - \frac{[N'(E_{\text{F}})]^2}{N(E_{\text{F}})} \right\} T^2, \quad (4.51)$$

where $N'(E_F) \equiv [dN(E)/dE]|_{E=E_{\text{F}}}$ and $N''(E_F) \equiv [d^2N(E)/dE^2]|_{E=E_{\text{F}}}$ are the first and second derivatives of the electron densities of states $N(E)$ with respect to the energy at $E = E_F$, N_A is the Avogadro constant, k_B is the Boltzmann constant, $\mu_{\text{eff}} = p\mu_{\text{B}}$ is the effective magnetic moment, and p is the number of Bohr magnetons μ_B. As opposed to usual expression for the Pauli paramagnetism of free electron gas (see [97], p. 174), the effective magnetic moment μ_{eff} is used in expression (4.51) instead of the magnetic moment of a free electron, which is equal to the Bohr magneton μ_B. It is caused by the presence of exchange electron-electron interaction in metals (especially in transition *d*-metals, including tungsten W) and in compounds of metals. In other words, metal contains not a free electron gas but a Fermi liquid of quasi-particles with arbitrary dispersion law. The electron-electron interaction increases the Pauli susceptibility. Taking this into account, the magnetic moment renormalization is made and the effective magnetic moment μ_{eff} of Fermi quasi-particle is used instead of the magnetic moment of a free electron.

It is clear from (4.51) that at 0 K the Pauli paramagnetism $\chi_p^-(0) = N_A\mu_B^2 p^2 N(E_F)$ depends only on the density of electron states at the Fermi level $N(E_F)$ and the temperature coefficient of the susceptibility is

$$B = (\pi^2 k_B^2 N_A \mu_{\text{eff}}^2/6)\left\{N''(E_F) - [N'(E_F)]^2/N(E_F)\right\} \equiv (\pi^2 k_B^2 N_A \mu_B^2 p^2/6)b, \tag{4.52}$$

where

$$b = N''(E_F) - [N'(E_F)]^2/N(E_F). \tag{4.53}$$

According to theoretical calculations of the electron energy spectrum of hexagonal tungsten carbide WC [99, 100], the Fermi level is located near a minimum in the *N*(*E*) curve. The density of states at the Fermi level $N(E_F)$ is small and is 0.3–0.4 eV^{-1} (4.1–5.4 Ry^{-1}) per unit cell. For the *N*(*E*) curves calculated in [99, 100] for carbides WC, the first derivative $N'(E_F)$ is negative and the second derivative $N''(E_F)$ is positive. Therefore, the temperature coefficient *B* can be positive or negative depending on the specific values of $N(E_F)$, $N'(E_F)$ and $N''(E_F)$. Numerical differentiation of the $N(E)$ curve from [99] in the vicinity of the Fermi energy yields $N(E_F) \approx 0.36\,\text{eV}^{-1}$, $N'(E_F) \approx -1.20\,\text{eV}^{-2}$ and $N''(E_F) \approx 6.10\,\text{eV}^{-3}$, from which we find that the temperature coefficient $b = 2.10 \pm 0.06\,\text{eV}^{-3}$. This value is in agreement with the positive experimental temperature coefficient *B* of the susceptibility of carbide WC. An analogous estimation using the *N*(*E*) curve from [100] yields a smaller, but also positive value of *b*.

Taking into account the experimental value $B \approx 2.55 \times 10^{-12}\,\text{cm}^3\text{·mol}^{-1}\text{·K}^{-2}$ and $b = 2.10\,\text{eV}^{-3} \equiv 2.1 \times (2.43 \times 10^{35})\,\text{erg}^{-3}$, we find from (4.52) that $p = [6B/(\pi^2 k_B^2 N_A \mu_B^2 b)]^{1/2} \approx 1.75 \pm 0.06$. The values $N_A = 6.0247 \times 10^{23}\,\text{mol}^{-1}$, $k_B = 1.38 \times 10^{-16}\,\text{erg·K}^{-1}$ and $\mu_B = 9.2741 \times 10^{-21}\,\text{erg·Gs}^{-1} \equiv 9.2741 \times 10^{-21}\,\text{cm}^{5/2}\text{·g}^{1/2}\text{·s}^{-1}$ are used for calculation of *p* (the number of Bohr magnetons). For $p = 1.75$ and $N(E_F) \approx 0.36\,\text{eV}^{-1}$, we obtain $\chi_p^-(0) = N_A\mu_B^2 p^2 N(E_F) = 6BN(E_F)/(\pi^2 k_B^2 N_A b) = (35.8 \pm 2.0) \times 10^{-6}\,\text{cm}^3\text{·mol}^{-1}$ and the Pauli paramagnetic contribution to the WC susceptibility is $\chi_p^-(T) = (35.8 + 2.55 \times 10^{-6}T^2 \pm 2.0) \times 10^{-6}\,\text{cm}^3\text{·mol}^{-1}$. If $m^* = m_0$, then $\chi_d^- = -\chi_p^-(0)/3 = -11.9 \times 10^{-6}\,\text{cm}^3\text{·mol}^{-1}$. The sum of following contributions $\chi_p^-(300) + \chi_d^- + \chi_{p\,\text{W}}^+ + \chi_{d\,\text{W}}^+ + \chi_{p\,\text{C}}^+ + \chi_{d\,\text{C}}^+ = (10.9 \pm 2.0) \times 10^{-6}\,\text{cm}^3\text{·mol}^{-1}$ is slightly smaller than the measured susceptibility $\chi(300) = (11.4 \pm 1.0) \times 10^{-6}\,\text{cm}^3\text{·mol}^{-1}$ of coarse-grained carbide WC. This means that the orbital paramagnetic contribution $\chi_{\text{VV orb}}$ in coarse-grained tungsten carbide WC is small and close to zero to within experimental error.

Let us consider the susceptibility of nanocrystalline tungsten carbide.

The typical dependence of the magnetic susceptibility χ of WC-plasm nanopowder on temperature is shown in Fig. 4.24. Over the entire temperature range studied, the nanopowder susceptibility is larger than that of pure coarse-grained carbide WC. In contrast to the smooth dependence $\chi(T)$ for coarse-grained WC carbide, the susceptibility curve for the WC-plasm nanopowder exhibits noticeable features. In the temperature, $\chi(T)$, and annealing-temperature, $\chi(300,T)$, curves for WC-plasm carbide, the susceptibility begins to increase sharply at temperature $T = 550-600$ K and reaches a maximum at ~820–870 K. Then the susceptibilities $\chi(T)$ and $\chi(300,T)$ of WC-plasm carbide decrease noticeably at 870–920 K. As the temperature increases further from 920–930 to 1220 K, the susceptibilities $\chi(T)$ and $\chi(300,T)$ decrease only slightly. The susceptibility $\chi(T)$ measured at $T > 600$ K is less than the susceptibility $\chi(300,T)$ measured at 300 K after the cooling of the carbide from temperature T. In a range of 300–600 K, the susceptibilities $\chi(T)$ and $\chi(300,T)$ coincide within the measurement errors (Fig. 4.24). When WC-plasm carbide is cooled from 1220 to 300 K and then is repeatedly heated to 1200 K, the susceptibility varies monotonically along curve *4* in Fig. 4.24 without any singularities.

Oxidation of WC-plasm nanopowder during susceptibility measurement could be one of the possible reasons of changes of magnetic susceptibility. According to [101, 102], in air the oxidation of WC nanopowders with the particle size of 20–60 nm begins at a temperature of 710–720 K and comes to an end at 770–790 K. The higher oxide WO_3 is a single oxidation product. However the temperature of the beginning of oxidation is much more than temperature ~600 K at which the sharp increasing χ is observed. Besides, tungsten oxides are diamagnetics with susceptibility about $-0.07 \times 10^{-6}\,cm^3{\cdot}g^{-1}$ ($-15 \times 10^{-6}\,cm^3{\cdot}mol^{-1}$), in this case magnetic susceptibility decreases during oxidation while experimental susceptibility increases. Susceptibility decreasing at temperature of 870–920 K also is not an effect of oxidation also. If the temperature 870 K was the temperature of the beginning of oxidation, the measured susceptibility would continuously be decreased at $T > 870$ K. In experiment sharp decreasing χ is completed at 920 K. Thus, oxidation does not cause the effects observed during susceptibility measurement of WC-plasm powder.

According to [103, 104], during vacuum annealing WC nanopowders start to lose carbon at temperature more than 870 K. Decarburization is caused by an interaction of carbon with impurity oxygen and leads to the variation in phase composition of WC nanopowders and to appearance of tungsten W in these powders. The quantity of precipitated tungsten changes from 0 to ~20 wt.% at vacuum annealing temperature of 870–1070 K and is equal to ~20 wt.% during vacuum annealing of WC-plasm nanopowder in the temperature range 1070–1470 K [103]. According to the generalized data [105–108], tungsten metal is a weak paramagnet with the temperature dependence of susceptibility $\chi(T) = \chi(0) + BT^2$ where $\chi(0) \approx 0.29 \times 10^{-6}\,cm^3 \cdot g^{-1}$ and $B \approx 1.6 \times 10^{-14}\,cm^3{\cdot}g^{-1}{\cdot}K^{-2}$. Taking this into account, at a temperature from 870 to 1070 K the contribution of tungsten to the susceptibility of WC-plasm nanopowder increases from 0 to $\sim 0.05 \times 10^{-6}\,cm^3{\cdot}g^{-1}$ and in the temperature range 1070–1250 K the presence of ~20 wt.% of tungsten metal in WC-plasm carbide gives an additional constant contribution $\sim 0.05 \times 10^{-6}\,cm^3{\cdot}g^{-1}$

to the magnetic susceptibility of the carbide. As a result, the magnetic susceptibility of WC-plasm carbide must reach $\sim(0.13–0.14)\times10^{-6}\,\text{cm}^3\cdot\text{g}^{-1}$ in the temperature range 1070–1250 K.

However, a spasmodic not continuous increase of susceptibility is observed in experiment. The sharp increasing χ starts at temperature ~600 K which is much less than temperature of the beginning of decarburization of WC nanopowders. Experimental value of a susceptibility jump is larger than that due to tungsten precipitation. Besides, decreasing susceptibility which is observed in experimental dependences $\chi(T)$ and $\chi(300, T)$ at temperature ~900 K can be connected with the presence of tungsten in no way. Therefore decarburization of WC-plasm and caused by this decarburization the subsequent tungsten metal precipitation are not the reasons of the effects observed during susceptibility measurement of this carbide.

Let us propose other possible reasons of the effects observed during susceptibility measurement of nanocrystalline tungsten carbide powder.

These effects can be associated with the changes in the magnetic properties and electronic structure of tungsten carbide or can be caused by relaxation of microstrains and presence of impurities. The diffraction data indicate that the size of coherent scattering regions of WC-plasm carbide remains almost unchanged during susceptibility measurements, while microstrains decrease almost to zero. It can be assumed that it is the relaxation annealing of microstrains that leads to the effects observed in the susceptibility. Note that the susceptibility of the initial nanocrystaline WC-plasm carbide (before χ measurement) even at 300 K is larger than that of coarse-grained carbide WC. The significant difference in the magnitudes of the susceptibility χ of pure coarse-grained (curve *1*) and annealed nanocrystalline (curve *4*) tungsten carbides (see Fig. 4.24) indicates a significant paramagnetic contribution in the susceptibility of annealed WC-plasm carbide.

According to the spectral analysis data, nanocrystalline WC-plasm carbide contains minor iron and cobalt impurities. However, measurements did not reveal a dependence of the susceptibility χ of the initial nanocrystalline tungsten carbide on magnetic field H. From this it follows that Fe and Co impurities are present in the initial WC-plasm carbide not in the ferromagnetic state. This is possible if these impurities are dissolved in the initial WC-plasm nanopowder. In this case, the contribution from these impurities to the susceptibility is paramagnetic rather than ferromagnetic. Precipitation of impurities as superparamagnetic particles also makes a paramagnetic contribution. However, at $T < T_C$ (T_C is the Curie temperature) this contribution will be larger than that from the same amount of dissolved impurities.

Let us discuss the following qualitative explanation of the susceptibility of nanocrystalline WC-plasm carbide.

The initial carbide WC-plasm has appreciable microstrains ε of up to 0.06 %. Any lattice strains reduce the symmetry of electron shells and increase the probability of electronic transitions from the ground to excited states. Therefore, in the presence of strains, quantities $\left|s|\hat{\mathrm{M}}|0\right| > 0$ and $\left[\left|s|\hat{\mathrm{M}}|0\right|\right]^2 > 0$ are positive and the susceptibility contains an additional positive contribution $\Delta\chi_{\text{orb}}$ of Van Vleck orbital paramagnetism. In addition, there is a small paramagnetic contribution from Fe and

Co impurities dissolved in the carbide lattice. With inclusion of these additional contributions, the susceptibility of nanocrystalline WC-plasm carbide is larger than that of pure coarse-grained carbide WC. It is well known that deformation of the lattice or a change in strains in it leads to precipitation of dissolved metal impurities. For example, precipitation of Fe particles from solutions of iron in copper or titanium was detected in works [93, 95, 109]. It can be assumed that annealing of microstrains, which begins in WC-plasm carbide at 570–600 K, initiates precipitation of Fe and Co impurities in the form of small superparamagnetic particles. As the temperature increases from 600 to 800–820 K, the number and volume of superparamagnetic particles increase, which is accompanied by an increase in the susceptibility up to a maximum value. At temperature ~820–870 K, precipitation of superparamagnetic particles terminates, but microstrain annealing continues up to ~920 K. The decrease in χ in the range 870–920 K is not an impurity effect but rather is inherent in carbide WC-plasm; namely, it is due to a decrease of the orbital paramagnetism $\chi_{\mathrm{VV\ orb}}$. Indeed, microstrains relaxation occurring at 870–920 K reduces distortions of the local symmetry of atomic environment and strain-induced distortions of the symmetric electron shells of ion cores. From this it follows that the probability of electron transition from the ground state to the excited state decreases to zero, therefore as a result of relaxation of microstrains at 870–920 K quantity $\left[\left|s|\hat{\mathrm{M}}|0\right|\right]^2$ vanishes, i.e., $\left[\left|s|\hat{\mathrm{M}}|0\right|\right]^2 \to 0$. As a result, the orbital paramagnetic contribution $\Delta\chi_{\mathrm{orb}}$ vanishes and the susceptibility decreases. The value of $\Delta\chi_{\mathrm{orb}}$ as estimated from the change in $\chi(300,T)$ in the temperature range 870–920 K (Fig. 4.24, curve *3*) is $\sim 3 \times 10^{-8}\,\mathrm{cm^3{\cdot}g^{-1}}$. At $T > 920$ K, there are no microstrains in WC-plasm carbide and the decrease in χ is caused only by the temperature dependence of impurity superparamagnetism.

Thus, the difference between the susceptibilities of the initial WC-plasm and coarse-grained carbide WC, $\Delta\chi(T) = \chi_{\mathrm{WC\text{-}plasm}}(T) - \chi_{\mathrm{WC}}(T)$, in the range 300–570 K (curves *2* and *1* in Fig. 4.24) is the sum of the paramagnetic susceptibility $\chi^{(m)}_{\mathrm{p\text{-}imp}}(T)$ of dissolved Fe and Co impurities and the orbital paramagnetism $\Delta\chi_{\mathrm{orb}}$, i.e., $\Delta\chi(T) = \chi^{(m)}_{\mathrm{p\text{-}imp}}(T) + \Delta\chi_{\mathrm{orb}}$. Hence, $\chi^{(m)}_{\mathrm{p\text{-}imp}}(T) = \Delta\chi(T) - \Delta\chi_{\mathrm{orb}}$, or

$$\chi^{(m)}_{\mathrm{p\text{-}imp}}(T) = [\chi_{\mathrm{WC\text{-}plasm}}(T) - \Delta\chi_{\mathrm{orb}}] - \chi_{\mathrm{WC}}(T). \tag{4.54}$$

The magnetization of a dissolved impurity is described by the dependence characteristic of Curie paramagnets. Going from the magnetization *M* to the susceptibility χ, the dimensionless paramagnetic contribution from one impurity at a temperature *T* in a magnetic field *H* can be written [97] as

$$\chi_{\mathrm{p\text{-}imp}}(T) = \frac{n\mu^{(\mathrm{imp})}_{\mathrm{eff}}}{H} L\left(\frac{\mu^{(\mathrm{imp})}_{\mathrm{eff}} H}{k_{\mathrm{B}} T}\right), \tag{4.55}$$

where n is the number of impurity atoms per unit volume, $\mu_{\mathrm{eff}}^{(\mathrm{imp})} = p_{\mathrm{imp}}\mu_{\mathrm{B}}$ is the effective magnetic moment of an impurity atom, and $L(x) = [\coth(x) - 1/x]$ is the Langevin function. The mass susceptibility (i.e., per unit mass) of an impurity is $\chi_{\text{p-imp}}^{(m)}(T) = \chi_{\text{p-imp}}(T)/\rho_i$. Therefore, the paramagnetic contribution from several different impurities (per unit mass) at a temperature T in a magnetic field H is

$$\chi_{\text{p-imp}}^{(m)}(T) = \sum_i \frac{n_i \mu_{\mathrm{eff}\,i}}{\rho_i H} L\left(\frac{\mu_{\mathrm{eff}\,i} H}{k_{\mathrm{B}} T}\right), \tag{4.56}$$

where ρ_i is the density of the ith impurity metal (subscript i specifies the impurity type). If the relative (by mass) impurity content is $c_i (c_i \ll 1)$, then the relative volume content is $c_{\mathrm{V}i} \approx c_i \rho_{\mathrm{WC}}/\rho_i$ and $n_i = c_{\mathrm{V}i} N_{\mathrm{A}} \rho_i / A_i = c_i N_{\mathrm{A}} \rho_{\mathrm{WC}}/A_i$ (A_i is the atomic mass of the ith impurity and $\rho_{\mathrm{WC}} = 15.8\,\mathrm{g\cdot cm^{-3}}$ is the density of carbide WC). Let us equate the right-hand sides of (4.54) and (4.56) and substitute for n_i its expression in terms of the impurity concentration, $n_i = c_i N_{\mathrm{A}} \rho_{\mathrm{WC}}/A_i$. As a result it follows the formulae (4.54) and (4.56) that the difference between the susceptibilities of the initial WC-plasm carbide (without orbital paramagnetism) and coarse-grained WC carbide, i.e. the mass susceptibility of an impurity at 300–570 K can be written as

$$\begin{aligned}\chi_{\text{p-imp}}^{(m)}(T) &\equiv (\chi_{\text{WC-plasm}}(T) - \Delta\chi_{\mathrm{orb}}) - \chi_{\mathrm{WC}}(T) \\ &= \frac{\rho_{\mathrm{WC}} N_{\mathrm{A}} \mu_{\mathrm{B}}}{H}\left[\frac{c_{\mathrm{Fe}} p_{\mathrm{Fe}}}{\rho_{\mathrm{Fe}} A_{\mathrm{Fe}}} L\left(\frac{p_{\mathrm{Fe}}\mu_{\mathrm{B}} H}{k_{\mathrm{B}} T}\right) + \frac{c_{\mathrm{Co}} p_{\mathrm{Co}}}{\rho_{\mathrm{Co}} A_{\mathrm{Co}}} L\left(\frac{p_{\mathrm{Co}}\mu_{\mathrm{B}} H}{k_{\mathrm{B}} T}\right)\right],\end{aligned} \tag{4.57}$$

where $p_{\mathrm{Fe}} = 2.22$ [110], $\rho_{\mathrm{Fe}} = 7.86\,\mathrm{g\cdot cm^{-3}}$, $A_{\mathrm{Fe}} = 55.85$, $p_{\mathrm{Co}} = 1.72$ [110], $\rho_{\mathrm{Co}} = 8.9\,\mathrm{g\cdot cm^{-3}}$, and $A_{\mathrm{Co}} = 58.93$. The experimental values of the difference between the susceptibilities of the initial nanocrystalline and coarse-grained tungsten carbides, $\chi_{\text{p-imp}}^{(m)}(T) = [\chi_{\text{WC-plasm}}(T) - \Delta\chi_{\mathrm{orb}}] - \chi_{\mathrm{WC}}(T)$, in the range 300–570 K are shown in Fig. 4.25a. Function (4.57) fits well the experimental values $\chi_{\text{p-imp}}^{(m)}(T)$ (Fig. 4.25a) for $c_{\mathrm{Fe}} = (3.6 \pm 0.5) \times 10^{-5}$ ($\sim$0.003–0.004 wt.%) and $c_{Co} = (7.3 \pm 0.5) \times 10^{-5}$ ($\sim$0.007–0.008 wt.%), which is in satisfactory agreement with the spectral analysis data.

If the sharp change in $\chi(T)$ of WC-plasm carbide at 600–820 K is caused by precipitation of impurity superparamagnetic particles, then the experimental contribution $\chi_{\mathrm{sp}}^{(m)}(T)$ to the susceptibility from the superparamagnetic phase is equal to the difference between the susceptibilities $\chi_{\text{WC-plasm}}^{(\mathrm{ann})}(T)$ of annealed WC-plasm carbide (Fig. 4.24, curve *4*) and $\chi_{\mathrm{WC}}(T)$ of coarse-grained WC carbide (Fig. 4.24, curve *1*), i.e.,

$$\chi_{\mathrm{sp}}^{(m)}(T) = \chi_{\text{WC-plasm}}^{(\mathrm{ann})}(T) - \chi_{\mathrm{WC}}(T). \tag{4.58}$$

On the other hand, the superparamagnetic contribution (per unit mass) at temperature T in a magnetic field H is

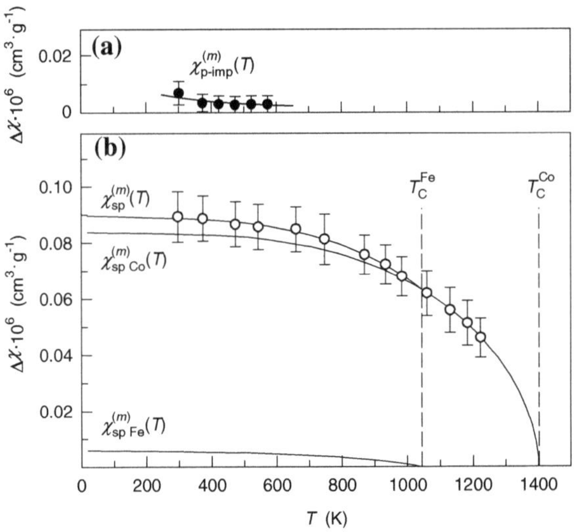

Fig. 4.25 Temperature dependences of the mass (i.e., per unit mass) magnetic susceptibility χ of impurity iron and cobalt phases in nanopowder WC-plasm. **a** experimental magnetic susceptibility (•) of dissolved Fe and Co, $\chi^{(m)}_{\text{p-imp}}(T) = [\chi_{\text{WC-plasm}}(T) - \Delta\chi_{\text{orb}}] - \chi_{\text{WC}}(T)$, is described by function (4.58) with $c_{\text{Fe}} = 0.000036$ and $c_{\text{Co}} = 0.000073$. (**b**) the experimental contribution (∘) from superparamagnetic Fe and Co impurity phases, $\chi^{(m)}_{\text{sp}}(T) \equiv \chi^{(m)}_{\text{sp Co}}(T) + \chi^{(m)}_{\text{sp Fe}}(T) = \chi^{(\text{ann})}_{\text{WC-plasm}}(T) - \chi_{\text{WC}}(T)$, to the susceptibility of carbide WC-plasm is fitted well by function (4.61). Approximating curves are shown by *solid lines*. *Vertical dashed lines* are Curie temperatures $T_C^{(\text{Fe})} = 1043$ K and $T_C^{(\text{Co})} = 1403$ K of iron and cobalt

$$\chi^{(m)}_{\text{sp}}(T) = \sum_i \frac{n_{\text{sp}\,i}\mu_{\text{sp}\,i}}{\rho_{\text{M}_i} H} L\left(\frac{\mu_{\text{sp}i} H}{k_{\text{B}} T}\right), \tag{4.59}$$

where $n_{\text{sp}\,i}$ is the number of superparamagnetic particles of type i per unit volume. If the volume $V^{(i)}_{\text{sp}}$ of a superparamagnetic particle of type i is independent of temperature, then $\mu_{\text{sp}\,i} = V^{(i)}_{\text{sp}} M^{(i)}_{\text{s}}(T)$. The saturation magnetization of the ith ferromagnet at temperature T is given by

$$M^{(i)}_{\text{s}}(T) = M^{(i)}_{\text{s}}(0)\tanh\left(\frac{M^{(i)}_{\text{s}}(T) T^{(i)}_{\text{C}}}{M^{(i)}_{\text{s}}(0) T}\right), \tag{4.60}$$

where $M^{(i)}_{\text{s}}(0)$ is the saturation magnetization at 0 K, $T^{(i)}_{\text{C}}$ is the Curie temperature of the i-th ferromagnet (for crystalline iron and cobalt, $M^{(\text{Fe})}_{\text{s}}(0) = 1740$ Gs, $T^{(\text{Fe})}_{\text{C}} = 1043$ K, $M^{(\text{Co})}_{\text{s}}(0) = 1400$ Gs and $T^{(\text{Co})}_{\text{C}} = 1403$ K [97, 110]).

Using the relation $\chi^{(m)}_{\text{sp}}(T) \equiv \chi^{(m)}_{\text{sp Co}}(T) + \chi^{(m)}_{\text{sp Fe}}(T) = \chi^{(\text{ann})}_{\text{WC-plasm}}(T) - \chi_{\text{WC}}(T)$ and (4.59), we can write that

$$\chi^{(\mathrm{ann})}_{\mathrm{WC\text{-}plasm}}(T) - \chi_{\mathrm{WC}}(T) = \sum_i \frac{n_{\mathrm{sp}i} V^{(i)}_{\mathrm{sp}} M^{(i)}_{\mathrm{s}}(T)}{\rho_{\mathrm{M}_i} H} L\left(\frac{V^{(i)}_{\mathrm{sp}} M^{(i)}_{\mathrm{s}}(T) H}{k_{\mathrm{B}} T}\right). \quad (4.61)$$

Summation in (4.61) is carried out over two species of impurity atoms, Fe and Co. The dependence of the saturation magnetization $M^{(i)}_{\mathrm{s}}(T)$ on temperature is a transcendent function; therefore, in order to simplify minimization of expression (4.61), the dependences $M^{(\mathrm{Fe})}_{\mathrm{s}}(T)$ and $M^{(\mathrm{Co})}_{\mathrm{s}}(T)$ were preliminarily calculated numerically. The calculated temperature dependences of the iron and cobalt magnetizations were used to minimize expression (4.61) and thereby determine the number and volume of superparamagnetic iron and cobalt particles. Fitting the experimental dependence $\chi^{(\mathrm{ann})}_{\mathrm{WC\text{-}plasm}}(T) - \chi_{\mathrm{WC}}(T)$ by function (4.61) made it possible to determine the contributions of each superparamagnetic phase to the impurity susceptibility (Fig. 4.25b) and to estimate the volumes $V^{(\mathrm{Fe})}_{\mathrm{sp}}$ and $V^{(\mathrm{Co})}_{\mathrm{sp}}$ of superparamagnetic iron and cobalt particles as 4.0×10^{-20} and $4.0 \times 10^{-19}\,\mathrm{cm}^3$; the corresponding linear particle sizes are ~3 and ~7 nm, respectively. The number of superparamagnetic iron and cobalt particles per unit volume is $n_{\mathrm{sp\ Fe}} = (5.9 \pm 1.0) \times 10^{12}$ and $n_{\mathrm{sp\ Co}} = (1.2 \pm 0.2) \times 10^{13}\,\mathrm{cm}^{-3}$, respectively. As seen in Fig. 4.25b, cobalt makes the largest and main impurity contribution to the susceptibility of nanocrystalline tungsten carbide. These calculations are only estimates, since they do not consider the Fe and Co solubility in tungsten carbide. Besides, the small constant paramagnetic contribution associated with the precipitation of tungsten W from WC-plasm carbide during susceptibility measurement at temperature more than 870 K does not take into account in these calculations.

Thus, hexagonal tungsten carbide WC in coarse-grained and nanocrystalline states is a weak paramagnet at temperatures of 300 to ~1250 K. The particle size in a nanocrystalline WC-plasm powder is ~55 nm and remains unchanged upon annealing in the temperature range 300 to ~1050 K, whereas total relaxation of microstrains occurs at 600–900 K. The jumps observed in the temperature dependences $\chi(T)$ and $\chi(300,T)$ of the magnetic susceptibility of nanocrystalline tungsten carbide are caused by relaxation annealing of microstrains. This annealing leads to decreasing the contribution from the orbital paramagnetism to the susceptibility of WC-plasm carbide and initiates precipitation of impurity iron and cobalt as superparamagnetic particles. The particle size of nanocrystalline tungsten carbide has no direct effect on the magnetic susceptibility, since the particle size (~55 nm) of WC-plasm nanopowder is not sufficiently small to lead to discrete electronic states in tungsten carbide. This effect could be expected, if the tungsten carbide particle size would not exceed 5 nm [14, 15]. During susceptibility measurement at temperature more than 870 K partial decarburization of WC-plasm nanopowder occurs. This decarburization is accompanied by the precipitation of tungsten W which makes small constant paramagnetic contribution to measured susceptibility at $T > 870$ K.

4.3.2 Oxidation of WC Powders in Air

Nanocrystalline powders of different substances can be used as separate materials and as precursors for nanostructured materials if in certain temperature intervals such powders possess stability of composition and properties in external media. For powders of transition metal carbides and especially tungsten carbide powders used in the production of hardmetals, it is necessary to know their stability against air oxygen. The oxidation of refractory carbides and nitrides, composite materials and alloys on their basis was studied repeatedly [111–116] and the obtained results even are generalized in handbooks [69, 117] but the overwhelming majority of studies is performed on compact sintered samples. The oxidation of tungsten carbide WC powders was investigated in [111, 113, 116]. According to [111], the oxidation of fine WC powder in air begins at temperature 790–800 K. According to Dufour and Simon [113], air oxidation of WC powder with a specific surface area of $0.37\,m^2\cdot g^{-1}$, corresponding to an average particle size of 1000 nm, begins at 770 K and occurs in the temperature range 790–920 K. Tungsten carbide oxidizes more rapidly than tungsten metal. Ribeiro et al. [116] investigated the oxidation of WC powder with an average particle size of 10 μm at several heating rates and determined a kinetic model for the oxidation process and the activation energy of oxidation. However, the dependence of oxidation of WC powders on the average particle size $\langle D \rangle$ in micrometer and nanometer ranges has not been studied up to 2010. For the first time, the effect of particle size on the air oxidation of hexagonal tungsten monocarbide powders was studied in works [101, 102]. The content of these articles with some changes and additions is described in this section.

Coarse-grained WC-coarse powder (Kirovgrad Plant of Hard Alloys, "KZTS", Kirovgrad, Russia) with the average particle size ~6000 nm, submicrocrystalline WC-submicro powder ($\langle D \rangle = 500$ nm, "Treibacher", Austria), and nanocrystalline WC-plasm and WC-mill powders obtained by plasma-chemical synthesis and high-energy ball milling, respectively, were used. The average particle size $\langle D \rangle$ for these nanopowders is equal to 55 and 20 nm, respectively. The procedures for plasma-chemical synthesis and high-energy ball milling used to prepare WC nanopowders are described in Sect. 4.2 (see also articles [59–64, 89–91]).

The composition of tungsten carbide powders was determined by chemical and spectral analyses. The oxygen and free (uncombined) carbon contents in tungsten carbide obtained by plasma-chemical synthesis are equal to 0.14 and 2.16 wt.%, respectively. In other WC powders under study, free carbon was absent and the oxygen content was up to 0.1 wt.%. The nitrogen content was less than 0.02 wt.% in all the WC powders. Powder WC-submicro contained alloying additions of VC (0.4 wt.%) and Cr_2C_3 (1.0 wt.%) carbides along with tungsten carbide WC.

The crystal structure and phase composition of WC powders with different particle sizes was examined using the X-ray diffraction (XRD) on DRON-UM1 and Siemens D-500 diffractometers. All the X-ray powder diffraction measurements were performed by the Bragg-Brentano method in the $CuK\alpha_{1,2}$ radiation in angle interval

2θ from 10° to 140° with scan step of 0.03° and 0.02° and a scanning time from 2 to 10 s in a point.

The particle size in WC powders was determined using JEOL 6310 and LEO DSM 982 Jemini scanning electron microscopes. Particle size distributions were obtained by the laser diffraction method on a HORIBA LA-950V2 Analyzer. Electron-microscopic examination showed that the average particle size of the coarse-grained WC-coarse powder was 6000 nm. The particles of WC nanopowders prepared by plasma-chemical synthesis and high-energy ball milling were agglomerated. According to electron microscopy data, the plasma-chemical WC-plasm nanopowder consists of large, very loose agglomerates ranging in size up to 10 μm. The agglomerate size in the ball-milled WC-mill nanopowder is equal to 100–150 nm as determined by the laser diffraction method and 200–400 nm according to electron microscopy data. The microstructure of studied tungsten carbide powders is described in detail in Sects. 4.2.1 and 4.2.2 and also in works [62, 118]. Since the WC nanopowders were highly agglomerated, the particle size (more precisely, the size of coherent scattering regions) was estimated by XRD method.

The size of the coherent scattering regions in WC nanopowders is determined from the broadening of the diffraction reflections. Method for determination of broadening, separation of the size and strain contributions to the measured broadening and estimation of average particle size is described in detail in Sect. 4.1.3 and works [14, 22, 23]. Our estimation showed that the average particle size $\langle D \rangle$ is equal to 55 ± 10 nm for the nanocrystalline WC-plasm powder produced by plasma-chemical synthesis and 20 ± 10 nm for the nanocrystalline WC-mill powder prepared by high-energy ball milling of the coarse-grained WC powder.

The effect of the dispersity of hexagonal WC carbide powders on their oxidation in air was studied by differential thermal and differential thermogravimetric (DTA-DTG) analyses on a Shimadzu DTG-60 thermoanalyzer. Measurements were carried out in the temperature interval 323–1173 K at a heating rate of 10 K·min^{-1}.

For quantitative evaluation of the heat of WC oxidation, we preliminarily studied the oxidation of hafnium (Hf) metal powder. Hafnium was chosen as the standard because Hf undergoes no phase transitions between 300 and 2100 K. Besides, only HfO_2 oxide having one structural modification is formed during Hf oxidation. Therefore the specific (per 1 g of hafnium) thermal effect $\Delta H = -7.90$ kJ·g^{-1} measured in the oxidation of Hf is due only to the formation of HfO_2. The experimental sample weight increment after oxidation of hafnium is equal to 18.01 % and is close to the theoretical value of weight increment $\Delta m = (M_{HfO_2} - M_{Hf})/M_{HfO_2} = 17.93\,\%$. X-ray phase analysis confirmed that the oxidation products contain only HfO_2 oxide. The experimental thermal effect was compared with the theoretical heat of oxidation $\Delta H_{HfO_2}(T) = -1123.7 + 0.01771T + 1.172 \times 10^{-6}T^2 + 1289.5T^{-1}$ kJ·mol^{-1} [72]. Taking the Hf oxidation temperature to be 946.8 K (the measured temperature of the DTA peak), we obtain $\Delta H_{HfO_2}(T = 946.8\text{K}) = -1104.5$ kJ·mol^{-1} (or $\Delta H_{sp\,HfO_2} = -6.18$ kJ·g^{-1} per 1 g of Hf). A comparison of the theoretical (−6.18 kJ·g^{-1}) and experimental (−7.90 kJ·g^{-1}) specific thermal effects of oxidation of hafnium allowed us to determine the coefficient $k = 0.78$ for recalculating the

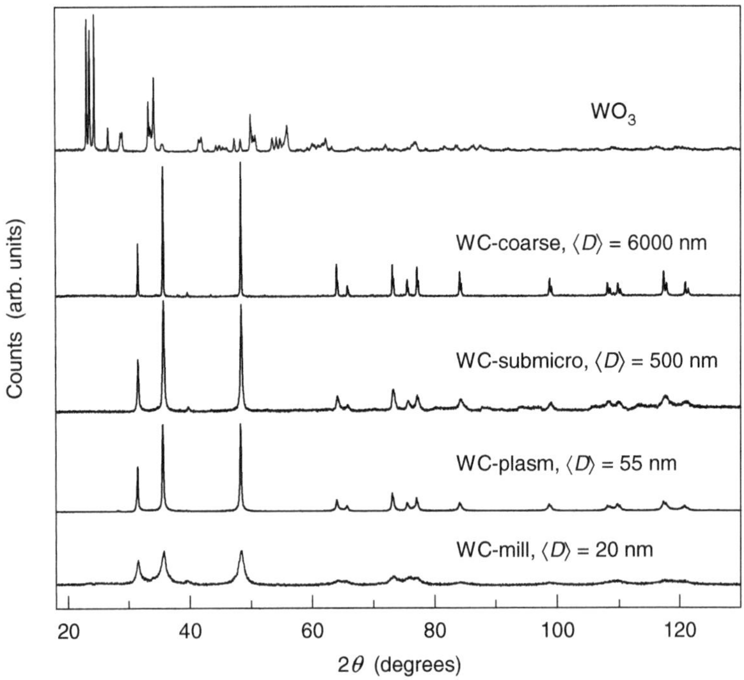

Fig. 4.26 XRD patterns of the hexagonal (space group $P\overline{6}m2$) WC tungsten carbide powders with different particle sizes and the monoclinic (space group $P12_1/n1(C_{2h}^5)$) higher WO_3 tungsten oxide formed during the oxidation of the tungsten carbide powders. The broadening of the diffraction reflections of the nanocrystalline WC powders is due to the small size of particles and microstrains. XRD measurements are performed using $CuK\alpha_{1,2}$ radiation

absolute value of the thermal effect in the oxidation of tungsten carbides powders, i.e. $\Delta H_{sp} = k \cdot \Delta H_{sp\ measured}$.

Figure 4.26 shows the XRD patterns in the 2θ range 20°–130° for four WC tungsten carbide powders with different particle sizes under study and the powder obtained after the oxidation. All the initial carbide powders contained only hexagonal (space group $P\overline{6}m2$) tungsten carbide WC with unit cell constants $a = 0.2906$ nm and $c = 0.2838$ nm. The diffraction reflections of the nanopowders obtained by ball milling and plasma-chemical synthesis are very broadened compared with the reflections of coarse-grained and submicrocrystalline tungsten carbides.

X-ray phase analysis (see Fig. 4.26) showed that the products of oxidation of the WC powders contain only tungsten oxide phase but no traces of tungsten W or lower tungsten carbide W_2C. According to the chemical analysis data, the oxidation products contain no carbon. Tungsten metal forms six oxide phases including monoclinic oxide WO_2 and five modifications of the higher oxide WO_3. Structural modifications WO_3 which are stable in different temperature intervals are low-temperature monoclinic phase stable at temperature $T < 230$ K, triclinic phase stable at $230 < T < 290$ K, monoclinic phase stable at $290 < T < 600$–700 K, orthorhombic phase stable in the temperature range $700 < T < 1000$ K, and tetragonal phase stable at temperature $T > 1010-1040$ K [119–125].

It follows from the results of XRD analysis that the product of oxidation of all four WC powders is the perovskite-like monoclinic (space group No 14 - $P2_1/c(P12_1/n1)$) higher oxide WO_3 [10], which is stable in the temperature range 290–700 K. The crystal structure of this tungsten oxide is described in work [125]. Thus WC was oxidized to WO_3 in all cases. For quantitative analysis of the oxidation and comparison of the calculated and experimental weight increments Δm and specific heats of oxidation ΔH_{sp}, we consider possible oxidation reactions of WC carbides to carbon oxides and WO_2 or WO_3:

$$WC + (3/2)O_2 = WO_2 + CO \tag{4.62}$$
$$WC + 2O_2 = WO_2 + CO_2 \tag{4.63}$$
$$WC + 2O_2 = WO_3 + CO \tag{4.64}$$
$$WC + (5/2)O_2 = WO_3 + CO_2 \tag{4.65}$$

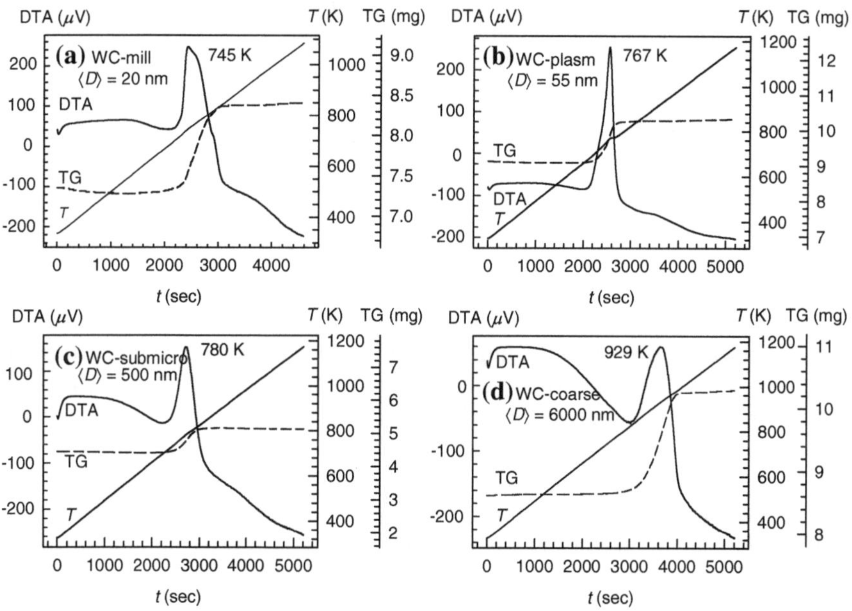

Fig. 4.27 Differential thermal and thermogravimetric analyses of the oxidation of tungsten carbide WC powders with different particle sizes $\langle D\rangle$: **a** WC-mill nanopowder with a particle size of ~20 nm, produced by ball milling of coarse-grained WC-coarse powder; **b** WC-plasm nanopowder with a particle size of ~55 nm, prepared by plasma-chemical synthesis; **c** submicrocrystalline WC-submicro powder ("Treibacher", Austria) with a particle size of ~500 nm; **d** coarse-grained WC-coarse powder ("KZTS", Kirovgrad, Russia) with an average particle size of 6000 nm

The experimental thermograms of the oxidation of WC powders with different particle size are shown in Fig. 4.27.

Table 4.3 lists the quantitative characteristics, including the sample weight increment after oxidation, Δm, and the specific heat of oxidation ΔH_{sp} (per 1 g of the oxidized WC carbide).

Measurements showed that the oxidation of tungsten carbide with any particle size is accompanied by a strong exothermal effect and an increase in the sample weight increment Δm (Fig. 4.27). The magnitude of Δm depends on the mass m of the oxidized carbide and the ratio of the molecular masses M of the final and starting solid reagents (WC and WO_2 for reactions (4.62) and (4.63) or WO_3 for reactions (4.64) and (4.65)). The theoretical weight increment is $\Delta m = m(M_{WO_2}/M_{WC}) - m = 0.102\,\mathrm{m}$ for reactions (4.62) and (4.63) and $\Delta m = m(M_{WO_3}/M_{WC}) - m = 0.184\,\mathrm{m}$ for reactions (4.64) and (4.65). As follows from Table 4.3, the experimental weight increments are closest to the calculated values corresponding to the oxidation by (4.64) and (4.65) reactions (i.e., to the higher oxide WO_3). This agrees with the X-ray phase analysis data according to which the solid products of oxidation contain only stoichiometric WO_3.

The oxidation rate $V = \Delta m/\Delta t$ (where Δt is the time interval between the start and end of oxidation) increased in going from coarse-grained WC-coarse powder with ~6 μm particles to submicrocrystalline WC-submicro powder with an average particle size of ~500 nm and the WC-mill nanopowder with $\langle D \rangle = 20$ nm, produced by ball milling (Table 4.3). The reason for this growth of the oxidation rate is the increase in the specific surface of powders due to the decrease in the particle size. If we assume, in the zero approximation, that all particles are spherical and have the same size, the specific surface will be $S_{sp} = s/m = 6/\rho\langle D \rangle$ (ρ is the density of a substance) for particles with the linear size D, mass $m = (\pi/6)\langle D \rangle^3 \rho$, and surface area $s = \pi \langle D \rangle^2$ [14, 15]. In other words, the specific surface of a powder is inversely proportional to the particle size and is larger for smaller size $\langle D \rangle$. In our case, the oxidation rate V of tungsten carbide powders should be proportional to $1/\langle D \rangle$, as is just observed experimentally (Fig. 4.28). The very high oxidation rate, ~0.0046 mg·s^{-1}, of plasma-chemical WC-plasm nanopowder disobeys this tendency because of the presence of a large amount of free (uncombined) carbon.

The thermal effect $\Delta H(T)$ of the oxidation of tungsten carbide (or heat of reaction) can be calculated using such reference data as the enthalpy of formation, ΔH°_{f298}, and the temperature dependence of enthalpy, $(H^{\circ}_T - H^{\circ}_{298})$, of the initial reagents and end products; i.e.,

$$\Delta H(T) = \Sigma[\Delta H^{\circ}_{f298} + (H^{\circ}_T - H^{\circ}_{298})]_{i\,\mathrm{final}} - \Sigma[\Delta H^{\circ}_{f298} + (H^{\circ}_T - H^{\circ}_{298})]_{i\,\mathrm{initial}}. \quad (4.66)$$

To calculate the thermal effect of reactions (4.62)–(4.65), we used the data on ΔH°_{f298} and $(H^{\circ}_T - H^{\circ}_{298})$ for tungsten carbide WC, oxygen O_2, tungsten oxides WO_2 and WO_3, and carbon oxides CO and CO_2 which were taken from handbook [72] (Table 4.4). For calculation, the reaction temperature was taken equal to the experimental temperature T_p of the peak of the exothermal effect.

Table 4.3 Dependence of the oxidation parameters on the average size $\langle D \rangle$ of WC powder particles

Powder WC	$\langle D \rangle$ (nm)	$m \pm 0.01$ (mg)	Δm (mg)			V (mg·s^{-1})	T_p (K)	$\Delta H \pm 1.0$ (J)	ΔH_{sp} (kJ·g^{-1}) per 1 g of WC				
			Experiment	Calculation for reactions					Experiment	Calculation for reaction			
			±0.05	(4.62), (4.63), (4.64), (4.65)					±0.3	(4.62)	(4.63)	(4.64)	(4.65)
WC-mill	20	7.39	1.036	0.754	1.360	0.0022	745	−54.7	7.4	−3.29	−4.74	−4.62	−6.07
[a]WC−plasm	55	8.28	1.484	0.845	1.524	0.0046	767	−126.6	−15.3	−3.29	−4.74	−4.62	−6.07
WC-submicro	500	4.43	0.690	0.452	0.816	0.0018	780	−35.0	−7.9	−3.29	−4.74	−4.62	−6.07
WC−coarse	6000	8.62	1.709	0.878	1.586	0.0017	929	−68.0	−7.9	−3.29	−4.73	−4.61	−6.05

[a] Plasma-chemical nanocrystaline WC-plasm powder contained up to 2 wt.% of free carbon

Note $\langle D \rangle$ is the average particle size, m is the powder mass, Δm is the weight increment, $V = \Delta m / \Delta t$ is the oxidation rate, T_p is the temperature of the DTA peak, ΔH is the thermal effect of reaction, $\Delta H_{sp} = \Delta H / m$ is the specific thermal effect (per 1 g of WC powder)

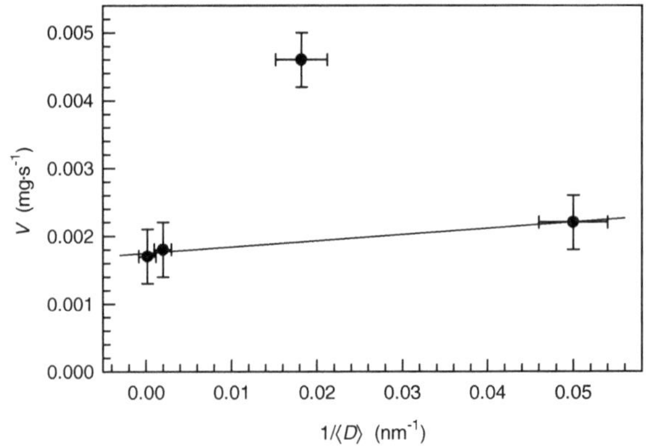

Fig. 4.28 Dependence of oxidation rate V of WC powders on quantity $\langle D\rangle^{-1}$ which is proportional to the specific surface S_{sp} of powders. Very high oxidation rate of the plasma-chemical WC-plasma nanopowder, which disobeys linear dependence $V(\langle D\rangle^{-1})$, is due to the large content of free (uncombined) carbon

Table 4.4 Enthalpy (heat) of formation ΔH°_{f298} and enthalpy $(H^{\circ}_{T} - H^{\circ}_{298})$ of substances involved in reactions (4.62)–(4.65) [72]

Reagent	Enthalpy of formation $\Delta H^{\circ}_{f298}(\mathrm{J\cdot mol^{-1}})$	Enthalpy $(H^{\circ}_{T} - H^{\circ}_{298})(\mathrm{J\cdot mol^{-1}})$
WC	−38100	$-10341 + 33.4T + 0.0045T^2$
O_2	0	$-9684 + 30.0T + 0.0021T^2 + 167472T^{-1}$
WO_2	−573592	$-16316 + 54.6T + 0.0046T^2$
WO_3	−840919	$-21624 + 73.9T + 0.0121T^2$
CO	−110599	$-8792 + 29.2T + 0.00205T^2 + 46055T^{-1}$
CO_2	−393777	$-16437 + 44.2T + 0.0045T^2 + 854107T^{-1}$

The experimental specific thermal effect of the oxidation of all WC powders (except plasma-chemical nanopowder) is approximately the same within in experimental error and is equal to $-7.7 \pm 0.3\,\mathrm{kJ\cdot g^{-1}}$ (Table 4.3). This means that all WC powders were oxidized to the same state. Indeed, according to XRD phase analysis data, only WO_3 tungsten oxide is observed after the oxidation. The specific thermal effect of the oxidation of the plasma-chemical WC-plasm powder is almost two times higher in magnitude and is equal to $-15.3\,\mathrm{kJ\cdot g^{-1}}$. This, as well as the high oxidation rate, is the consequence of the presence of 2.16 wt.% of uncombined carbon in plasma-chemical powder.

The calculated specific thermal effects $\Delta H_{sp} = \Delta H/M_{WC}$ for reactions (4.62)–(4.65) are given in Table 4.3. The absolute values of the calculated thermal effects of reactions (4.62)–(4.64) are markedly smaller than the absolute values of the experimental thermal effects. The theoretical estimates of the thermal effects of reaction (4.65) are the closest to the experimental values of ΔH_{sp}, although they are ~15 % smaller than the experimental thermal effects of the oxidation of WC carbides (see Table 4.3). A comparison of the experimental and calculated data in general indicates that the oxidation of WC carbide powders occurs by reaction (4.65). This agrees with the weight increments and XRD data of the oxidized powders. A decrease in the particle size $\langle D \rangle$ of WC powders leads to an increase in the oxidation rate and is accompanied by the reduction of the peak temperature T_p of the exothermal effect (Table 4.3), onset temperature and end temperature of oxidation.

Let us consider the reasons for the change in T_p in more detail using the methods developed by authors of [116, 126–135] for determining the kinetic parameters of heterogeneous chemical reactions from thermal analysis data. Quantitative kinetic analysis of these data is based on the general equation for the reaction rate suggested by Osawa [126] and refined in works [127–129]:

$$\frac{d\alpha}{dt} = f(\alpha)\frac{A}{\beta}\exp\left(-\frac{E}{RT}\right), \tag{4.67}$$

where α is the dimensionless degree of reaction; $f(\alpha)$ is the function that describes the kinetic model of reaction; A is the preexponential normalizing factor (frequency factor); β is the heating rate; and E is the activation energy. In (4.67) the kinetic parameters $f(\alpha)$, A, and E are interrelated and their separation is a rather complex problem. For solving this problem, mathematical procedures have been suggested [126, 129–134] that used the results of thermal measurements at several heating rates. The basic kinetic models used for describing the thermal analysis data are the one-parameter model $f(\alpha) = (1-\alpha)^n$, the two-parameter SB model $f(\alpha) = \alpha^m(1-\alpha)^n$ [135], the nucleation-growth JMA model $f(\alpha) = n(1-\alpha)[-ln(1-\alpha)]^{1-1/n}$, the two-dimensional 2D-diffusion model $f(\alpha) = 1/[-ln(1-\alpha)]$, and others [128, 129].

The oxidation of coarse-grained WC powder with a particle size $\langle D \rangle = 10\,\mu$m in air to higher oxide WO_3 was studied in [116]; the heating rate was 5, 10, 15, 17.5, and 20 K·min^{-1}. The analysis of obtained experimental data performed in work [116] was shown that the oxidation kinetics was well approximated by the SB and JMA models. According to [134],

$$A = \frac{\beta}{Q}\frac{E}{RT_p^2}\exp\left(\frac{E}{RT_p}\right), \tag{4.68}$$

where T_p is the temperature of the DTA or DTG peak, and Q is a constant for the found kinetic SB model of the reaction. At a fixed heating rate β, the frequency factor A is a constant. Therefore (4.68) can be transformed into

$$\frac{AQ}{\beta}T_p = \frac{E}{RT_p}\exp\left(\frac{E}{RT_p}\right) \text{ or } BT_p = \frac{E}{RT_p}\exp\left(\frac{E}{RT_p}\right), \tag{4.69}$$

where B is the proportionality coefficient.

Performed thermogravimetric measurements of the oxidation of WC powders with different average particle size $\langle D \rangle$ are shown that the DTA peak temperature T_p decreases nonlinearly with decreasing $\langle D \rangle$ (see Fig. 4.27). It is clear from (4.69) that this is possible at constant activation energy not depending on the particle size or when the activation energy is a certain function of particle size $\langle D\rangle$. With $T_p(D)$ and $E(D)$ dependences taken into account, (4.69) can be written in the form

$$BT_p(D) = \frac{E(D)}{RT_p(D)} \exp\left[\frac{E(D)}{RT_p(D)}\right]. \tag{4.70}$$

From the literature data on the oxidation of WC bulk samples and powders and our own experimental data it is clear that the $T_p(D)$ function should satisfy the following conditions: $T_p > 0$ at any $\langle D\rangle$; T_p should increase with increasing $\langle D\rangle$, should come up to saturation and reach the limiting value T_0 at $\langle D\rangle \to \infty$. The simplest function that satisfies these conditions has the form $T_pD = T_0[1 - c/(\langle D\rangle - d)]$, where $c$and d are the parameters of the function.

The activation energy $E(D)$ at $\langle D\rangle \to \infty$ should also go to saturation, reaching a certain limiting value E_0. Suppose $E(D) = E_0 f_E(D)$, $T_p(D) = T_0 f_T(D)$; then $E(D)/RT_p(D) \equiv (E_0/RT_0)f(D) = e_0 f(D)$, where $e_0 = E_0/RT_0 = const$ and $f(D) = f_E(D)/f_T(D)$. Therefore expression (4.70) is transformed into

$$T_p(D) = (e_0/B)f(D)\exp[e_0 f(D)], \tag{4.71}$$

for which a self-consistent solution is possible with numerical determination of the parameters of the $f(D)$ function if it is specified in the general form.

For calculation we used the temperatures T_p for four WC powders under study and the literature data of [111, 113, 116]. The temperature of the start of oxidation of the fine WC powder with an average particle size of ~3000 nm is ~800 K [111]. Since the temperature of the DTA peak is, on average, 40 K higher than the temperature of the start of oxidation, we can set that $T_p = 840 \pm 15$ K. According to [113], the oxidation of WC powder with a specific surface of $0.37\,m^2{\cdot}g^{-1}$, corresponding to an average particle size of 1000 nm, begins at temperature 770 K and hence T_p can be adopted equal to 810 ± 15 K. In the oxidation of WC powder with a particle size $\langle D\rangle = 10\,\mu m$ [116] at a heating rate of $10\,K{\cdot}min^{-1}$, T_p was equal to 990 K. Performed calculations showed that the description of the literature [111, 113, 116] and experimental T_p values by the function (4.71) is better if the model function has the general form $f(D) = 1 + a/(\langle D\rangle + b)$, where $a = -452$ nm, $b = 11816$ nm, $B = 14524\,K^{-1}$ and $e_0 = 14.15$ (for these parameter values particle size $\langle D\rangle$ in formula (4.71) is taken in nanometers). Calculation by the equation $T_p(D) = T_0[1 - c/(\langle D\rangle - d)]$ with the estimated values of T_p gave that $c = 7200$ nm, $d = -16200$ nm, and $T_0 = 1366$ K. Since $e_0 = E_0/RT_0$, from the e_0 and T_0 values we find an activation energy $E_0 = e_0RT_0 = 161\,kJ{\cdot}mol^{-1}$. The dependence of the activation energy on the average particle size is described by the equation

$$E(D) = E_0 f(D) f_T(D) = E_0\{1 - (m\langle D\rangle + n)/[(\langle D\rangle + b)(\langle D\rangle - d)]\} \tag{4.72}$$

with the parameters $m = 7652$ nm and $n = 8.914 \times 10^7\,nm^2$. The estimated activation energy E for WC powder with the average particle size 10 μm is equal to $114\,kJ{\cdot}mol^{-1}$. This almost coincided with the activation energy of the oxidation

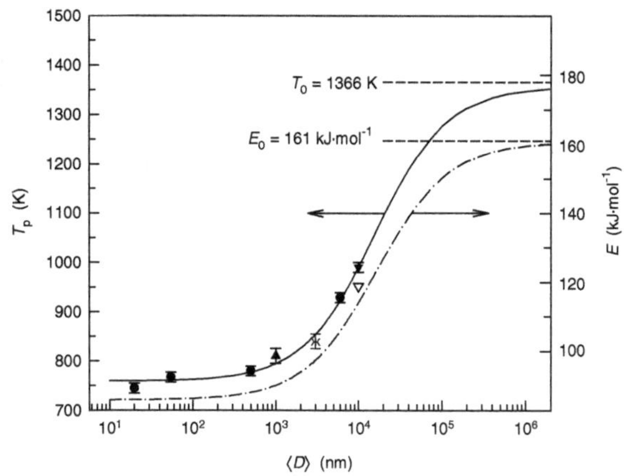

Fig. 4.29 Effect of the average size $\langle D \rangle$ of WC powder particles on the temperature T_p of the DTA peak in the oxidation curves and the activation energy E : (•) present T_p data; (×)T_p data from Newkirk [111]; (▲)T_p data from Dufour and Simon [113]; (▼) and (∇)T_p and E data, respectively, from Ribeiro et al. [116]. The *solid* and *dot-dashed lines* show the theoretical $T_p(D)$ and $E(D)$ dependences described by functions (4.71) and (4.72), respectively. Particle size $\langle D \rangle$ is shown in logarithmic scale

of this powder 120 kJ·mol^{-1} [116]. This confirms that the model used is correct. Figure 4.29 shows the experimental and literature values of T_p and E and the calculated dependences $T_p(D)$ (4.71) and $E(D)$ (4.72). As can be seen, the experiment and theory agree quite satisfactorily. For very coarse-grained powders having particles with hundreds and thousands micrometers in size, the $T_p(D)$ and $E(D)$ dependences become constant, approaching to the maximum limiting values of T_0 and E_0.

Thus the obtained data show that a decrease in the particle size $\langle D \rangle$ of WC powders leads to a decrease in the oxidation peak temperature T_p and growth of the oxidation rate.

4.3.3 Vacuum Annealing of Nanocrystalline WC Powders

Powders of WC tungsten carbide are widely used in the production of hardmetals for boring tools, high-speed cutting tools for machining of cast iron, bronzes, brass, and ceramics, as well as for application of wear-resistant fused coatings on metallic surfaces and for manufacture of reinforcing elements of coring bits and machinery parts in synthetic diamond production. Tungsten carbides are also used as catalysts of dehydration of alcohols, cyclohexane etc.

At present improving metal-cutting hardmetals is focused on the formation of these ultrafine-grained structure with carbide phase grains less than 200 nm. The use of nanocrystalline WC powders is regarded as the most promising way to produce nanostructured hardmetals with reduced sintering temperature, enhanced hardness, strength, and fracture toughness [1–4]. The size of WC particles is also important for tungsten carbide catalytic activity, which grows as the particle size decreases.

To date, different methods for production of nanocrystalline WC powder are known [14, 15, 140], among which two principally opposite techniques are of particular interest: high-energy ball milling of coarse-grained WC powder [64] relating to top-down nanotechnologies and plasma-chemical synthesis from tungsten oxide and hydrocarbon [141], i.e. a bottom-up nanotechnology. Each of the methods has its own advantages and drawbacks, but one drawback is common, namely, strong aggregation of obtained nanoparticles.

WC powders are used most extensively in the production of hardmetals, among which the most widespread materials are WC-Co hardmetals. The maximal sintering temperature of such hardmetals from microcrystalline powders by the conventional technology in vacuum or in protective or inert atmosphere ranges from 1650 to 1690 K (from 1380 to 1420 °C). Research into vacuum sintering of nanostructured hardmetals of the WC-Co system revealed a dependence between the average size of WC powder nanoparticles on the one hand and the optimal sintering temperature of hardmetals on the other hand [4, 118]. At the same time the authors [142] noted that the temperature of sintering affects the phase composition and microstructure of WC-Co hardmetals. According to [101, 102], the reduction of WC powder particle size leads to decrease in the onset temperature of oxidation and increase in the oxidation rate (see Sect. 4.3.2). Very large specific surface causes high chemical activity of nanocrystalline powders. As a result of storage in air, a surface of nanocrystalline carbide powders contains impurity oxygen in the chemisorbed form, in the surface oxide film and in adsorbed water [14, 143, 144]. The presence of impurity oxygen leads to changes of chemical and phase composition of WC nanopowders during annealing and sintering. Therefore it is important to know the thermal stability of composition and sizes of WC particles during annealing or sintering at temperatures below 1670 K (1400 °C).

The authors of works [103, 104, 145] examined how temperature influences the chemical and phase composition of nanocrystalline WC powders and size of WC nanoparticles during vacuum annealing in the temperature range from 400 to 1400 °C. These results are useful for the production of nanostructured hardmetals containing tungsten carbide nanopowders. Let us discuss the main content of articles [103, 104, 145] in more detail.

Vacuum annealing experiments were performed for three types of powders of higher hexagonal (space group $P\overline{6}m2$) tungsten carbide WC (δ-WC) with different average particle size: (1) microcrystalline powder WC-coarse made at the Kirovgrad Plant of Hard Alloys ("KZTS", Kirovgrad, Russia) with the average particle size 9 μm; (2) nanocrystalline powder WC-mill with the average particle size 20 nm produced by high-energy ball milling of WC-coarse powder in a planetary ball mill Retsch PM 200; and (3) nanocrystalline powder WC-plasm with the average particle size 60 nm, which was synthesized by the plasma-chemical method from tungsten oxide WO_3 and propane C_3H_8 in low-temperature hydrogen plasma flow in combination with thermochemical treatment (see Sect. 4.2.1). Milling conditions for WC powder are described in detail in Sect. 4.2.2 and articles [59–64], and the design of the electroarc plasmatron for plasma-chemical synthesis of tungsten carbide nanopowder is presented in Ref. [36]. Also thoroughly homogenized mixture of WC-mill nanopowder with 2.0 wt.% of carbon black MT-900 (purity 99.9 %) was used in one of the experiments; this mixture referred to as (WC-mill+C).

WC powders in the amount of 2–3 g were poured loosely into tantalum crucibles, which were placed into graphite crucibles and were annealed in 0.0013 Pa (10^{-5} mm Hg) vacuum in a furnace with tungsten lining and heaters. The annealing was performed in the following regime: slow heating (1–2 h depending on the maximal annealing temperature T_{ann}) to a temperature T_{ann}, exposure at this temperature for 1 h, and cooling with the furnace. Temperatures T_{ann} were 670, 870, 1070, 1270, 1470 and 1670 K. In each annealing experiment, only initial WC powders were used. Before vacuum annealing, all the initial WC powders were stored in air.

The phase composition of WC powders before and after annealing was examined using the X-ray diffraction (XRD) on DRON-UM1 diffractometer. All the X-ray powder diffraction measurements were performed by the Bragg-Brentano method in the Cu$K\alpha_{1,2}$ radiation in angle interval 2θ from 30 to 125° with a step of 0.03° and a scanning time of 2 s in point. The obtained XRD patterns were numerically analyzed by means of X'Pert Plus software (Version 1.0, Philips Analytical B.V.).

The particle size distribution in WC powders was determined using the laser diffraction method on a HORIBA-Laser LA-920 Analyzer. Besides, the average particle size of the powders was evaluated by means of a scanning electron microscope JEOL JSM 6390 LA. For nanocrystalline WC-mill and WC-plasm powders, dimensions $\langle D \rangle$ of X-ray coherent scattering regions (CSR) and microstrain ε values were determined by an X-ray diffraction method from reflection broadening (see Sect. 4.1.3). The average CSR dimension in the first approximation can be considered as the average size $\langle D \rangle$ of powder particles.

Chemical analysis of WC powders for total C_{total} and free (uncombined) C_{free} carbon content was performed with a METAVAC CS-30 analyzer (Russia). The content of impurity oxygen was determined by vacuum hot extraction in Balzers exhalograph EAO-201 equipment and by energy dispersion X-ray (EDX) analysis on an electron microscope JEOL-JSM LA 6390 with a JED 2300 analyzer. The content of adsorbed water in the powders was determined from the mass loss upon calcination of the powder for 1 h in vacuum 1×10^{-4} Pa at temperature 670 K. The adsorbed water content in the initial WC-coarse, WC-mill, and WC-plasm powders is equal to 0.1, 1.0, and 0.9 wt.%, respectively. In air, the mass of calcined powders increased rapidly (for 3–4 h) almost to the initial values owing to absorption of moisture from atmosphere. Spectral analysis of initial WC powders was carried out on a Perkin Elmer SCIEX-ELAN 9000 mass spectrometer. According to spectral analysis data, the content of impurity elements in the initial WC powders is almost the same and does not exceed 0.1 wt.% Na and 0.01 wt.% Fe, Ni, Mo, and Sn; the content of other impurity metals is less 0.005 wt.%.

Before vacuum annealing, all initial WC powders were certified for chemical composition and particle size.

The effect of vacuum annealing on nanocrystalline WC-mill and WC-plasm powders was studied by comparing them with microcrystalline tungsten carbide powder

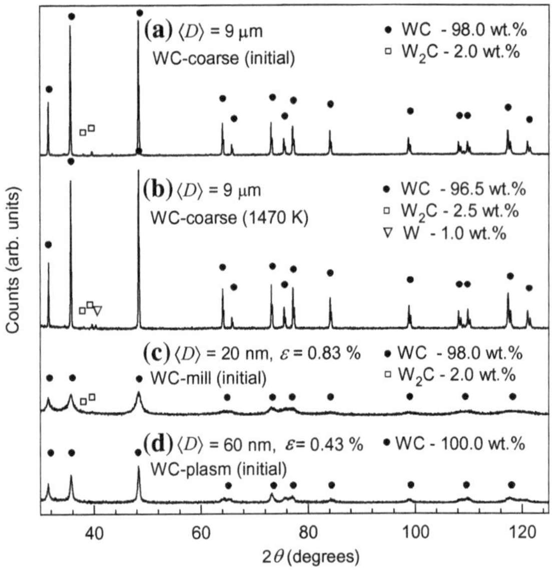

Fig. 4.30 X-ray diffraction patterns of WC powders recorded in Cu$K\alpha_{1,2}$ radiation: ***a*** initial microcrystalline pow-der WC-coarse with particles of average size 9 ± 1 μm; ***b*** WC-coarse powder after vacuum annealing at 1470 K; ***c*** initial nanocrystalline powder WC-mill with $\langle D \rangle = 20 \pm 10$ nm; ***d*** initial nanocrystalline powder WC-plasm with $\langle D \rangle = 60 \pm 10$ nm

WC-coarse used in commercial production of hardmetals with a microcrystalline structure.

Figure 4.30 displays XRD patterns of initial WC-coarse, WC-mill, and WC-plasm powders, as well as of WC-coarse powder upon vacuum annealing at 1470 K.

X-ray diffraction analysis showed that the initial microcrystalline powder WC-coarse contains a small content (~2 wt.%) of the lower tungsten carbide W_2C (Fig. 4.30a) in addition to the hexagonal (space group $P\overline{6}m2$) higher tungsten carbide WC. In the examined range of temperatures ≤1670 K, the trigonal (space group $P\overline{3}1m$) phase ε-W_2C is thermodynamically stable. According to [146, 147], it is the only ordered modification of the disordered lower tungsten carbide β-W_2C (W_2C) with the hexagonal (space group $P6_3/mmc$) structure. However, the disordered β-W_2C and ordered ε-W_2C modifications of the lower tungsten carbide are practically indistinguishable in the powder X-ray diffraction experiment. This is because they have the same hexagonal metal sublattice and the X-ray scattering amplitudes of W atoms is much higher than that of C atoms. Therefore, the lower tungsten carbide will be further designated as W_2C.

The initial nanocrystalline powder WC-mill was produced by high-energy ball milling of the microcrystalline powder, and therefore it contains a small amount of the W_2C phase (Fig. 4.30c). Spectral analysis showed that WC-mill powder contains also about 0.7 wt.% Co. This is due to small rubbing of the milling balls and cup lining made of WC-Co hardmetal, which occurs during grinding. The initial plasma-chemical WC-plasm nanopowder contains only the higher tungsten carbide WC as judged from its X-ray diffraction pattern (Fig. 4.30d).

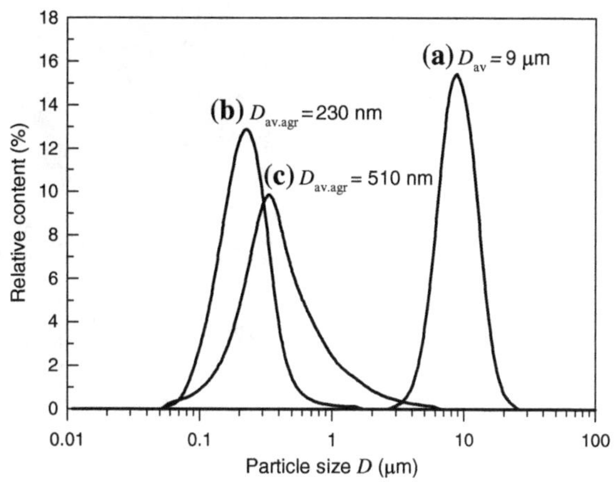

Fig. 4.31 Size distribution of particles in initial WC powders obtained by laser diffraction: ***a*** WC-coarse, $D_{av} = 9\,\mu m$; ***b*** WC-mill, $D_{av.agr} = 230\,nm$; ***c*** WC-plasm, $D_{av.agr} = 510\,nm$

Note that impurity phases in the WC nanopowders are difficult to determine from X-ray diffraction patterns (Fig. 4.30c, d) since reflections even of the basic phase WC are appreciably broadened, and in the region of angles $2\theta > 60°$ closely adjacent diffraction reflections overlap and merge. Analysis of diffraction reflection broadening for the nanopowders shows that it is due to a small size $\langle D \rangle$ of nanoparticles and the presence of microstrains ε in them. The values of $\langle D \rangle$ and ε were found from diffraction reflection broadening employing the technique described in Sect. 4.1.3 and works [14, 15, 22]. For WC-mill powder $\langle D \rangle = 20 \pm 10\,nm$ and $\varepsilon = 0.83 \pm 0.05\,\%$, and for WC-plasm powder $\langle D \rangle = 60 \pm 10\,nm$ and $\varepsilon = 0.43 \pm 0.05\,\%$. The average size $D_{av.agr}$ of aggregated particles determined by laser diffraction on a HORIBA LA-950V2 analyzer (Fig. 4.31) turned out to be almost an order of magnitude larger than the average dimension $\langle D \rangle$ of coherent scattering regions in WC nanoparticles. This is caused by aggregation of nanoparticles in large aggregates.

Scanning electron microscopy (Fig. 4.32) showed that the size of nanoparticles in WC-mill and WC-plasm powders is close to the average dimension of CSR, but the nanoparticles are actually aggregated in large (to several hundred nanometers) aggregates. According to well consistent data of laser diffraction (Fig. 4.31a) and electron microscopy (Fig. 4.32a), the average size $\langle D \rangle$ of particles in WC-coarse powder is 9000 nm. Indeed, the narrow diffraction reflections (Fig. 4.30a) are indicative of a coarse-grained powder and of the absence of microstrains.

In [14, 15], the specific surface of a powder is described by the dependence $S_{sp} = 6/\rho\langle D \rangle$, where ρ is density and $\langle D \rangle$ is the average size of powder particles. For WC, $\rho = 15.67\,g{\cdot}cm^{-3}$. Taking into consideration this value and the average size of particles of initial powders WC-coarse, WC-mill, and WC-plasm, their specific surface S_{sp} is 0.04, 19.2, and $6.4\,m^2{\cdot}g^{-1}$, respectively. The calculated value $S_{sp} = 6.4\,m^2{\cdot}g^{-1}$ for WC-plasm powder agrees well with the experimental value $S_{sp} = 6.0 - 9.5\,m^2{\cdot}g^{-1}$ determined for this powder by the BET method.

Note that the smallest content of adsorbed water (0.1 wt.%) is observed for WC-coarse powder having the smallest specific surface, whereas the greatest amount of

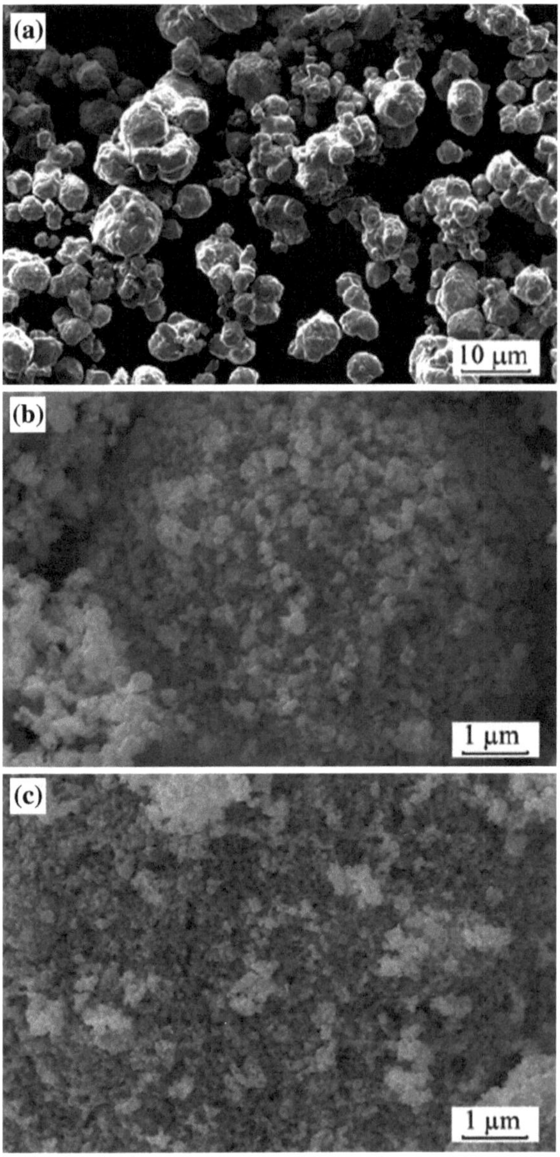

Fig. 4.32 Scanning electron microscopy of initial tungsten carbide powders: **a** WC-coarse, **b** WC-mill, **c** WC-plasm

adsorbed water (1.0 wt.%) is contained in the initial WC-mill nanopowder with the largest specific surface.

The data on chemical and phase analysis and other characteristics of initial and annealed tungsten carbide powders are listed in the Table 4.5.

Chemical analysis of the initial powders showed that the total and free carbon content in these powders is $C_{total} = 6.14 \pm 0.15$ wt.% and $C_{free} = 0.15 \pm 0.10$ wt.%

(see the Table 4.5), which corresponds to the tungsten carbide WC. It is noteworthy that C_{total} in the initial WC-mill nanopowder produced by milling of WC-coarse powder is by ~0.3 wt.% greater than that in WC-coarse powder. Most of this difference falls on free carbon, which is almost lacking in the initial WC-coarse powder. The possible reason for the presence of free carbon is milling the WC-mill powder in carbon-containing isopropanol $CH_3CH(OH)CH_3$: during high-energy ball milling 5 ml of isopropanol was added per every 10 g of the initial WC-coarse powder (milling conditions for WC powder are described in Sect. 4.2.2).

The content of oxygen in the initial WC-coarse, WC-mill, and WC-plasm powders is, respectively, 0.11, 0.91, and 1.23 wt.% according to chemical analysis, and 0.12 ± 0.05, 1.52 ± 0.05, and 1.80 ± 0.1 wt.% from EDX analysis (Fig. 4.33).

Figure 4.34 demonstrates as an example the results of EDX analysis of tungsten W, carbon C, and oxygen O distribution in an aggregated particle of the initial WC-mill powder. As is seen from Fig. 4.34, the particle have nearly the same ratio of tungsten W and carbon C along the whole scanning direction. EDX analysis of WC nanopowders revealed that impurity oxygen is distributed over the particle surface.

In addition, we performed EDX analysis of the WC-mill nanopowder, which was exposed in air for 20 h upon calcination (1 h in vacuum 1×10^{-4} Pa at temperature 670 K) to remove adsorbed water. According to EDX analysis, the content of oxygen in this powder is equal to 1.44 wt.%, which coincides with the initial value 1.52 wt.%. Thus, in air, the annealed WC nanopowders adsorb water and oxygen again. Taking into account EDX and thermogravimetric analysis data, as well as the literature data [143, 144, 148, 149], it can be supposed that impurity oxygen is contained in the adsorbed water and in the surface X-ray-amorphous oxide, hydroxide, and oxycarbide tungsten phases.

WC-coarse powder was annealed in vacuum at temperature 1470 K. Chemical and X-ray phase analysis results showed that annealing promoted decarburization of the powder, which, however, did not change its phase composition considerably. In Fig. 4.30a, b it is seen that the XRD patterns of WC-coarse powder before and after annealing are virtually identical, the only difference being the appearance of one weak tungsten line on the XRD pattern of the annealed powder. Quantitative phase analysis revealed that the content of the basic phase WC in WC-coarse powder upon annealing at 1470 K decreased by 1.5 wt.%, while the content of the W_2C phase increased by 0.5 wt.%; besides, a W phase in the amount of ~1.0 wt.% appeared (Table 4.5). From chemical analysis, the decarburization of WC-coarse powder as a result of annealing does not exceed ~0.01 wt.%.

After vacuum annealing of WC-mill nanopowder at temperature $T_{ann} \leq 870$ K, no changes in XRD patterns were observed. However, chemical analysis shows that with increasing T_{ann} up to 870 K, the powder is decarburized by almost 1 wt.% and at the same time it contains free carbon (see the Table 4.5). Analysis of diffraction reflection broadening shows that the particle size is stable and microstrains decrease from 0.83 to 0.64 % upon annealing at 870 k. Annealing of WC-mill nanopowder at 1070 k leads to almost 1.5 wt.% loss of total carbon as compared with the initial content and to the absence of free carbon, the XRD pattern (Fig. 4.35b) exhibiting clear reflections of metallic tungsten W with a body-centered cubic (space group

Table 4.5 Phase and chemical composition, average size $\langle D \rangle$ of nanoparticles, specific surface S_{sp}, and microstrain ε values for initial WC powders and the same powders after vacuum annealing at a temperature T_{ann}

Powder	$T_{ann} \pm 20$ (K)	$\langle D \rangle \pm 10$ (nm)	S_{sp} ($m^2 \cdot g^{-1}$)	$\varepsilon \pm 0.05$ (%)	Phase composition ± 0.1 (wt.%)			Chemical composition ± 0.05 (wt.%)		
					WC	W	W_2C	C_{total}	C_{free}	O^b
WC-coarse	300	9000 ± 1000[a]	0.04	–	98.0	0	2.0	6.00	0.05	0.11
WC-mill		20	19.2	0.83	98.0	0	2.0	6.27	0.24	0.91
WC-plasm		60	6.4	0.43	100.0	0	0	6.14	0.18	1.23
WC-mill	670	20	19.2	0.69	98.0	0	2.0	5.89	0.70	–
WC-plasm		60	6.4	0.42	100.0	0	0	5.86	0.14	–
WC-mill	870	20	19.2	0.64	98.0	0	2.0	5.31	0.58	–
WC-plasm		60	6.4	0.39	100.0	0	0	6.09	0.16	–
WC-mill	1070	30	12.8	0.60	80.5	19.5	0	5.15	Not found	–
WC-plasm		70	5.5	0.38	79.0	21.0	0	5.16	Not found	–
WC-mill	1270	50	7.7	0.59	77.5	22.5	0	4.84	Not found	–
WC-plasm		80	4.8	0.37	80.5	19.5	0	5.07	Not found	–
WC-coarse	1470	9000 ± 1000[b]	0.04	–	96.5	1.0	2.5	5.99	Not found	–
WC-mill		55	7.0	0.40	78.0	15.0	7.0	4.82	Not found	–
WC-mill+C[c]		110	3.5	0.05	100.0	0	0	6.28	1.28	–
WC-plasm		80	4.8	0.34	79.5	20.5	0	4.96	Not found	–
WC-mill	1670	77	5.0	0.26	64.0	0	36.0	4.83	Not found	–
WC-plasm		84	4.6	0.18	65.0	2.0	33.0	4.66	Not found	–

[a] Average size of particles in WC-coarse powder found from the size distribution (see Fig. 4.31a) measured by means of laser diffraction
[b] Chemical analysis for oxygen content determination was performed only for initial WC powders
[c] Mixture of initial powder WC-mill with 2 wt.% of carbon C

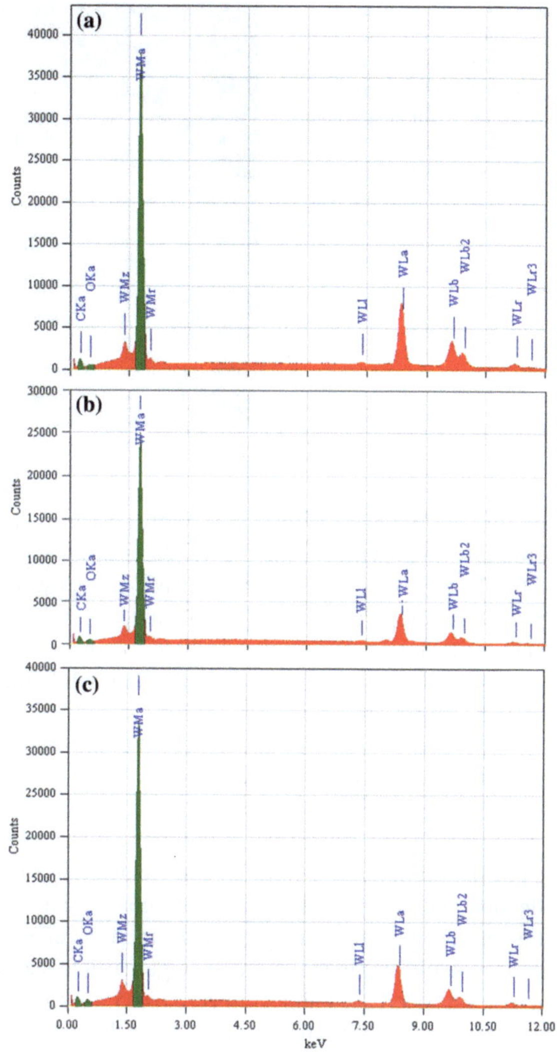

Fig. 4.33 EDX analysis of initial tungsten carbide powders: **a** WC-coarse, **b** WC-mill, **c** WC-plasm

$Im\overline{3}m$) structure. Observed narrowing of WC diffraction reflections corresponds to an increase of the average particle size to 30 nm and a decrease in microstrains to 0.6 %. The phase W_2C is absent although in the initial WC-coarse powder, which was milled to produce WC-mill powder, the content of W_2C was 2 wt.%. WC-mill powders annealed at 1070 and 1270 K have almost identical phase compositions, with the sole difference in the ratio of WC and W phases. As the annealing temperature is raised to 1270 K, the content of total carbon decreases and the content of W in WC-mill nanopowder increases. Moreover, WC nanoparticles grow to 50 nm. Upon annealing at 1470 K, WC-mill powder contains about 7 wt.% of the W_2C phase along

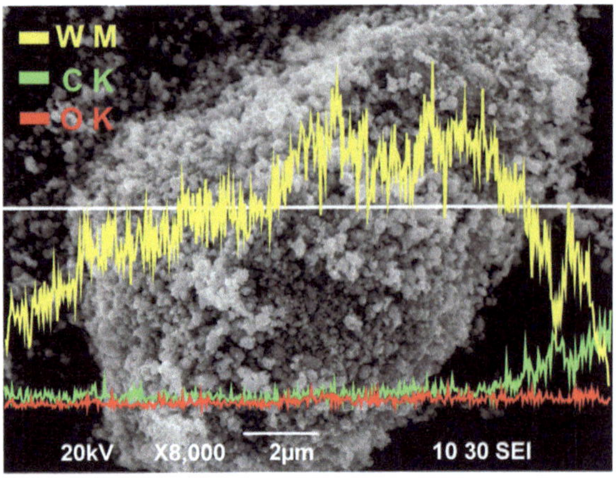

Fig. 4.34 EDX analysis of W, C, and O distribution in aggregated particle of initial powder WC-mill. Aggregate from 12 to 15 μm in size consists of many particles smaller than 100 nm. Aggregated particle was scanned along the *horizontal white line*

with WC and W phases (Fig. 4.35c). With a rise in T_{ann} to 1670 K, the content of W_2C increases to ~36 wt.%, while the W phase disappears (Fig. 4.35d). After annealing at 1670 K, the total carbon content in WC-mill powder decreased approximately by 1.5 wt.% as compared with the initial value.

During annealing the average particle size increased 4 times up to ~80 nm according to WC diffraction reflection broadening, while the microstrain value decreased to 0.26 % (Fig. 4.36).

In order to prevent decarburization of WC-mill nanopowder due to vacuum annealing and to retain it as a monophase powder, 2 wt.% of carbon black were added to the initial powder. The resulting mixture (WC-mill+C) was annealed in vacuum at 1470 K in the same way as WC-mill powder without carbon addition. The XRD pattern of the powder mixture (WC-mill+C) upon annealing (Fig. 4.35e) exhibits rather narrow reflections belonging only to one hexagonal (space group $P\bar{6}m2$) phase, namely, a higher carbide WC. Analysis of diffraction reflection broadening demonstrated that as a result of annealing the average particle size increased to 110 nm whereas microstrains decreased almost to zero (0.05 %). Chemical analysis revealed that the annealed mixture (WC-mill+C) contains 6.28 wt.% of total carbon and 1.28 wt.% of free carbon.

Scanning electron microscopy (SEM) shows that vacuum annealing of the initial nanopowder WC-mill at $T_{ann} \leq 1070$ K does not lead to considerable changes in the size and shape of the particles (Fig. 4.37a), which is indicative of their stability. At annealing temperatures 1470 K and higher, larger elongated dumbbell-like particles appear. Judging from the SEM images (Fig. 4.37b, c), most of them were formed by intergrowth of aggregated nanoparticles. According to SEM data, the mixture of the initial powder WC-mill with 2 wt.% of carbon C annealed in vacuum at 1470 K contained rather large well cut particles 1.0–1.5 μm in size (Fig. 4.37d) along with small-sized (~200 nm) particles.

Thus, owing to carbon adding WC nanopowder remains as a single-phase powder during annealing. However large addition of free carbon (≥1.0 wt.%) leads to an

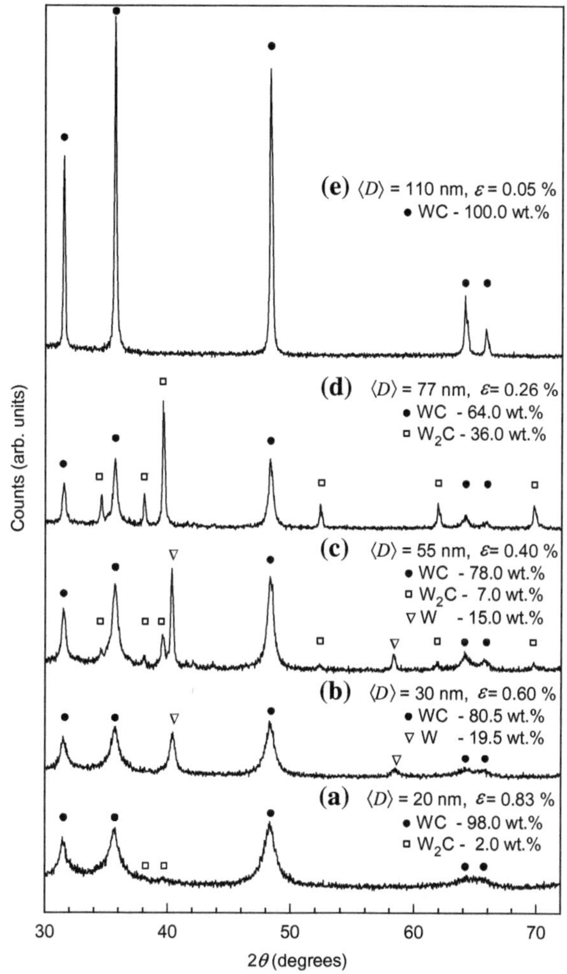

Fig. 4.35 X-ray diffraction patterns of WC-mill nanopowder recorded in Cu$K\alpha_{1,2}$ radiation: ***a*** initial powder; ***b***, ***c***, and ***d*** the same powder annealed in vacuum at 1070, 1470, and 1670 K, respectively; ***e*** mixture of initial powder WC-mill with 2 wt.% of carbon C annealed in vacuum at 1470 K

appreciable growth of carbide particles. If addition of free carbon into initial carbide nanopowder is small, the growth of WC nanoparticles during annealing is not great.

XRD patterns of WC-plasm nanopowders upon vacuum annealing up to 870 K exhibited no changes in the phase composition, average particle size, or microstrains. The stability of the powder composition follows also from chemical analysis showing that the powder contains a sufficient amount of total carbon for WC formation, as well as free carbon. Annealing at 1070 K resulted in a slight growth of particles to 70 nm and in an essential variation of the phase composition of the powder: in addition to the basic WC phase, it contains about 21 wt.% of metallic tungsten (Fig. 4.38b). Chemical analysis points to about 1 wt.% loss in total carbon and to complete disappearance of free carbon. As the annealing temperature rises from 1070 to 1470 K, the particles grow slightly to 80 nm, the microstrain value decreases a little, and the total carbon

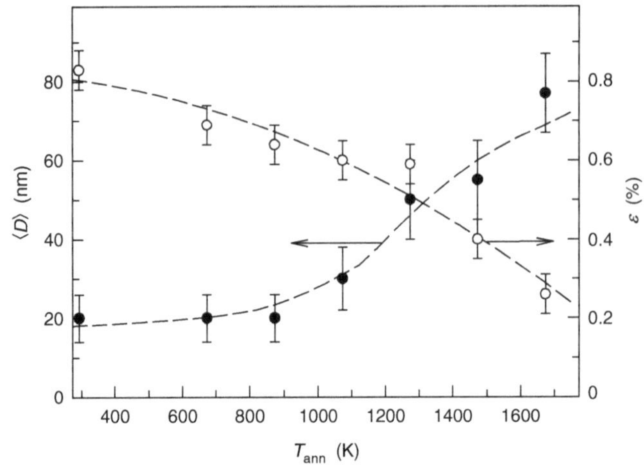

Fig. 4.36 The effect of vacuum annealing temperature on the average size $\langle D \rangle$ of powder particles and the microstrain ε values for the WC-mill nanopowder

Fig. 4.37 Scanning electron microscopy of WC-mill nanopowder annealed in vacuum at temperatures: **a** 1070 K; **b** 1470 K; **c** 1670 K; **d** mixture of initial powder WC-mill with 2 wt.% of carbon C annealed in vacuum at 1470 K

content lowers insignificantly that hardly affects the phase composition of the powder (79.5 wt.% WC and 20.5 wt.% W). Substantial changes are observed upon annealing of WC-plasm powder at 1670 K (Fig. 4.38d): ~33 wt.% of the W_2C phase appears

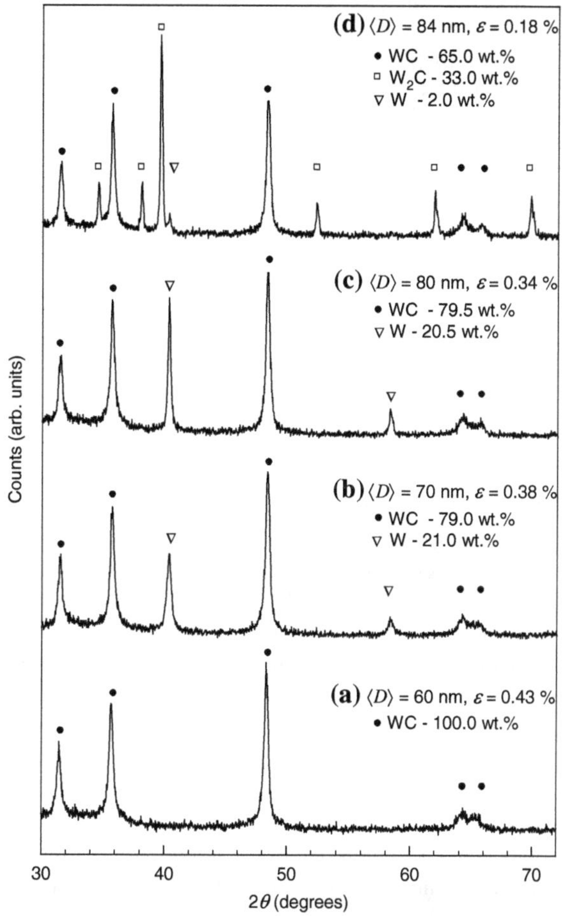

Fig. 4.38 X-ray diffraction patterns of WC-plasm nanopowder recorded in Cu$K\alpha_{1,2}$ radiation: ***a*** initial powder produced by plasma-chemical synthesis; ***b***, ***c***, and ***d*** the same powder annealed in vacuum at temperatures 1070, 1470, and 1670 K, respectively

in the powder, and the content of metallic tungsten drops abruptly to ~2 wt.% (only one weak reflection at $2\theta = 40.27°$ testifies is indicative of W).

Analysis of diffraction reflection broadening revealed that after annealing the size of the particles increased by 20 nm whereas the microstrain value decreased almost twice in comparison with the initial powder (Fig. 4.39). According to chemical analysis, the total carbon content decreased as a result of annealing approximately by 1.5 wt.% as compared with the initial WC-plasm powder.

Scanning electron microscopy of WC-plasm powders before and after vacuum annealing has shown the shape of particles remain nearly unchangeable during annealing although some coarsening of particles takes place (Fig. 4.40). This is confirmed also by diffraction data. Note that the growth of particles of the initial WC-plasm powder as a result of annealing is not as appreciable as for WC-mill powder, where the particle size upon annealing at 1670 K is 4 times greater than that

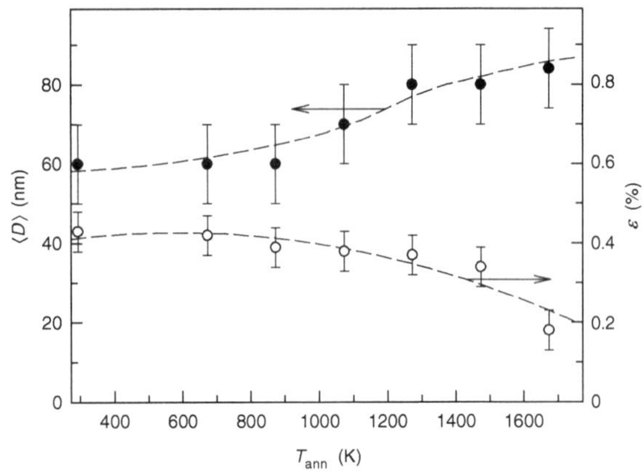

Fig. 4.39 The effect of vacuum annealing temperature on the average size $\langle D \rangle$ of powder particles and the microstrain ε values for the WC-plasm nanopowder

for the initial powder. Under annealing, WC-plasm powder particles retain the initial rounded shape.

Indeed, scanning electron microscopy revealed no intergrowing particles like those that formed in annealed WC-mill powder as a result of sintering. Nanoparticles of initial WC-plasm powder are strongly aggregated, and even under annealing at a temperature of 1670 K, which is a conventional temperature of hardmetal sintering, explicit intergrowth of nanoparticles (agglomeration or aggregation due to coalescence etc.) is not observed. It is the evidence of the poor sinterability of WC-plasm powder particles therefore application of WC-plasm powder can worsen the quality of the hardmetals.

As follows from the experiments, vacuum annealing of all the examined WC powders leads to their decarburization, which is most pronounced for WC nanopowders at annealing temperature 1070 K and higher. In turn, decarburization changes the phase composition of powders depending on the annealing temperature T_{ann} (Table 4.5), that is due to the peculiarities of the phase diagram of the system W-C [146]. According to [146, 150], the lower tungsten carbide W_2C emerges at temperature $\geq$1523 K; however, XRD patterns of WC-mill nanopowder annealed at lower temperature $\sim$1470 K contain diffraction reflection not only of WC and W, but also of the lower carbide W_2C.

The experimental data obtained suggest that the temperature T_{trans} of reversible transformation $W_2C \Leftrightarrow W + WC$ ($\varepsilon - W_2C \Leftrightarrow W + \delta$-WC) should be somewhat lower than 1523 K, namely, $\sim$1473-1503 K. In WC-mill nanopowder annealed in vacuum at temperature 1470 K close to T_{trans}, three phases were found: WC, W, and W_2C (see the Table 4.5). The presence of these three phases is probably due to the fact that the annealing temperature T_{ann} is close to T_{trans} and that fluctuations in T_{ann} are small (within $\pm$ 10 K): when WC-mill nanopowder is annealed at $T_{ann} \leq T_{trans}$, the phase W_2C, which is in equilibrium at a higher temperature, begins to decompose into W and WC, but eutectoid decomposition does not proceed to completion because of small overcooling relative to T_{trans}. A phase diagram of the W-C sys-

Fig. 4.40 Scanning electron microscopy of WC-plasm nanopowder annealed in vacuum at temperatures: **a** 1070 K, **b** 1470 K, **c** 1670 K

tem refined with allowance for the obtained experimental results is demonstrated in Fig. 4.41 for temperatures above 1000 K. As is seen from Fig. 4.41, variations in the phase composition observed experimentally during annealing of WC nanopowders are really possible in case of partial decarburization of WC powders.

In spite of similar particle sizes, WC-mill and WC-plasm nanopowders exhibit a number of distinctions which show up vividly in annealing experiments. While in

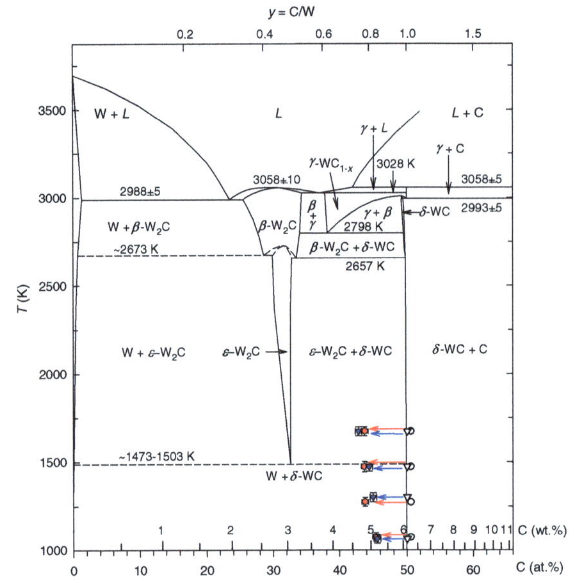

Fig. 4.41 The phase diagram of the system W – C and variation in the composition of WC nanopowders as a result of vacuum annealing at T_{ann}:(○) composition of initial nanopowder WC-mill, (●) composition of WC-mill nanopowder upon annealing at (▽) composition of initial nanopowder WC-plasm, (▼) composition of WC-plasm nanopowder upon annealing at T_{ann}

WC-mill powder nanoparticles growth and microstrain reduction occur rapidly with an increase in T_{ann}, these changes in WC-plasm powder are appreciably weaker. The same is observed for phase transformations. In WC-mill nanopowder, the phase W_2C appears already upon annealing at 1470 K whereas metallic tungsten disappears after annealing at 1670 K. As regards WC-plasm nanopowder, even traces of W_2C are lacking in it after annealing at 1470 K, while a small amount of tungsten remains upon annealing at 1670 K. These differences are due first of all to a greater specific surface of WC-mill nanopowder as compared with WC-plasm powder.

It is known that during vacuum evaporation WC loses mainly carbon and transforms to two-phase region (W + WC) or (W_2C + WC) depending on the evaporation temperature [69, 151–153]. However, partial carbon pressure over tungsten carbide even at a high temperature from 2100 to 2400 K is low and is equal from $\sim 3.7\times10^{-11}$ to $\sim 4.2\times10^{-8}$ bar (from $\sim 3.7\times10^{-6}$ to $\sim 4.2\times10^{-2}$ Pa) [151, 153]. In Ref. [151], the rate of carbon evaporation V_C from tungsten carbide at 2253 K is reported to be only 9.24×10^{-9} g·cm^{-2}·s^{-1}, while from data [152] the rate of evaporation in the high-temperature region from 2248 to 2923 K is described by the dependence $\lg V_C = 6.317 - 34310/T$. At temperature 1670 K and below, the rate of carbon evaporation from WC is hundreds thousand times smaller than V_C at ~2270 K and therefore cannot provide even a hundredth part of carbon loss observed experimentally during vacuum annealing of WC. Thus, the reduction of the carbon content as a result of vacuum annealing of WC at $T_{ann} \leq 1670$ K cannot be explained directly by carbon evaporation.

A more probable reason for the carbon content reduction during vacuum annealing of WC can be impurity oxygen present in the nanopowders. Indeed, 5 wt.% of impurity oxygen are found already in the nanopowder of vanadium carbide [154], which is the first nanocrystalline carbide powder examined comprehensively. Most of this oxygen was accounted for by adsorbed water and the surface oxide film, while the rest of oxygen was in the chemisorbed form on the surface of powder particles. The authors [143] studied plasma-chemical WC nanopowder and found ~2 wt.% of oxygen in it. More than 80 % of this oxygen enters X-ray-amorphous tungsten oxide phases, whereas the rest of oxygen is in water or in the adsorbed form. According to [148, 149], the surface layer on WC stored in oxygen-containing atmosphere is an oxide phase and is similar in the electronic state to WO_3. Auger-spectroscopy examination of the interaction of the WC(0001) surface with oxygen [149] revealed that surface tungsten oxycarbides and oxides WO and WO_3 are formed in high vacuum as a result of oxygen chemisorption.

It was established [144, 149] that heating of WC nanopowder at temperature from 1170 to 1370 K leads to removal of no less than 80 % of the total oxygen in the form of carbon oxide CO. Naturally, with a removal of oxygen in the form of CO, the content of carbon in the WC powder decreases, too. If the content of impurity oxygen is C_{O} wt.%, than in the process of vacuum annealing at $T_{\mathrm{ann}} \geq 1370$ K, $\Delta C_C = (A_C/A_O)C_O$ wt.% of carbon (A_C and A_{O} are atomic weight of carbon and oxygen) will be removed together with CO. Nanopowders of metals (W, Mo, V, Nb) and their carbides contain impurity oxygen chiefly on the surface [14, 143, 144, 148, 149, 154, 155]; that is why its content is proportional to the powder specific surface S_{sp}. Since the value of carbon loss ΔC_C in vacuum annealing is proportional to the impurity oxygen content C_O, ΔC_C should be also proportional to the specific surface S_{sp} of the initial powder. Note that annealing either in vacuum or in hydrogen of carbide nanopowders containing impurity oxygen is accompanied by a loss of carbon and by WC nanoparticle growth [156–158].

With allowance made for the unit cell sizes of the hexagonal (space group $P\overline{6}m2$) carbide WC, the number of W atoms per a surface unit, n_{s}, is about 1.4×10^{19} atom·m^{-2}. So, when a monolayer of the oxide phase WO_3 is formed on WC, mass of oxygen on 1 m^2 surface is equal to 3 $n_{\mathrm{s}} m_{\mathrm{a}} A_{\mathrm{O}}$ where $m_{\mathrm{a}} = 1.66 \times 10^{-24}$ g is the atomic mass unit and $A_{\mathrm{O}} = 16$ is the atomic weight of O. If the surface of particles is covered by p monolayers of the oxide phase, than in the first approximation the relative oxygen content in a powder with specific surface S_{sp} is

$$C_{\mathrm{O}} = 3pn_{\mathrm{s}}m_{\mathrm{a}}A_{\mathrm{O}}S_{\mathrm{sp}} = 0.00111pS_{\mathrm{sp}}, \tag{4.73}$$

hence, the number of the oxide phase monolayers is

$$p = C_{\mathrm{O}}/3n_{\mathrm{s}}m_{\mathrm{a}}A_{\mathrm{O}}S_{\mathrm{sp}} = C_{\mathrm{O}}/(0.00111S_{\mathrm{sp}}). \tag{4.74}$$

The specific surface S_{sp} of initial WC-mill and WC-plasm powders is respectively 19.2 and 6.4 m^2·g^{-1} (see Table 4.5), and the value of C_{O} from EDX analysis is 0.0152 and 0.0180 (1.52 and 1.80 wt.%). Therefore, the number of WO_3 monolayers p on the surface of initial powders WC-mill and WC-plasm is ~0.7 and ~2.5. This fits well with data [149] that the oxide film on WC is from 0.6 to 2.8 monolayers.

During heating, oxygen reacts with carbon and desorbs from WC mainly as carbon oxide CO [143, 148, 149], therefore, the relative loss of carbon is proportional to C_O:

$$\Delta C_C = (A_C/A_O)C_O = 3pn_s m_a A_C S_{sp} = 0.00083 p S_{sp}. \tag{4.75}$$

In conformity with (4.75), for the estimated values $p = 0.7$ and $p = 2.5$ the expected carbon loss ΔC_C from initial nanopowders WC-mill and WC-plasm is 0.011 and 0.013 (1.1 and 1.3 wt.%). With consideration for O and C content determination errors and the tentative character of the estimates, the obtained values of ΔC_C agree well with the maximal carbon loss ΔC_C from WC-mill and WC-plasm powders achieved upon annealing at 1670 K, namely, 1.4 and 1.5 wt.%.

In conclusion, we note the following.

The content of impurity oxygen in microcrystalline powder WC-coarse is very small. Therefore, vacuum annealing of WC-coarse powder at $T_{ann} \leq 1670$ K results in very small decarburization due to disappearance of free carbon and hardly affects the phase composition of the powder.

Nanocrystalline powders WC-mill and WC-plasm have a large specific surface and upon storage in air contain up to 1.8 wt.% of impurity oxygen. Vacuum annealing of these nanopowders at $T_{ann} \leq 1670$ K is accompanied by appreciable decarburization, variations in phase composition, and by coarsening of the powder particles due to intergrowth of aggregated nanoparticles. During annealing, the microstrain value decreases. To retain WC nanopowdes as monophase materials during annealing, a small excess of free carbon is needed. When more than 1 wt.% of excessive free carbon is added, WC nanopowders still remain single phases, but in this case appreciable particle growth occurs during annealing.

Thus, WC nanopowders used for vacuum or hydrogen sintering of nanostructured hardmetals WC-Co should be stored, mixed with Co powder and pressed in vacuum or inert atmosphere. This prevents the severe contamination of WC nanopowders with oxygen in preliminary process design of nanostructured hardmetals. Besides, in the manufacture of nanostructured WC-Co hardmetals, WC nanopowders with a slightly increased content of total carbon (compared with the stoichiomentric ratio C/W = 1) should be used. This will allow WC nanopowder to remain a single phase on the one hand and yet will lead to no appreciable growth of WC nanoparticles on the other hand.

References

1. Gleiter H. Materials with ultrafine microstructure: retrospectives and perspectives. Nanostruct Mater. 1992;1(1):1–19.
2. Birringer R, Gleiter H. Nanocrystalline materials. In: Cahn RW, editor. Encyclopedia of material science and engineering. vol. 1. Oxford: Pergamon Press; 1988. p. 339–49.
3. Siegel RW. Cluster—assembled nanophase materials. Ann Rev Mater Sci. 1991;21:559-78.
4. Siegel R. Nanostructured materials—mind over matter. Nanostruct Mater. 1993;3(1–3);1–18.

5. Schaefer H-E. Nanoscience. The science of the small in physics, engineering, chemistry, biology and medicine. Heidelberg: Springer; 2010. 772 pp.
6. Siegel RW. What do we really know about the atomic-scale structures of nanophase materials? J Phys Chem Solids. 1994;55(10):1097–106.
7. Morokhov ID, Trusov LI, Chizhik SP. Ultra-dispersed metallic substances. Moscow: Atomizdat; 1977. 264 pp. (in Russian).
8. Morokhov ID, Petinov VI, Trusov LI, Petrunin VF. Structure and properties of small metallic particles. Uspekhi Fiz. Nauk. 1981;133(4):653-92 (in Russian).
9. Morokhov ID, Trusov LI, Lapovok VN. Physical phenomena in ultra-dispersed substances. Moscow: Energoatomizdat:1984. 224 pp. (in Russian).
10. Petrov YuI. Physics of small particles. Moscow; Nauka: 1982. 360 pp. (in Russian).
11. Petrov YuI. Clusters and small particles. Moscow: Nauka: 1986. 368 pp. (in Russian).
12. Gusev AI. Effects of the nanocrystalline state in solids. Uspekhi Fiz. Nauk. 1998;168(1):55–83. (in Russian) (Engl. transl.: Physics - Uspekhi. 1998;41(1);49–76).
13. Gusev AI, Rempel AA. Nanocrystalline materials. Moscow: Nauka; 2000. 224 pp. (in Russian).
14. Gusev AI, Rempel AA. Nanocrysnalline materials. Cambridge: Cambridge Intern. Science Publ; 2004. 351 pp.
15. Gusev AI. Nanomaterials, nanostructures, and nanotechnologies. 2nd edn. Moscow: Nauka - Fizmatlit; 2007. 416 pp. (in Russian).
16. Krivoglaz MA. Theory of X-ray and thermal-neutron scattering by real crystals. New York: Plenum Press; 1969.
17. Rempel AA, Rempel SV, Gusev AI. Quantitative assesment of homogeneity of nonstoichiometric compounds. Dokl. Akad. Nauk. 1999;369(4):486–490. (in Russian) (Engl. Transl.: Doklady Phys. Chem. 1999;369(4–6):321–325).
18. Rempel AA, Gusev AI, Preparation of disordered and ordered highly nonstoichiometric carbides and evaluation of their homogeneity. Fiz Tverd Tela. 2000;42(7):1243–1249. (in Russian) (Engl. Transl.: Phys. Solid State. 2000;42(7):1280–1286).
19. Gusev AI, Rempel AA. Nonstoichiometry, disorder and order in solids. Ekaterinburg: Ural Division of the Rus. Acad. Sci. 2001. 580 pp. (in Russian).
20. Gusev AI, Rempel AA, Magerl AJ. Disorder and order in strongly nonstoichiometric compounds. Transition metal carbides, nitrides and oxides. Berlin: Springer; 2001. 608 pp.
21. Gusev AI. Nonstoichiometry, disorder, short-range and long-order in solids. Moscow: Nauka - Fizmatlit, 2007. 856 pp. (in Russian).
22. Kurlov AS, Gusev AI. Determination of the particle sizes, microstrains, and degree of inhomogeneity in nanostructured materials from X-ray diffraction data. Fiz Khim Stekla. 2007;33(3):383–92. (in Russian) (Engl. Transl.: Glass Phys Chem. 2007;33(3):276–82).
23. Gusev AI, Kurlov AS. Characterization of nanocrystalline materials by the size of particles (grains). Metallofizika i Nov. Tekhnologii. 2008;30(5):C.679–94. (in Russian).
24. Kozhevnikova NS, Kurlov AS, Uritzkaya AA, Rempel AA. Diffraction analysis of nanocrystalline particle size of lead and cadmium sulfides prepared by chemical deposition from aqueous solutions. Zh Strukt Khimii. 2004;45:147–53. (in Russian) (Engl. Transl.: J. Struct. Chem. 2004;45:S154–9).
25. Prabal D. On use of pseudo-Voigt profiles in diffraction line broadening analyses. Fizika A (Croatia). 2000;9(2):61–6.
26. Puerta J, Martin P. Three and four generalized Lorentzian approximations for the Voigt line shape. Appl Optics. 1981;20(22):3923–8.
27. Cagliotti G, Paoletti A, Ricci FP. Choice of collimators for a crystal spectrometer for neutron diffraction. Nuclear Instrum Methods. 1958;3(3):223–8.
28. Rietveld HM. A profile refinement method for nuclear and magnetic structures. J Appl Crystallogr. 1969;2(2):65–71.
29. Scherrer P. Bestimmung der Grösse und der inneren Struktur von Kolloidteilchen mittels Röntgenstrahlen. Nachr Ges Wiss Göttingen, Math-Phys. Kl. 1918;2:98–100.

30. Seljkow N. Eine röntgenographische Methode zur Messung der absoluten Dimensionen einzelner Kristalle in Körpern von fein-kristallinischem Bau. Z. Physik. 1925;31(5–6):439–44.
31. James RW. The optical principles of the diffraction of X-rays. London: G. Bell & Sons Ltd.; 1954.
32. Warren BE, Averbach BL, Roberts BW. Atomic size effect in the X-ray scattering by alloys. J. Appl Phys. 1951;22(12):1493–6.
33. Williamson GK, Hall WH. X-ray line broadening from filed aluminium and wolfram. Act Metal. 1953;1(1):22–31.
34. Gusev AI, Rempel SV, X-ray diffraction study of the nanostructure resulting from decomposition of $(ZrC)_{1-x}(NbC)_x$ solid solutions. Neorgan Mater. 2003;39(1):49–53. (in Russian) (Engl. Transl.: Inorg Mater. 2003;39(1):43–6).
35. Rempel SV, Gusev AI, Rempel AA, Method for determining the particle size in compacted and dispersed nanomaterials. In: Physical chemistry of ultradisperse (nano-) systems. Moscow: Moscow Institute of Engineering, Physics; 2003. p. 378–84. (in Russian).
36. Lebedev AV, Blagoveshchenskii YV, Moldaver VA. Creation of plasma-chemical equipment for manufacture of ultrafine powders of refractory metals and compounds. In: Physics and Chemistry of Ultradisperse Systems/Trans. 5th All-Russian Conf. (Yekaterinburg/Russia, October 9–13, 2000). Moscow: Moscow Engineering Physics Institute, 2000. p. 145–7. (in Russian).
37. Hall WH. X-ray line broadening in metals. Proc Phys Soc London. 1949. Sect. A;62, part 11(359A):741–3.
38. Hall WH, Williamson GK. The diffraction pattern of cold worked metals: I. The nature of extinction. Proc Phys Soc London. 1951; Sect. B:64, part 11(383):B.937–46.
39. Gusev AI, Nanocrystalline materials: synthesis and properties. In: Schwarz JA, Contescu C, Putyera K (editors). Dekker encyclopedia of nanoscience and nanotechnology (in 5 volumes). vol. 3. New York: Marcel Dekker Inc.; 2004. p. 2289–304.
40. Avvakumov EG, editor. Mechanical alloying in inorganic chemistry. Novosibirsk: Nauka; 1991. 305 pp. (in Russian).
41. Butyagin PYu. Problems in mechanochemistry and prospects for its development. Uspekhi Khimii. 1994;63(12):1031–1043. (in Russian) (Engl. Transl.: Russ Chem Reviews. 1994;63(12):965–76).
42. Balogh J, Bujdosó L, Faigel G, Gránásy L, Kemény T, Vincze I, Szabó S, Bakker H. Nucleation controlled transformation in ball milled FeB. Nanostruct Mater. 1993;2(1):11–8.
43. Teresiak A, Kubsch H. X-ray investigations of high energy ball milled transition metal carbides. Nanostruct Mater. 1995;6(5–8): 671–4.
44. Xueming MA, Gang JI. Nanostructured WC-Co alloy prepared by mechanical alloying. J Alloys Comp. 1996;245:L30–2.
45. Yavuz M, Maeda H, Vance L, Liu HK, Dou SX. Effect of ball milling materials and methods on powder processing of Bi2223 superconductors. Supercond Sci Techn. 1998;11(10):1153–9.
46. Coste S, Bertrand G, Coddet C, Gaffet E, Hahn H, Sieger H. High-energy ball milling of Al_2O_3-TiO_2 powders. J Alloys Comp. 2007;434–435:489–92.
47. Venugopal T, Prasad Rao K, Murty BS. Synthesis of Cu-W nanocomposite by high-energy ball milling. J Nanosci Nanotech. 2007;7(7):2376–81.
48. Wang Y, Li Y, Rong, C, Liu PJ. Sm-Co hard magnetic nanoparticles prepared by surfactant-assisted ball milling. Nanotechnology. 2007;18(46):465701-1–465701-4.
49. Zhurkov SN, Narzulayev BN. Time dependence of the strength of solids. Zh Tekhn Fiz. 1953;23(10):1677–1689. (in Russian).
50. Regel' VR, Slutsker AI, Tomashevskii EE. The kinetic nature of the strength of solids. Uspekhi Fiz. Nauk. 1972;106(2):193–228. (in Russian) (Engl. Transl.: Sov. Phys. Uspekhi 1972;15(1):45–65).
51. Yu BP. Mechanical disordering and reactivity of solids. In: Advances in mechanochemistry, physical and chemical processes under deformation. Harward Acad. Publ., 1998/Chemistry Reviews. 1998;23, Part 2: 91–165.

52. Putra IS, Suharto D (editors). Fracture and strength of solids VI (Proc. 6th Intern. Conf. on Fracture and Strength of Solids (FEOFS 2005), April 4–6, 2005, Bali, Indonesia). Key Eng. Mater. 2006;306–8. 1600 pp.
53. Sieradzki K. The fracture strength of solids near the percolation threshold. J Phys C: Solid State Phys. 1985:18(27):L855–6.
54. Fecht H-J. Nanostructure formation by mechanical attrition. Nanostruct. Mater. 1995;6(1–4):33–42.
55. Mohamed FA. A dislocation model for the minimum grain size obtainable by milling. Acta Materialia. 2003;51(14):4107–19.
56. Boiko VF, Verkhoturov AD. Model of grinding process of tungsten carbide. Perspektivnie Materialy. 2008;6:84–7 (in Russian).
57. Eckert J, Holzer JC, Krill CE, Johnson WL. Structural and thermodynamic properties of nanocrystalline fcc metals prepared by mechanical attrition. J Mater Res. 1992;7(7):1751–61.
58. Oleszak D, Shingu PH. Nanocrystalline metals prepared by low energy ball milling. J Appl Phys. 1996;79(6):2975–81.
59. Kurlov AS, Gusev AI. Effect of ball milling parameters on the particle size in nanocrystalline powders. Pis'ma v ZhTF. 2007;33(19):46–54. (in Russian) (Engl. Transl.: Technical Physics Letters. 2007;33(10):828–32).
60. Gusev AI, Kurlov AS. Production of nanocrystalline powders by high-energy ball milling: model and experiment. Nanotechnology. 2008;19(26). Paper 265302. P.265302-01–265302-08.
61. Gusev AI, Kurlov AS., Mechanical milling process modelling and making WC nanocrystalline powder. Neorgan Materialy. 2009;45(1):38–45. (in Russian) (Engl. Transl.: Inorganic Materials. 2009;45(1):35–42).
62. Kurlov AS, Gusev AI. Production of nanocrystalline powder of WC via ball-milling. In: Sigl LS, Rödhammer P, Wildner H (editors). 17 Plansee Seminar 2009: Proc. Intern. Conf. on High Performance P/M Materials (Reutte/Austria, May 25–29, 2009). Reutte, Austria: Plansee Group, 2009. V. 3. P.GT 24/1-GT24/11.
63. Kurlov AS, Gusev AI. Mathematical modeling of high-energy ball milling of powders. In: Zinigrad M (editor). Mathematical Modeling and Computer Simulation of Material Technologies/Proc. 6th Intern. Conf. MMT-2010 (Ariel, Israel, August 23–27, 2010). Ariel (Israel): Ariel University Center of Samaria, 2010. p. 1-105–1-113.
64. Kurlov AS, Gusev AI. Model for milling of powders. Zh Tekhn Fiz. 2011;81(7):76–82. (in Russian) (Engl. Transl.: Technical Physics. 2011;56(7):975–80).
65. Butyagin PYu, Streletskii AN. The kinetics and energy balance of mechanochemical transformations. Fiz Tverd Tela. 2005;47(5): C.830–5 (in Russian). (Engl. Transl.: Phys. Solid State. 2005;47(5):856–62).
66. Nazarov AA, Romanov AE, Valiev RZ. On the nature of high internal stresses in ultrafine grained materials. Nanostruct Mater. 1994;4(1):93–102.
67. Williamson GK, Smallman RE. Dislocation densities in some annealed and cold-worked metals from measurements on the X-ray Debye-Scherrer spectrum. Phil Mag. 1956;1(1):34–46. (Series 8).
68. Butyagin PYu. Physical and chemical ways of a relaxation of elastic energy in solids. Mechanochemical reactions in two-component systems. In: Avvakumov EG (editor). Mechanical Alloying in Inorganic Chemistry. Novosibirsk: Nauka; 1991. p. 32–52 (in Russian).
69. Ya Kosolapova T. Properties, Production and application of refractory compounds handbook. Moscow: Metallurgiya; 1986. 928 pp. (in Russian).
70. Gubicza J, Ribárik G, Goren-Muginstein GR, Rosen AR, Ungár T. The density and the character of dislocations in cubic and hexagonal polycrystals determined by X-ray diffraction. Materials Sci Eng A. 2001;A309–A310:60–3.
71. Bolton JD, Redington M. Plastic deformation mechanisms in tungsten carbide. J Mater Sci. 1980;15(12)3150–6.

72. Wicks CE, Block FE. Thermodynamic properties of 65 elements, their oxides, halides, carbides and nitrides/Bulletin No 605. Washington: US Government Printing Office; 1963. p. 240.
73. Glushko, VP. Thermodynamic properties of individual substances (handbook in 4 vol.). Moscow: Nauka; 1979 II, book 2, 340 pp.; 1982 IV, book 2; 560 pp. (in Russian).
74. Barin I. Thermochemical data of pure substance, 3rd edn. Weinheim: VCH, 1995. 1880 pp.
75. Warren BE. X-ray diffraction. New York: Dower; 1990.
76. Schaefer H-E. Interfaces and physical properties of nanostructured solids. In: Nastasi MA, Parkin DM, Gleiter H (editor). Mechanical properties and deformation behavior of materials having ultrafine microstructure. Netherlands, Dordrecht: Kluwer Academic Press; 1993. p. 81–106.
77. Gusev AI, Nanocrystalline materials: preparation and properties. Ekaterinburg: Ural Division of the Russ. Acad. Sci., 1998. 200 pp. (in Russian).
78. Edelstein AS, Cammarata RC (editors). Nanomaterials: synthesis, properties and applications. Baltimor: The Johns Hopkins University; 1998. 620 pp.
79. Gleiter H. Nanostructured materials: basic concepts and microstructure. Acta Mater. 2000;48(1):1–29.
80. Löffler J, Weissmüller J, Gleiter H. Characterization of nanocrystalline palladium by X-ray atomic density distribution functions. Nanostruct Mater. 1995;6(5–8):567–70.
81. Weissmüller J, Löffler J, Kleber M. Atomic structure of nanocrystalline metals studied by diffraction techniques and EXAFS. Nanostruct Mater. 1995;6(1–4):105–14.
82. Haubold T, Birringer R, Lengeler B, Gleiter H. EXAFS studies of nanocrystalline materials exhibiting a new solid-state structure with randomly arranged atoms. Phys Lett A. 1989;135(8–9);461–466.
83. Rempel AA, Gusev AI. Magnetic susceptibility of palladium subjected to severe plastic deformation. Phys Stat Sol (b). 1996;196(1);251–60.
84. Rempel AA, Gusev AI, Mulyukov RR, Amirkhanov NM. Microstructure, microhardness and magnetic susceptibility of submicrocrystalline palladium. Nanostruct Mater. 1996;7(6):667–74.
85. Korznikov A, Dimitrov O, Korznikova G. Thermal evolution of the structure of ultrafine grained materials produced by severe plastic deformation. Ann Chim France. 1996;21(6–7):443–60.
86. Kizuka T, Nakagami Y, Ohata T, Kanazawa I, Ichinose H, Murakami H, Ishida Y. Structure and thermal-stability of nanocrystalline silver studied by transmission electron-microscopy and positron-annihilation spectroscopy. Philos Mag A. 1994;69(3):551–63.
87. Ishida Y, Ichinose H, Kizuka Y, Suenaga K. High-resolution electron microscopy of interfaces in nanocrystalline materials. Nanostruct Mater. 1995;6(1–4):115–24.
88. Klemm W, Schüth W. Magnetochemische Untersuchungen. 3. Über den Magnetismus einiger Carbide und Nitride. Z Anorg Allgem Chemie. 1931;201(1):24–31.
89. Kurlov AS, Nazarova SZ, Gusev AI, Magnetic susceptibility of tungsten carbide: relaxation and impurity effects. Pis'ma v ZhETF. 2005;82(8):567–72. (in Russian) (Engl. transl.: JETP Letters. 2005;82(8):509–14).
90. Kurlov AS, Nazarova SZ, Gusev AI, Magnetic susceptibility and thermal stability of nanocrystalline tungsten carbide. Doklady Akad. Nauk. 2005;405(2):218–23. (in Russian) (Engl. transl.: Dokl. Phys. Chem. 2005;405(1):229–34).
91. Kurlov AS, Nazarova SZ, Gusev AI, Magnetic susceptibility and thermal stability of particle size of nanocrystalline tungsten carbide WC. Fiz Tverd Tela. 2007;49(9):1697–703. (in Russian) (Engl. transl.: Phys Solid State. 2007;49(9):1780–6).
92. Rempel AA, Gusev AI, Nazarova SZ, Mulyukov RR. Extrinsic superparamagnetism in plastically deformed copper. Doklady Akad Nauk. 1996;347(6):750–4. (in Russian) (Engl. transl.: Physics - Doklady. 1996;41(4):152–6).
93. Rempel AA, Nazarova SZ. Magnetic properties of iron nanoparticles in submicrocrystalline copper. J Metastable Nanocrystal Mater. 1999;1:217–22.

94. Rempel AA, Nazarova SZ, Gusev AI. Iron nanoparticles in severe-plastic-deformed copper. J Nanoparticle Res. 1999;1(4):485–90.
95. Rempel AA, Nazarova SZ, Gusev AI. Magnetic properties and defects in palladium, titanium and copper after severe plastic deformation. In.: Burhanettin A (editor). Severe plastic deformation: towards bulk production of nanostructured materials. New York: Nova Science Publishers; 2006. p. 415–30.
96. Gusev AI, Nazarova SZ. Magnetic susceptibility of nonstoichiometric compounds of transition d-metals. Uspekhi Fiz. Nauk. 2005;175(7):681–704. (in Russian) (Engl. Transl.: Uspekhi-Physics. 2005;8(7):651–74).
97. Vonsovsky SV. Magnetism. (2 vols.). New York: Wiley; 1974. 1032 pp.
98. Selwood PW. Magnetochemistry. 2nd ed. New York: Interscience; 1956. 435 pp.
99. Mattheiss LF, Hamann DR. Bulk and surface electronic structure of hexagonal WC. Phys Rev B. 1984;30(4):1731–8.
100. Liu AY, Wentzcovitch RM, Cohen ML. Structural and electronic properties of WC. Phys Rev B. 1988;38(14):9483–8.
101. Kurlov AS, Gusev AI, Particle size effects on the oxidation of tungsten carbide powders. Zh Fiz Khimii. 2010;84(12):2291–7. (in Russian) (Engl. Transl. Russ. J. Phys. Chem. A. 2010;84(12):2095–101).
102. Kurlov AS, Gusev AI. Effect of particle size on the oxidation of WC powders during heating. Neorgan Materialy. 2011;47(2);173–8. (in Russian) (Engl. Transl.: Inorg Materials. 2011;47(2);133–8).
103. Kurlov AS, Gusev AI. Pecualirities of vacuum annealing of nanocrystalline WC powders. Int J Refr Met Hard Mater. 2012;32(5):.51-60.
104. Kurlov AS, Gusev AI. Vacuum annealing of nanocrystalline WC powders. Neorgan Materialy. 2012;48(7):781–91. (in Russian) (Engl. Transl.: Inorg Materials. 2012;48(7):680–90).
105. Kriessman CJ. The high-temperature magnetic susceptibility of V, Nb, Ta, W, and Mo. Rev. Modern Phys. 1953;25(1):122–6.
106. Asmussen RW, Potte-Jensen J. The magnetic properties of the elements molybdenum and tungsten. Acta Chem Scand. 1957;11(7):1271–2.
107. Kojima H, Tebble RS, Williams DEG. The variation with temperature of the magnetic susceptibility of some of transition elements. Proc R Soc A. 1961;A260(1301):237–50.
108. Kohlhaas R, Wunsch KM. Der Paramagnetismus der Übergangsmetalle Nb, Ta, Mo, W und Re als Funktion der Temperatur. Zeitschr Angew Physik. 1971;32(2):158–63.
109. Tammann G, Oelsen W. Die Abhängigkeit der Konzentration gesättigter Mischkristalle von der Temperatur. Ztschr Anorg Chemie. 1930;186(3):257–88.
110. Kittel C. Introduction to solid state physics. 7th ed. Brisbane: Wiley; 1996. 673 pp.
111. Newkirk AE. The oxidation of tungsten carbide. J Am Chem Soc. 1955;77(17):4521–2.
112. Webb WW, Norton JT, Wagner C. Oxidation studies in metal-carbon systems. J Electrochem Soc. 1956;103(2):112–7.
113. Dufour LC, Simon J. Cinetique d'oxidation sous faible pression d'oxygene d'echantillons pulverulents de monocarbures de zirconium et de tungstene et d'hemicarbure de molibdene. Bull Soc Chim France. 1968;9:3644–51.
114. Voitovich RF, Pugach EA. Features of high-temperature oxidation of transition metal carbides of VI groups. Poroshk Metallurgiya. 1973;4:59–64. (in Russian). (Engl. Transl.: Powder Metallurgy Met Ceram. 1973;12(4):314–8).
115. Voitovich VB, Sverdel VV, Voitovich RF, Golovko EI. Oxidation of WC-Co, WC-Ni and WC-Co-Ni hard metals in the temperature range 500–800°C. Intern J Refr Met Hard Mater. 1996;14(4);289–95.
116. Ribeiro CA, Souza WR, Crespi MS, Gomes Neto JA, Fertonani FL. Non-isothermal kinetic of oxidation of tungsten carbide. J Therm Anal Calorimetry. 2007;90(3):801–5.
117. Voitovich RF. Oxidation of carbides and nitrides. Kiev: Naukova Dumka; 1981. 192 pp. (in Russian).
118. Kurlov AS, Rempel AA. Effect of WC nanoparticle size on the sintering temperature, density, and microhardness of WC-8 wt.% Co alloys. Neorgan Materialy. 2009;45(4):428–433. (in Russian) (Engl. Transl.: Inorg Materials. 2009;45(4):380–5).

119. Kehl WL, Hay RG, Wahl D. The structure of tetragonal tungsten trioxide. J Appl Phys. 1952;23(2):212–4.
120. Salje E. The orthorhombic phase of WO_3. Acta Crystallogr. B. 1977;33(2):574–7.
121. Woodward PM, Sleight AW, Vogt T. Ferroelectric tungsten trioxide. J. Solid State Chem. 1997;131(1):9–17.
122. Diehl R, Brandt G, Salje E. The crystal structure of triclinic WO_3. Acta Crystallogr. B. 1978;B34(4):1105–11.
123. Salje EKH, Rehmann S, Pobell I, Morris D, Knight KS, Herrmannsdörfer T, Dove MT. Crystal structure and paramagnetic behaviour of ε-WO_{3-x}. J Phys: Condens Matter. 1997;9(31):6563–77.
124. Airdy A, Domeneghetti MC, Mazzi F, Tazzoli V, Salje EKH. Sheet superconductivity in WO_{3-x}: crystal structure of the tetragonal matrix. J Phys: Condens Matter. 1998;10(33):L569–74.
125. Woodward PM, Sleight AW, Vogt T. Structure refinement of triclinic tungsten trioxide. J Phys Chem Solids. 1995;56(10):1305–15.
126. Ozawa T. Kinetic analysis of derivative curves in thermal analysis. J Therm Anal 1970;2(3):301–24.
127. Ozawa T. Kinetics of non-isothermal crystallization. Polymer. 1971;12(3):150–8.
128. Šesták J. Thermophysical properties of solids, their measurements and theoretical thermal analysis. Amsterdam: Elsevier; 1984, 440 pp.
129. Málek J, Šesták J, Rouquerol F, Rouquerol J, Criado JM, Ortega A. Possibilities of two non-isotermal procedures (temperature- or rate-controlled) for kinetical studies. J Therm Anal. 1992;38(1–2):71–87.
130. Kissinger HE. Reaction kinetics in differential thermal analysis. Anal Chem. 1959;29(10):1702–7.
131. Freeman ES, Carroll B. The application of thermoanalytical techniques to reaction kinetics: the thermogravimetric evaluation of the kinetics of the decomposition of calcium oxalate monohydrate. J Phys Chem. 1958;62(4):394–7.
132. Criado JM, Ortega A. Non-isothermal transformation kinetics: remarks on the Kissinger method. J Non-Cryst Solids. 1986;87(3):302–11.
133. Criado JM, Málek J, Ortega A. Applicability of the master plots in kinetic analysis of non-isothermal data. Thermochim Acta. 1989;147(2):377–85.
134. Málek J. A computer program for kinetic analysis of non-isothermal thermoanalytical data. Thermochim Acta. 1989;138(2):337–46.
135. Šesták J, Berggren G. Study of the kinetics of the mechanism of solid-state reactions at increasing temperatures. Thermochim Acta. 1971;3(1):1–12.
136. Berger S, Porat R, Rosen R. Nanocrystalline materials: a study of WC-based hard metals. Progr Mater Sci. 1997;42(1–4):311–20.
137. McCandlish LE, Kear BH, Kim BK. Processing and properties of nanostructured WC-Co. Nanostruct Mater. 1992;1(1):119–24.
138. Milman YuV, Chugunova S, Goncharuck V, Luyckx S, Northrop IT. Low and high temperature hardness of WC-6 wt.% Co alloys. Int J Refr Met Hard Mater. 1997;15(1–3):97-101.
139. Kurlov AS, Gusev AI, Rempel AA. Vacuum sintering of WC - 8 wt.% Co hardmetals from WC powders with different dispersity. Int J Refr Met Hard Mater. 2011;29(2):221–31.
140. Rempel AA. Nanotechnologies. Properties and applications of nanostructured materials. Uspekhi Khimii. 2007;76(5):474–500. (in Russian) (Engl. Transl.: Russian Chem Reviews. 2007;76(5):435–61).
141. Tsvetkov YV, Panfilov SA. Low-temperature plasma in reduction processes. Moscow: Nauka, 1980. 360 pp. (in Russian).
142. Kurlov AS, Rempel AA. Effect of sintering temperature on the phase composition and microhardness of WC-8 wt % Co cemented carbide. Neorgan Materialy. 2007;43(6):685–91. (in Russian) (Engl. Transl: Inorganic Materials. 2007;43(6):602–7).
143. Krasovskii PV, Blagoveshchenskii YV, Grigorovich KV, Determination of oxygen in W-C-Co nanopowders. Neorgan Materialy. 2008;44(9):1074–9. (in Russian) (Engl. Transl.: Inorganic Materials. 2008;44(9):954–9).

144. Ribeiro FH, Dalla Betta RA, Guskey GJ, Boudart M. Preparation and surface composition of tungsten carbide powders with high specific surface area. Chem Mater. 1991;3(5):805–12.
145. Kurlov AS. Effect of vacuum annealing on the particle size and phase composition of nanocrystalline WC powders. Zh Fiz Khimii. 2013;87(42):664–71. (in Russian) (Engl. Transl.: Russ J Phys Chem A. 2013;87:4:654–61).
146. Kurlov AS, Gusev AI. Phase equilibria in the W - C system and tungsten carbides. Uspekhi Khimii. 2006;75(7):687–708. (in Russian) (Engl. Transl.: Russian Chem Reviews. 2006;75(7):617–36).
147. Kurlov AS, Gusev AI. Neutron and X-ray diffraction study and symmetry analysis of phase transformations in lower tungsten carbide W_2C. Phys Rev B. 2007;76(17). Paper 174115. P.174115-01–174115–16.
148. Warren A, Nylund A, Olefjord I. Oxidation of tungsten and tungsten carbide in dry and humid atmosphere. Int J Refr Met Hard Mater. 1996;14(5–6):345–53.
149. Brillo J, Kuhlenbeck H, Freund H-J. Interaction of O_2 with WC (0001). Surface Science. 1998;409:199–206.
150. Kurlov AS, Gusev AI, Tungsten carbides and W - C phase diagram. Neorgan Materialy. 2006;42(2):156–63. (in Russian) (Engl. Transl.: Inorganic Materials. 2006;42(2):121–7).
151. Hoch M, Blackburn PE, Dingledy DP, Johnston HL. The heat of sublimation of carbon. J Phys Chem. 1955;59(2):97–9.
152. Bolgar AS, Turchanin AG, Fesenko VV. Thermodynamic properties of carbides. Kiev: Naukova Dumka; 1973. p. 170–1. (in Russian).
153. Kulikov IS. Thermodynamics of carbides and nitrides. Chelyabinsk: Metallurgiya; 1988. 320 pp. (in Russian).
154. Rempel AA, Gusev AI, Nanostructure and atomic ordering in vanadium carbide. Pis'ma v ZhETF. 1999;69(6):436–42. (in Russian) (Engl. Transl.: JETP Letters. 1999;69(6):472–8).
155. Levinsky YV, Blagoveshchensky YV, Vol'dman GM, Mel'nik YI. Mechanism of enlarging the nanoparticles of niobium, molybdenum, and tungsten during annealing in vacuum and hydrogen. Materialovedenie. 2006;10:38–47. (in Russian).
156. Klyachko LI, Fal'kovskii VA, Khokhlov AM, Hard alloys based on tungsten carbide with fine structure. Moscow: Ruda i, Metally; 1999. 48 pp. (in Russian).
157. Korolev YM, Polukarov YM, Zanozin VM. Decarburization of tungsten carbide during electrochemical application of cobalt coatings in aqueous solutions. Poroshk Metallurgiya. 1992;6:20–3. (in Russian) (Engl. transl.: Powder Metallurgy Met Ceram. 1992;31(4):476–9).
158. Blagoveshchensky YV, Isaeva NV, Mel'nik YI, Blagoveshchenskaya NV. Sintering of nanostructured hard alloys. In: Nano-2011/Trans. 4th All-Russian Conf. on Nanomaterials (Moscow, March 1–4, 2011). Moscow: Institute of Metallurgy and Materials Science of the Russian Academy of Sciences, 2011. p. 15. (in Russian).

Chapter 5
Hardmetals WC–Co Based on Nanocrystalline Powders of Tungsten Carbide WC

5.1 Sintering of WC–Co Hardmetals

The ternary system W–Co–C and its pseudobinary section WC–Co have been the subject of extensive studies because "tungsten carbide-cobalt" composites are widely used in industry as hard metals with superior mechanical properties and cutting performance. Sintering of ultrafine WC–Co hardmetals was studied repeatedly [1–3]. Nevertheless there are no literature experimental data on the influence of a particle size of tungsten carbide powders on the phase transformations which take place in hardmetals of the WC–Co system at various sintering temperatures. There are no data on the features of chemical interaction between nanocrystalline WC powders and cobalt. Such data are necessary for producing of fine-grained and, particularly, nanostructured hardmetals. The potential advantages of these hardmetals can be realized only at preservation of least size of grains [4].

Hexagonal tungsten carbide WC is hard and brittle substance and can only be sintered at very high temperatures. For this reason, a metallic binder is commonly added to WC carbide, which insures the required strength of the material without reducing its hardness to value not lower than the acceptable level. A binder metal should form no carbides and melted metal binder should wet WC grains well. In the case of WC, best of all these requirements are met by cobalt Co. Hardmetals based on WC and Co are used successfully in metal cutting, stamping, pressing, rolling, and mining. The wide use of WC–Co hardmetals is due to their high hardness (about 12–18 GPa), the thermal stability of their mechanical properties at heating up to $\sim$1270 K [5], their high elastic modulus $E \approx 600$ GPa, and small thermal expansion coefficient which is equal to $(5-6) \times 10^{-6}\text{K}^{-1}$ [6, 7].

The W–Co–C system contains binary and ternary compounds. For the first time the literature data on the W–Co–C system were systematized by Rautala and Norton [8], who constructed the projection of liquidus surface of this system, isothermal section of its phase diagram at a temperature of 1673 K, and also pseudobinary c-T phase diagrams along the sections C–$Co_{43}W_{57}$ and WC–Co. The influence of carbon content on the phase equilibria and structure of alloys in the pseudobinary section

A. S. Kurlov and A. I. Gusev, *Tungsten Carbides*,
Springer Series in Materials Science 184, DOI: 10.1007/978-3-319-00524-9_5,

© Springer International Publishing Switzerland 2013

WC_y–Co containing 16 wt.% Co was studied by Gurland [9]. Metallography and X-ray diffraction investigation of a crystal structure of cubic η-carbides of the M_6C-type which form in the W–Co–C system was performed by Pollock and Stadelmaier [10]. In particular, they constructed two isothermal sections of this system at 1273 and 1673 K and identified nonstoichiometric cubic η-carbide $Co_2W_4C_y$. Lattice constant of the unit cell of this ternary carbide depends on carbon content and varies from 1.107 to 1.125 nm. The isothermal sections of the W–Co–C system at 1373 and 1573 K were constructed by Ettmayer and Suchentrunk [11]. An experimental study of the phase equilibria in the W–Co–C system at a temperature of 1423 K in combination with thermodynamic calculations was performed by Johansson and Uhrenius [12].

According to results of many studies, the cubic (space group $Fd\overline{3}m$) ternary η-carbides Co_2W_4C [10] and Co_4W_2C [13] of the M_6C-type exist in the W–Co–C system. These carbides are deficient in carbon content and are often thought to be a single phase, $Co_3W_3C_y$, which is a pseudobinary cubic substitutional solid solution. An exhaustive review of the thermodynamic properties of binary and ternary compounds and the phase equilibria in the W–Co–C system was given by Guillermet [14]. He calculated phase boundaries for several isothermal sections of the W–Co–C system at temperatures 2073, 1773, 1698, 1673, 1548 and 1443 K. Calculated isothermal sections are presented in the handbook [15]. All of the isothermal sections contain a cubic (space group $Fd\overline{3}m$) carbide of approximate composition $Co_6W_6C_y$ ($M_{12}C$-type carbide) and $Co_3W_3C_y$ solid solution. In recent studies, the phase boundaries of the high-temperature cubic tungsten monocarbide δ-WC, hexagonal monocarbide WC and, especially, lower hexagonal tungsten carbide W_2C which formed in the W–C system were refined [16–18].

The WC–Co pseudobinary section is the most important section for hardmetals and contains hexagonal tungsten carbide WC and face-centered cubic (fcc) solid solution of WC in cobalt, β-Co(WC) at temperature from 1000 to 1560 K. A liquid phase appears in the WC–Co system at a temperature of 1560–1593 K. According to the data [14], the maximum WC solubility in solid cobalt is equal to 19.3 wt.% and does not exceed ~9 wt.% when temperature decreases to 1270 K. Cubic solid solution of WC in cobalt, β-Co(WC), forms as a result of diffusion dissolution and appears in the WC–Co system at sintering temperatures between 1420 and 1570 K. In hardmetals WC–Co containing no more 10 wt.% Co, the cobalt that is not incorporated into the β-Co(WC) solid solution melts in the presence of WC at a temperature $T \geq 1593$ K [14] and ensures wetting of the carbide grains.

Although the phase equilibria in the W–Co–C, WC–Co, and W–C systems have been studied in detail [14, 15, 18, 19], no reliable experimental data on the sequence of phase transformations in the WC–Co system at different sintering temperatures have been reported in the literature. The sequence of phase transformations has been described only for the WC–Co hardmetals produced from coarse-grained WC powder with average grain size of 1–2 μm [20]. Meanwhile these data are necessary for determination of optimal sintering temperature providing the most dense and homogeneous microstructure of the forming hardmetal.

Systematic study of the effect of sintering temperature and particle size of WC powders on the phase composition, microstructure, and some properties of a hardmetals identical in composition to the most widespread hardmetals WC–8 wt.% Co is performed by authors [21, 22]. The optimal sintering temperature of hardmetals WC–8 wt.% Co is also determined in these works. Sintering of hardmetals WC–6 wt.% Co and WC–10 wt.% Co with ultrafine structure is investigated in work [23]. Let us discuss the results of works [21–23] in more detail.

5.1.1 Sintering of WC–8 wt.% Co Hardmetals

Investigation of perfect pseudobinary WC–Co section does not allow to determine a sequence of the phase transformations taking place in the WC–Co hardmetal at different sintering temperatures because its qualitative phase composition (β-Co(WC) + WC) does not vary at given Co content in the hardmetal and at a temperature from 1000 to 1830 K. At small change of hardmetal composition from the pseudobinary WC–Co section towards the smaller carbon content the hardmetal displaces in three-phase region. Existence of such phases as WC, W_2C, β-Co(WC), Co_6W_6C, Co_3W_3C in this region is determined also by temperature. Comparison of temperatures of one or another phase appearance in the hardmetal, on the one hand, and a particle size of WC powder from which hardmetal is sintered, on the other hand, allows to find out whether size of WC particle influences on the temperature and sequence of phase formation. It is known that vacuum sintering of WC–Co hardmetals is accompanied by the loss of a small amount of carbon and slight shift of the hardmetal composition to the field of formation of fragile ternary Co_6W_6C and Co_3W_3C phases. Carbon loss is appreciable especially at sintering of hardmetals from fine-grained WC powders with a grains less than 300 nm in size. Therefore sintering of powder mixtures WC and Co containing carbon ~0.1–0.3 wt.% less than it is in perfect composition (WC + 8 wt.% Co) is investigated in works [21, 22].

WC–8 wt.% Co hardmetals are prepared using four tungsten carbide powders differing in particle size: (1) coarse-grained tungsten carbide WC-coarse with an average particle size of 6 μm; (2) submicrocrystalline tungsten carbide WC-submicro with an average particle size of about 150 nm; (3) nanocrystalline carbide WC-plasm with an average particle size of 60 nm, produced by plasma-chemical synthesis; and (4) nanocrystalline carbide WC-mill with an average particle size of 20 nm, produced by high-energy ball milling of coarse-grained tungsten carbide powder in a PM-200 Retsch planetary ball mill during 10 h (conditions of ball milling are described in detail in [24–29]). Coarse-grained, submicrocrystalline and ball-milled nanocrystalline powders of tungsten carbide contain 92–95 wt.% of the higher carbide WC and 5–8 wt.% of the lower carbide W_2C. The content of combined carbon in coarse-grained, submicrocrystalline and nanocrystalline powders of tungsten carbide is equal to ~6.05, ~6.10 and ~6.00 wt.%, respectively. The nanocrystalline tungsten carbide powder synthesised by plasma-chemical method contains WC phase only; the carbon content is equal to ~6.1 wt.% in this powder.

Powder mixtures are prepared by grinding one of mentioned above tungsten carbide powders and microcrystalline Co powder with average particle size of 2 μm for 2 h. Hereinafter, the powder mixtures and the hardmetals produced will be designated for brevity as TC8-1 (powder mixture of coarse-grained WC-coarse powder with 8 wt.% Co powder and the hardmetals sintered from this powder mixture), TC8-2 (powder mixture of submicrocrystalline WC-submicro powder with 8 wt.% Co powder and sintered hardmetals), TC8-3 (mixture of plasma-chemical WC-plasm powder with 8 wt.% Co powder and sintered hardmetals), and TC8-4 (mixture of ball-milled nanocrystalline WC-mill powder with 8 wt.% Co powder and sintered hardmetals). Note that the initial (starting) TC8-1 powder mixture used for sintering hardmetals has an industrial origin and was produced at the Kirovgrad Plant of Hard Alloys ("KZTS", Kirovgrad, Russia).

Green bulk samples were prepared by cold-pressing the powder mixtures in a cylindrical steel die with small addition of ethanol. For all of the samples, the compaction pressure was 0.7 GPa. After pressing, the samples had the same cylindrical shape and were close in dimensions. The green samples were sintered in a vacuum of 10^{-2} Pa ($\sim 10^{-4}$ mm Hg) at different temperatures from 1070 to 1870 K.

The sintering process involved slow heating the sample from room temperature to 970 K for 3 h, holding the sample for 1 h at 970 K, heating from 970 K to the chosen maximum sintering temperature $T_{\text{sint}}^{\text{max}}$ during 1 h, holding at thistemperature $T_{\text{sint}}^{\text{max}}$ for 10 min, and cooling to room temperature. The maximum sintering temperature was varied from 1070 to 1570 K with a step of 100 K and between 1610 and 1870 K with a step of 10 K. The used sintering conditions are represented in Fig. 5.1. Using of a many-stage thermal theatment conditions of sintering allows to take into account indirectly the sintering kinetics at different temperatures. Indeed, the time of heating from temperature 970 K to the maximum sintering temperature $T_{\text{sint}}^{\text{max}}$ (1070–1870 K) is the same. Therefore the less temperature $T_{\text{sint}}^{\text{max}}$ is the more slowly heating in the range from 970 K to $T_{\text{sint}}^{\text{max}}$ is. It ensures completion of phase transformations. Thus, completed phase transformations are attained by a change of heating rate and a small

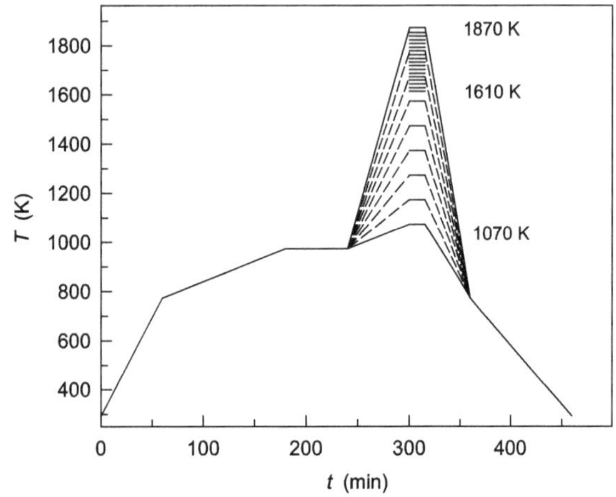

Fig. 5.1 Thermal theatment conditions used for sintering of WC-8 wt.% Co hardmetals

holding time at the sintering temperature $T_{\text{sint}}^{\max}$ is sufficient for fixation of these transformations.

The samples sintered at temperatures from 1070 to 1470 K had high open porosity, therefore their density ρ_{exp} was evaluated by measuring their mass and geometric volume on an air. The density of samples sintered at $T \geq 1570$ K was determined by means of the standard hydrostatic Archimedes method. The porosity $p = (\rho_{\text{theor}} - \rho_{\text{exp}})/\rho_{\text{theor}}$ of the sintered samples was evaluated from their measured density ρ_{exp} and the theoretical density $\rho_{\text{theor}} = 14.77\,\text{g} \cdot \text{cm}^{-3}$ of the WC-8 wt.% Co hardmetal.

The phase composition of the initial WC–Co powder mixtures and sintered hardmetals was studied by X-ray diffraction (XRD) on Shimadzu XRD-7000 and DRON-UM1 diffractometers. All the X-ray powder diffraction measurements were performed by the Bragg–Brentano method in the CuK$\alpha_{1,2}$ radiation at angles 2θ from 20 to 100° with a step of $\Delta 2\theta = 0.03°$ and a scanning time of 2 s at a point.

The average size of the coherent scattering regions in fine WC powders and nanostructured hardmetals was determined by an X-ray method from broadening of diffraction reflections [4] (*see* also Sect. 4.1.3). The size of the particles was evaluated also using the scanning electron microscopy (SEM). The particle size distribution in the WC powders was determined by laser diffraction method using a HORIBA partica LA-950V2 Analyzer.

The microstructure of the initial WC–Co powder mixtures and sintered hardmetals was examined by scanning electron microscopy in JEOL-JSM LA 6390 electron microscope equipped with EDX analyzer JED-2300. Some SEM images were obtained in LEO Gemini DSM982 (FELMI-ZFE, Graz) and JEOL 6310 (SCK-CEN) electron microscopes.

The composition of initial powders and sintered samples was certified by chemical and spectral methods. Chemical analysis for carbon and nitrogen was carried out by Dumas gas chromatography in a Carlo Erba CHN 1108 analyser. The oxygen content was measured by vacuum hot extraction in Balzers exhalograph EAO-201 equipment. The metal impurity content of the initial powder mixtures and of the sintered samples was determined by inductively coupled plasma-mass spectrometry in a Spectromass 2000. The total impurity content of the samples determined from the results of chemical and spectral analysis did not exceed 0.25 wt.%. The surface composition of the sintered samples was studied additionally by EDX method.

Metallographic cross sections for examining the microstructure and measuring the microhardness of the sintered hardmetal samples were prepared with a metallographic BUEHLER setup including PNEUMET-II, MOTOPOL-8, and MICROMET-1 modules. The sintered samples were cut on an electrical discharge machine and were then embedded in EPOMET molding compound on the PNEUMET-II module in automatic mode for subsequent polishing on the MOTOPOL-8 module. Coarse and mirror polishing were carried out with diamond suspensions with particles 30 to 1 μm in size. For exposure of grain boundaries and studying the microstructure, the polished sections were etched by the the saturated solution of iron chloride $FeCl_3$ in concentrated hydrochloric acid HCl, then the sections were washed by water and were etched by the solution consisting of 10 g sodium hydroxide NaOH, 10 g potassium ferricyanide $K_3Fe(CN)_6$, and 100 ml of H_2O.

The Vickers microhardness H_V was measured on polished samples using a MICROMET-I microhardness testing machine with automatic loading at a load of 50 g for a load time of 10 s. The average value of microhardness H_V was found as the average of 20 measurements on each sample.

Sequence of Phase Transformations During Sintering

The microstructure of the tungsten carbide powders used in [21, 22] is shown in Fig. 5.2.

The coarse-grained WC-coarse powder consists of rounded and irregularly shaped agglomerates with an average size of 6 μm; these agglomerates consist of fine particles about 2 μm in size (Fig. 5.2a). There are many individual fine particles not larger than 1 μm in size; content of these particles in coarse-grained WC powder reaches 10–15 vol. %. The average particle size of the submicrocrystalline WC-submicro powder (Fig. 5.2b) is equal to ~150 nm. Particles are strongly bonded to one another as if they were fused together. The WC-plasm powder produced by plasma-chemical synthesis has a microstructure (Fig. 5.2c) with well-observed large (up to 10 μm) loose agglomerates composed of smaller particles. The WC-mill nanopowder prepared by ball milling has a microstructure (Fig. 5.2d) similar to plasma-chemical WC powder. However, the agglomerates in the ball-milled powder have a smaller

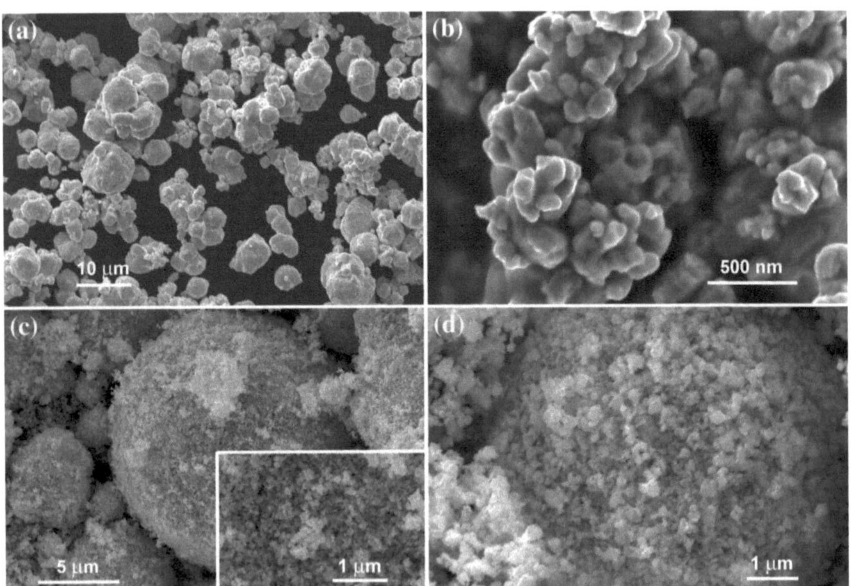

Fig. 5.2 SEM micrographs of tungsten carbide powders differing in particle size: **a** coarse-grained WC-coarse powder, **b** submicrocrystalline WC-submicro powder (the micrograph is obtained in cooperation with the FELMI-ZFE Centre, Graz), **c** nanocrystalline WC-plasma powder produced by plasma-chemical synthesis, the inset presents a SEM image of this nanopowder with large magnification, **d** nanocrystalline WC-mill powder produced by high-energy ball milling of coarse-grained WC-coarse powder in a PM-200 Retsch planetary ball mill

size which varies from 100 to 400 nm. According to XRD data, the average size of the coherent scattering regions is equal to 55 ± 10 nm in the plasma-chemical WC-plasm powder (*see* Sect. 4.2.1) and 20 ± 10 nm in the WC-mill powder prepared by ball milling during 10 h (*see* Sect. 4.2.2, Fig. 4.18). Thus, the coarse particles, up to several micrometers in size, observed by SEM are agglomerates of small particles with tens of nanometers in size.

According to the particle size distributions the average WC particle size is equal to 10 μm for the coarse-grained powder, 1 μm for the submicrocrystalline powder, 510 nm for the plasma-chemical powder and 310 nm for the ball-milled powder (Fig. 5.3). It means that the particles of powders are agglomerated. Observations in a scanning electron microscope (Fig. 5.2) also pointed to agglomeration of the WC powders. Note that the particle size determined by the XRD method is the smallest because the size estimated by the SEM and Horiba analyzer is a size of agglomerates rather than separate particles.

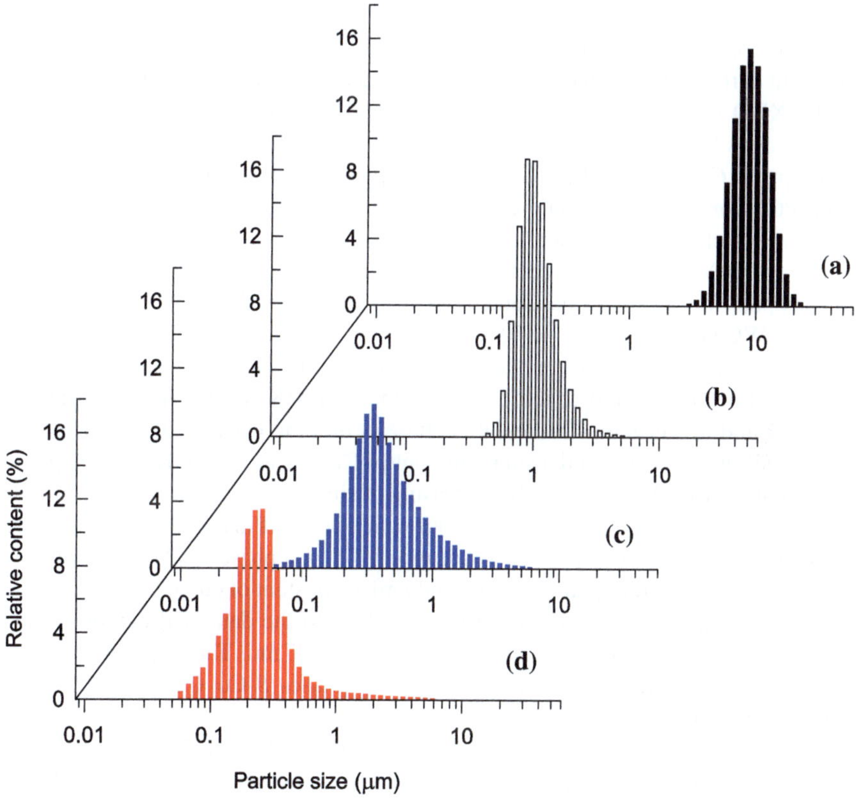

Fig. 5.3 Particle size distributions for the different WC powders: **a** coarse-grained WC-coarse powder, **b** submicrocrystalline WC-submicro powder, **c** nanocrystalline WC-plasm powder, **d** nanocrystalline WC-mill powder produced by ball milling of WC-coarse powder

Coarse-grained WC powder is stable during vacuum annealing. His chemical composition does not change after annealing in a vacuum at temperatures 1070 and 1270 K during 1 h.

Let us consider the variation of phase composition of the studied hardmetals during sintering at different temperatures.

The XRD patterns of the initial TC8-1 mixture of coarse-grained WC powder with Co powder and WC-8 wt.% Co hardmetals sintered from this powder mixture at temperatures $T_{sint}^{max} = 1070$, 1170, 1370, 1570, and 1870 K are shown in Fig. 5.4.

Chemical analysis shows that the initial powder mixture contains 7.9 wt.% Co in addition to tungsten and carbon; total content of carbon, C_{total}, is equal to 5.6 wt.% and includes 0.3 wt.% of free (uncombined) carbon. It should be noted that in the phase diagram of the W–Co–C ternary system the composition of the initial powder mixture is slightly displaced from the pseudobinary section WC–Co to lower carbon content. The XRD pattern of the initial WC-8 wt.% Co powder mixture contains diffraction reflections of three phases: the higher hexagonal (space group $P\overline{6}m2$) tungsten monocarbide WC, the lower hexagonal (space group $P6_3/mmc$) tungsten carbide W_2C, and cobalt α-Co with hexagonal (space group $P6_3/mmc$) structure. The contents of these phases are equal to ~85 wt. % WC, ~7 wt.% W_2C, and ~8 wt.% Co and are in agreement with the chemical analysis data within the experimental errors.

The XRD patterns of hardmetals sintered at 1070 and 1170 K (Fig. 5.4) show no reflections from α-Co, but there are reflections from the high-temperature cubic (space group $Fm\overline{3}m$) cobalt, β-Co, and a new phase Co_3W with a hexagonal (space group $P6_3/mmc$) structure. The cubic β-Co appears due to α-Co $\rightarrow$ β-Co phase transition which takes place at a temperature of 690 K(~420 °C) [30, 31]. The formation of Co_3W phase is connected with interaction between cobalt and tungsten which appears due to solid-phase eutectoid decomposition of the lower W_2C carbide to W and WC [17–19] at $T < 1270$ K. The lattice constants of the hexagonal Co_3W phase are $a = 0.514 \pm 0.001$ nm and $c = 0.413 \pm 0.001$ nm. This phase is present in the hardmetals sintered at temperatures up to 1170 K. In the samples sintered at 1370 K and higher temperatures, this phase is missing because of its peritectoid decomposition into β-Co and a small amount of Co_7W_6 at ~1370 K (see the phase diagram of the Co–W system [32]). Increasing of sintering temperature up to 1170 K leads to decreasing of content of W_2C phase which presents in initial powder mixture and to appearance of ternary cubic (space group $Fd\overline{3}m$) phase Co_6W_6C in addition to WC, β-Co, and Co_3W phases. Measured lattice constant of cubic Co_6W_6C phase is equal to 1.088 ± 0.001 nm and it is very close to the data [10] demonstrating the lattice constant of Co_6W_6C phase to be $a = 1.0897$ nm. The Co_6W_6C phase is formed through interaction of the Co-containing phases β-Co, Co_3W, and Co_7W_6 with W_2C and free carbon.

The XRD patterns of the samples sintered at 1270 and 1370 K show no reflections of the lower W_2C carbide, but there are well-defined reflections from WC, Co_6W_6C, and cubic solid solution of tungsten carbide in cobalt, β-Co(WC). The found phase composition is in good agreement with experimental data [10] and results of thermodynamic description of phase equilibria in the ternary W–Co–C system [14, 33]. In the XRD patterns of the hardmetals sintered at temperatures from 1270 to 1870 K,

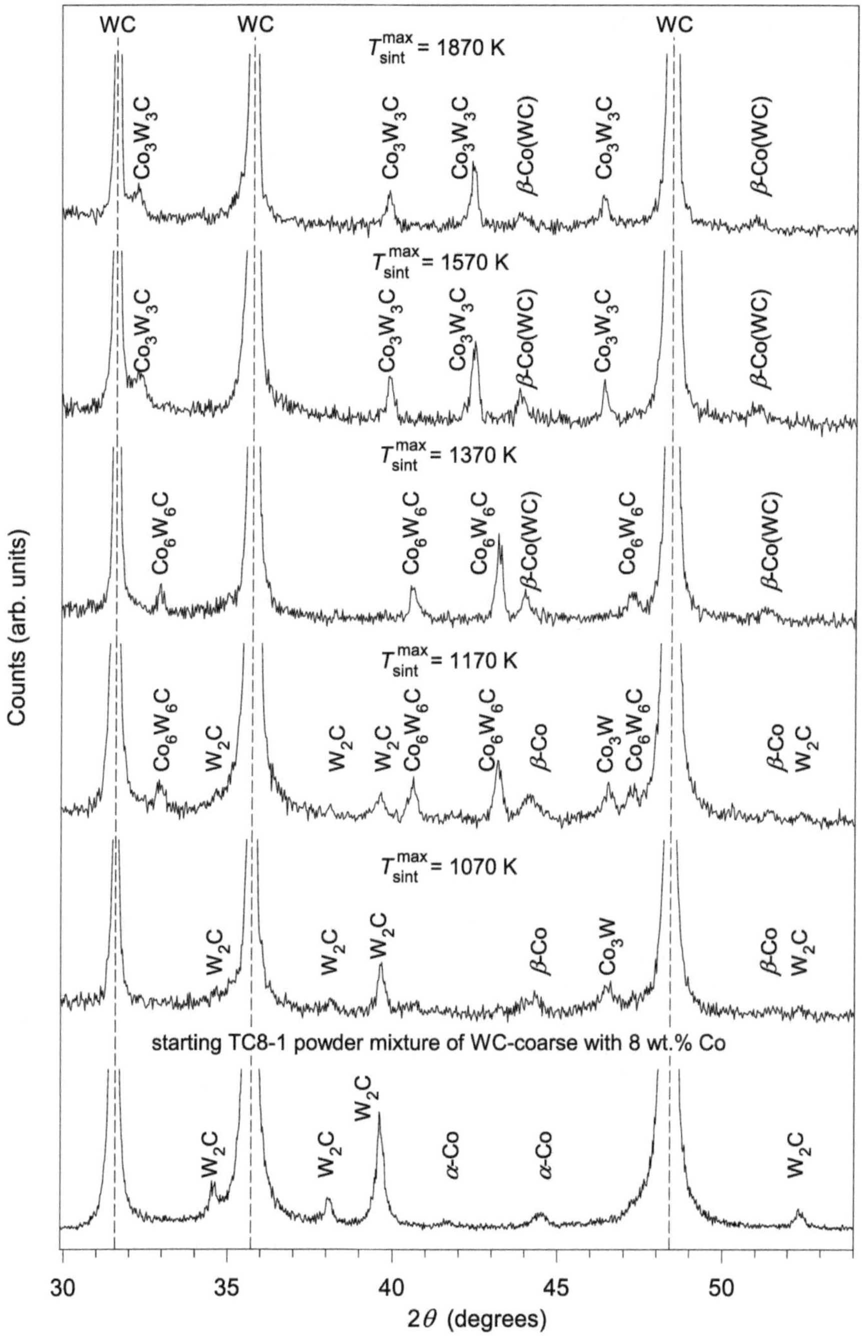

Fig. 5.4 XRD patterns of the initial TC8-1 mixture of coarse-grained WC-coarse powder with Co powder and WC-8 wt.% Co hardmetals prepared by sintering this powder mixture at different sintering temperatures T_{sint}^{max} with holding at T_{sint}^{max} for 10 min. All the XRD patterns are recorded in the CuK$\alpha_{1,2}$ radiation

the (111) and (200) diffraction reflections from the β-Co(WC) solid solution are slightly shifted to smaller diffraction angles in comparison with these from β-Co which are observed in the XRD patterns of the samples sintered at 1070 and 1170 K. This indicates that WC dissolution in β-Co is accompanied by increasing of its lattice constant. According to Guillermet [14], the tungsten carbide content in β-Co(WC) solid solution at temperatures from 1470 to 1870 K reaches 12–15 wt.%.

As the sintering temperature is raised from 1470 to 1570 K, the content of the carbide phase Co_6W_6C gradually decreases. In the sample sintered at 1570 K, the ternary Co_6W_6C phase disappears completely, but the sample contains a new cubic (space group $Fd\overline{3}m$) ternary phase Co_3W_3C which remains up to the highest sintering temperature 1870 K (Fig. 5.4). Thus, the Co_6W_6C phase is an intermediate sintering product and reacts with carbon forming the ternary Co_3W_3C phase. The lattice constant $a = 1.108 \pm 0.001$ nm of the cubic ternary Co_3W_3C phase is in good agreement with earlier data [10].

The XRD patterns of the TC8-1 hardmetals sintered in the temperature range 1570–1870 K have the set of reflections which are the same in the position and are close in the intensity. These diffraction reflections correspond to hexagonal (space group $P\overline{6}m2$) carbide WC with lattice constants $a = 0.2906 \pm 0.0002$ nm and $c = 0.2837 \pm 0.0003$ nm, cubic (space group $Fm\overline{3}m$) solid solution of tungsten carbide in cobalt, β-Co(WC), and cubic (space group $Fd\overline{3}m$) ternary Co_3W_3C phase. The insignificant changes in the relative intensity of the diffraction reflections from these phases indicate that the phase composition of the hardmetals sintered in the temperature range 1570–1870 K remains almost unchanged. Estimation of phase composition from XRD reflection intensities indicates that the hardmetals sintered at temperatures from 1670 to 1870 K contain ~87 wt.% WC, ~7 wt.% β-Co(WC), and ~6 wt.% Co_3W_3C. The effect of maximum sintering temperature T_{sint}^{max} with holding at T_{sint}^{max} for 10 min on the qualitative phase composition of TC8-1 samples investigated is illustrated in the Table 5.1.

Performed investigations have shown that the chosen holding time ($t = 10$ min) of a sample even at the lowest temperature $T_{sint}^{max} = 1070$ K is enough for achievement of such hardmetal phase composition which does not change qualitatively at increasing in holding time. Indeed, the increasing of a sintering time from 10 min up to 1 h and more at any studied sintering temperature $1070\,K \leq T_{sint}^{max} \leq 870$ K does not influence on a qualitative phase composition of a sample and changes slightly quantitative ratio of phases. As an example the XRD patterns of the WC-8 wt.% Co hardmetals sintered from TC8-1 powder mixture at $T_{sint}^{max} = 1070$ and 1650 K with holding for 10, 30 and 60 min are shown in Fig. 5.5. It is evident that increasing of the sintering time has not led to change of qualitative phase composition of hardmetals.

The XRD patterns of the initial (starting) TC8-2 mixture of submicrocrystalline WC powder and Co powder and the hardmetals sintered from this powder mixture at temperatures 1070, 1170, 1270, 1370, 1570, and 1870 K are shown in Fig. 5.6. Let us remind that the average particle size of the submicrocrystalline WC powder is equal to 150 nm.

The initial TC8-2 powder mixture contains three phases: the higher hexagonal (space group $P\overline{6}m2$) tungsten carbide WC, a small (<5 wt. %) amount of the lower

Table 5.1 Phase composition and porosity p of WC-8 wt.% Co hardmetals produced by vacuum sintering of mixtures of various WC powders with cobalt at different temperatures T_{sint}^{max} with holding at these temperatures for 10 min

$T_{sint}^{max} \pm 5$ (K)	TC8-1 (WC-coarse+Co)		TC8-2 (WC-submicro+Co)		TC8-3 (WC-plasm+Co)		TC8-4 (WC-mill+Co)	
	$^*p \pm 3\,\%$	phases	$^*p \pm 3\,\%$	phases	$^*p \pm 3\,\%$	phases	$^*p \pm 3\,\%$	phases
** initial	66	WC, W_2C, α-Co	51	WC, W_2C, α-Co	53	WC, α-Co	49	WC, W_2C, α-Co
1070	39	WC, W_2C, β-Co, Co_3W	50	WC, W_2C, β-Co, Co_6W_6C	52	WC, β-Co, Co_3W, Co_6W_6C	43	WC, β-Co, Co_3W, Co_6W_6C
1170	39	WC, W_2C, β-Co, Co_3W, Co_6W_6C	50	WC, W_2C, β-Co, Co_6W_6C	52	WC, β-Co, Co_3W, Co_6W_6C	41	WC, β-Co, Co_3W, Co_6W_6C
1270	37	WC, β-Co(WC), Co_6W_6C	49	WC, β-Co, Co_6W_6C, Co_3W_3C	46	WC, β-Co, Co_6W_6C, Co_3W_3C	38	WC, Co_6W_6C, Co_3W_3C
1370	32	WC, β-Co(WC), Co_6W_6C	48	WC, β-Co(WC), Co_3W_3C	35	WC, β-Co(WC), Co_6W_6C, Co_3W_3C	33	WC, Co_6W_6C, Co_3W_3C
1470	22	WC, β-Co(WC), Co_6W_6C	38	WC, β-Co(WC), Co_3W_3C	17	WC, β-Co(WC), Co_6W_6C, Co_3W_3C	24	WC, Co_6W_6C, Co_3W_3C
1570	7	WC, β-Co(WC), Co_3W_3C	25	WC, β-Co(WC), Co_3W_3C	10	WC, β-Co(WC), Co_6W_6C, Co_3W_3C	17	WC, Co_3W_3C
1670	3	WC, β-Co(WC), Co_3W_3C	17	WC, β-Co(WC), Co_3W_3C	9	WC, β-Co(WC), Co_3W_3C	14	WC, Co_3W_3C
1770	2 – 3	WC, β-Co(WC), Co_3W_3C	16	WC, β-Co(WC), Co_3W_3C	6	WC, β-Co(WC), Co_3W_3C	12	WC, Co_3W_3C
1870	1	WC, β-Co(WC), Co_3W_3C	9	WC, β-Co(WC), Co_3W_3C	3	WC, β-Co(WC), Co_3W_3C	9	WC, Co_3W_3C

* Porosity

* Initial green sample before sintering

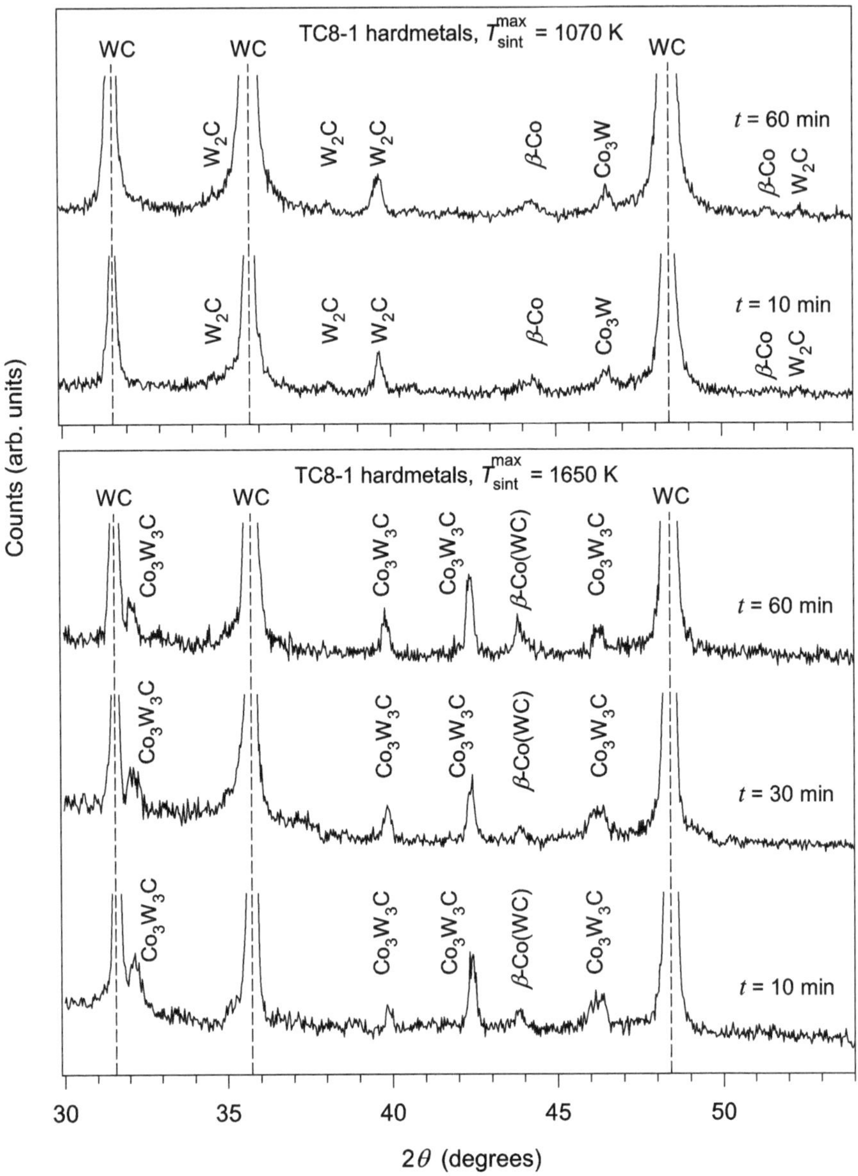

Fig. 5.5 XRD patterns of the WC-8 wt.% Co hardmetals prepared by sintering initial TC8-1 powder mixture of coarse-grained WC-coarse powder with Co powder with holding for 10, 30, and 60 min at maximum sintering temperatures $T_{\rm sint}^{\rm max}$ = 1070 and 1650 K. All the XRD patterns are recorded in the CuK$\alpha_{1,2}$ radiation

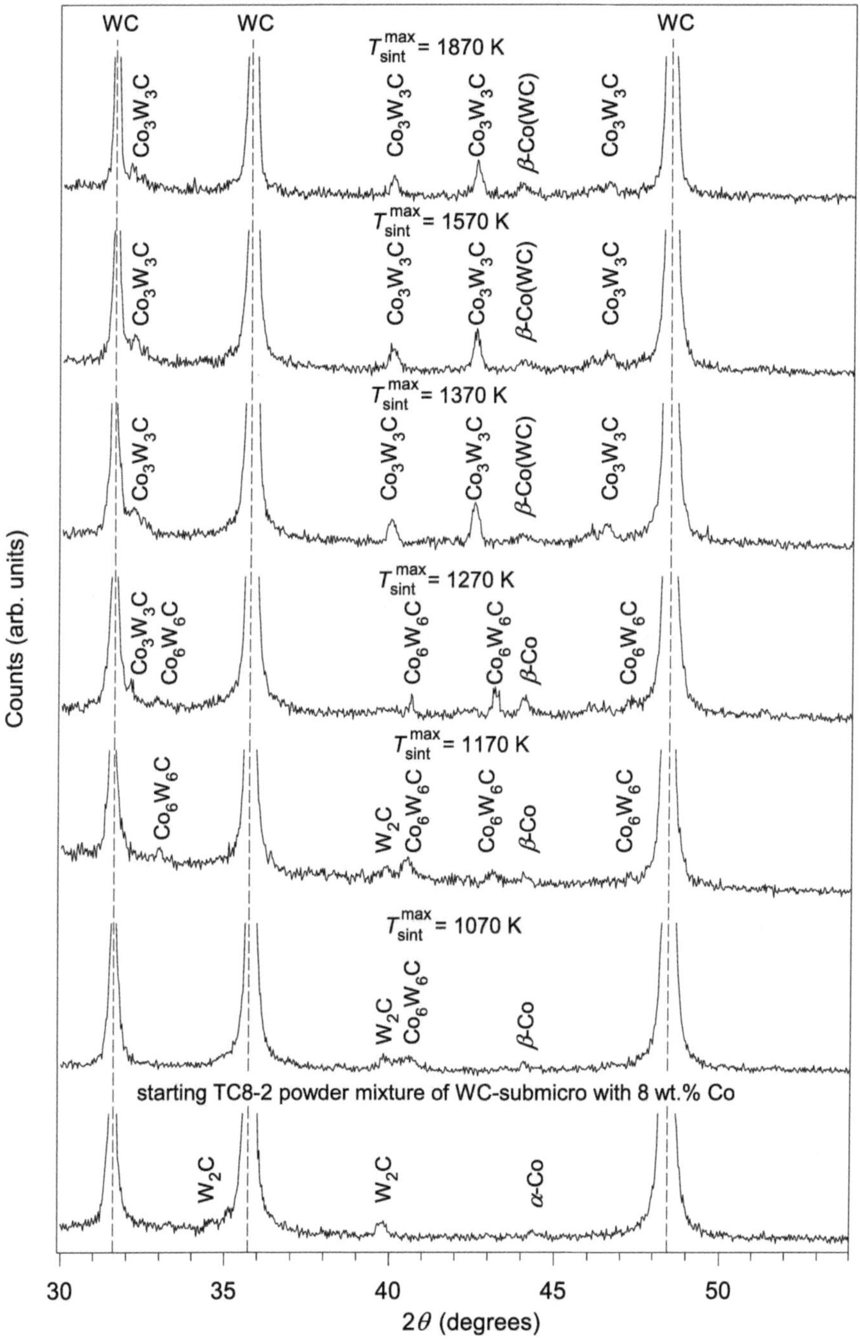

Fig. 5.6 XRD patterns of the initial TC8-2 mixture of submicrocrystalline WC-submicro powder with Co powder and the WC-8 wt.% Co hardmetals sintered from this powder mixture at different temperatures T_{sint}^{max} with holding at T_{sint}^{max} for 10 min. All the XRD patterns are recorded in the CuK$\alpha_{1,2}$ radiation

tungsten carbide W_2C, and α-Co with hexagonal (space group $P6_3/mmc$) structure. The XRD patterns of other powder mixtures (WC-plasm+Co) and (WC-mill+Co) do not contain diffraction reflections from W_2C carbide. It is a consequence of the low content of this phase and appreciable broadening of its diffraction reflections caused by the small size of the powder particles.

Diffraction reflections of hexagonal α-Co are absent in the XRD patterns of TC8-2 hardmetals sintered at temperatures 1070, 1170, and 1270 K (Fig. 5.6). The reflections of high-temperature cubic (space group $Fm\overline{3}m$) cobalt, β-Co, and reflections of the cubic (space group $Fd\overline{3}m$) ternary Co_6W_6C phase appear instead of α–Co reflections. The intensity of W_2C diffraction reflections decreases in comparison with the XRD pattern of initial powder mixture and these reflections are absent at a temperature of 1270 K. Ternary Co_6W_6C phase is formed owing to interaction between cobalt Co and lower tungsten carbide W_2C. Sintering at 1270 K leads to the formation of the new cubic (space group $Fd\overline{3}m$) ternary Co_3W_3C phase. The XRD pattern of the hardmetal sintered at 1370 K shows reflections of three phases: WC, Co_3W_3C, and a cubic solid solution of tungsten carbide in cobalt, β-Co(WC).

As the sintering temperature is raised from 1070 to 1270 K, the content of the Co_6W_6C phase decreases. In the samples sintered at 1370 K and higher temperatures, this phase is absent (*see* Table 5.1). The decreasing of the content of the ternary Co_6W_6C phase is accompanied by an increase of Co_3W_3C content. The ternary Co_3W_3C phase remains in hardmetal up to the highest sintering temperature $T_{sint}^{max} =$ 1870 K (Fig. 5.6). The XRD patterns of the TC8-2 hardmetals sintered at temperatures from 1470 to 1870 K contain the same set of reflections of hexagonal tungsten carbide WC, a cubic solid solution of tungsten carbide in cobalt β-Co(WC), and a cubic phase Co_3W_3C.

It is seen from Figs. 5.4 and 5.6 that change of phase composition of the TC8-1 and TC8-2 hardmetals at sintering temperatures $T_{sint}^{max} \geq 1370$ K is almost the same.

The XRD patterns of the initial (starting) TC8-3 mixture of plasma-chemical WC powder with Co powder and hardmetals sintered from this powder mixture at temperatures $T_{sint}^{max} = 1070$, 1170, 1270, 1370, 1570, and 1870 K are shown in Fig. 5.7. The initial TC8-3 powder mixture contains the higher hexagonal (space group $P\overline{6}m2$) tungsten carbide WC and cobalt α-Co with hexagonal structure. The WC diffraction reflections are broadened strongly due to small size of WC particles. Preliminary investigation has shown that vacuum annealing of one-phase plasma-chemical tungsten carbide powder leads to appearance of metallic W due to minor decarbonizing of WC. Indeed, content of bound carbon decreased from ~6.1 to ~4.9 wt.% after annealing of plasma-chemical WC powder in a vacuum at 1070 K for 1 h. According to X-ray data annealed plasma-chemical nanocrystalline powder contains ~79 wt.% WC and ~21 wt.% W. It is clear that the loss of carbon takes place at sintering of mixture of plasma-chemical WC powder and Co powder too. Therefore the composition of TC8-3 powder mixture shifts from the pseudobinary WC–Co section to the field of existence of ternary phases. As a result TC8-3 hardmetal sintered at a temperature of 1070 K contains ternary Co_6W_6C carbide together with WC, β-Co, and Co_3W (Fig. 5.7). Intermetallic Co_3W phase is formed due to interaction of Co with precipitating W.

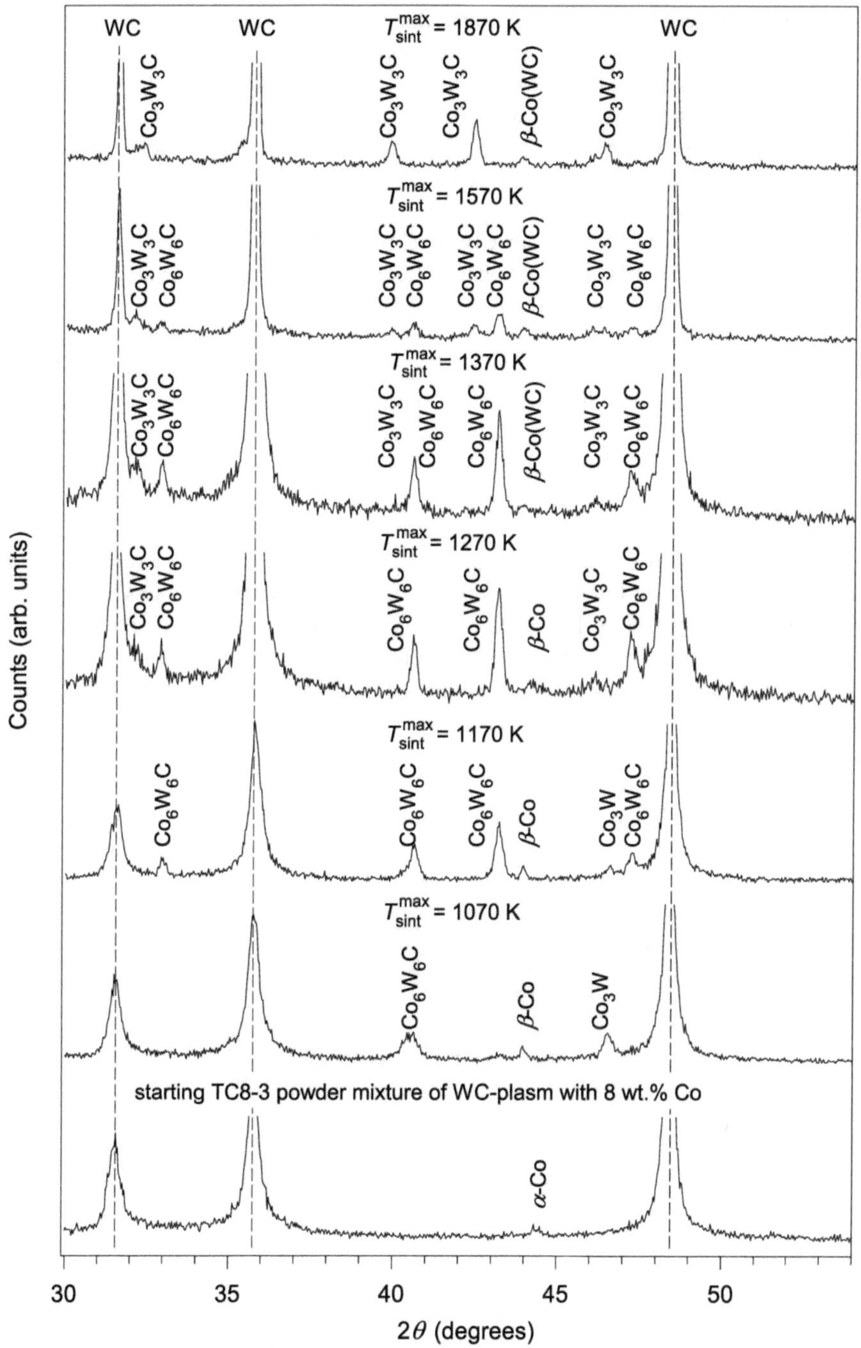

Fig. 5.7 XRD patterns of the initial TC8-3 mixture of plasma-chemical WC-plasm powder with Co powder and the WC-8 wt.% Co hardmetals sintered from this powder mixture at different temperatures T_{sint}^{max} with holding at T_{sint}^{max} for 10 min. All the XRD patterns are recorded in the CuK$\alpha_{1,2}$ radiation

The XRD pattern of the TC8-3 hardmetal sintered at 1370 K shows the diffraction reflections of WC, Co_6W_6C, Co_3W_3C and a cubic solid solution of WC in cobalt, β-Co(WC). When the sintering temperature T_{sint} is increased from 1370 to 1770 K, the relative content of the ternary Co_6W_6C phase decreases and it is absent in the sample sintered at 1770 K. The XRD patterns of the hardmetals sintered at a temperature of 1870 K have the set of diffraction reflections of hexagonal (space group $P\overline{6}m2$) tungsten carbide WC, cubic (space group $Fm\overline{3}m$) solid solution β-Co(WC) and cubic (space group $Fd\overline{3}m$) ternary Co_3W_3C phase. Judging by the intensity of the diffraction reflections, TC8-3 hardmetal sintered at 1870 K contains about 87 wt.% WC, 8 wt.% β-Co(WC)Co_3W_3C.

To determine the way of interaction between a fine-grained carbide phase and a metallic binder, we investigated a microstructure of TC8-3 hardmetals sintered at temperatures 1570 and 1670 K within 10 min. The initial powder mixture has been prepared from plasma-chemical WC-plasm nanopowder and Co powder with an average particle size of 2 μm. Feature of this WC-plasm nanopowder is aggregation of its nanoparticles into large agglomerates 5–10 μm in size (*see* Fig. 5.2c). Figure 5.8a represents the metallographic cross section of the sample sintered at 1570 K. This figure shows the interface between large (10–20 μm) cobalt inclusion (*on the right*) and grains of a hard carbide phase (*at the left*). The carbide phase grains are formed in a cobalt deficient field. In this field carbide phase grains with a size less than 1 μm are well connected, and boundaries between grains are almost not visible. On interface WC–Co there are well-faceted rectangular carbide grains with a length up to 3–4 (μm) on WC–Co interface. These large WC grains are formed due to contact with cobalt inclusion.

The presence of large cobalt inclusion in microstructure of the sample sintered at 1570 K means that the liquid-phase sintering has not occurred at this temperature yet. This is in agreement with results [1] according to which a sintering of fine-grained

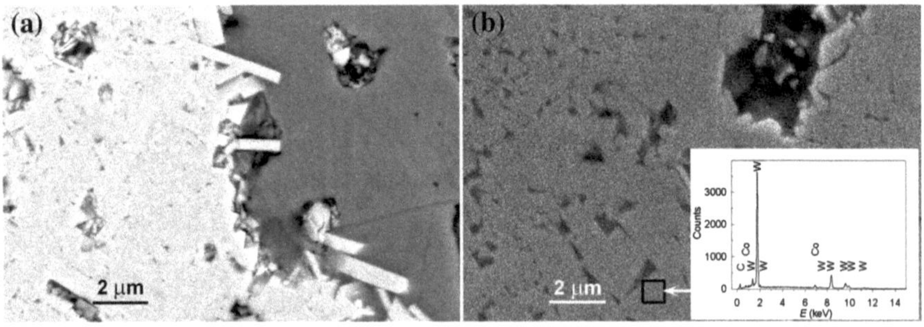

Fig. 5.8 Microstructure of WC-8 wt.% Co hardmetals sintered from TC8-3 mixture of plasma-chemical WC powder and Co powder: **a** $T_{sint}^{max} = 1570$ K, **b** $T_{sint}^{max} = 1670$ K. The inset in Fig. 5.8b presents an EDX spectrum obtained from marked section of the surface of the hardmetal. According to EDX-data content of tungsten W, carbon C and cobalt Co is equal to 46.6, 47.1, and 6.3 at%, respectively

WC–Co powder mixtures at temperature from 1470 to 1570 K takes place in a solid state not in a liquid one.

Sintering at a temperature of 1670 K occurs with a participation of a liquid phase and leads to disappearance of large cobalt inclusions and appearance of pores instead of these inclusions (Fig. 5.8*b*). This microstructure consisting of larger carbide grains has formed as a result of liquid Co binder infiltration in a carbide matrix. The presence of tungsten W, cobalt Co and carbon C in sintered hardmetal is confirmed by EDX method. EDX spectrum of the surface of TC8-3 hardmetal sintered at a temperature of 1670 K is shown in Fig. 5.8*b* (*inset*). According to the results of quantitative analysis, content of W, C and Co is equal to 90.1 ± 0.5, 6.0 ± 0.2, and 3.9 ± 0.6 wt.% (or 46.6, 47.1, and 6.3 at.%), respectively. Oxygen content does not exceed experimental measurement error equal to 0.1 wt.%.

Let us consider sintering of TC8-4 powder mixture containing cobalt and nanocrystalline WC-mill tungsten carbide powder produced by high-energy ball milling of coarse-grained WC powder. The average particle size of the ball-milled WC-mill nanopowder is equal to 20 nm. We prestudied whether vacuum annealing influences on chemical and phase composition of ball-milled nanocrystalline WC powder. Powder is annealed in a vacuum of 10^{-2} Pa at temperatures 670, 870, 1070, and 1270 K with soaking at annealing temperature for 1 h. Annealing at 670 and 870 K has not led to change of phase composition and average size of particles of WC nanopowder. Average particle size of WC nanopowder has not changed after vacuum annealing at 1070 K but phase and chemical compositions of a powder have varied. Content of bound carbon decreases from ~6.05 to ~4.95 wt. %, powder becomes two-phase and contains ~19 wt. % W in addition to ~81 wt. % WC. It is clear that sintering of mixture of ball-milled WC nanopowder and Co powder is accompanied by the decarbonizing too but to a less degree because a sintering duration is only 10 min.

The XRD patterns of initial (starting) TC8-4 powder mixture and the hardmetals sintered from this mixture at temperatures 1070, 1170, 1270, 1470, 1570, and 1870 K are represented in Fig. 5.9. The XRD pattern of initial TC8-4 powder mixture contains diffraction reflections of three hexagonal phases: WC and W_2C tungsten carbides and cobalt β-Co. The diffraction reflections of tungsten carbide WC are very strongly broadened owing to a small size (about 20 nm) of its particles. Coarse-grained tungsten carbide powder which was milled for preparing nanocrystalline WC powder, contained about 2 wt.% of the lower carbide W_2C (*see* Table 3.5). The only very weak diffuse reflection of W_2C carbide is presented on the XRD pattern of the initial TC8-4 powder mixture at $2\theta \approx 39.6°$.

In the XRD pattern of the TC8-4 hardmetal sintered at a temperature of 1070 K, the WC diffraction reflections are still diffused, but there are reflections of cubic (space group $Fd\overline{3}m$) ternary Co_6W_6C carbide, cubic (space group $Fm\overline{3}m$) cobalt β-Co and intermetallic hexagonal (space group $P6_3/mmc$) Co_3W phase (*see* Fig. 5.9). This phase is formed due to interaction of cobalt with tungsten W.

The diffraction reflection of ternary cubic Co_6W_6C phase is observed clearly. Apparently, interaction of cobalt with W_2C and WC occurs even at a temperature of 1070 K due to a very small size of the carbide particles. In the sample sintered at

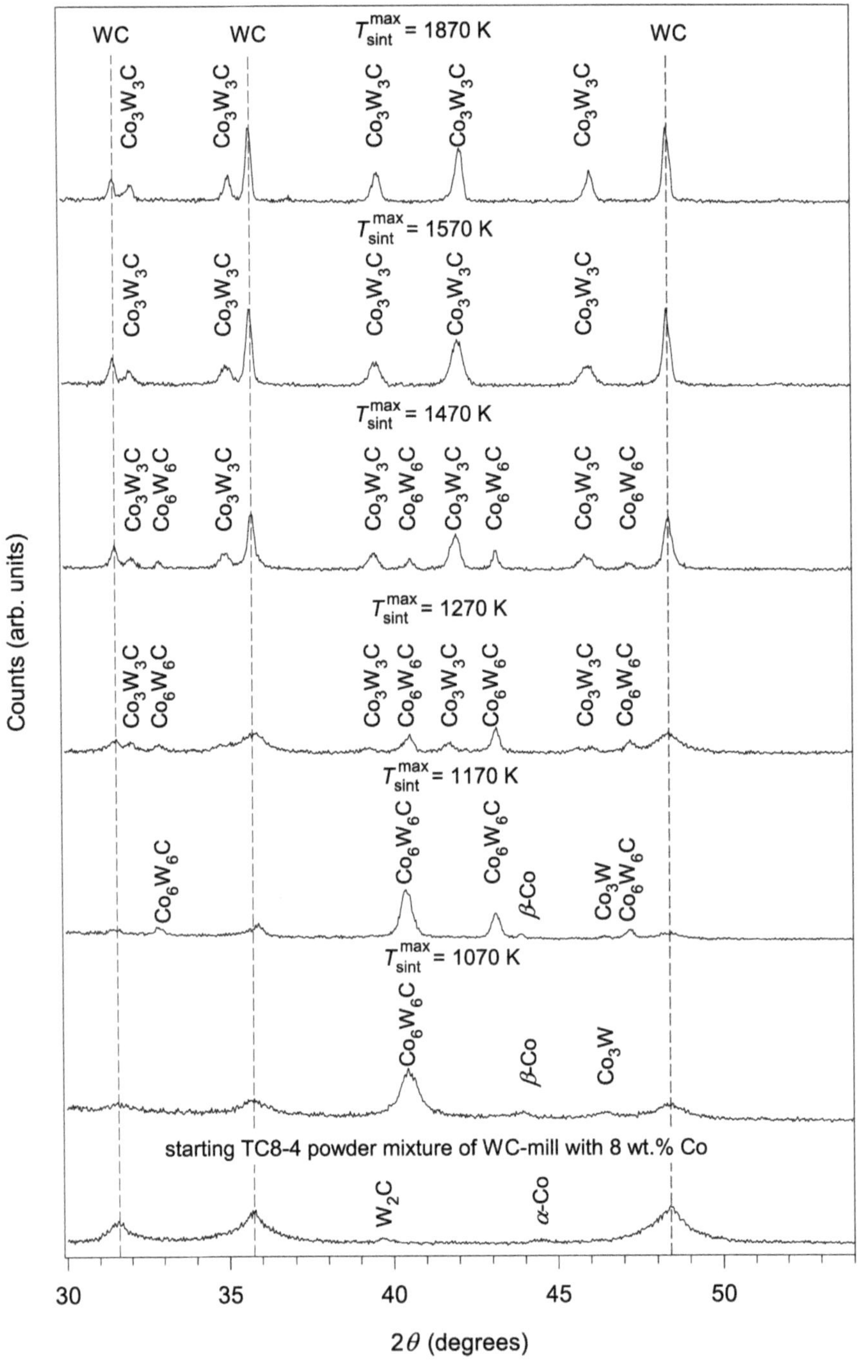

Fig. 5.9 XRD patterns of the initial TC8-4 mixture of ball-milled WC-mill powder with Co powder and the WC-8 wt.% Co hardmetals sintered from this powder mixture at different temperatures T_{sint}^{max} with holding at T_{sint}^{max} for 10 min. All the XRD patterns are recorded in the CuK$\alpha_{1,2}$ radiation

1170 K, there are the same four phases WC, Co_6W_6C, β-Co and Co_3W, but their diffraction reflections are narrowed owing to coarsening of grains at a sintering.

A sintering at $T > 1170$ K leads to disappearance of Co_3W phase and formation of ternary Co_3W_3C cubic phase. The diffraction reflections of three phases WC, Co_6W_6C and Co_3W_3C are visible in XRD patterns of TC8-4 hardmetals sintered at temperature from 1270 to 1470 K. Hardmetal sintered at 1570 K and higher temperatures contains WC and Co_3W_3C phases only (Fig. 5.9). As sintering temperature T_{sint} increases, diffraction reflections of ternary Co_3W_3C phase are shifted gradually towards greater angles 2θ because of decreasing the lattice constant of unit cell of this phase. While sintering temperature rises the broadening of the WC diffraction reflections is decreased owing to coarsening of WC grains.

On a whole, the study of the dependence of phase transformations on the sintering temperature of WC-8 wt.% Co hardmetals produced from WC powders with different average grain size has shown the following.

Samples of all studied hardmetals sintered at temperatures 1070 and 1170 K during 10 min have multiphase structure (*see* Table 5.1). It means that the sintering temperatures 1070 and 1170 K is not sufficient for the homogenization of chemical composition of hardmetal on all volume of the sample.

Judging by the XRD data obtained the influence of sintering temperature on the phase composition of the TC8-2, TC8-3 and TC8-4 hardmetals produced from three different fine-grained WC powders is almost the same (Table 5.1).

Owing to the smaller size of WC particle in these mixtures, the formation of the ternary phases Co_6W_6C and Co_3W_3C and β-Co(WC) solid solution in the TC8-2, TC8-3, and TC8-4 hardmetals begins at temperature approximately 100 K less than it is for TC8-1 hardmetal sintered from coarse-grained WC powder. For example, Co_6W_6C phase appears after sintering at 1170 K in case of coarse-grained TC8-1 powder mixture whereas this phase is observed in hardmetals from fine tungsten carbide powders sintered at 1070 K. Ternary Co_3W_3C phase forms in coarse-grained TC8-1 hardmetal after sintering at a temperature of 1570 K, and it arises in TC8-2, TC8-3 and TC8-4 hardmetals after sintering at 1270 K (*see* Table 5.1). Apparently, large area of interfaces in fine-grained hardmetals facilitates solid-phase interaction and reduces phase formation temperature. Studying of dependences of phase composition of the WC-8 wt.% Co hardmetals studied on maximum sintering temperature T_{sint}^{max} shows that using of fine-grained WC powders allows to sinter hardmetals at smaller temperatures.

Similar study has been made on WC-6 wt.% Co hardmetals sintered from tungsten carbide WC powders with different particle size. These results do not differ essentially from the results obtained for WC-8 wt.% Co hardmetals.

Effect of Sintering Temperature on Density, Microhardness, and Microstructure of Hardmetals

The density ρ of different WC–8 wt.% Co hardmetals as a function of maximum sintering temperature T_{sint}^{max} is depicted in Fig. 5.10.

At first let us consider sintering of green samples compacted from coarse-grained TC8-1 powder mixture. It is seen from Fig. 5.9 that sintering at temperatures from

1070 to 1270 K has an insignificant effect on the size of green sample. No densification occurs at these sintering temperatures in spite of the formation of the ternary Co_6W_6C phase and the intermetallic Co_3W phase. Considerable densification begins at sintering temperatures when Co binder becomes liquid. Indeed, a liquid Co appears at $T \geq 1470$ K, as follows from the phase diagram of the ternary W–Co–C system [14, 15]. The most active densification takes place at temperatures from 1470 to 1650 K (Fig. 5.10). It is in good agreement with experimental results [1] according to which the most intensive densification of hardmetal WC-15 wt.% Co produced from coarse-grained WC powder occurs at sintering temperature $T_{sint} \geq 1600$ K when a Co-binder is molten completely. As the sintering temperatures increased further, the density of TC8-1 hardmetals asymptotically approaches some limiting value which is constant within the experimental errors. When the maximum sintering temperature is raised from 1070 to 1650 K, the porosity of the sintered samples drops from 39 to 3 %. At higher sintering temperatures T_{sint}^{max}, up to 1870 K, the porosity of the sintered TC8-1 hardmetals varies from 3 to 1 % (*see* Table 5.1).

Variation of the density ρ of the TC8-2, TC8-3, and TC8-4 hardmetals with sintering temperature T_{sint}^{max} is shown in Fig. 5.10. Sintering at temperatures from 1070 to 1270 K has little effect on the sample dimensions. It means that no densification takes place at $T_{sint} \leq 1270$ K, in spite of the formation of the ternary phase Co_6W_6C. Densification begins at sintering temperatures above 1370 K. The most intensive densification occurs at temperatures from 1370 to 1570 K (Fig. 5.10). This is especially true of the TC8-3 samples: their porosity drops from 35 to 10 % in this temperature range. It means that sintering of fine-grained powder mixtures begins and takes place mostly in a solid state. It is in agreement with the conclusion of authors [1] that the WC–Co nanopowder densifies in the solid state at a sintering temperature from 1470 to 1540 K whereas a metallic binder melts completely at a temperature of 1600 K. At sintering temperatures T_{sint}^{max} above 1670 K, the density of all the hardmetals studied remains constant within the experimental errors (Fig. 5.10). At the

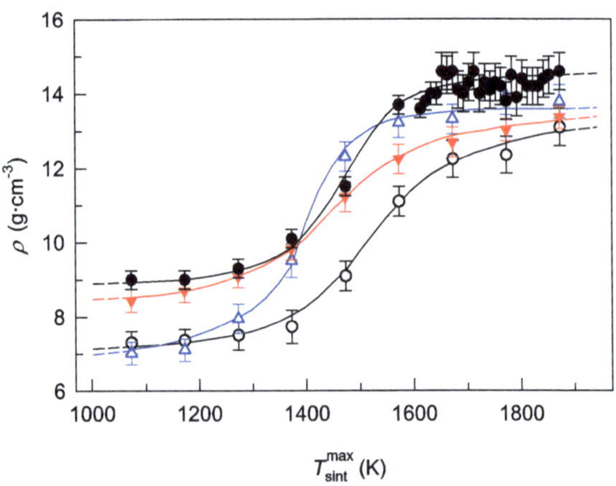

Fig. 5.10 Effect of maximum sintering temperature T_{sint}^{max} on the density ρ of the WC-8 wt.% Co hardmetals produced from different powder mixtures: (•) TC8-1, (○)TC8-2, (△) TC8-3 and (▼) TC8-4

highest sintering temperature 1870 K, the porosity of TC8-2 and TC8-4 hardmetals decreases to 9 %,and for TC8-3 hardmetal it is as low as 3 %.

As is seen from Fig. 5.10, the TC8-3 hardmetal has the highest density and TC8-2 hardmetal has the highest porosity among hardmetals obtained from fine-grained powder mixtures in the temperature range 1370–1870 K. The shape of the $\rho(T_{\text{sint}})$ curve for TC8-2 hardmetals is almost identical to that for the TC8-1 hardmetal but the density of the former hardmetal is lower in the entire temperature range examined.

Thus, it follows from the presented density data as well as from the XRD results that the use of fine-grained TC8-2, TC8-3 and TC8-4 powder mixtures allows one to reduce the sintering temperature by approximately 80–100 K compared to coarse-grained TC8-1 powder mixture.

The microhardness H_V versus sintering temperature $T_{\text{sint}}^{\text{max}}$ in the range 1270 to 1870 K for WC–8 wt.% Co hardmetals studied is shown in Fig. 5.11. The microhardness rises most rapidly with increasing maximum sintering temperature from 1420 to 1620 K and varies little at higher sintering temperatures. As seen in Fig. 5.11, TC8-2 hardmetal has the highest microhardness 19–20 GPa at $T_{\text{sint}}^{\text{max}} > 1670$ K.
The microstructure of WC-8 wt.% Co hardmetals studied is shown in Fig. 5.12.

The microstructure of the hardmetals sintered from coarse-grained TC8-1 powder mixture at 1610, 1660, and 1740 K was examined by scanning electron microscopy. The results indicate that sintering only begins at 1610 K because some of the grains remain rounded, as in the initial powder. This finding correlates with the densification and microhardness data: density ρ and microhardness H_V increase markedly at $T_{\text{sint}} > 1470$ K. The most active sintering takes place near 1660 K: the SEM micrograph (Fig. 5.12a) shows well-faceted tungsten carbide grains rather uniform in size (2–3 μm), without rounded grains, characteristic of the initial TC8-1 powder mixture. It should be mentioned especially that the size of grains of carbide WC phase in sintered hardmetal has not increased in comparison with the size of WC particles in the initial powder mixture. The same tendency is observed in the sample sintered at 1740 K; however, its microstructure is less uniform. It follows from the present

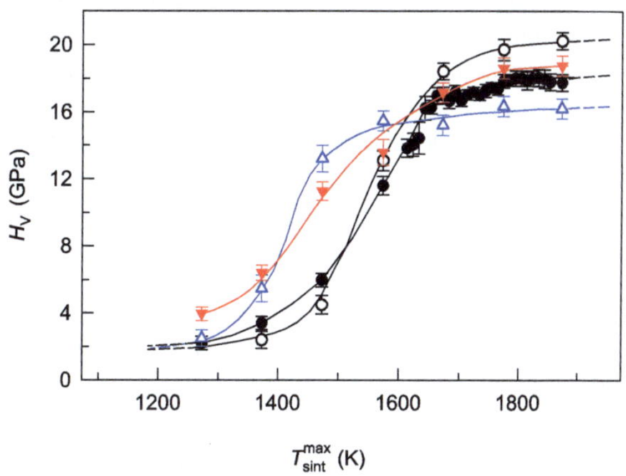

Fig. 5.11 Effect of maximum sintering temperature $T_{\text{sint}}^{\text{max}}$ on the microhardness H_V of the WC-8 wt.% Co hardmetals produced from different powder mixtures: (●) TC8-1, (○) TC8-2, (△) TC8-3 and (▼) TC8-4

Fig. 5.12 Microstructure of WC-8 wt.% Co hardmetals studied: **a** TC8-1 hardmetal sintered from coarse-grained powder mixture at 1660 K (the micrograph is obtained in cooperation with the FELMI-ZFE Centre, Graz), **b** TC8-2 hardmetal sintered from the powder mixture of submicrocrystalline WC with cobalt at 1670 K, **c** TC8-3 hardmetal sintered from powder mixture of cobalt and WC-plasm at 1670 K, **d** TC8-4 hardmetal sintered from mixture of cobalt powder and WC-mill nanopowder at 1650 K, the inset presents a SEM image of fracture surface of this hardmetal (the micrograph of fracture surface is obtained in cooperation with SCK-CEN, NMS Institute, Mol, Belgium)

experimental data that the highest density equal to 97–98 % of theoretical density and the lowest porosity of the hardmetals produced from coarse-grained TC8-1 powder mixture are reached by sintering at temperatures from 1650 to 1720 K. The most uniform (homogeneous) microstructure and the highest microhardness 17.0–17.5 GPa of the hardmetal studied corresponds to the same sintering temperatures.

The TC8-2 hardmetal sintered from mixture of Co powder and submicrocrystalline WC powder at a temperature of 1670 K has uniform microstructure with grains 300–400 nm in size (Fig. 5.12b). All grains are strongly bonded to one another. Sintered hardmetal contains large quantity of small pores. The grain size of the WC carbide phase in sintered hardmetal is approximately two times as high as a size of WC particles of initial powder mixture.

SEM micrograph of the TC8-3 hardmetal sintered from mixture of plasma-chemical WC powder with Co powder at 1670 K is presented in Fig. 5.12c. About 70 % of all grains of this hardmetal have a size no more than 500 nm, remaining grains have a size from 500 nm to 1.0–1.5 μm, and average grain size is ~600 nm. For the same hardmetals sintered at smaller temperature of 1570 K, an average grain size is equal to 400 nm.

The microstructure of the TC8-4 hardmetal sintered from mixture of Co powder and ball-milled nanocrystalline WC powder at a temperature of 1650 K is shown in Fig. 5.12d. The hardmetal contains well-faceted rectangular and triangular WC grains with a size 1–2 μm and a lot of fine grains with a size less than 500–600 nm. SEM fracture surface micrograph also shows that this hardmetal has enough uniform microstructure with WC grains about 700–900 nm in size (Fig. 5.12d, inset).

Studying of sequence of the phase transformations which take place at sintering of WC-8 wt.% Co powder mixtures and also microstructure, density and microhardness of the sintered hardmetals has shown that use of fine-grained and nanocrystalline WC powders allows to decrease sintering temperature $T_{\rm sint}^{\rm max}$ in comparison with hardmetals from coarse-grained WC powder approximately on 100 K. Thus, large density and uniform (homogeneous) microstructure are reached in nanostructured hardmetals at smaller sintering temperature than these in the hardmetals produced from coarse-grained WC powder.

Average size of grain in hardmetals sintered depends on the dispersity of initial powder mixture. In general the less a size of particles of initial WC powder is the less an average size of grains of carbide phase in hardmetals is. The size of grains of hardmetals sintered from fine-grained and nanocrystalline WC powders is one third - one forth as much as that of hardmetals produced from coarse-grained WC powder.

Investigation is shown that vacuum sintering of WC–Co hardmetals produced from nanocrystalline WC powders is accompanied by their partial decarbonizing (*see* also Sect. 4.3.3). The loss of carbon is observed already at minimum sintering temperature 1070 K and leads to an appearance of ternary carbide phases in hardmetals.

No carbon loss is observed at sintering of hardmetals from coarse-grained and submicrocrystalline WC powders. In this case ternary phase formation is caused by the carbon deficiency in initial carbide powders.

5.1.2 Sintering of Hardmetals WC-6 wt.% Co and WC-10 wt.% Co

To date, numerous various methods for producing nanocrystalline WC powders have been developed [34–36]. A most widely applied and productive chemical method for synthesizing fine-grained carbide powders is plasma-chemical synthesis, by which nanopowders with particles 10 to 100 nm in size can be produced.

In this section, results of investigation [23] are represented and technological problems that occur in using the conventional sintering method and affect the phase composition, microstructure, and mechanical properties of hardmetals produced from nanocrystalline WC-Co powder mixtures are discussed.

The initial materials for producing WC-6 wt.% Co hardmetal were nanocrystalline WC tungsten carbide powder with an average particle size of 35 nm that was synthesized by the plasma-chemical method with thermochemical treatment [37], microcrystalline Co cobalt powder with an average particle size of 2.2 μm, and nanocrystalline $VC_{0.87}$ vanadium carbide powder prepared by high-energy ball milling of coarse-grained $VC_{0.87}$ powder. A powder mixture TC6-nano containing 93 wt.% WC, 6 wt.% Co, and 1 wt.% $VC_{0.87}$ was prepared by multistage mixing in a Retsch PM 200 planetary ball mill. Owing to such a procedure, all the components were uniformly distributed throughout the mixture [38].

Hardmetal WC-10 wt.% Co was synthesized by sintering of a nanocrystalline powder mixture TC10-nano produced by the plasma-chemical method with thermochemical treatment and subsequent chemical deposition of cobalt [37]. A powder mixture TC10-nano contained inhibiting additives of nanocrystalline carbides $VC_{0.87}$, Cr_3C_2, and TaC in number 0.3–0.5 wt.% along with the main components (tungsten carbide WC and cobalt Co).

Usually, to obtain strong green bulk samples (compacts), a plasticizer is added to hardmetal powder mixtures. However, we failed to mix the nanopowder mixtures with a 9 % rubber solution in gasoline because the solution did not wet these mixtures. Therefore, samples were pressed from the nanopowder mixtures TC6-nano and TC10-nano without plasticizer. At a compaction pressure of ~200 MPa, the compacted samples were sufficiently strong and had very smooth and hard surface with glassy luster. However, the density of these compacts was ~95 % of the density of compacts made from coarse-grained powder mixtures of the same composition.

To avoid cracking and warping, the compacted samples were sintered in two stages. The first stage was slow annealing in a hydrogen atmosphere with a gradual increase in temperature to 1420 K. The second stage was the final sintering in a vacuum compression furnace with graphite lining. The treatment time at the maximal sintering temperature 1670 K was 1 h, for the first 30 min of which the sintering was performed in a vacuum at 5 mbar (500 Pa) and for the last 30 min of which it was done in an inert argon atmosphere at a pressure of 5 MPa (50 atm). After the sintering, the hardmetal samples got cold with the furnace.

Figure 5.13 presents the scanning electron microscopy images of particles of the nanopowders used for sintering of the hardmetals. The micrographs were taken at a magnification of 50 000 times and show individual particles, about 100 nm in size.

The powders are similar in morphology but differ in origin and composition. For example, WC-plasm nanopowder (Fig. 5.13a) and nanocrystalline powder mixture

Fig. 5.13 Scanning electron microscopy micrographs of nanopowders: **a** nanocrystalline tungsten carbide WC powder, **b** powder mixture TC6-nano, and **c** powder mixture TC10-nano

TC10-nano (Fig. 5.13c), which are produced by plasma-chemical method, have particles identical in shape and similar in size. The powder mixture TC6-nano was obtained by another method namely by mixing of the WC nanopowder and the microcrystalline Co powder with the inhibiting vanadium carbide additive. However in this mixture, scanning electron microscopy at various magnifications also failed to detect large cobalt particles. This is indicative of the fact that, in the course of the preparation of the TC6-nano mixture, microcrystalline cobalt was ground and uniformly distributed throughout the mixture owing to the intense interaction with hard carbide particles. Microscopic studies at various magnifications showed that most particles in all the three nanopowders are combined to form large loose aggregates reaching several micrometers in size. The aggregation of nanoparticles was also confirmed by the dispersion analysis of the powders which was performed by the laser diffraction method on a HORIBA LA-950V2 Analyzer. The dispersion analysis demonstrated that, in all the three nanopowders, particles are distributed in size within the range of 50 nm to 9 μm. Such a wide distribution is indicative of the fact that the powders contain both individual (disconnected) nanoparticles and large aggregates with average sizes of 800, 900, and 1300 nm in the WC, TC6-nano, and TC10-nano powders, respectively.

Quite a different result was obtained by analyzing the angular dependence of the broadening of X-ray diffraction reflections of WC carbide in the XRD patterns of various powders (Fig. 5.14). According to the X-ray diffraction data, the average size of the coherent scattering regions and microstrains in the WC, TC6-nano, and TC10-nano nanocrystalline powders are equal to 35 and 0.24, 32 and 0.33, and 40 nm and 0.24 %, respectively. The XRD analysis of the WC nanopowder showed that it consists of a single phase namely hexagonal (space group $P\bar{6}m2$) tungsten carbide WC (Fig. 5.14a). The XRD pattern of the powder mixture TC10-nano (Fig. 5.14c) exhibits both reflections of hexagonal tungsten carbide and pronounced reflections of cobalt. As judged from the positions of the reflections, cobalt is present in this mixture

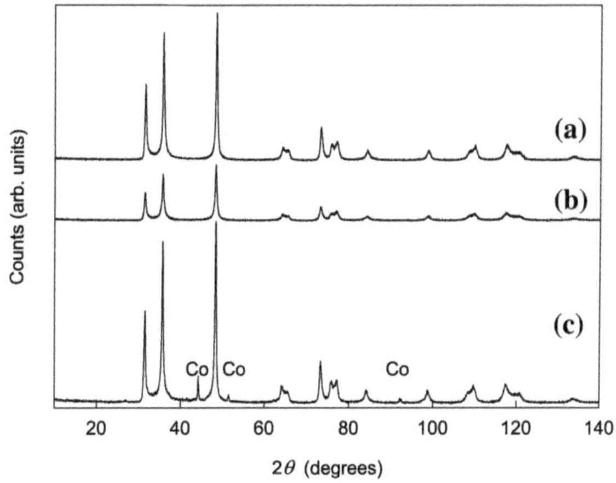

Fig. 5.14 XRD patterns of nanopowders: **a** nanocrystalline tungsten carbon WC powder, **b** powder mixture TC6-nano, and **c** powder mixture TC10-nano. In XRD pattern of powder mixture TC10-nano, cobalt reflections are marked. In other XRD patterns, cobalt reflections are not seen. All the XRD patterns are recorded in the $CuK\alpha_{1,2}$ radiation

as the high-temperature cubic (space group $Fm\overline{3}m$) modification. The XRD pattern of the powder mixture TC6-nano (Fig. 5.14b) shows only broadened reflections of hexagonal tungsten carbide; there are no cobalt reflections. This is because the powder mixture TC6-nano contains less cobalt than the powder mixture TC10-nano. Besides, the diffraction reflections of cobalt are highly broadened as a result of small size of particles and their deformations caused by intense attrition during preparation of the mixture. In other words, the reflections of deformed nanocrystalline cobalt particles are so broadened that they blend into the general background of the XRD pattern.

As was noted, a specific feature of the produced powder mixtures is the fact that they are hardly wetted by a 9 % rubber solution in gasoline, unlike coarse-grained mixtures of the same compositions. This forced us to compact the powder mixtures TC6-nano and TC10-nano without plasticizer. Nonetheless, the compacts made from the nanopowders had much higher strength, hard and smooth surface, but lower density than the samples compacted from coarse-grained mixtures. However, the single-stage vacuum sintering of the nanopowder compacts caused cracking and warping. Therefore, for sintering the nanopowder compacts, it was necessary to retain the stage of preliminary slow annealing in a hydrogen atmosphere at a temperature to 1420 K, which, in the process of sintering of samples from coarse-grained powders, is used for removing the plasticizer.

The samples of hardmetals WC-6 wt.% Co and WC-10 wt.% Co sintered from the powder mixtures TC6-nano and TC10-nano had no external defects.

The XRD analysis of the WC-6 wt.% Co and WC-10 wt.% Co hardmetals showed that the sintering did not change the phase composition (Fig. 5.15). The XRD patterns recorded from the surfaces of cross sections of the samples exhibit only reflections of hexagonal tungsten carbide WC and very weak reflections of the high-temperature cubic (space group $Fm\overline{3}m$) solid solution β-Co(WC).

A chemical analysis of the initial nanocrystalline WC powder used for producing the hardmetals demonstrated that it contains 6.14 wt.% total carbon and 0.41 wt.% free (uncombined) carbon. It is because of the excess free carbon content that we

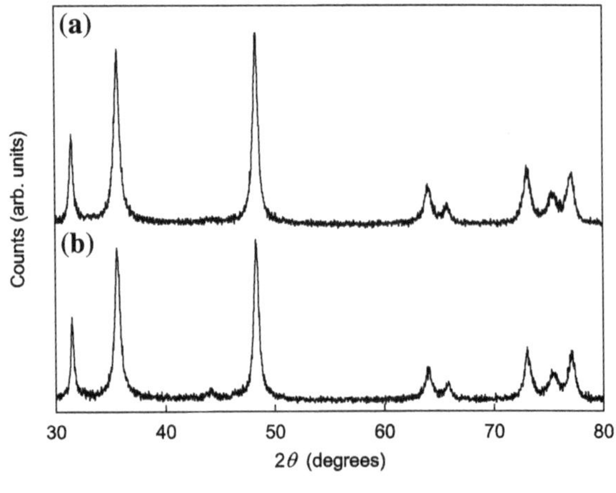

Fig. 5.15 XRD patterns of nanostructured hardmetals WC-6 Bec.% Co (**a**) and WC-10 Bec.% Co (**b**), sintered from powder mixtures TC6-nano and TC10-nano, respectively. All the XRD patterns are recorded in the CuK$\alpha_{1,2}$ radiation

managed to synthesize the hardmetals the phase composition of which perfectly corresponds to the quasi-binary section WC–Co of the phase diagram of the W–C–Co ternary system [14]. Indeed, it was shown [21, 22] that, in the sintering of WC–Co hardmetals from nanocrystalline WC powders, special attention should be paid to the carbon content of the WC nanopowder. It turned out that the standard carbon content of coarse-grained WC powders (5.80–6.12 wt.% total carbon and 0.08–0.10 wt.% free carbon) is too low when using nanocrystalline powders. The point is that nanopowders contain much oxygen adsorbed on the surface. As a result, the sintering is accompanied by partial decarburization of WC nanoparticles. This leads to a shift of the phase composition of hardmetals toward a three-phase region [17], in which the ternary η-phases $Co_3W_3C_y$ and $Co_6W_6C_y$ emerge, causing embrittlement of the sintered hardmetals. By using the WC nanopowder with higher carbon content, we avoided a shift of the phase composition of hardmetals toward a three-phase region.

By scanning electron microscopy with backscattered electron imaging, the micrographs of the surfaces of cross sections of the WC-6 wt.% Co and WC-10 wt.% Co nanostructured hardmetals were taken (Fig. 5.16). The micrographs show that the structure of each of the hardmetals is quite nonuniform and contains needle inclusions. The nanostructured hardmetals WC-6 wt.% Co and WC-10 wt.% Co contain both small grains with a size about 300–400 nm and elongated grains several micrometers long. Nevertheless, the produced hardmetals have minimal porosity A 0.02 % according to BS EN 24505:1993 (ISO 4505:1978). In other words, volume content of pores with a size less than 10 μm is mo more than 0.02 vol %. This is because the intergrain voids are filled with smaller grains. This is more clearly seen in the micrograph of the microstructure of WC-6 wt.% Co hardmetal (Fig. 5.16a). Indeed, the microstructure of WC-6 wt.% Co hardmetal is finer and more uniform than that of WC-10 wt.% Co hardmetal (Fig. 5.16b). Statistical processing of data on grain sizes using several microstructure micrographs showed that the average grain sizes in WC-6 wt.% Co and WC-10 wt.% Co nanostructured hardmetals are 0.6 and 0.8 μm, respectively. Analysis also demonstrated that the hardmetal grain size ranges from 80 nm to 4 μm, but more than 70 % of the grains are within the size range of 200–700 nm.

The measurements of the hardness and microhardness of the sintered nanostructured hardmetals showed that their hardness is higher than that of coarse-grained analogues [38, 39]. The average values of the hardness and microhardness of the hardmetal samples obtained from the nanocrystalline powder mixtures TC6-nano and TC10-nano are 93.9 HRA ($H_V = 2100\,\mathrm{kg{\cdot}mm^{-2}}$) and 91.4 HRA ($H_V = 1800\,\mathrm{kg{\cdot}mm^{-2}}$), respectively.

Conversely, the bending strength of the sintered hardmetals is low and is equal to 115 ± 5 and $125 \pm 5\,\mathrm{kg{\cdot}mm^{-2}}$ for nanostructured WC-6 wt.% Co and WC-10 wt.% Co hardmetals, respectively. This is likely to be caused by the lower density of the compacts made from the nanopowder mixtures.

Thus, for sintering nanostructured WC–Co hardmetals, it is necessary to use WC nanopowders with increased total carbon content, which ensures the correspondence of the phase composition of the sintered hardmetal to the quasi-binary cross section WC–Co of the phase diagram of the W–C–Co ternary system and rules out the formation of ternary η-phases impairing the mechanical properties of the hardmetals. However, the hardmetals sintered from WC nanopowders are inferior in strength to conventional hardmetals. Probably, only using nanopowders is insufficient to produce hardmetals with improved mechanical characteristics. Changes in the entire technology of hardmetals are required.

5.2 Microstructure of Nanocrystalline WC Powders, Micro-Structure and Bending Strength of WC–Co Hardmetals

The use of nanocrystalline WC powder as a component of hardmetals is the most perspective way for the production of nanostructured hardmetals having fine-grained microstructure, reduced sintering temperature, high hardness, strength and fracture toughness. Therefore a lot of investigations are devoted to the comparison of the mechanical properties of nanostructured WC based hardmetals with respect to properties of conventional hardmetals with micrometer grain sizes. Nevertheless, up to now it is not clear whether nanostructured tungsten carbide based hardmetals can replace the standard material in a wide application field. There is no literature data on interrelation of the particle size of WC powder used for sintering of hardmetals with the grain size and mechanical properties of the hardmetal produced. This problem is considered partially in works [40, 41] on an example of nanocrystalline WC powders and WC-8 wt.% Co hardmetals.

In the literature there is large number of papers which are devoted to the nanostructured WC-Co hardmetals. However the results of these works cannot be compared because the hardmetals were prepared from different precursors and more over by different technologies. Thus it is not clear whether the positive or negative effects on the hardmetal properties achieved in these works are results of fine microstructure or the result of technological parameters used in these works. That is why in this section

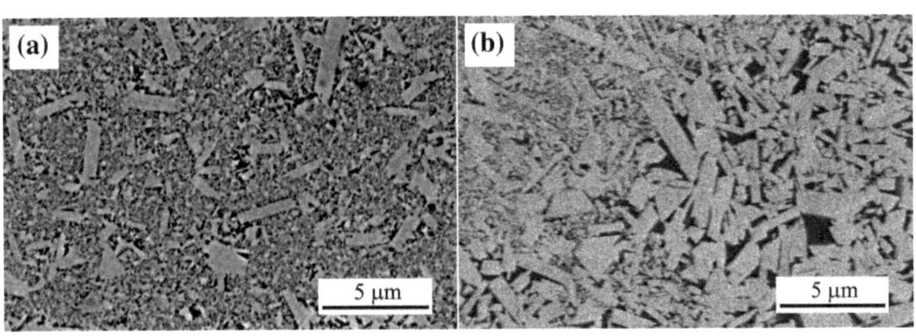

Fig. 5.16 Scanning electron microscopy micrographs of nanostructured hardmetals WC-6 wt.% Co (**a**) and WC-10 wt.% Co (**b**), sintered from powder mixtures TC6-nano and TC10-nano, respectively

the attempt to get comparable results on nanostructured WC–Co hardmetals is undertaken. For this reason the identical regime of sintering is applied for the preparation of WC-8 wt.% Co hardmetals starting from powder precursors with different size of particles.

This section is devoted to investigation of the influence of particle size of WC powder in micro- and nanoscale on the microstructure, hardness and bending strength of sintered TC8 hardmetals identical in composition to conventional widespread WC-8 wt.% Co hardmetal.

Four initial tungsten carbide powders WC-coarse, WC-submicro, WC-plasm, and WC-mill with different particle size were used for producing WC-8 wt.% Co hardmetals. Powder mixtures TC8-1, TC8-2, TC8-3, and TC8-4 were prepared by grinding one of mentioned above tungsten carbide powders with Co powder for 2 h. Initial tungsten carbide powders and composition of the powder mixtures are described in the beginning of Sect. 5.1.1. One more TC8-5 powder mixture was produced by high-energy ball milling of the standard coarse-grained TC8-1 powder mixture in a PM-200 Retsch planetary ball mill during 10 h at angular rotation speed 8.33 rps.

Green bulk rods with lengths of 40 mm and diameters of about 7 mm were prepared by cold-pressing the listed powder mixtures in a cylindrical steel die with small addition of ethanol. For all of the powder mixtures, the compaction pressure was 0.7 GPa.

The green rods were sintered in a vacuum of 10^{-2} Pa ($\sim 10^{-4}$ mm Hg). Figure 5.17 presents thermal treatment conditions used for sintering of hardmetal rods. Maximum sintering temperature T_{sint} is equal to 1550 K. It should be noted that the sintering temperature 1550 K is smaller than the eutectic temperature $T_e = 1590$ K of the pseudobinary WC–Co system [8, 14]. On the one hand, sintering at a temperature $T < T_e$ should provide the formation of homogeneous microstructure of hardmetals. On the other hand, sintering at $T < T_e$ allows to avoid strong recrystallization of carbide phase because recrystallization takes place in the presence of a liquid phase,

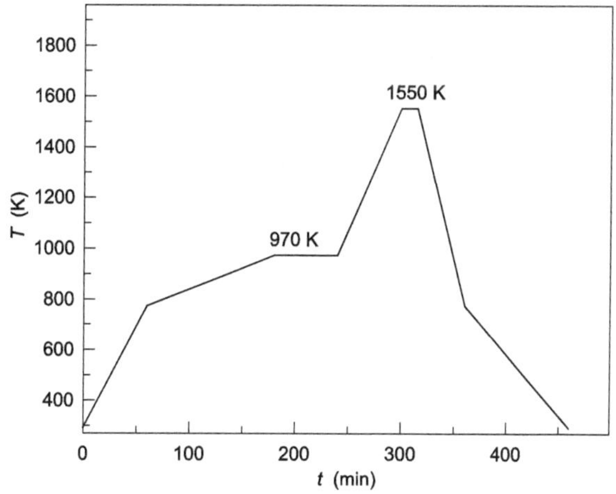

Fig. 5.17 Thermal treatment conditions used for sintering of hardmetals cylindrical rods from different powder mixtures of tungsten carbide with 8 wt.% cobalt

i.e. at $T > T_e$. For comparison a rod from the nanocrystalline WC-mill powder without Co binder was sintered.

The microstructure of the initial WC–Co powder mixtures and sintered hardmetals were examined by SEM in JEOL 6310 (SCK-CEN) and LEO Gemini DSM982 (FELMI-ZFE, Graz) electron microscopes. Some SEM images are obtained in JEOL-JSM LA 6390 electron microscope equipped with EDX analyzer JED-2300. The crystal structure and phase composition of the initial WC powders, WC–Co powder mixtures and sintered hardmetals were studied by XRD method. The average size of the coherent scattering regions in fine-grained WC powders and nanostructured hardmetals was determined from broadening of X-ray diffraction reflections (*see* Sect. 4.1.3). The size of the particles and grains is evaluated also using the scanning electron microscopy. The particle size distribution in the WC powders is determined using a HORIBA LA-950V2 Analyzer.

Bending strength was measured by means of three point bending tests with 18 mm distance between the support points using an INSTRON-1362 device. The cylindrical rods with length of 40 mm and diameter of about 7 mm are used for the measurements. The microstructure of the tungsten carbide powders used for producing WC-8 wt.% Co hardmetals cylindrical rods is shown in Fig. 5.18.

The coarse-grained WC-coarse powder consists of rounded and irregularly shaped agglomerates with average size of 6 μm; these agglomerates consist of fine particles about 2 μm in size (Fig. 5.18a). There are many individual fine particles not larger than 1 μm in size; content of these particles in coarse-grained WC powder reaches 10–15 vol. %. The average particle size of the submicrocrystalline WC-submicro powder (Fig. 5.18b) is equal to ∼150 nm. Particles are strongly bonded to one another as if they were fused together. The WC-plasm powder produced by plasma-chemical synthesis has a microstructure (Fig. 5.18c) with well-observed large (up to 10 μm) loose agglomerates composed of smaller particles. The nanocrystalline WC-mill powder prepared by ball milling has a microstructure (Fig. 5.18d) similar to plasma-chemical WC-plasm powder. However, the agglomerates in the nanocrystalline WC-mill powder have a smaller size from 100 to 400 nm. According to the SEM data, the average particle sizes in WC nanopowders obtained by plasma-chemical synthesis (Fig. 5.18c) and by ball milling (Fig. 5.18d) are equal to 100 and 180 nm, respectively. The mean particle size of the conventional TC8-1 powder mixture is equal to 2 μm (Fig. 5.18e).

Analysis of diffraction reflection broadening is shown that average size of the coherent scattering regions is equal to 55 ± 10 nm for the plasma-chemical nanocrystalline WC-plasm powder and 20 ± 10 nm for the nanocrystalline WC-mill powder produced by ball milling. The discrepancy in average sizes observed by SEM and XRD is evidence for a high degree of agglomeration of the nanopowders. Thus, the coarse particles, up to several micrometers in size, observed by SEM are agglomerates of small particles with tens of nanometers in size.

A similar situation is observed for the TC8-5 powder mixture produced by ball milling of the standard TC8-1 powder mixture. According to SEM data (Fig. 5.18f) the average particle size of the TC8-5 powder mixture is equal to about 180 nm and XRD analysis gives 20 ± 10 nm.

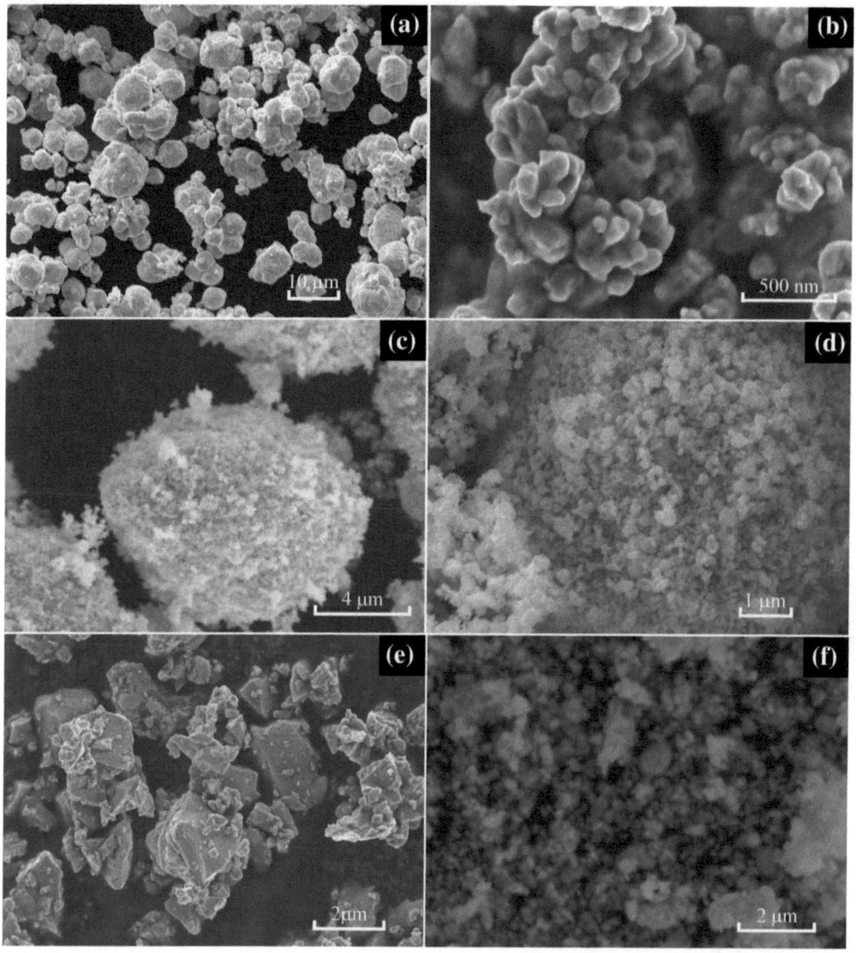

Fig. 5.18 Scanning electron microscopy micrographs of initial tungsten carbide WC powders and WC–Co powder mixtures: **a** coarse-grained WC-coarse powder, **b** submicrocrystalline WC-submicro powder, **c** nanocrystalline WC-plasm powder, produced by plasma-chemical synthesis, **d** nanocrystalline WC-mill powder, produced by high-energy ball milling of coarse-grained tungsten carbide powder, **e** coarse-grained TC8-1 powder mixture of tungsten carbide with 8 wt.% cobalt, **f** nanocrystalline TC8-5 powder mixture WC–Co, prepared by high-energy ball milling of coarse-grained TC8-1 powder mixture

SEM studies of a microstructure of sintered hardmetals (Fig. 5.19) show that the grain size for all WC-8 wt.% Co hardmetals is larger in comparison with the WC particle size for initial tungsten carbide powder and initial powder mixtures. SEM images of the fracture surface of all sintered hardmetals after the bending tests present in Fig. 5.19.

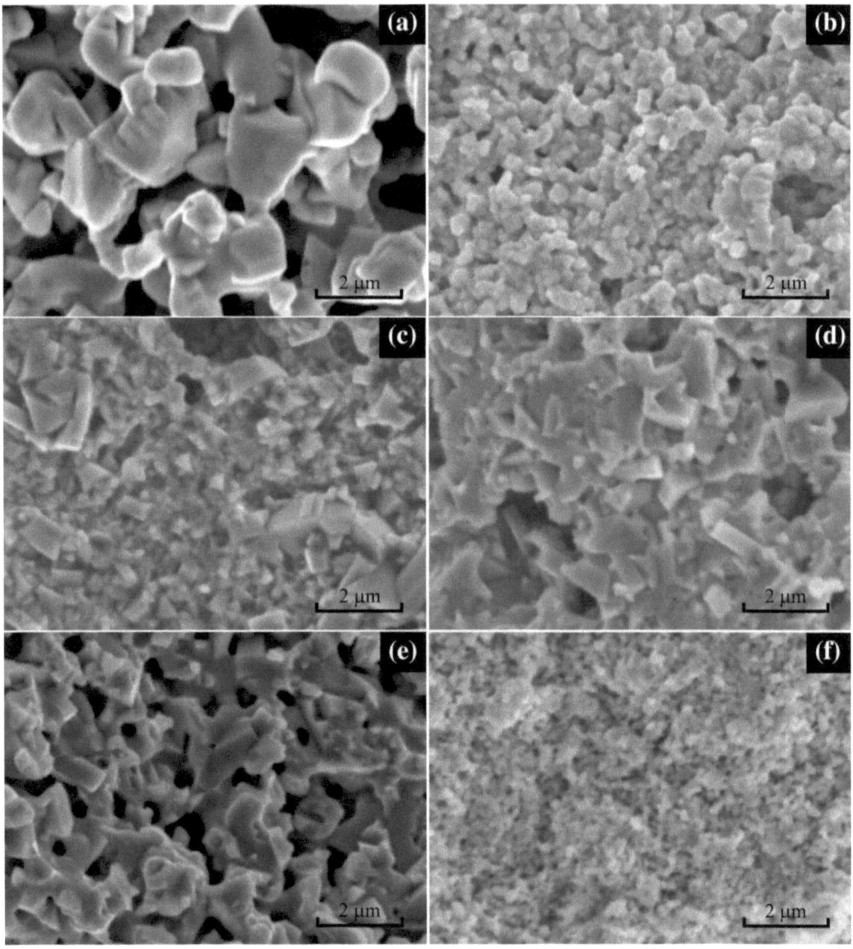

Fig. 5.19 Scanning electron microscopy micrographs of sintered hardmetals WC-8 wt.% Co and sintered pure tungsten carbide WC: **a** hardmetal TC8-1 sintered from coarse-grained WC-8 wt.% Co powder mixture: **b** hardmetal TC8-2 sintered from powder mixture of Co and submicrocrystalline WC-submicro powder: **c** hardmetal TC8-3 sintered from powder mixture of Co and nanocrystalline WC-plasm powder: **d** hardmetal TC8-4 sintered from powder mixture of Co and nanocrystalline WC-mill powder: **e** hardmetal TC8-5 sintered from nanocrystalline WC–Co powder mixture prepared by ball milling of coarse-grained TC8-1 powder mixture: **f** sample WC-mill sintered from nanocrystalline WC-mill powder without addition of cobalt Co

Insignificant grain growth up to 3 μm is observed for TC8-1 hardmetal (Fig. 5.19a) sintered from conventional coarse-grained powder mixture. The grain size of TC8-2 hardmetal is about 450 nm (Fig. 5.19b) and increases in comparison with a particle size of initial powder WC-submicro. Hardmetals TC8-3 and TC8-4 prepared from plasma-chemical and ball-milled nanocrystalline WC powders have grains about 450 and 850 nm in size, respectively (Fig. 5.19c and d). This is larger than the particle size for initial WC powders but is one second—one third as much as grain size of TC8-1 hardmetal sintered from conventional coarse-grained powder mixture. Average grain size of TC8-5 hardmetal is ~900 nm (Fig. 5.19e). Thus the grain size of hardmetals is larger in comparison with the particle size of initial powders. In particular, this is the case for TC8-3, TC8-4 and TC8-5 hardmetals prepared from fine-grained powders (Fig. 5.19c–e). Apart from this, the grain size in sample WC-mill sintered from nanocrystalline WC-mill powder without cobalt addition is practically the same as the particle size in initial powder (Fig. 5.19f).

Figure 5.19 shows that the microstructures of all sintered hardmetals are homogeneous. As is seen from Table 5.2, most fine-grained hardmetals are sintered from more fine-dispersed powders.

Loading plots $z(F)$ recorded during the mechanical testing are presented in Fig. 5.20 for all six rods. In this figure z is displacement of neutral axis of rod (i.e. flexure) and F is load which is normal to the neutral surface. Recorded loading curves $z(F)$ are typical for brittle materials. Indeed, the failure of rod is observed at critical loading without preliminary plastic deformation. The fracture begins when a critical value of bending stress equal to bending strength σ_{bend} is attained. Modulus of elasticity in bending is proportional to $\Delta F/\Delta z$ ratio where Δz is a change of flexure value at load changing by ΔF value.

Dependences $z(F)$ after convertation with mutual replacement of a position of z and F axes are presented in Fig. 5.21; the value of the breaking load (force) F_{break} at which the rod breaks is shown by a dashed line. As an example, dependences $z(F)$ for cylindrical rods from TC8-1 and TC8-3 hardmetals are shown in Fig. 5.21.

It is seen from Fig. 5.20 that the rods from TC8-1, TC8-2 and TC8-3 hardmetals are more elastic in comparison with TC8-4, TC8-5 rods and rod from WC-mill tungsten carbide because flexure z values of the former rods are smaller than that of the latter rods at the same load F.

SEM images of fracture surface of the rods produced from TC8-1 and TC8-3 hardmetals are shown in Fig. 5.22. Both hardmetals have porosity about 3 %. At a magnification of 10 times the fracture surface of the rod from TC8-1 hardmetal (Fig. 5.22a, top) represents a set of flat sections with small steps between them. Such element of a relief is named a "river pattern" and is characteristic for a macroscopic brittle failure [42]. In contrast to TC8-1 hardmetal, at the same magnification a fracture surface of the TC8-3 hardmetal rod is smooth (Fig. 5.22b, top). It means that the TC8-3 hardmetal have at least microplasticity traces. Average grain size of TC8-1 hardmetal is equal to ~2 μm (Fig. 5.22a, bottom). Average grain size of TC8-3 hardmetal is equal to 400–500 nm but there are separate large grains 1–2 μm in size (Fig. 5.22b, bottom).

According to [43], bending strength σ_{bend} of a cylindrical rod can be calculated using the formula

$$\sigma_{\text{bend}} = \frac{F_{\text{break}} L D}{8J}\left(1 + \frac{4z^2}{L^2}\right) = \frac{F_{\text{break}} L}{\pi R^3}\left(1 + \frac{4z^2}{L^2}\right), \tag{5.1}$$

Table 5.2 Properties of starting powders and sintered WC-8 wt.% Co hardmetals (average particle size or size of coherent scattering regions are given for the WC phase in the materials)

Starting WC–Co powder mixture or WC powders			Sintered hardmetals				
Powder	Average particle size according to SEM (nm)	Average particle size according to XRD (nm)	Hardmetal	Average grain size according to SEM (nm)	Vickers hardness $H \pm 0.5$ (GPa)	Microhardness $H_V \pm 0.5$ (GPa)	Bending strength $\sigma_{bend} \pm 10$ (MPa)
WC-coarse+Co	2000	-	TC8-1	2500	12.5	11.6	480
WC-submicro+Co	150	-	TC8-2	500	-	13.1	280
WC-plasm+Co	100	60	TC8-3	450	-	15.5	490
WC-mill+Co	180	20	TC8-4	850	-	13.6	90
ball-milled powder mixture WC–Co	180	20	TC8-5	900	-	16.6	190
WC-mill	180	20	WC-mill	180	16.8	16.0	70

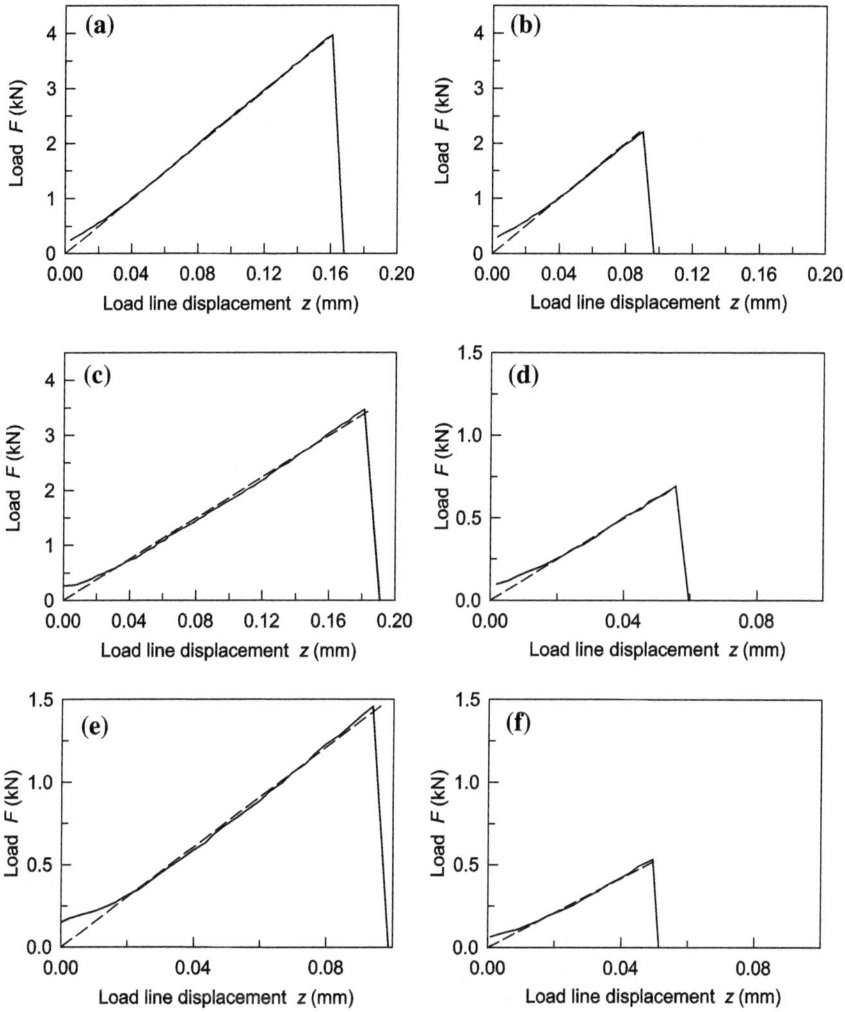

Fig. 5.20 Loading plots $z(F)$ of the bending tests for cylindrical rods of WC-8 wt.% Co hardmetals and of nanocrystalline tungsten carbide sintered without Co addition: **a** TC8-1, **b** TC8-2, **c** TC8-3, **d** TC8-4, **e** TC8-5, **f** WC-mill. Dependences $z(F)$ are recorded during the mechanical testing for measurements of bending strength

where F_{break} is the breaking load (force), $R = D/2$ is the radius of the transversal section of the rod, z is the flexure value, L is the distance between the support points, and $J = \pi R^4/4$ is the moment of inertia of the transversal section of a cylindrical rod. For brittle non-plastic materials value of flexure is negligibly small in comparison with L distance therefore the correction $(1 + 4z^2/L^2)$ can be neglected. In this case bending strength is

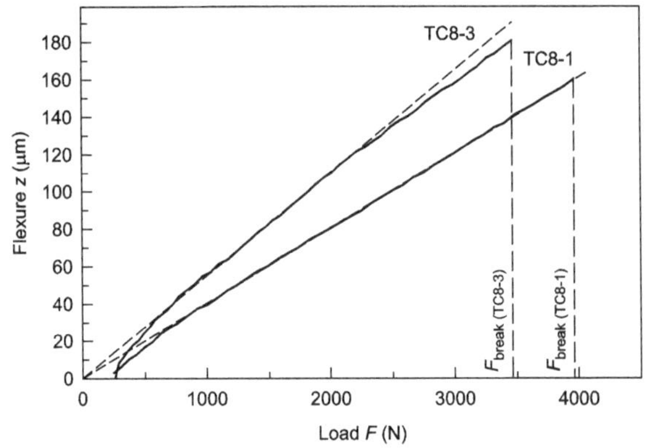

Fig. 5.21 Dependences $z(F)$ for cylindrical rods from TC8-1 and TC8-3 hardmetals after convertation with mutual replacement of a position of z and F axes. The value of the breaking load (force) F_{break} at which the rod breaks is shown by a *dashed line*

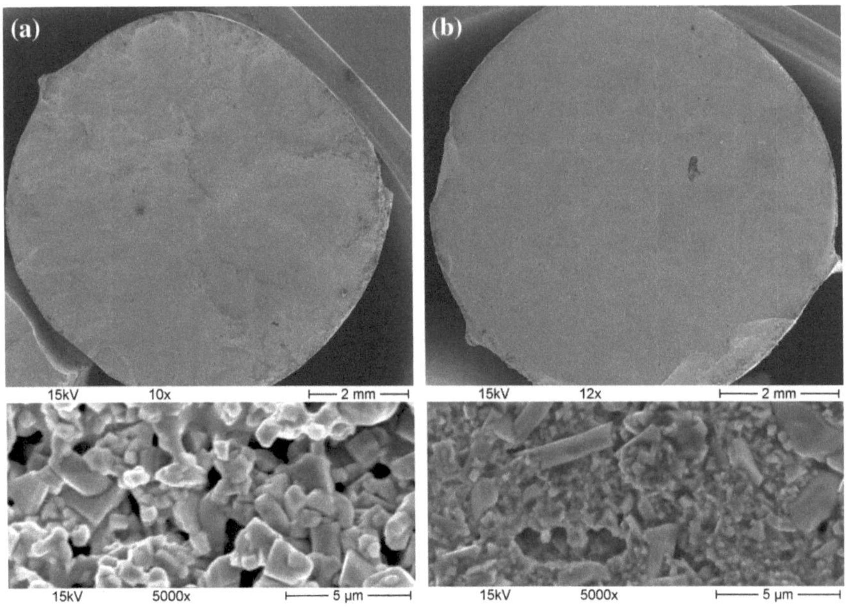

Fig. 5.22 Optical (*top*) and SEM (*bottom*) images of fracture surface of the rods produced from different WC-8 wt.% Co hardmetals: **a** the TC8-1 hardmetal, **b** TC8-3 hardmetal

$$\sigma_{bend} = F_{break} LD/8J \equiv F_{break} L/\pi R^3. \tag{5.2}$$

The calculated σ_{bend} values for the six measured rods produced from all hardmetals are given in the Table 5.2.

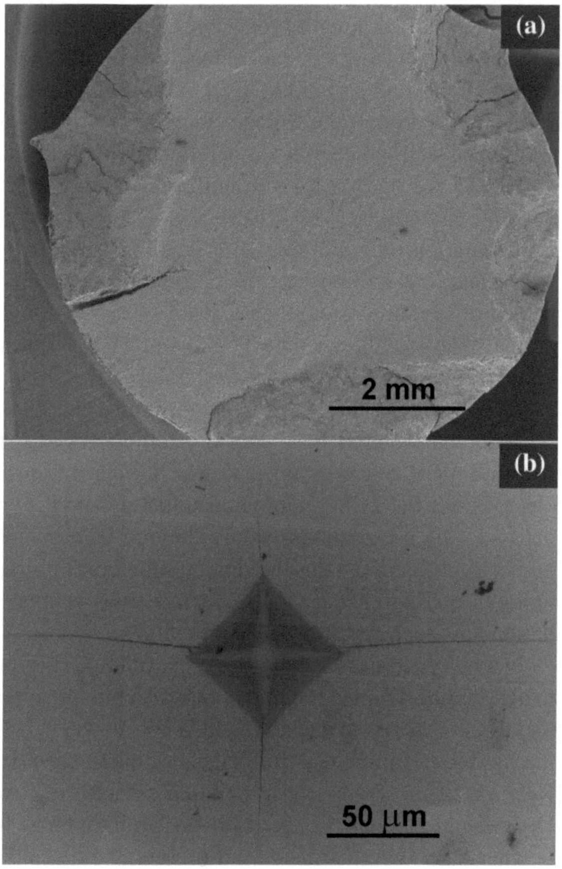

Fig. 5.23 Micrographs of the rod sintered from nanocrystalline WC-mill powder without addition of cobalt Co: **a** SEM image of fracture surface of the rod, **b** optical image of indentation obtained by Vickers hardness test under load of 50 N

Fine-grained hardmetals have enough smooth fracture surfaces and at a large magnification one can only see that destruction of hardmetals occurs mainly along the grain interfaces. The rod sintered from nanocrystalline WC-mill powder without Co addition has the lowest bending strength 70 MPa (*see* Table 5.2). This is because tungsten carbide itself is a very brittle material. It is confirmed by the formation of cracks in the vertices of indentations obtained at measurements of hardness and the presence of chips and cracks on the fracture surface (Fig. 5.23).

Measuring hardness H at large load P allows estimating value of fracture toughness K_{1c} of sintered samples from the hardness value and the length of cracks appearing in the indentation vertices. According to [44] fracture toughness is

$$K_{1c} = 0.016(E/H)^{1/2}P/l^{3/2}, \tag{5.3}$$

where E is Young's modulus and l is crack length measured from the center of indentation to the crack tip. We measured hardness H under load $P = 50$ N. The hardness

H and mean crack length l for the specimen sintered from nanocrystalline WC-mill powder without Co-binder are equal to 16.8 GPa and 105 μm (see Fig. 5.23), respectively. The Young's modulus E of WC carbide is equal to ~700 GPa. Taking these values into account, the fracture toughness K_{1c} of sample sintered from nanocrystalline WC-mill powder at temperature 1550 K is equal to about 5.0 MPa·m$^{1/2}$.

The TC8-3 hardmetal with uniform fine-grained microstructure has the highest bending strength σ_{bend} = 490 MPa closed to the σ_{bend} value of TC8-1 hardmetal. Confrontation of grain size with bending strength of TC8-2, TC8-3, TC8-4, and TC8-5 hardmetals finds out the correlation between these properties. The smallest values of bending strength σ_{bend} are observed for the TC8-4 and TC8-5 coarse-grained hardmetals whereas TC8-2 and TC8-3 hardmetals with smaller grain size have a larger bending strength (*see* Table 5.2).

As is seen from Table 5.2, the highest microhardness H_V is observed for the TC8-5 hardmetal sintered from nanocrystalline WC–Co powder mixture produced by ball milling of coarse-grained TC8-1 powder mixture. The lowest microhardness is observed for the TC8-1 hardmetal sintered from the coarse-grained WC–Co powder mixture. All WC–Co hardmetals sintered from nanocrystalline WC powders have higher microhardness than hardmetal sintered from coarse-grained WC powder. Even though the difference is small, this fact shows the need for further investigations of hardmetals prepared from nanocrystalline powders.

In whole, decreasing the particle size of initial WC powders down to nanosize leads to a more fine-grained homogeneous structure of sintered hardmetals. WC–Co hardmetals obtained from nanocrystalline WC powders have a higher microhardness than the hardmetals prepared from coarse-grained or submicrocrystalline WC powders. Relation of the grain size to bending strength of hardmetals is not so ambiguous. For example, cobalt free sample WC-mill sintered from nanocrystalline WC-mill powder has higher hardness and preserves the small grain size about 180 nm even after high-temperature sintering at 1550 K, but it is very brittle and has low bending strength. It means that the small grain size itself is not a sufficient condition for good mechanical properties of hardmetals.

Thus, while retaining the conventional technology of annealing and sintering, the replacement alone of coarse-crystalline powder mixtures of WC and Co by nanocrystalline powder mixtures allows one to obtain hardmetals with more fine-grained structure and increased hardness. However, the hardmetals sintered from WC nanopowders have less uniform microstructure than analogous standard hardmetals produced from ordinary tungsten carbide powders. Because of lower density of the initial compacts and nonuniform microstructure, the hardmetals made from nanopowder are inferior in strength to conventional hardmetals. Probably, only using nanopowders is insufficient to produce hardmetals with uniform fine microstructure and improved mechanical characteristics. It is necessary to make changes in the entire process line from plasticizer addition to nanocrystalline powder mixtures to their compaction and subsequent sintering of hardmetals.

5.3 Effect of the Morphology of Cobalt Powders on the Microstructure of WC–Co Hardmetals

Cobalt powders are an important component of hardmetals. Indeed, there are many hardmetals of the WC-Co system, which differ in the amount of cobalt, the dispersion of the carbide phase, the presence of hardening and inhibiting additives etc. All of them contain two components: tungsten carbide WC used as a "hard phase" and cobalt Co as a binder. The combination of high mechanical properties of tungsten carbide with plasticity and considerable impact toughness of cobalt provides increased hardness, strength, and wear-resistance of the hardmetals.

The most promising direction of improving WC–Co hardmetals is the formation of a fine-grained structure. This requires application of close-grain powders not only of tungsten carbide, but also of cobalt, as well as special sintering technologies [22, 34, 36, 41, 45–47] and consideration of the morphology of the initial cobalt powders. Let us consider how the morphology of cobalt powders affects the microstructure of WC–Co hardmetals containing 6 to 10 wt.% cobalt.

To produce test samples of WC-Co hardmetals, four different cobalt powders were used: thermochemical cobalt powders Co(th08) and Co(th14) with the average particle size 0.8 and 1.4 μm, respectively, electrolytic cobalt powder Co(el), and microcrystalline cobalt powder Co(kzts). Thermochemical powders were obtained by a one-step thermochemical method of decomposition and reduction of cobalt carbonate $CoCO_3$ in hydrogen atmosphere at temperature up to 890 K. Electrolytic powder Co(el) was produced by electrochemical deposition from an aqueous solution of cobalt sulfates and ammonium, in which ground plates of metallic cobalt were an additional source of Co^{2+} ions. Thermochemical and electrolytic Co powders were produced at the Electrochemical Converters Plant of the Urals Electrochemical Integrated Plant (ECP of the JSC UEIP). Microcrystalline cobalt powder Co(kzts) with the average particle size 2.2 μm produced at the Kirovgrad Plant of Hard Alloys (KZTS, Kirovgrad, Russia) was used for comparison. Also microcrystalline WC powder with the average particle size 7.6 μm produced at the KZTS was used for preparing test samples of WC–Co hardmetals.

All the initial powders of cobalt and tungsten carbide were certified using X-ray and laser diffraction, scanning electron microscopy, spectral and chemical analysis methods. The analysis procedure is described in Sect. 4.3.3. Chemical analysis of the initial WC powder showed that the total content of carbon C_{total} in this powder is 6.06 wt.% including 0.20 wt.% free (uncombined) carbon C_{free}. According to spectral analysis, the total content of impurity elements in the cobalt and tungsten carbide powders was less than 0.01 wt.%. The exception is the Co(el) powder containing 0.03 wt.% Cr and 0.35 wt.% Fe and the Co(th08) powder containing 0.01 wt.% Mg and 0.09 wt.% Ni.

The phase composition of the initial powders was studied by XRD method in the angular range $2\theta = 10$–$140°$ with a step 0.03° and scanning time 2 s at each point on a Shimadzu XRD-7000 diffractometer in $CuK\alpha_{1,2}$ radiation. The particle size distributions in the Co and WC powders were recorded by laser diffraction method on a HORIBA partica LA-950V2 Analyzer with the use of two sources of laser radiation with wavelengths 650 and 405 nm. In addition, powders of cobalt and tungsten carbides and also sintered WC–Co hardmetals were studied by the scanning electron microscopy (SEM) method on electron microscopes JEOL JSM 6390 LA.

Twelve homogeneous powder mixtures WC-6 wt.% Co, WC-8 wt.% Co, and WC-10 wt.% Co were prepared from tungsten carbide WC and one of the cobalt powders (Co(th08), Co(th14), Co(el), and Co(kzts)) in a planetary ball mill Retsch PM 200. Mixing was carried out in isopropanol for 30 min at the angular speed of rotation 5 rps and the ball to powder mass ratio 2:1. Upon mixing, the powders were dried at temperature 370 K for several hours. Then a plasticizer (a solution of rubber in benzene) was added to the powder mixtures and they were mixed again thoroughly till complete wetting. After that the powder mixtures were dried and sifted through a fine 120-mesh sieve. From these powder mixtures cylindrical green samples with diameter 7.5 mm and height ~8.0 mm were pressed. Axial static compacting pressure was 950 MPa for all the samples. The pressed green samples were sintered in $\sim 10^{-2}$ Pa vacuum at temperature conditions as presented in Fig. 5.24.

The phase composition of the hardmetal samples was examined by X-ray diffraction in the angle interval $2\theta = 30$–$80°$ with step $\Delta(2\theta) = 0.03°$ on a DRON-UM1 automated diffractometer using the Bragg-Brentano method in $CuK\alpha_{1,2}$ radiation. The density of the sintered samples was measured by hydrostatic weighing, and by comparing it with theoretical density the samples' porosity was determined.

Metallographic cross sections for examining the microstructure and measuring the microhardness were prepared by method which was described in Sect. 5.1.1. The Vickers microhardness H_V was measured on polished hardmetal samples using a MICROMET-I microhardness testing machine with automatic loading at a load of 100 g for a load time of 10 s.

The X-ray phase analysis showed that along with the basic WC phase the tungsten carbide powder contained about 2 wt.% of the lower tungsten carbide W_2C. All cobalt powders were two-phase and contained almost equal amounts of the low-temperature

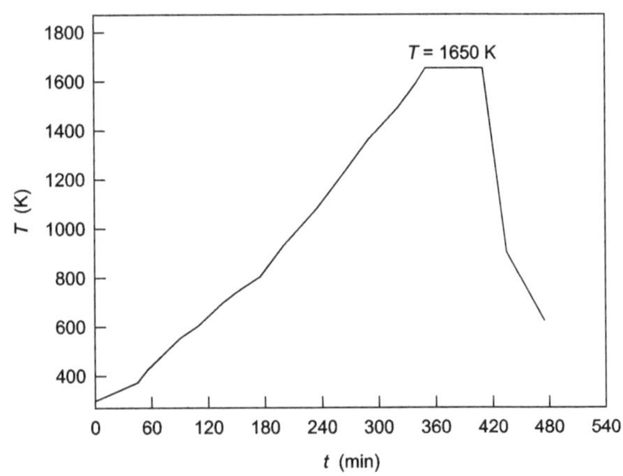

Fig. 5.24 Thermal treatment conditions used for sintering of WC-6 wt.% Co, WC-8 wt.% Co and WC-10 wt.% Co hardmetals

phase α–Co with a hexagonal close-packed structure and the high-temperature phase β–Co with a face-centered cubic structure. The WC–Co hardmetals with cobalt content from 4 to 16 wt.% are made by sintering at a temperature above 1270 K [22, 33], which is much higher than the temperature of the phase transition α–Co $\leftrightarrow$ β–Co which is equal to 700 K. That is why the two-phase character of the cobalt powders used as a metallic binder does not affect the structure and properties of produced WC–Co hardmetals, since already at the initial stages of sintering cobalt becomes monophase and contains only the cubic β–Co [20–22].

According to scanning electron microscopy, all initial powders are microcrystalline with the particle size ranging from 1 to 10 μm.

Scanning electron microscopy of the thermochemical powders Co(th) was carried out at a 1000× to 50000× magnification. Already at a 10000-power magnification it is visible that in the Co(th14) powder the extended ~1 μm long and ~0.5 μm wide particles are not bound with each other and form no dendritic microstructures (Fig. 5.25).

The powders Co(th08) and Co(kzts) have the same morphology, with particles from 300 nm to 5 μm, which are aggregates of smaller particles (Fig. 5.26).

The powder Co-el has a different morphology: at a 1500× magnification it is seen that its particles are 2 to 5 μm in length and ~1 μm in width and form a branched dendritic structure (Fig. 5.27a). The size of the dendrites is about 10 × 20 μm. At a magnification of 7000× it becomes clear that the large extended particles are agglomerates of particles having comparable (~ 1 μm) dimensions in all directions (Fig. 5.27b). The rounded aggregated particles of the WC powder have the average size of about 6 μm and consist of smaller particles with a size 100 nm or larger (Fig. 5.28).

The size distributions obtained with the use of laser diffraction exhibited wide dispersion of particle sizes (Table 5.3). The average size $\langle D \rangle$ of the powder particles was determined to be 8 to 20 μm, which is slightly larger than the values obtained by scanning electron microscopy. This is indicative of appreciable particle agglomeration in the powders.

Fig. 5.25 SEM images of thermochemical cobalt powder Co(th14): **a** 2500×, and **b** 10000× magnification

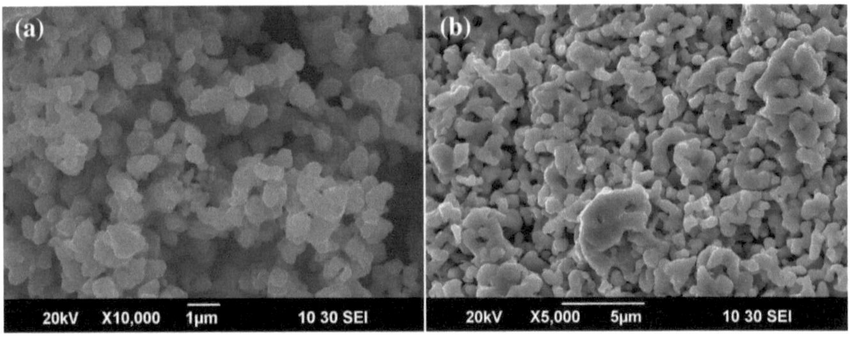

Fig. 5.26 SEM images of cobalt powders: **a** Co(th08), 10000×, and **b** Co(kzts), 5000× magnification

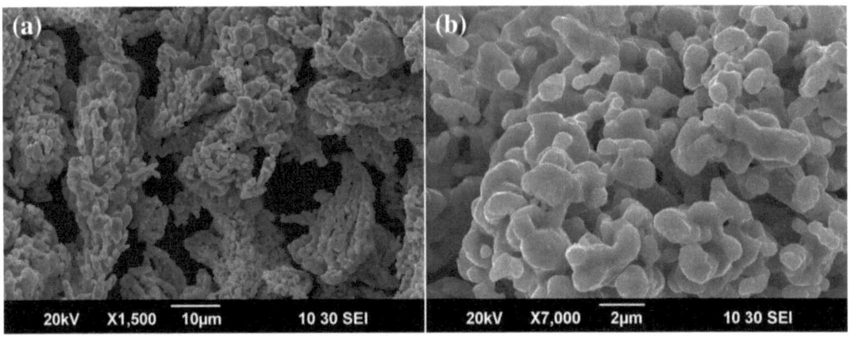

Fig. 5.27 SEM images of electrolytic cobalt powder Co(el): **a** 1500×, and **b** 7000× magnification

Fig. 5.28 SEM image of WC powder (5000× magnification)

Table 5.3 Particle size D for the initial powders according to laser diffraction data

Powder	$\langle D \rangle$ (μm)	D_{min} (μm)	D_{max} (μm)	*$D(50\,\%)$ (μm)	**$D(95\,\%)$ (μm)
Co(th08)	08.5	3.0	20.0	8.0	14.0
Co(th14)	18.5	2.0	90.0	15.0	45.0
Co(el)	19.2	4.0	50.0	18.0	30.0
Co(kzrs)	13.0	1.5	100.0	10.0	40.0
WC	5.8	0.8	17.0	5.5	11.0

* 50 % of all the particles have a size D less than $D(50\,\%)$; **95 % of all the particles have a size D less than $D(95\,\%)$

The morphology, crystal-chemical and size characteristics of the thermochemical and electrochemical cobalt powders are described in detail in works [48, 49].

The samples of the hardmetals WC-6 wt.% Co, WC-8 wt.% Co, and WC-10 wt.% Co containing one of the cobalt powders (Co(th08), Co(th14), Co(el) or Co(kzts)) as a binder were sintered in vacuum under the same temperature conditions (*see* Fig. 5.24).

Electron microscopic studies of a microstructure of the sintered hardmetals revealed that judging by the size of the hard phase grains all the hardmetal samples are microcrystalline and have a homogeneous and dense microstructure, except the alloys with the electrolytic cobalt powder Co(el) binder. Figure 5.29 shows as an

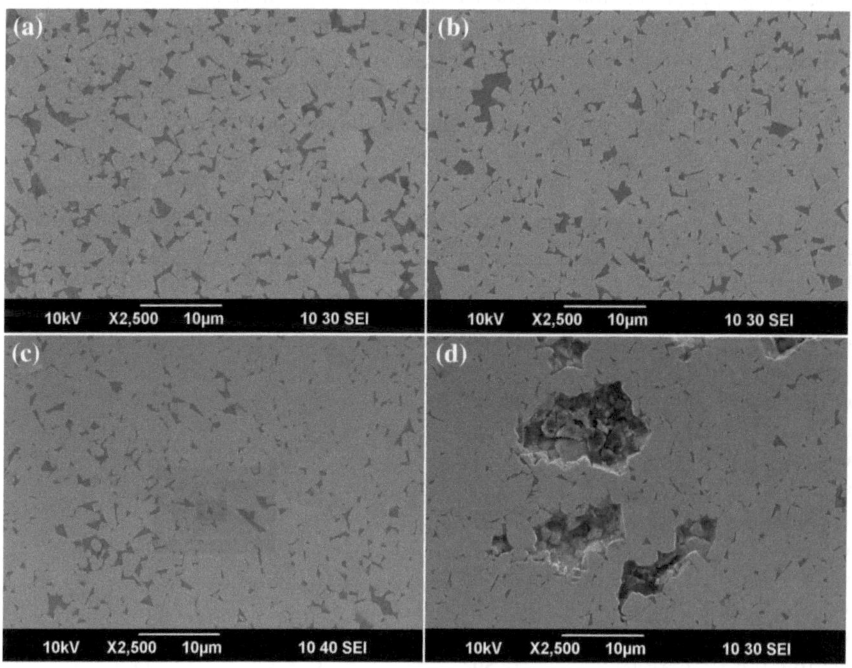

Fig. 5.29 Microstructure of WC-6 wt.% Co hardmetals, prepared with the use of different cobalt powders, at a magnification of 2500×: **a** Co(th08), **b** Co(th14), **c** Co(kzts), **d** Co(el)

example a microstructure of the WC-6 wt.% Co hardmetals obtained with the use of different cobalt powders. It is well visible that the pores are practically absent in the WC-6 wt.% Co hardmetals sintered using Co(th08), Co(th14), or Co(kzts) powders; the size of some pores is not greater than 1–2 μm. By contrast, the hardmetal sintered with the use of the electrolytic powder Co(el) has appreciable porosity with pore sizes to 10 μm (Fig. 5.29d).

The SEM data agree well with the results of density and porosity measurements: the hardmetals sintered with the use of the electrolytic powder Co(el) have the lowest density and the largest porosity (Table 5.4).

On the whole, the samples of the WC-Co hardmetals produced with the Co(th14) powder turned out to be the hardest and most dense species. The hardmetals sintered with the cobalt powder Co(th08) were slightly less harder (Table 5.4), which is clear since the range of the size distribution of the Co(th14) powder particles is almost twice as large as that for the Co(th08) powder. In this case, a closer packing is due to the presence of particles with different sizes in the powder mixture, when smaller particles fill the voids between larger particles.

Among the hardmetals produced with the use of Co(kzts) powder, only the samples of WC-8 wt.% Co were slightly inferior in hardness to the analogous hardmetals obtained using the cobalt powders Co(th14) and Co(th08).

Energy dispersion X-ray (EDX) analysis of the hardmetal surfaces revealed that the cobalt binder is uniformly distributed between the carbide grains, which bears witness to good compaction and good sinterability of the initial powders of tungsten carbide WC and thermochemical Co. The presence of large pores in the hardmetal WC-6 wt.% Co with the electrolytic cobalt powder Co(el) (Fig. 5.29d) points to poor morphological compatibility of the initial powders WC and Co(el). Judging from the pore size, it can be supposed that during sintering large dendritic particles of the initial

Table 5.4 Density ρ, porosity p and microhardness H_V of WC–Co hardmetals prepared with the use of different cobalt powders

Hardmetal	Co powder	Density $\rho \pm 0.04$ (g cm^{-3})	Porosity $\rho \pm 0.1$(%)	Microhardness $H_V \pm 0.5$ (GPa)
WC-6 wt% Co	Co(th08)	14.98	0.1	17
WC-6 wt% Co	Co(th14)	14.97	0.1	18
WC-6 wt% Co	Co(el)	14.48	3.4	16
WC-6 wt% Co	Co(kzts)	15.03	0	17
WC-8 wt% Co	Co(th08)	14.69	0.5	16
WC-8 wt% Co	Co(th14)	14.80	0	16
WC-8 wt% Co	Co(el)	14.58	1.3	16
WC-8 wt% Co	Co(kzts)	14.72	0.3	15
WC-10 wt% Co	Co(th08)	14.57	0	14
WC-10 wt% Co	Co(th14)	14.62	0	16
WC-10 wt% Co	Co(el)	14.62	0	16
WC-10 wt% Co	Co(kzts)	14.59	0	14

powder Co(el) melted and the resulting cobalt melt was redistributed on the grain boundaries of tungsten carbide. When a large dendritic particle of Co(el) melted, the volume it occupied inside the sample in the solid state was set free, and in the process of sintering it was not filled; as a result, large pores emerged. Thus, the use of the cobalt powder with a dendritic microstructure as a binder led to the formation of a porous hardmetals. On the contrary, under the same sintering conditions, the cobalt powders Co(th08), Co(th14), and Co(kzts), whose particles are not aggregates and have a rounded shape, gave rise to dense homogeneous hardmetals.

Thus, the morphology of the cobalt powders used as binders affects the microstructure of the sintered WC–Co hardmetals. For the preparation of hardmetals it is better to use the cobalt powders, whose particles are not aggregates and have a rounded shape. The dendritic microstructure of the cobalt powder impedes homogenization of the WC–Co powder mixture, sample compaction, as well as production of nonporous WC–Co hardmetals.

References

1. Arato P, Bartha L, Porat R, Berger S, Rosen A. Solid or liquid phase sintering of nanocrystalline WC/Co hardmetals. Nanostruct. Mater. 1998;10(2):245–255.
2. Sommer M, Schubert W-D, Zobetz E, Warbichler P. On the formation of very large WC crystal during sintering of ultrafine WC-Co alloys. Int. J. Refract. Hard. Mater. 2002;20(1):41–50.
3. Shao GQ, Duan XL, Xie JR, Yu XH, Zhang WF, Yuan RZ. Sintering of nanocrystalline WC-Co composite powder. Rev. Adv. Mater. Sci. 2003;5(4):281–286.
4. Gusev AI, Rempel AA. Nanocrystalline materials. Cambridge: Cambridge International Science Publications; 2004. p. 351.
5. Westbrook JH, Stover R. Carbides for high-temperature materials. In: High-temperature materials and technology. Campbell IE, Sherwood EM, editors. New York: Wiley; 1967. pp. 312–348.
6. Upadhyaya GS. Nature and properties of refractory carbides. New York: Nova Science Publ; 1996.
7. Gusev AI. Nonstoichiometry, disorder, short-range and long-range order in solids. Moscow: Nauka-Fizmatlit; 2007. p. 856. (in Russian).
8. Rautala P, Norton JT. Tungsten-cobalt-carbon system. Transact. AIME. 1952;194(4):1045–1050.
9. Gurland J. A study of the effect of carbon content on the structure and properties of sintered WC-Co alloys. Transact. AIME. 1954;200(3):285–290.
10. Pollock CB, Stadelmaier HH. The eta carbides in the Fe-W-C and Co-W-C systems. Metallurg. Transact. 1970;1(4):767–770.
11. Ettmayer P, Suchentrunk R. Über die thermische stabilität der eta-carbide. Monatsch. Chemie. 1970;101(4):1098-1103.
12. Johansson T, Uhrenius B. Phase equilibria, isothermal reactions, and a thermodynamic study in the Co-W-C system at 1150° C. Metal Sci. 1978;12(1):83–94.
13. Adelsköld V, Sundelin A, Westgren A. Carbide in kohlenstoffhaltigen Legierungen von Wolfram und Molybdän mit Chrom, Mangan, Eisen, Kobalt und Nickel. Zeitsch. Anorg. Allgem. Chemie. 1933;212(4):401–9.
14. Guillermet AF. Thermodynamic properties of the Co-W-C system. Metallurg. Transact. AIME A. 1989; 20A(5):935–956.
15. Villars P, Prince A, Okamoto H, Handbook of ternary alloy phase diagrams. Metals park. Ohio: ASM Publication; 1995. Vol. 5. pp. 6585–6605.

16. Kurlov AS, Gusev AI, Tungsten carbides and W-C phase diagram. Neorgan. Mater. 2006;42(2):156–163. (Engl. Transl. Inorganic Materials. 2006;42(2):121–27). (in Russian).
17. Kurlov AS, Gusev AI. Phase equilibria in the W-C system and tungsten carbides. Uspekhi Khimii. 2006;75(7):687–708. (Engl. Transl.: Russian Chem. Rev. 2006;75(7):617–636). (in Russian).
18. Kurlov AS, Gusev AI. Neutron and x-ray diffraction study and symmetry analysis of phase transformations in lower tungsten carbide W_2C. Phys. Rev. B. 2007;76(17):174115-01–174115-16. (Paper 174115).
19. C-W (carbon-tungsten). In: Binary alloy phase diagrams. Massalski TB, Subramanian PR, Okamoto H, Kasprzak L, editors. 2*nd* edition. Metals park. Ohio: ASM International Publication;1990. Vol. 1. pp. 895–896.
20. Kurlov AS, Rempel AA, Effect of sintering temperature on the phase composition and microhardness of WC-8 wt. % Co cemented carbide. Neorgan. Mater. 2007;43(6):685–691. (Engl. Transl. Inorg. Mater. 2007;43(6):602–607). (in Russian).
21. Kurlov AS, Rempel AA. Effect of WC nanoparticle size on the sintering temperature, density, and microhardness of WC-8 wt. % Co alloys. Neorgan. Mat. 2009;45(4):428–433. (Engl. Transl. Inorg. Mater. 2009;45(4):380–385). (in Russian).
22. Kurlov AA, Gusev AI, Rempel AA. Vacuum sintering of WC-8 wt. % Co hardmetals from WC powders with different dispersity. Int. J. Refr. Met. Hard Mater. 2011;29(2):221–231.
23. Kurlov AS, Rempel AA, Blagoveshenskii Yu V, Samokhin AV, Tsvetkov Yu V, Hard alloys WC-Co (6 wt. %) and WC-Co (10 wt. %) based on nanocrystalline powders. Doklady Akad. Nauk. 2011;439(2):215–220. (Engl. Transl. Doklady Chem. 2011;439(1):213–218). (in Russian).
24. Kurlov AS, Gusev AI, Effect of ball milling parameters on the particle size in nanocrystalline powders. Pis'ma v ZhTF. 2007;33(19):46–54. (Engl. Transl. Tech. Phys. Lett. 2007;33(10):828–832). (in Russain).
25. Gusev AI, Kurlov AS. Production of nanocrystalline powders by high-energy ball milling: model and experiment. Nanotechnology. 2008;19(26):265302-01–265302-08. (Paper 265302).
26. Gusev AI, Kurlov AS, Mechanical milling process modelling and making WC nanocrystalline powder. Neorgan. Mat. 2009;45(1):38–45. (Engl. Transl. Inorg. Mater. 2009;45(1):35–42). (in Russian).
27. Kurlov AS, Gusev AI. Production of nanocrystalline powder of WC via ball-milling. In: 17 Plansee Seminar 2009: Proceedings internnational conference on high performance P/M materials. Sigl LS, ödhammer PR, Wildner H, editors. Reutte: Plansee Group, 2009;3. p. GT 24/1-GT24/11. (25–29, May, 2009).
28. Kurlov AA, Gusev AI. Mathematical modeling of high-energy ball milling of powders. In: Proceedings 6*th* international conference on mathematical modeling and computer simulation of material technologies (MMT-2010). Zinigrad M, ed. Ariel: Ariel University Center of Samaria; 2010. p. 1-105–1-113. (23–27, Aug, 2010).
29. Kurlov AS, Gusev AI, Model for milling of powders. Zh. Tekhn. Fiz. 2011;81(7):76–82. (Engl. Transl. Tech. Phys. 2011;56(7):975–980). (in Russian).
30. Krebs H. Grundzüge der Anorganischen Kristallchemie. Stuttgart: Ferdinand Enke Verlag; 1968.
31. Okamoto H. Co-WJ. Phase Equilib. 2002;23(2):193–194.
32. Nagender Naudi SV, Sriramamurty AM, Rama Rao P. Co-W (cobalt-tungsten). J. Alloy Phase Diagrams. 1986;2(1):43–52.
33. Markström A, Sundman B, Frisk K. A revised thermodynamic description of the Co-W-C system. J. Phase Equil. Diff. 2005;26(2):152–160.
34. Klyachko LI, Fal'kovskii VA, Khokhlov AM. Hard alloys based on tungsten carbide with fine structure. Moscow: Ruda i, Metally; 199948. pp. 48. (in Russian).
35. Gusev AI. Nanomaterials, nanostructures, and nanotechnologies. 2*nd* edition. Moscow: Nauka-Fizmatlit; 2007. p. 416. (in Russian).

36. Fang ZZ, Xu W, Ryu T, Hwang KS, Sohn H.Y. Synthesis, sintering, and mechanical properties of nanocrystalline cemented tungsten carbide-a review. Int. J. Refr. Met. Hard Mater. 2009;27(2):288–299.
37. Blagoveshchenskiy YV, Samokhin AV, Tsvetkov YV, Alexeev NV, Isaeva NV, Kornev CA, Melnik YI. In: 17 Plansee seminar 2009: Proceedings international conference on high performance P/M materials. Sigl LS, Rödhammer P, Wildner H, editors. Reutte: Plansee Group; 2009. Vol. 3. p. GT 23/1-GT 23/6. (25–29, May, 2009).
38. Powder metallurgy and sprayed coatings. Mitin BS, ed. Moscow: Metallurgiya; 1987. pp. 184–186.
39. Schedler W. Hartmetall für den Praktiker: Aufbau, Herstellung, Eigenschaften und industrielle Anwendung einer modernen Werkstoffgruppe. Düsseldorf: VDI-Verlag; 1988. p. 558.
40. Kurlov AS, Rempel AA, Leenaers A, van den Berghe S. Bending strength and microhardness of WC-8 % (wt) Co hardmetals based on WC powders with various dispersity. Materialovedenie. 2009;4:18–21. (in Russian).
41. Kurlov AS, Leenaers A, van den Berghe S, Scibetta M, Schröttner H, Rempel AA. Microstructure of nanocrystalline WC powders and WC-Co hard alloys. Rev. Adv. Mater. Sci. 2011;27(2):165–172.
42. McCall JL, Steele H. Practical application of quantitative metallography. Philadelphia: ASTM International; 1984. p. 185.
43. Timoshenko S. Strength of Materials. Part 1: Elementary theory and problems. Second edition. Princeton: D. Van Nostrand Company; 1940. p. 359.
44. Anstis GR, Chantikul P, Lawn BR, Marshall DB. A critical evaluation of indentation techniques for measuring fracture toughness. I. Direct crack measurements. J. Am. Ceram. Soc. 1981;64(9):533–538.
45. Kurlov AS, Gusev AI. Vacuum annealing of nanocrystalline WC powders. Neorgan. Mat. 2012;48(7):781–791. (Engl. Transl. Inorg. Mater. 2012;48(7):680–690). (in Russian).
46. Kurlov AS, Gusev AI. Pecualirities of vacuum annealing of nanocrystalline WC powders. Int. J. Refr. Met. Hard Mater. 2012;32(5):51–60.
47. Kurlov AS. Effect of vacuum annealing on the particle size and phase composition of nanocrystalline WC powders. Zh. Fiz. Khimii. 2013;87(42):664–671. (Engl. Transl. Russ. J. Phys. Chem. A. 2013;87(4):654–661). (in Russian).
48. Kurlov AS, Rempel AA, Matrenin VI, Stikhin AS. Morphology and crystal-chemical characteristics of cobalt and nickel nanopowders prepared by thermochemical and electrolytic methods. Neorgan. Mat. 2013;49(2):144–150. (Engl. Transl. Inorg. Mater. 2013;49(2):153–158). (in Russian).
49. Kurlov AS, Gusev AI, Rempel AA. Morphology of ultrafine cobalt and nickel powders. Rev. Advans. Mater. Sci. 2012;32(1):52–60.

Index

A. S. Kurlov and A. I. Gusev, *Tungsten Carbides*,
Springer Series in Materials Science 184, DOI: 10.1007/978-3-319-00524-9,
© Springer International Publishing Switzerland 2013

MIX
Papier aus verantwortungsvollen Quellen
Paper from responsible sources
FSC® C105338

If you have any concerns about our products,
you can contact us on
ProductSafety@springernature.com

In case Publisher is established outside the EU,
the EU authorized representative is:
Springer Nature Customer Service Center GmbH
Europaplatz 3, 69115 Heidelberg, Germany

Printed by Libri Plureos GmbH
in Hamburg, Germany

Karin Peschel (Ed.)

Infrastructure and the Space-Economy

Essays in Honor of Rolf Funck

With 58 Figures

Springer-Verlag Berlin Heidelberg New York
London Paris Tokyo Hong Kong

Professor Dr. Karin Peschel
Institut für Regionalforschung
Christian-Albrechts-Universität, Kiel
Olshausenstraße 40
D-2300 Kiel 17, FRG

ISBN-13: 978-3-642-75573-6 e-ISBN-13: 978-3-642-75571-2
DOI: 10.1007/978-3-642-75571-2

This work is subject to copyright. All rights are reserved, whether the whole or part of the material is concerned, specifically the rights of translation, reprinting, reuse of illustrations, recitation, broadcasting, reproduction on microfilms or in other ways, and storage in data banks. Duplication of this publication or parts thereof is only permitted under the provisions of the German Copyright Law of September 9, 1965, in its version of June 24, 1985, and a copyright fee must always be paid. Violations fall under the prosecution act of the German Copyright Law.

© Springer-Verlag Berlin · Heidelberg 1990

Softcover reprint of the hardcover 1st edition 1990

The use of registered names, trademarks, etc. in the publication does not imply, even in the absence of a specific statement, that such names are exempt from the relevant protective laws and regulations and therefore free for general use.

PREFACE

This *Festschrift* is dedicated to ROLF FUNCK on occasion of his 60th birthday on February 7, 1990. It was a wholehearted desire of the authors who all are obliged to ROLF FUNCK in their research work to please and to honour him and to express their deep gratitude.

I gladly took the initiative of preparing this *Festschrift* since I have been closely associated with ROLF FUNCK for many years of his academic life. We first met in 1958 when he was Assistant Professor (Wissenschaftlicher Assistent) and I was student at the University of Münster. A few years later, when ROLF FUNCK, still very young, moved to the University of Karlsruhe as Professor of Economics and Director of the Institute of Economic Policy and Research, he offered me the opportunity of working with him and this started a long period of successful scientific collaboration. Nowadays we meet only occasionally at conferences, in committees etc., but I always recall with pleasure the warm-hearted and stimulating atmosphere he created at his Institute in Karlsruhe.

I express my thanks to the authors and to the publisher, the *Springer Verlag*, for their cooperation as well as to the sponsors who provided the necessary funds for the publication of this *Festschrift*. For her enthusiasm in compiling and processing the articles on the PC and producing the final draft I am indebted to *Mrs. Ilona Löhr*.

Kiel, November 1989 Karin Peschel

For ROLF FUNCK

by

Karin Peschel

When sitting at my desk and ready to do what is expected from the editor of a *Festschrift*, namely praising the laureate and his academic achievements, I imagined ROLF FUNCK seated in front of me. Over the rim of his glasses pulled down on his nose, he glanced at me, amused and sceptical, and he looked so young and impish that I realized: it is much to early for a definitive appraisal of his achievements. There is more to come and ROLF FUNCK should know that we all look forward to it! Hence, let us wait with such an appraisal for at least another 20 years!

From his impressive curriculum vitae we know of a large number of his activities we would certainly miss if ROLF FUNCK would indeed retire. Let me only mention a small sample of his present activities here: Director of the Institute of Economic Policy and Research of the University of Karlsruhe, President of the International Regional Science Association, Member of the Board of Trustees (Kuratorium) of the Fraunhofer Institute of Systems Technology and Innovation Research, Member of the Economic Advisory Board to the Federal Minister of Transportation, and Member of the August Lösch Commitee of the City of Heidenheim.

But ROLF FUNCK never lived in the academic world only. He has always been, and still is, eager to apply theoretical thinking for solving practical problems. Therefore he engaged himself in local politics in Karlsruhe where he is a long-standing member of the City Council, of several of its subcommittees and of some other bodies like the Board of Trustees (Stiftungsrat) of the Center of Arts and Media Technologies.

ROLF FUNCK has many friends. Hence it is not surprising that my suggestion of presenting to him a *Festschrift* on occasion of his 60th birthday met with unanimous and enthusiastic approval. That the papers collected in this *Festschrift* deal with rather heterogeneous topics is due to ROLF FUNCK's broad range of interest and research activities. Every of the contributors can justly claim that he is referring to ROLF FUNCK's work. When complaining to former and present staff members of his Institute at Karlsruhe that not even they did adhere to my proposed list of subject matters, I have been rightly reminded that this was merely the consequence of ROLF FUNCK's liberal attitude of allowing everyone to follow his or her own interests. I had to admit that any attempt of narrowing down the range of topics would have been against ROLF FUNCK's basic beliefs and against the spirit of his Institute.

Nevertheless, the core of this *Festschrift* is formed by papers dealing with a topic which is most prominent in ROLF FUNCK's own work: the material infrastructure and its regional impact. Moreover, the *Festschrift* reflects ROLF FUNCK's strong interest in purely methodological issues as well as in empirical studies.

To shape the world and therefore to link regional science to the requirements of practice became even more important for ROLF FUNCK during the course of his life. This view is clearly evident in his presidential address to the 29th European Congress of the Regional Science Association International, given in Cambridge UK in August of this year. He asked:

> Has regional science really overcome "the sweet and intoxicating flavor of l'art pour l'art ... and discovered the real world? And, if so, what are the characteristics of the real world today? What will they be tomorrow? And in which way do regional scientists cope with today's world, and tomorrow's?"

ROLF FUNCK is certainly correct, when pointing to the abundance of new problems which during the last few decades have emerged on a global scale although with different intensities in the various parts of the world. He is equally correct when reminding us of the diverse social, economic, or environmental problems which provoke different individual, social or political reactions depending on the structural characteristics of the particular regional systems that are confronted with them. Nearly every reader will agree with him, I believe, that the most important problems today are: exponential incrase in environmental pollution; limited reserves of fossil engergy resources; the stagnation or even decline of population in many western industrialized countries; radical change in production technology; continuing changes in the international division of labor; the aging of the infrastructure; the re-orientation of the spatial, social, economic, and communicative function of cities, the increasing demand for cultural, educational, and non-directional leisure activities, the rapidly increasing share of public activities, et cetera.

Can regional science cope with this challange? ROLF FUNCK's answer given in the presidential adress mentioned above is:

> "Often, regional scientists have sufficiently proved to be able to adjust their analytical tools to new developments, to adapt theories, developed in other fields of science, to their special, the spatial, perspective. What is needed, however, is anticipation rather than reaction, striding ahead rather than stumbling behind."

Nevertheless, due to his optimistic attitude ROLF FUNCK believes that progress in the right direction is possible, basically, because most changes "did not really come over us as a big surprise ... We simply refused to read the signs. *Perception* is what came as a lightning from heaven." (Presidential adress)

I know that ROLF FUNCK himself has never lacked imagination and vision. For example, as early as in 1965 we recommended to the Federal Minister of Transportation that the social costs of individual traffic should be internalized as far as possible by appropriate taxation. Presently this idea, under the label "ecological taxes", is widely and intensively discussed while ROLF FUNCK's attention has turned to new topics. Out of the foreseeable future changes he is i.a. interested in the regional implications of the Greenhouse problem and its potential sea level-rise.

Dear ROLF, do not loose your deep faith in humanity, your powerful imagination, your extraordinary ability to stimulate creative thinking and to awaken optimism! We all wish you many more healthy, productive and successful years!

Yours disciples and friends.

Financial support for publication was provided by

Daimler-Benz A.G., Stuttgart
Dambach-Werke GmbH, Gaggenau
Deutsche Bundesbahn, Frankfurt
Deutsche Verkehrs-Kreditbank A.G., Karlsruhe
Karlsruher Lebensversicherung A.G., Karlsruhe
Karlsruher Versicherung A.G., Karlsruhe
J.M. Voith GmbH, Heidenheim

C o n t e n t s

Spatial Patterns of Resource Utilization

by

Martin J. Beckmann

1. Resources are either localized at isolated points or along one dimensional lines, or they are dispersed over two-dimensional areas. Which resources are utilized depends on accessibility and proximity to markets. Accessibility determines utilization costs, and distance from markets determines transportation costs. Sometimes either the resource itself or the product may be transported, and comparative transportation cost determines which is more economical. All this is well known to location theorists who have treated this problem under the heading of orientation (cf. LÖSCH, PALANDER, ISARD, BECKMANN).

   What is less well understood are the patterns of resource use in the context of a two-dimensional setting when both demand and supply are widely dispersed. A special case is VON THÜNEN's "isolated state" where a single city obtains its agricultural products from a surrounding region with land of uniform utility (1826, 1921). In the following we show how the general situation may be analysed by means of a continuous flow model. A basic assumption made throughout is that the industry utilizing the resource is a constant returns industry. The following model applies in fact to all constant returns industries, even those that do not use localized resources.

2. In its classical form, location theory makes the following assumptions:
   - resources are localized;
   - they may not be substituted;

- hence they are used in fixed proportions;
- demand is insensitive to price;
- costs rather than production functions are considered as given data;
- resources are allocated to minimize costs;
- unless production costs depend on location, this reduces to minimizing transportation costs;
- an industry may be studied in isolation;
- partial equilibrium analysis is sufficient.

We do not intend to critizise these assumptions. In view of the complexity of spatial problems, such simplifications can be defended as necessary in a first approach. We mean instead to bring out the full implications of these assumptions by applying them in the context of a modern framework. This is the framework of the continuous transportation model which has been championed by PUU, TOBLER, and this author (1952).

3. We propose to consider in a two dimensional continuum the location of any industry with constant returns to scale. This will include the classical VON THÜNEN problem and, as a degenerate case, the WEBER problem of industrial location. With appropriate adjustments it also includes the location of residential housing, when a finite number ob housing types is considered.

These location problems are land use problems when several activities compete with each other. To begin with we consider a single industry, however.

This problem has been described in BECKMANN and PUU (1985, pp. 82-84) as follows:
At location $(x_1, x_2) = x$ an amount $z(x)$ of a commodity may be produced up to a limit $a(x)$ at constant unit costs $c(x)$. The demand for this commodity in location x is $q(x)$.
All variables are densities, i.e. quantities per unit area.

The unit costs of transportation at x in all directions is $k(x)$. We wish to allocate production z and transportation ϕ so as to minimize total costs

$$\text{Min} \int\int [k|\phi| + cz]\, dx_1\, dx_2 \tag{1}$$

The vector ϕ indicates the direction of commodity flow and its length equals the quantity of this flow per unit cross section per unit of time.

Minimization (1) is subject to the requirement that demand q is met in all locations. This problem is feasible if and only if total production capacity is not less than aggregate demand

$$\int\int q\, dx_1\, dx_2 \leq \int a\, dx_1\, dx_2 \tag{2}$$

The condition implied by meeting demand can be shown to be (BECKMANN and PUU, p. 15)

$$\text{div}\, \phi \leq z - q. \tag{3}$$

The production capacity constraint is

$$0 \leq z \leq a \tag{4}$$

Consider the Lagrangian

$$\int\int [- k|\phi| - cz + \lambda(z - q - \text{div}\, \phi) + \mu(a - z)]dx_1\, dx_2.$$

Maximization with respect to ϕ yields

$$k \frac{\phi}{|\phi|} = \text{grad}\, \lambda \tag{5}$$

the familiar gradient law of the continuous transportation problem (BECKMAN and PUU, p. 16–17). It states that goods move in the direction of the steapest increase of price, and that the rate of price increase is just equal to transportation costs.

Minimization with respect to output z yields

$$z \begin{Bmatrix} \leq 0 \\ \geq 0, \quad \leq a \\ = a \end{Bmatrix} \text{ if } \quad \lambda \begin{Bmatrix} < \\ = \\ > \end{Bmatrix} c \tag{6}$$

This means that production is feasible only where the commodity price λ is at least equal to unit production cost c, and that production is optimal at the maximum level a when price exceeds unit costs $\lambda > c$.

These statements, which are intuitively appealing, do in fact indicate where production of a good is economicaly optimal and where not. Production locations occupy two dimensional closed sets with continuous boundaries characterized by $c = \lambda$.

Where the commodity is not produced but consumed importation is cheaper than local production, $\lambda < c$.

4. This (conceptually) simple classical location problem becomes a land or resource use problem when several industries are considered in their competition for locations, i.e. for land use. (In BECKMANN and PUU this problem is studied in a neoclassical framework using utility and production functions rather than in the purely classical framework with fixed demands and constant unit costs).

Let q_i and c_i be the demand and unit costs for product i, respectively, and let the capacity limit a representing available land apply to the sum of production activities. No budged constraint is assumed for the total consumption of the various commodities i. This means that there are other products and services in the economy besides those considered explicitly as $i = 1,\ldots,n$.

A cost minimizing allocation of available land to the economic activities i requires

$$\min_{z_i,\phi} \int\int \sum_{i=1}^{n} [k_i |\phi^i| \, c_i \, z_i] \, dx_1 \, dx_2 \tag{7}$$

subject to the constraints

$$\operatorname{div} \phi^i \leq z_i - q_i \tag{8}$$

and

$$\sum_{i=1}^{n} z_i \leq a. \tag{9}$$

Consider the Langrangian

$$\iint \left\{ -\sum_{i=1}^{n} [k_i |\phi^i| + c_i\ z_i] + \sum_{i=1}^{n} \lambda_i [z_i - q_i - \operatorname{div} \phi_i] + \mu [a - \sum_{i=1}^{n} z_i \right\} dx_1\ dx_2 .$$

As before maximization with respect to ϕ implies

$$k_i \frac{\phi^i}{|\phi_i|} = \operatorname{grad} \lambda_i \tag{10}$$

Transportation cost of each product in the optimal direction must be covered by price gains.

Maximization with respect to output levels z_i yields

$$z_i \left\{ \begin{matrix} = \\ \geq \end{matrix} \right\} 0 \leftrightarrow \lambda_i - c_i - \mu \left\{ \begin{matrix} \leq \\ = \end{matrix} \right\} 0. \tag{11}$$

Moreover

$$\mu \left\{ \begin{matrix} = \\ \geq \end{matrix} \right\} 0 \leftrightarrow \sum_{i=1}^{n} z_i \left\{ \begin{matrix} \leq \\ = \end{matrix} \right\} a. \tag{12}$$

Here $\mu(x)$ is the rent of land, and the condition (11) describes the bidding for land by the various production activities i.

The condition (11) may in fact be rewritten

$$\mu = \max_i\ [\lambda_i - c_i]. \tag{13}$$

It is only on sets of measure zero that this maximum is achieved by more than one activity i (cf. BECKMANN and PUU, p. 57–58). It thus follows from (13) that land use will be spezialized. This is, of course, well-known in the VON THÜNEN model where zones of common land use form concentric circular rings. It is also true in urban housing and in fact applies to industrial land use, as well, provided the industry has constant returns to scale.

A WEBERIAN model of industrial location results, when the production activities include resource extraction and the combination of resources in fixed proportions to produce an output. Rather than single production locations, entire zones of production result, one production location for each market location. This is in fact an alternative idealization to the point locations that emerge in the classical LAUNHARDT and WEBER model.

The models developed here permit in fact a unified approach to the location of all constant returns activities, whether they are land intensive or not. In particular this approach can serve to bring out the comparative advantage of regions due to their resource endowment. Moreover it can account for the specialization of land use which has been observed particularly in land intensive economic activities.

Where it fails is in regard to economic activities which exhibit significant economies of scale. It is these of course that are the ultimate reason for the formation of cities. The foundation of urban economics and of central place systems must thus be thought elsewhere.

In conclusion we remark that the resource using constant returns industries studied here include as a prime example all types of agriculture and forestry. But they include also important energy resources such as oil fields and coal fields. Some of these are known to exist in locations where at this time utilization is uneconomical. As resource prices rise more and more of these distant resource locations will come within the reach of spatial markets. An understanding of this process requires a dynamic approach. This has not been attempted here, it remains the task for the future.

Location theory and its modern successor spatial economics are not to be taken as "L'art pour art". Rather they are meant as instruments to provide the understanding of spatial relationships that is needed for a rational economic policy. It is in this field of Regional Economic Policy that ROLF FUNCK has been a prime mover. We salute him today and wish him success in his many undertakings.

LITERATURE

Beckmann, M.J.; Puu T., 1985: "Spatial Economics: Density, Potential, And Flow", Amsterdam, North-Holland.

Funck, R., 1976: "Entscheidungshilfen für die Regionalpolitik", Heidenheimer Schriften, Verlag Stadt Heidenheim an der Brenz/Kulturamt, pp. 8-29.

Isard, W., 1956: "Location and Space-Economy", MIT Press, Cambridge, Massachusetts.

Isard, W.; Liossatos, P., 1979: "Spatial Dynamics and Optimal Space-Time Development", North-Holland, Amsterdam.

Launhardt, W., 1882: "Der beste Standort einer gewerblichen Anlage", Zeitschrift des Vereins deutscher Ingenieure, 26: 106-115.

Lösch, A., 1954: "The Economics of Location"; translation from the German "Die räumliche Ordnung der Wirtschaft", Yale University Press, New Haven, Connecticut. (German editions 1940, 1944).

Palander, T.F., 1935: "Beiträge zur Standorttheorie", Almqvist and Wiksell, Uppsala.

Predöhl, A., 1925: "Das Standortproblem in der Wirtschaftstheorie", Weltwirtschaftliches Archiv, XXI, 294-321. The principle is stated on p. 306.

von Thünen, H., 1921: "Der isolierte Staat", Waentig, Jena.

Weber, A., 1909: Über den Standort der Industrien", J.C.B. Mohr, Tübingen: translated into English as Weber, A. (1929) Alfred Weber's "Theory of Location of Industries", University of Chicago Press, Chicago, Illinois.

Regional Output and Price Effects of Spatial Price Discrimination*

by

John Greenhut, Melvin L. Greenhut, Hiroshi Ohta

1. Introduction

ROLF FUNCK's interests in recent years (or should we say decades) have centered more and more on applied economics. These interests have extended from the taxing of motor traffic and the determining of how to make mass transit systems more efficient to forecasting and implementing regional economic growth. One might wonder how, given his practicality, the authors of this paper decided to avoid such subjects as urban transit, regional development, benefit-cost analysis and instead to focus on what can easily be referred to as being the opposite side of that coin: namely the *theory* of regional outputs and prices. We have two answers for our deviations: (i) Professor FUNCK has also worked often with the other side of the coin[1], and (ii) our own interests have run the gamut from the one side to the other with our *recent focus* having centered on the hope that *spatial microeconomic theory per se* will take a giant leap forward during the 1990's. We trust that this paper will not only reflect recent advances but promote interest in a side of spatial microeconomics that must be moved forward rapidly in the immediate future. If such movement can (and does) take place in the years to come, applications of theory will, in turn, be more penetrating and important than in the past.

* This is a revised and extended version of a paper presented by one of the authors at the Pacific Regional Science Conference, Molokai, 1985, and published as Chapter 5 in Ohta (1988). Early financial support was providedby the Ganjoh Kosaka Memorial Scholarship of Aoyama Gakuin University, and is acknowledged as well as thanks to Yeung-Nan Shieh, Gary Anders, and Martin Bronfenbrenner for helpful comments.

By way of further background, let us note that PIGOU (1929) and ROBINSON (1933), in their classic analyses of regional price discrimination, considered two, and only two, separate markets as being served by a monopolist. M. L. GREENHUT and OHTA (1972) extended the PIGOU-ROBINSON analysis by introducing the concept of a variable space, i.e., variable market areas, the market spaces of which would be served under alternative pricing techniques. Of course, the basic endogenous variables would depend on the price policy itself. Going beyond mere analytical generalization, GREENHUT and OHTA arrived at the definitive conclusion that monopoly output tends to be strictly greater under spatially discriminatory pricing than under nondiscriminatory f.o.b. pricing.

The analysis of spatial price discrimination was later extended and widened in scope broadly by the same writers (1975a,b), HOLAHAN (1975), and NORMAN (1981), among others. Analytical deepening took place along the lines of generalizing the functional forms of the individual demands and the competitive market structures (J. GREENHUT, HWANG, and OHTA, 1974; J. GREENHUT and M. L. GREENHUT, 1975). This widening of scope, including the impacts of discrimination, were evaluated with respect to outputs, as detailed and extended in OHTA and WAKO (1988), as well as from a welfare criterion based on consumer surplus and profit, each under varying conditions of spatial competition and monopoly.

The robustness of the GREENHUT-OHTA findings was limited, however, by the algebraic form of the assumed individual demand, although the form utilized was more general than the conventional linear demand. Actually, the "basic" demand assumed by GREENHUT-OHTA belongs to the *class of demands* which are "relatively less convex" than the basic (local) demand whose elasticity with respect to net mill price (i.e., the price net of freight cost) remains constant; note, this constant elasticity applies regardless of consumer location (OHTA, 1980, 1981).[2] Let us stress the fact that the class of demands which is referred to herein as relatively less convex has also been referred to in the literature as less convex than the negative exponential demand.[3]

Local demands (in terms of f.o.b. mill prices), which relate to the relatively convex demand, are known to become increasingly elastic as buyer distances from the seller increase (as in OHTA, 1981). It follows that the spatial aggregate demand becomes more elastic than the basic demand, and this results

in lower prices and greater output. The question which must be addressed is whether the output effects of spatial price discrimination remain positive under these limited conditions alone? What happens to the GREENHUT-OHTA conclusion if the basic demand happened to be "relatively more convex"?
The present paper answers the questions raised above. It shows that the GREENHUT-OHTA conclusion remains unchanged even under the "relatively more convex" class of basic demands. However, it is important to note that even if aggregate output and overall welfare levels are greater under conditions of spatial price discrimination than under f.o.b. mill pricing, some consumers may gain more while others lose. Thus the present paper must also consider the question *who gains, who loses* under what conditions, if a seller is permitted to switch freely from a nondiscriminatory f.o.b. mill price system to a spatially discriminatory price pattern. May we suggest ala ROLF FUNCK that regional economic growth issues relate closely to pricing practices and freedoms.

Section 2 of the paper presents our basic assumptions. Though not all of them are retained throughout the paper, they constitute a departure point for later analysis while allowing a basic model of spatial monopoly to be set forth. Section 3 extends our earlier analysis of the output effects of spatial price discrimination under the more general model set forth in Section 2. Section 4 turns its attention to the price effects of spatial price discrimination, and provides a generalization as well as summary of the applicable, relevant findings that were provided in GREENHUT-OHTA-SAILOR (1984). Section 5 concludes this paper by presenting a few general propositions.

2. The Basic Model of Spatial Monopoly

Assume the following:

Assumption a. A monopolistic firm is located at an arbitrary point in an unbounded linear market.

Assumption b. The basic demand (density) q is constantly elastic in terms of the c.i.f. or delivered price. That price consists of the mill price m plus the freight rate x per quantity unit which, in turn, is a monotonically increasing function of distance from the firm.

Assumption c. The basic demand is uniformly and continuously distributed along the linear market in terms of distance cost x, as defined further below.

Pursuant to Assumption b, the basic demand q is given by:

$$q = a(m + x)^{-\alpha}, \quad a > 0, \ \alpha > 1 \qquad (1)$$

where, for any given positive coefficient a, the parameter α is the elasticity of the basic demand, e_B, i.e., $e_B = \alpha$. Note that from the seller's standpoint, elasticity based on the seller's price (i.e., the mill price) rather than on the delivered price to the buyer equals $\alpha m/(m+x)$, which decreases as distance x between the buyer and seller increases. Thus, demand of constant elasticity is of the form that can be referred to as the "more than convex" demand.

Pursuant to the above assumptions, equation (1) can be integrated into the market demand Q_M for a spatial monopolist who practices f.o.b. mill pricing in either direction. This provides:

$$Q_M = 2 \int_0^{\infty} a(m+x)^{-\alpha} \, dx$$

$$= \frac{2a}{\alpha - 1} m^{-(\alpha-1)}, \quad \alpha > 2 \qquad (2)$$

Several notes on equation (2) are in order:

(i) For any given mill price, the firm's market area extends without bound, approaching infinity as in (2) unless technically or otherwise restricted.[4]

(ii) The elasticity e_M of the spatial market demand, conventionally called the free spatial demand, is given by $e_M = \alpha - 1$. Thus, the market demand is *less* elastic than the basic/individual demands. Somewhat paradoxically, this means that even if econometric estimates of the market demand happen to be inelastic, individual consumers are more responsive to changes in mill price than in the aggregate.

(iii) It follows from (2) that if a good is to be supplied by a local monopolist *at a finite price*, the consumers' basic demand is required to be sufficiently elastic. Specifically, since e_M (= $\alpha - 1$) is required to be greater than unity, it follows that $e_B = \alpha > 2$.

(iv) As mentioned above, under constant elasticity of the basic demand, local demands in terms of mill price m becomes *increasingly elastic as the distance cost x increases* and the net demand schedule for the monopolist, i.e., net of freight rate x, shrinks. This implies that the elasticity of the spatial market demand which derives from aggregating the shrinking net demands tends to *increase* under conditions of competitive spatial entry, such as stressed by LÖSCH (1946). This follows because the greater is spatial competition, the smaller is the market area and the fewer are the total number of diminished net demands. With the subject net demands having monotonically decreasing elasticities, the smaller number of demands aggregated under increased competition necessarily yields greater elasticity of aggregate demand.

3. The Output Effects of Discrimination: The Constant Demand Elasticity Case

We are now in position to derive simple f.o.b. marginal revenue SMR and discriminatory marginal revenue DMR from the basic assumptions and equations set forth above. In doing this, we readily ascertain the output effects of spatial price discrimination.

SMR is derivable from the spatial market demand (2) by differentiating total revenue TR with respect to Q:

$$TR = m(Q)Q = \left(\frac{\alpha-1}{2a}\right)^{\frac{1}{1-\alpha}} Q^{\frac{1}{1-\alpha}+1}$$

$$\therefore \quad SMR = \frac{d(TR)}{dQ} = \frac{2-\alpha}{1-\alpha}\left(\frac{\alpha-1}{2a}\right)^{\frac{1}{1-\alpha}} Q^{\frac{1}{1-\alpha}}, \quad \alpha > 2 \qquad (3)$$

Note that the spatial monopolist subject to the constraint of f.o.b. mill pricing equates SMR with marginal production cost to obtain the optimal unique mill price *and* output (m^*, Q_s^*).

If the spatial monopolist is allowed to set different mill prices at different buying points (submarkets), the optimal policy for the firm is to separate each and every buying point. This signifies that every marginal revenue curve is aggregated horizontally and equated with marginal production cost, MC. The horizontal aggregate of these "local" marginal revenues establishes the spatial discriminatory marginal revenue cure, DMR. Because every local marginal revenue MR(x) is equated with MC, the optimal mix of mill prices and the quantity of output shipped to each buying point x, i.e., ($m^*(x)$, $q^*(x)$) is obtained. In particular, MR(x) derives from (1) by differentiating the *local* total revenue TR(x):

$$TR(x) = m(x)q(x) = a^{\frac{1}{\alpha}} q(x)^{1-\frac{1}{\alpha}} - xq(x)$$

$$\text{and } MR(x) = \frac{\alpha - 1}{\alpha} a^{\frac{1}{\alpha}} q(x)^{-\frac{1}{\alpha}} - x \qquad (4)$$

Equation (4) is then inverted to obtain DMR by holding all MR(x)'s equal (=DMR), as q(x) is aggregated over the entire market area:

$$q(x) = a\left(\frac{\alpha}{\alpha - 1}\right)^{-\alpha}(MR(x) + x)^{-\alpha}$$

$$\therefore Q = 2\int_0^{\infty} a\left(\frac{\alpha}{\alpha - 1}\right)^{-\alpha} (DMR + x)^{-\alpha}dx$$

$$= \frac{2a}{\alpha - 1}\left(\frac{\alpha}{\alpha - 1}\right)^{-\alpha}(DMR)^{1-\alpha}$$

$$\therefore \quad DMR = \left(\frac{\alpha}{\alpha - 1}\right)^{\frac{\alpha}{1-\alpha}} \left(\frac{\alpha - 1}{2a}\right)^{\frac{1}{1-\alpha}} Q^{\frac{1}{1-\alpha}} \qquad (5)$$

Equation (5) involves a one-to-one correspondence between equal MR(x)'s (=DMR) and the aggregate quantity Q that is transported. This relation holds if the firm handles local shipments q(x) in such way as to equate each local MR(x) with the common DMR, equated in turn with marginal production cost which therefore provides maximum global profits.

Comparing equations (3) and (5), we now demonstrate the strict relation that DMR > SMR when $\alpha > 2$. For proof, note that a sufficient condition for DMR > SMR is $[\alpha/(\alpha-1)]^{\alpha/(1-\alpha)} > (2-\alpha)/(1-\alpha)$. The right-hand side of this inequality can be rewritten as $y = 1 - \Omega$ and the left-hand side as $g = (1 + \Omega)^{-(1+\Omega)}$, where $\Omega = 1/(\alpha-1)$. Since α is assumed to be greater than two, the range for Ω is limited to $1 > \Omega > 0$. Over this domain, y and g have the following characteristics: $y(0) = g(0) = 1$, $y' = -1$, $g' < 0$, $g'(0) = -1$, and $g'' > 0$. Thus

$$g > y\,, \quad 1 > \Omega > 0 \tag{6}$$

which establishes a sufficient condition for DMR > SMR. It follows that regardless of the shape of the marginal production cost MC(Q), the level of output produced under spatial discriminatory pricing Q_D is, in equilibrium, greater than that under simple f.o.b. mill pricing Q_S even if the form of the basic demand happened to be relatively more convex.[5] Combined with previous knowledge for less convex demands, as in GREENHUT-OHTA (1975a), the present result thus leads to the following general proposition.

Proposition I: Spatial price discrimination tends to yield greater output than f.o.b. mill pricing under fairly comprehensive cost and demand conditions, i.e., regardless of the forms of the basic demand and cost functions.

4. The Price Effects of Spatial Price Discrimination: Who Gains under What Conditions?

The c.i.f. prices are presumed to depend on the pricing policies of the firm. Typically, discriminatory prices are higher for nearby customers and lower for distant customers than are the nondiscriminatory c.i.f. prices. (See GREENHUT, 1956 and BECKMANN, 1976.) This relation, however, derives from particular demand and cost conditions, where the required conditions are (i) a basic demand that is "relatively less convex," and (ii) there are at the most negligible economies of scale. Contrasting results are derivable if the initial condition is reversed and the latter (scale economies) made significant. The present section provides the foundation for these changes.

Consider as a departure point the standard case of constant marginal production cost and a "relatively less convex" basic demand. Since the local demand in this case in contrast to that of a constantly elastic basic demand *becomes increasingly elastic as distance from the firm increases,* the aggregate demand is unambiguously more elastic than the local demand of a buyer located next to the seller. In the event of constant MC, it follows that the price at the seller's mill under f.o.b. mill pricing stays strictly below the comparable price under discriminatory pricing, the former being related to the more elastic aggregate demand and the latter derived from the less elastic local demand (the basic demand). Moreover, though the lower mill price remains constant under f.o.b. mill pricing, the initially higher mill price under discriminatory pricing decreases as the elasticity of local demands increases with distance. The same freight rates being added to these mill prices under the two alternative pricing techniques causes the c.i.f. prices as a schedule (in terms of distance costs from the seller) to become flatter under discriminatory pricing than it is under f.o.b. mill pricing (see Figure 1).

Figure 1: Alternative Spatial Price Schedules under Alternative Cost Conditions

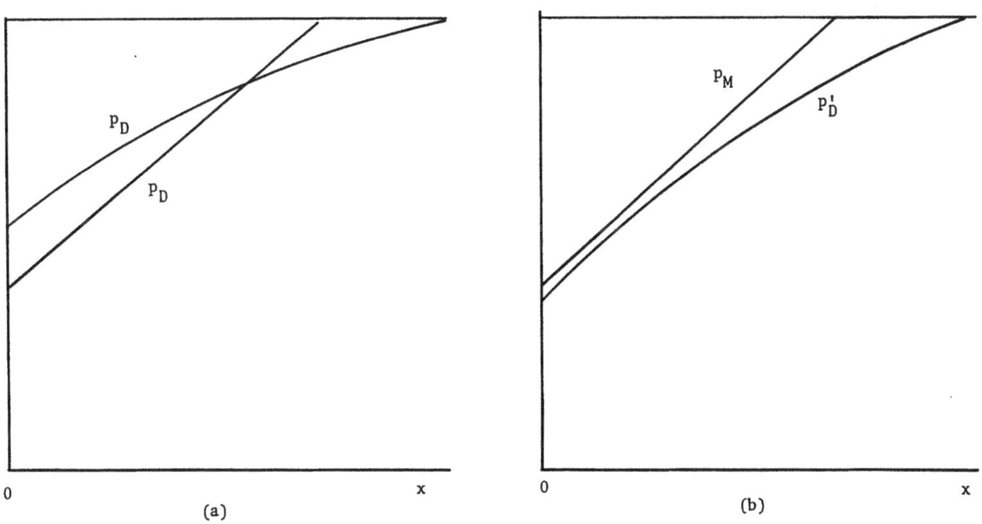

In Figure 1(a) the straight line P_M depicts a nondiscriminatory c.i.f. price schedule over economic space while P_D, which is everywhere flatter than P_M, represents a discriminatory c.i.f. price schedule. The P_M–P_D intersection in Figure 1(a) is no coincidence. Inexorably related to it are the facts that while the P_D is everywhere flatter than P_M and $P_D > P_M$ at the seller's site, the P_D is strictly lower than P_M at the market boundary point applicable to f.o.b. mill pricing. The reason for this, to repeat, is that distant local demand (in terms of mill price) is more elastic than the aggregate demand, given the assumed basic demand convexity. In passing, while P_M must be a straight line if transport costs are linear, the flatter line P_D need not be linear under varying demand conditions.

The welfare loss to nearby customers due to a price policy change from f.o.b. to discriminatory pricing is greater if marginal costs of production increase with output. Note in this case that P_D in Figure 1(a) would be shifted upward. So, distant buyers still benefit from price discrimination in the case of increasing MC insofar as P_D intersects P_M from above.

What is the effect of the pricing policy when MC decreases with increased output? In such case, the price schedule P_D shifts downward instead of upward. It is thus possible for everybody to benefit from spatial price discrimination under conditions of substantial scale economy. This result requires that the MC curve slope downward and intersect with DMR at a level substantially lower than with SMR. Substantial reduction in the equilibrium MC, along with a significant expansion in output as a consequence of discriminatory pricing, is not only a theoretical possibility, but also implies a substantial downward shift in the discriminatory price schedule P_D' below the nondiscriminatory counterpart P_M, as in Figure 1(b).[6]

The foregoing findings are based on Assumptions a and c (but not Assumption b) of Section 3. Let us summarize as follows:

Proposition II: Provided that the local demands become increasingly elastic as distances from the firm increase, the most distant customers, at the least, will benefit from discriminatory pricing regardless of the assumed cost conditions; moreover, everybody can benefit from discrimination under conditions of substantial scale economies.

The analysis of the present section has ignored to this point more convex basic demand. In fact, that form of basic demand is often avoided in the literature on spatial pricing, this avoidance for two reasons: first, it predicts discriminatory c.i.f. prices which increase more rapidly with distance than freight rates, the empirical verification of which is claimed by some writers, such as HOOVER (1937), to be unlikely on grounds of resale possibilities by the nearer buyers to more distant buyers. Second, any demand specification which does not have a finite price intercept value can be considered to be empirically irrelevant, especially with respect to a space economy.[7] However, the first point was rejected by DEWEY (1955) and OLSON (1972) on grounds of monopolistic power and/or substantial repackaging costs. Regarding the second point, a finite price intercept can be easily imposed by acting forth a constraint that (simply) involves the so-called consumer threshold concept (see DEVLETOGLOU, 1971). Then the *a priori* basis for rejecting that demand form no longer applies. Need, therefore, arises to reevaluate the basic assumptions of Section 2 in the particular context of spatial price discrimination and related price effects.

Now define the basic demand (1) as:

$$q = ap^{-\alpha}, \qquad p_0 \geqq p \geqq 0, \qquad p = m + x,$$
$$= 0, \qquad \text{elsewhere} \qquad (7)$$

The market demand QM under f.o.b. mill pricing is accordingly given by:

$$Q_M = 2\int_0^X a(m + x)^{-\alpha}\,dx\,, \qquad X = p_0 - m$$
$$= \frac{2a}{\alpha - 1}(m^{1-\alpha} - p_0^{1-\alpha}) \qquad (8)$$

Note that the average revenue curve, or the inverse of (8) currently has the finite price intercept value p_0. Correspondingly, the simple marginal revenue, SMR $\equiv d(mQ_M) / dQ_M$, which is derivable from (8) is:

$$SMR = m - \frac{Q_M}{2a}m^{\alpha}, \qquad m = \left(\frac{\alpha-1}{2a}Q_M + p_0^{1-\alpha}\right)^{\frac{1}{1-\alpha}} \qquad (9)$$

and has the same vertical intercept value p_0. Moreover, we observe $\partial(SMR)/\partial p_0 > 0$ for any $Q_M < 2ap_0^{1-\alpha}$. In the event of substantial threshold-sensitive consumer reactions, the SMR is shifted downward such that the lower the p_0, the lower is SMR.

To evaluate the impacts of consumer thresholds, if any, on the equilibrium f.o.b. mill price, initially assume that marginal production cost is constant. Then, the first order condition for profit maximization for the firm pricing f.o.b. mill is given by:

$$m - \frac{Q_M}{2a} m^{\alpha} = k, \tag{10}$$

where k stands for the marginal production cost MC, and Q_M is defined via (8) as a function of mill price m. It follows from (8) and (10) that:

$$(\alpha - 1)k - (\alpha - 2)m = p_0^{1-\alpha} m^{\alpha}, \tag{11}$$

where, for purposes of analytical continuity, the assumption of $\alpha > 2$ is preserved though no longer needed. The optimal solution for m^* is obtainable diagrammatically. Figure 2 depicts both the left-hand side (LHS) and the right-hand side (RHS) of (11). Siginificantly, the RHS curve alone shifts upward when p_0 is decreased. This causes a decrease in the equilibrium (optimal) mill price from m^* to $m^{*'}$.

Figure 2: Consumer Thresholds and the Optimal F.O.B. Mill Price: When MC is Constant

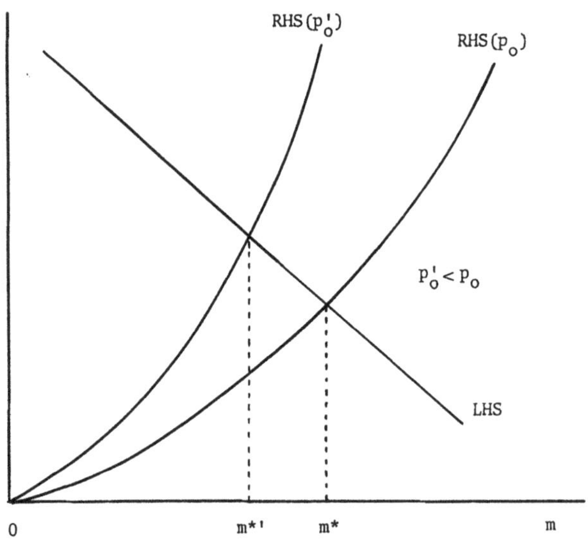

The price lowering effects of a consumer threshold are reinforced when the marginal cost curve slopes upward. This follows because given the downward shift in SMR as a consequence of p_0, the equilibrium MC decreases (leftward)

with the equilibrium output along the upward sloping MC curve labeled as $k'(Q) > 0$ in Figure 3. Without consumer thresholds in the figure, the reference point of equilibrium is given at (Q_M^*, k^*) whereas after the imposition of p_0, the new equilibrium is given at $(Q_M^*(p_0), k^*(p_0))$.

Figure 3: Simple Monopoly Equilibria under Alternative Cost Conditions and Consumer Thresholds

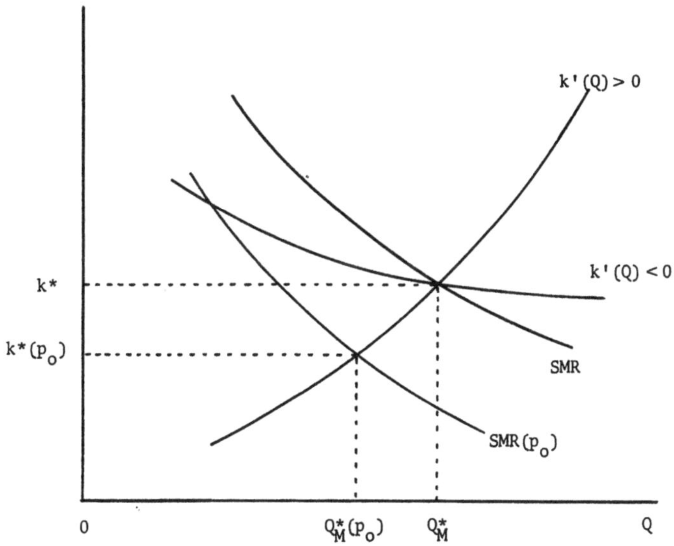

Thus, we have established:

Proposition III: When the basic demand defined in Section 3 is of constant elasticity, and the marginal production cost k remains constant, the optimal f.o.b. mill price m^* depends on the consumer threshold level p_0; and m^* is higher (lower) depending on whether p_0 is higher (lower).

If substantial scale economy exists in the sense that not only does the average variable cost curve slope downward, but so too the marginal cost curve, it is possible, though not necessary, that the equilibrium mill price will rise because of the increase in MC; of course, the equilibrium output is reduced along the negatively sloping MC curve, as given by $k'(Q) < 0$ in Figure 3.

Consider next the marginal revenue at location x. From (7) we have:

$$MR(x) = a^{\frac{1}{\alpha}} \left(\frac{\alpha - 1}{\alpha}\right) q^{-\frac{1}{\alpha}}, \qquad q \geq a p_0^{-\alpha}$$

$$= p_0, \qquad a p_0^{-\alpha} > q > 0 \qquad (12)$$

We know the discriminatory monopolist maximizes profit by equating each local marginal revenue MR(x) to marginal costs k plus transportation cost x. However, since equation (12) has a discontinuity at $q = a p_0^{-\alpha}$, the true optimality conditions include the inequality

$$MR(x) \geq k + x , \qquad (13)$$

where k is, as before, a function of the total output Q.

From (7), (12) and (13), the spatial price schedule under discriminatory pricing is then:

$$P_D = \frac{\alpha}{\alpha - 1} (k + x), \qquad \frac{(\alpha - 1) p_0}{\alpha} - k \geq x \geq 0$$

$$= p_0, \qquad p_0 - k \geq > x > \frac{(\alpha - 1) p_0}{\alpha} \qquad (14)$$

The slope of this price schedule $\alpha/(\alpha-1)$ is strictly greater than unity and of course steeper than the f.o.b. slope. The absence of threshold sensitivity along with Section 3's finding of a positive output effect resulting from price discrimination combine to provide the basis for comparing the two alternative spatial price schedules. We find in Figure 4(a) that the discriminatory price schedule P_D cuts the non-discriminatory schedule P_M *from below*. This establishes:

Proposition IV: When the basic demand is of constant elasticity, customers located proximate to the seller benefit from discriminatory pricing in the absence of threshold-sensitive reactions.

Figure 4: Impacts of Threshold Sensitive Consumer Reactions on Alternative Spatial Price Schedules: When MC is Constant

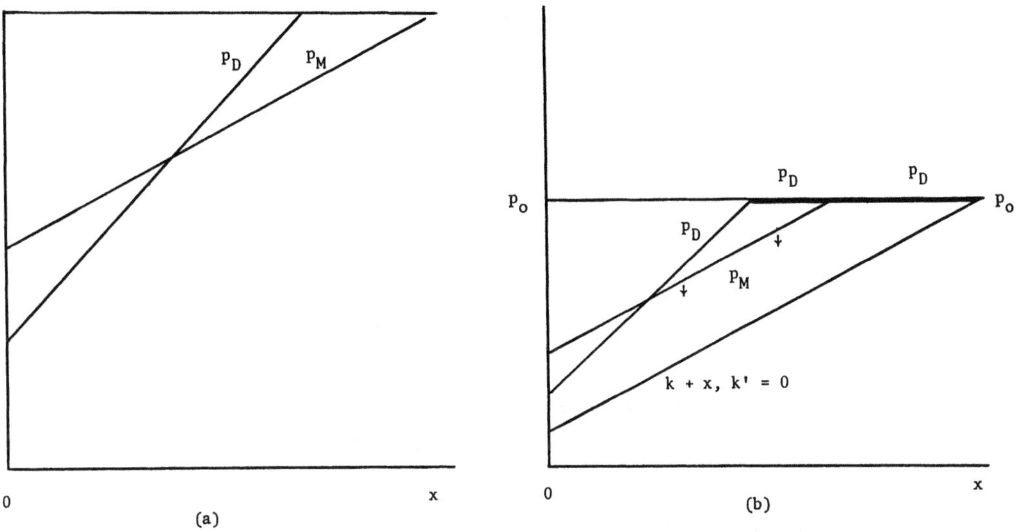

When a price ceiling is imposed on the delivered price schedules P_M and P_D as a consequence of threshold sensitive consumer reactions, the P_M schedule is shifted downward (unless the MC curve is decreasing sharply). The question arises, what happens to the P_D schedule? Consider the following:

(i) If marginal cost is constant, the price schedule P_D remains basically unaltered. However, as P_D approaches the price ceiling p_o, that price p_o becomes the c.i.f. price up to the point where the firm's marginal cost k plus transportation cost are at the threshold level. The impacts of threshold sensitivity on P_D and P_M are illustrated in Figure 4(b).

(ii) If marginal cost increases as output increases, the equilibrium MC will also increase as the firm switches its pricing policy from f.o.b. mill to discriminatory pricing. Recall in this regard that the price ceiling p_o shifts SMR downward. In contrast, DMR can be shown to remain unchanged, as it is completely independent of threshold sensitive price ceilings. This asymmetric and seemingly puzzling result stems directly from solving (12) and (13), via which the local sales quantities q(x) are obtained; then summing (i.e., integrating) these sales over the relevant spatial market domain, as given by (14), establishes

$$Q_D = \int_0^{x_1} a\left(\frac{\alpha}{\alpha - 1}\right)^{-\alpha}(k+x)^{-\alpha}dx + \int_{x_1}^{x_2} a p_o^{-\alpha}dx$$

$$\text{where } x_1 = \frac{(\alpha - 1)p_o}{\alpha} - k, \qquad x_2 = p_o - k$$

$$= \frac{\alpha}{\alpha - 1}\left(\frac{\alpha}{\alpha - 1}\right)^{-\alpha}k^{1-\alpha} \tag{15}$$

Equation (15) provides the output Q_D produced by the discriminating monopolist; of course, this output derives from equalizing the net marginal revenues at each market point with the marginal cost k.

Two other notes on (15) are in order. First, while Q_D is defined as a function of k *in the equilibrium*, the k itself is a function of total output. In other words, equilibrium k is determined at its intersection with (15). Second, the Q_D in (15) is a function of equilibrium k alone. It follows that DMR is completely independent of the price ceiling p_o.

Table 1: The Price and Output Effects of Alternative Pricing Policies under Threshold Sensitive Consumer Reactions

	P_M	P_D	Q_M	Q_D
$k'(Q)>0$	Down	Up	Down	Up
$k'(Q)=0$	Down	No Change	Down	Up
$k'(Q)<0$	Ambiguous	Down	Down	Up

Figure 5: Impacts of Threshold Sensitive Consumer Reactions on the SMR, DMR, and Related Equilibria under Alternative MC Conditions

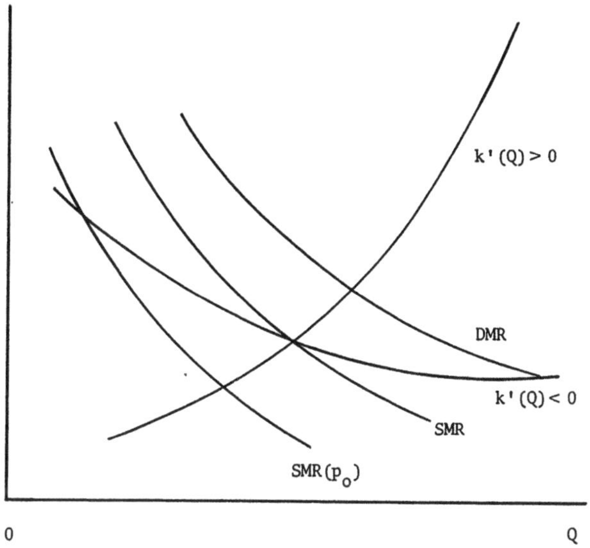

All the above findings and implications are summarily illustrated by Figure 5. Thus, compared with reference point E in the f.o.b. mill price equilibrium, the related equilibrium MC is shown to be lower or higher in the event of threshold-sensitive reactions on the part of consumers, depending on whether MC slopes upward or downward. (See and compare the intersection points generated by the curves SMR(p_o), $k'(Q) > 0$ and $k'(Q) < 0$.) On the other hand, the discriminator's equilibrium MC is shifted oppositely. (See the intersection points generated by DMR, $k'(Q) > 0$ and $k'(Q) < 0$.) The related price and output effects of discrimination *vis-a-vis* nondiscriminatory pricing, when subject to threshold sensitive consumer reactions, can be easily observed as summarized in Table 1 and illustrated by Figure 6 besides Figures 4 and 5.

Of particular interest is the relation exhibited in Figure 6(a). It can be stated in the form of the next proposition.

Figure 6: Impacts of Consumer Threshold Sensitive Reactions on the Alternative - Spatial Price Schedules under Alternative MC Conditions

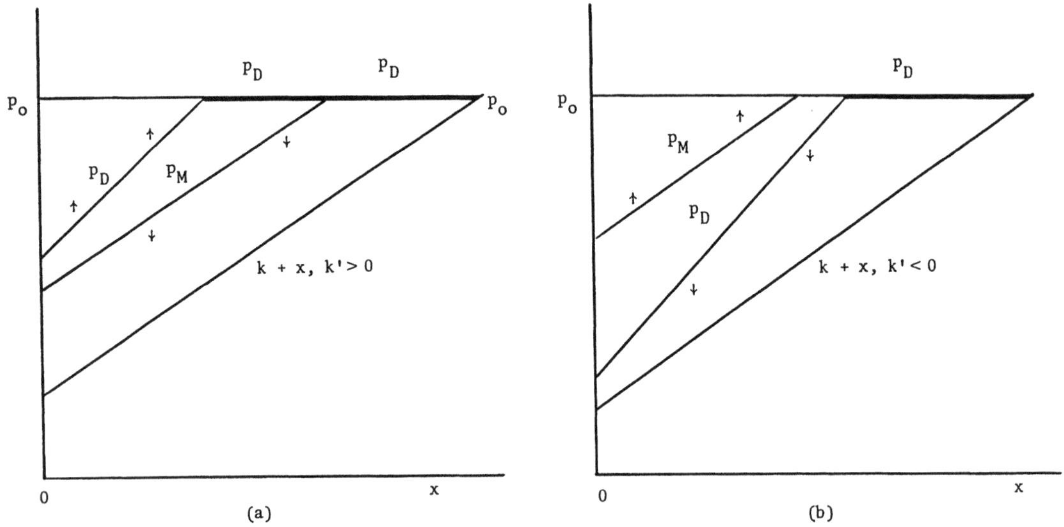

Proposition V: Given constant elasticity and threshold sensitivity, the number of nearer buyers who benefit from discrimination decreases unless the MC's negative slope is substantial. If, indeed, the MC slopes upward substantially and/or the consumer threshold level p_0 is sufficiently low, it becomes possible that virtually all consumers are better off under f.o.b. mill pricing than under discriminatory pricing. This is not to say, however, that the last result in Proposition V invalidates the latter part of Proposition II. According to the earlier finding, everyone is better off under discriminatory pricing whereas the present result provides a strict superiority of f.o.b. mill pricing over discriminatory pricing. Underlying the conflicting results are opposite technological conditions: substantial economies of scale versus diseconomies of scale. In fact, the superior result of discriminatory pricing proposed by Proposition II is reinforced when scale economies combine with consumer threshold reactions. This follows because a scale economy shifts P_D downward while it shifts the P_M upward when threshold sensitivity applies.[8]

5. Summary and Conclusion

This paper has considered the overall welfare effects of spatial price discrimination. Spotlighted are not only the aggregate outputs produced under discriminatory pricing *vis-a-vis* f.o.b. mill pricing, but also local c.i.f. prices. This is done so that we may determine who gains under what conditions. Conditions for evaluation including alternative production technologies and consumer tastes are represented by alternative cost and revenue curves. Impacts on consumer welfare of their own threshold-sensitive reactions are also examined. Basic results are summarized in the form of five propositions. We conclude that while the output effect of spatial price discrimination is generally positive, some buyers may gain and others may lose from discrimination, depending on where they are located. The relations reverse under different conditions. It is also possible that every consumer can benefit from discrimination (Proposition II) or on the contrary from f.o.b. mill pricing (Proposition V) depending on well-defined demand and cost conditions. Our findings point to the need to be particularly careful in establishing industrial policies and favored pricing practices. And while we mentioned at the outset to this paper that ROLF FUNCK has played with this (our) side of the coin, a natural extension from our findings would apply to his interest on the other side of the coin, as in FUNCK and BLUM (1981).

FOOTNOTES

1 See Funck and Kowalski (1984). Note that information costs, as with time costs, is counterpart to the cost of distance in spatial price theory. Or also see Funck (1986), which paper also provides counterpart insights into the subject of impacts of the costs of distance on the input-output efficiency of the firm in economic space.

2 A local consumer demand which is "relatively more convex" includes the local demand whose elasticity with respect to the net mill price, (i.e., the price net of freight cost) remains constant regardless of consumer location (Ohta, 1980, 1981).

3 See Greenhut, Norman, and Hung (1987).

4 The market area may be bounded if, for example, there exists a threshold-sensitive level of c.i.f. price, above which the basic demand may vanish. Section 4 will reveal some implications of such threshold sensitivity in consumer behavior.

[5] The profit maximizing market areas for the two alternative pricing policies under the present demand convexity conditions, i.e., $e_M > e_B$, turn out to be unbounded. This means that the positive output effect of price discrimination is in no way related to the market area "expansion" effect of discrimination observed under different demand convexity conditions as in Greenhut and Ohta (1972, 1975).

[6] As a straightforward example of this result, consider a linear basic demand with the price intercept value of a. Assume that MC in equilibrium under f.o.b. mill pricing happened to be a/2 while the equilibrium MC under discriminatory pricing decreased radically to approach zero. In this event, discriminatory price starts with a/2 at the mill whereas f.o.b. mill price clearly exceeds this level along with a slope throughout that is steeper than the former schedule.

[7] There is a third, and quite important reason for rejecting demands of the subject order; namely, as was pointed out in GREENHUT, NORMAN, HUNG, (1987) and detailed by M.L. GREENHUT in 1986, the elasticity requirements of each member of the set are extremely unlikely to exist. However, because this paper evaluates *welfare* impacts of spatial f.o.b. vis-à-vis spatial discriminatory pricing under conditions of spatial monopoly *and* spatial competition, the simple possibility that a member of this set might exist in the real-world requires the generality proposed above, and hence our relegating this third reason for rejecting demands of the subject order to footnote status only.

[8] And see M.L. GREENHUT, OHTA, SAILORS (1984). Note also that it can be readily shown that the elasticity of spatial market demand is increased by imposition of a threshold price ceiling regardless of the form of the basic demand. Since a concomitant downward shift in SMR will yield a constant or lower equilibrium MC in the absence of scale economy, it follows that the equilibrium f.o.b. mill price will be lower and so will the nondiscriminatory spatial price schedule P_M. However, when economies of scale exist, equilibrium MC will rise due to a downward shift in SMR! Thus, while the net effect of consumer threshold sensitivity upon equilibrium mill price is ambiguous when economies of scale exist, the more substantial the scale economy, the more likely is it that the effect will be positive: i.e., the mill price tends to rise due to threshold consumer reactions.

LITERATURE

Beckmann, M.J., 1976: "Spatial Price Policies Revisited," Bell Journal of Economics, 7, 619-630.

Devletoglou, N.E., 1971: Consumer Behavior, London.

Dewey, D., 1955: "A Reappraisal of F.O.B. Pricing and Freight Absorption," Southern Economic Journal, 1955, 22, 48-54.

Funck, R.H.: "August Lösch and the Concept of Region," Space-Structured Economy, R.H. Funck and A. Kuklinski, editors, Karlsruhe Papers in Economic Policy Research, 3, 55-56.

Funck, R.H.; Kowalski, J.S., 1984: "The Role of Information in Regional Urban Development," presented at the 11th meeting of the International Association for Regional and Urban Statistics, Copenhagen, I, 53-89.

Funck, R.H.; Blum, U.: "Urban Policy Analysis," in Cities in Transition: Problems and Policies (Alphen (Sijthoff & Noordhoff), P. Nijkamp and P. Rietveld (eds), pp. 243-256.

Greenhut, J.; Hwang, M.; and Ohta, H., 1974: "Price Discrimination Over Space - Theory Assumptions, and a Case Example," Annals of Regional Science, 8, 70-81.

Greenhut, J.; Greenhut, M.L., 1975: "Spatial Price Discrimination, Competition, and Locational Effects," Economica, 42, 401-419.

Greenhut, M.L., 1956: Plant Location in Theory and in Practice, 4th printing (Westport, Conn.: Greenwood Press, 1972, originally University of North Carolina Press.

Greenhut, M.L. 1986: "On Demand Curves and Spatial Pricing," chapter in Spatial Pricing and Differentiated Markets (Pion Press: London Papers in Regional Science Vol. 16) ed. G. Norman.

Greenhut, M.L.; Norman, G.; Hung C.S., 1987: The Economics of Imperfect Competition, London.

Greenhut, M.L.; Ohta, H., 1972: "Monopoly Output under Alternative Spatial Pricing Techniques, "The American Economic Review, 62, 705-713.

Greenhut, M.L. 1975a: "Discriminatory and Nondiscriminatory Spatial Prices and Outputs under Varying Market Conditions," Weltwirtschaftliches Archiv, 111, 310-332.

Greenhut, M.L., 1975b: Theory of Spatial Pricing and Market Areas, Durham, N.C.

Greenhut, M.L.; Sailors, J., 1985: "Reverse Dumping: A Form of Spatial Price Discrimination, "Journal of Industrial Economics, 34, 167-182.

Holahan, W.L., 1975: "The Welfare Effects of Spatial Price Discrimination," The American Economic Review, 65, 498-503.

Hoover, E.M., 1937: "Spatial Price Discrimination," Review of Economic Studies, 4, 182-191.

Lösch, A., 1954: Economics of Location, New Haven.

Norman, G., 1981: "Spatial Competition and Spatial Price Discrimination," Review of Economic Studies, 48, 97–111.

Ohta, H., 1980: "Spatial Competition, Concentration and Welfare," Regional Science and Urban Economics, 1980, 10, 3–16.

Ohta, H., 1981: "The Price Effects of Spatial Competition," Review of Economic Studies, 48, 317–325.

Ohta, H., 1988: Spatial Price Theory of Imperfect Conpetition, College Station, Texas.

Ohta, H.; Wako, T., 1988: "The Output Effects of Spatial Price Discrimination Revisited," Journal of Regional Science, 28, 83–87.

Olson, J.E., 1972: "Price Discrimination by Regulated Motor Carriers," American Economic Review, 62, 395–402.

Pigou, A.C., 1929: The Economics of Welfare, London, 3rd ed.

Robinson, J., 1933: The Economics of Imperfect Competition, London.

Smithies, A., 1941: "Monopolistic Price Policy in a Spatial Market," Econometrica, 9, 63–73.

Transport Flows in Tinbergen-Bos Systems

by

J. Hans KUIPER, Jean H.P. PAELINCK, Kenneth E. ROSING

1. Introduction

Work on Tinbergen-Bos systems has proven to be fruitful in discovering spatial properties concerning the organisation of activity.[1] In the paper just referred to a research programme is suggested, in which a portion is devoted to the study of metrics and transport flows in TBS. Some of these threads have already been taken up[2]; this paper however is a more fundamental approach. It makes explicit and justifies some hypotheses under which, it is hoped, more progress can be made in this field of research.

2. Axiomatics

We will base the analysis upon the following assumptions:

A1: the area to be studied is organised around a rectangular network with unit mazes;

A2: the area limits are located around a "Manhattan-circle"[3] with integer "radius";

A3: the area is structured by unit Manhattan-circles (squares) centered on the network nodes; this implies that the area's border is located outside the Manhattan-circle stated in A2[4];

A4: the spatial economy functions according to TBS-assumptions.[5]

Figure 1 hereafter represents the most elementary area.

Figure 1: elementary study area

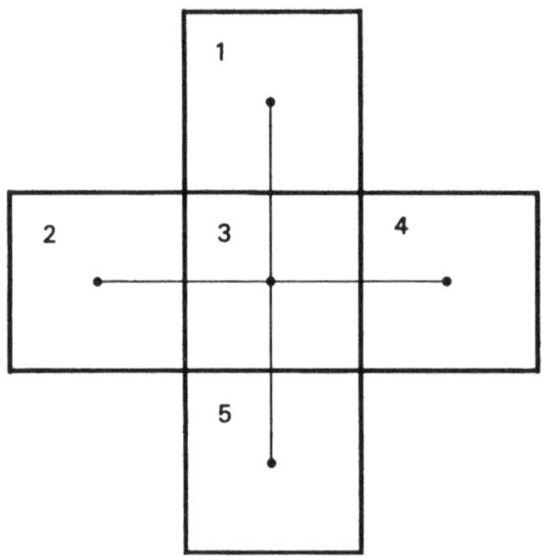

It is easy to compute that the number of elementary areas, n, is equal to

$$n = 2r\ (r+1) + 1 \qquad (1)$$

where r is the radius of A2.

Assumption 3 is based on the property of squares having, for equal surfaces, the smallest average Manhattan-distance from their centres, of all of the four classical figures in spatial analysis, to wit, circles, hexagons, squares and triangles; the last three regular polygons are the only area-covering ones[6] (Proposition 1).

Proof: one starts with the reference square as pictured in figure 2 as a 'Manhattan-circle', and some lemma's.

Figure 2: reference square

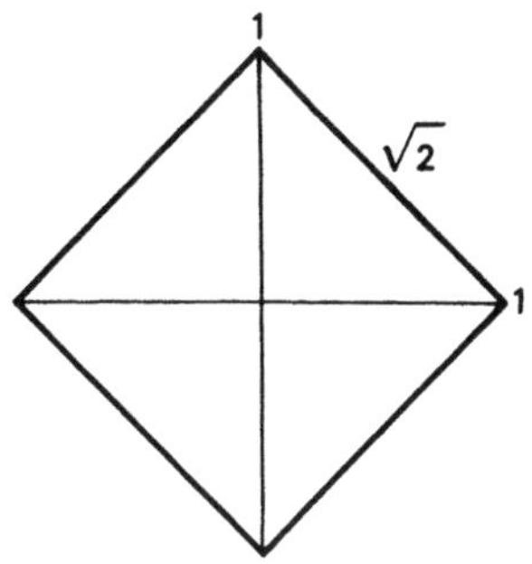

Lemma 1: the average Manhattan distance over an a (basis) – b (height) rectangle is (a+b)/2

Proof:

$$(\int_0^b \int_0^a x\,dx\,dy\;(\int_0^b \int_0^a dx\,dy)^{-1} = \frac{a^2b}{2}\,(ab)^{-1} = \frac{a}{2} \qquad (2)$$

and the same for y.

Lemma 2: the average Manhattan distance over an a (basis) – b(height) rectangular triangle is (a+b)/3

Proof:

$$(\int_0^a \int_0^{b(1-a^{-1}x)} y\,dy\,dx\;(\int_0^a \int_0^{b(1-a^{-1}x)} dy\,dx)^{-1} =$$

$$\frac{b^2}{2} \int_0^a (1-a^{-1}x)^2\,dx\,(\frac{ab}{2})^{-1} = \frac{ab^2}{6}\,(\frac{ab}{2})^{-1} = b/3 \qquad (3)$$

and the same for x.

Applications:

(i) Reference square

Figure 3: the reference square again

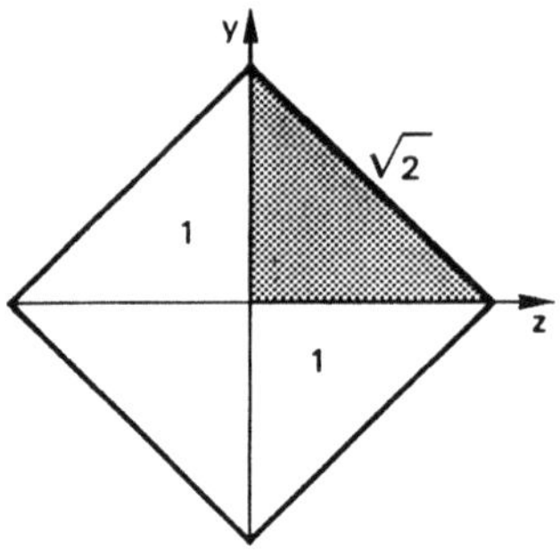

$\bar{d}$ = 2/3; S(surface of the total square) = 2 (4)

(ii) Hexagon

Figure 4a: the hexagon

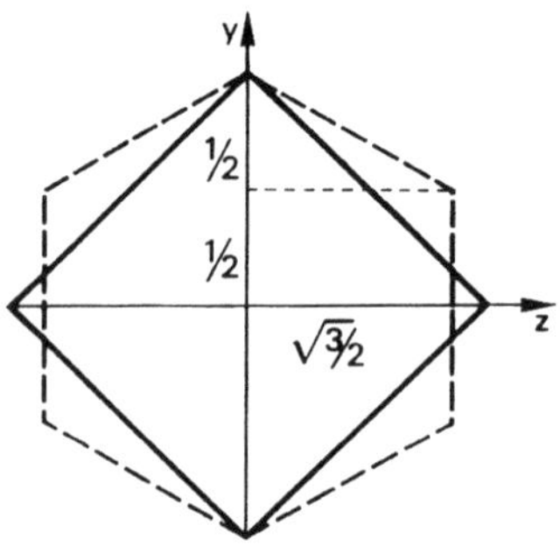

89

Figure 4b: a unit slice

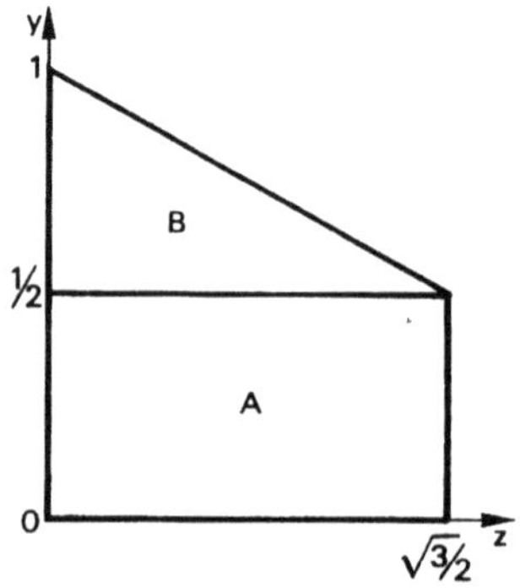

$$\begin{array}{llllr} & x & y & & \\ \bar{d}\ A = & \sqrt{3}/4 & 1/4 & w(=\text{weight}) = 2 & (5a) \\ \bar{d}\ A = & \sqrt{3}/6 & (1/2 + 1/6) & w(=\text{weight}) = 1 & (5b) \end{array}$$

$$\bar{d}_x = \frac{1}{3}\left(\frac{2\sqrt{3}}{4} + \frac{\sqrt{3}}{6}\right) = \frac{(3 + 1)\sqrt{3}}{3 \times 6} = \frac{2}{9}\ \sqrt{3} \qquad (6a)$$

$$\bar{d}_y = \frac{1}{3}\left(\frac{2}{4} + \frac{4}{6}\right) + \frac{1}{3}\ \left(\frac{3}{6} + \frac{6}{4}\right) = \frac{7}{18} \qquad (6b)$$

$$\bar{d} = \frac{1}{18}\ (7 + 4\sqrt{3}) = .7738 \qquad (6c)$$

the correcting factor for equal surfaces being computed as

$$2 = S = r^2\ \frac{3\sqrt{3}}{2} \qquad (7a)$$

whence

$$r = \sqrt{\ }\ 4(3\sqrt{3})^{-1} = .8774 \qquad (7b)$$

and

$$\bar{d}(\text{corr}) = .7738 \times .8774 + .6789 > 2/3 \qquad (8)$$

(iii) Triangle

Figure 5a: the triangle

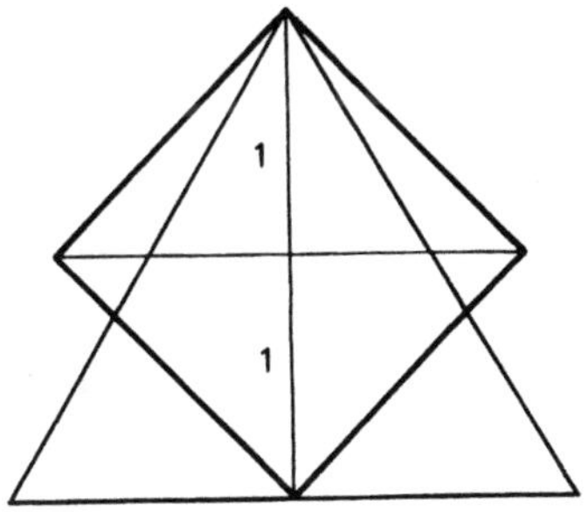

Figure 5b: a unit slice

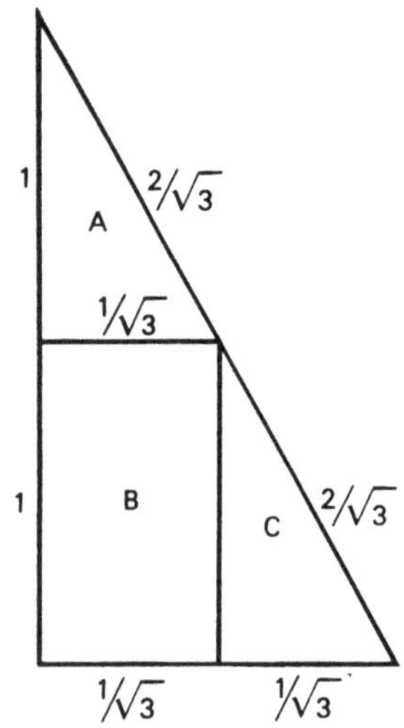

The calculation proceeds as follow:

	x	y	w	
$\bar{d}_A$	$\frac{1}{3\sqrt{3}}$	$\frac{1}{3}$	1	(9a)
$\bar{d}_B$	$\frac{1}{2\sqrt{3}}$	$\frac{1}{2}$	2	(9b)
$\bar{d}_C$	$\frac{1}{\sqrt{3}} + \frac{1}{3\sqrt{3}}$	$\frac{1}{3}$	1	(9c)

$$\bar{d}_x = \frac{1}{4}\left(\frac{1}{3\sqrt{3}} + \frac{3}{3\sqrt{3}} + \frac{3}{3\sqrt{3}} + \frac{1}{3\sqrt{3}}\right) \quad (10a)$$

$$\bar{d}_y = \frac{\frac{1}{3} + 1 + \frac{1}{3}}{4} = \frac{5}{12} \quad (10b)$$

$$\bar{d} = .3849 + 4167 = .8016 \quad (11)$$

the correcting factor being computed as

$$S = 2 = r\,\frac{1}{\sqrt{3}} \times \frac{r}{2} \times 8 = \frac{4r^2}{\sqrt{3}} \quad (12a)$$

whence

$$r = \frac{\sqrt{3}}{\sqrt{2}} = .9306 \quad (12b)$$

and

$$d\ (corr) = .7460 > 2/3 \quad (13)$$

If one now *exinscribes* a circle, one can compute the average y-distance as

$$\bar{d}_y = \int_0^{\pi_2} \int_0^{\sin\psi} y \sin\psi \, dy d\psi \quad \left(\int_0^{\pi_2} \int_0^{\sin\psi} \sin\psi dy \, d\psi\right)^{-1} \qquad (14a)$$

$$\int_0^1 \int_0^{\sqrt{1-x^2}} y \, dy \, dx \, \left(\int_0^1 \int_0^{\sqrt{1-x^2}} dy \, dx\right)^{-1} \qquad (14b)$$

$$= 4/3\pi = .4244 \qquad (14c)$$

and similarly for $\bar{d}_x$, the correcting factor for equal surfaces being

$$2 = S = \pi r^2 \qquad (15a)$$

whence

$$r = \surd 2 \pi = .7979 \qquad (15b)$$

so

$$\bar{d}(corr) = .8488 \times .7979 = .6773 > 2/3 \qquad (16)$$

x

x x

3. Tinbergen-Bos Mathematical Programme

Figure 1 can be extended to larger Manhattan-circles; we will denote, according to (1), the number of nodes by n.

Concerning the internal transportation costs in each elementary square, as long as no more than one plant is located in each node[7], and agricultural population is supposed to be equally spread over each elementary square, the 'internal' transportation cost (internal to each square) can be neglected, as the same quantities (or values) of every good have to be transported inside each square. If two or more plants are present in that square, average Manhattan distances within a square will still not be effected, and can be neglected according to Proposition 2 to follow.

Proposition 2: the average distance from the centre over an arbitrary slice of a Manhattan-circle is equal to 2/3 r.

Proof: start from the expression for the average y-distance of a slice of a Manhattan-circle with unit radius

$$\bar{d}_y = [\ \int_{a_2}^{a_1} \int_{(1-a_2)a_2^{-1}x}^{(1-a_1)a_1^{-1}x} y dy\, dx + \int_{a_1}^{a_2} \int_{(1-a_2)a_2^{-1}x}^{1-x} y dy\, dx\ ]$$

$$[\ \int_{0}^{a_1} \int_{(1-a_2)a_2^{-1}x}^{(1-a_1)a_1^{-1}x} dy dx + \int_{a_1}^{a_2} \int_{(1-a_2)a_2^{-1}x}^{1-x} dy\, dx\ ]^{-1}$$

$$= \frac{2}{3} - \frac{1}{3}(a_1 + a_2) \qquad (17)$$

and similarly for x:

$$\bar{d}_x = \frac{2}{3} - \frac{1}{3}\ [(1-a_1) + (1-a_2)] \qquad (18)$$

which leads to

$$\bar{d} = \bar{d}_x + \bar{d}_y = 2/3$$

x^x x

Thirdly, in TBS only one plant of the highest rank is allowed in each potential location; we will now relax this assumption and allow two or more.

In this paper we will restrict ourselves to only two goods, an agricultural product and an 'industrial' one, indexed i (i=0,1); the TBS is then very simple, and can be written out as follows:

$$y^* = y_0 + y_1 \qquad (19)$$

$$y_0 = \alpha_0 y^* \qquad (20a)$$

$$y_1 = \alpha_1 y^* \qquad (20b)$$

$$n_1 = n_1^* \qquad (21)$$

Indexing modes by j and k (j, k = 1, ... n), one computes easily

$$y_{1j} = x_{1j} \; y_1 n_1^{-1} \qquad (22a)$$

$$x_{1j} = x_{1j}^2 \qquad (22b)$$

The symbols used up to now have the following meaning:

y^* = national product (exogenous);

y_i = sectoral product;

α_i = consumption propensities ($\sum \alpha_i = 1$);

n_i = number of i-type plants;

y_{ij} = sectoral product i of center j;

x_{ij} = a zero-one (binary) variable.

There is no need in this case for the classical trade balance equilibrium equalities as agriculture has

$$y_0^d = (1-\alpha_0)\alpha_0 \; y^* \qquad (23a)$$

and industry

$$y_1^d = \alpha_1(1-\alpha_1) \; y^* \qquad (23b)$$

as disposable quantities or values to trade, and those are equal, given $\sum\alpha_i = 1$, the same being true of each node or elementary area.

Proposition 3: if $n_1 = 1$, the location is central; if $n_1 \geqq n$, the locations are completely deconcentrated.

Proof: for $n_1 = 1$, a central location minimises total transportation costs, as the average distance goods 0 and 1 have to be transported over, is minimized; for $n_1 \geqq n$, deconcentration leads to only 'internal' transportation costs, which we have discussed above.

x

x x

For $1 < n_1 < n$ different spatial patterns could be obtained, including multiple optima; in fact they can be derived from the following mathematical programme:

$$\min_{y_{ijk}, x_{ij}} \sum_{jk} t_0 y_{0jk} d_{jk} + \sum_{jk} t_1 y_{1jk} d_{jk}$$

$$= \sum_{jk} (t_0 y_{0jk} + t_1 y_{1jk}) d_{jk} \tag{24}$$

subject to

$$\sum y_{0jk} = y_{0j}^d \triangleq (1-\alpha_0)\ y_0 n^{-1}, \qquad \forall_j \tag{25a}$$

$$\sum_j y_{1jk} = (1-\alpha_0)\ y_0 n^{-1}, \qquad \forall_k \tag{25b}$$

$$\sum_k y_{1jk} = y_{1j}^d \triangleq x_{1j}(1-\alpha_1)\ y_1 n_1^{-1}, \quad \forall_j \tag{26a}$$

$$\sum_j y_{0jk} = x_{1k}(1-\alpha_1)\ y_1 n_1^{-1}, \qquad \forall_k \tag{26b}$$

$$\sum_j x_{1j} = n_1 , \tag{27a}$$

$$y_{ijk} \geq 0 \qquad \forall_{ijk} \tag{27b}$$

This programme is operationalised in the following section.

4. Obtaining Optimal Solutions

The optimal spatial arrangement of facilities to satisfy the requirements of a TBS, in a Manhattan-circle, can be found by applying a modified p-median model to the matrix of rectangular distances. The p-median model was developed to solve the classical plant location problem; in this problem one locates plants (here referred to as facilities) to serve demand located at points (here referred to as demand nodes) on a network. HAKIMI (1964, 1965) showed that an optimal solution existed in which the facilities are located at a selection of the demand nodes. WENDELL and HURTER (1973) and LOVE and MORRIS (1975) have shown that when dealing with rectangular distances only intersections of the coordinates of the "X" and "Y" coordinates need be considered since the objective function is separable into "X" and "Y" components; they further prove that only intersections inside the rectangular hull need be considered. ROSING (1989) notes that in the case of collinear points all but one of the equal coordinates can be eliminated. In the Manhattan-circle all points are collinear in either "X" or "Y" or both. Demand nodes can be used as potential facility sites; HAKIMI's theorem applies and the problem can be solved as a p-median problem.

Two principle lines of model specification are available for the p-median problem. The first is the MANNE (1964), BALINSKI (1965), REVELLE (1967), REVELLE and SWAIN (1970) linear programming formulation where the integer constraint is relaxed and the Simplex algorithm (Dantzig, 1963) used to solve the resulting continuous programme; REVELLE and SWAIN's definitive formulation is:

$$\min_{x_{Ij}} z = \sum_{i=1}^{n} \sum_{i=1}^{n} x_{ij}\, a_{ij} \qquad (28)$$

subject to:

$$\sum_{j=1}^{n} x_{ij} = 1, \qquad \forall\, i; \qquad (29)$$

$$x_{jj} - x_{ij} \geq 0 \qquad \forall\, i,\, j,\, i \neq j \qquad (30)$$

$$\sum_{j=1}^{n} x_{ij} = p^* \qquad (31)$$

$$x_{jj} = x_{jj}^2 \qquad \forall\, j; \qquad (32)$$

where:

- n = the number of demand nodes;
- a_{ij} = the weighted distance on a network;
- x_{ij} = the decision variable, one if demand node i is served by a facility at j, zero otherwise;
- p^* = an exogenous variable specifying the number of facilities.

Experience has shown the this formulation terminates fully integer in nearly all cases.

The second principle formulation is that of EFROYMSON and RAY (1966) which they solved by the, then new method, of branch and bound. Their formulation is similar to REVELLE's but replaces the n(n−1) constraints of type (30) with n constraints of type (33) (notation modified for consistency).

$$n\, x_{ij} - \sum_{i=1}^{n} x_{ij} \geq 0 \qquad \forall\, j,\ i \neq j \qquad (33)$$

This formulation can be successfully solved because their procedure of branch and bound enforces the integer constraint while relaxing constraint (33) and using it as a test of feasibility. Constraint (33) is, obviously, the sum of i constraints of type (30). Constraints of type (30) are "tight" constraints MORRIS, (1978) nearly guaranteeing integer solutions while constraints of type (33) are loose constraints, which, in a simplex solution, will nearly always be fractional. ROSING, REVELLE and ROSING-VOGELAAR (1979) take advantage of this difference to reduce the size of the fully specified REVELLE formulation by writing only limited sets of constraint (30), for nearby demand nodes, for each j and relying on constraints of type (33) to prevent assignments of more distant demand nodes; this formulation allows the solution of much larger problems than was previously feasible.

A requirement of the TBS is that each facility serves an equal demand. Neither of the formulations places any restriction on the amount of demand to be served by any facility. ROSING (1979) showed how constraints of type (33) can be used as capacity constraints thus limiting the demand served from each facility; utilizing this result only two steps remain before being able to solve the TBS as a modified p-median model. First, more than one facility must be able to be located at a single demand node. Constraint (29) can be changed from an equality constraint into a greater than or equal constraint to allow this. Each demand node will still be required to assign to some facility; demand nodes which are also facility sites will be able to have multiple facilities, and - one is minimising - no demand node will assign more than once. Second, the i ≠ j must be removed from constraint (33) in order to correctly count the assigned demand in cases of multiple facilities at a single demand point and thus ensure that all facilities serve exactly equal demand. Note that the only element which requires further development is the location of multiple facilities at one demand node.

This problem can now be reformulated as follows:

$$\min_{x_{ij},\, y_j} z = \sum_{i=1}^{n} \sum_{j=1}^{n} x_{ij}\left(|c_{1i} - c_{jj}| + |c_{2i} - c_{jj}|\right) \qquad (34)$$

subject to:

$$\sum_{j=1}^{n} x_{ij} \geq 1 \qquad \forall\, i \qquad (35)$$

$$y_j - x_{ij} \geq 0 \qquad \forall\, i,\, j,\, i \neq j \qquad (36)$$

$$d\, y_j - \sum_{i=1}^{n} x_{ij} \geq 0 \qquad \forall\, j \qquad (37)$$

$$\sum_{j=1}^{n} y_j = p^* \; ; \qquad (38)$$

$$y_j = y_j^2 \qquad \forall\, j \qquad (39)$$

where:

c_{1i} , c_{2i} are the first and second coordinates of the ith demand node respectively;

c_{1j} , c_{2j} are the first and second coordinates of the jth potential facility site;

y_j is a decision variable for the location of facilities, and;

d is $n/p^* + 0.00001$

The objective function (34) states that the rectangular matrix distance between each chosen facility and the demand nodes is counted if and only if that demand node is (fractionally) assigned to it. Constraint (35) requires that each demand node assign to chosen facility sites for at least 100%; minimisation will insure that no demand node assigns for more than 100%, but at the same time this constraint will allow more than one facility at a demand node. Y_j is a decision variable for the location of a plant in constraint (36). x_{ij} in the same constraint expresses the proportion of demand of its own elementary square satisfied by its own plant. This can be complete satisfaction (1.0), but a priori other cases are possible. Constraint (36) is the exact equivalent of constraint (30); constraint (37) however is more complex. In the TBS each facility can serve exactly $1/p^{*th}$ of the demand; here there is equal demand so d is calculated as n/p^*, however for rounding errors (which would result in an infeasible programme) 0.00001 is added to each capacity where necessary. Obviously the results are accurate only to the level of 0.0001, but a higher level of accuracy would test the single precision accuracy of the computer and be rejected by the programme employed (7 accurate digits). Constraint (37) is the requirement that the total number of facilities be the exogenously determined number. Constraint (38) is the requirement of fully integer facility locations.

Tables 1 through 3 hereafter present the results of computations for n = 5, 13 and 25 respectively. The nodes have been numbered from 1 to n, in each case from top to bottom and from left to right; this is the meaning of the figures appearing in Figure 1. The value of p is the number of facilities, the F.V. is the associated total cost of that system. Each different service area is enclosed in accolades; each node is given by a number followed by a comma, and then the proportion of elementary area served by the facility in that group. For example in table 1, p = 3, the total cost is 3.0; there is a facility in elementary region 1, which serves itself for 100%, and that facility also serves elementary regions 2 and 3 each for 33.33%.

Table 1, 5 element rectangular circle (bold numbers are facilities)

p	assignments
1	F.V. = 4.0000 {1,1.0; 2,1.0; **3,1.0** 4,1.0; 5,1.0}
2	F.V. = 4.0000 {1,1.0; 2,1.0; **3,0.5**} {**3,0.5**; 4,1.0; 5,1.0}
3	F.V. = 3.0000 {**1,1.0** 2,0.3333; 3,0.3333} {3,0.6667; **4,1.0**} {2,0.6667; **5,1.0**}
4	F.V. = 1.0000 {**1,1.0**; 3,0.2500} {**2,1.0**; 3,0.2500} {3,0.2500; **4,1.0**} {3,0.2500; **5,1.0**}
5	F.V. = 0.0000 {**1,1.0**} {**2,1.0**} {**3,1.0**} {**4,1.0**} {**5,1.0**}

Table 2, 13 element rectangular circle (bold numbers are facilities)

p	assignments
1	F.V. = 20.0000 {1,1.0; 2,1.0; 3,1.0; 4,1.0; 5,1.0; 6,1.0; **7,1.0;** 8,1.0; 9,1.0; 10,1.0; 11,1.0; 12,1.0; 13,1.0}
2	F.V. = 17.0000 {1,1.0; 2,1.0; **3,1.0**; 4,1.0, 7,1.0; 8,1.0, 9,0.5000} {5,1.0; 6,1.0; 9,0.5000; 10,1.0; **11,1.0;** 12,1.0; 13,1.0}
3	F.V. = 12.6667 {1,1.0; 2,1.0; **3,1.0**; 4,1.0; 7,0.3333} {7,0.6667; **8,1.0;** 9,1.0; 12,1.0; 13,0.6667} {5,1.0; 6,1.0; **10,1.0;** 11,1.0; 13,0.3333}
4	F.V. = 9.0000 {1,1.0; 2,0.2500; **3,1.0**; 4,1.0} {2,0.7500; 5,1.0; **6,1.0**;; 10,0.5000} {7,1.0; **8,1.0**; 9,1.0; 12,0.2500} {10,0.5000; **11,1.0**; 12,0.7500; 13,1.0}
5	F.V. = 9.4000* {1,1.0; 2,1.0; **3,0.5000**; 13,0.1000} {**3,0.5000**; 4,1.0; 7,1.0; 13,0.1000} {5,1.0, **6,1.0**; 10,0.6000} {**8,1.0**; 9,1.0; 12,0.6000} {10,0.4000; **11,1.0**; 12,0.4000; 13,0.8000}
6	F.V. = 9.0000* {1,1.0; 2,0.4167; **3,0.5000**; 7,0.2500} {2,0.4167; **3,0.5000**; 4,1.0; 7,0.2500} {2,0.1667; 5,1.0; **6,1.0**} {**8,1.0**; 9,1.0; 12,0.1667} {7,0.2500; 10,1.0; **11,0.5000**; 12,0.4167} {7,0.2500; **11,0.5000**; 12,0.4167; 13,1.0}

7	F.V. = 8.0000 {**1,1.0**; 3,0.8571} {2,0.2857; 3,0.1429; **4,1.0**; 8,0.4286} {2,0.7143; **5,1.0**; 6,0.1429} {7,0.2857; 8,0.5714; **9,1.0**} {6,0.8571; **10,1.0**} {11,0.8571; **12,1.0**} {7,0.7143; 11,0.1429; **13,1.0**;}
8	F.V. = 6.0000 {**1,1.0**; 3,0.6250} {**2,1.0**; 6,0.6250} {3,0.250; **4,1.0**; 8,0.3750} {**5,1.0**; 10,0.6250} {3,0.1250; 6,0.3750; **7,1.0**; 11,0.1250} {8,0.6250; **9,1.0**} {11,0.6250; **12,1.0**} {10,0.3750; 11,0.2500; **13,1.0**}
9	F.V. = 4.0000 {**1,1.0**; 3,0.4445} {**2,1.0**; 3,0.3334; 6,0.1112} {3,0.2222; **4,1.0**; 8,0.2223} {**5,1.0**; 6,0.4445} {6,0.4445; **7,1.0**} {8,0.4445; **9,1.0**} {**10,1.0**; 11,0.4445} {8,0.3333; 11,0.1112; **12,1.0**} {11.0.4444; **13,1.0**}
10	F.V. = 3.8000† {**1,1.0**; 3,0.3000} {**2,1.0**; 3,0.3000} {3,0.4000, **4,0.9000**;} {**5,1.0**; 7,0.3000} {**6,1.0**; 7,0.3000} {7,0.3000; **8,1.0**} {4,0.1000; 7,0.1000; **9,1.0**; 11,0.1000} {**10,1.0**; 11,0.3000} {11,0.3000; **12,1.0**} {11,0.3000; **13,1.0**}
11	F.V. = 3.0908† {**1,1.0**; 8,0.1817} {**2,1.0**; 6,0.1818} {**3,1.0**; 8,0.1818} {**4,1.0**; 8,0.1818} {**5,1.0**; 6,0.1818} {6,0.1818; **7,1.0**} {8,0.1818; **9,1.0**} {6,0.4545; **10,0.7273**} {10,0.2727; **11,0.9091**} {8,0.2728; **12,0.9090**} {11,0.0909; 12,0.0910; **13,1.0**}
12	F.V. = 1.6667† {**1,1.0**; 7,0.832} {**2,1.0**; 7,0.0833} {**3,1.0**; 7,0.0833} {**4,1.0**; 7,0.0833} {**5,1.0**; 6,0.0833} {**6,0.9167**; 7,0.1667} {7,0.0833; **8,1.0**} {7,0.0833; **9,1.0**} {**10,1.0**; 11,0.0833} {7,0.2500; **11,0.8333**} {11,0.0833; **12,1.0**} {7,0.0833; **13,1.0**}
13	F.V. = 0.0000 {**1,1.0**} {**2,1.0**} {**3,1.0**} {**4,1.0**} {**5,1.0**} {**6,1.0**} {**7,1.0**} {**8,1.0**} {**9,1.0**} {**10,1.0**} {**11,1.0**} {**12,1.0**} {**13,1.0**}

Table 3, 25 element rectangular circle (bold numbers are facilities)

p	assignments
1	F.V. = 56.0000 {1,1.0; 2,1.0; 3,1.0; 4,1.0; 5,1.0; 6,1.0; 7,1.0; 8,1.0; 9,1.0; 10,1.0; 11,1.0; 12,1.0; **13,1.0**; 14,1.0; 15,1.0; 16,1.0; 17,1.0; 18,1.0; 19,1.0; 20,1.0; 21,1.0; 22,1.0; 23,1.0; 24,1.0; 25,1.0}
2	F.V. = 45.0000 {2,1.0; 5,1.0; 6,1.0; 10,1.0; 11,1.0; **12,1.0**; 13,1.0; 17,1.0; 18,1.0; 19,1.0; 22,1.0; 23,1.0; 25,0.5} {1,1.0; 3,1.0; 4,1.0; 7,1.0; 8,1.0; 9,1.0; **14,1.0**; 15,1.0; 16,1.0; 20,1.0; 21,1.0; 24,1.0; 25,0.5}

3 F.V. = 36.6667
{1,1.0; 2,1.0; 3,1.0; 5,1.0; **6,1.0**; 7,1.0; 10,1.0; 11,1.0; 12,0.3333} {4,1.0; 8,1.0; 9,1.0; 12,0.3333; **14,1.0**; 15,1.0; 16,1.0; 20,1.0; 24,1.0} {12,0.3333; 13,1.0; 17,1.0; 18,1.0; **19,1.0**; 21,1.0; 22,1.0; 23,1.0; 25,1.0}

4 F.V. = 26.0000
{1,1.0; 2,1.0; **3,1.0**; 4,1.0; 6,0.5; 7,1.0; 13,0.75} {5,1.0; 6,0.5; 10,1.0; **11,1.0**; 12,1.0; 17,1.0; 18,0.75} {8,1.0; 9,1.0; 13,0.25; 14,1.0; **15,1.0**; 16,1.0; 21,1.0} {18,0.25; 19,1.0; 20,1.0; 22,1.0; **23,1.0**; 24,1.0; 25,1.0}

5 F.V. = 24.0000
{1.1.0; 2,1.0; **3,1.0**; 4,1.0; 8,1.0} {5,1.0; 6,1,0; 10,1.0; **11,1.0**; 17,1.0} {7,1.0; 12,1.0; **13,1.0**; 18,1.0; 19,1.0} {9,1.0; 14,1.0; **15,1.0**; 16,1.0; 21,1.0} {20,1.0; 22,1.0; **23,1.0**; 24,1.0; 25,1.0}

6 F.V. = 21.0000
{1,1.0; 2,1.0; **3,1.0**; 4,1.0; 7,0.1667} {5,1.0 10,1.0; **11,1.0**; 17,1.0; 18,0.1667} {6,1.0; 7,0.8333; **12,1.0**; 13,0.5; 18,0.8333} {8,1.0; 13,0.3333; **14,1.0**; 20,1.0; 24,0.8333} {9,1.0; 13,0.1667; **15,1.0**; 16,1.0; 21,1.0} {19,1.0; 22,1.0; **23,1.0**; 24,0.1667; 25,1.0}

7 F.V. = 19.1428
{1,1.0; 2,1.0; **3,1.0**; 7,0.5714} {4,1.0; 7,0.4286; **8,1.0**; 9,1.0; 14,0.1429} {5,1.0; 10,1.0; **11,1.0**; 17,0.5714} {6,1.0; **12,1.0**; 13,0.5714; 18,1.0} {14,0.5714; **15,1.0**; 16,1.0; 21,1.0} {13,0.4286; 14,0.2857; 17,0.4286; **19,1.0**; 20,1.0; 25,0.4286} {22,1.0; **23,1.0**; 24,1.0; 25,0.5714}

8 F.V. = 17.1250
{1,1.0; **3,1.0**; 4,1.0; 7,0.125} {2.1.0; 5,0.25; **6,1.0**; 7,0.875} {5,0.75;10,1.0; **11,1.0**; 12,0.375} {8,1.0; 13,1.0; **14,1.0**; 21,0.125} {9,1.0; **15,1.0**; 16,1.0; 21,0.125} {12,0.625; 17,1.0; **18,1.0**; 19,0.5} {19,0.375; **20,1.0**; 21,0.75; 24,1.0} {19,0.125; 22,1.0; **23,1.0**; 25,1.0}

9 F.V. = 16.0000
{1,1.0; **3,1.0**; 4,0.7778} {2,1.0; 5,0.7778; **6,1.0**} {4,0.2222; 7,0.3333; **8,1.0**; 9,0.2222; 14,1.0} {5,0.2222; 10,1.0; **11,1.0**; 12,0.3333; 17,0.2222} {7,0.6667; 12,0.6667; **13,1.0**; 19,0.4444} {9,0.7778; **15,1.0**; 16,1.0} {17,0.7778; **18,1.0**; 22,1.0} {19,0.5556; **20,1.0**; 21,1.0; 24,0.2222} {**23,1.0**; 24,0.7778; 25,1.0}

10 F.V. = 16.0000*
{1,1.0; 2,0.5; **3,1.0**} {2,0.5; 5,0.5; **6,1.0**; 12,0.5} {4,1.0; 7,1.0; **8,0.5**} {**8,0.5**; 9,1.0; 14,1.0} {5,0.5; 10,1.0; **11,1.0**} {12,0.5; **13,1.0**; 19,1.0} {**15,1.0**; 16,1.0; 21,0.5} {17,1.0; **18,1.0**; 22,0.5} {**20,1.0**; 21,0.5; 24,1.0} {22,0.5; **23,1.0**; 25,1.0}

11 F.V. = 16.0000*
{1,1.0; **3,1.0**; 4,0.2727} {2,1.0; **6,1.0**; 12,0.2727} {4,0.7273; 7,0.0909; **8,1.0**; 14,0.4545} {5,0.5; 10,1.0; **11,0.5**; 12,0.2727} {5,0.5; **11,0.5**; 12,0.2727; 17,1.0} {7,0.9091; **13,1.0**; 19,0.3637} {9,0.5; 14,0.2727; **15,0.5**; 16,1.0} {9,0.5; 14,0.2727; **15,0.5**; 21,1.0} {12,0.1818; **18,1.0**; 19,0.0909; 22,1.0} {19,0.2727; **20,1.0**; 24,1.0} {19,0.2726; **23,1.0**; 25,1.0}

12 F.V. = 16.0000*
(1,1.0; **3,1.0**; 4,0.0833) (2,0.5; 5,1.0; **6,0.5**; 7,0.0833) (2,0.5; **6,0.5**; 7,0.0833; 12,1.0) (4,0.9167; **8,1.0**; 9,0.1667) (10,1.0; **11,1.0**; 17,0.0833) (7,0.8333; **13,1.0**; 19,0.2500) (9,0.4167; 14,1.0; **15,0.5**; 21,0.1667) (9,0.4167; **15,0.5**; 16,1.0; 21,0.1667) (17,0.9167; **18,1.0**; 22,0.1667) (**20,1.0**; 21,0.6667; 24,0.4167) (19,0.375; 22,0.4167; **23,0.5**; 24,0.2916; 25,0.5) (19,0.375; 22,0.4167; **23,0.5**; 24,0.2916; 25,0.5)

13 F.V. = 15.0000
(**1,1.0**; 3,0.3077; 7,0.6154) (**2,1.0**; 0.9231) (4,0.6923; **4,1.0**; 8,0.2308) (**5,1.0**, 6,0.0769; 10,0.6154; 11,0.2308) (8,0.7692; **9,1.0**; 16,0.1538) (10,0.3846; **12,1.0**; 13,0.5385) (**14,1.0**; 15,0.0769; 16,0.8462) (11,0.7692; **17,1.0**; 18,0.1538) (7,0.3846; 13,0.4615; **19,1.0**; 20,0.0769) (15,0.9231; **21,1.0**) (18,0.8462, **22,1.0**; 23,0.0769) (20,0.9231; **24,1.0**) (23,0.9231, **25,1.0**)

14 F.V. = 13.0000
(**2,1.0**; 3,0.4286; 6,0.3570) (1,0.2143; 3,0.5714; **4,1.0**) (**5,1.0**; 6,0.6430; 11,0.1428) (1,0.7857; **7,1.0**) (8,0.7857; **9,1.0**) (**10,1.0**; 11,0.7857) (**12,1.0**; 13,0.7857) (8,0.2143; 13,0.2143; **14,1.0**; 15,0.2143; 20,0.1429) (15,0.7857; **16,1.0**) (11,0.0715; **17,1.0**; 18,0.7142) (18,0.2858; **19,1.0**; 23,0.4999) (20,0.7857; **21,1.0**) (20,0.0714; 22,0.2143; 23,0.5001; **24,1.0**) (22,0.7857; **25,1.0**)

15 F.V. = 11.0000
(**1,1.0**; 3,0.6667) (**2,1.0**; 6,0.6667) (3,0.3333; **4,1.0**; 8,0.3333) (**5,1.0**; 6,0.3333; 11,0.3333) (**7,1.0**; 8,0.3333; 19,0.3333) (8,0.3333; **9,1.0**; 15,0.3333) (**10,1.0**; 11,0.6667) (**12,1.0**; 13,0.3333; 18,0.3333) (13,0.6667; **14,1.0**) (15,0.6667; **16,1.0**) (**17,1.0**;18,0.6667) (20,0.6667; **21,1.0**) (**22,1.0**; 23,0.6667) (20,0.3333; 23,0.3333; **24,1.0**) (19,0.6667, **25,1.0**)

16 F.V. = 9.0000
(**1,1.0**; 3,0.5625) (**2,1.0**; 3,0.1875; 6,0.3750) (3,0.2500, **4,1.0**; 8,0.3125) (**5,1.0**; 6,0.1250; 11,0.4375) (6,0.5; **7,1.0**; 13,0.0625) (8,0.1250; **9,1.0**; 15,0.4375) (**10,1.0**; 11,0.5625) (**12,1.0**; 13,0.5625) (8,0.5625; **14,1.0**) (15,0.5625; **16,1.0**) (**17,1.0**; 18,0.5625) (13,0.375; 18,0.1875; **19,1.0**) (20,0.5625; **21,1.0**) (18,0.2500; **22,1.0**; 23,0.3125) (20,0.4375; 23,0.1250; **24,1.0**) (23,0.5625; **25,1.0**)

17 F.V. = 8.0588
(**1,1.0**; 3,0.4706) (**2,1.0**; 3,0.0588; 6,0.4118) (3,0.4706, **4,1.0**) (**5,1.0**; 6,0.1176; 11,0.3530) (7,0.4706; **8,1.0**) (**9,1.0**; 15,0.4706) (**10,1.0**; 11,0.4706) (6,0.4706; **12,1.0**) (7,0.4706; **13,1.0**) (**14,1.0**; 20,0.4706) (15,0.4706; **16,1.0**) (11,0.1764; **17,1.0**; 18,0.2942) (7,0.0588; 18,0.4118; **19,1.0**) (15,0.0588; 20,0.4118; **21,1.0**) (18,0.2940; **22,1.0**; 23, 0.1766) (20,0.1176; 23,0.3530; **24,1.0**) (23,0.4704; **25,1.0**)

18 F.V. = 7.4444
(**1,1.0**; 3,0.3889) (**2,1.0**; 3,0.3889) (3,0.2222; **4,1.0**; 7,0.1667) (**5,1.0**; 11,0.3889) (**6,1.0**; 7,0.3889) (7,0.3889; **8,1.0**) (**9,1.0**; 15,0.3889) (**10,1.0**; 11,0.3889) (7,0.0556; 12,0.3333; **13,1.0**) (**14,1.0**; 20,0.3889) (15,0.3889; **16,1.0**) (11,0.2222; 12,0.1667; **17,1.0**) (12,0.3889; **18,1.0**) (**19,1.0**; 20,0.3889) (15,0.2222; 20,0.1667; **21,1.0**) (12,0.1111; **22,1.0**; 23,0.2778) (20,0.0556; 23,0.3333; **24,1.0**) (23,0.3889, **25,1.0**)

19 F.V. = 6.6314†
{**1,1.0**; 3,0.3158} {**2,1.0**; 3,0.3158} {3,0.3158; **4,1.0**} {**5,1.0**; 11,0.3158} {**6,1.0**; 12,0.3158} {3,0.0526; **7,1.0**; 14,0.2632} {**8,1.0**; 14,0.3158} {**9,1.0**; 15,0.3158} {**10,1.0**; 11,0.3158} {12,0.2106; **13,1.0**; 14,0.1052} {15,0.3158; **16,1.0**} {11,0.3684; **17,0.9474**} {12,0.3158, **18,1.0**} {**19,1.0**; 23,0.3158} {14,0.3158; **20,1.0**} {15,0.3684; **21,0.9474**} {12,0.1578; 17,0.0526; **22,1.0**; 23,0.1054} {21,0.0526; 23,0.2632; **24,1.0**} {23,0.3156; **25,1.0**}

20 F.V. = 6.0000
{**1,1.0**; 3,0.2500} {**2,1.0**; 3,0.2500} {3,0.2500; **4,1.0**} {**5,1.0**;11,0.2500} {**6,1.0**; 13,0.2500} {3,0.2500; **7,1.0**} {**8,1.0**; 15,0.2500} {**9,1.0**; 15,0.2500} {**10,1.0**; 11,0.2500} {11,0.2500; **12,1.0**} {13,0.2500; **14,1.0**} {15,0.2500; **16,1.0**} {11,0.2500; **17,1.0**} {**18,1.0**; 23,0.2500} {13,0.2500; **19,1.0**} {13,0.2500; **20,1.0**} {15,0.2500; **21,1.0**} {**22,1.0**; 23,0.2500} {23,0.2500; **24,1.0**} {23,0.2500; **25,1.0**}

21 F.V. = 4.9522†
{**1,1.0**; 3,0.1905} {**2,1.0**; 3,0.1905} {3,0.1905; **4,1.0**} {**5,1.0**; 11,0.1905} {3,0.1903; **6,1.0**} {3,0.2383; **7,0.9522**} {**8,1.0**; 15,0.1905} {**9,1.0**; 15,0.1905} {**10,1.0**; 11,0.1905} {11,0.2381; **12,0.9524**} {7,0.0478; 12,0.0476, **13,1.0**; 23,0.0951} {**14,1.0**; 15,0.1905} {15,0.1905; **16,1.0**} {11,0.3810; **17,0.8095**} {17,0.1905; **18,1.0**} {**19,1.0**; 23,0.1905} {15,0.0476; **20,1.0**; 23,0.1429} {15,0.1905; **21,1.0**} {**22,1.0**; 23,0.1905} {23,0.1905; **24,1.0**} {23,0.1905; **25,1.0**}

22 F.V. = 5.0000†
{**1,1.0**; 3,0.1364} {**2,1.0**; 3,0.1364} {**3,0.7273**; 7,0.4091} {**4,1.0**; 8,0.1364} {**5,1.0**; 6,0.1364} {**6,0.8636**; 12,0.2727} {7,0.4091, **8,0.7273**} {8,0.1364, **9,1.0**} {**10,1.0**; 11,0.1364} {**11,0.8636**; 12,0.2727} {7,0.0909, 12, 0.0455; **13,1.0**} {7,0.0909; **14,1.0**; 20,0.0455} {**15,1.0**; 20,0.1364} {**16,1.0**; 20,0.1364} {**17,1.0**; 18,0.1364} {12,0.4091; **18,0.7273**} {**19,1.0**; 20,0.1364} {20,0.1634; **21,1.0**} {18,0.1364; **22,1.0**} {**23,1.0**; 24,0.1364} {20,0.4091; **24,0.7273**} {24,0.1364; **25,1.0**}

23 F.V. = 3.9130†
{**1,1.0**; 3,0.0870} {**2,1.0**; 3,0.0870} {**3,0.8261**; 7,0.2609} {**4,1.0**; 7,0.0870} {**5,1.0**; 6,0.0870} {**6,0.9130**; 7,0.1739} {7,0.1739; **8,0.9130**} {7,0.2609; **9,0.8261**} {**10,1.0**; 11,0.0870} {**11,0.9131**; 19,0.1739} {**12,1.0**; 19,0.0870} {7,0.0435; **13,1.0**; 19,0.0435} {8,0.0870; **14,1.0**} {9,0.0870; **15,1.0**} {9,0.0870; **16,1.0**} {**17,1.0**; 19,0.0870} {**18,1.0**; 19,0.0870} {19,0.0870; **20,1.0**} {19,0.0870; **21,1.0**} {19,0.0870; **22,1.0**} {19,0.0870; **23,1.0**} {19,0.0870; **24,1.0**} {19,0.0870; **25,1.0**}

24 F.V. = 2.3333†
{**1,1.0**; 3,0.0417} {**2,1.0**; 13,0.0417} {**3,0.9583**, 13,0.0833} {**4,1.0**; 14,0.0417} {**5,1.0**; 13,0.0417} {**6,1.0**; 13,0.0417} {**7,1.0**; 13,0.0417} {**8,1.0**; 14,0.0417} {**9,1.0**; 14,0.0417} {**10,1.0**; 13,0.0417} {**11,1.0**; 13,0.0417} {**12,1.0**; 13,0.0417} {13,0.2500; **14,0.7917**} {14,0.0417; **15,1.0**} {14,0.0417; **16,1.0**} {**17,1.0**; 19,0.0417} {**18,1.0**; 19,0.0417} {13,0.3750; **19,0.6667**} {19,0.0417; **20,1.0**} {19,0.0417; **21,1.0**} {19,0.0417; **22,1.0**} {19,0.0417; **23,1.0**} {19,0.0417; **24,1.0**} {19,0.0417; **25,1.0**}

25 F.V. = 0.0000
{**1,1.0**} {**2,1.0**} {**3,1.0**} {**4,1.0**} {**5,1.0**} {**6,1.0**} {**7,1.0**} {**8,1.0**} {**9,1.0**} {**10,1.0**} {**11,1.0**} {**12,1.0**} {**13,1.0**} {**14,1.0**} {**15,1.0**} {**16,1.0**} {**17,1.0**} {**18,1.0**} {**19,1.0**} {**20,1.0**} {**21,1.0**} {**22,1.0**} {**23,1.0**} {**24,1.0**} {**25,1.0**}

Notes to tables 1 through 3:

* indicate solutions which have one or more multiple facilities; † indicate solutions in which facilities do not cover all of their own territory.

One notices the following features:

- if, with respect to a given n/p situation joint, location takes place for p* = p + 1, the objective function never decreases, and sometimes even increases (see, e.g. the 13/4 and 13/5 cases);
- sometimes the own location region has to be split up in order to guarantee full covering of the area, with again the possibility of rising transportation costs with rising p (see e.g. the 25/21 and 25/22 case);
- in the majority of the cases no joint location takes place; firms act as local monopolists along the LÖSCH separable market area philosophy and this is in fact also true of the joint location case; the reason is that in some sense "barter" takes place of agricultural products for industrial products, and vice versa.

Finally it should be noted that the solutions presented are in fact a subset of multiple optima which can be obtained generally by rotation.

5. Conclusions and Directions for Further Research

Of primary importance is the rigorous proof of some characteristics of rectangular space in general and the elementary regions of the Manhattan-circle in particular. From the first work on TBS, where a discrete distance measure was used, it could be derived that optimal systems most of the time showed complete concentration of facilities; introducing a rectangular distance measure showed that concentration was not always optimal and also sometimes not allowed. Conclusions from different papers on TBS pointed towards the need of introducing some more spatial structure; a very special pattern, a square divided in subsquares, was used in KUIPER 1988. In this paper, introducing Manhattan-circles, a more fundamental approach is used. The location pattern of mono-industrial centres shows a dominant pattern of dispersed location, but for the time being one has to be careful in generalising this result for reasons mentioned below.

The methodological extension to an operational mixed integer programme has yielded results, some expected, some unexpected, some which should have been expected but were not. Facilities when required to supply equal demand spatially spread themselves, even in a TBS. Certain locations are geometrically favoured and with low numbers of facilities are included in nearly all solutions. With high numbers of facilities these same locations are again favoured but now by being in demand to be served i.e. Table 3; $p = 5$ and $p = 17$ are near mirror images. Two facilities in one elementary region never improve the cost of service, and may, even, increase it - the only divergence from a monotonic decrease in cost with increase in facilities except for situations when p is very high and facilities share their own elementary regions.

Several extensions are - quite obviously - envisaged in the research programme. First cases $n_1 > n$ have to be studied; second radius, r, has to be enlarged to 4 and more. But the strategic extension is to raise the number of non-agricultural goods to 2 and more. The development of the capacitated p-Median formulation here reported has immediate application in a number of real world location problems and is norm independent.

The authors hope to be able to report regularly on new results along the lines sketched above.

FOOTNOTES

[1] For a recent exposition on Tinbergen-Bos Systems - to be abbreviated hereafter as TBS - see J. Paelinck, 1988.

[2] See J.H. Kuiper, 1987, 1988, 1989, a,b, and c; H.J. Kuiper and N.C.H.M. Mares, 1989.

[3] This means that a Minkowski-metric with $p = 1$ is used; see J. Paelinck, 1983, pp. 44-50.

[4] In fact the area's borderline is fractal; see H.O. Peitgen and D. Saupe, 1988.

[5] Again J. Paelinck, 1988.

[6] For Euclidean distance, see J.H.P. Paelinck and P. Nijkamp, 1975, pp. 71–77; compare also J.H. Kuiper, 1986.

[7] Use is made here of the Hakimi-Levy property with concave transportation-costs as a function of distance; see J.-F. Thisse and P. Hanjoul, 1984.

LITERATURE

Balinski, M., 1965: Integer programming, methods, uses and computation, Management Science Series A, 12, 253–313.

Dantzig, G.B., 1963: Linear Programming and Extensions, (Princeton NJ: Princeton University Press).

Efroymson, M.; T. Ray, 1969: A branch-bound algorithm for plant location, Operations Research, 14, 361–368.

Hakimi, S.L., 1964: Optimum distribution of switching centers and the absolute centers and medians of a graph, Operations Research, 12, 450–459.

Hakimi, S.L., 1965: Optimum distribution of switching centers in a communications network and some related graph theoretic problems, Operations Research, 13, 462–475.

Kuiper, J.H., 1986: Distribution of Distances in Pre-geographical Space, Gower, Aldershot.

Kuiper, J.H., 1987: Transportation Costs in Tinbergen-Bos Systems, Erasmus University Rotterdam, Working Paper, presented at the Deltaseminar, Louvain-la-Neuve.

Kuiper, J.H., 1988: Location Patterns in Tinbergen-Bos systems, Revue d'Economie Rgionale et Urbaine, no. 1, pp. 29–50.

Kuiper, J.H., 1989a: Cots de transport dans certains modles d'quilibre conomique spatial, to appear in Les Cahiers Scientifiques du Transport.

Kuiper, J.H., 1989b: Transportation costs in some theoretical spatial equilibrium models, to appear in F. Dietz (ed.), Location Considerations of Industrial Organisations: Conditions and Effects,, Herzlia, Israel.

Kuiper, J.H., 1989c: The transportation costs in a Tinbergen-Bos model, to appear in Revista Urbana.

Kuiper, J.H.; Mares, N.C.H.M., 1989: Locational behaviour in a spatial economy: equilibrium model according to Tinbergen and Bos, Erasmus University, Faculty of Economics, Department of Theoretical Spatial Economics, Working Paper.

Love, R.F.; J.G. Morris, 1975: Computation procedure for the exact solution of location-allocation problems with rectangular distances, Naval Research Logistics Quarterly, 22, 441-453.

Manne, A.S., 1964: Plant location under economies-of-scale - decentralization and computation, Management Science, 11, 213-235, 1964.

Morris, J.G., 1978: On the extent to which certain fixed charges depot location problems can be solved by linear programming, Journal of Operational Research Society, 29, 71-76.

Paelinck, J.H.P.; Nijkamp, P., 1975: Operational Theory and Method in Regional Economics, Saxon House and Lexington, Farnborough and Lexington.

Paelinck, J.H.P., 1983: (with the assistance of J.-P. Ancot and J.H. Kuiper), Formal Spatial Economic Analysis, Gower, Aldershot.

Paelinck, J.H.P., 1988: L'quilibre gnral d'une conomie spatiale, in Cl. Ponsard (ed.), Analyse Economique Spatiale, Paris, P.U.F., Ch. 7, pp. 277-319.

Peitgen, H.O.; Saupe, D., 1988: The Science of Fractal Images, Springer-Verlag, Berlin.

ReVelle, C.S., 1967: Central Facilities Location, (Ithaca: Center for Environmental Quality Control, Cornell University) Report No. 30.

ReVelle, C.S.; R. Swain, 1970: Central facilities location, Geographical Analysis, 2, 30-42.

Rosing, K.E., 1979: A modified capacitated p-Median formulation for the reorganization of the ambulance service in the Rotterdam/-Rijnmond region, L' Association de Sciences Regionales de Langue Francaise, Analyse Spatiale et Services Publiques, Louvain-la-Neuve, Belgium, 10-12 September.

Rosing, K.E., 1989: Solutions to the Multi Weber Problem on the L (euclidean) norm via the L (rectangular) norm, E.G.I. Discussiestukken 89-8, (Rotterdam: Erasmus Universiteit).

Rosing, K.E.; C.S. ReVelle; H. Rosing-Vogelaar, 1979: The p-Median and its linear programming relaxation: an approach to large problems, Journal of the Operational Research Society, 30, 815-823.

Thisse, J.-F.; Hanjoul, P., 1984: Localisation de la firme sur un rseau, in J.M. Pillu and R. Guesnerie (eds), Modles conomiques de la localisation et des transports, E.N.P.C., Paris.

Wendell, R.E.; A.P. Hurter, Jr., 1973: Location theory, dominance, and convexity, Operations Research, 21, 314-320.

Economics of Multi–Habitation

by

Noboru Sakashita

1. Introduction

In the Fourth Nation–wide Comprehensive Regional Development Plan worked out by the Japanese government in 1987, there were a concept and policy recommendation of "multi–habitation".[1] People were encouraged to possess the second residence apart from their original residence partly because it may activate local areas which were losing their population as a consequence of reconcentration of population and economic activities to the Tokyo Metropolitan Area in 1980s. The aim of the present paper is to explore rationale of this multi–habitation policy from the viewpoint of regional economic theory.

A clue to the matter is the optimal allocation of living time for each household or individual. The allocation of living time decides the weighted average level of utility for each household as well as productive capacities of different regions. Maximizing the average utility under the constraints of regional productive capacities yields certain optimizing conditions in the setting of multi–habitation. It is important to know that the average utility maximized in the multi–habitation model is higher than the utility at the social optimum of single habitation with free mobility because the former does not have the equal utility constraint which is indispensable in the latter.

In section 2 which follows, a social optimum model of single habitation is formulated as a reference model. Introduction of a Benthamite social welfare function changes some of optimizing conditions which give us a hint to for–

mulate multi-habitation models. In section 3, the simplest model of multihabitation is introduced and the results of optimization are compared with those obtained in section 2. Section 4 explores a market mechanism model with policy intervention which can attain the equivalent conditions to the social optium. In section 5, the existence of incentives for individuals to switch from single habitation behaviour to multi-habitation behaviour is discussed as well as the possibility and desirability of "initial" subsidization to promote multi-habitation behaviour. Concluding remarks are given in section 6.

2. Social Optimum with Single Habitation

We utilize an extremely simplified version of David WILDASIN's general model (WILDASIN, 1987) as a starting point of discussion. Consider a two-region economy in which free migration between the two regions is assured. In region i (i=1,2), a resident enjoys his(her) utility U_i by the consumption of a private good C_i and a public good G_i, i.e. $U_i=U_i(C_i,G_i)$. Region i's productive capacity is expressed by its specific production function F_i the sole input of which is the residents' labour L_i. The total population of the two-region economy is fixed as $\bar{L}$, i.e. $\bar{L}=L_1+L_2$. U_i, F_i and $\bar{L}$ complete the description of socio-economic environment in which this two-region economy operates.

Let us consider the optimizing problem for a social planner who wishes to maximize the level of people's utility under the constraint of free migration, i.e. equal utility between the two regions. The problem is formulated as follows:

$$\text{Maximize} \quad U_1(C_1,G_1)$$
$$\text{w.r.t.} \quad C_i,G_i,L_i(i=1,2)$$
$$\text{subject to} \quad U_2(C_2,G_2) - U_1(C_1,G_1) = 0 \tag{1}$$
$$F_1(L_1) + F_2(L_2) - C_1L_1 - C_2L_2 - G_1 - G_2 = 0 \tag{2}$$
$$\bar{L} - L_1 - L_2 = 0 \tag{3}$$

The optimizing conditions for the above problem are:

$$\frac{\partial U_i}{\partial C_i} = \frac{\partial U_i}{\partial G_i} L_i, \qquad i = 1, 2 \tag{4}$$

$$F_1' - C_1 = F_2' - C_2 \tag{5}$$

Equation (4) implies a famous Samuelsonian condition for the supply of (local) public good in each region, and equation (5) is sometimes called 'social net benefit condition of migration'.[2]

As HARTWICK pointed out, the constraint of equal utility embodies the philosophy of Rawlsian social welfare.[3] If we introduce the concept of a Benthamite social welfare without equal utility constraint instead, we have the following different problem:

$$\text{Maximize} \quad L_1 \cdot U_1(C_1, G_1) + L_2 \cdot U_2(C_2, G_2)$$
$$\text{w.r.t. } C_i, G_i, L_i (i=1,2)$$

subject to (2) and (3)

In this case we have a little different set of the optimizing conditions:

$$\frac{\partial U_i}{\partial C_i} = \frac{\partial U_i}{\partial G_i} L_i, \qquad i = 1, 2 \tag{4}$$

$$\frac{\partial U_1}{\partial C_1} = \frac{\partial U_2}{\partial C_2} \tag{6}$$

$$U_1 + \frac{\partial U_i}{\partial C_i} (F_1' - C_1) = U_2 + \frac{\partial U_i}{\partial C_i} (F_2' - C_2) \tag{7}$$

We have Samuelsonian conditions again (equation (4)), and there is a stronger condition of equal marginal utilities of private good between regions (equation (6)) which implies equal amount of consumption of the same if the utility function is separable in C_i and G_i. Equation (7) replaces equation (5) of the previous case. Later the same equation will be interpreted as "social net benefit condition of multi-habitation" in a different context.

If we add the equal utility constraint (equation (1)) to the Benthamite problem, it becomes completely equivalent to the previous Rawlsian problem so that it is clear that the average utility ($L_1U_1 + L_2U_2/\bar{L}$) in the Benthamite optimum is higher than the utility ($U_1 = U_2$) in the Rawlsian optimum. This fact gives us a strong hint to consider an economy where multihabitation is encouraged.

In reality, however, it is almost impossible for us to dispense with equal utility constraint even in a planned economy, because it will cause a great social strain if the utility levels of different regions are kept unequal for long time. We must have, therefore, some device to realize a Benthamite-like social optimum *without* strain and this is the background of our multi-habitation model. Finally we wish to make it clear that the Rawlsian social optimum can be achieved by a market economy with modest intervention by the goverment. What is needed is a combination of interregional income transfer by the central goverment and optimal supply of local public good by the local government in each region.[4]

3. Social Optimum with Multi-Habitation

Let us consider that each household in the economy spends 100p percent of time in region 1 and 100(1−p) percent in region 2 every year. We assume further that movement cost between two regions is negligible. If the total population L is large enough, region 1 can maintain labour input of approximately (pL) always and region 2 can do that of ((1−p)L). Per capita consumption of the private good is p-and-(1−p)-weighted average of C_1 and C_2, but local public good must be supplied constantly at the levels of G_1 and G_2.

Social planner's problem in this case is:

$$\text{Maximize} \quad \{pU_1(C_1, G_1) + (1-p)U_2(C_2, G_2)\} \tag{8}$$

w.r.t. $C_i, G_i, (i=1,2), p$

$$\text{subject to} \quad F_1(p\bar{L}) + F_2\{(1-p)\bar{L}\} - \{pC_1 + (1-p)C_2\}\bar{L} - G_1 - G_2 = 0 \tag{9}$$

In mathematical sense this problem is perfectly equivalent to the problem of Benthamite optimization in the previous section so that we have the following four optimizing conditions two of which are exactly the same as those in the Benthamite optimization, i.e.[5]

$$\frac{\partial U_1}{\partial C_1} = \frac{\partial U_1}{\partial G_1}(p\bar{L}) \tag{10}$$

$$\frac{\partial U_2}{\partial C_2} = \frac{\partial U_2}{\partial G_2}((1-p)\bar{L}) \tag{11}$$

$$\frac{\partial U_1}{\partial C_1} = \frac{\partial U_2}{\partial C_2} \tag{6}$$

$$U_1 + \frac{\partial U_1}{\partial C_1}(F_1{}' - C_1) = U_2 + \frac{\partial U_2}{\partial C_2}(F_2{}' - C_2) \tag{7}$$

Equations (10) and (11) are variants of the Samuelsonian condition in the multi-habitation model, and equation (6) means the equality of marginal utilities of private good with the same interpretation as in the Benthamite model. Equation (7) means the following. If an individual (or household) shifts marginal length of time from the living in region 1 to that in region 2, region 1 loses one person's utility and the gap between his(her) marginal productivity and consumption of private good which is his(her) contribution to the regional economy. All values are evaluated by the dimension of utility and are calculated in annual term. To the contrary, region 2 gains the same set of values as region 1 loses. At the optimum, therefore, the loss in region 1 is equal to the gain in region 2, and there is no *social* incentive for an individual to shift the marginal time. We call this condition "social net benefit condition of multi-habitation" which is a generalization of the "social net benefit condition of migration" (equation (5)) in the Rawlsian optimum (RO).

Although mathematically equivalent, the multi-habitation optimum (MHO) has quite different implication when it is compared with the Benthamite optimum (BO). In BO the total population $\bar{L}$ is divided into region 1 (L_1) and region 2 (L_2) as *different* groups of the residents so that U_1 and U_2 are utilities of different people. In MHO, however, the same individual divides his(her) living time into the two regions by the ratio of p vs (1−p) so that $\{pU_1 + (1-p)U_2\}$ is the total utility of *one* specific individual. Therefore, there is no risk of social strain between regions in spite of the difference between U_1 and U_2. This is a very peculiar characteristic of MHO. Since MHO is mathematically equivalent to BO, its utility level is of course higher than that in RO.

It is important to know that we may be able to have an MHO even if one of two regions does not have productive capacity. If there is no production activity in region 2, F_2' in equation (7) vanishes. Even so we can have an MHO if U_2 is intrinsically much higher than U_1. The two utility functions, U_1

and U_2, are basically the same function of C_i and G_i and comparable with each other. If, however, there is the third argument in each utility function which is region-specific, the two will become different functions on the surface but will be still comparable with each other. More concretely, we make comparison between $U(C_1, G_1, R_1)$ and $U(C_2, G_2, R_2)$ where R_1 and R_2 are region-specific factors (regional amenity for instance). If R_i's have positive effect on U_i and if R_2 is much larger than R_1, we may have an MHO in which people do not make productive activities in R_2. Region 2 is nothing other than a pure resort area. (Perhaps, however, one may do some business activities, for instance, writing articles or composing music, even in resort areas.)

It might be appropriate to say something more concerning equation (6). If the variable C_i is separable from other arguments in utility function and if its marginal utility is monotonically decreasing, equation (6) implies the same amount of consumption (in annual term) by an individual in different living places. It is a very natural conclusion. Usually, however, C_i is *not* separable from, say, region-specific factors which enter the utility function. People drink cold beverages more in hotter region and prefer hot dishes in colder region. In that case, equation (6) does not imply equal consumption in different regions.

4. Multi-Habitation Market Equilibrium

The next question for us is to explore some system of the market economy with government intervention by which the equivalent situation to MHO can be realized. First we analyze a market system supported by the central government. Secondly we consider another system in which local governments perform certain rôle in the working of the system.

4.1 Market Equilibrium Supported by the Central Government

In this system an individual (household) maximizes his(her) weighted-average utility subject to budget constraint. His(her) income consists of wage in region $1(w_1)$ and the same in region 2 (w_2) plus an additional income which does not depend on his(her) residential location (Y). All variables are expressed in annual term.

$$\underset{\text{w.r.t. } C_1, C_2, p}{\text{Maximize}} \quad \{pU_1(C_1,G_1) + (1-p)U_2(C_2,G_2)\}$$

$$\text{subject to} \quad \{pw_1+(1-p)w_2 + Y\} - pC_1 - (1-p)C_2 = 0 \tag{12}$$

The following optimizing conditions are derived.

$$\frac{\partial U_1}{\partial C_1} = \frac{\partial U_2}{\partial C_2} \tag{6}$$

$$U_1 + \frac{\partial U_i}{\partial C_i}(w_1-C_1) = U_2 + \frac{\partial U_i}{\partial C_i}(w_2-C_2) \tag{13}$$

Equation (13) resembles "social net benefit condition of multi-habitation" (equation (7)) but we need an additional condition in order to state complete equivalence of these two equations.

Next we assume that a representative firm with a production function $F_i(L_i)$ maximizes its profit in each region under potentially competitive situation.

$$\underset{\text{w.r.t. } L_i}{\text{Maximize}} \quad \{ F_i(L_i) - w_iL_i \} \ , \ i = 1,2$$

The optimizing condition is:

$$F_i{}'(L_i) - w_i = 0 \ , \quad i = 1,2 \tag{14}$$

The labour market in each region is cleared by the following equation.

$$L_i = p_i\bar{L} \ , \quad i = 1,2 \tag{15}$$

Combining equations (13), (14), and (15), we can have the exactly same relation as equation (7) here.

Behaviour of the central government is formulated as follows. If it takes p as given, it can indirectly determine the volume of total consumption of private good Q by deciding the levels of G_1 and G_2, i.e.

$$Q = F_1(p\bar{L}) + F_2\{(1-p)\bar{L}\} - G_1 - G_2 \tag{16}$$

With an *arbitary* conjecture concerning the pattern of allocation of Q between the two regions, α vs $(1-\alpha)$, the central government can formulate its optimization problem as follows:

$$\underset{\text{w.r.t. } G_1,G_2}{\text{Maximize}} \left[pU_1\left\{ \frac{\alpha(F_1+F_2-G_1-G_2)}{p\bar{L}}, G_1\right\} + (1-p)U_2\left\{ \frac{(1-\alpha)(F_1+F_2-G_1-G_2)}{(1-p)\bar{L}}, G_2\right\} \right]$$

Then the following condition is derived.

$$\alpha\frac{\partial U_1}{\partial C_1} + (1-\alpha)\frac{\partial U_2}{\partial C_2} = \frac{\partial U_1}{\partial G_1}(p\bar{L}) = \frac{\partial U_2}{\partial G_2}\{(1-p)\bar{L}\} \tag{17}$$

Since $\frac{\partial U_1}{\partial C_1}$ and $\frac{\partial U_2}{\partial C_2}$ are always equalized with each other owing to the behaviour of (homogeneous) individuals (equation (6)), equation (17) becomes equivalent to equations (10) and (11) at MHO.

Finally the residence-free non-wage income of an individual can be defined as follows if there is an adequate reimbursement system.[6]

$$Y \overset{\text{def}}{=\!=} \frac{F_1+F_2-(G_1+G_2)-\{pw_1+(1-p)w_2\}\bar{L}}{\bar{L}} \tag{18}$$

Combining equation (12) and equation (19) we can derive the basic supply and demand relation in MHO (equation (9)).

Accordingly we could derive all conditions which characterize MHO for this market equilibrium under the general supervision of the central government. In this scheme the supply level of local public good in each region is determined in a synchronized manner by the central goverment. To some extent this is not an imagination but a reality in Japan. The Japanese governments eager to subsidize resort development in local areas selectively probably guided by some different principle from MHO. We cannot, however, find the rôle of local government in the scheme of this subsection, and it is unrealistic to disregared local autonomy in the world of multi-habitation. In the next subsection we proceed to the model with local governments.

4.2 Introduction of Local Governments - A Try and Error Process

Let us assume that the level of G_i is tentatively determined by the local government of region i as $\bar{G}_i$, i=1,2. An individual determines his(her) time allocation which maximizes his(her) utility under the supply of $\bar{G}_1$ and $\bar{G}_2$ and related budget constraint knowing that all individuals will take the same behaviour. Then similar conditions as equations (6) and (13) are derived.[7] By these conditions and the budget constraint, the "optimal" p^0 is determined as a function of $\bar{G}_1$ and $\bar{G}_2$ i.e. $p^0(\bar{G}_1, \bar{G}_2)$. Next, a local government, say that of region 1, determines the new optimal level of G_1 taking $(p^0\bar{L})$ and $((1-p^0)\bar{L})$ given and under a subjective presumption of autarky. Then there will be the new level of $\bar{G}_1(p_0)$ and the level of $\bar{C}_1(p_0)$ presumed by the local government with a Samuelsonian condition between $\bar{G}_1$ and $\bar{C}_1$. Similarly there will be $\bar{G}_2(p_0)$ and $\bar{C}_2(p_0)$ determined by the local government of region 2. Now there will be another process of adjustment by individuals and the new level of p, i.e. p^1, will be determined as follows:

$$p^1\{\bar{G}(p_0),\ \bar{G}_2(p^0)\} = \emptyset(p^0) \qquad (19)$$

Therefore we observe a repeated process of $p^{t+1} = \emptyset(p^t)$ which may result in a stationary equilibrium.

There are, however, two difficulties in this try and error process. The first difficulty is that it is not clear whether the above process is stable or not. A sufficient condition for the stability is combination of $\emptyset > 0$ and $\emptyset'(p)<1$ for all possible p, but the function $\emptyset$ may be too complicated to be checked for these properties. The second difficulty is that $C_1{}^*$ and $C_2{}^*$ which are actually determined by consumers at the equilibrium (p^*) are generally different from $\bar{C}_1(p^*)$ and $\bar{C}_2(p^*)$ which are fictitiously determined by the local government at the same equilibrium. Therefore, there is no guarantee that we can have Samuelsonian conditions at the equilibrium.

One possibility to overcome these difficulties is an adequate intervention by the central government to break autarky presumtion by the local governments. It imposes positive and negative subsidy S on both of the local governments, i.e.

$$F_1(p\bar{L}) - pC_1\bar{L} - G_1 + S = 0 \tag{20}$$

$$F_2\{(1-p)\bar{L}\} - (1-p)C_2\bar{L} - G_2 - S = 0 \tag{21}$$

The resultant equilibrium values $p^*(S)$, $C_1{}^*(S)$, $C_2{}^*(S)$, $\bar{C}_1\{p^*(S)\}$, and $\bar{C}_2\{p^*(S)\}$ can be induced to make $C_1{}^* = \bar{C}_1$ and $C_2{}^* = \bar{C}_2$ and to realize Samuelsonian conditions by selecting appropriate S.

Is it possible to conceptualize a system of this sort in which local autonomy is reestablished? Considering the existence of fiscal externality and inefficiency of free migration in the single habitation system, we think, the possibility is not so promising.[8]

5. Incentive to Switch from Single Habitation System to Multi-Habitation System

People are usually living in a single habitation system in which equal utility between regions is realized. Is there strong enough incentive for them to switch from the current system to that of the multi-habitation autonomously? Let us begin with RO in the two-region single habitation economy in which $U_1\ (C_1{}^*, G_1{}^*) = U_2(C_2{}^*, G_2{}^*)$. Each individual supposes that he(she) can obtain disposable incomes of $C_1{}^*$ and $C_2{}^*$ in each region. The problem to be solved is:

$$\underset{\text{w.r.t. } C_1, C_2, p}{\text{Maximize}} \quad \{p_1U_1(C_1,\ G_1{}^*) + (1-p)U_2(C_2,\ G_2{}^*)\}$$

$$\text{subject to } pC_1{}^* + (1-p)C_2{}^* - pC_1 - (1-p)C_2 = 0 \tag{22}$$

This problem can be approached by a step-wise manner. We first maximize the objective function with a fixed p, then the following optimizing condition is obtained:

$$\frac{\partial U}{\partial C_1} = \frac{\partial U}{\partial C_2} = \lambda \tag{23}$$

where λ is a Lagrangean multiplier for the constraint (22).

Next we maximize the maximized objective function with respect to p. By the envelope theorem we have:

$$\frac{d\Phi}{dp} = \frac{\partial\Phi}{\partial p} = U_1 - U_2 + \lambda\{(C_1{}^*-C_2{}^*) - (C_1-C_2)\} \tag{24}$$

where Φ is the related Lagrangean form. The second-order derivative is:

$$\frac{d^2\Phi}{dp^2} = \frac{\partial U_1}{\partial C_1}\frac{dC_1}{dp} - \frac{\partial U_2}{\partial C_2}\frac{dC_2}{dp} + \frac{d\lambda}{dp}\{(C_1{}^*-C_2{}^*) - (C_1-C_2)\}$$

$$+ \lambda\left(\frac{dC_2}{dp} - \frac{dC_1}{dp}\right)$$

$$= \frac{d\lambda}{dp}\{(C_1{}^*-C_2{}^*) - (C_1-C_2)\} \tag{25}$$

by equation (23). We can calculate $\frac{d\lambda}{dp}$ by comparative static solution to the first stage optimization problem, i.e.

$$\frac{d\lambda}{dp} = \frac{1}{Q}\{(C_1{}^*-C_2{}^*) - (C_1-C_2)\}\left(\frac{\partial^2 U_1}{\partial C_1{}^2}\right)\left(\frac{\partial^2 U_2}{\partial C_2{}^2}\right) \tag{26}$$

where $Q = p\frac{\partial^2 U_2}{\partial C_2{}^2} + (1-p)\frac{\partial^2 U_1}{\partial C_1{}^2}$ (< 0). From equations (25) and (26) we can conclude that:

$$\frac{d^2\Phi}{dp^2} < 0 \tag{27}$$

It has become clear that the objective function is concave with respect to p. However, continuity of $C_1(p)$ and $C_2(p)$ at p = 0 or p = 1 is unclear because we have switches of the phase there. If the utility function is separable between C_i and G_i, we can say that:

$$\lim_{p\to 0} C_i(p) = C_2{}^*\ ,\quad \lim_{p\to 1} C_i(p) = C_1{}^* \tag{28}$$

by equations (22) and (23) and the continuity is established.

In the continuity case we can safely say that there is enough incentive for individuals to switch from single habitation behaviour to multi-habitation behaviour (see figure 1). An individual adopts a time allocation of $\hat{p}$ and $(1-\hat{p})$ with given $G_1{}^*$ and $G_2{}^*$. Then local or central governments start to change G_is in response to $\hat{p}$ and the whole adjustment process described in the previous section begins.

Figure 1: Switch from Single Habitation to Multi-habitation

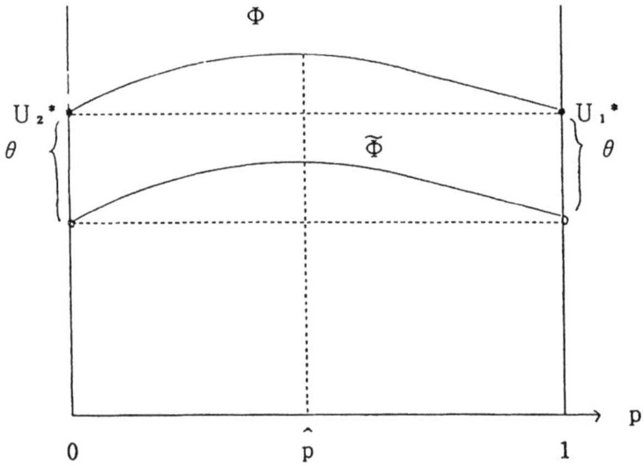

If there exists some "set-up" cost or movement cost θ in order to shift from single habitation to multi-habitation such as construction cost of the second house for instance, the maximized $\tilde{\Phi}(\hat{p})$ net of the set-up cost could be lower than $U_1^*(U_2^*)$ in single habitation. In this case the switch process does not start (see figure 1). Even if $\tilde{\Phi}(\hat{p})$ net of the set-up cost is higher than $U_1^*(U_2^*)$, an individual needs some "finance" to start multi-habitation behaviour so that it does not begin automatically when the financial market is not perfect. In the latter case there is a necessity for the government to give a subsidy to individuals to initiate switches from single habitation to multi-habitation.

6. Concluding Remarks

There are several things to be discussed more concerning multi-habitation behaviour. First, rise in labour productivity in the form of upward shift of production functions may give different impacts on the maximized utility levels in RO and MHO. Therefore, a switch from single habitation to multihabitation may start at the certain level of technical progress in production, i.e. people become affordable to begin multi-habitation behaviour at certain stage of economic growth.

Second, the local or central governments may attempt to improve the amenity of the regions by certain investment in natural environment etc. to attract people there. This action changes the position of region-specific utility functions. Some benefit-cost analysis can be in order for the public activities of this sort utilizing the model developed here.

Introduction of heterogeneity among individuals will be another important generalization particularly to make comparison of the present model with a Tiebout type model of separation. On the other hand, introduction of private capital etc. will not change the conclusion obtained here essentially.

Appendix: Numerical Example of Rawlsian Optimum and Multi-habitation Optimum

A numerical calculation has been done to deduce the values of several variables at RO and at MHO. The specification of related functions is given below table 1. In this example, RO gives smaller C and larger G in rgion 1 compared to them in region 2 and assures the same utility level between the two regions. On the other hand, the level of C is equalized for the two regions, but G_1 is bigger than G_2 implying higher U in region 1 than region 2 at MHO. Therefore, p is much bigger than (1−p) and it is also bigger than $(L_1/\bar{L})$ at RO. The gap (U_1-U_2) at MHO is offset by the same in $(\frac{\partial U_i}{\partial C_i} F_i')$ because of equation (7), i.e. $F_2'>F_1'$.

Table 1: Numerical Examples of RO and MHO

	(i) Rawlsian Optimum	(i) Multi-habitation Optimum
U_1	0.83	0.90
U_2	0.83	0.57
p	—	0.83
AVERAGE U	—	0.84
C_1	0.17	0.18
G_1	3.37	3.75
C_2	0.22	0.18
G_2	1.18	0.74
L_1	78.8	—
L_2	21.2	—

Specification of Functions

$$\begin{cases} U_i = 2 + 0.8 \log C_i + 0.2 \log G_i , \; i=1,2 \\ F_1 = 1.2 L_1^{0.6} , \; F_2 = L_2^{0.6} \\ \bar{L} = L_1 + L_2 = 100 \end{cases}$$

Although the difference between U_i at RO and $(U = pU_1+(1-p)U_2)$ at MHO is very small, the two optima give quite different allocation of local public goods between the two regions.

FOOTNOTES

[1] National Land Agency (3, p. 57).

[2] See Wildasin (6, Chapter 2).

[3] See Hartwick (2).

[4] See Sakashita (4).

[5] The secondary condition for the maximum is not so easy to be worked out. One possibility is to make a two-step approach. First solve the problem with a fixed p, then show that the maximized objective function, say $\hat{\Phi}$, is concave with respect to p. In this case, if the derivative $\frac{d\hat{\Phi}}{dp}$ becomes zero for some $p \varepsilon (0,1)$, then we can have an interior maximum. Calcula= tion of $\frac{d^2\hat{\Phi}}{dp^2}$ may be messy, but it becomes much easier with an assumption of separable utility.

[6] Notice, however, that the value of Y could be negative which means a lump-sum charge to finance the supply of local public goods.

[7] A consumer's problem is as follows:

$$\underset{\text{w.r.t. } p, C_1, C_2}{\text{Maximize}} \; \{pU_1 (C_1, \bar{G}_1) + (1-p) U_2 (C_2, \bar{G}_2)\}$$

$$\text{subject to } \; F_1(p\bar{L}) + F_2\{(1-p)\bar{L}\} - (\bar{G}_1 + \bar{G}_2) - \{pC_1 + (1-p)C_2\} \bar{L} = 0 \tag{F.1}$$

The optimizing conditions are:

$$\frac{\partial U_1}{\partial C_1} = \frac{\partial U_2}{\partial C_2} = \lambda \bar{L} \tag{F.2}$$

$$U_1 + \lambda \bar{L}(F_1' - C_1) = U_2 + \lambda \bar{L}(F_2' - C_2) \tag{F.3}$$

[8] See the following literatures: Boadway and Flatters (1), Sakashita (4), and Sakashita (5).

LITERATURE

Boadway, R.; Flatters, F., 1982: Efficiency and Equalization Payments in a Federal System of Government: a Synthesis and Extensions of Recent Results, Canadian Journal of Economies, Vol. 15, pp.613–633.

Hartwick, J.M., 1980: The Henry George Rule, Optimal Population, and Interregional Equity, Canadian Journal of Economics, Vol. 13, pp.695–700.

National Land Agency, 1987: The Fourth Nation-wide Comprehensive Regional Development Plan, Ministry of Finance (Japan) Press, (in Japanese).

Sakashita, N., 1988: Spatial Interdependence and Externalities, ISEP Discussion Paper Series (University of Tsukuba), No. 391, (Presidential Address, The 35th North American Meetings of Regional Science Association, 11th Nov.

Sakashita, N., 1989: Market Equilibrium and Optimum Social Welfare in a Two Region Economy, R. Sato and T. Negishi ed.: Developments in Japanese Economics, Tokyo, Harcourt Brace Jovonovich Japan, Chapter 3.

Wildasin, D.E., 1987: Urban Public Finance, Fundamentals of Pure and Applied Economics, Vol. 10, Chur: Harwood Academic Publishers.

Operationalizing the Trade Variable in a Funck-type Multi-Region Global Model*

by

Walter Isard

1. Introduction

My association with ROLF FUNCK goes back many years. Our first real interaction took place at the Summer Institute in Regional Science, the University of California (Berkeley), 1962. There, and ever since, I have been much influenced by his emphasis on regional policy development within a conceptual framework of a comprehensive multi-regional (interregional) system. Perhaps the one contribution of his which has most influenced my work was his paper "A Policy-Oriented Model of Regional Economic Growth" and the associated materials he presented at the Advanced Studies Institute on Recent Developments in Regional Science which he organized at Karlsruhe. As most of us know, the Institute was exceedingly fruitful and led to the inauguration of the subsequent set of stimulating European Summer Institutes of the Regional Science Association.

The FUNCK paper was published in the first volume of the *Karlsruhe Papers in Regional Science*, 1972, London, Pion, which he edited. Since then many developments have taken place. Yet, one sees the emphasis of FUNCK and others on the need for policy development within a broad interregional

* This manuscript draws heavily upon materials presented at the Workshop on Large Scale (Global) Social Science Models, September 1989, Urbana, Illinois jointly sponsored by the National Center for Supercomputing Applications of the University of Illinois and the Cornell National Supercomputing Facility.

framework still very much alive in and basic to the recent spate of partially or fully integrated multi-regional models which are operational. (E.g. see ISARD and ANSELIN, 1982; ISSAEV et al., 1982; ISARD,BOYCE et al. (1981) and most recently SMITH (1989).

But this emphasis and the associated ideas are disseminating even more widely. Since that first Summer Institute at Karlsruhe, the international system has become much more interdependent, and we also have come to recognize much more fully this interdependence. The emphasis of ROLF FUNCK and others on policy development within a broad interregional framework has simply been extended to cover the global multiregion system, wherein each region may be a nation (a political region) or other meaningful spatial unit. I can testify to this influence in our current global model work at Cornell, the University of Illinois, the University of Pennsylvania and elsewhere.

Let me be specific, and in doing so I shall report at some length on progress in making the basic trade equations of FUNCK's paper operational.

First, in one version of our operational global model, which can easily be viewed as an operational FUNCK scheme dealing with nations as regions, we may begin with projection of Keynesian-type magnitudes: investment, consumption, government expenditures, exports and imports as dealt with in standard national econometric models. This is as FUNCK does in the figure in his paper. For a world system, however, each nation's exports and imports must be consistent with those of other nations, since total global imports and total global exports by definition must be equal. Hence each nation's econometric model (constructed in isolation, that is without reference to the econometric models of other nations) must be viewed as yielding preliminary estimates. These preliminary estimates, as is done in the LINK model (and also in LEONTIEF's World Input-Output model), are made consistent with each other through a series of adjustments making use of a constant shares approach, projected prices, and other mechanisms. We then obtain a consistent set of total national exports and imports which are fed back into each nation's econometric model to yield a new set of projected GDP's, and exports and imports. The projected exports and imports again may be tested for consistency and so forth. This process can proceed until discrepancies and inconsistences become small enough to justify bringing the iteration to a halt.

Once a set of consistent projections by nations is achieved, we are able to proceed through typical input-output and SAM (Social Accounting) procedures to generate projections of relevant magnitudes. In a forthcoming paper, my colleague SMITH (1989a) has developed the procedures to link input-output and SAM operationally to a set of consistent econometric models by nations.[2]

As FUNCK's paper indicates, at all times welfare policy must be of primary concern. In our current work using the supercomputer facilities at Cornell, we are particularly concerned with not only arms reduction but also environmental policy (e.g. managing the greenhouse effect and the acid rain problem) via projecting pollution generation. From such projection we go to multiregion policy analysis (which could be added as a box in FUNCK's figure) ... and then to targets, goals and constraints which define various scenarios whose implications we can establish for welfare (as measured by GDP, employment, per capita income, acid rain level, greenhouse effect, agricultural productivity by region, and so forth).

When the FUNCK framework is *extended to a system of nations* (as a system of political regions) we find that one place at which major improvement is required is, as already alluded to, in projecting trade among nations, consistently and in a *theoretically meaningful way*. While for a system of regions within a nation there are no exchange rate variables, there are in a global system. There are of course rigorous models that deal with exchange rates. Unfortunately, however, these models have not demonstrated an ability to do significantly better than a random walk when they are used to project magnitudes outside their sample periods (ISARD, P., 1988). Since we cannot project exports and imports using conventional price variables if exchange rates are not specified, and since exchange rates as variables cannot be projected on a non-random basis, resort must be had to other approaches. One such is the constant shares approach (described below) used by LEONTIEF, CARTER and PETRI (1977) in their World Input-Output model and KLEIN, PAULY and VOISIN (1982) in their 79 nation LINK model. However, the use of a constant shares approach (whether or not in combination with questionable projections of relative prices) leaves much to be desired. Hence, we have explored the possible use of a gravity model approach - one which has had a modicum of, if not great, success in travel and transportation studies.

2. Projecting Trade in a FUNCK-type Global Model

There are several problems in the use of a gravity model. The first concerns the measure of mass. Clearly, when we define interaction among nations as trade, we can, for a start, use GDP as a good proxy for mass. Next, how define distance? To make a long story short, the procedure we judged to be best, after going thoroughly through the literature, is as follows. Where the major fraction of trade between two nations is via water, we take the distance to be the nautical mileage between the major port of each plus the equivalent nautical mileage of associated land shipment. Each mile of land shipment is taken to be equivalent to two nautical miles of shipment, and an "on the average" mile of land shipment is taken to be associated with each shipment arriving at or departing from the main port of shipment. Where trade between two nations is primarily by land, we take the distance to be the equivalent nautical mileage of the actual land mileage separating the major industrial node (or equivalent) of each plus in many cases an "on the average" land distance to points of destination and origination in a node's hinterland.[1]

A third problem concerned the choice of a functional form of the gravity model. While theoretically and in certain empirical studies an exponential form has been shown to be superior (SMITH 1978, WILSON 1970), we have employed the basic Newtonian form for our initial work, namely

$$T_{ij} = G_{ij} a_i M_i^{\alpha_i} c_j M_j^{\alpha_j} / d_{ij}^b \qquad i, j = A, \ldots, U \quad i \neq j \qquad (1)$$

where T_{ij} = interaction as represented by trade

M_i, M_j = the masses, respectively of nation i and j, taken to be GDP_i and GDP_j, respectively

d_{ij} = distance between i and j

G_{ij} = a gravitational constant re nations i and j

b = a power to which distance is raised

α_i, α_j = powers to which the respective masses are raised and

a_i and c_j = weights, to be discussed below.

A fourth problem is related to the number of nations to be covered in our operational model. The basic set of data employed was provided on a tape by Fred CAMPANO of the United Nations - the best set of data available. It is a consistent set in that it eliminates the numerous inconsistences in the published data of the United Nations, the International Monetary Fund and other organizations. (A typical inconsistency exists when a nation i reports imports from nation j which differs from nation j's report of exports to i). The data were for years 1965-86, a period wherein many nations experienced major inflation. Hence, because we needed to make adjustments for inflation, we confined our empirical analysis to twenty-one countries for which GDP estimates in constant prices were available for at least most of this time period.

A fifth problem relates to the extent to which the disaggregations of the trade data are to be used. Currently, the United Nations Campano tape reports bilateral trade data in total, and disaggregated by some standard commodity classes[3] and aggregates of them. Typically, an analyst might judge it more desirable to explain and understand trade in each category of goods, utilizing all the information available on that category of goods as well as information on total trade. Such implicitly assumes at the minimum that the analyst has available testable theory for explanation of trade in each or most of the categories. However, when such theory is lacking or has been found to be statistically invalid, and when on the other hand theory regarding trade totals alone has been found to be empirically reasonable, one may justifiably set aside the disaggregated trade data for use pending further adequate theory development. As discussed elsewhere (ISARD, 1989) we in essence do this in what follows when we use a gravity-type model, where we by and large confine ourselves to the use of a macro-type theory regarding total trade.[4]

A sixth problem concerns the derivation of the parameters of equation (1). Early in our exploratory work we confronted the problem. At that time we were using a set of 24 nations. Hence we had 552 observations for each of nine years (1975, 1978-85). We undertook a number of simple regression analyses. From among several distance conventions, we found the highest R^2 for that one described above. For that convention we found that $b = 0.9999$, its estimate being significant at the one percent level. Hence, we set $b = 1.0$ for all later computations. Further, we had no theoretical basis whatsoever for expecting α_i to be different from α_j. So we set $\alpha_i = \alpha_j = \alpha$. Our estimate for α was 0.9305, significant at the one percent level. Other regression explorations led to somewhat similar results, but did not rule out exploring the use of the very simple gravity model, consistent with the strict analogy to the definition of mutual energy in Newtonian physics where α is set equal to unity as well as b. The determination of the a_i and c_j will be discussed below.

Further, note that our framework is, of necessity, a comparative statics framework. There exists no framework having the comprehensiveness that we seek which also has that dynamic character for which FUNCK's framework aspires.

3. A Run-down of Some Preliminary Findings and the Many Unresolved Questions

Because of data inadequacies and other problems, we took year 1975 as a base year from which to project trade for year 1985. GDP data, nominal and in constant prices, were available from U.N. publications and consistent trade data were provided in the U.N. Campano tape. However, there were problems in using the data. As already noted the reported trade data (stated in a common unit, namely the U.S. dollar) reflected inflation effects. But so, too, did the data on GDP_i and GDP_j. Now in eq. (1), the left side deals with a bilateral trade estimate that is to be affected by inflation only once to be comparable to the actual bilateral trade; on the other hand, the right side involves the product of two magnitudes (GDP_i and GDP_j), each inflated once. Thus the right side contains a "double inflation" effect, and hence the right side must be "deflated once" to be usable. Therefore, in deriving any estimated bilateral T_{ij} we divided the right side by the arithmetic average rate of inflation for nations i and j.[5]

Once we computed the bilaterals, we had to derive an f factor to reduce each estimated T_{ij} so that their sum equals the sum of the total imports (exports) of the 21 nations. The f factor thus is an element of each G_{ij} parameter that will emerge. Next, as is well known in the gravity model literature, consistency needs to be established with the marginals, that is the actual 1985 total imports by nation and total exports by nation. Typically, the standard RAS method[6] is used to achieve the required consistency. In this method, all the elements of each row are adjusted by a factor so that their total equals the actual 1985 total exports (the marginal) of that row; next all the resulting elements of each column are adjusted so that their total equals the actual 1985 total imports (marginal) of that column; then a second adjustment of all realized elements in each row is made, and a second adjustment of all resulting elements in each column and so forth until, within an acceptable deviation, convergence to the marginals is attained. Ultimately the derived adjustment a_i by rows and c_j by columns may then be viewed as the weights to be applied to the GDP_i and the GDP_j, respectively. As a consequence, each nation k then comes to have two effective masses, a_k GDP_k, *its effective exporting* (product supplying) *mass*, and b_k GDP_k, *its effective importing* (product consuming) *mass*.

With these adjustments, we next compared the estimated bilaterals for 1985 and the actual bilaterals for our 21 nations. Employing the total of absolute deviations as one relevant measure of fit, we obtained the sum of 226,204 in millions of U.S. dollars. How well did the gravity model project? To answer the question, we compared its total of absolute derivations with that yielded by another procedure used in LEONTIEF's World Input-Output model, KLEIN's LINK model and others, namely the constant shares approach, already alluded to. In one version of this approach, the imports IM_j of goods by a nation j are taken to be a fixed share k_j of the nation's domestic consumption D_j of goods.[7] That is,

$$\frac{IM_j}{D_j} = k_j \qquad j = A,\ldots,K. \quad \text{(importing countries)}$$

Imports come from a world pool of goods, to which pool each exporting nation contributes a fixed percentage k_i. That is,

$$E_i = \bar{k}_i \sum_i E_i \qquad i = L,\ldots,U \quad \text{(exporting countries)}$$

where over all importing countries the total of imports equals over all exporting countries the total of exports. That is,

$$\sum_j IM_j = \sum_i E_i \ .$$

This pool-type constant share approach thus implies that whenever a nation increases (decreases) its imports, each exporting nation will provide a fixed share of that increase (decrease), regardless of which nation's imports increase (decrease), this fixed share being the share that the exporting nation has contributed to the world pool. In our use of this approach, we introduce much more flexibility. We allow the imports of each nation to be met by a different fixed pattern of exports from supplying nations. Put otherwise, we take the percentage of a nation's imports that comes from a given exporting nation to continue to be the same in future years as in the base year, although this fixed percentage for an exporting nation may vary from one importing nation to another. In particular we take the percentages of our base year 1975 and apply them to each nation's total imports in 1985 to derive the bilateral trade between pairs of nations in 1985. Thus, given the actual 1985 total imports of each of our 21 nations, we estimate the exports from each of the other 20 nations as exporting nations. Taking the total of absolute deviations from the actual 1985 bilaterals of the bilaterals obtained from the use of the constant shares approach, we obtain the figure of 199,907 in millions of U.S. dollars. Thus the constant shares approach seems to be superior to the gravity model.

At this point, however, we need to make explicit the theoretical bases of each approach. The constant shares approach captures the inertial forces stemming from institutional ties, past practices and trading relations that have worked well, reluctance to incur the costs and risks of trying out new relationships, and so forth. The gravity model through its mass variables captures the demand, the productivity factor, the scale economy and other agglomeration and related effects, as they change over time. In actuality both sets of factors are at play, and therefore a better model than either of the above alone would be one which combines the two approaches. We thus applied different weights to the two, and found the best result when a weight of 0.4 was applied to the estimated bilaterals of the gravity model and a weight of 0.6 to those of the constant shares approach. The total of the resulting deviations from actual was 163,852.

Likewise, when we projected trade for each of the years 1983 and 1984, instead of 1985, where the base year is 1975, we found that the use of a combination of the two to be better than the use of one alone.

The next problem arises, however, when we attempt to project bilateral trade for a year in the future, when either (i) total exports or total imports for each nation are not projected (so that marginals do not exist), or (ii) they are projected but are questionable. How then to proceed?

If we face a situation where only GDP's are projected, then we must be satisfied with the use of gravity model projections alone, where we employ the f, a_i and c_j of a base year in making projections.[8] The constant shares approach cannot be used at this point, since it requires exogenous projections of total imports by nation. Conceivably, however, the column sums of the projected gravity model bilaterals can be used to yield the required projections of total imports by nations to which the constant shares approach can be applied. Then we can proceed to use a combination as described above - a combination that thus introduces the important inertial aspect of reality into the gravity model projections. This direction of research remains to be explored.

When, as with LINK, or with other models such as the computable general equilibrium ones[9], we have projections for total imports and total exports as well as GDP by nation - where these total imports and exports and their GDP's may have been iteratively adjusted to achieve consistency - we may maintain that the procedures used in the adjustments did not embody the best theoretical structure, and in particular did not embody as good a rationale as the use of a combination of a gravity model and a constant shares approach. We may take this position particularly if the use of import-adjusted input-output coefficients in the operation of say a fused LINK-Input-Output, or a fused LINK-SAM operational model generates total imports by nation which are significantly different from the total imports by nation generated by the use of say LINK alone. Our current thinking suggests that a fused LINK-Input-Output or a fused LINK-SAM framework should be used to generate a first round estimate of total imports by nation, using the LINK magnitudes on investment, consumption, government expenditures and exports as the bases for the first-run final demand magnitudes. Once these magnitudes are premultiplied by the inverse of an import-adjusted coefficients matrix, total output by sector and total imports from import coefficients by sector can then be obtained for a nation. With these totals by nation, a constant shares approach can then be used to determine a first-run set of bilaterals. This set then yields *"constant shares" total exports* by nation when the bilaterals are summed by rows. Simultaneously, the projected GDP's and the a_i and c_j of a base year can be used to project gravity model bilaterals. By columns these bilateral can then be multiplied by a g_j factor so that the total of each column j equals the total imports of nation i that was obtained with the use of the imported-adjusted I-O coefficients matrix.[10] Summing these adjusted bilaterals by row yields first-run *"gravity model" total exports* by nation. We then obtain a new set of total exports (to be designated second-run total exports) and bilaterals by combining the above two sets of total exports with the weights found best in combining constant shares and gravity model bilaterals for replicating trade flows in the base year. These second-run total exports are then introduced as exogenous elements in a second run of LINK. From this second run of LINK, we obtain the basis for the second run final demand magnitudes. And so we continue in iterative fashion until convergence within negligible differences is attained.

4. Conclusions

In bringing this paper to a close, I want to reiterate that what we have accomplished in operationalizing the trade equation in the FUNCK-type conceptual framework for a global multi-region system represents results from preliminary thinking only. There remain a number of research questions which must be attacked much more effectively. These concern: (i) the most appropriate functional form; (ii) the proper representation of distance and mass; (iii) appropriate exponents (the α_i and α_j) to which the masses must be raised; (iv) the appropriate weights a_i and c_j to be applied and the meaning of these weights; (v) the treatment of inflation/deflation effects in deriving parameters; (vi) the proper weighting or treatment of the constant shares and gravity model projections in defining a best combination (the theoretical basis for some combination cannot be denied); (vii) the introduction of the comparative cost (delivered price) approach, the constant elasticity of substitution approach (as a factor that tempers the constant shares rationale) and others into the combination so as to yield the best form of a multi-method approach to trade projection (ISARD and DEAN, 1987); and (viii) possibly the introduction of exchange rate variables, relative prices and dynamic elements.

Nonetheless, while a tremendous amount of further research is essential, we do judge that we have been able to take a step forward. Hopefully, with the use of the ever-increasing capacities of our computer networks our new generation of regional scientists will forge rapidly ahead in this research area.

FOOTNOTES

[1] Let me note that we are operating this model using old data that exist in appropriate form. Once the required operations are routinized, we anticipate the effective use of more recent data.

[2] This brief statement of the treatment of the distance variable fails to note the numerous exceptions that were taken. In some cases, we had to represent a nation by two major ports. In some cases we had to consider two water routes. In other cases we had to take into account the extent of common borders and contiguity of industrial structures. In still other cases, we had to use a major port or rail node of a third country. And so forth. The convention that one land mile of shipment was equivalent to two nautical miles of shipment was employed to reflect the higher costs of a mile of land shipment. We failed to take account of scale economies in transport, terminal charges, back-haul rates, transport networks and numerous other factors. Nor were we able to go much beyond some initial empirical testing of our convention. Nonetheless, we feel we reached a much more satisfactory procedure for measuring distance than had hitherto been employed, and can only urge scholars to develop an even better procedure. See LINNEMANN (1966), GRUBER and VERNON (1970), and BERGSTRAND (1985) for other treatments of the distance variable.

[3] These classes are: SITC 0 and 1 Food, Live Animals, Beverages and Tobacco; SITC 2 and 4 Crude Materials, Oils and Fats (Fuels excluded); SITC 3 Mineral Fuels, Lubricants and Related Material; SITC 5 Chemicals; SITC 7 Machinery and Transport Equipment; and SITC 6 and 8 Other Manufactured Goods.

[4] Note, however, that for certain commodities whose shipments are very large in internation trade, e.g. wheat, rice, coal, oil etc., or even commodities whose shipment may not be large, there may exist a firm basis for projection (as in aluminum, coal, steel, etc.) via comparative cost (delivered price) analysis or investigation of institutional ties and agreements. Hence, one may take these specific items out of the bundle of commodities whose aggregate is to be projected with a gravity-type model, and project them on this firm basis. Such does weaken the rationale for the use of the gravity model that stems from its overall "averaging-out" characteristic. When proceeding in this direction there is in essence a trade-off: the gains from improvement in the projection of individual commodities, or group of commodities, should outweigh the decrease in the validity of gravity-model projection

[5] It would involve too long a footnote to go through the various experiments we conducted and the long-drawn out thinking we carried on with regard to this question. See ISARD (1989). In no way do we remain satisfied with what we did.

[6] In this method, a square matrix is constructed with the nations listed in the same order along the rows and columns. The bilaterals (trade between pairs of nations) are entered in the cells such that the sum of the bilaterals along the rows represent total exports by nations and the sum of the bilaterals along the columns represent total imports by nation.

[7] This approach may also be applied with regard to individual goods h, $h=1,\ldots,\ell$ instead of to the total value of bilateral imports and bilateral exports over all traded goods.

[8] Should future prices among nations be projected in this situation, along with GDP's, we still question the use of projected prices since such prices depend on future exchange rates, and since, as already noted, it has been shown that the best exchange rate models cannot be said to do significantly better than a random walk when they are applied to a time period outside their sample period (P. ISARD, 1988).

[9] This type of model has only recently been developed and exists for only a few nations, the best being the ORANI model of Australia. Both its relative advantages (e.g. in replicating more fully the market system) and disadvantages are yet to be fully explored.

[10] We could, of course, doubly constrain the gravity model operation with the use of total exports by nation (sums of rows) as well as total imports by nation generated by the constant shares approach. A combination of the gravity model bilaterals and the constant shares bilateral might then be taken to project a superior set of bilaterals. However, in this operation the gravity model would not have any effect on projecting a more reasonable set of total exports by nation and hence a more reasonable set of bilaterals as well.

LITERATURE

Bergstrand, Jeffrey H., 1985: "The Gravity Equation in International Trade: Some Microeconomic Foundations and Empirical Evidence", Review of Economics and Statistics, 67:474-81.

Gruber, W.H.; R. Vernon (eds.), 1970: The Technology Factor in International Trade, New York, National Bureau of Economic Research.

Isaaev, B., et al. (eds.) 1982: Multiregional Economic Modelling: Practice and Prospects, North-Holland, Amsterdam.

Isard, P., 1988: "Exchange Rate Modelling: An Assessment of Alternative Approaches" in Bryand, R.C., et al. Empirical Macroeconomics for Interdependent Economies; Washington, D.C., Brookings, pp. 183-201.

Isard, W.; L. Anselin, 1982: "Integration of Multiregion Models for Policy Analysis", Environment and Planning A, Vol. 14, pp. 359-376.

Isard, W.; D. Boyce, et al., 1983: "Integration of Multiregion Models", submitted to the U.S. National Science Foundation to be published in Smith, C. (ed.), forthcoming, Practical Methods of Regional Science and Empirical Applications, (Selected Papers of Walter Isard), Macmillan, London.

Isard, W.; W. Dean, 1987: "The Projection of World (Multiregional) Trade Matrices", Environment and Planning A, Vol. 19, pp. 1059-1066.

Isard, W., 1989: "On Bilateral Trade Matrix Projection", Workshop on LargeScale (Global) Social Science Models, September, mimeo.

Klein, L. R.; P. Pauly; P. Voisin, 1982: "The World Economy: A Global Model", IBM Perspectives in Computing, Vol. 2(2), pp. 4-17.

Leontief, W.; A.P. Carter; P.A. Petri, 1977: The Future of the World Economy;, Oxford University Press, New York.

Linnemann, Hans, 1966: An Econometric Study of International Trade Flows, Amsterdam, North-Holland.

Smith, C., 1989: Integrated Multiregions Models for Policy Analysis: An Australian Perspective. Amsterdam: North Holland.

Smith, C., 1989a: "On the Integration of Econometric, Input-Output, Social Accounting Models and Frameworks", Workshop on LargeScale (Global) Social Science Models, September.

Smith, T.E., 1978: "A Cost-Efficiency Principle of Spatial Interaction Behavior", Regional Science and Urban Economics, Vol. 8, pp. 313-338.

Wilson, A.G., 1970: Entropy in Urban and Regional Modeling: London, Pion.

The Importance of Infrastructure in Regional Evolution

by

Dieter Bökemann

1. Foreword

In more than twenty years my friendship with ROLF FUNCK has long grown to become more than a professional relationship. In many discussions, some joint research projects and minor consulting contracts for policy-makers "the importance of infrastructure in regional development" time and again caused us distress. And at times, after having gained this insight or the other, it even instilled us with feeling of pleasure.

ROLF FUNCK, an economist with a politicizing and evaluating mind, stimulated me, the engineer with a penchant for theories, in our discussions in many ways: suggesting, for instance, to become a member of the Regional Science Association and to join - just as himself - as one of the founding fathers the "Gesellschaft für Regionalforschung", the Association's German-speaking group; suggesting I should gear my transport-system-centered thinking more to issues of location optimization and infrastructural technology. It was by no means a rare occurrence that ROLF FUNCK, with his preconceived idea that "an engineer of course understands how things work", used to prompt me into a more structured analytical thinking; and myself, whom it concerned, all of a sudden had to find my way through the new land of economic theory, unknown to me so far.

ROLF FUNCK has never ceased challenging me and - as I am aware, also many others of his colleagues and friends - to dare and break new scientific ground. For this I want to convey to him my heartfelt thanks and the thanks of all those, who - just as myself - have used some of Rolf FUNCK's challenges for their own benefit.

I hope, dear ROLF, that you will fulfil this all-important function of a catalyst in our regional science for yet a long time to come.

2. Introduction

Up to the present, regional science has focused on (and largely restricted itself to) (i) portraying, evaluating and optimizing (at rare instances in relation to political objectives and aims) the development of regional patterns in macroeconomic terms and (ii) on explaining individual behavior (in particular the choice of locations and decisions on utilization) in the overall context of a given settlement pattern in microeconomic terms. In formulating a set of theories within the framework of regional science to explain the dynamics of regional development, many of the insights gained outside the bounds of economics, have remained largely unheeded (v. inter alia FUNCK and KUKLINSKI, who deplored the lack of theories in regional science originating from outside economics at various panel discussions held by the Regional Science Association).

In order to analyze regional development under an evolutionary aspect, we have put together some arguments from the sociobiological theory of reproduction, the theory of technological infrastructure, from the economic theory of property rights and the sociological theory on the development of social rules, joining them with the location theory. Whereas the modern sociobiological theory of reproduction relates to the objective function of individual behavior in regional development, the assumptions underlying infrastructural technology and the economics of property rights try to define, how means of shaping individual options have proven effective throughout the process of settlement and how they are further developed. In their working and in their historical development, regional structures are based on a complex set of social rules. In this respect, the socio-philosophical theory on the development of social rules is of particular importance when it comes to analyzing regional development (von HAYEK, 1971; BUCHANAN,1971).

Against this scientific backdrop, the present contribution will discuss the following *hypothesis:*

In settlements, the individual is accorded location-related options through communication and safeguarding facilities. Individual needs for more communication opportunities and more security are subjected to the primary search for options. Their purpose is to ensure and heighten the chances of survival for their offspring. Infrastructural technologies and territorial structures, which satisfy individual needs for communication and security in social terms, develop as sets of standards after having stood the test of evolution.

3. Regional Patterns as a Product of Individual Interests

Regional patterns are considered as environments created by man according to social and technical standards. It is assumed that in regional development (i) social evolution, which is oriented towards the purposes of human reproduction and (ii) cultural evolution, which is oriented towards the means for shaping the environment condition each other. Whereas sociobiological and ethnological assumptions explain, how mankind develops a specific behavior in a given environment for energy-saving reproduction ("environment-determined reproduction strategies" as part of social evolution), socio-technological arguments explain the development of those institutional and material instruments, by means of which people shape their environment into options ("reproduction-linked shaping of the environment" as part of cultural evolution) if specific (reproduction-oriented) interests are pregiven. According to sociobiological and ethnological evolutionary theories (IMMELMANN/SCHERER/VOGEL/SCHMOOCK, 1988) specific reproductional forms and strategies become established in their environments (individual free scopes) by way of trial and error. In individual attempts to learn from experience and to reduce personal uncertainties, social rules develop and consolidate into reproductive behavior. According to socio-technological theories, technical standards develop as instructions to shape the environment in an evolutionary process of selective testing, whereby technical innovations tend to become accepted only, if they prove sufficiently compatible with existing material and institutional systems and after widespread approval by those concerned.

Regional patterns are a material expression of social structures; in the following they shall be dealt with as (i) individual options (settlement equaling a certain allocation of individual location-linked options) and as (ii) interpersonal communication opportunities and property rules (settlement equaling a system of locational utilities and of inter-locational infrastructural links and property boundaries). Again, this concept (BÖKEMANN, 1982) is first related according to sociobiological findings (inter alia HAMILTON, 1964; DAWKINS,1978) to the following underlying *hypothesis:*

Over the long term (here this means over generations) regional evolution follows a program for the survival of certain combinations of genes in the human gene pool. For this purpose, there are social rules working over the short term (here this means within generations) which regulate the meeting of individual needs for communication and security. In regional evolution, social rules are accepted according to how much they contribute to stabilizing the program for the individual's survival in a collective (BUCHANAN, 1971; AXELROD, 1987). Seen from the individual within one generation, the needs for communication and security - mostly regarded as original so far - are considered as secondary needs; derived from the primary search for options and development potential for the descendants or for securing the longest possible lineage (over time) (according to WEISMANN, v. WICKLER and SEIBT, 1977).

We know from sociobiological findings (HAMILTON, 1964; DAWKINS, 1978; WILSON, 1981; VOGEL, 1986; VOLAND, 1986, and others) that individual behavior is phenotypically and primarily geared towards preserving the identity of the individual genotype as long as possible. Hence, the individual is seen as a mere mediator who hands down genetic information ("reservoir for certain genetic combinations") to the next generation. His needs are mainly focused on those means, by which he may increase the chances for survival of his own genetic identity beyond his individual existence. This genetic chance of survival of the individual identity ("overall fitness" according to HAMILTON) onto the next generation depends:

(i) on the totality of descendants and

(ii) on the viability and assertive capabilities ("fitness") of each and every individual offspring.

As a primordial aim in all his activities, the individual thus will seek to maximize the chances for survival of his genetic identity onto the next generation ("overall fitness").

In an economic analysis, the genetic chance for survival may be interpreted as individual benefit in the form of a curve of indifference (see Figure 1) related to the number and qualification of the descendants. (The curve of indifference represents a certain level of utility; as in reality different people will start arguing at different levels of utility.)

In his reproduction strategy an individual may basically influence both variables determining his genetic chance of survival: the number of offspring by selecting and manipulating his partner(s); and the viability and assertive capability of each individual descendant by means of an adequate education and inheritance. Given the limited time and energy budget available to every individual, a larger number of progeny will imply reduced viability and assertiveness of the individual child, and, vice versa, better qualification of the individual child through investments in education and training (for higher skills) and savings to be passed on (wealth and power) will imply a smaller number of offspring. It thus seems obvious that indigent parents will procreate as many children as they can in the - statistically justified - hope that at least one of them will succeed in the next generation by virtue of its innate talents and strength. In contrast, affluent parents decide quite rationally in terms of the stated sociobiological assumptions, if they concentrate their available means on a smaller number of children, in order to secure them a head start and superiority in the competitive race against their fellows and peers through education and inheritance (from their own time and energy budget). More affluent parents therefore expect their children to use the options they are gaining from parental care (investments in education, capital resources and transferable relationships) by their own initiative, in order to transmit their common genetic program ("individual combination of genes") efficiently.

Figure 1: Reproductive trends of affluent and indigent parents

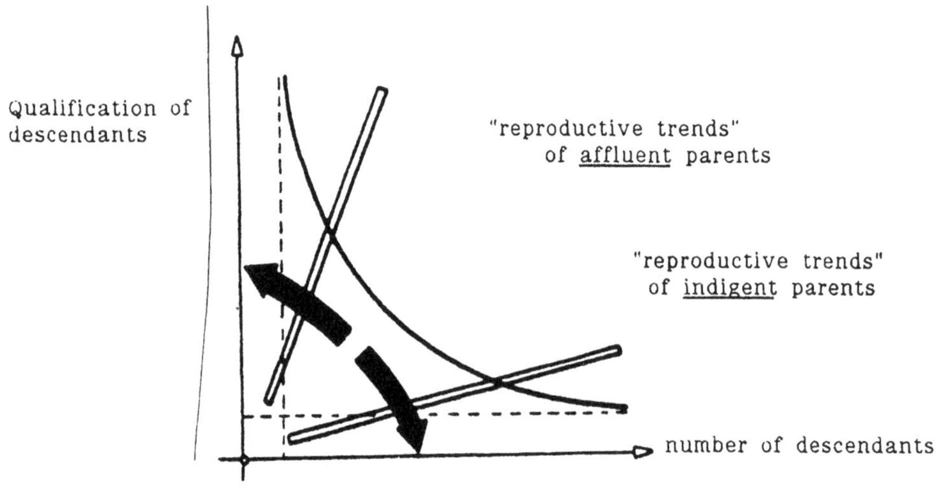

Comparing the social, economic and regional systems, we find that people who are living in the sociotope of a city tend towards concentrating their relative wealth on fewer descendants, thus a preference for higher qualification of children against numerous offspring. The colonization of new land, i.e. the early phases of settlement, was generally characterized by a reproduction strategy which aimed at "the largest possible number of children - while necessarily foregoing educational investments to ensure higher qualification and renouncing to pass on capital legacies."

Economically, these empirical findings can be explained only, if the individual's genetic chances of survival in the sociotope of the city are heightened in a more cost and energy-saving manner through a "higher qualification of a quantitatively reduced offspring". Though this proposition is only valid, if the different economic systems and settlement patterns entail different costs for producing and ensuring a higher qualification of offspring. There is much reason to believe that the costs to be borne by the individual for educating and training his descendants and for securing them other advantages in survival decrease with growing urbanization.

Figure 2: Settlement-related reproductive strategies of the population based on different costs for qualification and production of descendants.

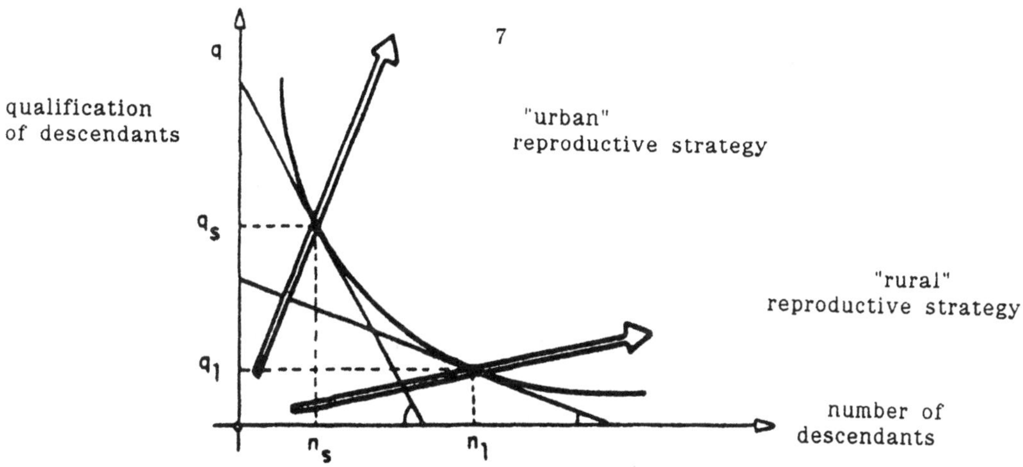

If, indeed, the urban population found the "higher qualification of a quantitatively reduced offspring" to be the most efficient reproduction strategy (see Figure 2), the city was bound to become that social habitat, where educational and training facilities for the descendants, but also hereditary - and thus institutionalized - power and protected monetary and real capital accumulated and specialized to a particularly large extent, compared to other social, economic and regional systems. It may be assumed that especially the individual's sociobiologically-rooted urge to hand down his power and capital resources, his personal experiences and social relationships to his descendants would give rise to the development of that set of rules and instruments we call urban organization and urban infrastructure. Whereas the individual may well be in a position to safeguard his own options through bilateral agreements and contracts, it is assumed that the need for a public area with publicly accessible facilities and generally applicable legal norms emerged as the urban population sought to transmit a maximum of options to its descendants. This is the mechanism underlying the development of public urban facilities, which, on the one hand, force the competition between individuals for more options into a system of rules and which, on the other, have fostered the advantage of cooperation for all urban citizens. The strategy of a "higher qualification of a quantitatively reduced offspring" thus largely accounts for that innovative force, which distinguishes the urban sociotope from other social, economic and regional structures. Also, the strategy of a "higher qualification of quantitatively reduced offspring" presumably favored the emergence of such rules in a phenotypical way, which, after all, enabled cooperative competition in the form of markets, time and energy-saving production methods and the procurement of living means (in the broadest sense of the word) based on a division of labor and cultivated administration of justice and government (bureaucracy). Furthermore, the city, with its numerous opportunities for individual development, lost nothing of its appeal and attractiveness for non-urbans as this strategy was implemented. In the past, this attractiveness of cities to the non-urban surroundings compensated (and in some cases even more than compensated) their traditional lack of progeny by (im)migration.

The sociobiological interpretation of the genetic chances for survival constitutes a new argument which has now been put forward in the theoretical discussion of regional development. The findings on genetic survival as part of social evolution go well together with reflections on the social regulation of individual communication and security needs, which are manifest in regional patterns as infrastructural links and borderline systems regulating property rights (as part of cultural evolution). The fact that conflicting interests, which arise from individual communication and security needs, are settled marginally by infrastructural and territorial measures, is basically founded on the following *hypothesis:*

In regional evolution, individual interests subordinated to the reproductive interest are conditioned by the technical and institutional means available for meeting these interests and vice versa. This is true in particular for the communication and security needs and how they relate to individual options or potentials for development. The desire to possess other peoples' goods is hidden by the individual need for communication, whereas the concern of losing one's own property is concealed by the individual need for security. Apparently, this individual need for communication creates the willingness to provide, together with others, opportunities favoring contact and interchange; be it by building publicly accessible roads or by keeping public spaces freely accessible. The individual need for security in turn manifests itself in endeavors to protect property by delineating, controlling and fortifying the territorial borders on the one hand, and on the other, by the individual's readiness to respect his neighbors property, if they are committed to reciprocal behavior. Such mutual commitments are based on the individual's interest in saving costs for protecting his property and also on the individual's expectation that others will sanction infringements on his property in a similar fashion as he himself would. In such a way, individual interests give rise to the formation of rules and institutions, which (according to BUCHANAN, 1971) reduce uncertainties in decision-making ("prisoner's dilemma" situations), while at the same time stabilizing existing property structures over the long term.

Infrastructural transport and supply systems serve the individual need for communication; the extent to which the individual may use such systems will largely determine his economic power and social position. Systems of property dedication and border fortifications serve the individual need for security: the degree to which the individual may act during a given period of time without interference will determine his personal security and the value of his property. One may assume that the individual needs for communication and security do not primarily evolve from personal qualities (traits), but are rather derived from the individual's social position, which is the resultant of a comparison between one's own options ("potential") and the options enjoyed by the members of one's "own" social group. Infrastructural communication facilities and territorial safeguarding installations complement one another in contributing towards individual options; thus one might say that the need for security grows with the number of extra communication opportunities available to the individual, just as the individual need for communication grows as additional security and property guarantees are being obtained. Hence, individual communication and security needs develop within the predetermined structural and institutional regional (urban) pattern; that pattern, in which the individual finds his options and which gives him a subjective feeling of restriction caused by a lack of communication facilities or by outside interference; and the same pattern, in which the concentration of power and relationships instills the lower-positioned individual with a feeling of wistfulness.

4. Organization of Settlements Through Infrastructure and Territorial Structure

We may deduct from the sociobiological hypothesis of the "egoistic gene" (DAWKINS, 1978) that individuals seek to heighten the chances for the survival of their genetic program over the long term by creating options for themselves and passing them on to their descendants. According to what has been said so far, also the means for shaping individual options ("environments") are subjected in terms of their technical differentiation to the evolutional criteria of selection and mutation (in this case cultural evolution). It is true in particular for the technical/material and legal/institutional means of shaping the environment, that, in principle, innovations must be economically compatible with existing material and institutional systems and acceptable for the majority they concern. The question therefore arises, how individual options emerge in regional evolution as hereditary and tradable goods; and secondly, how the technical and institutional means used to shape land-related options evolve.

Basically, individual options are determined socially: (i) by the need for communication with others and (ii) by the need to protect property from disturbing outside intervention.

The `individual's interest in his own options may thus be explained by the way he expects others to behave. Assuming such reciprocity in individual motives and expectations, the options existing for different persons in a settlement are mutually conditional.

To put it in abstract terms: a settlement is composed of an aggregate of individual land-related options, shaped by appropriate technical/material and legal/institutional means in a manner which ensures that the mutual communication and security needs are met reciprocally. Along these lines, the concepts of communication facilities and security of property are related to infrastructure and territorial structure as means of shaping regional patterns (BÖKEMANN, 1982).

Communication, we assume, fulfills the function of taking away from others a good one does not have (or only in limited amounts) in exchange for a good one does not need (or only to a limited extent). Communication is a quid-pro-quo relation. The willingness to communicate grows as the differential of goods owned by individuals becomes greater. It decreases with increasing transport costs.

Security of territorial property, just as communication, is considered an elementary human need. Seen from this angle, the individual is sure of his property to the extent it is recognized by others. A plot of land or a built-in space are seen as territorial property (no matter whether as freehold or leasehold). This territory (in the sociological sense) is protected and defended against disturbances and intruders with a degree of intensity that decreases with increasing area size as a function of the individual's power. The area occupied by a person (or also by a collective such as a family, an enterprise, a community) may be conceived of (according to HALL, 1966) as being divided into zones:
Around the center, which is characterized by a maximum need for intimacy and privacy, we find ring-shaped zones with a radially decreasing need for access control. The interest in access control and the readiness for defence vanish at a certain distance, the socalled orbit, where we find a growing interest for spontaneous encounter and communication and a growing interest for uncontrolled experience of the public-at-large ("urbanity"). In a "territory occupied" by a person HALL (1966) distinguishes between reach and access control:

- a core area of socalled "intimate distance."
  In this area the individual seeks love, consolation and protection through physical contact. He has developed powerful defence mechanisms against unwanted physical contact.

- a first ring-shaped zone of socalled "personal distance".
  Here the individual entertains relations with persons "close to him", yet without intimate contacts. Examples include talks between friends in an apartment or between colleagues at work. If these persons intrude into the

"arm's length distance", this is felt as disturbing and will be sanctioned accordingly.

- the second ring-shaped zone of socalled "social distance".
  This area is normally used for non-personal business (shopping, etc.)

- a third zone of socalled "public distance".
  Here we find a complete lack of commitment, escape and defense are possible.

5. A Model of Regional Evolution

Communication basically requires the existence of adequate traffic routes and public (market) places. Once the initial construction efforts have been deployed, the communication advantage for the individual user will grow with improving quality, which then in turn causes the number of users to rise. Property - which initially referred almost exclusively to land ("territory") - must at least be delineated at its borders (if others are to respect it); accordingly, a certain degree of security of territorial property requires the fortification of borders (by walls etc.).

In its imagined pristine state, the land was neither endowed with constructed roads, which would favor the exchange of goods, nor with border fortifications to secure the property owned by the individual or by collectives. There were many natural obstacles in the way of people meeting one another, and also, the individual's territory was open to intrusion from outside and had to be constantly guarded. Hence, man's shaping of the environment evidently aims at sparing himself recurring transportation efforts by the one-time construction of a road system and saving himself the constant troubles of protecting his property through the one-time erection of obstacles. Out of this motivation, additional and new installations are constantly set up in order (i) to improve transportation between different locations and (ii) to improve the protection of locational properties. In physical terms, such construction activities are designed to rationally reduce the link-specific (frictional) resistance for certain goods flows and to rationally heighten borderline obstacles against intruders and to expand the land-related potential (also frictional resistance) for absorbing diffuse pollutants and disturbances. While performing the same amount of transportation work it becomes possible to travel greater distances at a lower link-specific (frictional) resistance and/or to achieve higher transport speeds and shorter traveling times. Also, it becomes possible to neutralize polluting sources within a close range by increasing the land-related absorption potential.

For practical purposes new roads are built along the routes connecting various locations, whose users exchange goods and factors. The need to invest in the construction of new roads is generally justified with the relatively vast communication potentials existing in the locations being connected. Normally, one bottleneck will affect the performance of the whole connection. Rational investments for border fortifications are made between locations, whose users are likely to disturb one another in the enjoyment of their property. These investments are generally justified by invoking the existing potential for conflict (differentials in property, emission of pollutants, etc). Pollutants (just as intruders) propagate from the source of pollution by their own energy, as if seeking the place of maximum permeability. Seen from an economic perspective, investments in connecting links (new routes, such as roads, railroads, canals or improvements of ducts) which require a certain constructional effort (leveling, smoothing, overbridging, underbridging etc.) primarily spare the users of new or improved links (understood as part of a communication, supply and waste disposal or transport system) efforts they have to deploy for transportation. Investments in border fortifications to improve protection of locational property (fences, walls and other protective facilities) primarily reduce the risk of actual damage being caused and relieve the owner from permanently having to control and guard his property. For the owner, who is relatively unprotected, particularly these controlling and guarding efforts are highly time-consuming.

It is not wrong to assume that the development of territorial authorities went hand in hand with that of regional patterns. Our analytical interest mainly focuses on those facilities and installations, being collective goods by their nature, which open up and at the same time delimit individual options (freedoms and opportunities). Infrastructure and territorial structures thus become major factors in opening up and safeguarding individual options, though they may also be looked upon as primary instruments of rulership used by landlords; ruling instruments used by sovereigns to allocate options to their subjects. Here, infrastructure is defined as the aggregate of inter-locational linking systems which create procurement, production, selling and reproduction opportunities for the individual. The territorial structure, however, is defined as the aggregate of inter-locational borders which govern the individual's territorial disposal and property rights.

Reasoning in terms of "infrastructure" and "territorial structure", the cultivation of the land in recent settlement history was determined by agriculture. It may be traced back to two points of departure (BÖKEMANN, 1982):

- the reclamation of land by rural "colonists" or

- irrigation technology as a precondition of soil fertility, especially in dry areas, used by rural "cultivators".

Starting our line of reasoning with land reclamation, we assume as a precondition that the natural soil is already endowed with a minimum of fertility for the cultivation of crops. After having cleared the forest, the colonizing farmer will first remove boulders and stones from the plot of land he has occupied before he can start grading it. Removing materials hindering exploitation may thus be interpreted as an original form of land reclamation. "Removing" at the same time implies: taking away obstacles from the exploitable land area and depositing and stacking them at its edges. In order to rationalize transportation, these "edges" will generally be laid out in a centrifugal arrangement from the focus of the intended use, for instance a warehouse or tool shed, the farm building etc. Piling up obstructing materials at the edges of the useful plot "marks" the borderlines of an original territorial property held by an individual. In other words: the concrete activity (of "removing obstacles") in land reclamation and cultivation corresponds to defining and delimiting it as property.

There is an analogy if we start our argumentation on the cultivation of land with irrigation technology. As a precondition we must assume, however, that the land first had to be rendered arable by artificial irrigation, as is common in arid landstretches. The waste material from building irrigation drains was then used for grading and loosening the arable land. Arable land equals irrigated land, the reach of the linking irrigation system within the farmer's land property defines the border of his territory; his territorial property remains non-defined outside the irrigated area. Different territorial properties will come into contact, whenever the individuals use a common (collective) irrigation system: in such a case, the collective linking system - which may be attributed a land-reclaiming function - will at the same time separate the individual plots as different territorial properties.

Whereas under the theory of land reclamation obstacles (boulders, roots etc.) are first simply piled up to delimit the territorial borders, these fortified edges fulfil an additional function as dams in the theory of irrigation. Especially in arid areas land plots must be flooded before they can be exploited. In addition to their delimiting function, dams are designed to maintain a scarce commodity, i.e. water, (land as such is often not considered a scarce commodity) on the cultivable (and thus valuable) land. Hence, the concept of "land cultivation" stands for an activity, by which land is leveled, hemmed in by dams and taken possession of at the inside of these dams. In all this, the amount of water available via a linking aquifer will limit the yield produced by the imaginary primordial farmer more than the size of the plot under cultivation.

The fact that land-reclaiming (collective) linking systems tend to define property borders is added by an analogous finding: property borders marked by obstacles may at the same time function as primary (collective) land-reclaiming links or route systems. Under this aspect, we will continue our line of reasoning according to which land cultivation starts with "reclamation".

Reclamation implies that a plot of inherently fertile land is made arable by removing obstacles, whereas an other part of the land, compacted by debris, becomes unfit for agricultural exploitation. The impossibility of using the borderline for agricultural purposes either technologically (due to soil compaction) or in an economically rational manner (due to its marginality) will induce the original farmer, conceived as a colonist, to keep pedestrian, equestrian and vehicle traffic off his land and - amongst other reasons in anticipation of field damage - to direct it, if possible, along the borderline. Pursuing this theory, reinforced ways and roads developed on the fortified borderlines around cultivated plots of land, since the unbroken fortification of ways and roads (which reduces kinetic resistance) is an essential precondition for any kind of vehicle traffic, not least with a view to climatic and seasonal independence. Hence, the economic efforts deployed for protecting properties and creating ways, alongside the technical function, were - at least originally - united in one activity.

We may assume that the roads, having such been built between the territorial properties of individual private owners (located in a no-man's land, as it were), were subsequently taken into public care by landlords or territorial authorities without resistance, since the borderline areas lose their value as cultivable areas and traffic routes, if repair and further development of ways even beyond the individual property are not guaranteed. It may also be assumed, however, that during this phase of development the concerned individual recognized (maintenance and repair being guaranteed by a landlord or an ("original") territorial authority) the economic advantages arising from the dispossession of the private borderline, if it went hand in hand with a transformation into a public traffic route. In the process of regional evolution, the development of areas for public traffic owned by territorial authorities, especially by municipal communities, can be explained as the appropriation of no-man's land; because eventually, the borderline areas between private land plots may be interpreted as such. We can thus distinguish between two basically different phases in land reclamation:

- in the first phase, land is reclaimed by farmers and - as has been outlined - cultivated on a private basis;
- in the second phase, after the appropriation of no-man's land remaining between private properties through landlords, the now arable land is reclaimed and developed politically by subjecting the conditions of private interchange and property structures prevailing at the edges of the utilized land to sovereign rules.

Embodying the first quasi-governmental function of "regulating the conditions of interchange and property structures between individuals", the landlords took hold of an ecological niche within the regional pattern. By reducing the costs of interchange through investment in traffic routes and by reducing the costs for transactions via borderline guarantees, quasi-governmental functions create the basis for the subsequent functional and spatial division of labor. It is in particular the selling of sovereign privileges to private users, which was to prosper in that niche. The interplay of regional development and the continued formation of hierarchically structured political institutions is characterized by the ever-growing rationalization of production at public sites and by the voluntary political allocation of this product to parts of the population or the economy, selected more or less at will.

Looking at regional development in macroeconomic terms as an evolutionary process, we find that policy-makers and political bodies fulfil a catalytic function; they act as catalysts controlling - accelerating or also slowing down - interaction between private subjects.

The presented model of regional development roughly outlines rules according to which individual communication and security needs are coordinated and physically structured. Approximate as they may be, these rules reveal - at different degrees - settlement patterns which are similar in character: the spatial division of an individual building into zones of intimacy, private and quasi-public areas and into traffic, supply and waste disposal zones is thus comparable to the organization of intra-urban infrastructure and to building regulations and zoning laws; analogous structures are found in the hierarchical set-up of territorial authorities, in regional infrastructure and in central place systems.

6. Looking beyond the Frontiers of Regional Science

Regional development has been described as a model which is partly empirical and partly based on simple assumptions. The question now arises how the very rules which govern the physical or institutional shaping of individual communication and security needs became established. At the socio-philosophical level, it was mainly von HAYEK (1971) and BUCHANAN (1971) who found a general answer to that question. According to BUCHANAN, rules (such as those for defining and respecting private land property and for delineating, recognizing and using public traffic routes) may be characterized as tools, which reduce frequently recurring conflictprone uncertainties of individual behavior (according to the decisional theory of the "prisoner's dilemma"), while at the same time

(i) coordinating parallel individual interests (such as communication and security needs in the settlement area) and

(ii) reducing inequalities in the benefits the individual derives from collective decisions, which may arise as infrastructure and territorial structures develop (preventing in particular socalled "free-rider" situations).

Rules governing individual behavior may result from

(i) certain behavioral patterns which have been successfully tested at multiple instances in the evolutionary process of trial and error (von HAYEK, 1971) or

(ii) an agreement based on economic rationality (BUCHANAN, 1971; RAWLS, 1975).

It is assumed that the rules of individual behavior give rise to the emergence of social structures (and thus also settlements). On the other hand, the argument goes, consolidated social structures (and thus also existing regional patterns) give rise to the emergence of conflict-prone uncertainties in individual behavior; uncertainties, which in turn provoke the emergence or establishment of new rules.

AXELROD (1987) has convincingly proved that rules for the cooperation of individuals develop the very moment strangers try to build up trust, while signaling at the same time that they do not want to be exploited. For the emergence of behavioral rules it is therefore neither necessary for the individual to pursue his interests selfishly, nor to sacrifice himself altruistically to further the "public weal". A simulation model developed by AXELROD shows that sufficiently stable rules develop as soon as the individual starts consistently emulating an attitude of give and take ("tit for tat", as proposed by RAPAPORT).

The theory explaining the development of these rules is similarly applicable to the emergence of those technical standards, instructions and plans, according to which equipment, buildings and regional structures, and especially the infrastructural linking systems and territorial border systems, are created. The standards and plans required for producing any kind of goods may be seen as the product of a separate cultural evolution, of a historical process which, in essence, is subjected to similar criteria of selection and mutation as biological evolution. Because every technical innovation must be compatible

(i) with the interests and values upheld by direct users and those indirectly concerned at the social level and
(ii) with existing material systems, into which they are to be fitted functionally or materially as a novel piece of equipment or tool, as a new component or a new building.

Seen from this perspective, new technical standards invariably develop on the basis of existing operational systems (as specializations or "bridging links" between still incompatible systems).

Concluding the presented reflections one may state that men primarily use infrastructural linking systems and territorial border systems in cooperative action to create their settlements; the primary aim being (according to BUCHANAN) to reduce or avoid uncertainties in individual behavior ("prisoner's dilemma" situations), and the secondary aim being to improve their overall genetic fitness and to save reproduction costs for the individual genetic program.

LITERATURE

Albert, H., 1968: Traktat über kritische Vernunft, Tübingen.

Axelrod, R., 1987: Die Evolution der Kooperation, München, Wien.

Barker, R.G., 1968: Ecological Psychology, Stanford.

Bökemann, D., 1982: Theorie der Raumplanung, München, Wien.

Buchanan, J., 1971: The Limits of Liberty, Chicago.

Dawkins, R., 1978: Das egoistische Gen, Heidelberg.

Elias, N., 1969: Über den Prozeß der Zivilisation, 2.Band, Bern, (2. Auflage)

Hall, E.T., 1966: The Hidden Dimension, Garden City/New York.

Hamilton, W.D., 1964: The Genetical Theory of Social Behaviour, in: Journal of Theoretical Biology, 7.

von Hayek, F.A., 1971: Die Verfassung der Freiheit, Tübingen.

Immelmann, K.; Scherer, K.; Vogel, C.; Schmook, P., 1988: Psychobiologie, Stuttgart.

Linde, H., 1972: Sachdominanz in Sozialstrukturen, Tübingen.

McKenzie, R.B.; G. Tullock, 1984: Homo Oeconomicus, Campus, Frankfurt/ New York.

Molt, W., 1976: Raum- und Sozialverhalten, in: ATTESLANDER, P. (Hrsg.), Soziologie und Raumplanung, Berlin.

Rawls, J., 1975: Eine Theorie der Gerechtigkeit, Frankfurt.

Rogers, A., 1983: Sources of Urban Population Growth and Urbanization 1950-2000: A Demographic Accounting, in: Economic Development and Cultural Change.

Schmid, J. (Hrsg.), 1985:Bevölkerungswissenschaft, Die "Bevölkerungslehre" von Gerhard Mackenroth - 30 Jahre danach, Frankfurt, Campus.

Schüller, A., (Hrsg.), 1983: Property Rights und ökonomische Theorie, Vahlen, München.

Schumpeter, J., 1926: Theorie der wirtschaftlichen Entwicklung, Berlin.

Vogel, Ch., 1986: Populationsdichte - Regulation und individuelle Reproduktionsstrategien in evolutionsbiologischer Sicht, In: KRAUS, O., (Hrsg.), Regulation, Manipulation und Explosion der Bevölkerungsdichte, Göttingen.

Voland, E., 1986: Die Evolution der reproduktiven Selbstbeschränkung, In: KRAUS, O. (Hrsg.), Regulation, Manipulation und Explosion der Bevölkerungsdichte, Göttingen.

Weismann; v. Wickler; Seibt, 1977: Das Prinzip Eigennutz, Hamburg.

Wilson, E.O., 1981: Sociobiology: A Synthesis, Cambridge (Mass.).

Zimmer, D.E., 1983: Die Macht der Gene, Die Koevolution von Natur und Kultur muß sorgfältig entziffert werden; In: Die Zeit, Nr. 43.

Zipf, G.K., 1949: Human Behavior and the Principle of Least Effort, Addision Wisley Press, Reading.

Barriers to Technological Transfer: Infrastructural Difficulties in Nigeria

by

Nicholas W. Balabkins

1. Introduction

Dear ROLF: In your first major work, *Verkehr und volkswirtschaftliche Gesamtrechnung* of 1961, dealing with the transportation component of the material infrastructure, you declared clearly and precisely your commitment to economic theory as an indispensable part of economic inquiry or what Joseph A. SCHUMPETER called "scientific economic analysis". In that volume, you were a veritable captive of LEONTIEF's Input-Output analysis. $[I-A]^{-1} \cdot Y = X$ seemed like a lodestar to your future as a German academic economist.

When you entered academic life, the work of American economists radiated a powerful, almost magnetic attraction to young German economists. After having lost touch with developments in the field of economic theory in the U.S. and the United Kingdom during the 1930s and particularly during World War II, German economists wanted to catch up and join the prevailing Mainstream Economics. They found a profession caught in the sweep of a mathematical revolution, a revolution calling for new analytical techniques, new approaches, and new attitudes towards the subject. To become a competent academic economist almost meant becoming a mathematician. In America, many economists trained first as mathematicians and then switched fields in the graduate school. These changes required young German economists to shed the non-mathematical image of German economics, and follow their American cousins. In addition to mathematics, one had to become fluent in the English language. A few years residence in America became virtually a necessity. ROLF, you remember these *Zeitgeist* phenomena all too well.

Economic theory usually manifests itself in the form of a mathematical model, a hypothesis or an "As-If" statement. The math model of a purely competitive firm, for instance, is usually cast in terms of empirical equations, definitional equations, and necessary inequalities. You solve for X, and once you know your output, you can easily get your total revenue, cost, and profit or loss. The world is simple, precise, and elegant.

Alas, every economist knows the great difficulties one faces in doing theoretical work. For one,. there is the *recognition problem*, which means that an economist must accept the empirical nature of economic science and make statements about reality. Next, the *transformation problem* refers to the economist's attempts to transform subjectively observed reality into a set of scientific statements. Finally, the *representation problem* describes the economist's attempts to cast these scientific statements in the form of a theory (see DOPFER, 1988, pp. 553).

Because economic theorists forever simplify the almost infinite diversity and complexity of real life, logical rigor and conceptual precision in their models may yield neat and reasonably analytical substitutes for a not-so-neat real world. The trick is to remember the difference between economic theory and reality. Very often an economist who knows the score, so to speak, is not quite able to attain the highest level of theoretical sophistication. Thorough knowledge of prevailing reality and possession of the necessary mathematical tools of analysis would represent an ideal, of course.

The great economists of the past were also great practitioners and successful businessmen. John Maynard KEYNES comes to mind. Alas, as reported by the *London Economist* in 1987, only 1 of every 20 academic economists is interested in practical problems of the day. The other 19 prefer to remain captives of the logical order of classroom rigor, and sweep institutional, psychological, and technological problems under the rug. They yearn so much for the approval of the members of the economist's academic guild, with its commitment to *a priori* reasoning, that they ignore everything that does not lend itself to easy quantification.

In the field of infrastructure studies, however, *a priori* reasoning does not take you very far. Rolf, you have been interested in the above dichotomy for a long time. You are fully aware of the relative character of economic truth. You also know that life is messy and that the problems of infrastructure, your lifelong intellectual hobby, are too chaotic to be squeezed into a neat mathematical model.

This "messy" essay was written after the completion of my book, *Indigenization and Economic Development: The Nigerian Experiences*, published in 1982. As you know, I made six field trips to Africa. In 1988, I published an essay "Factors which have Influenced Nigeria's Industrial Development", in *Rivista Internationale.* Even though the material is a bit dated, the situation is worse today than it was before. Please accept this essay as a token of our long friendship. In my native Latvian, I wish you "Daudz baltu dieninu" in years to come, ROLF!

This essay deals with the barriers to technological transfer to Nigeria. Economists who favor such transfer assume that infrastructural deficiencies will not block such a transfer. This is a story how it actually works.

2. What is Technology Transfer All About?

Technology transfer is a huge topic. For the purposes of this paper, technology transfer will refer to proprietary knowledge conveyed through commercial transactions from the USA to Nigeria (see CHUDSON, 1982, § 37 p. 3). That is, Nigerian businesses, private or stateowned, seek certain technology and are willing to pay for it. They may want either production or engineering technology–hardware; or they may want financial, marketing and management skills–software.

For underdeveloped or less developed countries, technology transfer has been viewed as a potent catalyst in the modernization process. In operational terms, modernization means teaching tradition-bound people reationality; how to increase the productivity of their farms and factories; how to raise the living standards of their masses; how to bring about national consolidation.[1] Scores of essays, articles, books and technical reports have been written on technology transfer and modernization since the end of World War II, when Gret Britain, France, Spain, the Netherlands, and Belgium decolonized their overseas empires and a whirlwind of independence swept through the Third World.

The early optimism that decolonization would lead to successful technology transfer was not warranted (SPENCER AND WORONIAK, 1968). The prevailing social rigidities in the underdeveloped countries made the technology transfer impossible. In the classroom, the technology transfer idea is simple enough, but achieving it is extremely difficult.[2] Similarly, the optimistic Western decolonizers of the 1960s and 1970s were hoping unrealistically to transfer and firmly install Western-style democracies in the "new" countries. An American style democracy was installed with much fanfare in the fall of 1979 in Nigeria only to be eliminated by the military at the end of 1982.[3]

Technology transfer, as posited by Columbia's Professor Nathaniel H. LEFF, has three objectives. The first is to provide indigenous businesses with the requisite manufacturing know-how to reduce production costs and raise productivity (1980, p. 18). The second goal is to enable LDCs to possess their own advanced technology, rather than lease someone else's. LDCs are particularly anxious to reduce the duration of the contract periods during which they pay lincense fees in hard currency. These countries also object to restrictions imposed on the use of advanced technology by multinational corporations.

Professor LEFF's *Goal III*, looking at the problem from the recipient's end, refers to the desire of the less developed countries to build their own indigenous technology, so as to reduce or eliminate the "emotional dependency syndrome". Of course, intellectuals from the Third World countries are rawnerved on this issue and use pejorative language to describe the process of technology transfer from, let us say, the USA to Nigeria as "exploitation".[4] Most African intellectuals consider technology transfer as "unequal". They have accused the Western oil companies "of deliberately avoiding and resisting the transfer of technical know-how to Nigeria" (TURNER, 1980, p. 201).

3. Stages in the Technology Transfer Process

Technology transfer is not a single act but a process. One source described it as a series of steps: i. awareness, ii. interest, iii. evaluation, iv. trial, and v. adoption (PYTLIK, LAUDA and JOHNSON, 1978, pp. 112). Let us construct an example from the currently highly fashionable CAD/CAM technology. The awareness stage refers to the first inklings in the mind of a Nigerian manufacturer. The manufacturer knows that the application of CAD/CAM technology has led to significant increase in productivity elsewhere, but he lacks precise information. Most likely, he has been informed by his son in the USA, let's make him an industrial engineering student, of the wonders of this new technology. In the second stage the manufacturer obtains information on CAD/CAM equipment from field representatives of the XYZ Corporation. Trade publications, government reports, and the mass media can also offer information. The interested manufacturer has to *evaluate* the CAD/CAM equipment in the third stage. This is the YES or NO stage. He will ask such questions as, "By how much did the CAD/CAM technology increase output per man in the US plant over a year or two? Was it the most advanced plant in the industry? Or, was it the least advanced? An average facility? What was the average increase in productivity over a given period after the introduction of the CAD/CAM equipment?

The answers to these questions may come from salesmen visiting Nigeria or, preferably, from personal inspection of US facilities using the particular CAD/CAM equipment. A pilot study or a trial run simulating Nigerian factory conditions could also be useful in this stage. If the evaluation of the CAD/CAM equipment turns out to be positive, our Nigerian manufacturer will probably decide to acquire this equipment for his plant. Discounting all other problems, including that of foreign exchange, the process of technology transfer has now entered the final stage, namely, adoption.

4. Infrastructural Barriers to Technology Diffusion in Nigeria

A decade ago I was writing a book on the indigenzation process in Nigeria during the 1970s. Because insufficient authentic information was available in the United States, I made six extended trips to Africa where I lived like a Nigerian and was able to see beneath the surface. I lived with friends and acquaintances, mostly parents of my Ph. D. candidates at Lehigh University. I experienced the vexing problems of everyday Nigerian life, such as the regular electricity blackouts and brownouts, monumental traffic gridlock in Lagos, overflowing and stinking street sewers, poor to nonexistent telephone service, blatant official corruption, and food shortages. I met with government officials, and conducted interviews with Nigerian businessmen and bankers and, above all, with the "little people" or "forgotten people", who taught me a great deal about contemporary Nigeria. During my last visit in 1978/79, I also saw what the barriers to technology transfer were at the receiving end.

Professor Mark PERLMAN, of the University of Pittsburgh, has persuasively argued (1982, pp. 9) that technological change has often been presumed to be not only a new method but also the development of a new material product or service, intermediate or final. Such a perception confuses a necessary with a sufficient condition. Professor PERLMAN (1982, pp. 10 and 19) stressed the significance of the necessary general conditions that must precede the success of innovations. One of his preconditions was infrastructure. The study of infrastructure in developing countries is not exactly an exciting field for young Ph. D. students or academics. Such studies are considered "too sociological"; they do not command prestige among academics, and for that reason they are sorely neglected, omitted altogether, or relegated to the dustheap of the *ceteris paribus* clause. As one economist with considerable research experience in Nigeria noted, economist know little about infrastructure and its effects on the economy (SCHATZ, 1977, pp. 114). What then, is infrastructure?

5. The Concept of Material, Human and Institutional Infrastructure

If one takes the view that *material infrastructure* in the form of roads, canals, and railroads must be built before agricultural and industrial production can begin, then one would expect that investment in material infrastructure should lead to a rapid growth of production. In other words, material infrastructure can be a source of favourable or negative external economies. If the roads are in good repair, they generate favourable external economies, and industrial firms and farmers enjoy lower costs of production; if not, both industrial firms and farmers will have higher transportation costs.

The second component of infrastructure or social overhead capital, the *human infrastructure*, generally refers to the population of working age, its education, adaptability to change, nutrition, health, mortality, morbidity, and fertility. Human infrastructure is influenced by the economic, institutional and technical characteristics of the material infrastructure. Its role in the economic growth process raises the same host of questions as does the material infrastructure. Primary, secondary, and vocational education, the presence or absence of competent and motivated teachers and students alike, universities, and ongoing research are also components of human infrastructure and are indispensable for the sustained evolution of a country.[5] The nation's health includes such basic matters as the incidence of disease and of malnutrition (KAMARCK, 1976, pp. 57–80).

Again, although high or low levels of educational achievement and good or bad health do not produce anything by themselve, a literate, healthy, and well-fed population constitutes the development potential of a country. This part of infrastructure can also be called the "brain power" (the human capital) of a country. Without it, wealth can be neither created nor re-created; without it, industry, trade and agriculture are negatively affected (SINGER, 1964, p. 66). A country without sufficient brain power is reflected in low output *per man*, *per machine*, or *per ox*.

The third component, *institutional infrastructure*, consists of the country's legal framework (property rights, contracts, tax and corporate laws, and inheritance laws, among others), its law enforcement (or lack of it through a permissive corrupt judiciary), and its social phenomena-habits, customs, mores, traditions, the presence or absence of a caste system, social classes and social distance among classes, land tenure, local government, degree of corruption, the presence or absence of a "work ethic", the existence of extended families, down to such details as the tendency of villagers to fear all strangers (BREWSTER, 1967, p. 70 and pp. 79-80).

To contemporary economic theorists the concept of institutional infrastructure is too vague, too "sociological", too difficult to quantifiy, and thus not very meaningful. Yet, elusive as it may be, institutional infrastructure represents the essence of a nation's culture. It must be analyzed if economists want to understand the backwardness of Africa and Asia and propose viable solutions and reforms. But, specifically, what were the problems I noted in Nigeria? Let me cite some examples.

6. Characteristics of Nigeria's Infrastructure

Roads: Nigeria's infrastructure is, to put it kindly, inadequate. In the late 1970s, traffic congestion was monumental. Every Lagos resident knew the meaning of "Lagos driving". It took six hours to fly from London to Lagos and seven hours to get from the airport to downtown Lagos, according to one popular joke that was more than a little credible. Traffic jams and slowdowns made it almost impossible to drive to work on time. Billions of man-hours were lost each year. For a couple of weeks in 1978, I commuted by bus from Surulere to downtown Lagos. On the average, it took more than two hours to negotiate the six kilometers. People began lining up at the bus station at 6:00 a.m. When the bus arrived, the driver pulled close to an open sewer that was usually filled to the brim; it often had no lid. I saw people fall into the sewer as they fought their way onto the bus. For those who made it, the ride downtown cost ten kobo or 16 U.S. cents.

Lagos is usually hot and humid, and its buses are steamy and crowded. After a two-hour ride, a wash-up and some air-conditioning would help workers put in a productive day. But many offices and factories downtown had no electricity or running water. To compound their distress, most workers have no electricity or running water at home either. Thus they can not bathe or shave or have a cup of tea before setting off to work.

In the 1960s and 1970s, road building became a national obsession. Engineers planned a vast system of superhighways crisscrossing the country. Ninety thousand kilometers of hardtop highways were built (HARMAN, 1982, p. 1). But despite the massive construction of roads, extensions and access points, Nigeria's road network leaves much to be desired in terms of speed, safety and convenience.

To build roads is one thing, but to keep them in good shape is something else again. Road maintenance has been a veritable nightmare in Nigeria. In 1974, the World Bank Country Economic Report evaluated the conditions of the roads leading to rural communities. Federal roads varied from good to extremely poor, with surfaces generally poor and inadequately maintained. In 1978, another report detailed with dismay the rapid dilapidation of the newly built highways. Many of these new roads lacked adequate drainage systems and were becoming death traps in rainstorms. Indiscriminate dumping of garbage into drains and channels were causing constant flooding (ODEDE, 1978, p. 7).

Inadequately maintained roads literally destroy vehicles. The best trucks last only a year or two. But the number of heavy road vehicles has grown rapidly, largely as a result of another infrastructure problem - Nigeria's poor railway service. So autos and roads end up inflicting enormous wear and tear on each other. The new roads outside Kaduna and Zaria are good, but long stretches of other roads are marred by coffin-sized potholes.

Truck drivers face other problems. It is difficult or impossible to obtain spare parts for foreign-made trucks because of monumental tie-ups in the ports and erratic foreign exchange controls. It is impossible to operate vehicles at night, when robbers control the highways. And Nigerian drivers create accident hazards by abandoning broken-down vehicles smack in the middle of the road, without warning flares or lights. These factors raise truck driver's operating and maintenance costs and insurances rates, while decreasing utilization rates.

Another problem that plagues Nigeria's roads is the reckless speeds at which Nigerians drive. There is a saying that Nigerians are never in a hurry except when they are driving a car. Nigerians are much wilder than the all-out German *Autobahn* motorists or the Swiss on the *Autoroute*. They pass downhill, uphill and on curves. One reasons Nigerians speed is a genuine fear of being caught by armed robbers. But many drivers, it is said, have obtained driving licenses "even before they master how to insert the ignition key". From 1962 to 1972, traffic fatalities grew at a compound rate of 10 percent. The rate has remained high ever since.

The country's material infrastructure in the form of roads has generated negative external economies which have affected adversely Nigeria's industry and agriculture. Nigerians had ample petro-dollars in the 1970s. They also had the will to push for structural change in the country. But what they seems to have lacked was the competence to carry their fine ambitions through, as the London Economist has recently put it. In early 1983, the well-known Nigerian journalist, Gordon Tialobi, asked why Nigerians were asking their former colonial masters to run their airlines, ports, import-inspection system, railways and, recently, even the hotels. Was it because the Nigerians themselves could not do so efficiently (TIALOBI, 1983).

Energy: Electricity production and distribution facilities are another component of a country's material infrastructure. Reliable power production facilities can boost productivity in external economies. Such a public good usually is provided ahead of what economists call the final demand.

Generation of electric power in Nigeria dates from 1898, but expansion of the power net got into high gear only after the country gained independence (INTERNATIONAL BANK FOR RECONSTRUCTION AND DEVELOPMENT, 1955, pp. 56, 431). After the end of the Civil War (1967–1970), the former Electricity Corporation of Nigeria and the Niger Dams Authority were merged in 1972 to form the National Electric Power Authority (NEPA), which controls electricity supply (OLAYIDE, 1976, p. 87) Generating capacity grew rapidly in the 1970s, especially after the Kainji Dam completion (VISSER, 1970).

From 1960 to 1980, the average continuous growth rate in installed electricity generating capacity was 15.8 percent. However, this rate was quite inconsistent. From 1960 to 1970 the average continuous rate was 24 percent, whereas between 1970 and 1980 it fell to 7.7 percent. Given this slowdown, the projected capacity in 1990 is likely to be 3,758 megawatts. But will that be enough to maintain rapid economic growth? After all, generating capacity must be installed *ahead* of final demand. In the 1970s, when installed capacity grew at an annual rate of 7.7 percent, the oil boom caused electricity consumption rates to rise by more than 20 percent a year. Total installed capacity more than doubled during the decade, but available generating capacity still could not meet demand (UKPONG, 1975, p. 14). This divergence has been reflected in frequent load sheddings, electricity blackouts and brownouts (UKPONG, 1974, p. 5). The reasons most often cited were a dearth of executive talent in NEPA, grossly inefficient maintenance service because of an acute shortage of qualified help, ancient distribution circuits, and dependence on imported power equipment and supplies (OLAYIDE, 1976, pp. 89; UKPONG, 1975, p. 15). Damaged underground cables were also responsible for electricity blackouts.[6] Improvisation has become a way of life in coping with the malfunctioning infrastructure. As the London *Economist* put it recently, "The well-off have their own bore holes for water, their own standby generators for power, their own servants to run the endless errands." (HARMAN, 1982, p. 1).

As a strategic input in the development process, electricity supply must be regular and must increase with demand from industry and agriculture. The economic consequences of prolonged and continued electricity blackouts or brownouts were thus severe. Industrial output fell, production costs increased, supplies of consumer goods became irregular, unemployment rose, and national income declined. Electricity brownouts also destroyed sensitive electronic equipment, whose replacement required considerable outlays by industrial firms. The cause for the extra outlays originated outside the firm, but these exogenous factors raised operating costs nevertheless.

To cope with such uncertainty, Nigerian businessmen installed standby generating equipment which, again, raised their costs of operation for the simple reason that NEPA was unable to provide adequate energy. In other words, without an adequate, regular and inexpensive electricity supply, modernization and industrialization, even with plenty of petro-dollars, was very difficult.

7. Communications

Communications systems form an indispensable part of a country's material infrastructure. An extensive and well-functioning telephone network is not only a necessity for the business and administrative life of a country, but it also has a considerable cultural value and makes for closer contact between the country and the city. This part of the material infrastructure is a public good that generates external economies and thus affects the country's industry and agriculture. As a public good, it is also "lumpy" - indivisible - and must be provided "ahead of the final demand for goods and services", as economists always put it.

In the 1970s, Nigeria's telephone system, far from being a model of efficiency, was an enormous burden on commerce and industry. Telephones and telegraphs either were dead or erratic. In 1978, the president of the Nigerian Manufacturer's Association reported that in Lagos State alone some 1.500 telephones were out of order. Some manufacturer's telephones were out of use for more than five months; the manufacturers had installed telex machines but were unable to use them because the Post and Telecommunications Department lacked both the manpower and the spare parts to keep them operating (ODUTOLA, 1978, pp. 31).

In Nigeria, the post office and internal communications services are owned and operated by the federal government, whereas international communications are provided by Nigerian External Telecommunications Corporation, co-owned by the government and by a private company. Even before the onset of the oil boom, the demand for phone and cable services exceeded by far the existing capacity. For this reason, the Nigerian government invested heavily during the Third Plan Period (1975-1980) to revamp communication links.

Making a telephone call in Nigeria is a tricky business. During peak working hours it is next to impossible. In 1976, Nigeria adopted a communications system called TCOM, which is a balloon outfitted with electronic packages hanging above a mooring tower (THE EVENING SUN, 1976) Despite such improvements, telephone communications are slow and unreliable; thus it is necessary to employ an army of messengers using cars and motorcycles.

The inadequacy of Nigeria's communication systems in a milieu of rapidly growing Gross Domestic Product represents a serious drag on the country's industry, commerce and agriculture. Its communications systems, interpreted as a public good, generated negative external economies, which were very costly for the directly productive sector of the economy. These experiences suggested that, like roads and electricity-generating facilities, communications systems must be built ahead of the final demand. The functioning of the various components of the Nigerian material infrastructure in the 1970s was quite erratic. It imposed considerable financial costs on the country's commercial and industrial plants, and it angered millions of Nigerians. It gave overseas Nigerians fits of anger. One such person, after many years abroad, recorded his impressions as follows: You feel the shock as you get off the plane. You pick up a telephone and it doesn't work, you get on a bus and it doesn't move, you step on an elevator and the power goes off. All those little everyday things that you had taken for granted. You compare it with London and you feel a sense of shame that things should be this way. Shame and then rage.[7]

8. Summary

This paper focussed on the infrastructural difficulties which constitute considerable barriers to technological diffusion in contemporary Nigeria. Nigeria's infrastructure is "messy". One has to experience it to understand the frustration of everyday life and the frustrations for Westerners and Nigerians seeking to transfer technology into the country. The Westerners always overlook the inadequacy of the country's infrastructure.

FOOTNOTES

[1] For a discussion of modernization goals see Myrdal (1968, Vol. I, pp. 57–69).

[2] See: Baranson (1967). Another useful book on this problem is that by Goulet (1977).

[3] See: Winder (1984) and Blackborn (1984).

[4] See: Turner (1980, pp. 199).

[5] Myrdal (1968): Chapters 29-33 deal with such matters as "Investments in Man", health, education, literacy and adult education, pp. 1533-1828.

[6] See: Dark Days Ahead, NEPA Warns, Daily Times (1978, p. 1).

[7] See: Brain Drain Hurting Nigeria. And its New 'Self-Exiles', The New York Times (1976).

LITERATURE

Africa, Nr. 72, 1977: Growth of Radios, Television and Telephones in Africa, 1973, p. 101.

Balabkins, N., 1982: Indigenization and Economic Development: The Nigerian Experience, JAI Press, Greenwich, Connecticut.

Baranson, J., 1967: Technology for Underdeveloped Areas: An Annotated Bibliography, Pergamon Press, New York.

Blackborn, P., 1984: Nigeria Military Tries to do What Civilians Didn't, The Christian Science Monitor, January 6.

Brewster, J.M., 1967: Traditional Social Structure as Barriers to Change, in: Southworth, H.M.; Johnston, B.F., eds.: Agricultural Development and Economic Growth, Cornell University Press, Ithaca.

Chudson, W.A., 1982: Transfer of Proprietary Technology to Developing Countries, in: Walter, I.; Murray T., eds.: Handbook of International Business, John Wiley and Sons, New York.

Dopfer, K., 1988: How Historical is Schmoller's Economic Theory, in: Journal of Theoretical Economics (Zeitschrift für die gesamte Staatswissenschaft), vol. 144.

Funck, R., 1961: Verkehr und volkswirtschaftliche Gesamtrechnung, Vandenhoeck and Ruprecht, Göttingen.

Goulet, D., 1977: The Uncertain Promise: Value Conflicts in Technology transfer, New York, iDoc, North America.

Harman, N., 1982: The Most African Country, in: The Economist, January 23.

International Bank for Reconstruction and Development, 1955: The Economic Development of Nigeria, The Johns Hopkins Press, Baltimore, MD.

Karmarck, A.M., 1976: The Tropics and Economic Development, The Johns Hopkins University Press, Baltimore, MD.

Leff, N.H., 1980: Technology Transfer and the Developing Countries. Management in Nigeria, Vol. 16, 8.

Myrdal, G., 1968: Asian Drama. An Inquiry into the Poverty of Nations, New York, The Twentieth Century Fund.

Odede, J., 1978: Will Our Roads Ever Be Safe? in: Daily Times, December 10.

Odutola, T.A., 1978: The Sufferings of the Private Sector, in: Management in Nigeria, Vol. 14, No. 3, March/April, pp. 31–32.

Olayide, S.O., (ed.), 1976: Economic Survey of Nigeria (1960–1975). Aromolaran Publishing Company, Limited, Ibadan.

Perlman, M., 1982: Conflicting Aproaches to Technology Transfer, mimeo.

Pytlik, E.C.; Lauda, D.P.; Johnson, D.L., (eds.), 1978: The Transfer of Technology: An Overview in: Technology Change and Society, Davis Publications, Worchester, Massachusetts.

Schatz, S.P., 1977: Nigerian Capitalism, California University Press, Berkeley.

Singer, H.W., 1964: International Development: Growth and Change, McGraw-Hill, New York.

Spencer, D.L.; Woroniak, A., (eds.), 1968: The Transfer of Technology to Developing Countries, F.A. Praeger, p.V., New York.

Tialobi, G., 1983: To Sell or Not to Sell, in: The Punch, March 19.

The Evening Sun, November 3, 1976.

The New York Times, November 1, 1976.

Turner, T., 1980: Nigeria: Imperialism, Oil Technology and the Comprador State, in: Oil and Class Struggle, Nore, P., and Turner, T., (eds)., London ZED Press.

Ukpong, I.I., 1974: Economic Aid and the Development of Infrastructure in Nigeria, paper presented at the 17th Annual Meeting of the African Studies Association, Chicago, Illinois, mimeo.

Ukpong, I.I., 1975: Nigeria: Electrical Energy Development, in: Africa, No. 50.

Visser, S.A., (ed.), 1970: Kainji. A Nigerian Man-Made Lake, Nigerian Institute of Social and Economic Research, Vol. 1, Ibadan.

Winder, D., 1984: When the Going Gets Tough in Africa, the Tough Sometimes Stage a Coup. The Christian Monitor, January 4.

Telecommunications Infrastructure and Regional Development

by

Maria Giaoutzi

1. Introduction

The relationship between infrastructure and regional development is one of the most intriguing problems in regional policy for less prosperous areas. Since World War II the extensive literature on economic growth and development has emphasized the key role of "infrastructure" or "economic and social overhead capital" in national and regional development. There has been much discussion on the crucial role of infrastructure and the temporal order in which infrastructure and other forms of capital should be provided so as to stimulate economic development in less developed countries as well as in lagging regions of industrialized societies. The level and quality of services flowing from existing infrastructure is often low, in many cases falling below 'threshold' levels, so that their effects on economic output are negligible.

In the 60s and 70s research on the impact of telecommunication on economic development focused primarily on the implications of introducing voice telephone service in less developed countries and in rural and remote regions of developed countries. During this period, studies of innovation processes considered the implications of information technology in the manufacturing sector and less frequently in the service sector. In the 80s, though, the convergence of computer and telecommunications technologies supported by innovations in software and microelectronics has been reflected in virtually all research. The links between automated manufacturing processes, local area networking capability and public and private telecommunications infrastructures are now widely recognized. Much of the emphasis attached to telecommunications in a regional development perspective is based on the assumptions that:

(i) telecommunications will provide a uniform distribution of information over space; and

(ii) technologies and their applications will be adopted by significant shares of a growing information-dependent market.

The uniform validity of these arguments is, however, questionable. At present there is very little evidence supporting either the dispersion or the concentration hypothesis attached to telecommunications. In light of the above discussion, the focus of this paper will lie upon the implications of New Information Technologies (NIT) on Regional Development.

2. The Information Sector

Despite many uncertainties, it is clear that the information sector has grown in importance over the past decade. This importance was illustrated in an article appearing in BUSINESS WEEK (June 30, 1980, p. 102), which stated that:

> "The old industrial society that generated wealth in the form of capital goods and manufactured products is giving way to a new society valued in terms of intangible assets, such as knowledge and information processing. In fact, for 15 years more people have been working at processing information than any other type of job."

The increasingly important role of information will no doubt exert a profound impact on our future ways of living, working, travelling, recreation and maintaining social contacts. The scientific and commercial use and management of modern information systems (in terms of hardware, software and orgware) requires systematic and coherent collection, storage and transmission of knowledge (based on integrated network patterns) and will here be called *informatics*. The informatics sector is clearly one of the most rapidly growing sectors in the industrialized world.

Informatics includes both information technology (the hardware part, based on microelectronics) and the use and management of advanced information systems (the software and orgware part, based on data banks, modelling approaches, geocoding systems, decision support systems, etc.). Information systems are, in a norrower sense, often related to both the software part (design and use) of the informatics sector and the organizational-institutional part of managing information systems (e.g. multi-level and multimodular planning information systems, VAN EST et al., 1985; MASSER and WILSON, 1984; NIJKAMP, 1985; NIJKAMP and RIETVELD, 1984).

According to SOETE (1986), the information technology field includes the following sectors:

Sector	Description	Product example
1. Industrial Automation	Equipment used in specific factory and design automation applications	* Numerical controllers * Process control * Robotics
2. Office Automation	Equipment used in office automation	* Word processors * Executive Workstations
3. Telecommunications	Voice and data communication equipment currently handled by PTTs	* Central Office switching * Modems
4. Defense, aerospace and communication	Equipment used in defense, aerospace and communication	* Aviation * Guidance systems * Radar
5. Other industrial	Miscellaneous equipment sold to industrial users and equipment incorporated in other products	* Medical electronics * Automative electronics
6. Consumer	Consumer electronics	* Video equipment * Home information systems
7. Computer	General purpose processors and peripherals and complete dataprocessing systems	* Mainframe computers * Minicomputers * Terminals
8. Components	Components used in the manufacture of products in all other sectors	* Integrated circuits * Discrete semiconductors * Lasers
9. Software and services	Software and computer services sold independently of hardware	* Remote computer services * Package software

In recent years, the results of telecommunications have been spectacular due to the emergence of video transmission, telefax, teleshopping, etc. Informatics and communication have become key elements in industrial organization and location. A new "paperless" or "wired" society (MARTIN, 1978) is likely to emanate, so that the printing technology which has held a superior position for many centuries may lose ground. Talking and communication machines may be a next step, although it must be realized that - despite the technological potential - the human mind has in general only a limited capacity for adopting revolutionary innovation (TURKLE, 1985). In this framework, NAISHBITT argues that 'high tech' development evokes a countervailing 'high touch' development, as the alienation caused by technology forces people to be compensated for by increasing personal and social face-to-face contact.

Such new technologies, especially in the field of the information and telecommunications sector, may have a farreaching impact on the locational pattern of firms and hence on regional development. Modern technology also offers a large growth potential for traditional rural areas, provided that these areas are connected to an information and knowledge network. Thus, the adoptive capability of a region for this new technology will be decisive for its future economic position. Clearly, New Information Technlogy may also exert an impact on the mobility behaviour of people (e.g. teleshopping, teleconferencing). The use of modern telecommunications facilities technically allows for the substitution of physical commuting by telecommuting, so that more jobs may cease to require physical travel. However, the degree to which this will take place is still very much subject to speculation.

3. Telecommunications and Spatial Organization

Before embarking upon the likely impacts on future spatial organization, we shall start by sharing the widely accepted view that telecommunications is a friction-reducing technology (NIJKAMP et al., 1987), so that it may induce geographical dispersion of economic activities. It is, however, questionable whether this simple hypothesis may remain valid as a general rule, since the likely impact of telecommunications on spatial developments and more specifically on spatial deconcentration is a rather complex issue (GODDARD, 1980;

GIAOUTZI, 1988; GIAOUTZI et al., 1989).

A useful tool for identifying such impacts can be the systematic distinction among the three phases of implementing such new technologies (see STONEMAN, 1983):

- design and generation,
- diffusion and adoption,
- effects.

The first phase refers mainly to the invention of a new product, the production process, production systems and institutional systems, and eventually leads to the emergence of a new technological regime (NELSON and WINTER, 1982). Those places likely to generate telecommunications technologies, as in the case of the high tech sector as a whole, tend largely to be areas possessing a highly skilled labour force, advanced R&D centers or universities which provide the necessary conditions for such new telecommunications technologies to emerge (MALECKI, 1982; NIJKAMP, 1985; GIAOUTZI, 1987). The existence of such requirements initially provides certain areas with a relative comparative advantage over others competing in the same spatial context.

The second phase incorporates a wide range of diffusion paths with diversified temporal and geographical dimensions. The above paths are strongly conditioned by a number of factors, such as the initial conditions existing in the diffusion model, the parameters involved, the feedback structure of the evolutionary model and resistance factors on the part of potential adopters (HÄGERSTRAND, 1967). The conventional deterministic diffusion models of the previous decades are gradually losing their relevance, giving way to more behaviourally oriented models incorporating decision uncertainty, systemic resilience and environmental complexity (PRIGOGINE, 1976). Unlike conventional diffusion models, telecommunications is not mainly supply-driven, which means that the introduction of a specific technology in this area is very much determined by a number of constraints as well as by user needs and preferences. A useful distinction in the above set of 'barriers' to the diffusion and use of telecommunications may be made among: barriers caused by extended obstacles such as physical factors (e.g. mountains, lakes); cultural, political, institutional (e.g. language barriers) or socio-economic factors (e.g. welfare discrepancies); and internal obstacles resulting from a lack of capacity of the network concerned (e.g. congestion) (NIJKAMP, 1989).

Barriers essentially involve a certain degree of discontinuity in a network interaction pattern (or at least a rather sudden change in the cost of interaction). The existence of such barriers largely determines the spatial distribution of access to telecommunications, which in turn leads to a certain range of diversified sets of locational requirements and as such varying potential development profiles in space.

The third phase of new technology implementation refers to the likely effects of such technologies. These are usually multi-dimensional in nature and may relate to locational effects (deconcentration vs. concentration), employment effects (deskilling vs. skilling), environmental effects, land use effects and/or effects on production structure on varying geographical scales.

4. Telecommunications and Regional Development

The most highly emphasized property of telecommunications is its friction-reducing capability, which may induce geographical dispersion of economic activities. This has led to the belief that the following two assumptions can become the basis for promoting regional policies:

(i) that telecommunications will provide a uniform distribution of information overs space; and

(ii) that the technologies and their applications will be adopted by significant shares of a growing information-dependent market.

In this respect, this capacity appears as a challenging tool for policy-making aimed at reducing spatial inequalities. The uniform validity of these arguments is, however, questionable. Only one plausible generalization can be made in this respect: that firms and individuals may enjoy greater flexibility in their locational decisions, as a result of the major impact widely attributed to telecommunications technologies, viz. that they produce a different cost-distance distribution from that based exclusively on physical movement. While telecommunications rates are significantly less distance-sensitive than travel costs, they do exhibit greater sensitivity to the duration of communication (NIJKAMP et al., 1988).

There are numerous moderating variables affecting the spatial patterns of NIT-produced phenomena. NIJKAMP and SALOMON (1988) identify among others the following as most important:

- *the scale effect* (ranging from micro-scale through local, metropolitan and regional scales to the national and international scales), where a certain effect observed at one level (e.g. metropolitan scale) may not necessarily be repeated on other scales;
- *the prior geographical distribution effect:* the impact of telecommunications on the spatial distribution of activities is likely to depend on the distribution prevailing prior to the introduction of new technology);
- *information demand:* the demand for telecommunications will derive from the intensity of demand for information;
- *organizational structure:* spatial behaviour will vary in response to organizational structure, while the market composition will affect the extent of decentralization attributable to telecommunications;
- *logistics:* even highly info-based industries involved in production rather than R&D have a certain need for physical movement, which has a two-fold influence: in location decisions the costs of goods movement must also be considered; and total vehicle movement to remote locations increases as a result of the footloose nature of these firms, which can in itself create spatial impacts such as service centers of requirements for road improvements.

In the context of the above discussion, it must be mentioned that there is very little empirical evidence to support either the dispersion or the concentration hypothesis attributed to telecommunications, while on the other hand sound theoretical arguments exist concerning the impact of these technologies on equity grounds.

5. The Impact of NIT on the Spatial Development Process

The increasing importance attached to the role of NIT in the regional development context is mainly associated with its catalytic function for a set of structural transformations termed "Information Economy" (GODDARD, GILLESPIE, 1987). The paramount role played by NIT in light of the Information Economy perspective is mainly related to the fact that *knowledge-* and *information-oriented* activities are being considered more and more as the "strategic resources" upon which *competitiveness* and *efficiency* are based. These strategic resources have nowadays become the necessary requirements for the survival and flourishing of modern societies.

NIT may be characterized by its revolutionary impact on the regional development process, since the pervasiveness of its applications contributes to transforming many of the traditional features of present-day *production, exchange* and *distribution systems.* This transformational potential is considered the driving force allowing for the realization of a fundamental shift in the "locus of competitive advantage", thus leading to a changing geography of competition (GODDARD, GILLESPIE, 1987).

The economic significance of these technological developments to further the competitive strategies of regions is nowadays greatly appreciated. NIT developments are taking the form of *new products, distribution systems, production processes, transaction innovation, management and decisionmaking innovations, control innovations etc.*

The transformation of information into a *commodity form* and its distribution via modern telecommunication networks have considerably contributed to the diffusion of information, reducing the barrier of distance and thus facilitating market extension. Information availability is far less geographically constrained before.

In the context of *production processes*, NIT has significantly contributed towards the integration of these processes, which have been automated and subjected to computer control. Computer-integrated manufacturing (CIM) is nowadays a reality and has been considered by various sectors as a key element which will enhance their competitive position as to their ability to respond rapidly to changing market conditions and, most important, to exploit *economies of scope* - in contrast with *economies of scale* - offered by the new technological developments, which will additionally allow for the integration of all economic, organizational and technological functions within one organization, promoting in this sense the improvement of process predictability, timely delivery, quality and efficiency without losing flexibility and variety (NIJKAMP, 1987; GOLDHAR et al., 1985; GIAOUTZI, 1987).

One of the most important applications of new technological developments is the *just-in time* (JIT) production and distribution systems enabling a number of companies, which may be geographically dispersed, to establish a complex inter-organizational production and distribution system, thus creating chains of competitive advantage with considerable competitive strenght, insofar as the aggregation of the competitive advantage of each member company is achieved.

In the managerial context, NIT increases the ability of multi-site organizations to control and integrate their activities over space through the opportunities offered by computer networks operating via modern telecommunications channels. This leads to a separation of the *location of activity* from the *location of control*, thus reinforcing the centralization of strategic functions, while depriving local areas of their initial relative autonomy. In this respect, NIT is *facilitating* the creating of such organizational structures characterized by strong centralization patterns, and hence an even more complex spatial division of labour is emerging.

The availability of information and the interaction between decision-makers and information services available on the network market have considerably improved the knowledge of an organization or a region in the context of its environment, which - as a result of access to information networks - now refers to a greatly extended geographical space. A consequence of the above extension is that NIT - and more specifically its applications in the telecommunications field and access to information networks via telecommunications channels - will serve to facilitate the *internationalization of capital and goods markets* by making available distantly located opportunities (GODDARD, GILLESPIE, 1987). In the above context it is clear that NIT may play a steering role in reshaping spatial relationships, given its potential for altering the relative distances between places. This distance-shrinking effect is essentially considered on the basis of computer/telecommunications network innovations.

The impact of NIT upon spatial structures and its potential to produce a new interregional geography of competition is expected to radically affect core-periphery relationships in the political and social systems of national economies. As a consequence of NIT making available distantly located opportunities and extending geographical scope, new actors whose locational decisions are made according to global rather than national objectives are certain to emerge.

In the context of the globalization of the network economy, the core-periphery relationship will be dramatically transformed. *Core areas* are to a large extent maintaining their comparative advantage in terms of access to networks. Firms, on the other hand, in their effort to minimize risks, are orienting towards those locations which will provide them with a wide variety of polyvalent information, in other words towards the core regions. As a result, core areas in a multinodal network context are becoming strategic nodes with increasing control power owing to NIT, through not only the quantity but also the quality of the information they convey.

Peripheral areas, on the other hand, are faced with new prospects associated with the possibility of having direct access to information (GIAOUTZI, NIJKAMP, 1988), as well as threats stemming from the uncertainty of profound impacts upon our future ways of living, working, travelling, recreation and maintaining contacts, while their competitive position will be mainly determined by their ability to meet the locational requirements of the various agents at work.

The effects of NIT are usually multi-dimensional in nature and may be visualized as effects on employment, land use, industrial structure, etc. on different geographical scales (NIJKAMP, SALOMON, 1988). In light of the above discussion, a powerful issue may come to the fore: Will NIT increase the discrepancies between core and periphery or under certain conditions reverse the trends? The answer seems to rely more upon a smart regional development policy enabling a specific region to benefit from the use and allocation of the Information sector, since technological development incorporates a set of antagonistic forces towards both concentration and deconcentration (NIJKAMP, 1987).

6. New Perspective for Core and Peripheral Areas

The role of New Information Technology and its dynamic potential in the process of economic restructuring and revitalization taking place at an international level has already been recognized. People, firms and countries as a whole are confronted nowadays with new challenges as to their locational and mobility patterns, organizational structures, technological functions, skill requirement, ranges of expertise, educational orientation, etc. (GIAOUTZI, 1987).

Due to the profound restructuring observed in the world economies, an increasing competitive pressure is being exerted upon the various actors involved, leading them to a reorganization of their competitive strategies in order to take full advantage of the benefits to be gained from technological developments, as it has been realized that international competitiveness and efficiency are to a large extent related to the range of technological expertise.

In this context *access to and use of information* becomes a key force, as has been rather clearly reflected in the efforts of many countries when establishing either regional or national innovation policies in their territories. The increasing importance attached to information-intensive activities and the use of NIT as a key strategy in order to gain competitiveness at the international as well as the national level greatly support the idea that the 'space of places' is gradually being superseded by a 'space of flows', (CASTELLS, 1984). In this respect, the need for network advanced services, modern communication facilities and material and social infrastructure appear as essential factors for the emerging structural transformation in space.

Therefore, it becomes evident that the various actors are greatly inclined to locate their activities in those areas where all the above necessary infrastructure is available. This in turn implies that core areas, in light of their ability to offer the above requirements, are to a large extent maintaining their comparative advantage in terms of access to technological advances and networks to the detriment of peripheral areas.

Core and peripheral areas in an information economy perspective will compete on the grounds of the quality of their regional information environment. This quality is greatly determined by a number of factors such as the stock of knowledge of the region's labour force and enterprises, based essentially on the education and training system; the quality and intensiveness of the information networks established within the region, which will enable enterprises to take full advantage of the existing knowledge for their own benefit; and finally the nature and variety of information and communication links established between the region and the rest of the world, which will to a large extent determine the updating of the region's knowledge stock. In this respect, it becomes obvious that insofar as core areas are displaying a more structure information environment compared to peripheral areas, they will remain a more attractive location for the generation, adoption and diffusion of technological innovation as well as for the formation of new firms (incubators), which will continue to reinforce their central role while at the same time increasing regional disparities.

Many research efforts have been devoted to a better insight into the locational patterns of activities in such a context, the development of interregional trade patterns and the corresponding regional division of labour. In this framework, the issue of comparative advantage - associated with the potential of one region to specialize in the production and export of several commodities at a relatively competitive cost vis-a-vis others - needs to be enriched by the new dynamics induced by the properties of a 'space of flows', which has been created by the recent logic of NIT.

The product cycle approach, which disaggregates regional economic activities on the basis of their product life stage, provides a promising explanatory ground in this respect. It also offers a sound exploratory framework which may contribute to a better understanding of the steering role that core areas are likely to play on the global scene in a long-run perspective. According to the product life cycle theory, *core areas* - being characterized by a high level of R&D activities, better access to information and communication networks, high quality of scientific and technical labour force, academic and cultural activity, conditions promoting synergy and instability and an increasing capacity for geographical market extension - are more efficiently forming an environment highly supportive of the renewal and development of knowledge and competence, thus appearing as centers of innovative information-oriented activities and attractive locations for both a highly skilled scientific and technical labour force and high-tech activities. As a result of scientific and technical labour force and high-tech activities. As a result of all the above, the new technology sector tends to agglomerate around the core areas, in order to take advantage of the various opportunities there.

On the other hand, *peripheral areas,* lagging in terms of access to technological developments and networks and infrastructure in general, are chosen by different actors as production sites for their routine activities, inasfar as these activities require a substantial input of low labour skill profile and have routinized and simplified their technological production requirements. In this context, the gap separating core and peripheral areas is further widened, thus placing peripheral areas in a less advantagous position, mainly associated with the lack of innovative and technological expertise necessary for an active rather than a reactive role vis-a-vis technological developments. What we actually observe is that a clear spatial distinction exists between routine production and non-routine R&D and decision-making activities, promoting descrepancies between core and periphery hosts.

The above polarization, coupled with the complementary role attributed to peripheral areas, reflects to a great extent the inability of peripheral areas - supply - to gain access to a number of assets such as information networks, a highly skilled labour force, etc. required by various actors in order to satisfy their locational demands as imposed by the new technological regime. Such a situation obviously has undesirable effects on the economic base of peripheral areas in terms of technological expertise, sources of added value and employment opportunities, thus weakening even further their competitiveness and in a way predetermining their future trajectories.

The spatial consequences of the above restructuring stemming from NIT developments are associated with an international/interregional division of labour, particularly in high-tech industries. These allow for a disjunction of different operations across space, given the fact that communications networks are able to restore the necessary links among various production units. What follows is a considerable accelertion of uneven development and disequilibrium of the spatial structure.

Significant changes are also certain to occur within the large metropolitan areas, mainly associated with the growth of corporate services and high-tech manufacturing, the decline of traditional activities and the development of new economic sectors. The metropolises are acting as a virtual magnet at an international level, strongly attracting people, capital, brainpower, information, materials and energy and thus reinforcing their prosition vis-a-vis regional structures (ANDERSSON, JOHANSSON, 1984).

In light of the above discussion, it is evident that peripheral areas are facing considerable bottlenecks as to their development perspective, which they will only be able to overcome insofar as they manage to meet the locational requirements (e.g. educational facilities, knowledge networks and networks in general) enabling firms to profit from their environmental advantages.

On the other hand, core areas may lose the central role traditionally assigned to them due to the economies of scale which are mainly dictated by traditional technology requirements, to the extent that economies of scope will become the new emblem in the goals that corporate strategy must set in an information age (GIAOUTZI, 1987).

LITERATURE

van Est, J.J.P.M.; Maack, V.; Witiuk, S.; Broome, S., 1985: Information Systems for Urban and Regional Planning, Paper presented at SORSA-UNESCO Meeting on "Information Systems for Sub-National Planning", Visakhapatnam, India.

Giaoutzi, M., 1989: Information Technology and the SMEs: A Spatial Perspective, Annals of Regional Science (forthcoming).

Giaoutzi, M.; Nijkamp, P., 1988: Informatics and Regional Development, Gower, London.

Giaoutzi, M.; Stratigea, A., 1989: The Impact of New Information Technologies on Spatial Inequalities, Paper presented at the World Conference R.S.A., Israel.

Gillespie, A.; Goddard, J., 1981: Information and Communications Technology and Regional Development, Paper presented at the OECD Seminar on Information and Telecommunications Technology for Regional Development, Athens.

Goddard, J.B., 1980: Technology Forecasting in a Spatial Context, Futures, pp. 90-105.

Goddard, J.B; Gillespie, A., 1988: Advanced Telecommunications and Regional Development, in Giaoutzi, M. and P. Nijkamp (eds.), Informatics and Regional Development, Gower, London.

Hägerstrand, T., 1967: Innovation Diffusion as a Spatial Process, University of Chicago Press.

Malecki, E.J., 1982: Corporate Organization of R&D and the Location of Technological Activities, Regional Studies 14, pp. 91-110.

Martin, J., 1978: The Wired Society, Prentice-Hall, Englewood Cliffs.

Masser, I.; Wilson, T., 1984: Approaches to Information Management in County Planning Authorities in England and Wales, Urban Studies, Vol. 21, pp. 415-425.

Nijkamp, P., 1985: Information Technology and Urban Planning, in J. Brotchie, P. Newton, P. Hall and P. Nijkamp, (eds.), The Future of Urban Form, Croom Helm, London, pp. 206-220.

Nijkamp, P., 1987: The Strategic Role of High-Tech Firms in Peripheral and Border Areas, Research paper, Dept. of Economics, Free University of Amsterdam.

Nijkamp, P.; Rietveld, P. (eds.), 1984: Information Systems for Integrated Regional Planning, North-Holland Publ. Co., Amsterdam.

Nijkamp, P.; Salomon, J., 1988: Telecommunications and the Tyranny of Space, in Orishimo, I., G. Hewings and P. Nijkamp (eds.): Information Technology: Social and Spatial Perspectives, Springer, Berlin.

Prigogine, I., 1976: Order through Fluctuation: Self-Organization and Social Systems, Evolution and Consciousness (E. Jantsch and C.H. Woddington, eds.), Addison-Wesley, Reading, pp. 93-133.

Soete, L., 1986: Employment Policies with Regard to the New Technologies. Paper presented at IIASA Conference on "Technology Alternatives and Employment Policy Options to the Year 2000", Luxembourg (mimeographed).

Stoneman, P., 1953: The Economic Analysis of Technological Change, Oxford University Press, Oxford.

Turkle, S., 1985: The Second Self: Computers and Human Spirit, Canada Publ., London.

Health Care Modeling

A Potential Research Area in Regional Science

by

Manas Chatterji

1. INTRODUCTION

In 1987 the total health care cost for the United States was $500 billion. As compared to $94 billion in 1972, this growth rate is substantially higher than the general economic growth during the same period, and consistently higher than the general inflation rate of the country. Researchers estimate that by the year 1990, total health care costs in the U.S. will equal $700 billion, out of which governmental spending will be about $500 billion. Due mainly to medicare and medicaid, the federal government's share will be about 75%. Changes have occurred in the pattern of health care expenditures during the last two decades. With an average rate of $13% per year increase, recently hospital costs have risen to 45% of the total health care cost, compared to 39.9% in 1966. Nursing home costs have increased from 3% during the same period to about 10%. This increase in hospital and nursing home care has been offset by a slight decrease in the cost of physicians' services, dental care, and cost of medicine.

Consumers paid about 49% of personal health care expenditures in 1966, which dropped to 32% in 1978. On the other hand, the government's share increased from 25% to 39% during the same period. If we look at the health care systems in Western European countries, regardless of whether the system is a national health service or a private sector activity, serious problems of cost, equity and efficiency have arisen. Although a generalization cannot be made, it can be said that in those countries, there is more need for private initiative rather than national government programs. The scene in the developing countries however, is more complicated. Lack of affordable health care is linked with general developmental problems, nutrition and sanitation and, more importantly, the equity problem associated with an advanced level of health care facility.

There is considerable variation in health care expenditures per capita among different states in the U.S. There are a number of reasons why the hospital costs vary over space. Some of the factors are:

(i) Salaries of employees
(ii) Prices charged
(iii) Length of stay
(iv) Illness patterns
(v) Case mix in hospital
(vi) Per capita income
etc. etc.

Due to the factors mentioned above, we can expect wide variations in hospital costs among regions and states for different types and ownerships of hospitals. Any national policy for controlling health care cost should consider the geographical dimension. It should be an important parameter for a planning model, whether it be policy oriented or mathematical.

For quite some time, the health field was treated as quite technical, dealing with medical knowledge and techniques. But the situation has changed recently. It has been realized that it is linked to other sectors of human activity. The components of health care systems have grown independently of each other. But if we want to look for efficiency in the system, we have to consider it as a whole. There are at least two facets in health care where the mathematical approach can be used. One is the modeling of the components within the health care sector; namely, different departments within the hospital, the physician, etc.

Some of the methods are (i) inventory, (ii) queuing, (iii) simulation, (iv) allocation, competition and search.

(i) Inventory

In a hospital the decision to store items like blood, drugs, cleaning fluids, etc. or to decide how many beds to keep open, may be important to reduce the cost or avoiding the risk of not having time. For this purpose, the inventory control models of operations research can be applied. In the deterministic model, the demand is known and the problem is to choose the size of the order and the number of times the order is placed. In the buffer stock model the demand is assumed to follow normal distribution and the reorder point (level of inventory when the order will be given) is to be determined to minimize total cost. The order quantity is determined as in the deterministic model. In the mathematical model (probablistic inventory model), the reorder point and order-quantity are determined through iteration. Although inventory models have been used for hospital stores (SMITH, GREGORY, MAUIRE, 1975), in blood banks (STIMSON & STIMSON, 1972), mergers and contradiction (REINKE (1972)), the cost of conducting a sophisticated model sometimes exceeds the benefits.

(ii) *Queuing*

Queuing theory assumes that there are facilities like hospital or health centers where customers arrive and wait for service. If the level of the service is low (the facility is small) and/or the clients do not come in a specified order, they have to spend a long time in queues or they may not get the service at all. On the other hand, if the facility is large, there is a cost of idling the facility. The objective in the queuing problem is to provide the level of facility or form the queue in such a way that the overall costs are minimum. The arrival pattern may be constant, variable, bunched or random. In the case of random arrival, the probability of a certain number of arrivals can be computed from Poisson distribution. In addition to this probability, the expected number in queue and the expected waiting time in queues can also be computed. Queuing theory has wide application in inpatient and outpatient facilities (THOMPSON, 1963). The problem with this theory is that the arrival pattern may not follow the Poisson distribution, the behavioristic aspect of the doctors may be important and we may not get reliable estimates of service time. In this situation, the simulation technique may be used.

(iii) *Simulation*

When the mathematical process of a system, whether it is queuing or inventory is not known or difficult to obtain, simulation can be used most effectively. Simulation is built on the representation of reality and it tries to estimate the impact of the system due to the changes of policy variables. Changes for which an objective is optimized are made. This simple tool is usually used by reproducing the system by sampling distribution from a probability distribution (often normal). It should be clearly understood that simulation is not based on theory and the generality which mathematical approach brings can never be achieved by simulation. Nevertheless, authors such as STIMSON & STIMSON (1972), WREN (1974) and GORDON (1978), have effectively used this technique in health care.

(iv) *Allocation, Competition and Search*

Health care planning is a perfect example where the allocation problem is involved in face of limited resources. Linear, non-linear, dynamic, integer and goal programming techniques can be effectively utilized for this purpose. Presently the system is so complex, it is important to measure the impact of alternative decision processes for which the tools of Decision Science can be effectively utilized. Control Chart techniques of statistical quality control can also be used to identify real deviation (not random) of the system from the set standard for the introduction of corrective procedure. These micro tools of health care delivery can also be coordinated with concepts of medical geography, i.e., diffusion models PYLE (1986) and CLIFF et. al. (1986).

The other area of mathematical modeling is the relationship of the health sector with economic, political and social sectors of the country. The objective of this paper is to briefly summarize some theoritical research on health care economics and outline some empirical work on New York State I am doing.

There is considerable scope of developing theoretical models in health care economics. PAELINCK and CHATTERJI (1989) give such an abstract model. I shall briefly outline it here. The objective is to develop a theoretical and behavioral model of each of the following health sectors and link them. Later we intend to use empirical observations to estimate the parameters and test the validity of the model.

The health care sector here covers the following subsectors:

- physicians
- patients
- hospitals
- medical drugs
- insurance

2. Regression Models

The most widely used technique used in health care demand is the regression analysis. Here we define a dependent variable, say number of admission, or patient days per 1000 population, etc., and regress it with such variables as income, price, age, sex, etc. NEWHOUSE and PHELPS (1974), for example computed price and income elasticities for a cross section sample of n = 122 after regressing the dependent variable, "Hospital Days" to such variables as hospital coinsurance, wage income/week, non-wage income, education, age, family, size, sex, etc. I have also used the regression analysis to estimate cost and production function of New York State hospitals (CHATTERJI, 1983). For each Cost Center of the hospitals such as surgery, cafeteria, etc., I have also estimated the cost function. The data used are pooled time series-cross sections for all acute care hospitals for the time period 1974-86 with each hospital having a maximum 500 variables.

Presently I am embarking on a project to develop a socio-econometric model sketched in Fig. 1. As a part of the socio-economic model of health care of New York State counties we have collected data for large number of variables of the following types and have estimated regression equations.

PDTHOU	Patient days per 1000 population
RBDTHOU	Regular beds per 1000 population
SBDTHOU	State beds per 1000 population
VBDTHOU	Veterans beds per 1000 population
AMTHOU	Ambulances per 1000 population

MDTHOU	Doctors per 1000 population
NRBDTHOU	Nurses per 100 population
	etc.

Figure 1: A Fused Input-Output Programming Framework of Health Care in New York State

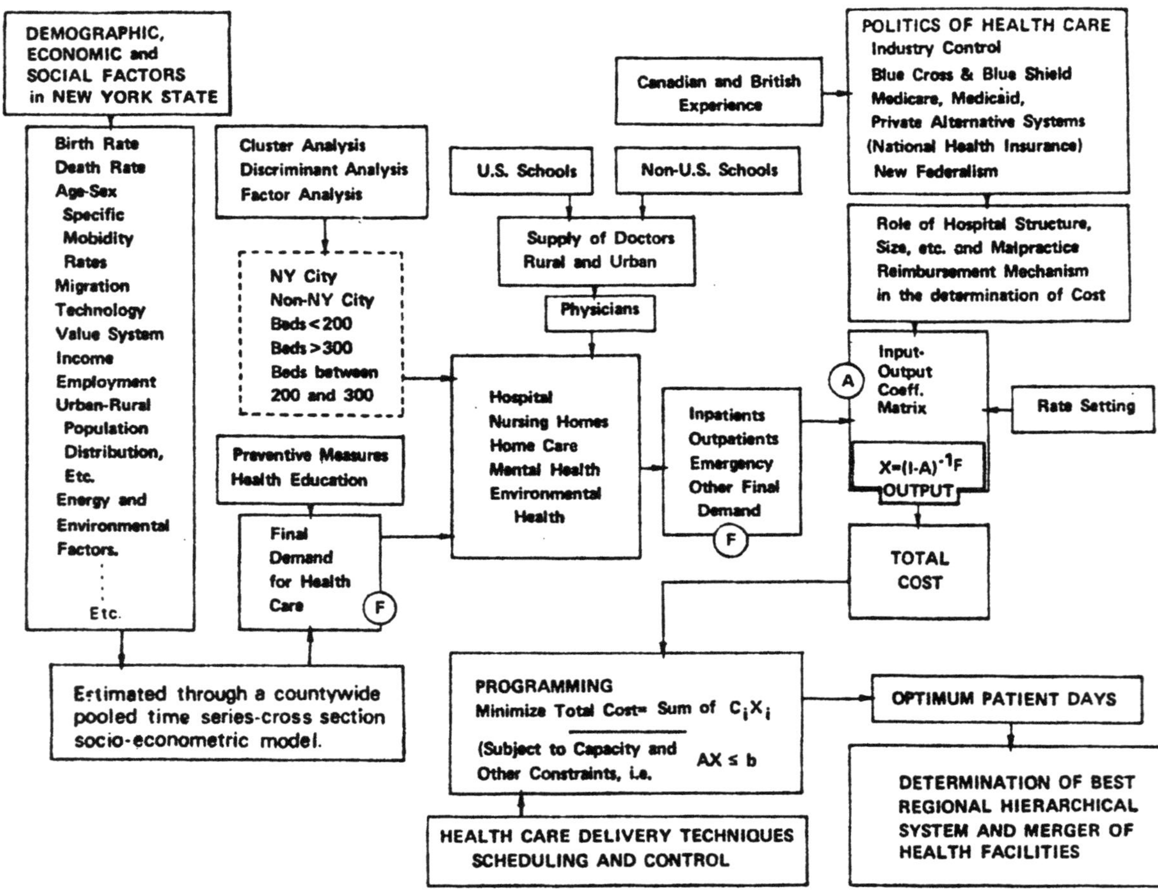

Most of the work done in health using regression analysis is of single equation ordinary least square (OLS) type estimated on the basis of time series data. Uses of these types of data help us to estimate price elasticity. But since time series data for many variables are not available and the significant changes took place in health field; namely, Medicare, Medicaid, etc., combined cross section and time series analysis appear to be more desirable rather than cross section or time series analysis alone. The most comprehensive structure of an econometric model is presented by FELDSTEIN and KELMAN (see CHATTERJI, 1983).

They divide the health care institutions into five sets; namely, hospitals, nursing homes, outpatient clinics, doctor's offices and patient's home. The total product of the health care sector is obtained as the weighted sum of services provided in the different sectors. The price level of health care is also defined as a Laspeyer's index based on previous year's price. In the second stage, the supply and demand of each type of institution are determined through a number of equations. Then, the resulting manpower demand and supply are estimated. Typical equations in each sector look like:

$$S_H^* = f(AGE,\ MAR,\ SEX,\ URB,\ \frac{Y}{CPI},\ INS_H,\ WEL.....) \qquad (1)$$

$$OCC_H = f(S_H^*,\ AVG_H,\ P_H^{'},\ \underline{POP}.....)$$

where

S_H^* = desired stock of hospital beds per thousand population
P_H^o = Price
Y = median family income
AVG_H = average size of short-term hospitals
WEL = public assistance for medical purposes
INS_H = hospital insurance
(All other variables are self-explanatory)

The model we are working on is similar to that of FELDSTEIN (see CHATTERJI, 1983) except that it is a pooled (N.Y. State as a whole) time series and cross section (with N.Y. counties). It will also be an integration of Feldstein type health care model and well known examples of econometric model including sociological variables.

The demand for health care delivery starts from the population growth and its spatial distribution with respect to age and marital status. So, before constructing any model for health care sector, a demographic submodel is essential. After population, income, disposable personal income, etc., origin of such income are important factors to consider. Other competing needs besides health care also affect the demand. This demand can be divided into separate categories like hospital care, nursing homes, outpatient services, etc. Within each component, as in hospital care, different types of submodels may be specified, cost functions for different size hospitals can be estimated. For such functions, the rising wage rates of hospital employees, particularly in larger SMSA's is another factor. Considerable resources are also being used for the purchase of expensive equipment, particularly in the teaching hospitals. It will be worthwhile to investigate whether expenditures on such assets have a positive relationship with the rising costs.

The hospital cost also varies with the size and type of ownership. The hospitals in the SMSA's are of a completely different character than those in more rural areas. The largest SMSA's are in many ways different from the small sized SMSA's. To avoid the problem of a small number of observations, we can divide the SMSA's into two population groups: (i) less than 200,000 population, and (ii) more than 200,000 population. Other factors which contribute to the variation of hospital costs are the supply of beds, utilization of inpatient and outpatient services, occupancy rate, case mix, and regulation.

A number of observations seems to be in order while describing demand functions in health care estimated through regression equations. One important point to remember is that health care is derived demand, i.e., the demand of the patient is determined by the doctor. Also, there should be a distinction between demand measured by actual use and the need as determined by the professionals. There is also a distinction between private and public demand. For example, water fluoridation, immunization, etc., are public demand. Usually, in health demand studies, private demand is considered. One should remember that there is a considerable amount of public expenditure on such items such as subsidies to hospitals, welfare of patients and veterans administration programs. Medical care is composed of other things such as drugs, hospital and physicians care and quality differences are important variables. Medical care is not merely hospital visits, it is good health and there are many other things involved relating to good health.

The basic factors involved in the demand for medical care are income and price. As income changes, demand also changes. It is not, however, clear whether different types of services demanded increase or the quality changes. The intensity of demand is measured by the variables: hospital admissions, patient days, length of stays, physician visits. We can divide health care into different services and construct demand equations in each sector separately as total expenditure is expressed as a function of price and quantity. The price can be taken as some sort of an index obtained by aggregating the prices for individual sectors. But the problem with the approach is that the quality has drastically changed over the period. So, it is well to use a cross section approach. As mentioned before, that the demand of health care is a derived demand just like the demand for electricity for residential use is derived demand coming through the use of appliances. So, the demand is a function of economic, cultural-demographic factors, prevailing medical practices as well as the incidence of illness.

Another factor which differentiates health care demand from other demand studies is the concept of choice. Here the choice depends on knowledge and the availability of substitutes. The physician makes the choice for the patient. The physician combines the components in such a way that it has least relative cost to the patient and himself.

In economics, we are interested to know what goods and services should be produced, how shall they be produced, and for whom. The answer is dictated by the marginality principle which states that last dollar worth of resources devoted to health care increased human satisfaction equal to that of last dollar worth devoted to other goods. The marginal principle leads to optimum allocation of resources if the perfect competition exists, i.e., (i) many well-informed buyers and sellers, (ii) buyers and sellers act independently, (iii) there is free entry and exit. In the market of hospital, physicians, etc., this is obviously not the case. Again, it is not clear what we are supposed to maximize since the perception of health care is different depending to whom and when and where it is made.

One factor often not taken into account while relating consumption of health care and income is the question of existing health status. People who are poor are more likely to have ill health than non-poor. So the elasticity does not represent the real situation. Therefore, income and expenditures are not the only variables. Other variables related to type of occupation, family condition, etc., should be considered. Another problem is the introduction of new programs. Medicare and medicaid have changed the financial situation. So it is difficult to compare 1963 date with 1968 data.

3. Input–Output Analysis

The input–output analysis first devised by W. LEONTIEF was applied to the macro–linear system. The total economy is divided into a number of sectors and a flow table is constructed (see CHATTERJI, 1983). In this input–output table, two kinds of sectors are identified: the intermediate sector, and the final demand sector. This model has been used (CHATTERJI, 1983) to estimate the multiplier effect when there is a change, say, in the level of a particular activity, i.e., surgery. It will increase facilities and personnel in the operating room, kitchen and cafeteria, laundry and sterile supply unit, etc. Thus feedback and multiplier effect cannot be neglected. Accordingly, the central thesis is that the direct input care for a surgical patient from a particular sector, say, housekeeping, requires inputs from the cafeteria, maintenance team, administration, etc. All these individual requirements of input should add up to the direct requirements, and this can only be done by the input–output method. The cost of allocation process can be further refined by using such methods as factor and cluster analysis, discriminant factor analysis and statistical cost functions (CHATTERJI, 1983).

It is to be noted here that the operation of the input–output model will depend upon the specification of the final demand, i.e., what different kinds of services from a hospital will be needed by the community at a future time. This, of course, depends on the demographic and socio–economic structure. For example, in many communities in New York State, the population is aging, requiring more and specific types of health care. Income is also changing, and the amount of reimbursement as Blue Cross and Blue Shield, Medicare, and Medicaid, is also going to be affected. If a national health care insurance plan is adopted in the U.S., then its impact on the final demand can be determined by a comparative cost analysis of the Canadian experience before and after the implementation of its national health care insurance, and then applying that observation to the U.S. situation.

4. Factor Analysis of New York State Hospitals

Previously we obtained some regression relationships of the total cost and its constituent parts in a hospital, with such independent variables as the number of beds, patient days, and the like. The usefulness of those relations lies in the fact that once the future levels of the independent variables are known or stipulated by policy decisions, the values of the dependent cost variables can be estimated. One difficulty with this method is that in a given case the number of independent variables may be quite large, and many of them may be correlated. In that situation the estimates of the regression coefficient will be biased with the error of multi-collinearity. In addition, for a given hospital in a particular location and offering a number of services, a question may be asked about the dimensions of its structure; that is, many intermediate services may be represented by a single dimension or structure. Thus instead of looking at the services separately, we need only pay attention to this dimension, or *factor,* as it is called, provided we know how this factor is interrelated with the variables. Once the different factors are identified through factor analysis, they can be related to the cost variables. This will reduce the necessity of using too many variables in the regression equation. It is then a method of aggregation by which a set of related variables is combined in a new variable termed a *factor* so that it is a representative of the total set of variables.

I have conducted a factor analysis for New York State hospitals. There are 91 variables that characterize the structure of N.Y. State hospitals (CHATTERJI, 1983). From the correlation matrix (91X91) of the variables, ten factors were extracted. They explain about 90 percent of the variability. In other words, these ten factors (hitherto unnamed) will approximately produce the same correlation matrix (91X91); that is, they are representative of the 91 variables. To decide which variables belong to what factors and the names of the factors, we have to refer to the varimax rotated matrix. This matrix (not shown here) gives the factor loadings of each varimax factor with the corresponding variable. Looking at each row, we can identify the strongest linkage of any variable with the corresponding factor. For example, we might match the highest coefficient with F_1 (factor one). We then group the variables in one class. It was found out that all these variables in the class refer to such items as available bed days, use of beds, and the like. Consequently, we identify F_1 as a demand or capacity factor. Similarly, it was found that all variables having the strongest relations with F_2 are related to radiology. So we term F_2 radiology. Similarly, factor 3 was termed length of stay, factor 4 operations, factor 5 maternity, factor 6 laboratory, and factor 7 pediatrics. We disregarded the other three factors.

Next we compute the factor scores for each hospital. If, for example, a hospital has a high score for factor 4, it means that the hospital has more than the average number of operations. With the scores of the seven factors as datafor seven independent variables, we computed the regression of the cost of each center.

5. Frontiers of Further Research

The current research can be improved in many ways and be made a part of a broad systemwide study. Without offering a detailed description, we shall indicate a number of fruitful directions.

5.1 Analysis of Variance and Covariance and χ^2 Test

We have used the analysis of varience (ANOVA) techniques to test the diversity of hospital structure to determine which classification is most reasonable. On that basis we took the following classes:

- Number of beds less than 200
- Number of beds between 200 and 300
- Number of beds more than 300
- All hospitals
- All New York City hospitals
- All hospitals except New York City

It is interesting to note that on the basis of analysis of variance, the cutoff points selected were 200 and 300. Nevertheless, this is not the only potential of ANOVA. It can be used to test a multitude of hypotheses after one has been cross-classifying the data by different classification variables. For example, does the operating cost per patient day very between different size hospitals and locations?

The analysis of covariance (ANCOVA) is even more important. We can easily identify the two independent variables that vary together, take out the effect of one variable from the other, and test whether the residual amount of the second variable is related to the cost.

In a similar fashion, we can also construct m × n contingency tables and make a χ^2 test given by

$$\chi^2 = \sum \frac{(O-e)^2}{e} \qquad (2)$$

and decide whether two classifications - say, size and cost - are independent. This regression analysis can be supplemented by χ^2, ANOVA, and ANCOVA tests. Obviously, the χ^2 test in many cases will give the same conclusion as the ANOVA test. It is true that ANOVA and ANCOVA cannot help us to forecast costs, but they can help us to detect the difference between the control and the experimental group. The regression analysis, in addition to these, can provide us with a forecasting relation.

5.2 Discriminant and Cluster Analysis

The first stage of the analysis of cost function is to classify the hospitals into several groups on the basis of some discriminating variables, such as those previously considered in the factor analysis study. The hospitals in the group are expected to differ in reference to the variables. On the basis of the values of the discriminating variables, a discriminate function of the following form can be constructed:

$$D_i = d_{i1}Z_1 + d_{i2}Z_2 111 + Z_{ip}Z_p \qquad (3)$$

Where D_i is the score on discriminate function for ith unit, the d's are weighted coefficients and the Z's are the standardized value of the p discriminating variables (in the case of factor analysis example, p = 91). The functions above is formed in such a way as to maximize the separation of groups. Once the discriminate functions have been estimated, they can be used for either analysis or classification. In the analysis stage, we can test whether the functions really discriminate. In the case of classification, future units of a given hospital with different values of variables can be put into appropriate groups.

The cluster analysis, like the discriminate analysis, serves similar purposes, except that it identifies several groups at a time. This technique has been used to set rates.

5.3 Programming Models

So far we have not been concerned with any optimal behavior. We are only trying to understand the effects on endogenous hospital variables when something happens in the exogenous system. We may be interested, however, to know, for example, the optimal combination of patient days in the different departments of the hospital for which either the total cost is minimum or the total charges are maximum, as long as the capacity is not exceeded and other restrictions are satisfied. In symbolic form, we may wish to maximize

$$Y = C_1X_1 + C_2X_2 \dots + C_nX_n$$

subject to

$$\begin{aligned} A_{11}X_1 + A_{12}X_2 \dots + A_{1n}X_n &\leq b_1 \\ A_{21}X_1 + A_{21}X_2 \dots + A_{2n}X_n &\leq b_2 \\ &\vdots \\ A_{m1}X_1 + A_{m2}X_2 \dots + A_{mn}X_n &\leq b_m \end{aligned} \tag{4}$$

and

$$\begin{aligned} X_1^L \leq X_1 &\leq X_1^U \\ X_2^L \leq X_2 &\leq X_2^U \\ &\vdots \\ X_n^L \leq X_n &\leq X_n^U \end{aligned}$$

In this formulation:

X_1 = Number of patient days in medical service

X_2 = Number of patient days in surgical service

X_3 = Number of patient days in intensive care

X_4 = Number of patient days in coronary

X_5 = Number of patient days in pediatrics

X_6 = Number of patient days in maternity...

.

.

.

X_n = Number of patient days in routine nursery

and C_1, $C_2 \ldots C_n$ are the charges received per n types of patient days; $b_1 \ldots b_m$ are the capacity restrictions (in dollars or in physical units) in m service departments. These departments can be the hospitas services or operative departments. They can be grouped into fewer numbers - say, 20 departments - or into primary, secondary, and independent sectors, as defined before. The coefficients A_{ij} are input coefficients that state how much of the services of each department i is needed per unit of patient day in the jth type of service - say, maternity. This information can be obtained from hospital statistics. If we decide to include the physical inputs, some information, but not all, will be available from published data. The variables (X_1^L, X_1^U,...X_n^L, X_n^U) denote the lower and upper allowable limits for each type of case. The limits can be obtained from legal guidelines, policy statements, and the like.

For a preliminary study, the base period values (1976) of the b coefficients and c coefficients and hypothetical values of upper and lower limit can be used. A simulation with different states of hypothetical values can also be conducted.

5.4 Other Techniques

There are a number of other techniques which have great potential. Due to the limitation of space and time, it is not possible for us to discuss them here (see CHATTERJI, 1983). Some of them are:

(i) Markov Chain approach as used by LAVE & LAVE (1974). To consider the health status of the patients and admission to different types of medical facilities.

(ii) Highly developed ones of location)allocation in geography and regional science.

(iii) Optimum aggregation of services provided by different facilities.

(iv) Hospital consolidation.

(v) Modeling emergency care within the hospital and the community.

(vi) Fuzzy system.

(vii) Consolidation of services or hospitals may lead to lower morale and less participation by the numbers in the organization. To devize an optimum system, gravity models and participation potential models may be of use.

(viii) It is not possible or necessary to provide all services (such as heart operations) in all locations. For devising a regional health care system with changing demographics and economic structures, techniques of hierarchical systems may be of value.

LITERATURE

Chatterji, M., 1986: Health Care Cost Containment in New York State: Problems and Possibilities (edited), The Nelson A. Rockefeller Institute of Government, Albany.

Chatterji, M., 1986: Towards a Hierarchical Model of Health Care Delivery in New York State. In Actes du Symposium "Analyse de Systaames et Modaales Mathaamatiques." Besanaaon Cahiers de Gaaographie.

Chatterji, M., 1986: Health Care Cost and Its Allocation. In Health Care Cost in New York State, edited by M. Chatterji, Nelson A. Rockefeller Institute of Government, Albany.

Chatterji, M., 1985: An Econometric Model of Health Care Delivery in New York State. In Actes des Journaaes Internationales d"Economaatrie de la Santaa, No. 2.

Chatterji, M., 1984: A Hierarchical Model of Health Care Delivery in NEw York State. In Proceedings of the Third International Conference on System Science in Health Care, International Society for Systems Science and Health Care, Munich.

Chatterji, M., 1983: Health Care Cost Containment Policy: An Econometric Study, Kluwer-Nijhoff, Boston.

Cliff, A.D.; P. Haggett; J.K. Ord, 1986: Spatial Aspects of INFLUENZA Epidemics, Pion Limited, London,

Frech, H.E.; P. Ginsburg, 1981: "The Cost of Nursing Home Care in the United States: Government Financing, Ownership, and Efficiency." In Health Economies, and Proceedings of the World Congress on Health Economics, Leiden, The Netherlands, September 1980. Ed. Jacques van der Gaag and Mark Perlman. New York: North-Holland Publishing Co..

Gordon, G., 1978: 'Simulation-Computation' in Moder, J.J. & Elmaghriby, S.E. (eds.) Handbook of Operations Research: Foundations & Fundamentals, Van Nostrand Reinhold, pp. 566-585.

Intriligator, M.D.; B.H. Kehrer, 1972: "A simultaneous Equations Model of Ancillary Personnel Employed in Physicians' Offices." Paper delivered at the Econometric Society Meetings, Toronto, December 29.

Intriligator, M.D.; B.H. Kehrer, 1974: "Allied Health Personnel in Physicians' Offices: An Econometric Approach." In Economics of Health and Medical Care, ed. M. Perlman, New York: Wiley.

Lave, J.R.; L.B. Lave; BS. Leinhardt, 1974: "Modeling the Delivery of Medical Services." In The Economics of Health and Medical Care, ed. M. Perlman, New York: Wiley, pp. 326)351.

Newhouse, J.P.; C.E. Phelps, 1974: "Price and Income Elasticities for Medical LLCare Services." In The Economics of Health and Medical Care, ed. M. Perlman, New York: Wiley, pp. 139)161.

Paelinck, J.H.P., 1985: (avec la collaboration de J.-P. Ancot, H. Gravesteijn, J.H. Kuiper et Th. ten Raa), Eléments d'Analyse Economique Spatiale, Genaave, ERESA.

Paelinck, J.H.P., 1984: Les difficultés de la médicométrie régionale, in A.S. Baily et M. Périat (éds), Médicomaatrie régionale, Editions Anthropos,Paris, pp.13-19.

Pyle, G. F., 1986: The Diffusion of Influenze: Patterns and Paradigms, Rowman &Littlefield, New Jersey.

Reinke, W.A., 1972: Health Planning: Qualitative Aspects & Quantitive Techniques,LLDepartment of International Health, School of Hygiene & Public Health, John Hopkins University.

Reisman A.; M. Kiley (eds), 1973: Health Care Delivery Planning, Gordon and Breach Science Publishers, New York.

Smith, A.G.; K. Gregory; J.D. Maguire: "Operational Research for the Hospital Supply Service", Operational Research Quarterly, Vol. 26, No. 2, ii pp. 375)388.

Stimson, D.H.; R.H. Stimson, 1972: Operations Research in Hospitals, H.R.E.T..

Thompson, J.D. et al, 1963: 'Predicting Requirements for maternity Facilities', Hospitals, February 16.

Wren, G., 1974: Modern Health Administration, Georgia U.P.

Yett, D.E.; L. Drabek; M.D. Intriligator; L.J. Kimbell, 1979: A Forecasting and LLPolicy Simulation Model of the Health Care Sector, Lexington Books, Lexington, Massachusetts.

Zweifel, P., 1982: Ein Oekonomisches Modell des Arztverhaltens, Springer, Berlin.

Pareto-Efficient Relation Between Infrastructure and Industry

by

Ryszard Domański

1. Introduction

We assume the hypothesis that the relation between infrastructure and industry exerts significant influence on the efficiency of urban and regional economy. There exist statistical evidences which allow for the acceptance of this hypothesis with high degree of confidence. Hence, it is purposeful to search methods enabling efficient shaping of this relation.

This paper suggests a method based on the idea of Pareto optimum. Such a method is used in discussing the relation between industry and natural environment. Its extension on infrastructure is justified since the infrastructure can be regarded as an element of the environment in its broader sense, i.e. the element of urban or regional environment embracing both natural as well as socio-economic components. Suggested method has been tested using data of the Polish regional statistics.

2. Methodological Approach

The industry uses various resources in its production. They can be divided in two groups: (i) own resources of industrial enterprises – capital, labor, organisation, research – and (ii) local resources – environment, infrastructure. Let us denote individual resources in the following way: K – capital of industrial enterprises, L - labor in industrial enterprises, E - environmental resources, R - instrastructure.

Industrial production, Q, is a function of individual production factors. With the above denotion it assumes the form

$$Q = f\,[K,\ L,\ E,\ R].$$

Let us consider the dependence of industrial production on infrastructure. In terms of marginal calculus, the part of industrial production obtained due to the use of infrastructure can be expressed as the partial derivative of function Q with respect to R, i.e.

$$\frac{\partial Q}{\partial R}.$$

Assuming that the marginal value product of infrastructure shows a decreasing tendency, the graphic presentation of the above derivative will take the form of curve

$$\frac{\partial Q}{\partial R}\ /\ (\text{Fig. 1}).$$

Figure 1: Pareto optimum for the relation between infrastructure and industry

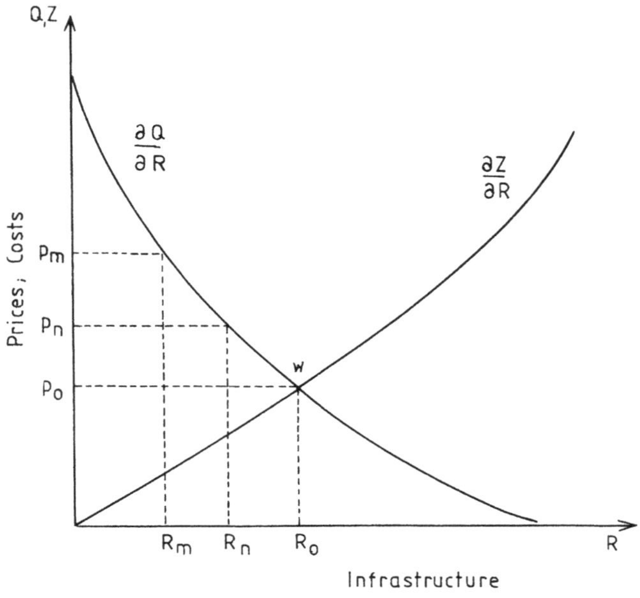

The use of infrastructure by industry bears costs in the local sector of urban and regional economy. Theses costs are a function of capital, labor and other resources layed out in infrastructure by the local sector. The industry is being charged for the use of these resources. Assuming that the marginal cost of the use of infrastructure supplied by the local sector shows an increasing tendency its graphic presentation will take the form of curve

$$\frac{\partial Z}{\partial R}.$$

We can read Fig. 1 from right to left. Let us start with the curve

$$\frac{\partial Z}{\partial R}.$$

It decreases with the decrease of infrastructure supplied by the local sector. Such an effect results in the decrease of local expenditures, and consequently local taxes. Hence, it can be considered as public benefit. However, the low level of infrastructure creates limitations for industrial growth. This, in turn, limits the revenues of local budget collected from industrial enterprises.

The curve

$$\frac{\partial Q}{\partial R}$$

read in the same direction increases with the decrease of infrastructural resources used by industry. This, however, means an increase of goods not produced and not delivered to consumers. Hence, it can be considered as public loss.

Theoretically, we can decrease the use of infrastructure by industry to zero level. This would mean, however, putting industry out of action, interruption of deliveries, and reduction of tax revenues. On the other hand, we can encourage industry to grow by developing infrastructure facilities. This would mean the growth of industrial production, but at the same time the growth of burdens of the local sector.

A question arises under what condition the relation between industry and infrastructure reaches its optimum. In terms of Pareto optimality, it occurs when the marginal value product of infrastructure is equal to the marginal cost of the use of infrastructure, i.e. when

$$\frac{\partial Q}{\partial R} = \frac{\partial Z}{\partial R} .$$

Optimal solution can be found graphically. In Fig. 1 this is the point of intersection of the curves

$$\frac{\partial Q}{\partial R} \text{ and } \frac{\partial Z}{\partial R} \text{, point w.}$$

Given the production function and the infrastructure cost function, its coordinates indicate the efficient size of infrastructure [point R_o] and the correspondent value of production gained by industry thanks to infrastructure [point P_o]. This gain creates the financial sources of fees and taxes payed by industry for the use of infrastructure. It should be then a determinant of the amount of these charges.

The usefulness of the method derived from Pareto optimum relies in that it ensures the connectivity - not necessarily the equality - between the value gained and the charges paid, and thus between industry and infrastructure. It can be useful also in solving other problems of this type, e.g. in the coordination of local and supralocal sectors of urban and regional economy.

3. Approximate Solution

In empirical investigation of infrastructure and industry one faces very often difficulties connected with the deficiency of information. Simple information can be processed by means of simple methods. If we take into account only linear dependences, then marginal calculus becomes irrelevant.

Let us consider a linear production function embracing several production factors. Such a function would be a sum of several functions relating the value of production and individual production factors. Their general form would be the following: $y = ax$. The derivative of such a function is equal to the coefficient a. Hence, the exogenous variable x, representing production factors, would disappear in mathematical analysis.

When considering only linear dependences it is sufficient to apply regression analysis. Below a simple procedure is presented, which allows for the determination of efficient relation between infrastructure and industry. It refers to the idea of Pareto optimum, although it does not yield a result which could be recognized as Pareto optimum in the proper sens of this term. This procedure was tested by means of statistical data relating to Poznań region 1975–1986. Lacking information were filled in by statistical estimates. This procedure consists of the following steps:

(i) Description of the dependence of industrial production on infrastructural and environmental factors in the form of regression equations; estimation of the parameters of equations.

(ii) Apportionment of the industrial production obtained due to the use of infrastructure and environment. Applying the above symbols we can denote respective portions of production in the following way: Q_{RK} – prouction apportioned to the capital of infrastructure, Q_{RL} – production apportioned to the labor employed in infrastructure, Q_W – production apportioned to the water, used and polluted by industry, Q_a – producion apportioned to the air, polluted by industry.

The apportioned parts of industrial production are functions, the arguments of which, i.e. exogenous variables are infrastructural and environmental factors. The value of coefficients expressing intensity of the use of factors were calculated by means of regression analysis. These functions are presented in Fig. 2–6 as the lines of regression.

(iii) Unit costs of the use of infrastructure as well as the costs of environmental protection were estimated and assumed as given. Total cost are represented in Fig. 2–6 by lines Z.

(iv) Symmetric rotation of regression lines. This is very important step in the procedure. The regression lines express the production obtained due to the use of infrastructure or environment and delivered to the market. However, we are interested in tracing the production not obtained and not delivered due to the infrastructural and environmental constraints. Therefore, we carry out the rotation. Now, we can relate the public loss the decrease of production and the public benefit [the decrease of infrastructural and environmental costs.] Thanks to the rotation, both lines can intersect. The point of intersection indicates the relation between infrastructure or environment and industrial production efficient in Pareto sense.

As mentioned before, the coordinates of this point indicate the efficient size of infrastructure and the correspondent value of production gained by industry. In case of environment they indicate ecconomically admissible pollution and correspondent industrial production.

Actual variability of statistical characteristics of infrastructure and environment is defined by broken lines on x-axis. We can relate the position of the point of efficiency to this range of variability. In this way we can determine which infrastructural resources are exploited above the point of efficiency, and which could be used in broader range. In case of environment the comparison indicates, whether actual pollution is higher or lower than the level of efficiency.

One can also compare the value of production gained by industry due to the use of infrastructure and environment on one hand, and the amount of fees and taxes paid by the industry on the other.

Fig. 2 and 3 characterize the relation between infrastructure and industry in central city of the region under investigation. The point of efficient use of infrastructural capital [Fig. 2] takes the position within the interval of actual uses. However, it places itself relatively closer to the lower limit. It means that in the period under investigation, taken as a whole, the actual uses exceeded the point of efficiency. This can be interpreted as an overload of infrastructure by industry. Similar position takes the point of efficient use of labor in infrastructural sector [Fig. 3]. The interpretation, however, is different. The employment, relatively higher than efficient one, means unproductive use of labor in infrastructural sector.

It is worth noticing that the productivity of labor in central city differs substantially from that in the region. Fig. 4 shows that the point of efficiency in the region places itself far beyond the upper limit of actual employment. It means that further increase of employment could be efficient. Such a hypothesis is plausible in the light of real processes undergoing in

<u>Figure 2:</u> **Efficient relation between infrastructural capital and industrial production**

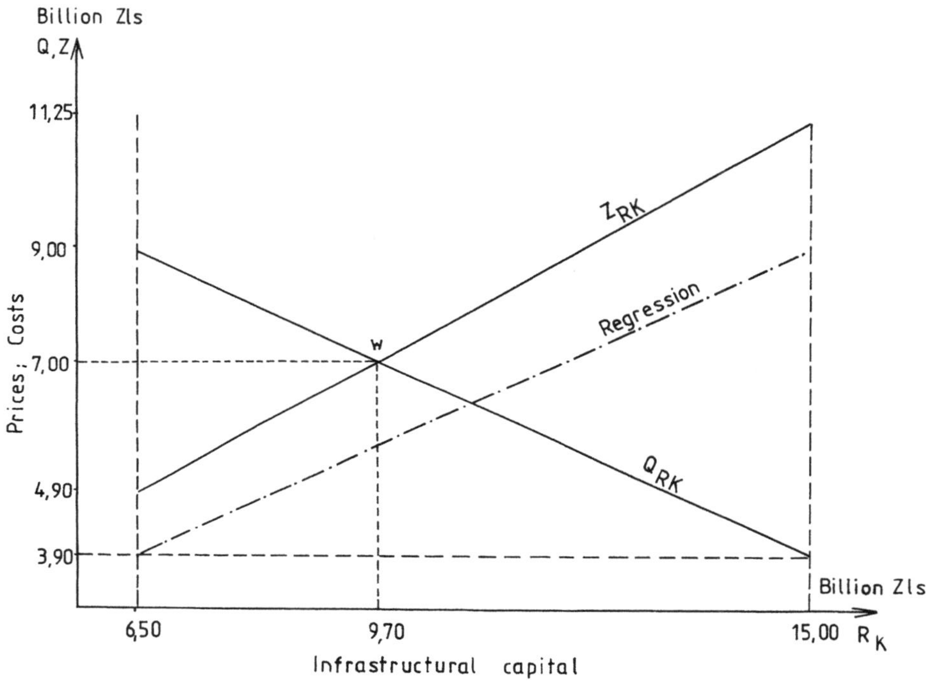

our region. The structure of regional economy, differently than the structure of urban economy, is formed by agriculture, food industry, services, small and middle-size manufacture firms. High level of agriculture, and harmonized development of industry and services favors the increase of economic efficiency in the region, whereas capital demanding industry, excessive employment in administration, and disequilibrium between industry and infrastructure decreases economic efficiency in central city.

Figure 3: Efficient relation between labor in infrastructure and industrial production: central city

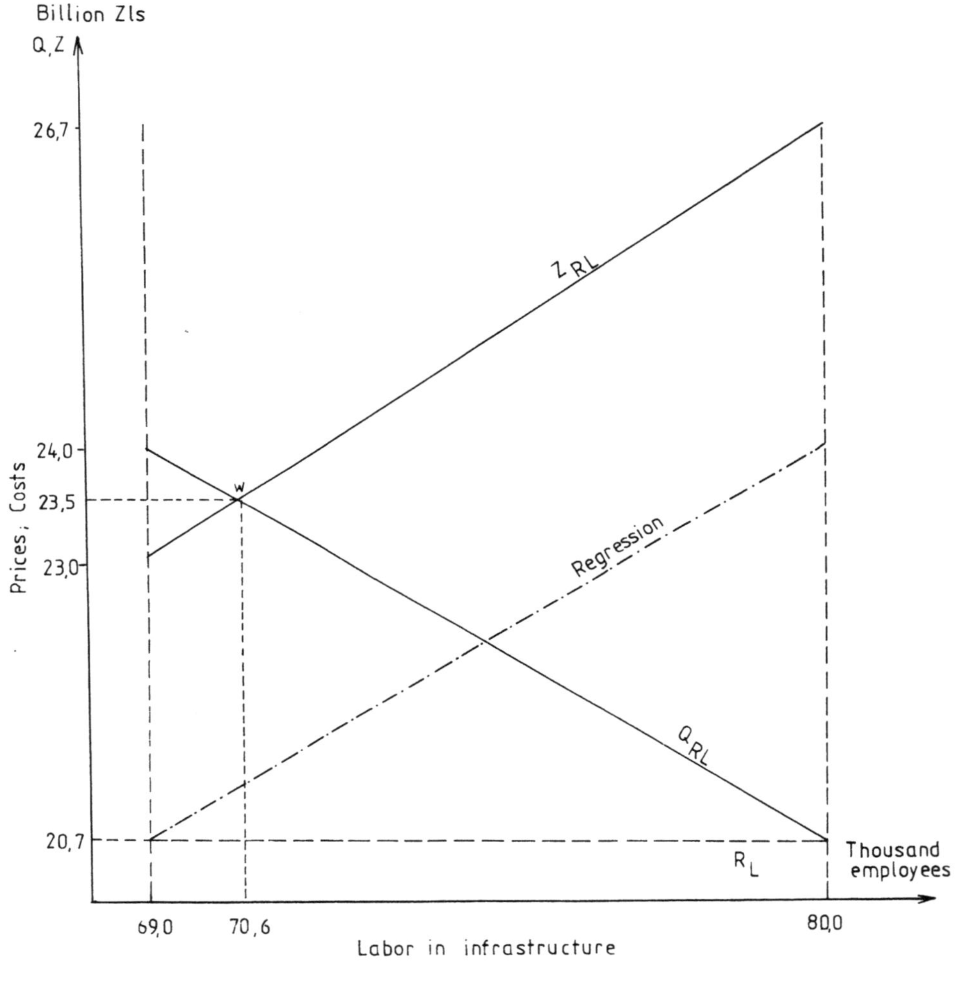

Figure 4: Efficient relation between labor in infrastructure and industrial production: region

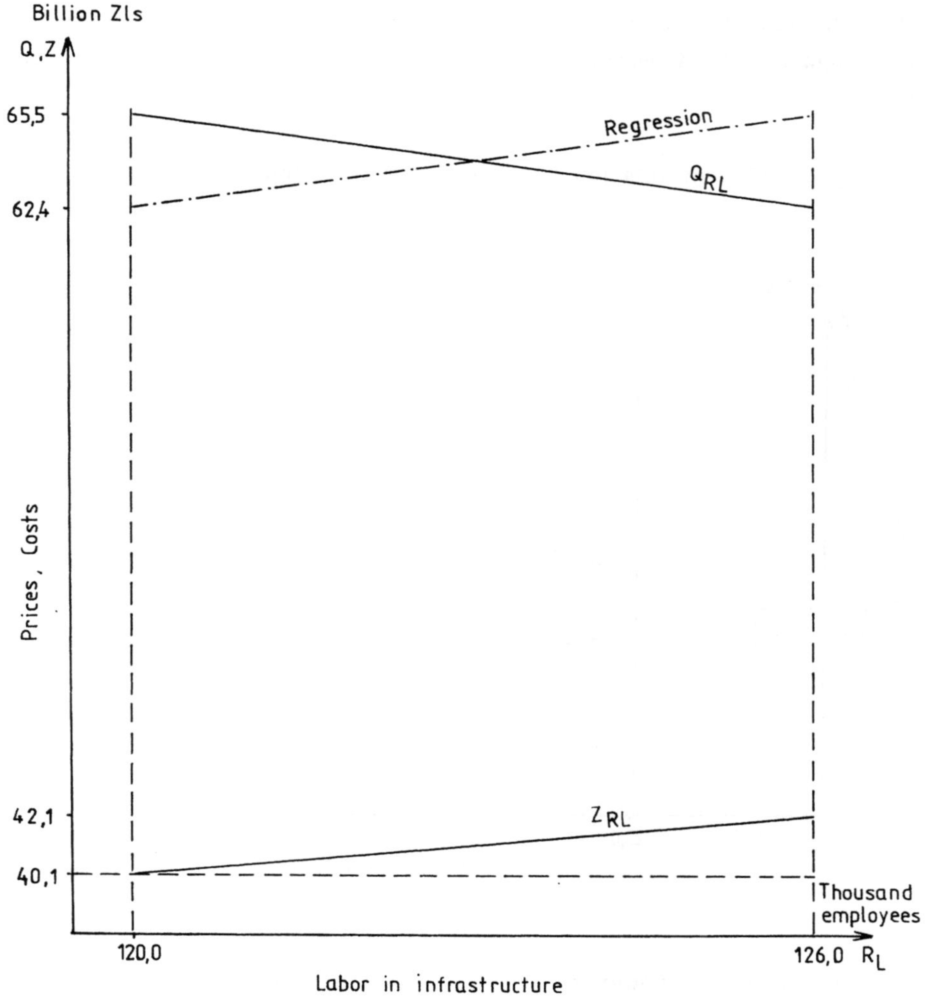

Fig. 5 and 6 characterize the relation between industry and environment. They demonstrate high level of pollution of water and air in the central city. The emission of pollutants exceeds considerably the level of efficiency. The point of efficiency places itself to the left of actual pollution, i.e. below the lowest actual pollution. Such a position of the point of efficiency suggests that the restoration of the quality of natural environment requires not only the engagement of marginal value product of environment but also extra funds.

Figure 5: Efficient relation between water pollution and industrial production

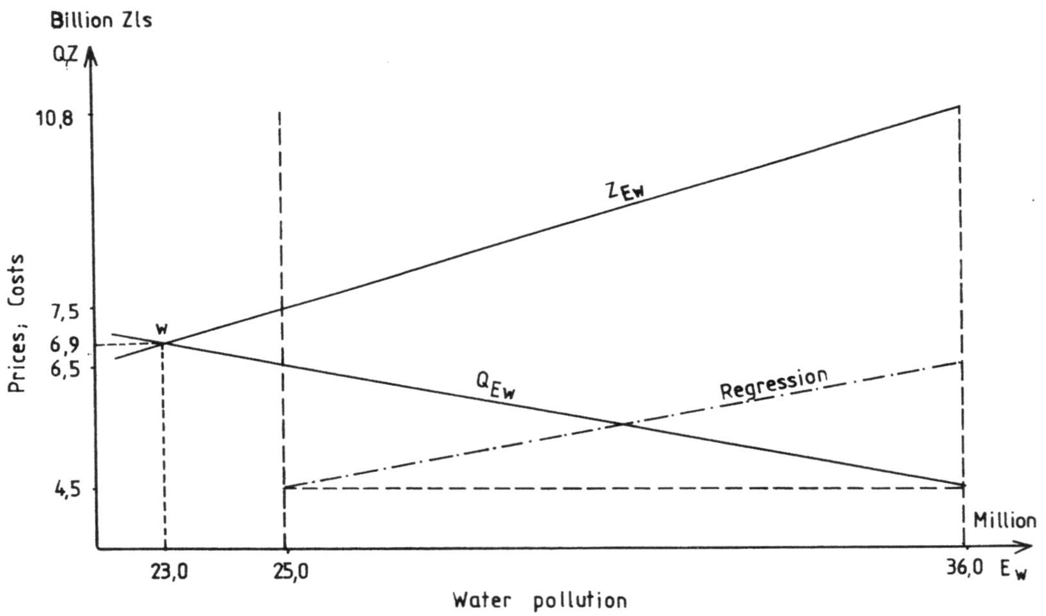

Real situation is, however, just opposite. Industry does not pay full equivalence of the value gained thanks to the use of local resources. It results from our estimates that the value of production gained by industry due to the use of infrastructure and environment is equal approximately to total revenues of local budget. However, the fees, taxes and other payments of industry to this budget constitute less than 50 % of total revenues. Hence, the industry pays less than it gains.

Figure 6: Efficient relation between air pollution and industrial production

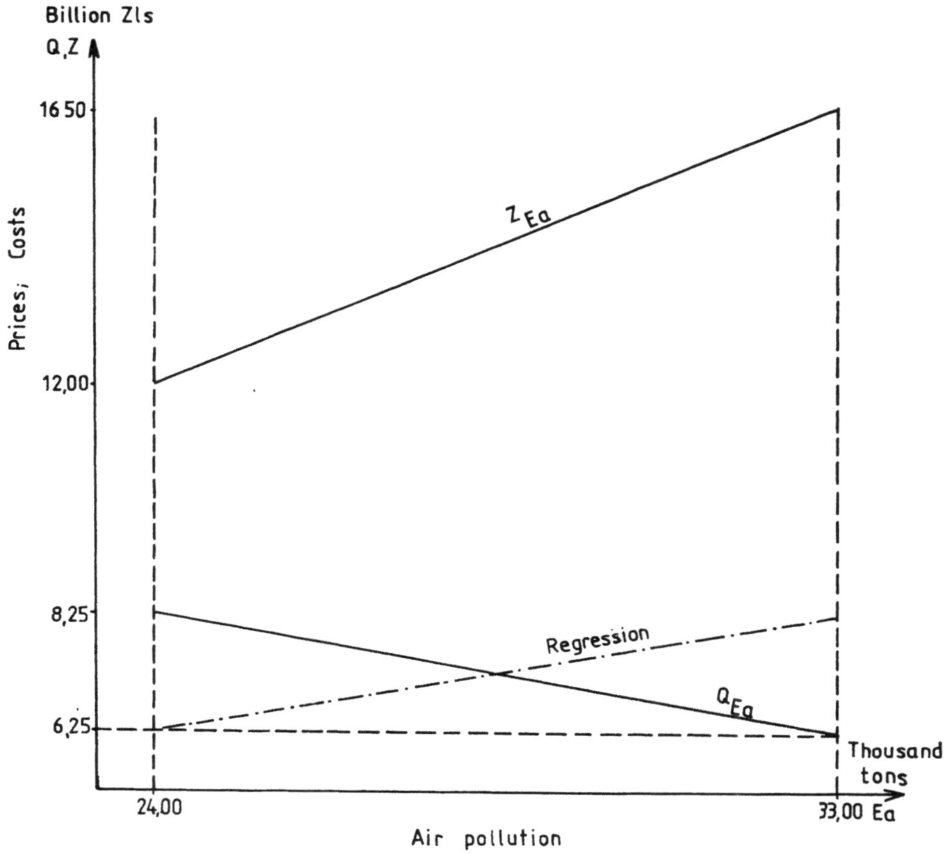

The difference is a contribution of local economy to the creation of extremal economies. Local authorities are interested in the increase of attractiveness of their area for industry. Industry creates jobs and activates socio-economic life. However, local goverment should harmonize the relation between infrastructure, environment and industry. The idea of Pareto optimum in its operational form can be a useful tool of such a harmony.

4. The Relation Local-Supralocal Sectors and the Efficiency of Economy

Efficient relation between infrastructure and industry exerts an influence not only on the functioning of these two sectors. This influence extends also on other sectors of economy. The above relation can be regarded as a special case of a broader problem, namely the problem of relation between local and supralocal sectors of economy.

We examined this broader relation using Polish statistical data. The examination was carried out on three levels: national, regional and urban. The two last levels were represented by the Poznań region, and the city of Poznań. National income was used as a measure of the efficiency of national economy. On regional and urban level, due to the lack of time series of regional and urban income, this measure was replaced by relevant indicators-representatives correlated with the income. The relation between local and supralocal sectors was presented in terms of fixed assets and employment in activities numbered among these sectors. We faced, of course, known difficulties in numbering these activities among sectors. Time series embracing the period of 1975-1986 were examined. Correlation was used to determine the strength of association between sectoral relation and economic efficiency. The significance of correlation was checked using standard statistics.

On national level we received very high negative correlation between national income and the dominance of supralocal over local sector [r = 0,815]. We can put forward the following hypothesis: The efficiency of national economy increases with the decrease of the dominance of supralocal sector, and decreases with its increase. The presumption is accepted that on certain level the curve of efficiency will reach the point of inflexion, after which the efficiency will decrease.

There are two possible explanations of this result: (i) Well developed local sector creates favorable environment for supraregional sector. If, however, the later grows fast, whithout parallel growth of the former, this environment deteriorates, which exerts unfavorable influence on the efficiency of supralocal sector. Only the implementation of new infrastructural projects establishes new sectoral equilibrium. (ii) Inefficiency of several large state-owned industrial enterprises constituting predominant part of supralocal sector.

It is worth noticing that the correlation of economic efficiency and the dominance of supralocal sector decreases in the period of recession. This can be explained by the fact that supralocal sector is more sensitive to economic fluctuations, whereas local sector, being more diversified and tied strongly with less elastic local demand, is more resistent to them.

If instead of time series for the whole national economy we use in calculations geographical [regional] series for one or two specific years the results in terms of correlation are quite different. The geographical series in contrast to time series do not show significant correlation between regional incomes and the relation local-supralocal sectors. The reason of this fact is that regions are not miniatures of the country. Similar level of regional income can be obtained with quite different sectoral structures.

When interpreting results obtained, we must take into account the influence of pricing system and the method of income estimation. Due to the differentation of subsidies and taxes, prices in various industries can diverge substantially from the value of their products. The activities of industrial, transportation, commercial and service enterprises often outflow the administrative boundaries of regions for which regional incomes are calculated. All these factors bear inevitable errors in the calculation of both regional and national income as well as the relation local-supralocal sectors.

On the level of individual region - Poznań region - some of the coefficients of correlation confirm the conclusions drawn from the calculations on national level, other do not. This is due to the reform of administrative division carried out in 1975. It changed the territory of the region and in consequence the sectoral structure formed in previous decades. The reform cut off relatively more supralocal than local activities. This disturbed former associations and equilibrium.

The strongest negative correlation between income and the dominance of supralocal sector occured on the level of city - city of Poznań. The coefficients oscillated from 0,873 to 0,976. This means that the city organism is the most sensitive to the disequilibrium between local and supralocal sectors. The following arguments can be brought to support this conclusion. The industry and other supralocal sectors located in big city require extensive infrastructure. If it does not meet these requirement, economic efficiency of industry decreases. On regional level some deficiencies in urban infrastructure can be filled in by the facilities located in neighbour cities. In calculations, however, we take the cities in their administrative boundaries. Within these boundaries the deficiencies of infrastructure mean sectoral disequilibrium and weaken correlation. Economic efficiency can be ensured only if the growth of supralocal sector is anticipated or at least accompanied by the growth of local sector.

5. Conclusions

The relation between infrastructure and industry as well as between local and supralocal sectors exerts significant influence on efficiency of urban and regional economy. Hence, it is indispensable to form this relation in optimal shape. This paper suggests a method which allows for approximation to this shape. The method refers to the idea for Pareto optimum.

Estimation Procedures for
Transportation Network Equilibrium Models*

by

David Boyce and Ashish Sen

1. Introduction

This paper examines the estimation of parameter values for a class of urban location and travel choice models in which the travel costs are endogenously determined. These models are variously known as combined trip distribution and assignment models (EVANS, 1976), network equilibrium models of urban location and travel choice (BOYCE et al., 1988), or models of congested spatial interaction behavior (SMITH, 1983). A synthesis and discussion of the results of research performed from 1976 to date by the first author and his students is given in BOYCE (1990). The spatial interaction, or trip distribution, model embedded in these models can take the form of the generalized gravity model, a major research interest of the second author (SEN and SOOT, 1981; SEN, 1986).

The question of parameter estimation for such models has arisen in the context of model implementation and evaluation studies (BOYCE et al., 1985; CATS and UIUC, 1987), and has been addressed by CHON (1982), Y.J. LEE (1985), C.K. LEE (1987), BOYCE and LUNDQVIST (1987), LENNERSTRAND and BOYCE (1988). In seeking to re-examine this question, the present paper inevitably has become concerned with the foundations of such models. Fortunately, a wide trail has been blazed through this territory by SMITH (1978, 1983, 1985, 1987) and ERLANDER (1977, 1980, 1985, 1989).

* An earlier version was presented at the Third World Congress of the Regional Science Association, Jerusalem, April 3–6, 1989.

The organization of this paper is as follows. First, the nature and possible interpretation of the residence-workplace data available to estimate these "combined models", as we shall refer to them here, is examined. Next, we consider alternative approaches to choosing values of model parameters. Two approaches are considered. One is the well-known maximum likelihood estimation method of classical statistics. The other involves the solution of the model as a nonlinear programming problem. In the case of fixed travel cost models, it is well known that the maximum likelihood and maximum entropy approaches are equivalent (e.g. ANAS, 1983); however, minimizing cost subject to an entropy constraint leads to a different solution. Extending the maximum likelihood approach to the combined model with endogenous travel costs leads to an important conjecture concerning the conditions on the maximum likelihood solution. We have unsuccessfully searched for a proof of this conjecture; indeed, because of the bilevel programming structure of the formulation, an analytical proof may not exist. Therefore, we undertook computational studies which do seem to confirm the conjecture. Moreover, the proposed conditions show how to improve the parameter estimates, as illustrated by an example drawn from studies of the Chicago region. We conclude the paper with a discussion of future research.

2. Nature of Urban Travel-to-Work Data for Parameter Estimation

Parameter estimation for many types of urban travel models involving origin-to-destination flows has been based on census of population data since 1970 in the United States and in several European countries. In the U.S. census, roughly one in twelve households are asked to report the workplace location and mode of travel for employed adults residing at that housing location (actually about one in six households were questioned but only half the responses were tabulated). From the responses, a residence-workplace-mode table is compiled for this sample, which may be interpreted as a 24-hour travel-to-work table. This table is typically expanded to a total trip table by multiplying each row of the table by essentially the ratio of total workers to the surveyed workers in the corresponding zone. If departure rates for work trips are available by departure time as in the 1970 census or from other surveys, this 24-hour table could be factored to a peak period representing the trips per hour from residence to workplace by mode during the peak commuting period. In Sweden, a similar residence-workplace-mode census was taken in 1975 for all employed persons. In succeeding Swedish censuses, workplace locations were obtained, but travel modes were not.

A characteristic of such census-based, travel-to-work tables is that they are very sparse. For example, in one version of the Chicago region table with about 300 zones (90,000 origin-destination pairs), 71 percent of O-D pairs have no observed auto trips, and 96 percent have no observed transit trips. How should such data be interpreted in the context of a probability model of origin-destination-mode choice which assigns a positive probability to each feasible travel opportunity? We now consider two sampling interpretations of such census data.

Our underlying hypothesis or model concerning origin-destination interaction and mode choice largely abstracts from the realities of individual attributes, tastes and even income. It posits that origin-destination interaction by mode depends only on workplace and residence totals by zone and generalized travel costs by mode. The hypothesis concerns the expected number of trips occurring during some commuting period, given these abstract characteristics. These expected numbers are positive if the travel costs are finite and are zero otherwise.

The number of residence to workplace by mode combinations enumerated for some sampled proportion of households (e.g. 1 out of 12 or 1 out of 1) may be regarded as corresponding to the situation on the census day. Six months later a somewhat different situation would be observed. Likewise, if we asked everyone to leave their homes, and repopulated the area with a different population, a different enumeration would occur.

Given the true probability of each origin-destination-mode combination, a random sample corresponding to 1/12 or 1/1 could be drawn. Since the expected number of actual trips per combination is relatively small (about six during the peak hour for the Chicago region table mentioned earlier), we can expect that many combinations would not be selected at all. Indeed the relative frequency of combinations with no, one, two, three,... trips should be similar to the relative frequencies observed in the census data; however, these frequencies bear only a rough correspondence to the probability matrix.

Similarly, consider some future year. Many realities could obtain in the future concerning the number of trips by origin-destination-mode combination that are consistent with the underlying probability model. In preparing a forecast of this future, say, for the purpose of evaluating a transportation plan, it is not meaningful to consider these many possible outcomes. What we desire is the expected value of these outcomes, i.e., the mean of the possible futures that are consistent with the model and the assumptions about the future.

To summarize, the observed number of trip combinations, say N_{ijm}, is a random variable, whether it is based on a partial enumeration or a complete one. Its expected value, say T_{ijm}, the expected number of trips, is a function of the number of origins and or destinations and the travel costs, depending on what specific model is postulated.

The form of the combined model with which we are concerned in this paper depends on the above account in a most important manner. Briefly, such models embody a constraint which depicts the dispersion of origin-destination-mode choices to higher cost modes, subject to that information assumed to be known (e.g. origin and destination totals). The measure of this dispersion which corresponds to our exponential models is the entropy function:

$$- \Sigma_{ijm} \; p_{ijm} \log[p_{ijm}] = S$$

where S is the dispersion (entropy) and p_{ijm} is the probability of a trip occurring between origin i and destination j by mode m. This function achieves its maximum value when all the probabilities are all equal and its minimum value when the probabilities are as clustered as possible, given the governing constraints. If unobserved origin-destination-mode combination are disregarded, or considered to be zero, assuming 0 log(0) = 0, then clearly, the expected values of the trip table will have a higher dispersion, i.e., a larger value of S, than will an enumeration, or random sample, based on the underlying actual probabilities. Thus the enumeration appears more clustered than the expected values. Of course, the expected values cannot be observed. Therefore, the question arises as to how to estimate the values of parameters corresponding to this dispersion constraint where the constant S cannot be directly observed. To propose an answer to this question that takes account of related results on models with fixed costs, we first consider the fixed cost model.

3. Fixed Cost Model

This section compares results obtained by three methods for the gravity model (or equivalently for the combined distribution and assignment model with fixed link costs) for three estimation methods: maximum likelihood (ML), minimum cost (MC) and maximum entropy (ME). The last two were originally used to derive the gravity model. However, some authors have modified them for estimation purposes. The key question we address is whether they give identical estimates, or at least close to identical ones.

For this purpose we need to distinguish three quantities: the random variable flows N_{ij}, the expected flows $T_{ij} = E[N_{ij}]$ and estimated flows $\hat{T}_{ij}$, which are estimates of T_{ij}. A parenthetic superscript f will denote the future, i.e., $T_{ij}^{(f)}$ is the future value of T_{ij}, and while we will not use it, $N_{ij}^{(f)}$ would be the future value of N_{ij}. Perhaps, the ideal forecast is $N_{ij}^{(f)}$; usually the best we can do is estimate $T_{ij}^{(f)}$.

First consider the following problem: Let O_i, D_j and C be positive numbers with $O_+ = D_+$ where replacement of a subscript by a + denotes summation over the subscript (e.g., $O_+ = \Sigma_i O_i$). Now maximize $-\Sigma_{ij} T_{ij} \log[T_{ij}]$ under the constraints

$$T_{i+} = O_i$$

$$T_{+j} = D_j$$

$$\Sigma_{ij} c_{ij} T_{ij} = C \qquad (1)$$

where c_{ij} is the cost of travel between i and j. If there is a feasible solution to this problem (there is not always one, obviously), it is

$$T_{ij} = a_i b_j \cdot \exp(-\beta c_{ij}) \qquad (2)$$

where a_i and b_j are constants associated with origin i and destination j whose values are unknown. Moreover, the solution is unique, i.e., there are unique values of a_i, b_j and β for which (2) gives a unique set of T_{ij}'s which satisfy (1). It follows that there is a one-to-one correspondence between C and β.

Now consider the following problem: for a given number S, minimize

$$\begin{aligned} & \Sigma_{ij}\ c_{ij}\ T_{ij} \\ \text{so that}\quad & T_{i+} = O_i \\ & T_{+j} = D_j \\ & -\Sigma_{ij}\ T_{ij}\ \log\ [T_{ij}] = S \end{aligned} \qquad (3)$$

If the solution exists, it is equation (2) with a unique value of β. Obviously then there is a one-to-one correspondence between S and β. Consequently, there is also a one-to-one correspondence between C and S.

The above equivalence breaks down if we base O_i, D_j, C and S on the random variables N_{ij}, and solve for a_i, b_j and β as numerical values of Lagrange multipliers.

For the maximum entropy solution of the Lagrange multipliers, replace (1) with

$$\begin{aligned} T_{i+} &= N_{i+} \\ T_{+j} &= N_{+j} \\ \Sigma_{ij}\ c_{ij}\ T_{ij} &= \Sigma_{ij}\ c_{ij}\ N_{ij} \end{aligned} \qquad (4)$$

and solve for the values of a_i, b_j and β that maximize

$$-\ \Sigma_{ij}\ T_{ij}\ \log\ [T_{ij}]$$

For the cost minimizing method, minimize

$$\begin{aligned} & \Sigma_{ij}\ c_{ij}\ T_{ij} \\ \text{s.t.}\quad & T_{i+} = N_{i+} \\ & T_{+j} = N_{+j} \\ & \Sigma_{ij}\ T_{ij}\ \log\ [T_{ij}] = \Sigma_{ij}\ N_{ij}\ \log\ [N_{ij}] \end{aligned} \qquad (5)$$

for numerical values a_i, b_j and β. In general, this procedure yields different numerical values since there is no reason for the correspondence between C and S in the model to hold for $\Sigma_{ij}\ c_{ij}\ N_{ij}$ and $-\Sigma_{ij}\ N_{ij}\ \log\ [N_{ij}]$. Indeed, these latter values would only have the one-to-one correspondence of C and S if $N_{ij} = T_{ij}$. This will rarely, if ever, be the case.

Now, consider the maximum likelihood estimates of a_i, b_j and β. As is well-known, these are obtained by maximizing

$$\Sigma_{ij}\ N_{ij}\ \log[T_{ij}] = \Sigma_{ij}\ N_{ij}\ \log[a_i b_j\ \exp\ (-\beta c_{ij})] \qquad (6)$$

The first order optimality conditions for this unconstrained maximization problem are equations (4), the constraints in the maximum entropy problem. This is not surprising since

$$E\ [\Sigma_{ij}\ N_{ij}\ \log\ (T_{ij})] = \Sigma_{ij}\ T_{ij}\ \log\ [T_{ij}]$$

This well known result was evidently not published for a long time; see ANAS (1983).

Since the ML and ME methods produce the same numerical values of the model parameters, we may attribute the statistical properties of ML to the ME result. These include consistency, asymptotic efficiency and asymptotic normality. It has been shown by Monte Carlo methods (SEN and MATUSZEWSKI, 1987) that when samples are even very small, the estimate of β is relatively unbiased, has a fairly small variance and is approximately normal. Moreover the estimate is quite robust. The latter properties are not surprising, since the observtions enter the computations only through the sufficient statistics N_{i+}, N_{+j} and $\Sigma_{ij} c_{ij}\ N_{ij}$, which are not likely to be too small even when individual N_{ij}'s are small.

In contrast, the ML and MC methods produce different results. Therefore, the value of the estimates of β computed based on equations (5) is likely to be very biased. If one were to estimate S by $\hat{S}$ where

$$\hat{S} = -\ \Sigma_{ij}\ \hat{T}_{ij}\ \log[\hat{T}_{ij}]$$

and $\hat{T}_{ij}$ are the ML estimates of T_{ij}, then the MC method would agree with the ML method. Evidently, the only way to do this is, in fact, to use maximum likelihood. Hence, we conclude that solving for the values of β using the MC approach with S estimated as $-\sum_{ij} N_{ij} \log[N_{ij}]$ is not advisable.

Now consider forecasts. Under the ML tradition, we assume β to be constant into the future, and generate $\hat{T}_{ij}^{(f)}$'s under the assumption that $\hat{T}_{i+}^{(f)}$ and $\hat{T}_{+j}^{(f)}$ are exogenously given. In the ME tradition we may assume that $\bar{C} = C/N_{++}$ is constant into the future or changes in some exogenously assumed way. The value of $\beta^{(f)}$ then depends on $T_{i+}^{(f)}$, $T_{+j}^{(f)}$, $c_{ij}^{(f)}$ and $C^{(f)} = \bar{C}T_{++}^{(f)}$. In general, this value of $\beta^{(f)}$ is different from the ML estimate of β. For the reasons above, the MC value of β computed from $S^{(f)}$ is also likely to be different from the other values.

4. Endogenous Cost Models

The above section on fixed cost models is essentially a review of past results for the purpose of considering the analogous estimation problem for endogenous cost models. For the purpose of this exposition, we use the combined gravity (trip distribution) and useroptimal route choice (assignment) model formulated by ERLANDER (1977), which has a mathematical equivalence to the formulation of EVANS (1976).

ERLANDER's formulation may be viewed as a generalization of the cost minimization formulation given in the above section:

$$\underset{(h_r,\ T_{ij})}{\text{minimize}} \quad \sum_{l \in L} \int_0^{f_l} c_l(x)\, d_x$$

$$\text{subject to: } \sum_{r \in R_{ij}} h_r = T_{ij}$$

$$T_{i+} = O_i$$

$$T_{+j} = D_j$$

$$-\sum_{ij} T_{ij} \log[T_{ij}] = S$$

$$h_r \geq 0$$

where $f_l \equiv \sum_{ij} \sum_{r \in R_{ij}} h_r\, a_{lr}$

Here, h_r with $r \epsilon Rij$, is the route flow on route r, an element of the set of routes, R_{ij} connecting i to j; f_l is the flow on link l, an element of the set of links L. The indicator variable a_{lr} = 1, if link l belongs to route r, and 0 otherwise.

The first order optimality conditions for this problem are the above constraints plus:

$$T_{ij} = a_i b_j \exp(-\beta u_{ij})$$

$$h_r \; [\sum_l c_l(f_l) \; a_{lr} - u_{ij}] = 0$$

$$\sum_l c_l(f_l) \; a_{lr} - u_{ij} \geq 0$$

The u_{ij} are Lagrange multipliers corresponding to the first constraint set listed above.

It should be noted that there also exists an entropy maximizing version of this formulation. In this case the cost constraint becomes

$$\sum_l \int_0^{f_l} c_l(x)\,dx = C'$$

where C' could be calculated from observed link flows. Since this paper is concerned with estimation of models based on census data, and because we are unaware of any suitable set of link flow data, we do not pursue this estimation problem here. A related problem has been studied by ANAS (1988).

As with the fixed cost MC model, the values of the Lagrange multipliers β, ,a_i, b_j can be determined by assuming that equations (5) hold. Just as in the fixed cost model, however, we may expect that the value of β determined in this way will be biased. For this reason, we consider a maximum likelihood approach to the estimation of β:

$$\sum_{ij} N_{ij} \log \, [a_i b_j \exp\,(-\beta u_{ij})]$$

with respect to a_i, b_j and β, where the u_{ij} are the endogenous travel costs that solve the above combined gravity and route choice model. The structure of this problem is a bilevel nonlinear programming problem; see BOYCE (1986). It can be written as an ordinary nonlinear programming problem by replacing the gravity and route choice lower problem with its first order conditions as constraints to the ML problem. In general, such problems are nonconvex; however, C.K. LEE (1987) computationally showed that for this problem a unique local maximum exists.

Since we have not been able to develop analytical first order conditions for the above ML problem, and we suspect that it is not possible to do so, we present the following conjectures by analogy with the fixed cost case:

$$a_i \;:\; T_{i+} = N_{i+}$$
$$b_j \;:\; T_{+j} = N_{+j}$$
$$\beta \;:\; \Sigma_{ij} u_{ij}\, T_{ij} = \Sigma_{ij} u_{ij} N_{ij}$$

There is little doubt that the first two sets of conditions hold, since they correspond to the constraints in the combined problem. The third condition defined on the endogenous travel costs u_{ij} has never been examined, until now, in our extensive computational analysis of this problem. In the past, we have simply chosen β, as well as related generalized cost parameters, so as to maximize the likelihood function. We did not employ the total cost statistic as is normally done in fixed cost case, either as a basis for choosing successive values of β or to monitor its convergence to the optimal value.

In the implementation of a multimodel network equilibrium model, one typically defines the endogenous travel cost u_{ij} as a linear function of a vector of endogenous and fixed costs. For example,

$$u_{ijm} = \gamma_0 \delta_m + \gamma_1 t^1_{ijm} + \gamma_2 t^2_{ijm} + \gamma_3 k_{ijm}$$

where δ_m is an indicator variable equal to 1 for transit and 0 for auto; $t^1{}_{ijm}$ is the in-vehicle travel time from i to j by mode m; $t^2{}_{ijm}$ is the out-of-vehicle travel time from i to j by m; and k_{ijm} is the out-of-pocket travel cost from i to j by m. The coefficients γ_0, γ_1, γ_2, and γ_3 are the weights implicitly attributed to these variables by travelers in choosing among mode and location combinations. The values of these coefficients need to be estimated together with the mode and location deterrence parameter values.

$$\gamma_0 \;:\; T_{++m} = N_{++m}$$
$$\gamma_1 \;:\; \Sigma_{ijm}\, T_{ijm}\, t^1_{ijm} = \Sigma_{ijm}\, N_{ijm}\, t^1_{ijm}$$
$$\gamma_2 \;:\; \Sigma_{ijm}\, T_{ijm}\, t^2_{ijm} = \Sigma_{ijm}\, N_{ijm}\, t^2_{ijm}$$
$$\gamma_3 \;:\; \Sigma_{ijm}\, T_{ijm}\, k_{ijm} = \Sigma_{ijm}\, N_{ijm}\, k_{ijm}$$

If the values of the cost variables are fixed, then the right-hand-sides of these equations are known. If the cost variables are endogenous, then the right-hand-side values can only be calculated after the model is solved for a trial set of values of the coefficients. In the case of the examples given below, t^1_{ij1} and k_{ij1} are endogenous, where mode 1 is automobile. All other values are fixed.

6. An Example

The values of these first order conditions were computed for three network equilibrium models summarized in BOYCE (1988). These models are alternative forms of nested logit models, as described by WILLIAMS (1977) and BOYCE et al. (1983).

1. Simultaneous Residential Location and Mode Choice (SLMR)

$$T_{ijm} = \frac{D_j \; O_i \; \exp\,(-\mu \; u_{ijm})}{\Sigma_{im} \; O_i \; \exp\,(-\mu \; u_{ijm})}$$

2. Combined Residential Location and Mode Choice (CLMR)

$$T_{ijm} = \frac{D_j \; O_i \; \exp\,(-\eta \; \tilde{u}_{ij})}{\Sigma_i \; O_i \; \exp\,(-\eta \; u_{ij})} \quad \frac{\exp\,(-\mu \; u_{ijm})}{\Sigma_m \exp\,(-\mu \; u_{ijm})}$$

where $\tilde{u}_{ij} \;=\; - \frac{I}{\mu} \log \,[\Sigma_m \exp\,(-\mu \; u_{ijm})\,]$

3. Combined Mode and Residential Location Model (CMLR)

$$T_{ijm} = \frac{D_j \; \exp\,(-\mu \; \tilde{u}_{jm})}{\Sigma_m \exp\,(-\mu \; u_{jm})} \quad \frac{O_i \; \exp\,(-\eta \; u_{ijm})}{\Sigma_i \; O_i \; \exp\,(-\eta \; u_{ijm})}$$

where $\tilde{u}_{jm} = \frac{I}{\eta} \log \,[\Sigma_i \; O_i \; \exp\,(-\eta \; u_{ijm})\,]$

In all three models, route choice (R) is governed by the following equilibrium conditions:

$$h_r \quad [\Sigma_l \; c_l \; (f_l) \; a_{lr} - u_{ijm}] = 0, \; r \in R_{ijm}$$

$$\Sigma_l \; c_l \; (f_l) \; a_{lr}' - u_{ijm} \geq 0, \; r \in R_{ijm}$$

In addition, the parameters a_i and b_j in the general formulation given in Section 4 are solved explicitly in these functions with

$$T_{+j} = D_j = N_{+j} \; , \; j = i,\ldots n$$

There is no origin constraint; instead $0_i = N_{i+}$ is used as a measure of the attractiveness of locating in residential zone i. One additional condition can be defined for Models 2 and 3. In Model 2 (CLMR), the value of the parameter μ cannot be distinguished from the values of Υ, since they always appear as a product:

$$\exp(-\mu \; u_{ijm}) = \exp \; [-\mu(\gamma_0 \; + \gamma_1 \, t^1_{ijm} \; + \gamma_2 \, t^2_{ijm} \; + \gamma_3 \, t_{ijm})]$$

Therefore we may set $\mu = 1$ without loss of generality. It may be more useful, however, to set $\mu = (\gamma_0+\gamma_1+\gamma_2+\gamma_3)$ and normalize the values of Υ by this sum. This procedure is followed in Table 1.

It then remains to define a first order condition for the parameter η. By analogy with the condition for μ and γ, it is:

$$\Sigma_{ij} \; T_{ij+} \; - \; \tilde{u}_{ij} \; = \Sigma_{ij} \; N_{ij+} \; - \; \tilde{u}_{ij}$$

or replacing $\tilde{u}_{ij}$ by its definition and simplifying,

$$\Sigma_{ij} \; T_{ij+} \; \log \; [\Sigma_m \; \exp(-\mu \; u_{ijm})] = \Sigma_{ij} \; N_{ij+} \; \log \; [\Sigma_m \; \exp(-\mu \; u_{ijm})]$$

In order to check the validity of these conditions computationally, we computed all of the sums, except the last one, for three models estimated by LEE (1987). The results are given in table 1. The top third of the table gives the estimated values of the coefficients. The middle third gives measures for evaluating the equality of the first order conditions:

$$\left(\frac{\text{estimated cost - observed cost}}{\text{observed cost}}\right) \times 100\ \%$$

If the condition is met exactly, this measure is zero. The bottom of the table gives some additonal evaluation measures, including the log likelihood and

$$e^2 = \frac{\text{(model likelihood - null likelihood)}}{\text{(perfekt likelihood - null likelihood)}} \times 100\ \%$$

These performance measures show that the three models are not estimated equally well. In particular, the CMLR model is poorly estimated. For this reason, the values are inconclusive measures of the model performance.

Table 2 shows the same performance measures for three estimates of the CLMR model varied with respect to the coefficient on out-of-pocket cost. According to these results, the best case is γ_3 = .00526 as judged by log likelihood. However, the estimates for γ_3 = .0060 have better performance measures for all coefficients except in-vehicle time. These results suggest that the values of all of the coefficients need to be adjusted, rather than just one value at a time. Various schemes for adjusting such values to obtain convergence have been suggested in the literature. One simple procedure is to set

$$\gamma_k(n) = \gamma_k(n-1)\ \frac{C_k^{est}}{C_k^{obs}}$$

where n is the iteration number, k is the coefficient index and C_k^{est} and C_k^{obs} are the estimated and observed costs for component k of the generalized cost function. Another possibility is to apply the method of POWELL (1970). The performance of these methods is being tested in our current research.

Table 1: Coefficient Estimates and Evaluation of First Order Conditions for Three Alternatives

	observed	Model SLMR	CLMR	CMLR
generalized cost coefficients				
in-vehicle time		0.2477	0.2809	0.2002
out-of-vehicle time		0.2767	0.5132	0.3975
out-of-pocket cost		0.0359	0.0377	0.0210
transit bias		-0.4398	0.1683	0.3812
dispersion parameters				
mode - μ		0.2928	0.1493	0.2500
location - η		------	0.2675	0.3811
evaluation of first order conditions (estimated-observed)/observed				
in-vehicle time		-0.105%	-0.100%	0.056%
out-of-vehicle time		0.597	0.835	4.879
out-of-pocket cost		1.711	2.229	3.363
transit share		-2.215	-0.883	-13.967
generalized cost		0.527	0.696	- 1.551
regional transit share	18.2%	17.8%	18.0%	15.7%
CBD transit share	67.3%	59.8	52.4	49.8
Likelihood		4483682.6	4485453.5	4506299.2
ρ^2		84.40%	84.30%	83.49%

Likelihood of null model - 6592219.7
Likelihood of perfect model - 4093809.3

Table 2: Comparison of Performance Measure for Alternative Coefficient Values γ_3

Out-of-Pocket Cost (change)	.00526 (-.00037)	.00563 (Table 1)	.00600 (+.0037)
Likelihood/minimum L	1.0	1.000069	1.000254
% diff. in estimated less observed transit share	-1.1603 %	-0.9371 %	-0.5325 %
% diff. in estimated less observed in-vehicle time	0.6936 %	-0.1004 %	-1.0128 %
% diff. in estimated less observed access/egress time	0.9454	0.8346	0.7775
% diff. in estimated less observed out-of-pocket cost	2.9303 %	2.2902	1.7015
% diff. in estimated less observed generalized cost	1.2259	0.6962	0.1373

5. Future Research Directions

We conclude this exposition of estimation issues with a brief research agenda. The first research issue concerns the statistical properties of maximum likelihood estimates of endogenous cost models. In particular, the bias associated with the estimate of μ and with associated generalized cost parameters not discussed here, should be determined using standard Monte Carlo methods. In addition, other statistical properties of these estimates should be examined.

Second, we would propose that a study be undertaken of the forecasting accuracy of these models based on census data. Following the tabulation of the 1990 census, similar origin-destination-mode data will be available for U.S. metropolitan areas for 1980 and 1990. We suggest that a doubly-constrained multimodal gravity model with endogenous travel costs be estimated on the 1980 data. Then using 1990 origins, destinations, and travel networks, these parameters should be used to predict 1990 flows. Alternately, 1990 values of the parameters can be calculated by holding S or C constant at their 1980 normalized values.

Hopefully, such a study will be conducted for several metropolitan areas. Once this is accomplished, the routine application of such models in planning practice will be warranted.

LITERATURE

Anas, A., 1983: Discrete choice theory, information theory and the multinominal logit and gravity models, Transportation 17B, pp. 13-23.

Anas, A., 1988: Statistical properties of mathematical programming models of stochastic network equilibrium, Journal of Regional Science 28, pp. 511-530.

Boyce, D.E., 1986: Integration of supply and demand models in transportation and location: problem formulations and research questions, Environment and Planning, A 18, pp. 485-489.

Boyce, D.E., 1990: Network equilibrium models of urban location and travel choices: a new research agenda, New Frontiers in Regional Science, M. Chatterji and R. Kuenne, eds., New York, Macmillan, forthcoming.

Boyce, D.E.; Chon, K.S.; Ferris, M.E.; Lee, Y.J.; Lin, K.T.; Eash, R.W., 1985: Implementation and evaluation of combined models of urban travel and location on a sketch planning network, research report to the Chicago Area Transportation Study, Chicago.

Boyce, D.E.; Chon, K.S.; Lee, Y.J.; Lin, K.T.; LeBlanc, L.J., 1983: Implementation and computational issues for combined models of location, destination, mode and route choice, Environment and Planning A 15, pp. 1219-1230.

Boyce, D.E.; LeBlanc, L.J.; Chon, K.S., 1988: Network Equilibrium models of urban location and travel choices: a retrospective survey, Journal of Regional Science 28, pp. 159-183.

Boyce, D.E.; Lundqvist, L., 1987: Network equilibrium models of urban location and travel choices: alternative formulations for the Stockholm region, Papers of the Regional Science Association 61, pp. 93-104.

Chicago Area Transportation Study and University of Illinois at Urbana-Champaign, 1987: The sketch planning ridership model for the Regional Transportation Authority strategic plan, report to the Regional Transportation Authority, Chicago, Illinois.

Chon, K.S., 1982: Testing of combined urban location and travel choice models, Ph.D. thesis in civil engineering, University of Illinois at Urbana-Champaign, Urbana.

Erlander, S., 1977: Accessibility, entropy and distribution and assignment of traffic, Transportation Research 11, pp. 149-153.

Erlander, S., 1980: Optimal spatial interaction and the gravity model. Lecture Notes in Economics and Mathematical Systems, 173, Berlin, Springer-Verlag.

Erlander, S., 1985: Remarks on spatial distribution modelling, Environment and Planning A 17, pp. 699-703.

Erlander, S., 1989: Efficient population behavior in networks, dispersed equilibrium and the combined distribution and assignment model, working paper, Linkoping University, Sweden.

Evans, S.P., 1976: Derivation and analysis of some models for combining trip distribution and assignment, Transportation Research 10, pp. 37-57.

Lee, C.K., 1987: Implementation and evaluation of network models of urban residential location and travel choice, Ph.D. thesis in civil engineering, University of Illinois at Urbana-Champaign, Urbana.

Lee, Y.J., 1985: Methods for implementing network equilibrium models of urban location and travel choice, Ph.D. thesis in civil engineering, University of Illinois at Urbana-Champaign, Urbana.

Lennerstrand, K.F.M.; Boyce, D.E., 1988: Remodeling commuting computation for the Stockholm region, paper presented at the 28th European Congress of the Regional Science Association, Stockholm.

Powell, M.J.D., 1970: A hybrid method for nonlinear equations, in: Numerical Method for Nonlinear Algebraic Equations, P. Rabinowitz, ed., London: Gordon and Breach.

Sen, A., 1986: Maximum likelihood estimation of gravity model parameters, Journal of Regional Science 26, pp. 461-474.

Sen, A.; Matuszewski, Z., 1987: Properties of maximum likelihood estimates of gravity parameters, paper presented at the North American Meetings of the Regional Science Association, Baltimore.

Sen, A.; Soot, S., 1981: Selected procedure for calibrating the generalized gravity model, Papers of the Regional Science Association 48, pp. 165-176.

Smith, T.E., 1978: A cost-efficiency principle of spatial interaction, Regional Science and Urban Economics 8, pp. 313-337.

Smith, T.E., 1983: A cost-efficiency approach to the analysis of congested spatial interaction behavior, Environment and Planning A 15, pp. 435-464.

Smith, T.E., 1985: Remarks on the most-probable-state approach to analyzing probabilistic theories of behavior, Environment and Planning A 17, pp. 688-695.

Smith, T.E., 1988: A cost efficiency theory of dispersed network equilibria, Environment and Planning A 20, pp. 231-266.

Williams, H.C.W.L., 1977: On the formation of travel demand models and economic evaluation measures of user benefit, Environment and Planning A 9, pp. 285-344.

Airline Network Structure and Regional Economic Development:
US Case Studies*

by

Aisling J. Reynolds and Geoffry J.D. Hewings

1. Introduction

Professor FUNCK has had a long interest in the role of infrastructure in shaping the path of development for regions at different spatial scales. In this paper, we provide some empirical evaluation of the effects of a major change in the regulation of a transportation industry on the supporting spatial infrastructure in the United States. Contrary to many popular statements, the processes of change were evident before deregulation; while the deregulated system has moved at a more rapid pace, the network structure was evolving over the last several decades.

The process of change in the airline industry in the United States has been heralded as the dawn of a new era in the management of transportation systems. The initial enthusiasm that greeted the Airline Deregulation Act was borne out in the increase in the number of certified carriers, increased frequency of service, lower fares and often dramatic changes in the number of carriers serving any given airport. Recently, there has been a decline in the number of carriers, increasing fares (though, in real terms, these are still below the fares in existence prior to deregulation) and substantial evidence

* Support for this work was provided in part by a grant from the Ameritech Foundation. Additional support was provided by the Regional Economics Applications Laboratory at the University of Illinois

of movement towards oligopoly in the industry as a whole and spatial monopoly at several key airports in the national system. In the space of ten years, the pleas for deregulation have given way to more persistent calls for some form of re-regulation. At the same time, one might note that the process of deregulation is proceeding apace in Europe and several other countries (such as Australia).

In the next section of this paper, the process leading to deregulation is described. The major focus is on describing and documenting network changes by the airlines, and their effects on US airports over a 20 year period. The trends in the network structure for the airline system are revealed through the use of Gini indices of concentration. In the fourth section, an examination of two major airlines, Continental and American, is provided. Finally, some reflections are provided in the context of regional development theory.

2. The Deregulation of U.S. Commercial Aviation

2.1 The Process and Rationale for Regulation

For a period of 40 years, from 1938 to 1978, U.S. commercial aviation was regulated by various parts of the U.S. federal government. Airline regulation was imposed in 1938 in part to preserve the existence of 16 airlines ("trunk" lines) in operation at that time; it was felt that the unregulated free market might result in many of these firms moving to a state of bankruptcy. Hence, it was the fear of market failure and the feeling that few air traffic markets could support more than one airline that provided this initial regulatory trend (PANZAR, 1980). The allocation of federal subsidies in mail carriage provided a major cushion for airlines unable to survive by just transporting passengers. Thus, the airline markets could be considered to have been joint product natural spatial monopolies. All responsibility for the determination of mail rates and passenger fares was consolidated into one agency, the Civil Aeronautics Authority (and later, in 1940, the Civil Aeronautics Board, hereafter the CAB). This institution was given authority over routes, entry into the industry and entry into markets; it was also able to restrict the number of existing carriers that would compete in a given market. The net result was that major long-haul markets were preserved for existing trunks, while new entrants were only able to serve regional markets. Hence, the system of control produced route and market specialization (GRAHAM, KAPLAN and SIBLEY, 1983). Until 1981, when CAB authority ended, an air carrier's certificate listed all routes it was authorized to serve. The CAB was required to apply a test of 'Public Convenience and Necessity' before awarding an interstate route authorization to a new or existing air carrier. The CAB selected the airline to serve the route from the applications tendered; the process was long and cumbersome, with the burden of proof falling on the applicant to demonstrate that other incumbent airlines would not be harmed in the process. Usually, two to three carriers were assumed to be sufficient on any route (CAB, 1984; GRAHAAM, KAPLAN and SIBLEY, 1983). The CAB made a conscious effort to preserve all existing carriers and to narrow the differences between the strong and weak.

It was claimed that regulation allowed for service to a much larger number of cities than would have been the case in a deregulated environment; this assertion was promoted on the basis of the implicit cross-subsidy from the high volume (and hence high profit) markets to the thinner markets. However, in the 1960s and 1970s, questions began to be raised about the level and degree of cross-subsidization actually taking place, especially after empirical evidence was provided that some commuter/air taxi carriers (unregulated by the CAB) managed to provide profitable, unsubsidized service on many routes over which even the subsidized local carriers had not been able to break even (KEELER, 1972).

2.2 The Process of Deregulation

In 1978, the U.S. Congress passed the Airline Deregulation Act (ADA) and provided for a six year transition period. The ADA allowed the CAB to grant operating rights to any carrier applying to serve a route on which a minimum level of service was not being provided; although other carriers had authorization to provide service, they had chosen not to do so. Essential air transportation services were provided in the thinner markets through the continuation of subsidies for a ten year period. Without CAB approval, carriers could decrease fares by up to 50% or raise them 5% above standard industry fares. CAB authority over routes ended in December 1981; a year later, authority over rates (fares), mergers and acquisitions ceased, with the CAB itself ceasing to exist in December, 1984. (MOORE, 1986; CAB, 1982, 1984; MEYER et al. 1981)

2.3 The Effects

How did an airline change its strategy in this post-regulatory environment ? Much depended on the niche held by the airline before regulation. Competition essentially focused on the battle between the existing trunks, characterized by high costs, union-dominated labor and the newer entrants that were low cost and non-union. The markets for which they competed contained a high percentage of discretionary passengers, many of whom found that flying could often prove to be cheaper than Greyhound Bus or AMTRAK rail. Competition focused first on fares and then on quality of service. Spatially, trunk airlines moved from point-to-point service (mainly non-stops flights) to hub-and-spoke oriented systems. The smaller trunk airlines contracted with their niche absorbed by the larger trunks or local service airlines. The latter carriers often entered into joint agreements with the trunks to provide service from smaller communities into the major hubs. It is interesting to note that the local carriers established the hub-and-spoke system because of the large proportion of passengers connecting to the trunk carriers. Their success led to experimentation by the trunks along the same lines (CAB, 1984; KORAN, 1983; MOORE, 1986; MEYER, 1986).

At first, there was an increase in the number of carriers - as predicted by the proponents of deregulation. However, this initial increase was not sustained with the result that the market under free entry and exit now resembles the market that existed under regulation in terms of the number of carriers and the market structure. Mergers and acquisitions in the airline industry are now treated not differently from similar processes in iron and steel or the automobile industries.

In the next section, some of the effects will be described in more detail for the whole U.S. system. Thereafter, the focus of attention will be on two large carriers, to provide specific examples of the patterns and processes of change in the last two decades.

3. Network Structure: Trends in US Commercial Aviation

3.1 Introduction

Economic assessments of the industry since 1978 have failed to focus on the extent and nature of specific network changes, and secondly, the effect of these changes on the national airport system. In this section, both of these issues will be addressed. Specifically, the aims are to (i) describe the changes taking place in the national airways system and attempt to link these with patterns of change among air carriers' route systems; (ii) to ascertain whether the air carriers' propensity to concentrate traffic in their system at a small number of points (i.e. develop and emphasize so-called 'hub-and-spoke' systems) has occurred only since 1978, or if this trend was in place before deregulation.

The next sections begins with a description of the national airways system and the pressing spatial questions are emphasized. The subsequent sections describe the airport system, the traffic distributions over the system in the last 20 years, and the supporting air carrier industry. These carriers are then examined closely; industry characteristics are set out, and two airlines' traffic patterns are carefully analyzed for a 20 year period. The Gini index of concentration is used extensively in the following sections, and justification for its use is provided.

3.2 The National Airways System

The Federal Aviation Administration (FAA), under the Department of Transportation control the airway system, and categorize airports and airfields into two groups: (i) those airports used for air carrier operations which serve the needs of scheduled air carriers and (ii) those for general aviation, serving all other civil air carriers (air taxis, business, agriculture etc.). In this paper it is the former category only, which is the focus of attention. For this category, the FAA developed its air traffic hub structure, and they use this classification scheme to measure the concentration of all civil air traffic, in their economic and operations research procedures (CAB, 1983). The FAA point out that social, economic and other factors influencing a community's ability to generate air carrier traffic are brought together in this one medium. The FAA air traffic hubs are not airports per se, rather, they are the cities and Standard Metropolitan Statistical Areas (SMSAs) requiring aviation services. These communities fall into a four hub classification scheme, depending on the community's percentage of the total enplaned passengers in all services and operations of U.S. certificated route air carriers within the 50 States and other designated areas. The hub classes are given below in Table 1.

Table 1:

FAA hub classification scheme

FAA hub type	Percentage of total US passenger enplanements
Large	1.00% or more
Medium	0.25 to 1.00%
Small	0.05 to 0.25%
Nonhub	less than 0.05%

Airline 'hubs', in comparison, usually refer to those key airports (typically two or three) around which the entire route system is organized. Airline hubs emphasize the transfer of passengers, and accordingly, at these airports, the airlines offer a large number of connecting flights to other areas. The airlines construct what they call 'complexes', which are large blocks of flights arriving at the hub airport within a very short period of time, and this is followed by an equally large number of departures, when passengers have made their connections. The hub airports usually have between 3 and 5 complexes in operation every day. The term 'hub' in the context of airline schedules is more loosely applied.

While most large cities began studies of new airports 20 or more years ago, very little progress has been made in the 1980s, in terms of designing and building new large airports. Since the completion of facilities at Dallas-Fort Worth airport over a decade ago, no major airport has been built in the U.S., though plans were recently outlined for a new 'superhub' at Denver, Colorado. Investment at the major airports in the recent period has been aimed more at upgrading facilities rather than expanding capacity.

At the lower end of the hierarchy, carrier network changes could involve a small community either loosing all of its service, or experiencing phenomenal growth in terms of demand for its facilities, but not being able to satisfy this demand. Given the volatile nature of the market and traffic patterns at smaller airports in particular, the investment and planning process aimed at maintaining and structuring the national airway system has been made considerably more risky. Some major carriers have been increasing their traffic feed to larger hubs by aligning themselves with small carriers, or by directly serving the smaller communities. The airlines have tended to select underused airports in expanding their systems, since these airports would be comparatively free of constraints that restrict many of the larger airport facilities, such as airside and groundside capacity, environmental concerns and airport policies that inhibit further growth. For the FAA, long term planning goals are difficult to formulate when projections of traffic need to be revised as carriers continue experimenting with new routes and connections, but fail to provide advance information to the FAA on their expansion or contraction plans. For the smaller airports being selected by the airlines, the grounds on which they may deny access are limited and difficult to establish now under deregulation. It is up to the community to make a case to the FAA, who are more likely to side with the airline. Deregulation allowed airlines to determine their own routes, rates and service frequency, but no mechanism was left for determining which airlines should get access when an airport has limited facilities (QUINONES, 1986).

The structuring of the air carrier's route system has wide ranging implications not just for the firm's profitability and competitive standing, but also for the nation's airways system. The stability and sustainability of this system depends on the air carriers' choice of airports, their service frequency and the possible connections that may be generated. The understanding of these relationships is important for federal agencies in order for future investments and planning to be effective and efficient.

3.3 Patterns in U.S. Commercial Aviation 1969–87

Since the early 1960s, U.S. passenger traffic has been growing steadily at an average annual rate of 5.8%, with minor declines in 1973/74 and 1980/81 due to the impact of the oil crises, and in the case of the 1980/81 drop, decline must also be attributed to a correction after the temporary but phenomenally high enplanement figures for 1978 and 1979, the deregulation year. Figure 1 shows the traffic trend for the period 1969–87, where traffic is indexed to 1977. International traffic has consistently accounted for between 3 and 5 percent of the total U.S. traffic over this period.

Figure 1: U.S. Passenger Enplanements
Traffic Index 1969–87

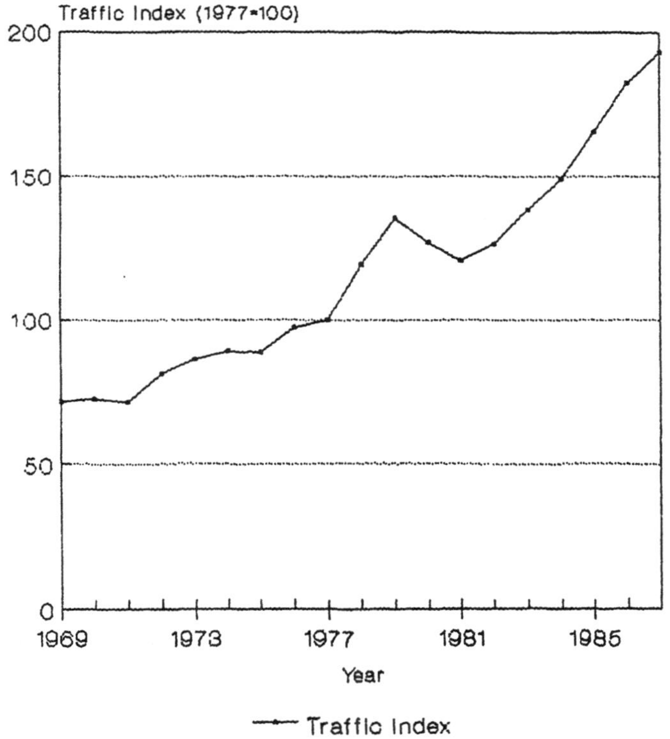

The increase in air traffic has not been equally distributed over the U.S. airport system. Some airports' traffic levels have grown at rates greatly exceeding the national average, while others have declined and indeed a substantial proportion have ceased serving the scheduled traffic market. The changing distribution of traffic can be very simply captured by computing rank correlation coefficients of the top 100 airport's ranks in 1987, with their ranks in 1985, 1981, 1977 and 1973. The vast majority of scheduled air traffic takes place at these top 100 'hubs'. Table 2 shows the percentage of U.S. traffic at these hubs in the four time periods mentioned above, and the proportion of all airports/communities certificated to receive scheduled service that the top 100 represent.

Table 2:

Proportion of traffic at top 100 U.S. hubs 1973-85, and correlation coefficients of the top 100 hubs' ranks in 1987 with other years.

Year	% Traffic	Proportion of CPs[1]	Total no. of CPs	Correlation with 1987 ranks
1969	96.9	27.9	524	-[2]
1973	92.5	13.0	772	0.707
1977	92.3	15.7	638	0.749
1981	94.7	15.9	627	0.848
1985	96.0	26.7	374	0.953
1987	97.7	26.2	477	1.000

Source: CAB/FAA *Airport Activity Statistics of the Certificated Route Air carriers*, Washington D.C..

[1] CPs refers to certificated points.

[2] The correlation coefficient was not computed for 1969, since it was not possible to determine the ranks of all hubs in that year.

The total number of FAA certificated points has dropped substantially since deregulation, so in effect, a smaller number of airports/communities are handling this substantially greater volume of traffic in the 1980s. Figure 2 shows the total numbers of large, medium and small hubs for the period 1969-87. In Table 3, the traffic proportions at all large, medium and small hubs for the period 1969-87 are summarized. The number of large and medium hubs in the system has remained relatively constant in the last 12 years, though the number of small hubs has fallen noticably, and accordingly so has the proportion of traffic at these nodes in a given year. US air traffic has become considerably more concentrated at a smaller number of centers. It is most likely that many of the smaller communities began receiving alternative service when scheduled service ceased, but these data are not as easily or readily available.

Figure 2: U.S. FAA Hubs 1969–87
Number of large, medium and small hubs

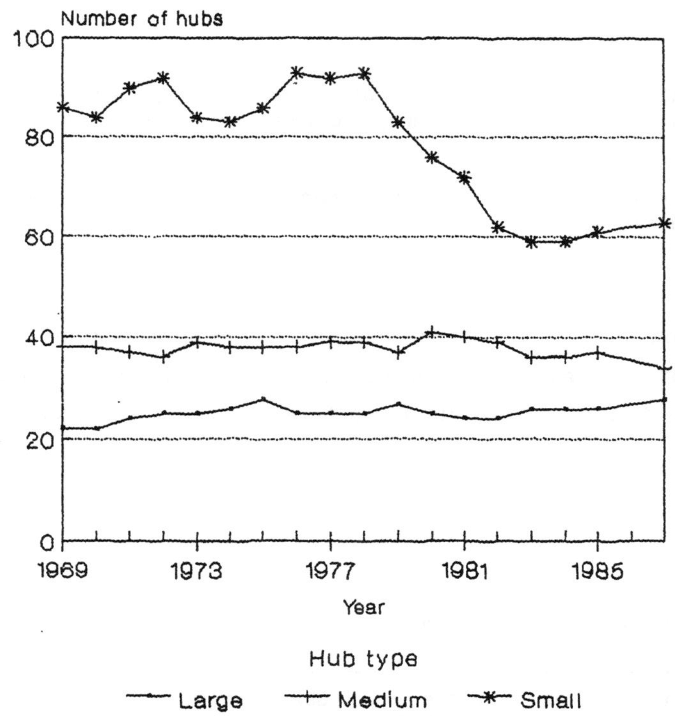

Table 3:

Traffic at large, medium and small FAA hubs 1973-85

Year	Large hubs No.[1]	Large hubs % traffic[2]	Medium hubs No.[1]	Medium hubs % traffic[2]	Small hubs No.[1]	Small hubs % traffic[2]
1969	22	69.7	38	20.4	86	9.9
1973	25	68.4	39	18.4	84	9.0
1977	25	68.1	39	18.4	92	10.2
1981	24	70.2	40	19.0	72	7.9
1985	26	72.8	37	18.1	61	6.7
1987	28	74.1	34	16.6	63	7.1

Source: CAB/FAA Data.

[1] Number of large, medium and small hubs (note: hubs refer to SMSAs and communities, *not* airports, so that there may be more than one airport at the FAA hub).

[2] % traffic refers to the proportion of U.S. domestic passenger enplanements at large, medium anf small hubs.

Along with the reduction in the number of centers receiving scheduled air services, the spacing of the remaining airports has also changed over the recent decades. In Table 4, the average distances between large hubs and

Table 4:

Mean distances between U.S. hub types in miles

Year	Large to large hubs	Large to large and medium hubs	Large to large, medium and small hubs
1965	258	137	68
1969	256	145	72
1973	235	143	80
1977	236	139	79
1981	240	130	90
1985	216	127	101

Source: CAB/FAA Data.

orther large, medium and small hubs for the period 1965–85 are presented. Since the 1960s, it can be observed that the mean distance separating large hubs has been reducing gradually. A less stable trend is noted for the large to large or medium hub figures. The average mileage between large and all other hubs has been increasing over the period.

3.4. Traffic Patterns at FAA Hubs, 1969–87

In the previous section, the tremendous increase in passenger enplanements was traced for the U.S. over an eighteen year period. In this section, the traffic distribution at a sample of those airports are examined so that the pre–1978 patterns of service by carriers may be compared and contrasted with the patterns observed in the deregulated years. In the federal and academic literature on deregulation, many authors often refer to the noticeable change in carriers' routing strategy from predominantly linear, or point-to-point networks, prior to 1978, to 'hub-and-spoke' structures in the 1980s. Most of the evidence for this comes from visual inspection of route schedules and from examining traffic proportions at carriers' busiest airports. The aim in this section is to present data for a longer time period (i.e. 1969–87), and measure this propensity by the airlines to use hub-and-spoke routing systems, and in addition, to examine the effects of this hubbing phenomenon on the U.S. air system.

A random sample of five large, five medium and five small FAA hubs were selected from the 1987 *Airport Activity Statistics of the Certificated Route Air Carriers*. The total domestic scheduled and nonscheduled passenger enplanements were recorded for each certificated carrier serving the hub, and these traffic distributions were detailed for the years 1969, 1973, 1977, 1981, 1985 and 1987, (five year periods, with 1977 being used as a base year, since it was the last fully regulated year; 1987 was the most recent year for which data were available, and is included for this reason).

For the five large hubs (serving between 1.65% and 5.30% of total U.S. traffic), the average number of carriers increased from 8.2 in 1969 to 15.2 in 1987. This increase in the number of carriers is quite marked after 1977, since the ADA allowed carriers to enter new markets freely after 1979. There was an average of 17 carriers serving the large hubs in 1985, though this number dropped in 1987, due most likely to the contraction in the total number of carriers serving the industry, because of mergers, takeovers/acquisitions or less typically, financial failure. The traffic distributions at one of the large hubs in the sample over the study period are illustrated in Figure 3.

Figure 3: Concentration at FAA Hubs
Traffic distribution at a sample large hub: Las Vegas, NV.

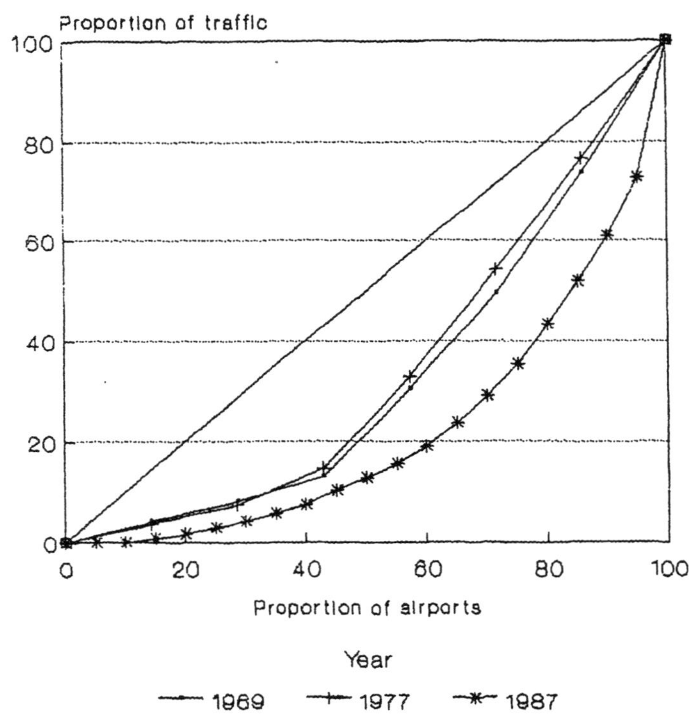

The Gini index, measuring the difference between actual traffic distribution and a traffic distribution where all carriers carried the same proportion of traffic, rose from an average of 0.526 in 1969, to an average of 0.701 in 1987 for FAA large hubs . A Gini index of 0.5 is quite high under any circumstances, so it must be stressed that almost a decade before deregulation, traffic patterns at large U.S. hubs exhibited a high degree of concentration, and after passage of the Act, this tendency became ever more marked. The Gini index is positively and significantly correlated with the number of carriers serving each airport (the R-squared is 0.46). The Gini index is also positively and significantly correlated with the proportion of traffic at each airport (R^2 = 0.44). Regressing the Gini index on both of these variables gives an R^2 of 0.63. The relationship between the proportion of traffic and the number of carriers is very weak, the R^2 value being 0.19. Thus, the busiest airports do not necessarily have a large number of carriers serving them, but these airports are likely to have a more concentrated traffic distribution in comparison to the smaller airports. Those airports being served by a larger number of carriers are also likely to have a more concentrated traffic distribution.

The dramatic increase in concentration, particularly at the largest hubs, reflects the large airlines' consolidation of their "hubs" at these airports in the deregulated environment. The concentrated nature of the traffic distribution prior to 1978, indicates that many carriers had already selected and grown around a small number of key nodes in their systems, and, deregulation allowed this growth to continue in a more streamlined and efficient manner. The air carriers would logically have had a small number of major maintenance facilities at the airports they served, and aircraft rotation schedules would be planned around these nodes. The problem for the airlines with the regulations enforced by the CAB was that the ability to add and cut routes was very limited, and so, traffic flows and aircraft movements could not be channeled through such a narrow subset of nodes. The ADA did allow this streamlining to take place, and the hub-and-spoke networks currently being observed, are the logical progression for the airlines.

For all FAA hubs, it must be remembered that huge growth was taking place in the volume of traffic at these airports, in order for them to maintain their status (i.e. for medium hubs, they consistently needed to carry between 0.25 and 0.99% of total U.S. traffic). As with the large hubs, the increase in the number of carriers serving these airports is quite dramatic after 1977, though for the medium hubs, the increase is at a slower rate, with this trend being consistent for the 10 years since 1977. The Gini index for the medium hubs is consistently lower than for the large hubs. The index began increasing substantially after 1977, rising from 0.361 in that year to 0.567 in 1987. The airlines select these nodes as obvious centers of growth and accordingly, have moved in and in many instances, used them as secondary hubs in their systems. Figure 4 plots the average Gini index for each hub type for the 18 year period, while Figure 5 shows the average number of carriers serving each hub type over the same period.

Figure 4: U.S. Commercial Air Traffic
Average Gini coefficient at each hub type

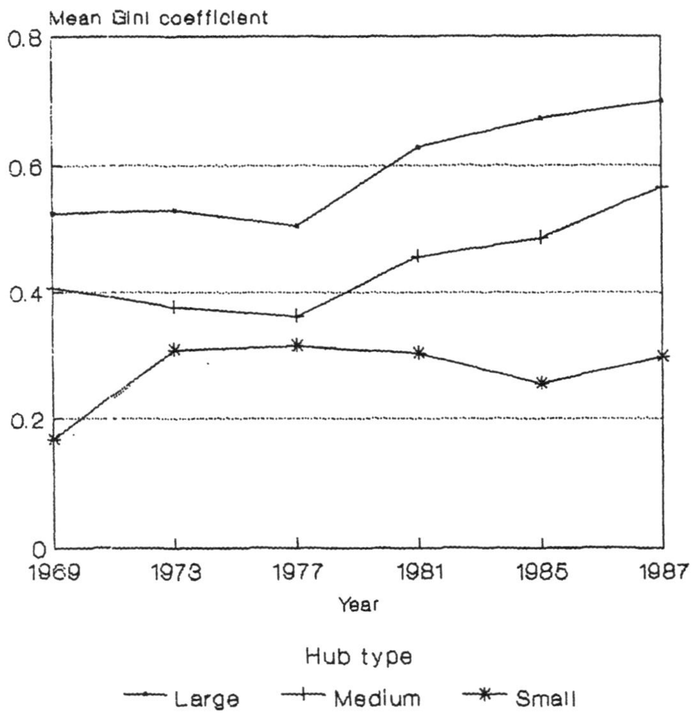

The number of carriers serving small hubs has increased also in the deregulated period. In 1969, there were an average of 2.4 carriers at the five small hubs in this sample. There were 2.6 on average in 1977 and 5.6 in 1987. However, looking at the traffic distributions at these airports, it is noted that only a small change has taken place in the concentration levels, in the 10 years since deregulation. Generally speaking, the Gini index continues to reflect a more equitable distribution of traffic among carriers serving small hubs. This is good for these communities, since if their traffic load is being relatively evenly spread over an increasing number of carriers, they are also spreading their risks, giving them greater stability.

Figure 5: U.S. Commercial Air Traffic
Average number of carriers at FAA hubs

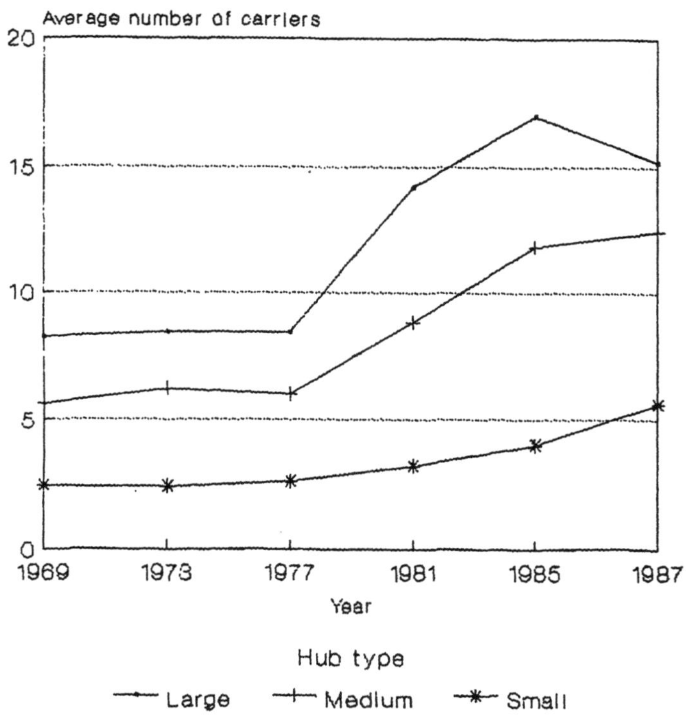

3.5. U.S. Air Carriers 1969–1987

Attention is now turned to the airline industry, and the size of that industry over the 18 year period, 1969–87.

Figure 6 plots the total number of certificated carriers for the selected time period. Between 1969 and 1978, the number of carriers fluctuated between 33 (1974) and 37 (1970), dramatically rising to 52 in 1979, the first fully deregulated year. The industry continued to expand in the subsequent years. In 1983, the total count dropped by 9 carriers, and the general trend recently, has been a gradual fall-off in the number of carriers in the industry. The decline does not necessarily reflect financial failure of carriers, but rather the extent of mergers, takeovers and acquisitions by other airlines in the industry.

Figure 6: U.S. Air Traffic 1969–1987
Number of Certificated Carriers

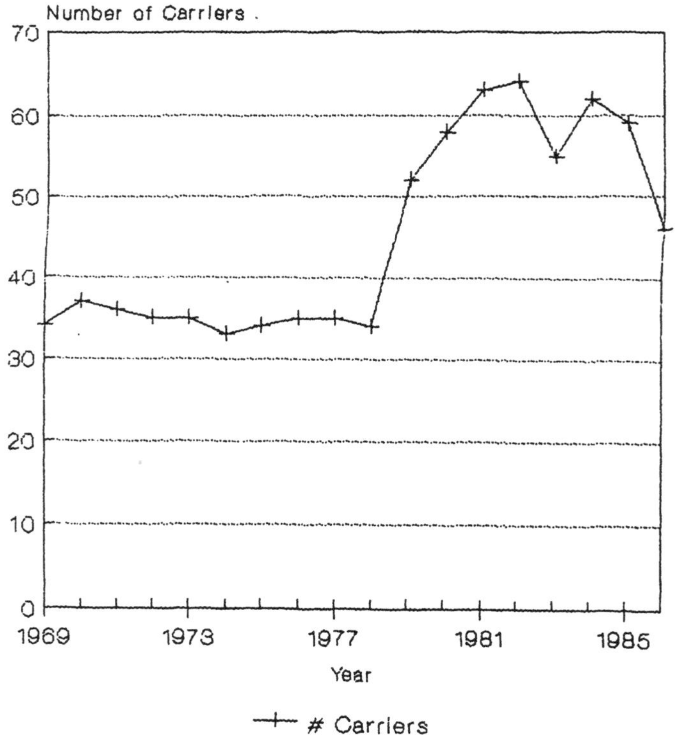

Air carriers are allocated to one of four groups: The majors (with annual operating revenues of at least $1 billion); the nationals (operating revenues of between $100 million and $1 billion); the large regionals (operating revenues of between $10 and $100 million); and the medium regionals (with operating revenues of less than $10 million). The number of majors has remained relatively constant over the period 1969–87. The greatest changes have occurred in the medium and large regionals categories. The number of medium regionals fell sharply from 26 in 1982 to just 4 in 1987, while the number of large regionals has varied up and down from a high of 23 in 1984, to 15 in 1987. Looking at the trends for the earlier period (1969–80), there were consistent numbers in all categories up until 1979. The categories were different to those listed above, focusing on carriers certificates rather than traffic volume; this older scheme broke down after passage of the Act. The major carriers served 77.4% of the total U.S. passenger traffic in 1987, and attention is now turned to examining two of these carriers in more detail.

4. Case Studies: Continental Airlines and American Airlines

4.1. Introduction

Both Continental and American existed before regulation was first imposed in 1938, and as was pointed out above, airlines in this position were given special attention and protection under CAB regulation. The economic status of air carriers in the current period was, to a large extent, predetermined by their status under regulation. The trunk carriers emerged in the late 1970s with by far the largest share of the air passenger traffic market, the most extensive aircraft fleet and gate ownership levels and with long standing public reputations. In the ten years since 1978 however, three of the 11 trunks of 1978 have declared bankruptcy and two were acquired by other trunks. In 1988/89, 9 of the 11 trunks were flying, many under different management and corporate frameworks. Table 5 summarizes the main changes taking place between 1978 and 1989.

Table 5:

Major changes in the U.S. trunk carriers 1977-1989.

1977 Trunks	Changes since 1977
American Airlines	[1]
Braniff Airways Inc.	Bankruptcy declared 1982. Re-entered as a national carrier 1984
Continental Air Lines Inc.	Acquired by Texas International 1981. Bankruptcy declared 1983. [3]
Delta Air Lines Inc.	
Eastern Air Lines Inc.	Acquired by Texas International 1986 Bankruptcy declared 1989. [3]
National Airlines Inc.	Acquired by Pan American 1979.
Northwest Airlines Inc.	[2]
Pan American World Airways	
Trans World Airlines Inc.	Acquired Ozark Airlines
United Air Lines Inc.	
Western Air Lines Inc.	Acquired by Delta Air Lines 1986.

[1] American Airlines acquired Air California 1987

[2] Northwest acquired Republic in 1986 (Republic consisted of North Central, Southern and Hughes Southwest)

[3] Texas Air (formerly Texas International) now consists of Continental Airlines (of which People Express (a post 1978 entrant), New York Air, Frontier, Rocky Mountain Airlines, Bar Harbor Airways, Britt Airways and Provincetown-Boston Airlines Inc are subsidiaries) and Eastern Airlines.

4.2. Continental Airlines and American Airlines Traffic Patterns, 1969-87.

In this section, two major U.S. airlines are focused, the aims being to (i) examine the effects of deregulation on the size and share of two carriers' operations, (ii) trace the characteristics of these carriers' networks for a roughly 20 year period, and examine the changes in these carriers' networks in terms of their size, features of the airports served, traffic, and, distribution of traffic and levels of concentration in the overall network system.

Table 6 shows a traffic index for Continental and American from 1969 to 1987 (1977, the base year, =100) and also the share of total domestic traffic handled by these airlines over the same period. American's volume of traffic has increased 240% since 1977, though it's share of the passenger market is almost exactly the same in 1987 as it was in 1969. This airline's market share dropped in the 1970s and continued to decline initially under deregulation. American has managed to recapture its earlier share of the market via extension of its network and purchase of smaller feeder airlines (e.g. Air California Inc., and American Eagle). Continental Airlines was a smaller trunk, and its market share varied from 3.0% of total traffic in 1969 to 3.4% of traffic in 1973. The volume of traffic handled by Continental rose slowly after 1978. As seen in table 5, Continental was acquired by Texas Air (formerly Texas International) in 1981, and declared bankruptcy in 1983. After re-organization in 1984 and 1985, the company purchased People Express (and its subsidiary, Frontier), New York Air and Rocky Mountain Airways Inc., from its parent company. Continental rapidly increased its volume of traffic, 530% over the 1977 volume. Its share of the U.S. market grew from 3.04% in 1981 to 9.1% in 1987.

Table 6:

Continental and American Airlines' traffic indices and share of domestic U.S. traffic market 1969-87

Year	Continental Airlines		American Airlines	
	traffic index	share of U.S. market	traffic index	share of U.S. market
1969	63.68	3.009	86.99	12.642
1973	90.03	3.443	92.48	10.874
1977	100.00	3.293	100.00	10.125
1981	111.49	3.048	107.92	9.071
1985	208.50	4.153	181.98	11.145
1987	534.82	9.099	242.29	12.675

Source: CAB/FAA *Airport Activity Statistics of the Certificated route air carriers*, various editions.

In Tables 7 and 8, the number of domestic airports served by both carriers are recorded for the period 1969-87, and these are subdivided into the FAA hub classifications. Also given in this table, are the proportions of traffic at each hub type over the period. Both airlines have greatly expanded the size of their domestic networks since 1977, Continental increased its numbers from 33 airports in 1977 to 122 in 1987, while American increased its numbers from 50 in 1977 to 119 in 1987. Looking at the emphasis within the networks, American Airlines has consistently focused their operations on the FAA large hubs. The proportion of traffic at these large hubs increased in the deregulated period, when all FAA large hubs were served by this airline.

The proportion of traffic at medium hubs has declined even though the airline served a larger proportion of FAA medium hubs in 1985 and 1987 then it did in 1977.

Table 7:

Continental Airlines: traffic distributions and domestic system characteristics, 1969–1987

Year	Number of airports served	Number of hubs served				Proportion of traffic at each hub type			
		L	M	S	N	L	M	S	N
1969	28	10	9	8	1	63.35	23.37	13.09	0.19
1973	32	10	10	9	2	65.01	22.11	12.68	0.20
1977	33	12	9	9	2	67.81	20.49	11.92	0.07
1981	48	19	17	9	1	71.73	22.06	5.64	0.57
1985	54	23	16	6	3	83.37	11.84	3.89	0.89
1987	122	27	28	33	23	78.81	14.16	4.89	2.13

Table 8:

American Airlines: traffic distributions and domestic system characteristics, 1969–1987

Year	Number of airports served	Number of hubs served				Proportion of traffic at each hub type			
		L	M	S	N	L	M	S	N
1969	44	14	21	3	1	72.27	25.31	0.76	0.09
1973	47	15	22	6	0	71.47	25.46	1.43	0.00
1977	50	16	21	9	1	74.17	23.02	2.73	0.08
1981	67	23	28	14	0	79.85	17.94	2.22	0.00
1985	102	26	34	30	3	78.44	15.93	5.41	0.23
1987	119	28	32	39	10	72.69	20.98	6.01	0.32

Continental Airlines has significantly increased its traffic proportion at the large hubs it serves. In 1977, 67.8% of its traffic passed through large hubs, while in 1987, 78.8% of its traffic was channelled through these communities, when all but one of the FAA large hubs were served. As with American, the proportion of traffic at medium hubs has declined, along with a decline in the proportion of traffic at small hubs also. In 1987 however, this airline dramatically increased the number of small hubs and nonhubs that it served: much of this increase in the network size is explained by the mergers and acquisitions mentioned above.

The traffic distributions over the airlines' systems were examined for each of the time periods selected. Scheduled and nonscheduled domestic passenger enplanements for each airport in the system were recorded for the six years. A series of summary statistics were computed for these traffic distributions for both carriers, and these are detailed in tables 9 and 10. The two most popular series of measures used in the economics literature to summarize traffic distributions are the Herfindahl index and the proportion of traffic at the top 1,2 and 3% most utilized airports. In Tables 9 and 10, these figures are recorded for both airlines.

The Gini index, Theil measure and coefficient of variation all have one obvious major advantage over the proportion of traffic measure, since they utilize the full distribution, and therefore take into account the increases in the number of airports served by both airlines. Looking at one example, the fact that 16.4% of Continental Airline's traffic was at one airport in 1969, and 18.1% of it's traffic was at its busiest airport in 1987, masks the 436% increase in the size of the system, as measured by the number of nodes. 18.1% of an airline's traffic at one airport, when it serves a total of 122 nodes, is intuitively more concentrated than having 16.4% of its traffic at one airport when only 28 nodes are served. In addition, any changes in the traffic distributions at the lower ranked airports, will not effect the summary statistics usually computed (i.e. the Herfindahl index and proportion of traffic indices).

Table 9:

Measures of concentration for Continental Airlines traffic distributions 1969-87

Year	C.O.V.	Theil index	Gini index	Traffic at top 1,2,3 & 4 airports				Herfindahl index
1969	1.185	-0.831	0.565	0.164	0.328	0.411	0.491	0.086
1973	1.220	-0.840	0.553	0.164	0.318	0.401	0.464	0.078
1977	1.255	-0.838	0.555	0.173	0.314	0.413	0.468	0.078
1981	2.252	-0.737	0.674	0.313	0.408	0.501	0.560	0.127
1985	2.382	-0.729	0.659	0.241	0.478	0.531	0.561	0.124
1987	2.763	-0.730	0.741	0.181	0.305	0.422	0.472	0.071

Table 10:

Measures of concentration for American Airlines traffic distributions 1969-87

Year	C.O.V.	Theil index	Gini index	Traffic at top 1,2,3 & 4 airport				Herfindahl index
1969	1.276	-0.844	0.589	0.139	0.244	0.350	0.417	0.061
1973	1.387	-0.829	0.603	0.142	0.253	0.353	0.423	0.062
1977	1.525	-0.816	0.618	0.160	0.278	0.372	0.446	0.066
1981	2.157	-0.748	0.705	0.198	0.356	0.430	0.504	0.084
1985	3.255	-0.704	0.713	0.278	0.448	0.492	0.523	0.114
1987		-0.731	0.718	0.235	0.377	0.424	0.459	0.085

The coefficient of variation, Theil's entropy measure and the Gini index show gradual increases in both American and Continental Airlines' traffic distributions between 1969 and 1987. These three measures are generally consistent in the trends they show. The Herfindahl index shows a different trend however, the change in this index is much smaller over the period. It suggests very slight increases in concentration of traffic for both airlines from 1969-85, and a drop in concentration between 1985 and 1987 (even though, in the space of two years, there were 126% and 17% increases in the number of airports served by Continental and American Airlines respectively). The proportions of traffic at the top 1,2,3 and 4 airports served by both airlines, show the same trend as the Herfindahl index.

The Gini index is proposed as a superior measure to those previously used in the economic literature, because it takes account of the entire distribution and makes unit free comparisons. In addition, it satisfies the axioms proposed by SEN (1976) which capture several intuitive properties of a good measure of concentration. In the context of air traffic for example, the Gini index is superior in measuring the effects of passenger transfer from one of the top ranked airports to one of the lower ranked airports in an air carriers system (REYNOLDS, 1989). The Gini index is also easily illustrated, and it is used now to highlight changes in both Continental and American' straffic patterns in the last c.20 years.

Figures 7 and 8 illustrate the changing nature of Continental and American Airlines' traffic distributions between 1969 and 1987. The high level of concentration in both airlines' traffic patterns in 1969 and 1977 indicate that the recent phenomenon of hub-and-spoke networking is not so drastic a change in organization, but rather a refinement of a general trend among larger carriers, to concentrate flows from an increasing number of nodes in their system, through a small number of key airports or carrier hubs. The same comments were made earlier in section 3.4, when airport patterns were examined. These trends have become exaggerated since 1977, but were very definitely in place prior to deregulation.

Figure 7: Continental Airlines
Traffic distribution 1969-87

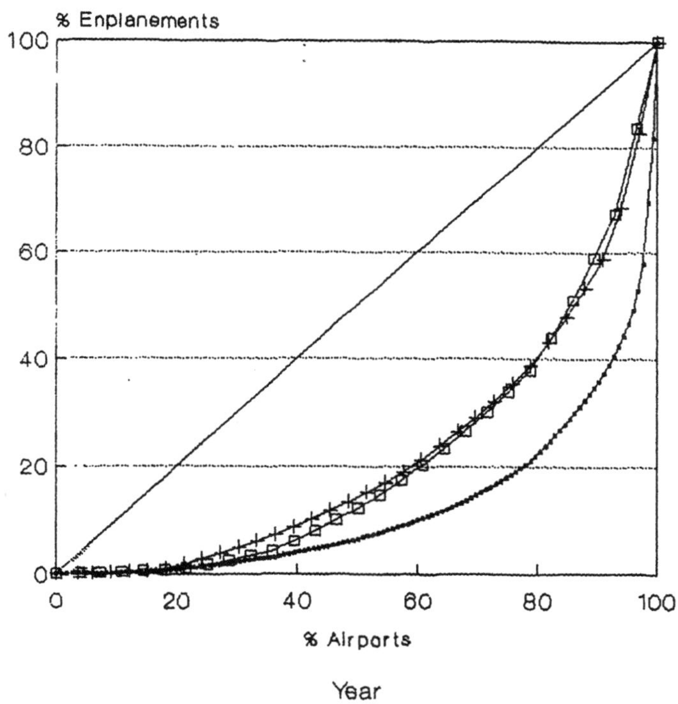

Figure 8: American Airlines
Traffic Distribution 1965–85

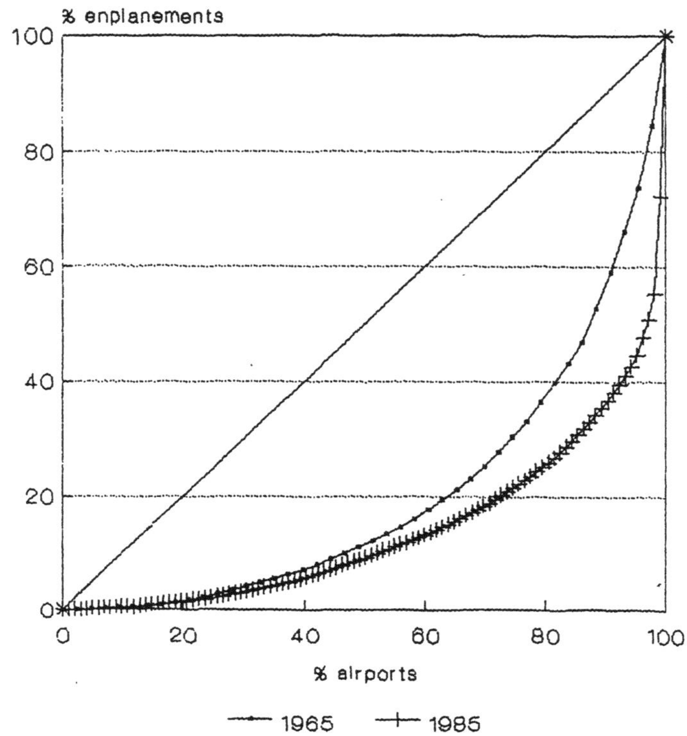

To summarize then, the major U.S. carriers examined here, have greatly expanded their networks (be it through natural growth or merger with/ acquisition of other carriers) since 1977, and have maintained or increased their market shares by serving a very high proportion of the large and medium FAA hubs, and by streamlining their services through a small number of key hub-airports. The size of the national airways system serving the scheduled traffic market has contracted considerably since deregulation. This smaller national network which is handling a tremendously increased volume of traffic, is now being served by a greater proportion of the major carriers. The 11 trunk carriers in 1977 accounted for 77.86% of all U.S. traffic in that year. In 1987, the 12 major carriers accounted for 77.40% of U.S. traffic:

change in the industry structure has occurred among the smaller carriers: the large and medium regionals were the classes of scheduled carriers experiencing the greatest change, in terms of new entrants and subsequent exits.

In the final section of this paper, the implications of these changes for regional economic development and issues in regional development theory are outlined. Suggestions for further avenues of research are offered.

5. Implications for Regional Development Theory

What role do transportation systems play in conditioning the path, direction and pace of regional economic development? In particular, can one document a clear link between transportation improvements and economic change? How do changes in the organization of transportation systems affect regional development. The first two questions are ones that have occupied a great deal of attention in the literature at a variety of spatial scales as authors have struggled to relate transportation improvements to overall levels of national and regional welfare. The last question raises an issue that is far more difficult to answer since the effects of changes in spatial organization are often more difficult to trace. In the case of the airlines, the differences in organization have often come about in two important ways. First, there are the types of changes described in sections 3 and 4, the movement from point-to-point to hub-and-spoke systems. Secondly, and often complementing this reorientation are changes in the nature and character of the service provided. For example, the changes in the organization may increase the frequency of departures from an airport; the involvement with a hub in which interchange among several connecting flights might be possible will improve the indirect connectivity of the airports at the spokes with the rest of the system. Finally, one may note that the capital embodied in the service may change; greater frequency of service might be offered but in smaller aircraft and perhaps involving a substitution of non-jet for jet or narrow-bodied jet for wide-bodied jet aircraft. While these changes can be documented, the problem remains one of tracing the effects of these changes on other indicators within the regional economy.

The process of industrial restructuring that has characterized the economies of almost all western nations has resulted, in SCOTT's (1988) terms in the articulation of "flexible production spaces." This process has been accentuated by substantial, sustained vertical disintegration in the production process. Improvements in telecommunications and transportation in general have created opportunities for firms to expand the choice set of location possibilities. Factors that in traditional locational analysis appeared at the lower ends of the ranking now would appear to have been given greater prominence; the character of other factors, especially their form, would appear to have changed. This is particularly true in the case of access to transportation, first seen in terms of railroad access, then to highway/freeway access and most recently access to major airports.

As more and more commodities and inputs (especially human capital) with high value-to-weight ratios are being transported by air, proximity to major airports would appear to have become an important consideration. Yet, to date there would appear to be little documented evidence of the impact that the designation of community as a hub has had on its economic development. As with most transportation impact analysis, there remains the important distinction between cause and effect in a synergistic process of development; designation of transportation improvement as the cause or effect of economic development continues to prove intractable.

The process of hub creation has a major impact on a community in terms of jobs created and sustained over a period of incredible change and uncertainty in traditional, manufacturing enterprises. But to what degree have Raleigh-Durham (American Airlines), Dayton, Ohio (Piedmont, now USAir) or Denver (United and Continental) benefitted from the development of hubs at their airports in contrast to places like Columbus, Ohio or Indianapolis that are non airline hub airports?

It is interesting to note that there have been some serious discussions about the development of "Wayports," large airports located in the central part of the U.S. away from major urban concentrations. These airports would serve the role of transfer points rather than final destinations. Needless to say, the major airlines, with substantial infrastructure investments in major urban airports are not very excited about such propositions. The construction of wayports would involve disembodiment of the airport from the urban system and represent a phenomenon that has not occurred in other transportation systems.

What has been the effect of changes in the spatial structure of the airline system on the process of regional development? The available evidence would seem to suggest that the airlines have responded to rather than anticipated economic development. The decrease in the number of carriers, through attrition but mainly mergers and acquisitions, may serve to further neutralize the potential impacts of the post regulated environment on regional development.

There are however, some major changes that are likely to require close monitoring. The first is the movement towards joint ventures between US airlines and carriers from other countries (e.g., SAS and Texas Air; Delta and Swissair, United and British). Closely associated with this is the degree to which US airlines are positioning themselves to take advantage of the increasing internationalization of the productive process. Penetration of the Pacific market has yielded substantial profit for those carriers fortunate enough to have authorization to serve points such as Tokyo, Hong Kong and Seoul. Across the Atlantic, US airlines are rapidly completing arrangements to serve European cities before 1992 when the fear of dealing with an unknown European Community agency might limit market penetration.

In the final analysis, the ability to offer substantial evidence for a clear link between changes in network structure and regional development remains elusive. Available data are not in a form to render the kinds of analysis and interpretation that would be necessary. On the other hand, the degree to which communities benefit from the presence of an airline hub may revolve more around the characteristics of the individual airline's total network rather than the presence of absence of the hub per se. In the United States, the absence of major developments of new airports over the next decade is likely to create additional opportunities for the specification of alternative, integrated systems of transportation. The spokes in the hub and spoke system may be served by high speed rail rather than short-haul jet and non-jet aircraft; if the development of magnetic levitation rail systems can be brought into economically viable reality, offering service at speeds up to 300mph, the character of the airline system will undergo a transformation that might be even more dramatic than the changes that have accompanied the movement to the present hub-and-spoke system.

LITERATURE

CIVIL AERONAUTICS BOARD, 1982: "The Consequences of Airline Mergers Since Deregulation", CAB and DOT Report, Washington D.C.

CIVIL AERONAUTICS BOARD, 1984: Implementation of the Provisions of the Airline Deregulation Act of 1978, Report to Congress, Washington D.C.

CIVIL AERONAUTICS BOARD; U.S. DEPT. OF TRANSPORTATION, 1965-1987: Airport Activity Statistics of the Certificated Route Air Carriers, Washington D.C.

Graham, D.R.; Kaplan, D.P.; Sibley, D.S., 1983: "Efficiency and Competition in the Airline Industry", Bell Jour. Econs. 14(1), pp.118-138.

Keeler, T.E., 1972: "Airline Regulation and Market Performance", Bell Jour. Econs.3, pp.399-424.

Koran, D.W., 1983: "The Welfare Effects of Airline Fare Deregulation in the United States", Jour Transport, Econs. and Policy Vol. 17, pp.177-188.

Meyer, J.R.; Oster, C.V.; Morgan, I.P.; Bergman, B.A.; Strassmann, D.L., 1981: Airline Deregulation - The Early Experience, Auburn House, Mass.2

Meyer, J.R., 1986: "Book Review of Bailey, Graham & Kaplan's 'Deregulating the Airlines'", Rand Jour. Econs. 17(3), pp.461-466.

Moore, T.G., 1986: "U.S. Airline Deregulation: Its Effects on Passengers, Capital and Labour", Jour. Law and Econs Vol. XXIX, pp.1-28.

Panzar, J., 1980: "Regulation, Deregulation and Economic Efficiency: The Case of the CAB", American Economic Review, 70, pp.311-315.

Quinones, W., 1986: "Airports battle airlines on right of access to community airfields" Airport Services Management, January, pp. 8-11.

Reynolds, A.J., 1989: "Measuring the impact of network changes by the airlines in the deregulated environment", unpublished Ph. Dissertation, University of Illinois, Urbana.

Scott, A.J., 1988: Metropolis: From the Division of Labor to Urban Form. Berkeley, University of California Press.

Sen, A., 1976: "Poverty: an Ordinal Approach to Measurement", Econometrica, 44(2), pp.219-231.

Employment Impacts of Infrastructure Investments

A Case Study for the Netherlands

by

Frank Bruinsma, Peter Nijkamp, Piet Rietveld

1. Introduction

The interest in infrastructure - from both a scientific and a political angle - is exhibiting a wave-like pattern. In the period of the sixties a broad interest in infrastructure expansion emerged as a result of the unprecedented economic growth necessitating the construction of large infrastructure networks. In the seventies much more emphasis was placed on a more efficient use of existing network capacity, mainly as a result of the limits to growth discussion (including environmental externalities and resource scarcity). In the eighties the attention shifted towards infrastructure and economic stagnation, as a result of a particular interest in the relationship between public investment and employment, whilst in the second part of the eighties also the restructuring effects of new infrastructure (e.g., informatics, telematics) came to the fore (see e.g. GIAOUTZI and NIJKAMP, 1988). Very recently also the relationship between infrastructure and international trade (and competition) - from the viewpoint of the unification of the European market - has been given due attention (including the interest in high speed transport systems).

Unfortunately, various contributions to infrastructure research have been either only theoretical in nature or weak in empirical foundations. Speculations on the impact of infrastructure have been more prevalent than solid scientific analyses. Apparently, the influence of infrastructure on the behaviour of economic actors (e.g., the (re)location of firms) is not a very well researched issue. For instance, in a critical overview of the research on the structural impacts of infrastructure on regional or urban development, VAN GENT and NIJKAMP (1988) conclude that in answering this question, the following elements have to be taken in consideration:

(i) Regional development is a latent concept, which cannot be unambiguously measured. Observable indicators such as gross regional product or income per capita do not capture all elements of economic development. A similar problem applies to infrastructure: this is not an unambiguous concept either.

(ii) The contribution of infrastructure to regional development depends also on its uniqueness. An increase in an ubiquitous infrastructure category does not exert a driving progress on a region. For instance, road expansion in an industrial area with a highly developed infrastructure network will have lower effects than that in an underdeveloped area (decreasing marginal benefits).

(iii) Infrastructure is a conditio sine qua non, but certainly not a sufficient condition for growth. First, infrastructure policy requires a comprehensive and tailor-made supply of all relevant infrastructure categories (due to synergetic effects). Second, infrastructure will only have a positive impact if the region at hand has already a favourable existing potential for new development. The implementation of new infrastructure in an economically weak region may even run the risk that the region at hand suffers from strong competition of enterprises in more distant regions. Thus infrastructure has to be considered in relation to the whole locational profile of an area. Third, infrastructure investments will only have a discriminating effect on regional development, if the competitive position of a region is enhanced. Fourth, the impacts of infrastructure are also co-determined by the general economic situation: in case of a less favourable economic situation, the probability of surpassing a threshold level for survivorship is much lower.

(iv) The impacts of infrastructure are also determined by technological renewal. The improvement of the regional competitive position also requires flexibility and resilience in terms of infrastructure investments.

Altogether, network infrastructure is indispensable for regional development but the extent to which it will have a decisive influence on regional growth is not unambiguous. But in any case, it is evident that regions or countries with a poor infrastructure network ('missing links') run the risk of staying behind in the national and international economic restructuring.

A major problem in most infrastructure analyses is that long-term impact inferences have to be drawn from a static and short-term framework (LAKSHMANAN, 1989). This is also illustrated in an interesting study of MILLS and CARLINO (1989) who conclude that the long-term employment effects of infrastructure investments may be several times larger than the corresponding short-term effects, although their model with distributed lags is essentialy based on a comparative static production function approach. Apart from specification problems (e.g., the separation of infrastructure effects from agglomeration effects), a major flaw in infrastructure impact analysis is caused by a severe lack of reliable and up-to-date statistical information (see also WIGREN, 1984).

In this context, an interesting contribution has been provided by BLUM (1982), who employs a Cobb-Douglas function in which the production factors related to infrastructural stocks are separately defined as input potentials characterized by spatial immobility for a medium-to short-term period. Such input potentials may refer to the transportation network, the telecommunication network or the knowledge or information network.

In any case, it is evident that in a more competitive international environment accessibility is a sine qua non for economic efficiency, i.e., that the development of appropriate infrastructure is a critical success factor for countries and regions in Europe. In this context, it is also important to call attention for the timing of infrastructure investments. For instance, Van DER ZWAN (1986) claims that - beside conventional long waves in economics - also infrastructural developments display cyclical movements associated with new technologies and economic upswings and downswings. Coordinated infrastructural investments (e.g., in the area of communication and transport technology) might reinforce economic growth, an idea already advocated by HIRSCHMAN (1958) in his theory on unbalanced growth (including the tuning of private and public investments).

In this paper we will pay particular attention to the measurement and impacts of infrastructure investments. Therefore, in the next section we will first define infrastructure in a more operational sense, whilst next a brief review of various infrastructure impact models will be given. Then we will present in more detail results from an infrastructure impact study in the Netherlands, with a particular view on the assessment of its economic importance in terms of investment amounts and (direct and indirect) employment effects. This study was undertaken in order to provide a more solid empirical basis for the question whether (public) infrastructure investments would significantly contribute to an improvement of the employment situation in the Netherlands. The difficulties of analysing the spin-off effects of investments in telecommunication infrastructure on the Dutch economy will be discussed using the results of a multisectoral macro-economic model.

2. Methodology of Infrastructure Impact Analysis

2.1 Identification of Infrastructure

The measurement of infrastructure is far from easy. Infrastructure is usually defined as the stock of all social overhead capital that is (directly or indirectly) necessary for a proper functioning of all direct productive capital. Usually it has the following features (see NIJKAMP, 1986):

- immobility (in geographical sense)
- indivisibility (in terms of 'lumpiness')
- non-substitutability (in the sense of limited replacement possibilities)
- polyvalence (in terms of multiple purpose use for other activities)

In a cross-national empirical research project for the European Community a broad view of infrastructure was adopted (see BIEHL, 1986). The following main infrastructure categories (subdivided into more than 100 subclasses) were distinguished:

- transportation
- communication
- energy supply

- water supply
- environment
- education
- health
- special urban amenities
- sports and tourist facilities
- social amenities
- cultural amenities
- natural environment

Clearly, in various other studies different definitions have been adopted (see for a review also LAKSHMANAN, 1989). A general subdivision which will also be used in this paper is into economic infrastructure (as a necessary condition for increasing productivity and efficiency) and social infrastructure (as a necessary condition for increasing individual and social welfare). Examples of the first category are roads, railways or airports, whilst examples of the second category are schools, museums or community centres.

In our study the following infrastructure items were distinguished, according to a typology made by the DUTCH SOCIAL & ECONOMIC COUNCIL (1987). It is noteworthy that infrastructure does not explicitly include publicly financed network projects. Clearly, an unambiguous classification is very hard to reach; for instance, universities belong to social infrastructure, but contribute indirectly to economic efficiency (and hence may be regarded as partial economic infrastructure). Furthermore, infrastructure is exhibiting drastic qualitative changes, for instance, from conventional physical networks to teleports and logistic chains linked to various transport modes. The following categories are dealt with in our study:

- TRANSPORTATION
  * roads
  * railroads
  * waterways
  * airports
  * harbours
  * information transmission
  * pipelines

- WATER AND ENERGY SUPPLY
  * water supply
  * natural gas supply
  * electricity supply
  * district heating

- WATER CONTROL SYSTEMS
  * water quantity control
  * water quality control

- INFORMATION CONTROL AND KNOWLEDGE DEVELOPMENT
  * general information networks (i)
  * specific information networks (ii)
  * knowledge development (iii)

- CITY STRUCTURE
  * sewage
  * industrial sites
  * waste disposal
  * urban layout

- LAND (RE)STRUCTURING PROJECTS
  * new land (iv)
  * land restructuring (v)

(i) Information networks for public use, for instance, a public library in a computer system with which people can connect their own home computer;

(ii) Information systems for a special group of users, for instance, logistic transportation systems wich shorten time needed for customs formalities;

(iii) large technological institutes;

(iv) Projects whereby, for instance land is reclaimed from the sea;

(v) Landrestructuring projects whereby landscapes are adapted to technological development in agriculture.

For each of these categories an assessment of the order of magnitude (in terms of average annual investments) and related employment impacts will be provided in section 3.

2.2 Infrastructure Impact Analysis

Economic infrastructure will evidently lead to various economic effects in terms of value added, productivity and employment. Clearly, infrastructure may also incorporate embodied technological progress and contribute to productivity increase via a rise in capital productivity. It may also be a substitute or a complement to various production factors.

Infrastructure impact analysis for assessing the effects on employment can be undertaken in two complementary ways, *direct* effects related to design, construction and building aspects of infrastructure provisions and *indirect* effects related to the derived (second order) consequences of the creation of infrastructure. The first category may be seen as the conventional (Keynesian) expenditure effects, whereas the second category refers to all effects generated by intermediate deliveries (as multiplier effects). In both cases, the pay-back effects of infrastructure investments in the framework of employment policy may play an important role (see also KUIK and NIJKAMP, 1987). Furthermore, we may distinguish *programme* effects (or 'structuring' effects) associated with the supply of infrastructure. One group of programme effects consists of long-term structural employment effects based on maintenance and management. Another group of programme effects consists of spin-off effects caused by changes in the relative locational attractivenes of places or regions for new enterprises.

Three different modelling approaches can be distinguished in the analysis of spin-off effects of infrastructure (cf. RIETVELD, 1989).

Firstly, improvement of infrastructure leads to an increase in the productivity of the private production factors labour and capital. This can be modelled by means of *production functions* where both private and public production factors are taken into account. A sample of models following this approach has been summarized in table 1 (derived from RIETVELD, 1989). It appears that it is usually very difficult to carry out such an analysis at a high level of detail in terms of economic sectors and types of infrastructure.

This is a pity, since the infrastructure requirement may vary strongly among economic sectors.

Table 1: Examples of the Production Function Approach to Infrastructure Modelling

Author	Country	Number of sectors	Number of types of infrastructure	Presence of: labour	Presence of: private capital	Form of production function
Biehl	E.C.	1	1	yes	no	Cobb-Douglas
Blum	F.R.G.	3	8	no	no	Cobb-Douglas
Andersson et al.	Sweden	1	7	yes	yes	Cobb-Douglas (with modification)
Snickars and Granholm	Sweden	21	5	yes	yes	Leontief
Nijkamp	The Netherlands	1	3	yes	no	Cobb-Douglas
Fukuchi	Japan	3	3	yes	yes	Cobb-Douglas
Kawashima	Japan	8	1	yes	no	linear

Secondly, improvement of infrastructure in a region or a country leads to a spatial reallocation of labour or capital. This leads to a *factor movement approach* to infrastructure impacts. Most of the models in this group introduce accessibility as a location factor for employment or private capital. Improvement of network infrastructure leads to an improvement of accessibility which in its turn may lead to an increase of employment or investment. We use the word 'may' on purpose, since from a theoretical viewpoint an improvement of accessibility may also lead to a decrease of employment in a region. This may be the case when a local industry, formerly sheltered by isolation, cannot stand the competition of industries located elsewhere after the improvement of accessibility. However, in empirical research it is an exception when such a negative impact is found (cf. BOTHAM, 1983).

A third way of analyzing the effects of network infrastructure is the *trade flow approach*. Improvement of infrastructure leads to a reduction of transportation costs and hence to a reorientation of trade patterns. For this purpose one needs a detailed multi-regional model of the impact of infrastructure on transportation costs, as well as estimates of the sensitivity of trade flow parameters to changes in transportation costs. Examples of such models can be found in AMANO and FUJITA (1970) and LIEW and LIEW (1985).

3. A Case Study for the Netherlands

In this section a general review will be given of the development of the investments in the infrastructure network in the Netherlands. The development of the infrastructure categories and the amounts of investments in these categories throughout the last fifteen years will be shown with some common trends in infrastructural development and investment patterns in the Netherlands. It appears that officially published statistical information is not always sufficient as a data source. For various infrastructure categories a deeper investigation of raw data - and some minor estimations - had to be made in order to trace the total direct and indirect employment effects of infrastructure investments in the Netherlands in the year 1985.

3.1 General Evolution of Infrastructure Investments

In the Netherlands nearly 80 % of the total expenditure in infrastructure is public investment, of which some 75 % is invested by lower governments. The projects not financed by the government are mostly semi-state company projects such as energy-supply projects.

The total amount of infrastructure investments in the Netherlands rose sharply after the Second World War. The expenditures needed to restore the war damage were not yet completed, when a great part of the south-west of the country was flooded in 1953. To defend the nation against the sea an extensive system of walls and dikes was initiated, the so-called Deltaplan. These works were finished only recently. The economic upswing in the sixties made huge investments in public infrastructure not only necessary but also possible. In the seventies it became clear that growth has its limits. The saturation point is reached and the financial position of the government could not bear any more high infrastructural investments. Figure 1 shows over the period 1977-1985 that, although the investment in the civil engineering works in current prices rose slightly, the investments in constant prices of 1980 demonstrates a sharp decline (from 9.0 billion in 1977 towards 6.9 billion in 1985). As a share of the national production, the decline is even more pronounced.

Figure 1: Investments in Civil Engineering Works in the Netherlands in Current Prices and Constant Prices of 1980 for the Period 1977-1985 (in Billion Dutch Guilders)

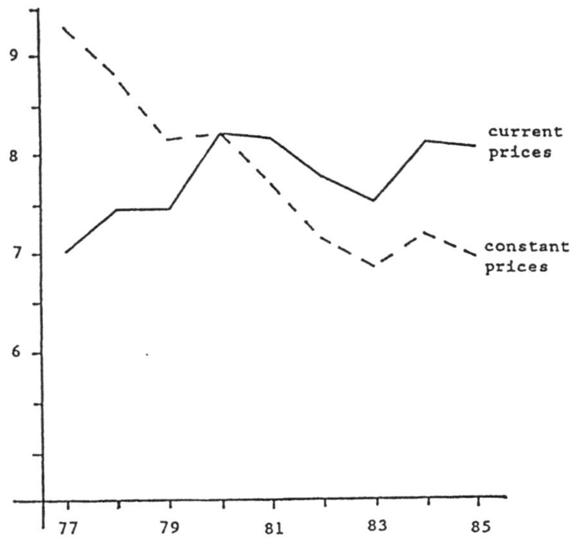

3.2 Common Trends in the Growth Path Investments in Various Infrastructure Categories

The first common trend in the development of most infrastructure categories with a netwerk structure is a growth in terms of the length of the total network in the sixties and early seventies. After this period there is a tendency that the *length* of the network remains constant, but that the *capacity* of the existing links increases. For instance, in the early period the total amount of miles of highway rose rather fast, whilst in the latter period there is a tendency for growth to take place mainly in the form of an increase in the number of lanes on the highways. In the period 1970-1988 the total amount of travel miles in the Netherlands doubled, so that the need for a further increase of the road capacity is clear. There are two infrastructure categories with a netwerk structure that demonstrate different development patterns, i.e., waterways and railroads. Both exhibit a declining network in the sixties and seventies because of the closure of unremunerative goods-lines. Remarkable is that the railroad network expands again in the eighties due to the construction of new commuter lines in or into the metropolian area 'the Randstad'.

It is rather difficult to trace the yearly investments of certain infrastructure categories over a longer period of time. Specific problems arise when different governments (e.g., local, regional or national) subsidize each others investments. For example, the State subsidizes road investments of the provinces which in their turn subsidize local road investments. This problem does not arise when there is only one prime investor, such as the railroad company, the telephone company and the energy-supply companies.

The investment pattern of the electricity-supply company (see figure 2) is a typical example of the pattern which can be observed for most infrastructure categories. There is a tendency of a small rise of the amounts of investment

<u>Figure 2:</u> Investment Pattern Electricity Companies in the Netherlands in Current Prices and Constant Prices 1980 for the Period 1970-1987 (in Million Dutch Guilders)

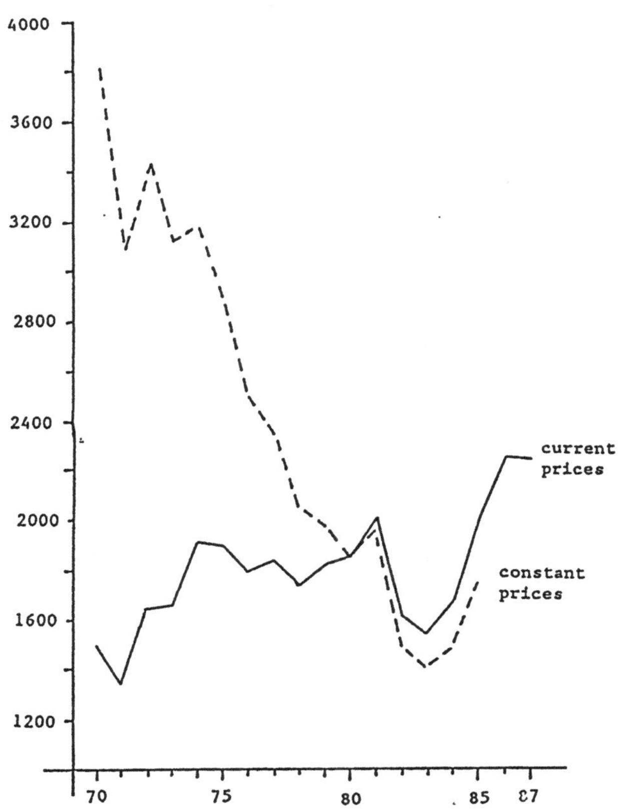

in current prices but, on the other hand there is a rather strong decline in constant prices of 1980. Again the railroads show a rather different pattern. Both current prices as well as constant prices demonstrate a rising tendency. This is due to the government policy to stimulate public transport systems as a substitute for private car ownership. This reorientation of investment policy away from the road system towards more advanced railway systems can be observed in several other European countries (cf. E.C.M.T., 1988). The European railway system is showing signs of a logistic saturation path for its conventional infrastructure, but is - in view of high speed railway transportation - certainly not at the end of its life cycle as is sometimes believed (cf. ANDERSSON and BATTEN, 1988).

3.3 Direct and Indirect Employment Effects

In this section the direct and indirect employment effects of the investments in the different infrastructure categories are put together to provide an overview of the total employment effects (excluding programme effects) of infrastructure investments in the Netherlands in 1985. For each category the direct and indirect employment effects per million of investment are calculated. The direct employment effect per million investment varied from 5.5 towards 9.0 manyear. The indirect employment effect ranged from 3.5 towards 5.4 manyear. It was almost impossible to calculate figures for the rather new infrastructure categories of 'Information control and Knowledge development'. The assessment of all figures is presented in table 2.

Obviously, the figures on indirect employment are most difficult to obtain. They can be computed in principle by means of employment multipliers generated in the context of input-output analysis. However, an input-output matrix with this degree of sectoral detail is not available. Therefore the indirect employment effects had to be estimated on the basis of raw data on the input structure of various segments of the construction sector.

Altogether, with a total investment in infrastructure of roughly 9.2 billion Dutch guilders, nearly 88.000 manyears were involved. This implies roughly that an extra investment of about 105.000 guilders in infrastructure would create a new job. It must be admitted that the volume of labour created by infrastructure is relatively small compared to the total number of unemployed infrastructure is relatively small compared to the total number of unemployed (about 500.000 persons). Nevertheless, an increase of government expenditure in this direction would be a welcome contribution to the decrease of unemployment. Such an increase of public investment may also have negative effects on employment in the private sector, however. Crowding out at the

Table 2: Investment in Economic Infrastructure and the Direct and Indirect Employment Effects (1985)

Category	Investment	Employment	
		Direct	Indirect
TRANSPORTATION			
roads	2.600	14.300	12.740
railroads	585	3.510*	2.106*
waterways	130	1.040	624
airports	95	570	342
harbours	164	966	573
information transmission	978	4.172	2.503
pipelines	7	42	23
WATER AND ENERGY SUPPLY			
water supply	419	2.514*	1.466*
natural gas supply	429	3.432	2.145
electricity supply	1.990	11.110	5.000
district heating	290	1.619	730
WATER CONTROL SYSTEMS			
water quantity control	132	3.590 (both)	2.166 (both)
water quality control	487		
CITY STRUCTURE			
sewage	460*	3.640	2.116
industrial sites	51*	306	183
waste disposal	15*	90*	60*
urban lay-out	nn	nn	nn
LAND (RE)STRUCTURING PROJECTS			
new land	370 (both)	2.479 (both)	1.480 (both)
land restructuring			
TOTAL	9.202	53.380	34.257

* = estimation
nn = figures unknown

capital market may lead to a decrease of private investments, e.g. due to higher interest rates. One must be aware that there may also be an opposite effect of public investments on private investments, however, i.e., when spin-off effects occur. Such effects will be discussed in the next section.

3.4 Spin-off Effects of Infrastructure

The empirical analysis of spin-off effects of infrastructure is a difficult task. As an example of the problems one usually encounters, we discuss the results of a study on the impacts of investments in telecommunication infrastructure on the Dutch economy. For this impact study, a multisectoral macro-economic model has been used (SECMON). See DRIEHUIS (1987) for a discription of the main features of the model. SCHRIJVER and VAN WESTRENEN (1987) present simulation results of a dfl. 500 million increase in yearly investments in communication infrastructure during a five year period. Six different types of effects are distinguished:

(i) An exogenous investment impulse in infrastructure leads to an increase in domestic production and employment because of input-output relationships. A countervailing effect is that the rate of interest rises, wich leads to a decrease in other investments.

(ii) A change in consumption patterns occurs (more videotex, less leisure activities), which has differential effects on employment in the various sectors.

(iii) Improved telecommunication leads to an increase of intermediate deliveries of this sector to other sectors.

(iv) An increase in telecommunication infrastructure leads to a higher labour productivity. This results in a combination of a price decrease, higher production and less employment.

(v) In a similar way one may also expect a higher capital productivity, which leads to a decrease in prices and investments, and an increase in production.

(vi) The price decrease mentioned in (iv) and (v) implies also increased competitiveness at the international market, so that export increase.

Model simulations show that the six effects are very different in size and sign. As far as the employment impact is concerned, the largest effects are the labour productivity effect (iv) which is negative, and the export effect (vi) which is positive. The overall effect on employment of the infrastructure impulse is scarcely larger then zero. However, the effect on the aggregate production level is clearly larger than zero (a yearly increase of about 0.5 %, the export effect being the dominant component).

In the simulations, the situation in the rest of the world is assumed to be constant. If there would be similar investment programs in the other countries, the effects on the Dutch economy would be clearly smaller. In that case the Dutch infrastructure program would only help to maintain the existing market shares: the export effect would be zero.

It is helpful to analyse this result in terms of generative versus distributive growth. Distributive growth occurs when a region grows at the expense of other regions. Generative growth occurs when the overall system of regions grows. The present simulation results suggest that a considerable part of the possitive effect on production in the Dutch economy is in fact a redistribution from elsewhere. This conclusion holds true at the national level. At the regional level an even larger share of the total effect would consist of redistribution effects.

Although SECMON is a well-developed economic model, it needs several additional assumptions in order to arrive at the simulation results. It seems that the simulation results depend as much (or even more) on these additional assumptions as on the parameters of the SECMON model. For example, additional assumptions are needed on changes in consumption patterns, input-output coefficients, labour productivity and export growth in order to use the SECMON model for simulating the employment impact of the investment programme. The assumptions have to be made at an ad-hoc basis, which underlines how difficult it is to generate the spin-offs of specific infrastructure investments.

4. Concluding Remarks

Infrastructure is an indispensable factor for economic development. Yet its role is often poorly understood. A major problem concerns causality: almost all modelling approaches assume that regional development follows infrastructure improvement. The reverse may be equally true, however, as was already pointed out by HIRSCHMAN (1958). Thus, correlations between the two elements say nothing about the effectiveness of infrastructure policies as a tool for regional development.

There are clear decreasing returns to scale in infrastructure. Adding another highway in an already well developed region will not give a large economic response. Investments in new types of infrastructure may have much larger effects.

Direct and indirect employment effects of infrastructure investments may be quite large, but they only exist in the short run. Input-output analysis is a suitable tool for this purpose, but standard input-output tables are not always directly applicable, since technical coefficients may be rather different for particular infrastructure projects.

Spin-off effects of infrastructure take place in the long run. Some modelling approaches have been discussed in this paper. It is seldomly recognized that spin-off effects of network infrastructure may be negative for some of the regions concerned. For instance, negative effects may occur when domestic producers cannot stand the competition from other producers after an increase in accessibility.

Many regional studies do not clearly distinguish between distributive and generative effects of infrastructure. In such studies substantial employment effects may be found, but these may be mainly distributive by nature. By neglecting the jobs lost in other regions, one may easily arrive at exaggerated effects of infrastructure on employment.

LITERATURE

Amano, K.; Fujita, M., 1970: A Long Run Economic Effect Analysis of Alternative Transportation Facility Plans - Regional and National, Journal of Regional Science, vol 10, pp 297-323.

Andersson, A.E.; Batten D.F., 1988: Creative Nodes, Logistical Networks and the Future of the Metropolis, Transportation, vol 14, pp 281-293.

Biehl, D., 1986: The Contribution of Infrastructure to Regional Development, Regional Policy Division, European Community, Brussels.

Blum, U., 1982: Effects of Transportation Investments on Regional Growth, Papers of the Regional Science Association, pp 151-168.

Botham, R., 1983: The Road Programme and Regional Development: The Problem of the Counterfactual, in: K.J. Button and D. Gillingwater (eds.), Transport, Location and Spatial Policy, Gower, Aldershot, pp. 23-56.

Driehuis, W., 1979: An Analysis of the Impact of Demand and Cost Factors on Employment, De Economist, vol 127, 1979, pp 255-286.

E.M.C.T. (European Conference of Ministers of Transport), 1988: Investment in Transport Infrastructure in E.M.C.T. - countries, Paris.

Gent, H.A. van; Nijkamp P., 1987: Mobility, Transportation and Development, I.A.T.T.S. Research, vol 11, no 1, pp 62- 68.

Giaoutzi, M.; Nijkamp, P., 1988: (eds), Informatics and Regional Development, Gower, Aldershot.

Hirschman, 1958: The Stategy of Economic Development, Yale University Press, New Have.

Kuik, F.C.; Nijkamp, P., 1987: Relevance of Pay-back Effects in Public Investment Appraisal, Project Appraisal, vol 2, no 3, pp. 153-156.

Lakshmanan, T.R., 1989: Infrastructure and Economic Transformation, Advances in Spatial Economic Theory and Dynamics, (A. Andersson, D. Batten, B. Johnsson and P. Nijkamp, eds), North-Holland Publ. Co., Amsterdam, pp 241-262.

Liew, C.K.; Liew, C.J., 1985: Measuring the Development Impact of a Transportation System: A Simplified Approach, Journal of Regional Science, vol 25, pp 241-257.

Mills, E.S.; Carlino, G., 1989: Dynamics of County Growth, Advances in Spatial Economic Theory and Dynamics, (A. Andersson, D. Batten, B. Johnsson and P. Nijkamp, eds.), North Holland Publ. Co., Amsterdam, pp 195-206.

Nijkamp, P., 1986: Infrastructure and Regional Development: a Multidimensional Policy Analysis, Empirical Economics, vol 11, pp 1-21.

Schrijvers, F.J.; Van Westrenen, C.M., 1987: De betekenis van de telecommunicatie-infrastructuur voor de ontwikkeling van de Nederlandse economie, O.S.A.-Voorstudie V21, The Hague.

Social & Economic Council, 1987: Advies Economische Infrastructuur, The Hague, 1987.

Rietveld, P., 1989: Infrastructure and Regional Development: A Survey of Multiregional Economic Models, Research Memorandum, Dept. of Economics, Free University, Amsterdam.

Wigren, R., 1984: Measuring Regional Efficiency: A Method Applied to the Manufacturing Industry in Sweden 1973-1975, Regional Science and Urban Economics, vol 14, pp 363-379.

Zwan, A van der, 1986: Infrastruktur en Economische Groei, Tijdschrift voor Vervoerswetenschappen, vol. 22, no. 1, pp. 19-30.

Deregulation and the Problem of Capacity/Price Planning for the Transport and Communication Infrastructure

by

Werner Rothengatter

1. Introduction

Transport and communication are sectors in which the efficiency of the price mechanism is traditionally questioned and government regulation is assumed to be necessary in addition to the normal rules of competition. In all countries roads and waterways are very largely publicly owned. Railways mostly belong to government-controlled companies, and the same holds for the telecom sector.

However, experience was also acquired early on with the privatisation of sub-sectors, as in the case of the railways in the United States. In addition, more and more countries have gone over to financing and operating particularly expensive parts of roads infrastructure, such as motorways, tunnels and bridges, according to private enterprise criteria.

Because of the high costs involved in new and replacement investment, the authorities' ability to speed up infrastructure development in keeping with economic requirements is increasingly overtaxed. For this reason it is now a topical issue in many countries as to whether the interfaces between private and public investment in transport should be redefined.

2. Transport Infrastructure as a Pure Public Good or a Club Good

2.1 Network Infrastructure Characteristics

(i) *General characteristics*

The characteristics of network infrastructure are as follows:

- Sub-additivity

As a rule, doubling of capacity costs less than double the expenditure and doubling the size of a network frequently gives more than double the utility. This is not, however, a natural law applicable to all network sizes. It implies that only networks of certain sizes reach the breakeven point. But if they are of a high standard and are closely knit, the sub- additivity characteristic no longer necessarily exists.

- Irreversibility and high fixed costs of facilities

Fixed infrastructure costs are high compared with variable costs. In addition, expensive preliminary financing of fixed costs is required because of the high breakeven point. This means a high risk for an investor, especially if the assets are exclusively intended for the provision of certain types of goods and cannot be put to other uses.

Sub-additivity and irreversibility provide the basis for a natural monopoly. Because of the economies of scale, the strongest supplier will ultimately prevail on the market. The high fixed costs are a market access barrier for competitors, while the irreversibility of the investment forms a market exit barrier for the monopolist. But a natural monopoly is still not a sufficient reason for government regulation. It may still be possible for strong competition from substitute goods to exert enough pressure (contestable market) for the monopolist to implement a market-conforming price and supply policy. Regulatory action is called for only if there is any threat of competition from substitute goods being eliminated by the natural monopolist.

(ii) *Special characteristics of transport and communication networks*

There is a whole series of markets for which goods are provided or distributed by means of utility networks: electricity, gas, water, sewage disposal, the telephone, broadcasting, passenger and goods transport, information and data transmission.

In addition to the general characteristics of network infrastructure, the following are also to be noted in the case of transport and communication networks:

- Joint production of services

Different services are provided on transport networks, such as the carriage of goods and passengers in short and long-distance transport operated by vehicles of different types. Transport operations with a different economic purpose can be carried out technically in exactly the same way, as for example passenger transport with a minibus and mixed goods transport with a delivery vehicle.

Joint production is also performed on communication networks. ISDN communication networks for example transmit speech, pictures and data by digitalized signals simultaneously through the network. The information is finally decoded by the receiving stations.

- Public interest

Transport and communication networks are subject to strong public interests which concentrate on two issues:

- First the infrastructure should contribute to a just personal and regional income distribution; this leads to public influences on the spatial allocation of investment funds and to particular requirements for pricing and operating such as obligatory services and equal prices over space.

- Secondly the control of external diseconomies becomes a problem of increasing importance. Transport induces accidents, exhaust emissions, waterproof surfaces and undesired spatial effects like urban sprawl.

Efficient communication networks can be abused for criminal activities.

- congestion

In many transport fields demand is not co-ordinated on an optimum basis, so that capacity overloading temporarily occurs. Sub-optimality in road transport as a result of non-co-ordinated private decisions to travel has been highlighted by PIGOU (1951), KNIGHT (1924), and WARDROP (1952). In this case the market error is due to incomplete information and the wrong cost allocated to those generating the costs.

(iii) *Basic problems in the organisation of transport networks*

The central issue in the transport network market system is whether services should be provided by public or private enterprises in view of the substantial public interests involved. According to SAPPINGTON and STIGLITZ (1987), the considerable difference between the private and public provision of services is due to intervention possibilities. In the case of public enterprises, the authorities can intervene directly in corporate decisions, i. e. in capacity, programme, quantity and pricing policy, whereas only limited control can be exercised in the case of private production. The transaction costs of control are generally lower in public than in private enterprises. However, this is offset by the disadvantages of the lack of incentive since compensation is not based on performance. In deciding on the type of firm, the frequency of the government intervention required in running the firm if public goals are to be achieved must therefore be considered.

If business activity can be defined in a way which keeps government intervention to a modest level, privatisation is the best solution. Privatisation is also advantageous if the government can obtain with a few basic standards what it considers to be the optimum result from production decisions without being familiar with the production technology itself.

In principle, three possible types of system - depending on the level of transaction costs for the supervision of government objectives - can therefore be used in the transport field.

Public enterprises

Infrastructure and operation are set up and organised by public corporations. This type of system is to be found in most railways in Western Europe. It leads back to the period of about 100 years ago, when the railways were an important instrument in regional economic development, and when protection against exploitation from monopolies and issues concerning the regional and personal distribution of income justified strict government control over the railways. It is increasingly doubted whether the order of preference prevailing at that time is still appropriate today.

Private enterprises with limited monopoly powers

In the United States and, in a recent development, in Japan too, the rail supply side (network and operation) is shared among regionally or sectorally defined (long-distance traffic, regional passenger traffic, goods traffic) private monopolies. Special problems obviously arise at network interfaces and in the event of joint utilisation. Economies of scale and scope are not fully used. Instead greater efficiency is achieved through the pressure of competition on identifiable, manageable segments. Strict price and quantity regulation has been abandoned (in the USA since 1980, Staggers Act).

Separation of the network responsibility from operation

With the separation of the network responsibility from operation, the natural monopoly on the operator's side is eliminated as soon as the central co-ordinated facilities are also included with the network (e.g. a railway operating control system, or air traffic control in the case of commercial aviation).

This organistion system has been implemented practically everywhere in road transport; the infrastructure is provided by the government or a government-controlled company (in the case of toll roads). Users pay directly through tolls or indirectly through taxes, with utilisation being regulated to a varying extent: very moderately in the case of passenger transport (exceptions are to be found in densely-populated areas of Japan, Hong Kong and Singapore, but also increasingly in European towns such as Turin or Zurich), but closely in the case of road haulage since the railways are exposed to competition from substitute modes. Such a system is required for German Railways, i. e. by its management board or also by political parties (see, for example, the proposals for the amended Federal Railways Act of the Social Democratic Party).

It would be possible to go a step further and open the competition on the railway network, since a monopolistic form of railway operation is not an imperative from the viewpoint of the operational technique. This assumes that operational control would be supervised by the authorities so that no monopoly can be exercised through ownership of the control facilities. Mixed solutions, such as the centralisation of basic services inclusive of operational control within a public enterprise, and decentralisation of supplementary services (e. g. regionalisation of short-distance, traffic elimination of branch lines without economies of scope), are also conceivable. An example for such possibilities is given by the Airport-Express service which is provided between Frankfurt, Cologne and Dusseldorf by Lufthansa. Such forms of operation are also conceivable in goods transport. Since a considerable proportion of goods wagons are now privately owned, the formation of entire trains by private carriers should not be ruled out either.

The advantage of separating the network from operation is that the economies of scale and scope can be fully used. This might be partly offset by the problems of undesired market entry (cream skimming) when private competitors are allowed into parts of the market, while the public sector supplier has to serve the entire market under defined conditions (e. g. legal obligation to accept contracts, uniform tariffs in the region). Basic principal/agent problems also arise, which will be discussed in the following section.

2.2 Separation of the Network Responsibility from Operation: Principal/Agent Problems

Let us take a market system in which the network infrastructure and the central co-ordination facilities are provided by the government. The government decides on the scope of infrastructure supply and on the pricing system imposed on users of the network. The users decide on the volume and structure of market activities and, if they provide services for third parties, on their prices.

If there is no natural monopoly in network operation and if there has been no competition from substitute modes, there is firstly no ground for further regulation in the form of market entry, quantity and price rules. Quantity rationing in the form of entry restrictions and output conditions are then required only if prices are not perfectly efficient as a regulator, whether for reasons concerning incomplete information or false signalling, and when prices have stopped acting as the indicator for excess demand in other sectors of the economy as well. It is also conceivable that government objectives cannot be fully expressed in costs and thereby in prices. An example of this is given by the environmental and safety aspects, in which no market economy cost prices can be worked out in cases of environmental effects or risks that endanger life (death of woods, threat to wildlife, health hazards).

This results in two decision levels: as the principal, the government decides on the creation of infrastructure and on the prices for its use, possibly supplemented by rationing. It can then consider an order of preference that covers the entire spectrum of public welfare, i. e. one that includes the external effects regarded as relevant, distribution goals and the provision of public or merit goods. The users' preferences may be a constituent part of the principal' s entire preference structure. The users, i. e. private firms and consumers, utilise the infrastructure as agents and therefore try to maximise their individual preferences. They also signal to the principal the need for further infrastructure, which he should anticipate as far as possible in his long-term planning.

Here the principal's task is to find a capacity/price/rationing scheme which meets his priority objectives, without endangering the individual economic interest and the effectiveness of competition. There is naturally a limit of the inclusion of private market forces in investment decisions and funding with such a system.

3. Basis for Investment and Pricing Theory in Transport

Investment and pricing theory in transport is in general independent of the type of organisation. Because of its special interest in transport structure, the government is, in any case, required to form an opinion about the desirable development of this sector. But the type of system selected determines the form in which the government takes action to impose its views with regard to the private sector.

At first sight, it seems unusual that the unity of investment and pricing should be emphasized here. But it can be seen that the financing side of investment activity, which defines, in the long term, the limits to the share in transport infrastructure costs to be met by the private sector, is considered simultaneously with pricing. In the case of the prices to be set in conjunction with investment planning, the reference can, of course, be only to long-term strategic prices, and therefore not to short-term peak load prices for adjusting demand to current capacity. Since the beginning of the 1970s (e.g. in LELAND, 1972), similar ideas have been expressed in specialist publications concerning corporate investment/price strategies in the event of monopolistic competition.

3.1 Short-term Marginal Cost Theory and the Ramsey Pricing Technique

Simple marginal cost theory involves pricing at the short-term marginal costs of infrastructure. But this rule is subject to limits in the case of networks with economies of scale and scope, because degressive cost patterns occur. Therefore allocation at marginal costs would always lead to deficits that would have to be covered through withdrawal of resources from other fields. The optimality proof for a taxation/subsidy system required for this purpose can be given only by means of premises that have little to do with the real world.

The same applies to the Ramsey pricing technique. Subject to an otherwise similar set of premises, the costs for the network complex in question must be covered out of the revenues obtained. The result obtained is that the percentage deviation of prices from marginal costs should at best be inversely proportional to demand price elasticities.

In judging this approach, we shall mention only the most important premises (BÖS, 1986):

(i) All non-controlled prices correspond to marginal costs.
(ii) The private sector is organised on a fully competitive basis.

(iii) The transfer distribution to offset the deficit is optimal in the sense of compensated demand.

(iv) Quality of service is at a given, constant level. Demand must be met.

(v) The (compensated) demand curves must be known.

In addition, unwanted side effects on distribution and dynamic efficiency occur, if the Ramsey prices exceed marginal costs:

- Sections of the population with low incomes and no substitution possibilities (captive riders) could be especially price inelastic and therefore be taken into account for the deficit coverage in an undesirable way from the distribution policy viewpoint.

- Groups which are in, or move into higher income classes are encouraged to reduce their "captivity" and obtain substitution possibilities.

3.2 Long-term and Dynamic Marginal Costing

If capacity extension over time is taken into account, pricing at short-term marginal costs is no longer a meaningful principle. Starting with BOITEAUX (1956), the French marginal cost school has developed a long-term marginal cost concept that includes capacity development costs.

Boiteaux discusses the non-convex problem of railway investments. Under the usual marginal costing rules, the price would have to remain at zero as long as the wagon was not fully loaded and then move up to the level of the costs for providing an additional wagon as the capacity limit is exceeded. Since with a price of zero a high demand exceeding capacity, and with the marginal cost price a low demand less then the capacity limit, may prevail, no equilibrium point can be defined. Boiteaux then introduces the time dimension, demand increasing with time and new investment frequency in order to define an optimum price and investment policy combination. This leads to the "longterm marginal revenue/marginal cost rule", according to which the revenues generated by the last wagon acquired must correspond to this wagon's marginal development costs for optimum capacity to be provided.

Apart from the question of the Boiteaux-model's internal consistency (WILLIAMSON, 1966; STARRETT, 1978), it should be realised that the model for the long-term marginal costs can be used only in the case of constant adjustment of capacity, with capacity necessarily broken down into small units. But as soon as the capacity change is indivisible and - related to a longer time span - a one-off development, the French marginal cost theory is overtaxed. In principle, it is wrong to speak of long-term marginal costs if a long-term cost curve cannot be worked out.

This applies to all in transport infrastructure, such as the construction of new motorway routes, high-speed rail sections and information transmission facilities like the ISDN. A decisive factor is that such new investments are accompanied by a change in superstructure technology (i. e. in operation).If, in addition, this form of infrastructure quality improvement is not continuous, but takes the form of investment surges over time, marginal cost theory loses its entire empirical application basis.

This fact is worth noting insofar as marginal theory has also made deep inroads within the European Community - due to the influence of Anglo Saxon and French publications - and is regarded as a scientific basis for the distribution of infrastructure costs according to the territoriality principle.

3.3 Principles of an Operational Investment and Price Theory

3.3.1 Basic Model

An empirically meaningful theory of investment and price planning in the transport sector must be firstly freed from the narrow efficiency concept of classical welfare economics. It is not only the provision of material goods which is decisive for the welfare of a society, but also the quality of the environment and the solution of distribution conflicts within and between generations.

It must be also be freed from the concept that prices have only market clearing functions and could also always carry them out. Just as the prices of goods may have quite different functions in the normal economic process, such as the indication of product qualities, infrastructure utilisation prices also have, in addition to the short-term task of market clearance (e. g. peak load pricing), long-term functions such as the appropriate share of users in the investor's risk and the creation of fair distribution arrangements for various user groups or carriers. Taking these factors into account, we arrive at the following basic structure for an investment/pricing model for transport infrastructure:

$$\max_{p_t,\, x_{tr},\, y_{tr}} \left\{ \sum_{r=1}^{R} \sum_{t=1}^{T} u_{tr} * y_{tr} \right. \qquad (1)$$

where

$$u_{tr} = \int_{\tau=t}^{T_r} [\, b_\tau \, (x_\tau \, (y_1, \ldots, y_\tau \, ; \, p_1, \ldots, p_\tau)) - k_\tau \, (y_1, \ldots y_\tau))\,] \exp(-\gamma\tau) d\tau \qquad (2)$$

b_τ: utility for the project r within period

x_τ: vector of transport or communication flows in period τ

$x_\tau = (x_{\tau 1}, \ldots, x_{\tau r}, \ldots x_{\tau N}); \; x_{\tau r} \geq 0$

y_τ: vector of decision variables

$y_\tau = (y_{\tau 1}, \ldots, y_{\tau r}, \ldots y_{\tau R}); \; y_{\tau r} \in \{0,1\}$

N: number of links in the network

T_r: utilisation time for project r

γ: discount rate

P_τ: price in period τ

Unlike the Ramsey-Boiteaux model, b is not evaluated by means of the consumer's surplus concept but with a direct approach. The following are defined as utility elements, for example, in federal transport network planning of W. Germany:

- lower transport costs;
- lower infrastructure maintenance costs;
- contributions to transport safety;
- contributions to accessibility improvement;
- spatial effects;
- contribution to environmental protection; and
- effects from non-transport functions.

The indivisibility assumption is called for in the case of larger investments from which considerable changes in quality of service can be expected:

$$\sum_{t=1}^{T} y_{tr} \leq 1 \qquad \forall r \tag{3}$$

Additional constraints can be defined for introducing budget and technical requirements.

On the second decision level the operators decide on the activities which they wish to carry out on the network provided. This decision is described by means of an activity vector that shows the strength of the transport or communication flows on the network segments (sections).

The following points are to be noted in this context:

(i) There may be several categories of users who influence one another.

(ii) Substitution or complementary relationships covering the entire network may arise

(iii) Demand may develop in surges, as certain threshold values are reached (e. g. surge in goods traffic by night, critical audience ranges in the case of cable TV).

(iv) The change in demand on the network in question may consist ofinduced or transferred demand. This is relevant for the utility assessment.

As a result of the second decision level, we therefore obtain the forecast of demand behaviour in the form of section loading. This is included as an important secondary condition in the calculation for the first decision level:

x is a policy-sensitive forecast of the pattern of activity on
the transport network with substitution and complementary
relationships being taken into account. (4)

Because of the many interactions on and between networks, a full description of all relevant factors in a simultaneous approach is hardly possible. x can, in fact, cover several decision levels in turn. This applies to all forms of transport in which the consignors and carriers are not identical, as in goods transport or bus, rail and air passenger transport. The possibility of controlling network activities varies: while the railway can make an attempt at central co-ordination of flows, road transport flows are based on decentralised, unco-ordinated individual decisions. Accordingly, there are differing reference concepts which can be expressed in terms of Wardrop's first and second principles (WARDROP, 1952). The first principle of user equilibrium, based on non-co-ordinated decentralised decisions, is thus a behaviour description with a non-optimum result. The second principle of system equilibrium is a normative status description for an optimally controlled system.

3.3.2 Decomposition by Submodels

The comprehensive version of the model (1) .. (4) looks simple. But it summarizes a series of submodels which are large and complex if they are intended to represent parts of the reality. These submodels provide the following informations:

(i) policy sensitive forecasting of transportation demand for passengers and freight and for all relevant transportation modes;

(ii) design of networks, definition of projects and project appraisal;

(iii) priority setting and budgeting of investment activities;

(iv) standard cost accounting and price setting for the transportation demand or for the carriers, respectively;

(v) bottleneck analysis, rationing and regulation.

Note that all submodels are highly interdependent. Such the price setting in submodel (iv) is an important input parameter for the forecasting model (i) because prices influence traffic behavior. An appropriate pricing policy might decrease the necessity for regulatory measures (model (v)) which is needed to achieve the objectives of improving the income distribution or reducing the external diseconomies.

It is not possible to even describe the basic characteristics of all submodels. Therefore I shall restrict the discussion to one of them, the cost accounting and pricing model.

3.3.3 Standard Cost Accounting and Price Setting

The problem of cost accounting is very similar for the railway and telecommunication sector. In both cases the planned regulation regimes comprise a public control of the infrastructure, a designated carrier for the network and potentially further carriers for value added services. Such in both cases there are two levels of pricing: first the government has to decide on the share of infrastructure costs which should be covered by the designated carrier. This is not necessarily 100 % because political objectives like income distribution or control of external diseconomies may lead to infrastructure investments which are not profitable from the private business point of view.

Secondly the designated carrier has to share the network with and to provide the operation system for further companies supplying value added services. It would be fair and reasonable if these companies would contribute to covering the infrastructure and operation control costs in the same way as the designated carrier. These two issues imply specific requirements for the accounting scheme. The aggregate cost volume has to be allocated flexibly to transportation activities characterized by time, location, OD-pair and mode.

(i) *Volume of aggregate costs*

The volume of costs which are to be allocated comprises
- costs of infrastructure minus the cost for providing general public services (e.g. defense, general communication)
- external diseconomies minus external economies.

The first term has been discussed for decades in the literature. Many computations have been performed on the national level (e.g. road track costs) which unfortunately are in many cases not comparable because of different definitions and basic assumptions. The DIW has prepared a national accounting for the federal transportation infrastructure since 1972 and every three years. This accounting is based on a very detailled data base listing all major elements of investment activities over time. This data base allows for a reliable estimation of capital costs (depreciation and interest on capital), costs of repair and improvement and costs of operation. Note that this "business type" cost accounting comes up with costs values exceeding the actual expenditures for the infrastructure. But this is the correct way and helps to make the cost figures comparable to that of private business activities which are substitutive or complementary to transport such as holding inventories or applying logistic principles, e.g. "just-in-time" service.

Although the calculation of social costs (second term) cannot be based on the solid base of damage cost/traffic activity relationships it should be considered in the cost allocation to avoid a general subsidisation of transport. A rich literature has grown on the features of the problem and its welfare implications and also some estimations have been done which are summarized by QUINET (1989).

The DIW (1987) has suggested cost estimation method by using avoidance costs for additional accident and environmental costs. This method in particular is appropriate in the context of a standard cost approach based on future cost figures derived from investment plans and future traffic activities. This approach starts from defined limits of accidents and pollution, continues by deriving the optimal techniques for satisfying the derived levels and ends up with estimating the related costs.

(ii) *Allocation of aggregate costs*

The most difficult step of cost accounting is to allocate cost elements to the users of the transport system. Basically there are two approaches

- allocation according to technical relationships between transport activities and the related costs,
- allocation according to club principle.

The first approach has been widely used and thus a broad variety of key functions is known which are based on experimental results or purely on unproved hypotheses (e.g. AASHO-road-test, ton miles, axle miles, static or dynamic space consumption, flow-capacity indicators). The problem of applying such relationships is that, once the transport infrastructure has been realized, the major part of costs consists of fixed and common costs which don't vary with a single activity.

The club principle is more appropriate if one can assume that for the time span of cost calculation the decision on all infrastructure elements can be taken independently of all decisions made before. As each member of the club of users has specific requirements as to the dimensions of the infrastructure one can identify stand alone and common investment costs for different user classes. This is the base of club allocation principles, two of which are explained in more detail in the following section.

3.3.4 Cost Allocation and the Club Principle

From the formulation of the model 1) to 4), it emerges that price and investment policy are closely connected and that both refer to the future. The basis for pricing is accordingly the forecast costs which will arise in the future as a result of investment and utilisation decisions. The planned long-term or strategic prices are the focal point in the study.

Over a period of time a transport network is used by various categories of users and at different periods by several generations of users. Accordingly, it is a matter of defining a price system which corresponds to the extended concept of efficiency (by objective functions (1)) and which is fair in both horizontal and vertical (intergenerational) terms. In this connection, aspects concerning long-term costing possibilities and the adjustment of the competitors' start up chances on the network are to be taken into account.

For this reason, a distribution principle is selected which is based on fairness concepts and takes into account the club characteristics of a large part of network capacity. Procedures based on game theory are particularly appropriate for this purpose. Without going into a detailed discussion of the procedures, which is provided for example by RAUHUT, SCHMITZ and ZACHOW (1979) or SCHOTTER and SCHWÖDIAUER (1980), let us take two possibilities - the Shapley value and the nucleolus concept - which can be considered as basic principles for fair distributions and have already been tested in applications (LITTLECHILD and THOMSON, 1978; LOEHMANN et al., 1979; ROTHENGATTER, 1983).

The type of cost pool on a network can be summed up by the concept of the "club principle" devised by COASE (1946). This says that the decision to provide an infrastructure to be used jointly can be conceived in such a way that the user groups form a company for the construction and maintenance of infrastructure and share with one another the accruing, one-off and current costs according to certain effeciency and fairness rules. The club principle's efficiency rules are:

(i) The club management should have the incentive and financial power to set the optimum dimensions for the infrastructure. It should aim precisely for cost coverage.

(ii) There should be no incentive for any coalition of users to break away and set up equipment for themselves.

(iii) The cost allocation scheme should leave any user who is in a position to meet his share of costs with the incentive to use the infrastructure.

The efficiency conditions thereby simulate the result of a free negotiation process in a system with exclusively private property rights. As many cost distribution schemes can be efficient in terms of the three conditions, a choice has to be made. Fairness rules such as those defined by LITTLECHILD and THOMSON (1977) can be introduced for this purpose.:

- Lower capacity use should also lead to lower costs.
- The difference in the costs allocated between any two user groups should not be greater than the cost difference which would occur if each group provided the infrastructure required for its own needs.

Even with the addition of these rules, there can still be several solutions that must be reduced to a cost distribution scheme through further axiomatic concepts relating to fairness, like those developed in cooperative games theory. Here only two possibilities are considered, which are described in the application of games theory as the cornerstones of fair distributions, i. e. the "Shapley value" and the "nucleolus concept".

Shapley value

Let us start with a cost allocation game, in which each user group causes lower additional costs for the provision of infrastructure by joining the club than it would have incurred by providing it individually on its own account. The question is, therefore, how the costs saved by creating a club should be distributed over the user groups.

As a solution, the Shapley value assigns to each user group the expected value of the costs saved by it, with the assumption that the order of entry into the club is quite accidental. To simplify the picture, it is assumed in this definition that the utility side can be neglected. A change in this assumption also modifies the Shapley value.

The practical result of this allocation principle is very simple. It corresponds to incremental costing which was already discussed in the Supplementary Report on the Highway Cost Allocation Study in the United States (1981): the infrastructure costs which are incurred by the user groups with the smallest/lightest vehicles are allocated evenly to all the groups, the cost growth incurred by the next smallest/lightest group of vehicles is allocated to all groups up to the first, etc., and the cost growth incurred by the biggest/heaviest group of vehicles is finally allocated only to this group (Figure 1).

Figure 1: INCREMENTAL COST ALLOCATION AMONG USER GROUPS

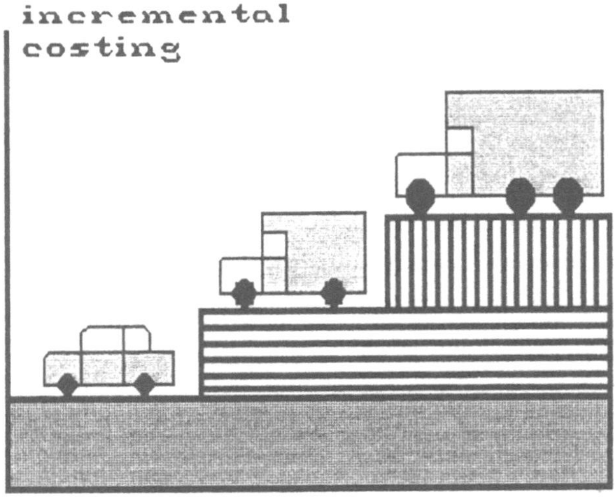

The Shapley value allocation principle is also found in the proposals for the further development of the Swiss road transport calculation with the "concept of the minimal road" (EPEL, 1985).

An applied version: The Generalized Shapley Value (LOEHMAN and WHINSTON, 1976):

The cost allocation by Shapley's value implies that each coalition can be formed and each ordering of arrival of members within a coalition is equally likely. This may not be true for many practical planning problems.

It is necessary, to modify the Shapley value to introduce this consideration into an allocation formula. Then, the resulting generalized Shapley value is defined by

$$x_i \ (v,N) = \sum_{S \subset N} P \ (S, S \setminus \{i\}) \cdot (v(S) - v(S \setminus \{i\})) \ , \ i = 1,\ldots,n,$$

$$\sum_{i \in S} \ \sum_{S \subset N} P \ (S, S \setminus \{i\}) = 1 \ .$$

The terms P (·) can be interpreted as weights, or prohabilities for player i to join an existing coalition $S \setminus \{i\}$. So the generalized Shapley value assigns to each player his expected marginal contribution, if there are different probabilities for forming coalitions.

Note that - contrary to the original Shapley value - the order of entrance into the coalitions now plays an important role.

Notation:

Notation:	x_i :	pay-off for partner i
	v :	characteristic function
	v(S):	worth of coalition S
	P :	weights
	N :	number of partners in the game .

The nucleolus concept

By nucleolus is meant a cost allocation scheme that places the user coalition benefiting the least from the creation of a club in the best possible position.

The main idea in this concept is similar to Rawls's fairness concept. It goes further in that it applies not only to the user groups themselves but to any coalition within these groups. With regard to infrastructure cost applications, it appears particularly advantageous that the allocation result depends entirely on costs and therefore remains unaffected by the utility side (proof in LITTLECHILD and OWEN, 1976). This is an important asset in view of the difficulties of quantifying utility in the infrastructure user sector.

On the other hand, the computation process is relatively expensive (lexicographical minimisation by mean of a series of linear programmes). In simplified terms, the computer operation can be so designed that the user groups are classified by increasing size (weight) and, in turn, are allocated the highest possible amount of overheads on the basis of the two fairness conditions, with total overheads being distributed. This is equivalent to the opposite procedure, according to which the user groups are classified by decreasing size (weight) and are, in turn, allocated the least possible amount of overheads subject to the same conditions.

An applied version: The excess of "(n - 1)"-coalitions

Computation of nucleoli may be tedious, because a series of linear programs must be solved. Therefore, we approximate the nucleoli by simple formulas, which can easily be calculated and are intuitively understandable for decision makers.

If the excesses for the coalitions are equal in size, no coalition will feel being treated in an unfair way with respect to another one. Instead of equalizing the excesses of all coalitions we only take the excesses of the "(n - 1)"-coalitions $N \setminus \{i\}$. These excesses show, in absolute terms, the additional costs occuring for a coalition $N \setminus \{i\}$, if player i leaves the grand coalition, and thus can be interpreted as bargaining powers of the players i. The approximation formula for the nucleolus is:

$$\tilde{N}u\ (v,N) = \{x \in X\ (v,N)\ \mid\ e\ (x, N \setminus \{i\}) = e\ (x, N \setminus \{j\}),\ i,j = 1,\dots,n, i=j\}.$$

Result of this operation is a situation for which the propensity of excluding one member from the grand coalition is equal for all "(n - 1)" - coalitions.

4.3 Infrastructure Costs

The infrastructure cost account of the DIW consist of

- capital costs (depreciation according to section 3.2 and interest)
- costs of major repair and maintenance
- yearly expenditures for current repair, maintenance and operation.

The principles of acounting these costs have been developed in the late sixties and applied by the DIW in cost accounts which have been performed every three years from 1972 onwards for road, rail, inland waterways and air traffic (airport landing tracks and flight control).

Because of continuously employing the same accounting system the results from 1972 until 1984 are comparable to each other. The infrastructure costs can be compared with the public income from specific taxes, such as vehicle taxes, fuel taxes and parking fees (note that in W. Germany no road tolls are raised). Figures 4 and 5 show the development of the income/cost- ratio for rail and road.

Figure 2:

Gross Infrastructure Investments and Replacement
Deutsche Bundesbahn
Mill. DM,base 1983

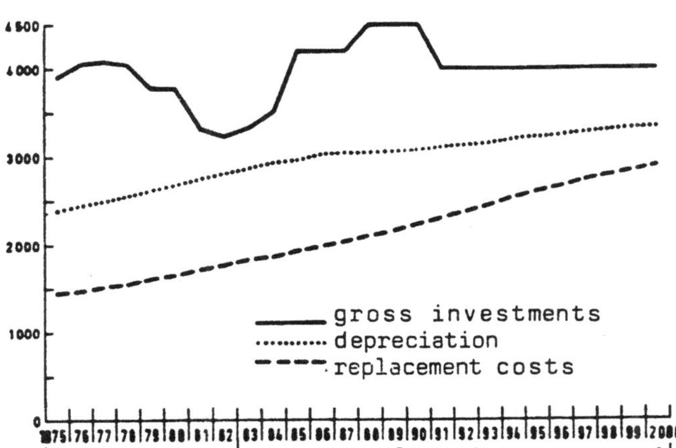

Notation:

e : excess (complaint)

$$e(x,S) = v(S) - \sum_{i \in S} x_i .$$

4. Some Figures on Investment Decisions and Costs, Replacement and Infrastructure Costs in W. Germany

4.1 Investment for the Federal Transport Infrastructure Master Plan of 1985

The Federal Transport Infrastructure Master Plan (BVWP/85) comprises all investment expenditures of the Federal Minister of Transport from 1986 until 1995. These activities have been selected by means of benefit cost analysis as indicated in section 3.3.

The sum of all investment costs is 126 Bill. DM which is allocated as follows:

* Rail	35.0
* Roads, Interstate	50.1
* Inland Waterways	8.0
* Air	2.3
* Local Public Transport	13.9
* Local Roads	13.9.

4.2 Replacement Cost

The DIW has calculated the replacement costs for federal and local infrastructure investments according to the perpetual inventary concept which is applied in the most national accounting systems of the OECD countries. Two figures are given here which exhibit the results for the infrastructure of the Deutsche Bundesbahn (figure 2) and of the federal roads (figure 3).

Figure 3:

Gross Investments for Federal Expressways

Mill. DM, base 1983

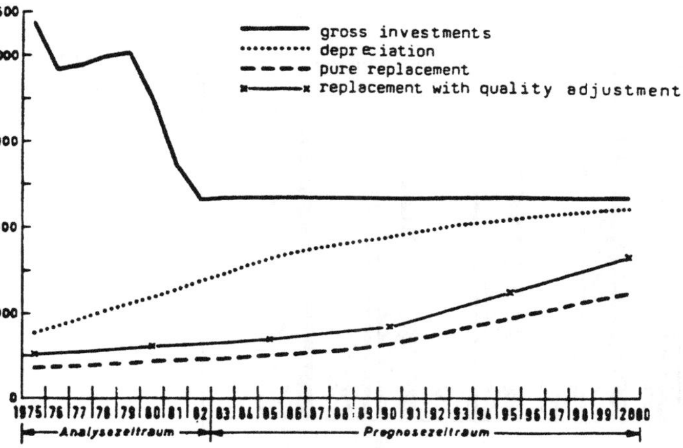

Figure 4:

Income/Cost-Ratio - Rail

1966 - 1984

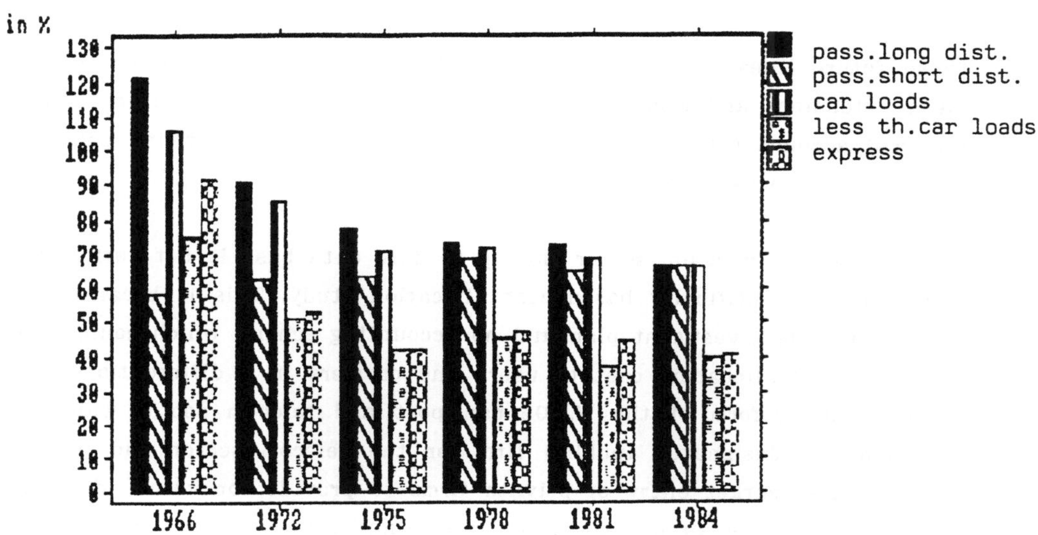

Figure 5:

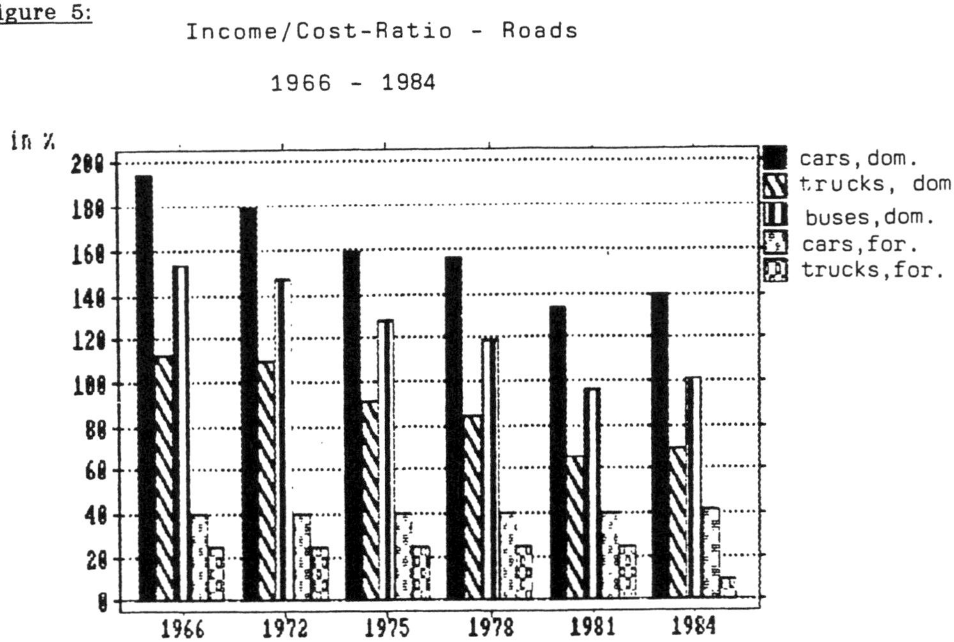

The cost account can be extended by the components

- standard costing based on future investments and costs
- social cost calculation
- cost allocation by employing the club principle, which has not been performed yet.

These future lines have been explained in section 2 and 3. After finishing these extensions and modifications the results of the infrastructure and social cost account could be taken as a base for the public pricing policy towards users and carriers on the networks.

In the telecommunication sector the up to date data base is not sufficient to carry out a scientifically based cost allocation study. This is because there is no long term investment plan and no accounting scheme which could provide a base for allocating cost to users and carriers in a differentiated way. As after the reform of the W. German post and telecom sector which will establish one designated (public) firm for the telecom sector and plans to allow for the market entry of private firms to provide for value added services there will be an urgent need for solving cost allocation problems. The designated public telecom firm is expected not to discriminate private carriers by imposing to them the user charges which correspond to the costs of its own use (see Regierungskommission Fernmeldewesen, 1987). This presupposes that the firm applies a detailed cost accounting scheme.

5. Summary and Conclusion

In many countries of W. Europe the problems of deregulating the transport and telecom sectors are planned to be solved by a separation of the responsiblility for the network infrastructure from the operation. In this case major problems of investment planning and price policy occur. These problems can be solved by applying sophisticated forecasting, evaluation and cost allocation models. One can argue that deregulation should make the world more simple and that a deregulation scheme inducing such complex problems for adjusting the behavior of the public leads to a dead end. However there are good arguments for this type of deregulation because the synergetic effects of an integrated network can be used and the external diseconomies can more easily be controlled.

LITERATURE

Bös, D., 1986: Public Enterprise Economics. Amsterdam.

BVWP/85, 1985: Gesamtwirtschaftliche Bewertung von Verkehrsinvestitionen. Bewertungsverfahren für den Bundesverkehrswegeplan 1985. Schriftenreihe des Bundesministers für Verkehr. Heft 69. Bonn.

Coase, R.-H., 1946: The Marginal Cost Controversy. In: Economica. XII. 169-182.

DIW, 1985: Vorausschätzung des Ersatzinvestitionsbedarfs für die Bundesverkehrswege. Beiträge zur Strukturforschung. Heft 83. Verf.: B. Bartholmai, H. Enderlein und J. Niklas.

DIW, 1987: Berechnung der Kosten für die Wege des Eisenbahn-, Straßen-, Binnenschiffs- und Luftverkehrs in der Bundesrepublik Deutschland für das Jahr 1984. Gutachten im Auftrage des Bundesministers für Verkehr. Bearb.: H. Enderlein und H. Rieke. Berlin.

DIW, 1987: Erweiterung methodischer Ansätze zur Wegekostenrechnung und Erarbeitung eines Konzeptes für eine Gesamtkostenrechnung des Verkehrs. Gutachten im Auftrage des Bundesministers für Verkehr. Bearb.: J. Niklas, H. Rieke, W. Rothengatter und U. Voigt. Berlin.

EPFL, 1985: Institute des Transports et de Planification (Hrsg.): Compte Routier de Repartition des Couts. Rapport Final. Lausanne.

Honkapohja, S., 1980: A Note on Monopolistic Quantity Rationing. In: Economic Letters. 6. 203-209.

Knight, F.H., 1924: Some Fallacies in the Interpretation of Social Costs. Reprinted in: Irwin, R.D. (ed.): Readings in Price Theory. Chicago.

Leland, H.E., 1972: Theory of the Firm Facing Uncertain Demand. In: American Economic Review. 62. 278-29122.

Littlechild, S.C.; Owen, G., 1976: A Further Note on the Nucleous of the "Airport Game". In: International Journal of Game Theory. 5. 91-95.

Littlechild, S.C.; Thompson, G.F., 1977: Aircraft Landing Fees: A Game Approach. In: The Bell Journal of Economics. 8. 186-204.

Loehman, E.; Orlando, J.; Tschirhart, J.; Whinston, A., 1979: Cost Allocation for a Regional Wastwater Treatment System. In: Water Resources Research. 15.2. 193-202.

Pigou, A.C., 1952 (1st edition 1920): The Economics of Welfare. London.

Quinet, E., 1989: The Social Costs of Land Transport. Paper prepared for OECD. Paris.

Rauhut, B.; Schmitz, E.; Zachow, W., 1979: Spieltheorie. Stuttgart.

Rees, R., 1986: Indivisibilities, Pricing and Investment: The Case of the Second Best. In: Zeitschrift für Nationalökonomie. Suppl. 5. 195-210.

Regierungskommission Fernmeldewesen, 1987: Neuordnung der Telekommunikation. Heidelberg.

Rothengatter, W., 1983: Allocating Joint Costs of Public Investments. Discussion Paper. Universität Ulm.

Rothengatter, W., 1987: Scheduling of Interstate Road and Railway Investments. In: Environment and Planning A. 18. 465-483.

Sappington, D.E.M.; Stiglitz, J.E., 1987: Privatization, Information and Incentives. Working Paper No. 2196. National Bureau of Economic Research. Cambridge.

Schotter, A.; Schwödiauer, G., 1980: Economics and the Theory of Games: A Survey. In: Journal of Economic Literature. 18.2. 479–527.

Starret, D.A., 1987: Marginal Cost Pricing of Recursive Lumpy Investments. In: Review of Economic Studies. 45. 215–227.

Stiglitz, J.E., 1987: The Causes and Consequences of the Dependence of Quality on Price. In: Journal of Economic Literature. XXV. 1–48.

Wardrop, J., 1952: Some Theoretical Aspects of Road Traffic Research. In: Proceedings Institute of Civil Engineers. Part II. Vol. 1.

Meso-Economics and Organizational Ecology

by
Jan G. Lambooy

1. Introduction

Meso-economics can be defined as the intermediary level between macro-economics and micro-economics. The field of mesoeconomics is subdivided into the study of regional economics and that of sectoral developments or the industrial organization. Naturally, this latter subdiscipline is basic to understanding regional economic developments. In many cases the reverse relation is also true (MASSEY, 1984; Van DUIJN and LAMBOOY, 1982). Anyway, regional economics has to reckon especially with two aspects of sectoral developments: (i) the dynamics and the structure of the market; and (ii) the ecology of organisations, or the relational structure with the socioeconomic environment. The first aspect is connected with the dynamic nature of competition in spatially and technologically developing markets; the second one concerns the spatial nature of relations between firms, more specifically the network structure.

In order to be able to understand the development of regions, the meso-economic approach has proved to be increasingly useful. The complexity of markets and the problems with regard to the effective use of macro-economic indicators have caused us to look more intensively to the mesoeconomic level. However, it has to be stressed that the mesoeconomic approach does not render the other approaches useless. The body of economic theory can be enriched by adding new avenues to examining reality.

2. The Complexity of Markets

Markets are very different in their structure and in their dynamics. In fact, 'competition' is the very heart of capitalism and includes continuously changing markets, effectuated by competition itself. Markets are like fluids or amoebes, adapting perpetuously to altered conditions. If markets work, it is impossible to accept neoclassical statics as realistic; only in exceptional cases of stability and equilibrium this might be possible. The same goes for one of the basic neoclassical assumptions: that of complete information. If that would be a realistic assumption, it would also be possible to accept the success of central planning. Why would an entrepreneur know more than a central planner?

The success of the 'free market' is that it allows a *trial-and-error-process,* and that, as a system, it accepts many failed entrepreneurs in the competitive society. The main reason why they fail is that they have not understood the complexity of the markets and of the wider environment. Information is not complete or free; information is incomplete, often fuzzy, and costly to acquire. It is difficult to adopt the right trends.

Another problem with traditional neoclassical theory is that it quite often assumes that all entrepreneurs and firms are identical and behave as if they were each others clones. Firms were viewed as rational units with mechanistic behavioural patterns.

Mesoeconomics accepts many results and methods of the neoclassical approach, but adds *complexity, incomplete information,* and a *differentiation of behavioural strategies.* It uses the theory of industrial organization and regional economics to examine the effects on economic development. The central concept remains the market.

Markets are mechanisms by which parties attempt to acquire strategic positions, profits, income, or satisfaction, by the buying and selling of products and/or services. The possibility to reach one's goal is constrained by various factors, such as a lack of information, technology, capital, labour, or income. The complexity of the environment is such that an optimal result is acquired if the actor has the impression that what he did was successful, or that others have that impression. Imitation and psychology are very important for economic actors. Nevertheless, some have better results than others – when measured by some common measure, like profit, or market share. Certain entrepreneurs 'behave' more efficiently, because they had a better insight, or simply because they had more luck in acting. Innovative behaviour is no guarantee for success, however. Success – or failure – depends also on the structure of the market. Success means that a certain behaviour in a certain environment leads to survival and/or growth.

'Structure' denotes stability, interrelations, and integration, but where markets are concerned, dynamic changes result in a certain relativity of positions: not rigid, but with certain limitations as to flexibility and speed of response. The concept of the 'product life cycle', for instance, provides us with a tool to show the dynamism of a competitive structure. During a product's course through the market, the producing firm has to change its behaviour according to the development stage of the product.

In general: technological, demographical, institutional, and competitive developments are strong reasons for accepting the fact that markets have to be examined in the context of a complex environment. Firms – especially the larger corporations – are actively monitoring the environment and are devising tools and strategies to survive. For example, multinationals continuously adapt their geographical organizational structure.

The complexity of markets relates to uncertainty and the possibilities and the costs of obtaining information. These factors are even more important, because the markets are not the only significant parts of the environment. Besides the markets, the relevant environment encompasses two other highly important parts, viz. (i) *institutional relations,* such as politicial parties, religious organisations, labour unions, and governments; and (ii) *infrastructure and land,* which is a very important relationship for firms, conditional for them to function. Schematically:

```
              <-------> markets (other firms and consumers)
THE FIRM      <-------> institutions
              <-------> infrastructure and land
```

Contrary to EUCKEN (1950), these latter two factors cannot be regarded as given for decisions at the micro level. They form an integral part of the information structure and they influence markets; and markets, in turn, influence these parts of the relevant socioeconomic environment.

Technology and demography have different, indirect influences on firms. Technologies are developed within firms and are transferred through the market. Demographic changes influence consumer markets or the labour market, or they can have influence via institutions.

For firms, it may be contended that the markets are the most important part of the socioeconomic environment. But the institutional environment is increasingly important, because governments, political parties and various pressure groups are attempting to enforce rules on market behaviour, e.g. in the case of mergers or the introduction of new pharmaceuticals, or to acquire more inside information, e.g. through legislation on accounting rules. A new - but vigorous - influence is that of the ecological movements.

Each firm, whether a small one or an international corporation, tries to design a location and a market strategy for optimizing its goals, whether this is to survive, to earn higher profits, to gain a larger market share, or to successfully introduce a new product. Because the socioeconomic environment is so complicated, quite often the strategies developed along these lines are of a 'trial and error'-nature, based on the available information, on resources and on entrepreneurship.

3. Complexity and Strategy

The relationships between firms and the environment are complex and they differ according to the size, the products, the location, the organizational structure, et cetera, of the firm concerned. It is possible to define this complexity with the following equation:

$$S = W, E, R_{WW}, R_{eW};$$

in which: S = the system;
W = the set of elements within S;
E = the set of elements not in W;
R_{WW} = the set of relations internal to W; and
R_{eW} = the set of relations between W and E.

Elements can be persons, or departments, or divisions, or behavioural rules, et cetera. It is almost impossible to fully describe and analyse all of the relevant information. In theory and in practice, reductions are made before decisions can be made. The relevant question is: to what extent is such a reduction permitted in order to obtain the relevant answers necessary to formulate theoretical propositions or to make practical decisions? In many cases, the far reaching reduction of neoclassical economics is not sufficient to respond to the dynamic reality.

In recent years, a large body of new literature was published on the industrial organization, the market structure, and strategic behaviour (De JONG, 1985; SHEPHERD, 1970; PORTER, 1986) and on the relationship with environmental factors (LAMBOOY, 1984; MASSEY, 1984; KAMANN, 1988; DAVELAAR, 1989; HANNAN and FREEMANN, 1989). Firms are widely differing organisations, and they also react quite differently. Some of the main strategies to overcome the challenges of the complex environment are concerned with solving spatial problems, others are focussed on selecting the right products and processes.

Strategies can be examined as being *consciously chosen*, by more or less informed decision-makers, or as *induced*, by environmental (structural) forces. In its most extreme position, the latter viewpoint is that of (Neo-)Marxist theory. A less extreme position is taken by geographers and by regional economists; they adhere to the incubation theory, to the theory of agglomeration economies and to the theory of competitive advantages. But this last mentioned group will not accept a deterministic or mechanistic approach.

The main classical authors from the industrial organization theory, like BAIN (1956), accepted a rather static approach with some deterministic and mechanistic flavour in it. Their model was based on the SCP-base (SCP stands for Structure-Conduct-Performance). A firm's performance is more or less determined by its structural position. The most logical and efficient outcome of the firm's behaviour in the market could be predicted if the market structure was known. This model was strongly attacked by De JONG (1973) and PORTER (1986): both authors approached strategic behaviour from a more dynamic angle. PORTER argued that firms should be viewed as actively searching for information about the relevant environment, and then consciously taking decisions as to what strategies would be likely to be the most appropriate in *their specific case*.

An entrepreneur (or a manager) can choose from many optional strategies; these are mainly related to:

(i) the product (price, kind, quality);
(ii) the technology and the processes;
(iii) the organisation (internal, external); and
(iv) the location.

Within each category (or in a combination), one or more strategies can be developed. Strategies are aimed not only at market parties, but at other relevant persons or groups (or governments) as well. Lobbying is one instrument, especially suitable to influence political decisions. It depends on the kind of firm involved, what possibilities there are to influence the environment. The global corporation is, of course, more powerful than are smaller firms. Yet, if a small firm is innovative, it can acquire a 'niche' on the market, and with that a virtual monopoly for a certain product and for a certain time - until others have imitated it and succeed in getting larger market shares.

4. Networks

The 'network concept' is used, or can be used, as:

(i) *infrastructural systems* of interconnected computers, telecommunication equipment, or for transportation;
(ii) *spatial systems*, such as relations between central places in an urban hierarchy;
(iii) *social structures* which do not possess the more stable characteristics of 'groups'; and
(iv) *economic relations* between agents in the market process, such as supplying or purchasing.

For the development of regional economics it might be useful to combine these concepts. Particularly the concepts of 'unequal market relations' and 'agglomeration economies' can be connected with that of 'networks'.

The complexity of the economy can be diminished by contracts and by a division of labour. One of the expressions of the complex economy is the network structure of activities, for instance in the relationships between multinational corporations (MNC's) and suppliers. A central issue is how the **structure** of a network is set up. The following suggestions for a classification are here conveyed for further discussion. This typology is an adjusted form of the one given by STOKMAN, ZIEGLER and SCOTT (1985).

First of all, a differentiation can be made according to the *nature of the links:*

(i) personal relations (formal, informal);

(ii) institutional relations (firm/firm, firm/government, firm/ university); and

(iii) contractual relations (occasional, structural).

A second classification is based on the *contents* of the relationship:

(i) relations based on natural resources;

(ii) relations based on financial resources;

(iii) relations based on management control;

(iv) relations based on property rights of licenses;

(v) relations based on a personal union of directors or share-holders;

(vi) relations based on complementarity/cohesion; and

(vii) relations based on consultancy.

The third classification can be based on the *position* in the network:

(i) centre of the network;

(ii) periphery of the network; and

(iii) intermediary in the network.

These classifications may be important to understand, for instance, the mechanism of transfers of innovations, and the forces behind the increased vitality of the small and medium-sized enterprises (SME's) in the recent economic restructuring process.

An economic approach of networks has to focus on their influence on the structure of markets and on the behaviour of the market parties. By the concept of 'market structure' we mean the set of positions of market parties, the set of regulations and the set of rules accepted by or enforced upon the parties. Factors such as the level of incomes, the position in the labour market, and the differences in technological knowledge and other information, belong to the relevant aspects of the market structure. It delimits the possible arena of agents in a market and denotes the 'starting positions' in the 'dynamic market'. The quintessence of modern market theory is that competition does not only occur between equal partners, but quite often also between unequal parties. The market is almost always divided by entrepreneurs and firms with unequal access to finance, resources, knowledge, and politicians. Many relations between firms, or between firms and consumers, are fixed in contracts which quite often express the unequal positions of the participants (HOLLAND, 1987, Ch.5).

In regional economics, the facts of unequal relations and of unequal regional development were already recognized by PERROUX (1950; 1970) and in development economics. The theory of Perroux is most intricately interwoven with the concept of networks. His basic idea was that market relations could be described as networks, using the mathematical language of topology. A 'growth pole' was to be considered as a dominant ('oligopolistic firm') centre within a network system. The topological network space was called the 'economic space'. The growth poles were 'key enterprises' within these economic spaces.

Growth poles are dominant and they have a determining power for the market structure. In their networks, the suppliers and buyers are dominated by the 'spider in its web'. According to PERROUX, growth oriented investments by the government should concentrate on these growth poles. The growth effects would then effectively trickle down to the other market participants.

In later publications, PERROUX himself and many other authors (BOUDEVILLE, 1966; VANNESTE, 1967; LAMBOOY, 1969) have stressed the double use of the concept of the growth pole. Although the original meaning was meant to be explicitly topological, many authors have used it as a spatial or topographical concept, especially in situations in which the concept was used as applicable for regional policies (KUKLINSKI, 1972). In that approach, a growth pole was considered to be a centre in an underdeveloped region, where investments should be concentrated in order to acquire a maximum effect by creating agglomeration economies. With the shift of approach, the network theory and the agglomeration theory were connected, but not consciously so. The basic relevance of both (i) market relations between unequal parties in networks, and (ii) the extra dimension of spatial proximity, encouraging non-market and market relations, remained central issues during the last 30 years of research in regional economics.

Agglomeration economies are the results of a mutual enforcement of market and non-market relations; both forces are activated in and through networks. Von BORRIES (1969) argued that the complex environment of agglomerations creates 'Fühlungsvorteile' (advantages of atmosphere and of informal relations).

Another kind of economic network is based on the concept of the 'filière' as used by French and Swiss researchers (BAILLY and MAILLAT, 1986). The idea is very familiar to regional economists and is related with the theory of industrial complexes (CHARDONNET, 1953; ISARD, 1960). The 'filière' stresses the fact that markets can be conceived as networks of suppliers and purchasers. The principal line in this network is to be selected by starting from the final product, for instance a watch. From that point, a system of suppliers can be traced back, not only in one line (as in a production column), but in all relevant directions, i.e. backward, forward, and diagonal linkages.

The main relevance of using the concept of the 'filière' is to display the spatial and functional interdependence within a region. If the full **network configuration** is located within the region, the development potential of that region is quite different from that of a region containing only a (small) part of it.

Figure 1: **A simple 'filière'**

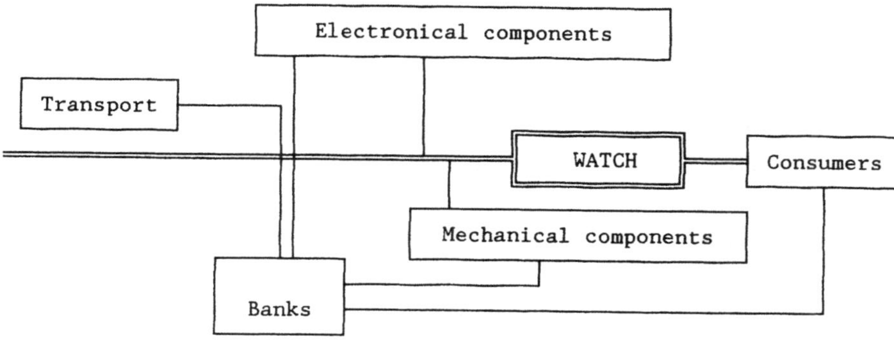

Banks

This configuration provides us with a picture of the degree of openness of a region with regard to its main products. It does not show the ownership, nor does it show the network of information streams.

In many studies (PRED, 1979; Van der KNAAP, 1978) a relationship is assumed to exist between the development of regions and a third kind of network, the hierarchy of cities and physical infrastructure. This kind of network is a framework for economic activities, rather than an economic network as such. However, it does condition the development of economic networks, and therefore it is very important to investigate this network in conjunction with economic and social networks.

Urban hierarchies focus on unequal spatial markets and unequal access to information. Incubation and diffusion theories support the idea of unequal parties in the structure of market relations. The issue of the hierarchical structure of diffusion has to be confronted with the impact of the new infrastructural networks established for using the new combinations of information and telecommunication: telematics and integrated networks.

5. Complex Environments and Networks

Agglomeration economies are commonly considered to be a very important subject in regional economic theory. But at the same time it is a phenomenon which is not yet clearly understood. RICHARDSON (1971, p.36) argued that the basic condition for creating agglomeration economies is proximity, enabling face-to-face contacts. He contends: "Urban location decisions are interdependent. This interdependence very often shows itself in agglomerations. For instance, agglomeration of similar establishments can create external economies - the ease of face-to-face contacts in the office zone or the fact that locating shops together minimizes commuting costs and attracts customers. Furthermore, complementary activities tend to agglomerate - offices and lunch bars, theatres and restaurants, wholesalers and transport firms. On the other hand, some activities repel each other; for example: nuisance creating manufacturers and high class residences".

The immense complexity of economic life, and especially of urban life, makes it necessary to develop specific approaches to clarify the working of the (urban) market. RICHARDSON (1971, pp.12/13) argues: "(...), the urban economy must, if it is to be understood, be conceived of in terms of more dimensions than is customary in most branches of economic analysis. We cannot comprehend it with static resource allocation models that determine the optimal combination of factors in a spaceless, timeless world".

This contention leads us to take into consideration the two mechanisms which influence the location decisions of investments: (i) the market, and (ii) the non-market decisions. Both can have a network character, on which we will elaborate here.

With the concept of *'market'*, the economist denotes a relation between at least two parties in order to transfer the ownership or possession of goods or services ('property rights') for a price. The level of the price depends on various factors, for example the preferences, the fact whether or not a certain competition exists, production costs, and so on.

Conventional economic theory starts with the assumption that quantities, production costs, incomes, and the transaction prices are decisive factors.

Environmental, or contextual, conditions are quite often assumed away, through the ceteris paribus formula. None of the economic schools deny the decisive importance of the market, but they differ when they consider the influence of contextual factors. Marxism, for instance, takes it for granted that the social and political circumstances make a really 'free market' - in which people are free to choose - impossible, because of the different starting positions of the market parties. Many participants are price-takers, and not price-makers. Many other economists accept the implicit value judgements about the set of conditions under which the market functions. This is illustrated by MILTON FRIEDMAN (1962, p.13) when he claims that: "fundamentally, there are only two ways of coordinating the economic activity of millions. One is central direction involving the use of coercion - the technique of the army in the modern totalitarian state. The other is the voluntary cooperation of individuals - the technique of the market place". FRIEDMAN denies that the market can be dominated by monopolistic forces providing different starting positions for the participants in the market. De JONG (1985, pp.6-9) distinguishes between three principal kinds of relationships between firms: (i) *competition* (i.e. the relations based on rivalry); (ii) *co-operation;* and (ii) *control.* These three are principles of coordination of economic agents. Not all activities can be explained by free competition.

The dominance of some market participants is enforced by the process of social networks, such as the 'Old Boys' network. This kind of social structure determines decisions - and does so quite differently from the theoretical market theory. This kind of network can be defined as "a club without premises, constitutions, or the life membership. Not simply a clique. Not quite an élite. Not exactly a trade union. But with some of the qualities of all these alliances" (TOMALIN, cited in HEALD, 1983, p.13).

In this paper, this type of network is considered as only one kind of a broader species of networks, which are important for the functioning of the economy and for agglomeration in particular. Some of those networks can be integrated with conventional economic theory; some cannot.

It is argued here that networks and/or markets are instrumental in diminishing the complexity of the real world. Complexity and uncertainty are very basic phenomena and many actions are focussed on reducing these characteristics. People who are interacting with each other, create markets, bureaucracies, institutions, and networks for co-ordination and control. One of the purposes of this kind of activities is to set up a framework for acquiring regular and secure information, thus enabling certain market parties to be stronger and more effective than others. It secures a better starting point in a 'dynamic market' (De JONG, 1985).

There is a difference between networks and institutions. Institutions consist of (i) a set of rules and values; (ii) the decision-making mechanism; and (iii) groups or organisations. By creating institutions to confine the free functioning of markets, or to organize them, the market process becomes a social process. The market is not an institution by itself, but its functioning is defined by institutions. The 'market' is confined within the limits of laws, prescriptions, and habits. The invisible hand is quite often the result of the 'marriage' between the search for reducing uncertainty and a set of institutionalized rules that determine the outcome of the interaction of seemingly independent decisions.

Not all institutions are created to facilitate exchanges in a market structure. Universities, churches, and tennis clubs are established to ensure the free exchange of information, or to facilitate religious or leasure activities. They may be linked with property rights and with market transactions, but their primary goal is different from that of common market participants. They are, however, interest groups similar to those active in market situations (LAMBOOY et al., 1982).

An institution is not always explicitly set up for its accepted functions. It may be inherited from the past, but if its basic goals could not be accepted, it would not be possible to maintain its existence without force.

Nobel-prize winner HERBERT SIMONS (1976, p.129) has presented the importance of institutions to understand the economic process, as follows: "Economics (...) is a description and explanation of human institutions, whose theory is more likely to remain invariant over time than the theory of bridge design. Decision processes, like all other aspects of human institutions, exist inside human heads. They are subject to change with every change in what human beings know, and with every change in their means of calculation. For this reason, the attempt to predict and prescribe human economic behaviour by deductive inference from a small set of unchallengable premises must fail and has failed". Nevertheless, human institutions themselves are devices facilitating the reduction of information costs and uncertainty and for channelling the lessons of past experiences.

For the heterogeneous economic world, information is an essential factor - so important, in fact, that it is a production factor in itself. Information about products, processes, or markets, acquired by R&D and marketing activities, affects a firm's capacity to enter or dominate a market: innovation, the product life cycle, diffusion, and other concepts can be used to indicate this fact. The necessity of institutions to confine the freedom of action of the heterogeneous market participants affected the organization of enterprises. Information and co-ordination required increasing resources devoted to the organisation itself.

The rise of the corporation has to be explained by scale economies and by 'institutional economies'. The technology of the 19th century was characterized by large fixed capital investments. It required financial resources, a mass of cheap labour, and an easy access to natural resources. The organisation became increasingly complex, and this led to specialization of the large corporations (NORTH, 1986).

The present information revolution leads to new forms of organisations. Investments in 'human capital' become more decisive than large capital assets. R&D, innovation, and an easy access to knowledge are considered to be basic inputs to the firm, in order to be able to survive (KAMANN, 1988). The market has become more heterogeneous, uncertain, and changes very quickly. Product life cycles of 'high tech' and consumer goods now are shorter than those of the products of previous periods. The volume and the importance of exchanges and transactions increase.

One of the most important processes of our time is the development of firms specialized in dealing with transaction and information, such as lawyers, brokers, insurance companies, banking, accounting, computer services, marketing analysts, data-banking, et cetera. At the same time, government has grown enormously. The Israelian economist OFER (1973) has shown that even the civil service, as a share of the total labour force, increased with complexity and the national income. Our complex societies have an inherent need for rules and co-ordination. Appeals made by the market participants to the government to act as a third party, have become natural, although very often this is regarded as a dangerous development (NORTH, 1986; OLSON, 1982). But contracts, subsidies, and tax deductions attract firms like bees looking for honey.

Another important phenomenon is the specificity of institutions as to national states or regions. This is one of the most decisive factors in explaining the behaviour of firms (LAMBOOY, 1986). For international trade, the possible conflict between the state and the multinational corporation is expressed by KADAR (1984, p.11): "One of the growing contradictions of our age is that, whereas the productive forces are increasingly becoming internationalized, the economic structure and the levels of advancement in the various countries are still determined by national institutions; it is in their framework that social, cultural and political progress is taking place, and the nation-states seem to be going to play a decisive historic role even in determining the patterns of economic development for a long time to come".

In general, the problem arises because states and regions attempt to keep control over transactions related to their territory, whereas multinational corporations and other exporters and importers have to deal with more territorial markets and institutions. One solution is to establish branch plants or marketing organizations specifically adjusted to national institutions.

The increasing complexity necessitates a reduction of uncertainty and heterogeneity. Professional services are one solution; a structured relationship (networks) between large enterprises and small and medium-sized enterprises is another one (ESB, 1987, various articles).

6. Organizational Ecology

Large corporations have developed from local to multidomestic to really global corporations. But, as PORTER (1986, p.5) contends: "There is no one pattern of international competition nor one type of global strategy ... Sometimes an international industry is nothing more than a collection of largely separate domestic industries. In other industries, competition is global, though in many different ways. Similarly, there is no archetypical global strategy, but numerous ways of competing globally that reflect different mechanisms through which a firm's international presence yields competitive advantages". And further on: "The particular forces driving globalization as well as the resulting patterns of competition vary widely from industry to industry". PORTER (1986, p.6) then concludes: "Coordination among increasingly complex networks of activities dispersed worldwide is becoming a prime source of competitive advantage", and he continues (p.7): "Implementing a global approach to strategy requires a difficult organizational reorientation for many firms. ... The solutions arise as much from attitudinal changes, education, and organizational processes as they do from formal reporting relationships".

Two points can now be made. First, that economic networks are one form or organization MNC's can develop. And secondly, that certain regions are likely to be better equipped to be a focus of headquarters, production units, and networks. As has already been emphasized earlier, certain regions are 'incubation regions' in which new ideas and products can be developed, whereas other regions are 'receptor regions', which benefit from the diffusion from the first type of region. This does not mean that the latter kind has activities with a lower value added. DAVELAAR (1989) has found that quite often regions of the latter type are better off, because they benefit from the diffusion of more mature products and have less costs in the development process. The organizational structure differs, however. The incubator regions convey a wide variety of professions, firms, and success rates. The headquarters of industrial corporations are located in the largest agglomerations with highly developed producer services. The geographical pattern of headquarters and the consulting professions is much more concentrated than that of production units and suppliers, although these are less concentrated in countries like the Netherlands, Germany, and Switzerland, because there the spatial structure of cities is much less centralized than is the case in the UK or in France.

SABEL a.o. (1987, p.38) have stressed that even regions such as Baden-Württemberg, without a strong concentration of large corporations with a 'leading industry' or big agglomerations, can be successful if they are: (i) "Industry structures which *coordinate specialization* among firms and provide infrastructural *services* which allow individual companies to run all the risks associated with renouncing self-suffficiency; a state *government* which seconds these efforts at coordinated specialization by shoring up the infrastructure of innovation with its own authority and resources", and (ii) "the conditions include *specialized, but flexible firms* that depend on the use of *skilled, versatile labor* and general purpose machines; *sub-contracting systems* which allow firms to spread the risks of research and development".

Thus, there are two sides to organizational ecology: firstly the character of regions, and secondly the character of firms and industrial sectors. Regions are rather stable in their characteristics, because of historical and social accumulation. It is very dificult to change infrastructure, education, political and labour attitudes over a short period. Corporations are more flexible. They can change their geographical organization quite easily, although even the largest among them have to behave 'socially'. Public media, governments, labour unions, and consumer organisations set certain limits as to relocation or the changing of prducts and production processes.

Organisations - even the large corporations - have to acknowledge the influence of the socioeconomic environments in formulating their geographical strategies. Organisations reflect their environments. Diversified, dynamic regions 'produce' other kinds or organizational response than do stable and undiversified regions. HANNAN and FREEMAN (1989, p.27) argue: "In fact, ecological-evolutionary theory suggests that uncertain, volatile environments will support diverse organizational forms and that the apparent winners will fluctuate from time to time as conditions change". That is precisely the case for MNC's in a global economy. In previous periods in which traditional Taylorist or Fordist management and technology prevailed, it was possible to have rather rigid and formalized hierarchical companies. Large-scale production and the geographic dispersion of the markets gave them a decisive advantage over smaller firms. In our time, modern technology, more educated labour, and political pressures have altered the picture. New and more flexible organisations are the new leaders. They are not always the largest firms. Perpetuously, new entrants in the market are able to break through, endorsed by new technologies, new tastes, new geographical markets or new institutional arrangements. Even the strongest corporations have to adjust continuously their organisation to the regional differentiated development patterns.

7. Conclusions

The mesoeconomic approach to (regional) economic development can be developed in co-operation with contributions from organizational theory and from industrial organization. Traditional neo-classical theory reduces too much, losing insights, because it assumes complexity away. Various studies by economic geographers (JANSEN, 1974; 1987; MASSEY, 1984) and regional economists (DAVELAAR, 1989) show that this is a useful approach to better understanding the dynamics of regional economic development.

LITERATURE

Bailly, A.S.; D. Maillat, 1986: Le secteur tertiaire en question. Paris: Editions Anthropos.

Bain, J.S., 1956: Barriers to new competition; their character and consequences in manufacturing industries. Cambridge, Mass.

Borries, H.W. von, 1969: Ökonomische Grundlagen der westdeutschen Siedlungsstructur. Hannover: Jnecke Verlag.

Boudeville, J.R., 1966: Problems of regional economic planning. Edinburgh: Edinburgh University Press.

Chardonnet, J., 1953: Les grands types de complexes industriels. Paris: Librairie Armand.

Davelaar, E.J., 1989: Incubation and Innovation; a spatial perspective. Amsterdam: VU Uitgeverij.

Duijn, J.J. van; J.G. Lambooy, 1982: Technological innovation and regional economic growth; a meso-economic analysis. Amsterdam: University of Amsterdam, Dept. of Economics. Research Memorandum Nr.8207.

Economisch-Statistische Berichten (ESB), Special issue of 7 October 1987: 'De Netwerkeconomie' (The Networks Economy), Rotterdam.

Eucken, W., 1950: Die Grundlagen der Nationalökonomie. Berlin/Heidelberg: Springer Verlag.

Friedman, M., 1962: Capitalism and freedom. Chicago: Phoenix Books.

Hannan, M.T.; J. Freeman, 1989: Organizational Ecology. Cambridge (Mass.): Harvard University Press.

Heald, T., 1983: Networks; who we know and how we use them. London: Coronet Books, Hodder & Stoughton.

Holland, S., 1987: The market economy; from micro- to mesoeconomics. Vol.I., London: Weidenfeld.

Isard, W., 1960: Methods and regional analysis. Cambridge, Mass.

Jansen, A.C.M., 1974: Elementen van een vestigingsplaatsmodel voor industriele 'multi-plant' ondernemingen (Elements of a locational model for industrial 'multi-plant' enterprises). In: Tijdschrift voor Sociale & Economische Geografie (TESG), Nr.3, pp.174-193.

Jansen, A.C.M., 1987: Bier in Nederland en Belgie; een geografie van de smaak (Beer in the Netherlands and in Belgium; a geography of taste). Amsterdam: KNAG/EGI, NGS Series, Nr. 39, (Ph.D.-thesis University of Amsterdam).

Jong, H.W. de, 1972: Dynamische concentratietheorie (Dynamic concentration theory). Leiden: Stenfert Kroese.

Jong, H.W. de, 1985: Dynamische markttheorie (Dynamic market theory) Leiden: Stenfert Kroese (3rd Edition).

Kadar, B., 1984: Structural changes in the world economy. London: Frances Pinter.

Kamann, D.F.J., 1988: Spatial differentiation in the impact of tech nology on society. London: Gower.

Knaap, G.A. van der, 1978: A spatial analysis of the evolution of an urban system. Utrecht: Ph.D.thesis Erasmus University, Rotterdam.

Kuklinski, A., 1972: Growth poles and growth centres. The Hague: Mouton.

Lambooy, J.G., 1969: Het geografisch systeem en de groeipooltheorie (The geographical system and the growth pole theory). Amsterdam: Free University, Inaugural lecture.

Lambooy, J.G., 1984: The regional econoly of technological change. In: J.G. Lambooy (ed.), New spatial dynamics and economic crisis, pp. 63-76. Tampere: Finnpublishers. IRPA Yearbook.

Lambooy, J.G., 1986: Information and internationalization. In: Revue d'Economie Rgionale et Urbaine, No.5, pp.719-732.

Lambooy, J.G.; P.C.M. Huigsloot; R.E. van de Lustgraaf, 1982: Greep op de stad?! (A hold on the city?!). The Hague: Staatsuitgeverij.

Massey, D., 1984: Spatial divisions of labour; social structures and the geography of production. London: Macmillan.

North, D.C., 1986: The new institutional economics. In: Journal of Institutional and Theoretical Economics, Vol.142, No.1, pp.230-238.

Ofer, G., 1973: The service sector in Soviet Economic Growth. Cambridge: Cambridge University Press.

Olson, M., 1982: The rise and decline of nations. Newhaven.

Perroux, F., 1950; 1970: Note on the concept of 'growth poles'. In: D.C. McKee et al., Regional Economics. New York.

Porter, M.E. (ed.), 1986: Competition in global industries. Boston (Mass.): Harvard Business School Press.

Pred. A.R., 1979: City systems in advanced economies. London: Hut chinson.

Richardson, H.W., 1971: Urban economics. London: Penguin Books.

Sabel, C.F.; G.B. Herrigel, R. Deeg; R. Kazis, 1987: Regional prosperities compared: Massachusetts and Baden-Württemberg in the 1990's. Berlin: Forschungsschwerpunkt Arbeitsmarkt und Beschäftigung.

Shepherd, W.G., 1970: Market power and economic welfare. New York.

Simon, H., 1976: From substantive to procedural rationality. In: S.J. Latsis (ed.), method and appraisal in economics. New York.

Stokman, F.N.; R. Ziegler; J. Scott (eds), 1985: Networks of corporate power. Cambridge: Polity Press.

Vanneste, O., 1967: Het groeipoolconcept en de regionaal-economische politiek (The concept of the growth pole and regional economic policy). Antwerpen: SWU.

On Concepts of Regional Economic Structure

by

Rodney C. Jensen, Guy R. West, John H.L. Dewhurst

1. Introduction

The world of economics is apparently one of endless diversity, both in the nature of economic activity and in the economic structure of those myriads of entities which we call "economies." The economic literature implies that this diversity is sufficiently extensive to be a dominant feature of economies, or at least intractable to successful economic analysis. Amidst this diversity, economists have been able to identify commonalities or regularities in economic structure only in the most general and non-operational terms, and have made virtually no progress towards either classifying economies into "like", "similar" or "dissimilar" groups, of identifying a basis for measures of similarity among economies.

This paper addresses the question of diversity and similarity among economies. Section 2 addresses first the implicit presumption of endless diversity in economics. Section 3 probes the notion of similarity in more depth, indicating that the science of taxonomy could provide some valuable insights into the classification of economies, and the expression of degrees of sameness or similarity. Section 4 suggests the use of input-output tables as a data base for measuring the degree of similarity and suggests a revised concept of Fundamental Economic Structure as one approach to this measurement. Section 5 provides some concluding comment.

The central thrust of this paper is the suggestion that regional science, and economists in particular, might profitably address regularities in economic structure, rather than diversity, and that such an approach has potentially significant rewards in terms of insights and research resources. While economists have been content to assume regularities in economic behaviour, they have generally not sought to address questions of regularities in economic structure. This paper suggests some concepts and approaches which may be relevant on these issues.

2. Diversity in Economics

We can interpret the often vaguely-used term "economy" simply as the aggregate economic activity occurring in a given physical space. These spaces could be defined in a regional science context, in terms of a spatial hierarchy of economies, from the world economy, to groupings of national economies, national economies (the most common unit), state or provincial, regional, sub-regional, urban, locality and community economies. These economies may or may not refer to areas which are continuous or contiguous. Most economies are defined spatially on political or administrative boundaries; few are defined more flexibly on more rationally-determined criteria. On this basis alone, analysts are faced with a vast array of economies with an apparently equally vast range of characteristics or differentiating features. Considering also the array of national economy types, the apparent degree of diversity in economies and economic structure could be considered to approach that degree of diversity occurring in the natural world.

The literature suggests that economists have presumed the diversity of the economic world, and of the structure of economies in particular, to be almost infinite, or at least sufficiently extensive to be intractable to organisation/classification in all but the most general and rudimentary terms. This is evident from the propensity of the economist to consider and analyse chosen single-unit (usually national or regional) economies, and to imply a high degree of individuality and uniqueness for that economy, to place a high value on that individuality in economic analysis, and to imply that the integrity of economic analysis is compromised unless these values are observed.

It is evident also from the small amount of effort devoted to the detailed empirical description of economies and the comparison of research results across a number of economies; the implication is that such description and comparison has little value in the high degree of diversity of the existing array of "types" of economies.

Within this apparently endless diversity, regional scientists have recognised that some regularities in economy-types can be expected to exist. It is accepted that as we move from smaller to larger economies, increasing levels of economic complexity will be observed, and that over time within any one region type, we might observe increasing levels of economic interdependence, if not increasing complexity. This reflects the decreasing marginal propensity to import with increased region size and development. While these regularities have been recognised (e.g. MULLIGAN, undated), even this simple concept of relating import propensities to region size has not been widely operational in regional economic analysis. Certainly systematic studies of differences in regional economic structure across the small-largenational continuum of economies, or at points in time for regional economies, have not progressed to the stage where regional economic structural change can be described in any but the most basic terms.

In summary, studies in regional economic structure, and theories of regional economic growth, appear to have taken us only to the point where we recognise some general regularities amid an apparently intractable degree of diversity in regional types and structures. It is the intention of this paper to suggest that this perception of diversity is inappropriate, that it is more fruitful to concentrate on the degree of sameness among economies than on the degree of diversity, and that potentially workable approaches to identifying this level of sameness or similarity can be developed.

3. The Notion of a Taxonomic Approach

A very large amount of quantitative information exists with respect to a variety of characteristics of the seemingly endless number of entities which we term economies. Much of this information is assembled at the national level in reasonably-standardised systems of national accounts, including input-output tables; no such systems operate efficiently at the sub-national level in most countries. Accounting systems such as the United Nations System of National Accounts are designed to allow understanding of the nature and structure of single national economies rather than groups of national or sub-national economies. A broader view of economic structure would consider each economy as one observation of economic structure in the array of economies which make up a population of observations, and seek to determine the extent of diversity or similarity among this array. To this end it is necessary to consider the question of classifying economies if we are to develop useful approaches to managing the large amount of loosely-related data which is available. Section 3 considers the notion of a taxonomic approach to classifying economies, along the lines of JENSEN et al (1987).

3.1 The Principles of Taxonomy.

"Faced with a vast array of diversity .... man instinctively classifies" (HEYWOOD, 1976). Dividing this diversity into smaller, more identifiable and manageable groups is a virtually instinctive, everyday process. Taxonomy is the scientific study of classification, of the ordering of units (plants, animals, in more established taxonomic studies) into groups termed taxa of higher and lower orders, forming a taxonomic hierarchy. The taxonomic problem is the arrangement of those units to reflect best the similarities and differences of members, and at the same time to organise this array of knowledge/data in the most convenient and useful manner at all levels of classification. The taxonomic problem is essentially that of developing a system of classification, and requires:

(i) determination of differentiating features or characters of the units to be classified, and
(ii) the process of identification - determining the extent to which two or more units can be described as "the same", or determining measures of "sameness" or degrees of "similar but different".

Classification provides important advantages:

(i) the summarising and organisation of a large amount of information/data/knowledge into manageable proportions, to reduce the processes of description and analysis,

(ii) provides an ability to make generalisations at different and known levels of generality,

(iii) provides an ability to make predictions for individual units, based on sets of collective identifiable characteristics.

Reference is conventionally made to two approaches to classification, namely natural and artificial. A natural classification is based on the sum total (or at least many) of the characters of the units, classifying together those units which are most alike in most ways. An artificial classification is based on one or a few chosen characters. Natural classifications have a high predictive value since predictions are based on, and convey, the most information. Artificial classifications, although easier to identify, have little predictive value.

Finally in this section, taxonomists refer to genotype, which is the basic genetic constitution of the unit, and which remains unchanged throughout the life-cycle and determines the general framework and limits to the existence of the unit, and phenotype which is the individual unit within a genotype; environmental conditions and management will ensure a range of phenotypes for each genotype. Taxonomists prefer to base classifications on the genetic structure represented by the genotype rather than the individual unit variations of the phenotypes.

Most modern systems of taxonomy are mixtures of various approaches, with a tendency to increased natural content. In recent years, taxometrics or numerical taxonomy has provided more rapid and comprehensive measures of similarities in respect of large numbers of characters previously beyond the capacity of the taxonomist, and has increased the level of objectivity in the process of taxonomy.

3.2 On Taxonomy and Economies

This section considers the principles of taxonomy outlined above, in the context of the acknowledged diversity in economies and economic structure which was discussed in Section 2. It is suggested that this diversity of economic data/knowledge could be amenable to organisation (at least in part) through a taxonomic approach, and that the advantages of such an approach would be significant. It is not suggested that we should envisage the sum total of universal economic data organised in internationally acknowledged taxonomic systems of the type which have been applied to the plant and animal kingdoms. It is simply suggested that a modest acknowledgement of the principles and practice of taxonomy in economics could lead to a more efficient use of research resources with improved insights into the universal view of economic structure. It is later suggested that the input-output table offers a useful and workable avenue to a modest beginning to a taxonomic view of economies.

Economists have invariably used artificial classification systems, usually based on one character, when classifying or describing economies. Economies have been classified on the criteria of geographical base, resource base, political system, political and economic community, racial or religious dominance, size of economy, income level, spatial level, level of development, and other special-purpose criteria. The level-of-development criteria tend to broad generalisations such as third and fourth world, developed and lessdeveloped or developing, north and south, and the well-known seven level-of-development classifications of KUZNETS (1957, 1963). These classifications have the real advantage of ease of application, but have the disadvantages of the artificial approach; they are based on and convey little information, and have virtually no predictive value because of the potentially large range of economy types within each group.

In principle, a more comprehensive and natural taxonomic system of classification/description of economies would provide some of the advantages to the economist which are currently enjoyed by counterparts in the natural sciences. These could be:

(i) the organisation, ordering and summarising of the large amount of data/knowledge on the nature and structure of economies, into smaller and manageable groups,

(ii) the ability to make generalisations, based on the structure/behaviour of groups of economies of known levels of similarity, and at various levels of generality; generalisations could be based on a sound empirical footing in the sense that known levels of closeness and similarity can be stated, and

(iii) improved predictive value with respect to individual economies which belong to a group of economies defined on a large number of similar characteristics. We would expect that the similarity of many structural characteristics would be a sounder base for prediction of other characters of behaviour patterns than the simplistic criteria of the artificial classification schemes.

Apart from these three obvious advantages, a significant fourth advantage would emerge. A taxonomic approach to considering the nature of economies would encourage economists to visualise any individual economy as one of a larger array, differing from some other economies in small details or unimportant characters, while differing from others in much more comprehensive ways. Economists would be encouraged to identify the inevitable commonalities, or degrees of commonness on structure and behaviour, rather than implicitly emphasizing the dissimilarities between the economies or implicitly assuming high levels of uniqueness for each economy. This suggests that substantial commonalities among economies can be recognised in practice and that degrees of commonality in terms of commonly-shared characteristics can be satisfactorily defined. The authors suggest that groups of economies can be recognised empirically which are "more the same than different", especially in spatial sub-national groupings, and that a great deal of research effort can be saved by recognising these groups. In particular, it could well be the case that the development of an economic model for an individual region adds little to an economic model for the group of which that region is a member, or to a model for a region which is quite similar in structure. A taxonomic approach would at least encourage economists to attempt to "type" the economies they study, to describe their characteristics and to be aware of possible similar and dissimilar economies, and the extent of variation in economy "types". The current state of organisation of economic knowledge does not allow a high degree of awareness of these matters.

Some discussion of the choice of differentiating characters, directions of classification, the process of classification and measures of similarity are provided by JENSEN et al. (1987). Clearly at this early stage of organising economic data, the parameters of any suggested system must be largely speculative. Before providing more specific suggestions on the role of the input-output table in such a system, we briefly visit the literature for a review of approaches which have some relevance to the description or classification of economies.

3.3 Assistance from the Literature

We visit the literature briefly to identify some of the potential sources of assistance in either the definition of the classification base or potential members of the character set. The issue was raised by STONE (1986) in the context of an input-output taxonomy, but this was focussed more on the direction of classification, definitions and treatment of the mechanics of producing tables. One such source lies in the studies which have applied triangulation to input-output tables (LEONTIEF, 1963; SIMPSON and TSUKUI, 1965; HELMSTADTER, 1969; LAMEL et al., 1971) in attempts to recognise patterns or blocks of transactions within the table. Although relatively ignored in recent years, the ideas of patterns and block recognition as integral parts of an economy are very forceful indeed, leading SIMPSON and TSUKUI (1965) to ponder the existence of a "fundamental structure of production". This concept was further developed into the notion of "fundamental economic structure" by JENSEN, WEST and HEWINGS (1988); this concept is developed in some detail later in this paper. DEUTSCH, SYRQUIN and URATA (1986) have examined a large number of input-output tables to try to uncover a relationship between economic development and the structure of production.

A second source are measures of holistic matrix description summarised by HEWINGS, JENSEN and WEST (1987), and, in particular, the concept of connectedness as represented by HAMILTON and JENSEN (1983) and SZYRMER (1985, 1986). Connectedness as a measure of "churning" in the table, and the critical-cells approach of SHERMAN and MORRISON (1950); HEWINGS and ROMANOS

(1981); and WEST (1981), provide some powerful measures of economic attributes of economies which are not clear from surface examination of transactions tables. These measures indicate the extent to which the economy churns and the more significant components in this process.

A third source lies in the work which has been undertaken mainly by regional scientists in the general area of analysis of hierarchies and ordering. For example a wealth of literature has developed on the combination of numerical taxonomy techniques, collectively termed cluster analysis, which has proven particularly useful where the taxonomic system of classification is unknown and is to be derived as part of the classification process. Several variations of the technique exist (ANDERBERG, 1973) although applications to input-output tables have been limited to clustering sectors within the tables as part of the study of aggregation (FISHER, 1969; BLIN and COHEN, 1976), rather than groupings of economies as a whole as suggested in this paper. Cluster analysis techniques have typically been used to summarise large tables, for example the French interdepartmental migration tables (SLATER and WINCHESTER, 1978), and the trading ties among nations (SLATER and SCHWARZ, 1979). Cluster analysts have, however, tended to demonstrate a stronger interest in the theoretical aspects of the technique and its application, than in the inherent logic of the classification systems. Other approaches tend to be less arbitrary by exploring more of the linkages within the table, for example, higher order factor analysis (O'HUALLACHIAN, 1985), causal modelling (ASHER, 1976; GETIS, 1985), and optimal ordering (GHOSH and SARKAR, 1970; BLIN, 1973). This literature on numerical ordering of classification systems pays little regard to economic logic in the classification process; it would seem important for such a regard to be a central feature of any approach to a classification of economies as suggested in this paper.

4. Input-output Tables and the Concept of Fundamental Economic Structure in Taxonomic/Classification Approaches

Previous sections of this paper have highlighted three basic issues, namely the diversity of economies and economic data, advantages to be gained from some system of classification of economies into groups of similar units, and the need to isolate the similarities among economies rather than implicitly assume that diversity is the dominant character of economies. This section directs the focus of the paper to the conventional input-output model and the role which this model might assume in expressing similarities or diversity among economies. It suggests that the input-output table , through analysis of the structure of the table, is a useful character set for the process of taxonomy, at least at this early stage of research in this area. More specifically it introduces revised notions of the concept of Fundamental Economic Structure (FES) as a logical base for classification of economies.

4.1 The Input-output Table as a Taxonomic Character Set

Perhaps the main raison d'etre for input-output tables has been their ability to represent in some detail the structure of economies in terms of interdependence among sectors, and in the analytical implications on this interdependence. It is therefore surprising that input-output tables have not been used more extensively in the study of economic structure, beyond the limited number of applications summarised in Section 3.3. We would argue that the input-output table offers perhaps the most promising source of data for economy classification for three reasons, (i) large numbers of input-output tables exist for a variety of economies at various spatial levels; this range will continue to increase as table compilation technology improves, (ii) it is possible to consider each cell of the input-output table as an observation of the level of a specific category of economic activity, and hence as a character in the classification process, (iii) the input-output table provides the possibility of identification of patterns of cell structure, in groups, blocks or layers which can collectively express important characteristics of the economy, and allow comparison with the patterns of other economies.

One might consider each input-output table as a phenotype, as an individual unit within a genotype to be identified, e.g. a table representing a regional economy within a group with similar "genetic" constitution defining a genotype of regions. The difficulty with such a suggestion is the need to isolate the characteristics which might define the genotype, or the basic genetic or structural nature of the economy in question. The following sections suggest the FES as one approach to this problem.

4.2 Fundamental Economic Structure – the Original or Partitioned Approach

The earlier work on taxonomy (JENSEN et al, 1987) offered the general concept of FES as an organising principle upon which a taxonomy of economies might be based. At that stage the earlier concept of FES (termed now the original or partitioned approach) had not been fully developed, and was largely an expression of empirical regularities derived from an analysis of input-output tables for the eleven regions of Queensland (JENSEN, WEST and HEWINGS, 1988). The approach simply sought to identify regularities in each cell across these regions, which ranged from a less-developed remote rural region to a modern metropolitann region; these analyses sought to identify also any patterns in cell behaviour. Those cells which showed transactions levels which were consistently present at predictable levels over this range of economies were termed "fundamental" meaning that category of economic activity was inevitably required at a predictable level.

The results of this empirical work with the Queensland tables, which were consistent with the results achieved later in a study of sets of regional tables for the Australian states of New South Wales, Victoria and South Australia, suggested that most (perhaps 75%) of the cells of Australian input-output tables showed regularities and were "predictable" in the sense that acceptable regression relationships were established. These regularities were shown to be present in the secondary-tertiary portion of the primary-secondary-tertiary continuum of the input-output table, which was considered to define the FES. Those cells where regularities were not identified occurred mainly in the mining and agricultural sectors, or those sectors which are more region-specific in nature, and were collectively termed the Non-Fundamental Economic Structure (NFES). Hence it was possible to identify both an FES and an NFES partition of a set of input-output tables.

The results obtained from these earlier studies have stimulated further research and evaluation of the FES concept. They provided the first evidence that regularities in economic structure occur in cross-sectional studies of a number of regional input-output tables, that therefore similarities in economic structure were significant in some parts of the table, and that it could be possible to provide empirical estimates of these regularities or relationships between tables and cells of tables. In this sense the results of the original paper (JENSEN, WEST and HEWINGS, 1988) were considered worthy of further development.

A review of the original or partitioned approach to FES (JENSEN, DEWHURST, WEST and HEWINGS, 1989) suggested that a more comprehensive development of the FES concept was desirable for three reasons, (i) the concept was basically an expression of observed empirical regularities, rather than an expression of clearly-defined theories and economic logic, and required further conceptual development, (ii) procedures used in the identification of regularities in the original study were considered to be unnecessarily restrictive in that some existing regularities may not have been identified, and (iii) it is possible that the lack of an established regularity by these methods could imply an irregular or unordered structure which may not be an appropriate conclusion. For these reasons, and the fact that several regional tables were required to test for regularities, the view was formed that the partitioned approach may be too restrictive to permit significant further development of the concept, and that it was a special case of a more general and flexible concept of FES outlined below and termed the tiered approach.

4.3 Fundamental Economic Structure - The Tiered Approach

This section outlines a new and more comprehensive approach to the FES concept, as described first in JENSEN, DEWHURST, WEST and HEWINGS (1989). It relies on the proposition that it is reasonable to consider any (regional) input-output table as the sum of two input-output tables, one of which can be termed the FES, and the other the NFES, and which have the following properties:

(i) the FES tier will be "predictable" within any economic system, in that each region within that economic system can be expected to exhibit an FES tier which is generally determined by the characteristics of that system as a whole, and is "predictably different" from the FES of other regions in that system.

(ii) The NFES tier, which will not be predictable within an economic system, but will be region-specific, or determined by the relationship of the individual region to the rest of the world.

This approach suggests that each cell of the input-output transactions table (and also the coefficient table) should be considered as consisting of two elements, one of which is fundamental or predictable, and one which is non-fundamental, region-specific or "random" in nature. We distinguish between these two tiers, or elements in each cell, according to two types of reaction paths:

(i) internally-oriented reaction paths, which originate within the region from the regional level of population and the demand for people-oriented goods, and are directly and indirectly determined endogenously within the region. These will represent the FES tier.

(ii) externally-oriented reaction paths, which originate outside the region, and will represent the NFES tier.

This suggests a simple functional model based on internally and externally-oriented reaction paths, of the type developed by CZAMANSKI (1963) and ISARD and CZAMANSKI (1965) and extended later by JENSEN (1981). This model specified three categories of economic activity:

(i) Geographically-oriented activity, the level of which is determined by forces external to the region, i.e. activities serving extra-regional markets, but located in the region due to regional advantages of location or resource availability. Included will be agricultural and mining industries, non-local tourism, and some defence and transportation facilities. (notated G_i)

(ii) Complementary activity, the level of which is determined by the G_i, either in the processing of output of the Gi, or providing inputs to these activities. Included will be agricultural and mining processing, and agricultural and mining supply industries. (notated C_i)

(iii) Urban-oriented activity, the level of which is determined by the level of regional population, through provision, directly or indirectly, of goods and services for personal consumption. (notated U_i)

Each cell, or category of economic activity in region r, T_{ijr}, may be considered as containing elements of the above, i.e.

$$T_{ijr} = G_{ijr} + C_{ijr} + U_{ijr} \tag{1}$$

Since the G_i and C_i are taken to represent the NFES, in terms of the tiered approach to FES, this may be rewritten as

$$T_{ijr} = NFES_{ijr} + FES_{ijr} \tag{2}$$

if we make the assumption that the level of the U_{ijr} can be predicted directly from population, given known levels of personal consumption.

In terms of the conventional approach to input-output, we need to recognise the FES and NFES elements of final demand, i.e. to distinguish those elements of final demand which may be considered to be determined by the level of population, and those determined extra-regionally. We can decompose final demand into a number of elements in the usual input-output identity:

$$X = (I-A)^{-1} [F_1 + F_2 + \ldots\ldots + F_m] \tag{3}$$

where, for the present A is the usual matrix of average input-output coefficients and X is a vector of gross outputs. It is possible to attribute a level of output to any one, or a combination, of final demand elements, as

$$X_i = (I-A)^{-1} F_i \tag{4}$$

where $\hat{X}_i$ is a vector of (decomposed) gross outputs such that $\Sigma X_i = X$, or

$$\Sigma X_i = \Sigma (I-A)^{-1} F_i \tag{5}$$

and to calculate that part of the input-output table attributable to F_i :

$$A \hat{X}_i = A [(I-A)^{\hat{-}1} F_i] \tag{6}$$

where X_i is a diagonalised matrix of gross outputs. Collecting together those elements of the F_i which may be considered to vary with household expenditure, the FES components F_F, and those determined exogenously, the NFES component F_N, we can calculate output levels attributable to the FES and NFES as

$$X_F = (I-A)^{-1} F_F \qquad \text{(FES)} \qquad (7)$$
$$X_N = (I-A)^{-1} F_N \qquad \text{(NFES)}$$

where $X_F + X_N = X$, and the FES and NFES tiers as

$$T_{ijrF} = A \hat{X}_F = A \ [(I-A)^{\hat{-}1} F_F] \qquad \text{(FES)} \qquad (8a)$$

$$T_{ijrN} = A \hat{X}_N = A \ [(I-A)^{\hat{-}1} F_N] \qquad \text{(NFES)} \qquad (8b)$$

The above incorporates an important and potentially controversial assumption that the standard A matrix is appropriate to each category of final demand. This assumption is common in input-output but may be unacceptable in an FES-NFES context. If different A matrices are considered to apply to different elements of final demand F_i, then sector specific coefficient (sub)matrices A_i would be necessary. In terms of the FES-NFES dichotomy, the two tiers may be defined as:

$$T_{ijrF} = A_F X_F = A_F \ [(I-A_F)^{-1} F_F] \qquad \text{(FES)} \qquad (9a)$$

$$T_{ijrN} = A_N X_N = A_N \ [(I-A_N)^{-1} F_N] \qquad \text{(NFES)} \qquad (9b)$$

In this case $T_{ijrF} + T_{ijrN}$ will not necessarily equal T_{ijr}, and some balancing term, presumably representing a synergistic component, will be necessary.

In practice the FES tier can be calculated from equation (8a) if the average A matrix is considered appropriate, or equation (9a) if the A_F matrix can be estimated. The NFES tier can be estimated by subtraction if the input-output table is known, or by survey methods if the table is under construction. In hybrid methods of table construction, the FES tier could possibly constitute the mechanical component, and the NFES tier constitute the superior data or prior information.

5. Some Implications and Limitations of this Approach

This paper has offered two general thoughts. First it has suggested that economists might be rewarded more by the study of similarities or regularities in structure among economies, than by the study of diversity. Secondly, it has suggested a revised concept of FES as one possible base for the study of these similarities or regularities in economic structure. This section contemplates some of the implications of these suggestions and some limitations on progress which appear at this early stage of development.

First, this approach is to a large extent the antithesis of the conventional eeconomic base (EB) approach. Although the EB approach has been frequently and widely criticised, it is probably true to suggest that the general precepts of EB (that the total level of economic activity of a region, and in particular the level of non-basic or residentiary activity is determined largely by export sales) remain broadly acknowledged as appropriate. The FES-NFES approach implies a quite different line of reasoning. It implies that the export base (NFES) and the remaining sectors of the economy are largely independent of each other, and that knowledge of the economic base provides no (or at least very little) information on the level of non-basic activity. It implies that the only link between the NFES and other sectors will be through the personal consumption of those households dependent on FES activities.

Second, we need to know the relative sizes of the FES and NFES tiers in various types of economies. It will be useful to know whether the "predictable" part of the economy is the minor tier; in this case the concept may only be interesting and of little analytical significance. If the FES is the dominant tier of the economy, and the NFES tier analytically insignificant, then large proportions of the table can be seen as predictable in a given set of regions. We have at this stage no evidence of the relative sizes of the FES and NFES tiers. We suspect the FES to be the larger tier in the general case, and that the relative size of the NFES tier will decrease as region size increases, and that the NFES will show variation in relative size from region to region.

Third, we would expect that the FES tier of the table, since it represents personal consumption, will be the more stable of the two over time, and be less liable to fluctuation than the NFES tier, which will be at least partly subject to national and international markets, including commodity markets.

These three points imply that the FES appears to combine predictability, stability and economic significance. As such the nature of the FES would be of central importance to approaches attempting to measure the degree of similarity among regional economies. Some questions may be posed:

(i) We need to examine in detail the relationship between FES and population. It is probable that this relationship is not linear, but will show non-linearities reflecting different mixes of consumer goods and services at different region (population) sizes.

(ii) It is possible that the definitional base of the FES should be re-examined; one further development could be the disaggregation of the FES into components of particular interest, or components which could be expected to exhibit different behavioral patterns.

(iii) Is it possible for these concepts to be developed to the stage of analytical operation? The potential savings in research resources could be enormous.

(iv) Again, we need to determine if similarity or diversity is the dominant feature of economic structure; if we have the answer to this question, what implications follow for the directions of research into economic structure in the future?

6. Concluding Comment

This paper has sought to stress the advantages of analysing the degree of sameness or similarity among regional economies, rather than accepting the current implied assumption that economies should be considered as wholly unique in structure, and together too diverse to be tractable to any detailed comparison. It suggests that economists should acknowledge the principles of the science of taxonomy which seeks to measure degrees of similarity and sameness among units and form groups of units with higher degrees of similarity. The suggestion is made that these groupings should be formed on natural classifications, i.e. considering a large number of characters, rather than the artificial classifications currently used in economics, which employ one or two characters; such a system would maximise our ability to generalise and predict from and within groups.

The input-output table would appear to offer a useful data set for a natural classification of economies, and the FES appears to be a useful starting point for the identification of similarities among regions. A revised or tiered approach to FES is suggested for this purpose.

LITERATURE

Anderberg, M.R., 1973: Cluster Analysis for Applications, Academic Press, New York.

Asher, H.B., 1976: Causal Modelling, Sage, Beverly Hills, California.

Blin, J-M., 1973: "A further procedure for ordering an input-output matrix: some empirical evidence", Economics of Planning, 13:121-9.

Blin, J-M; Cohen, C., 1976: Technological similarity and aggregation in input-output systems - a cluster analytic approach, Northwestern University, Center for Mathematical Studies in Economics and Management Science, Discussion Paper No. 163, Evanston, Illinois.

Czamanski, S., 1964: "A model of urban growth", Papers and Proceedings of the Regional Science Association, 13:177-200.

Deutsch, J.; Syrquin, M.; Urata, S., 1986: Economic Development and the Structure of Production, Bar Ilan University, Economics Research Institute, Discussion Paper 8608, Ramat Gan, Israel.

Fisher, W.D., 1969: Clustering and aggregation in economics, Johns Hopkins Press, Baltimore.

Getis, A. 1985: "The economic health of municipalities within a metropolitan region: the case of the Chicage SMSA", Paper presented to the 32nd North American Meeting of the Regional Science Association, Philadelphia.

Ghosh, A.; Sarkar, H., 1970: "An input-output matrix as a spatial configuration", Economics of Planning, 10:133-42.

Hamilton, J.R.; Jensen, R.C., 1983: "Summary measures of interconnectedness of input-output models", Environment and Planning A, 15:55-66.

Helmstadter, E., 1969: "The hierarchical structure of interindustry transactions" in International Comparisons of Interindustry Data, United Nations Industrial Development Organisation Report, Industrial Planning and Programming Series No.2, United Nations, New York.

Hewings, G.J.D.; Jensen, R.C.; and West, G.R., 1987: Holistic matrix descriptors of regional input-output systems, Working Paper No.60, Department of Economics, University of Queensland.

Hewings, G.J.D.; Romanos, M.C., 1981: "Simulating less developed regional economies under conditions of limited information", Geographical Analysis, 13:373-90.

Heywood, V.H., 1976: Plant Taxonomy, Edward Arnold, London.

Isard, W.; Czamanski, S., 1965: "Techniques for estimating local and regional multiplier effects of a change in the level of major government programs", Papers of the Peace Research Society, 3:19-44.

Jensen, R.C., 1981: "A model of regional economic growth and decline in agricultural regions", in: Buhr, W. and Friedrich, P. (eds) Regional Development under Stagnation, Nomos-Verlagsgesellschaft, Baden Baden.

Jensen, R.C.; Dewhurst, J.H.Ll.; West, G.R.; Hewings, G.J.D.,(1989": On the concept of fundamental structure", paper to The Burn Conference on Input-Output Modelling, The Burn, Scotland.

Jensen, R.C.; Hewings, G.J.D.; West, G.R., 1987: "On a taxonomy of economies", Australian Journal of Regional Studies, 2:3-24.

Jensen, R.C.; West, G.R.; Hewings, G.J.D., 1988: "The study of regional economic structure using input-output tables," Regional Studies, 22:209-220.

Kuznets, S., 1963: "Quantitative aspects of the economic growth of nation, VIII. distribution of income by size," Economic Development and Cultural Change, 11:1-71.

Lamel, J.; Richter, J.; Teufelsbauer, W., 1971: "Triangulation of inputoutput tables for OECD countries," Annex III to L. Gentile et al Preparation of the set of the ECE standardised input-output tables - first results of comparative analysis technigues, Palais des Nations, Geneva.

Leontief, W.W., 1968: "The Structure of Development" in Leontief, W. (ed), Input-output Economics, (2nd Ed.), Oxford University Press, New York.

Mulligan, G.F. (undated): "A general regression model for estimating economic base multipliers; the case of Arizona towns", Department of Geography, University of Arizona.

O'hUallachain, B., 1985: "Complementary linkages and the hierarchical structure of regional economies," Geographical Analysis, 17:130-42.

Sherman, J.; Morrison, W., 1950: "Adjustments of an inverse matrix corresponding to a change in one element of a given matrix," Annals of Mathematical Statistics, 21:124-7.

Simpson, D.; Tsukui, J., 1965: "The fundamental structure of input-output tables: an international comparison," Review of Economics and Statistics, 47:434-46.

Slater, P.B.; Schwarz, W., 1979: "Global trade patterns: scaling and clustering analysis", IEEE Transactions on Systems, Man and Cybernetics, 9:381-7.

Slater, P.B.; Winchester, H.L.M., 1978: "Clustering and scaling of transactions flow tables: a French interdepartmental migration example," IEEE Transactions on Systems, Man and Cybernetics, 8:635-40.

Stone, R., 1986: "Where are we now? a short account of the development of input-output studies annd their present trends" in I. Sohn (ed), Readings in Input-output Analysis, Oxford University Press, New York.

Szyrmer, J.M., 1985: "Measuring connectedness of input-output models: part I - survey of measures," Environment and Planning A, 17:1591-612.

Szyrmer, J.M., 1986: "Measuring connectedness of input-output models: part 2 - total flow concept," Environment and Planning A, 18:857

How Different Are Regions?
An Evolutionary Approach to Regional Inequality

by

Michael Steiner

1. Introduction

What is a region and how and why does it differ from others is one of the fundamental questions of most people related to regional science. At several occasions ROLF FUNCK (e.g. 1976, 1978, 1986) has emphasized that the definition of a region must meet a number of criteria the economic ones being only a part of them. But he also pointed to the fact that there are different ways of approaching these economic criteria (e.g. in interpreting the merits of A. LÖSCH in respect to this question): "Static equilibrium is the standard of reference ..., but there is plenty of room for dynamic, disequilibrium interpretation of reality." (FUNCK and KOWALSKI, 1986, p. 3).

Thus he reminded of an old dispute in German location theory: Is the spatial pattern of the economy predetermined by allocative mechanisms or is it the result of chance events. In the first half of the century especially ENGLÄNDER (1927), RITSCHL (1927) and PALANDER (1935) emphasized that the spatial distribution of economic activity is influenced by concrete historical events.[1]

Behind this distinction of a pre-ordained or a process-dependent interpretation of economic criteria influencing the spatial distribution of economic acitivity we can recognize two different views of the role of markets: The one view emphasizes the allocative function of the market, the other its creative function. In its creative function the market is an instrument for transmitting incentives for economic change (KALDOR, 1972). Both functions imply different behaviour of economic agents: They might behave in a "creative", "entrepreneurial" (in the sense of HAYEK) way or "adaptive", reacting to a given situation of the parameters of price and quantity.

These specific ways of behaviour demand certain economic environments with a special factor endowment. If we regard the "structure" of a region as formed by combinations of specific factors which enable certain kinds of economic behaviour (a behaviour which influences in a reciprocal process this factor endowment at the same time) we can differentiate regions according to these criteria. As a consequence regional inequality results from the fact that not all regions are capable of producing such "creative" incentives - because of their different structure - or capable of adapting to the consequences of such incentives. The differentiation of regional types then is a result of their varying capability to produce a behaviour necessary for economic survival. In the following we will briefly touch upon "structural" and static-allocative interpretations of an unequal distribution of economic activities in space (Ch. 2), try to develop a dynamic perspective of regional disparity (Ch. 3), deduce from this different types of regions (Ch. 4), underline the evolutionary aspects of this approach (Ch. 5) and finally draw some conclusions of this line of reasoning.

2. Regional Inequality as a "Structural" and/or Static-allocative Phenomenon?

One of the first approaches to differentiate between regions was (and still is) to differentiate according to their "structure". From this structure in the sense of composition of industries different development potentials or barriers to development were deduced: The economic potential of a region was defined by its structure - countless shift-share-analyses and normal growth patterns relied on this concept: To what extent causes the composition of sectors and industries a rate of growth that differs from other regions or from the whole national economy? Especially in the case of the socalled "old industrial areas" the monostructure - the concentration of a few declining industries - served as a dominant reason for their stagnation.

Yet this much used approach does not answer the question how a specific formation of regional structure comes about: Why is there a concentration of growing industries in one region, why a concentration of declining industries in another? Why was a growth-impeding structure not changed? The theoretical basis of this interpretation is mainly demand-oriented - the expansion or contraction of sectors or industries depends on the increasing or decreasing income elasticity of demand of their products. The method of analysis essentially consists of the isolation of the role of the "industrial structure"; the analysis of direction and result of structural change was dominating. This much-used explanation had to be dismissed: The empirical analysis did not attribute much weight to the structural component thus reducing the importance of structural change (see e.g. DENISON 1967, MADDISON 1972, STEINER, 1981). This also led now and then to wrong conclusions: The influence of worldwide structural change affects regional inequality only marginnally, there is even a decrease in this inequality as measured in terms of rates of growth of employment, of production and of sales.

This comparative-static approach does not only assume the existence, but also the knowledge of a "right" economic structure - structural problems consist of a deviation form this right structure. As such it is in contrast to a dynamic analysis, one that only assumes the existence of an optimal structure or optimal path of development but not (necessarily) its a priori recognition. According to this dynamic interpretation it is the task of the economic system itself to find this path of development and to realize it. As a consequence the problem does not consist of the given composition of industries which deviates from the desired one but the problem consists of the fact that the dominating economic institutions are not able to fulfill this task. Regional inequality according to this dynamic (in contrast to the comparative-static) approach results form the different presence of those factors the combination of which is better capable of mastering this search process and which is more able to cope with structural change.

This open question - why is there a specific composition of industry, a specific allocation of economic activity between regions and in space - is partly answered by location and agglomeration theory. They emphasize the individual decision of economic agents and their space-differentiating effect. Yet they also rely on static principles of regional distribution of economic acitivity. The "geography" of economic acitivity is the result of static allocation which stems from material linkages, from the existence of specific ressources by means of market interaction: Out of this regional hierarchies and regional specialisations come into existence.

This is dominated by the idea that an optimal spatial distribution is possible and that it is dependent of a few market factors; entrepreneurs with maximizing behaviour react to these factors and adapt through means of cost minimizing location decisions to this optimal distribution. This concept assumes maximizing behaviour, complete information of locational ressources and of the costs of distance, reversibility of decisions with the possibility of marginal adaption. From this results an adaptive behaviour of entrepreneurs - they are "adaptive" and "location-conscious" in the sense of paying attention to the calculation of locational costs; they are also technologically adaptive, their technique is led through the minimization of factor costs.

Altogether the emphasis lies on the allocative function of the market; under these assumptions market forces tend to an equilibrium, the result is an allocatively efficient distribution of economic activity in space. This equilibrium does not mean that all regions have to be "equal"; yet the causes of inequality may be reduced to a few factors, the extent of inequality is predetermined.

3. Regional Inequality as a Result of a Dynamic Process

3.1 The Extended Concept of the Market

Beside the allocative function of the market - and this is mostly less emphasized - there is also a further function: The market is an instrument which transfers incentives for economic change. This emphasis of the creative function of the market started with the Austrian School of economic thinking. Especially HAYEK underlined the explorative potential of the market - its capabilities do not rest so much in the allocation, but in its flexibility to adjust to new situations. The market is an optimal process of exploration, it is a development process. In the same sense SCHUMPETER always regarded the market as a process and not as a state: Economic competition is not price-taking behaviour under conditions of perfect competition, but a process with winners and losers. The procedual character of the market brings about - by its creative distruction - innovations. These innovations are not only of a technical kind, but also have organizational character thus opening new markets. Compared to these dynamic aspects the pure static, allocative advantages of the market are loosing importance (STREISSLER, 1980, p. 8): "The market is predominantly an optimal process of exploration, beyond this a process of development, a process of change of data".[2]

In the centre of this approach there is the creative function of the entrepreneur who has to create markets (and who does not exclusively react to a given set of data) and who has to master production under conditions of uncertainty (see CASSON, 1982, p. 23). This personal element is a dominant feature of the market economy: The entrepreneur is the one who explores hitherto unknown opportunities and who realizes them, who creates innovations connected with entrepreneurial risk and hazard. He is the one who explores future needs, the one whose informative function is of importance especially in the case of "thin" markets (where the number of participants in the market is too small to warrant a fairly stable intertemporal development of prices). The emphasize of the allocative function obstructs the understanding of the importance of the entrepreneur, the person of the entrepreneur does not exist.

What does this differentiation between allocative and creative functions of the market mean for the interpretation of the locational behaviour of entrepreneurs, what does it mean for the resulting differences in regional economic potentials? The static approach of location theory, which emphasizes the allocative function, demands the adaption to an efficient distribution of economic activity in space. The correct adaption leads to high growth rates and to high productivity. In contrast to this the creative dimension demands an adaption in the dynamic sense which we may call "adaptibility" and which requires creative ability: This contains ways of behaviour which are beyond maximizing (or which cannot be explained by this alone) or, in the above sense: entrepreneurial behaviour, especially innovative acting, or, more generally, "activities, which are linked with spontaneous activities and which as such do not represent a reflex of an adaptive behaviour leading to an equilibrium" (RÖPKE, 1980, p. 143). Such a behaviour eludes an allocative determinism: Entrepreneurs are "indecisive" in regard to their location, as far as the costs of location are concerned there are no decisive limits to the location decision. In contrast to the concept of a stongly limited "adaptive" locational behaviour, which is determined by the equilibrium model of cost minimizing, there is more room for "adaptable" entrepreneurs who are "market-creating"; at least more room which is not limited by locational costs. It is important to point to the different kinds of behaviour: In the one case entrepreneurs have room for strategic action, they can themselves induce change, whereas the strategic opportunities of "adaptive" entrepreneurs are considerably limited, they are forced to follow inflexibly the given developments, they have to adapt to the changes transferred by the market.

Regional inequality results - this is our hypothesis - from the unequal occurence of such entrepreneurial behaviour: It is the spatially different dominance of "adaptable" and "adaptive" behaviour. Location theory with its concentration on the allocative function of the market cannot explain that there is and why there is a spatially different occurence of such forms of behaviour.

3.2 The Evolutionary Concept

As soon as economic behaviour and its spatial consequence are not only "adaptive" and fixed through a few market data, as soon as the creative function of the market opens up room for strategic action, then the result is indeterminate, is dependent from chance events which happen in the course of history. Then each state is not independent of the antecedent state. The spatial distribution, the regional differentiation is a result of an evolutive process happening in real time.

We want to call an economic approach as "evolutive" - following NELSON and WINTER (1982) - if it explains the state of an economy or of economic sectors by the mechanisms of survival which are similar to the biological evolution: The competition of the market is seen in analogy to the biological competition, firms are exposed to a test of survival on the market. As search and selection have great importance in the biological development so firms have to "search" in the market and are subject to a process of selection. Yet this search is not a behaviour in the sense of maximizing but a behaviour which is partly influenced by routine, which is every now and then modified, a modification which is found through search. The behaviour of a firm can be explained by its history and it will be imitated by other firms. What drives the economic and technological development ahead is explained endogenously by the processes of search which are partly influenced by the routine activities of the firms. This development is smooth and continual, with the condition "in each period bearing the seeds of its condition in the following periode" (NELSON and WINTER, p. 19). These transitions and the probabilities of its results are not subject to a causal determinism.

This evolutionary perspective has several consequences:

- It considers innovations as the central phenomenon to be explained: how do innovations come about, what general influences do they have. The emergence of new opportunities and situations has to be explained endogenously.
- It requires a broader concept of economic rationality, a broader one than in orthodox economic theory. Economic decision making is thoroughly regarded as "rational": rational in the sense that human beings have objectives which they pursue with consistence and skill and foresight. But this rationality is not interpreted in the tight context of a maximizing behaviour with given objectives and known set of choices, but it is guided by past behaviour, by institutions and by routine. This rule-bound behaviour allows for a reduction of complexity in the face of uncertainty and of lack of information. So it is not so much the question if there is rational behaviour or not, but rather if decision theory of traditional economic theory represents the relevant form of rational behaviour.
- What happens economically is a process occurring in real time; economic events are to a large part a result of past happenings, of the experience of economic agents which is gained in time - learning by doing is an essential way of gaining knowledge which is based on experience.
- Economic action does not only know the parameter of price and quantity, it is more than allocative mechanism guided by prices. It also depends on economic, legal and social institutions. The influence but also the emergence of institutions is a legitimate field of economics. This leads to decision rules which differ form the maximizing behaviour of traditional economic theory.

- Economic processes are not subject to a predetermined causality. Each antecedent state influences the next one, yet the concrete form of it is open to several possibilities. The process of search and selection does not know stable terminal points and therefore no equilibrium to which it necessarly moves. The results of this search process possess a stochastic character.
- Economic agents are in an interplay with their surroundings. The concept of selection means that they either have to adjust to it or that they perish: non survival of the non fit. The manyfold processes of adaption are in analogy to variation or mutation. They represent deviations of hitherto unknown forms of behaviour. These performance of adjustment have to be differentiated. They can mean adaptability or adaption (PRISCHING, 1986, p. 66): Adaption is a reaction to the surroundings, it is an optimal use of the exterior facts. But not every effort of adaption serves in a positive respect the probability to survive - a too excessive adaption leads to specialisation and excludes any further adaption. Adaptability in contrast means to keep open structures and ressources which are not bound for specific purposes but which are at disposal for necessary reorganisations in times of new challenges.

3.3 The Cyclical Concept

Proceeding form the assumption that the allocative and creative functions of the market are reflected in different forms of behaviour also with regard to the location decision we may from an evolutionary perspective interpret them as no doubt different but not each other excluding forms of behaviour. We may regard these contrasting forms of spatial behaviour, the one creative, the other allocative, as being united in an evolutionary process (see ELLINGER, 1977). This behaviour is dependent of the age of the firm.[3] In the early stage firms are not limited in regard to their location decision. In the early stages of growth the combination of super profits, of imperfect competition (because of existing limits to capital mobility and because of price inelastic demand) allows for succesful developement without consideration of regional differences in costs of location. This is the consequence of their specific decision making behaviour: They are "adaptable" - i. e. they are either innovators themselves or quickly adopt innovations, they decide to start production soon after a major innovation. In later stages this behaviour changes: Firms become "adaptive" - they adjust to their economic and regional environment. Their behaviour becomes "locationally decisive" in the sense that they consider differences in locational costs.[4]

Both forms of entrepreneurial behaviour - the adaptable one and the adaptive one - thus follow a temporal sequence; the spatial consequences are part of an evolutionary process. Yet the basic freedom of location in the early phase is restricted in real space and in real time - the ressources which are necessary for an adaptable behaviour cannot whereever be attracted. From this there results an interrelationship between the locational needs of innovating firm, the manyfold factors of production which are necessary for adaptable behaviour; on the other side is the distribution of the regional supply of such factors not independent of the distribution of such firms - the location and the situation of many factors can be influenced by the decisions of firms in their innovative stage. Thus regional development becomes a "self fulfilling prophecy" for areas with innovating, adaptable firms. A region is distinguished by the mass of ressources which are demanded from the firms within this region. The kind of ressources is dependent form the stage of development of the dominant firms within a region. The distinction between different evolutionary stages implies a specific supply of factors which defines the regional economic potential and which also implies different types of regions.

The difference in regional economic potentials therefore lies in these dominating supply of factors which are characteristic for the specific evolutionary stages. We may also put them in relation to the different adjustment performance which these supplies of factors are able to produce:

- Adaptability as the potential to leave open ressources for new challenges and to put them into use if necessary relates to the "creative" behaviour in the early stages - "adaptable" regions are distinguished by firms which are able to create markets and which need factors necessary for such a behaviour.
- Adaption as a different form of an adjustment performance, one which tries to use the external situation in an optimal way, relates to a behaviour fulfilling an allocative function - "adaptive" regions are dominated by firms which react to locational costs and which demand factors where the price is a relevant element.

- There is of course an insufficient adjustment performance, deficits in adaption, which result in negative selection either because an "over adjustment" led into a dead end street of the evolution[5] or because a too fast change of the environment (to which there was no timely reaction) exceeded the adaptive potential; "non-adapted" regions stagnate, they have lost their potential to survive, their dominating firms were too stable; "complete stability results in stagnation" (RADNITZKY, 1984, p. 12).

4. Regional Types as a Consequence of Different Performances of Adjustment

The consequences of an evolutionary perspective are expressed by the different performances of adjustment which the behaviour that dominates the specific stages of the evolutionary cycle - and also the corresponding regions - are able to perform. An empirical test (which can here only very shortly be outlined) for Austrian districts showed results which support our interpretation (STEINER, 1989). Three basic types of regions (which can be differentiated with further subdivisions) could be extracted of a number of 99 districts, the sequence of which corresponds to the above ordering of performance of adjustment:

(i) First there were districts with an economic potential which may be regarded as adaptable. This adaptability results from a combination of indicators which can be deduced from the sort of behaviour dominating the early stages of the cycle: an "entrepreneurial" behaviour bringing about impulses for economic change, finding new opportunities. As far as possible with a limited set of data we tried to seize this behaviour (and the necessary factors for it) through the following combination of variables: a high qualification of the workforce, a low percentage of blue collar workers, a high possiblity for communication, diversification of industry, high rates of firm entry, a lack of dominance of big firms. And the method used - a combination of factor and cluster analysis - showed that the bundle of these variables (which we regarded as significant preconditions and as indicators for the entrepreneurial and creative element of the market) is concentrated in a limited number of districts (and not in others).

(ii) A different form of adjustment performance is achieved in regions dominated by standardized production. Here economic activities are adaptive. It is a behaviour which is rather oriented by the economic parameters of price and quantity, by a wellknown economic environment, it is a behaviour reacting to these parameters and to this environment. Low factor costs (cheap land, low wages) make these areas attractive, the management needs only limited capacity to process different kinds of information, the workers perform routine activities demanding low qualifications. All this represents still a performance of adjustment which enables survival. It is a survival within a world dominated by the allocative function of the market. This world has become more stable, needs less information, an information largely based on the usual parameters of the market. In our empirical test we described this type of regions with the following variables: a low qualification of the workforce, low wages, high female participation, high mobility of firms (high birth rate and death rate) - our method of grouping districts to regions resulted in a marked concentration of these indicators within a limited number of districts.

(iii) An insufficient performance of adjustment is produced in the so called "old industrial areas". They are neither "adaptable" nor "adaptive". In this type of region the failure of the creative function of the market is evident - no new opportunities are found, no hitherto unknown combinations are taken up, no markets are created. The proxies we used signal the lack of this kind of entrepreneurial behaviour: dominance of big firms with strong internal hierarchies and with a small number of managers with decision-making authority, a low challenge through a diversified industrial structure within close spatial limits, low vertical and horizontal disintegration. Low entry and exit rates indicate strong barriers, which are enforced by high wages. Yet this also impeded the allocative function of the market: a behaviour in response to price and quantity signals. The wages are too high for the low rate of return, the number of firms leaving the market is too small.

5. Evolutionary Aspects of our Interpretation of Regional Inequality

We call our interpretation "evolutionary" because several aspects of this theoretical approach are necessary parts of our regional differentiation.

(i) We regareded the role of innovations as an essential criterion of this perspective: "The real key to the understanding lies in the phenomenon of search, of trial and realization of the hitherto unknown, for which the economic agents are more or less able ... they are the decisive force behind socioeconomic changes" (WITT, 1986, p. 15). The source of regional inequality is economic change which is caused through innovations. These innovations appear with different intensity in the regional types. The gradation of performances of adjustment was essentially according to this criterion: to what extent is it probable that there are economic dynamism and innovations in regions with a certain economic potential; to what extent may we assume that entrepreneurial, creative behaviour may take place there. The empirical factors by which we tried to express differences in regional potentials (and which could only be hinted at here) can also be interpreted as an expression for the possibility of creative action. The claim was not to describe exactly the lags and leads in innovation or to describe where innovations exactly come about, but to list conditions which make innovations more or less probable and to localize geographically these conditions.

(ii) As a further criterion we emphasized the role of the historical process. The concrete materialization of regional inequality is not predictable, it is not explainable by a mechanism influenced by a few parameters and tending towards equilibrium. The fact that this process which differentiates regions is path-dependent excludes a determinism the result of which is exactly deducable. As mentioned above this question of determinism versus path-dependency was already a controversial one in the German location theory in the first half of this century. The regional distribution of economic activity was according to the one line of thought predictable, linked to a specific terminal state, according to the other view dependent of foregone distributions, to which new industries and firms add, which react to locational advantages but which are in interplay with these advantages and especially with each other.

ARTHUR (1986, 1987) could show that the evolution of locational patterns depends to a great extent of how far the model includes "chance" (as expressed by historical events) or "necessity". Yet this historical "chance" may be influenced by probabilities: "Chance, of course, ist not the only factor here. Regions that are geographically attractive to many firm types - regions that offer high economic benefits - will have a higher probability of being selected early on. And this will make them more likely to become the single region the dominates the industry. To use an analogy borrowed from genetics, chance events act to "select" the pattern that becomes "fixed"; but regions that are economically attractive enjoy "selectional advantage", with correspondingly higher probabilities of gaining dominance" (ARTHUR, 1987, p. 7). Where firms do locate depends of chance events the appearance of which may be more or less probable and influenced by the different attractiveness of regions.

(iii) Evolutionary approaches exceed the model of rational behaviour as postulated by orthodox economic theory. Economic behaviour is not (only) optimizing behaviour starting with exact objectives, possessing a clearly definded knowledge and including only maximization: objectives are not unequivocal, human rationality is bounded, the interrelationship between means and objectives is unclear, decisions are made in real time with incomplete knowledge about their consequences - all this makes other interpretations of behaviour more appropriate in modelling economic reality.

In our model of economic behaviour, a behaviour that changes in the cyclical process and which concentrates in specific types of regions, we assumed that in each case different forms of economic activities are of importance. The behaviour in the early stages is more influenced by uncertainty than a behaviour in later stages, the relationship between means and objectives is to larger extent undefined, the consequences of economic activity highly unpredictable. Because of this openness of the economic process there is no room for optimizing. But uncertainty is constantly decreasing. With growing maturity of production manufacturing processes became more standardized, more mechanized, decision making can rely on past experience, production as a whole gets routinized. There is greater stability on the demand side, the economic environment is better known thus reducing the incompleteness of information; the set of data becomes more stable. All this contributes to the possibility to think in terms of optimizing: to reach given objectives with known means. But this behaviour may under different conditions become inadequate: If routinized maximizing behaviour is not aware (or too late aware) of changes in the economic environment, it is maximizing with a wrong set of data: Economic activity is conforming to the traditional rules of behaviour without taking notice of a change in the economic environment.

We call these different stages of the dominant form of economic activity adaptable, adaptive and neither adaptable nor adaptive behaviour. This distinction corresponds to different functions of the market: Adaptable behaviour corresponds to spontaneous entrepreneurial economic activity, which is able to create markets. This stage is dominated by the creative function of the market where impulses for economic change are set which cannot be explained in terms of optimizing calculation. It is a state of distance to equilibrium where calculations for a situation marked by equilibrium conditions are inadequate. The adaptive behaviour in contrast presupposes a more or less known state of equilibrium in which the allocative mechanism is justified. In this situation it is sufficient to pay attention to the usual market signals of prices and quantities. Because this state is close to equilibrium these signals are able enough to reduce ignorance so that the traditional calculus of economic behaviour is sufficient and adequate. Thus we suppose a complementarity (or a sequence) of these market functions as formulated by KUNZ (1985, p. 22); "Allocation theory on the one side and evolutionary and coordination theory on the other side certainly are mutually exclusive approaches. But they confront each other in a complementary way. And only the use of both approaches can lead to a 'full' understanding of the market system." This thesis of complementarity assumes that evolutionary processes "only can start under conditions of distance to equilibrium" (KUNZ, p. 26); this distance to equilibrium is created by the phenomenon of constitutive ignorance. An essential characteristic of the distance to equilibrium is the fact that the knowledge necessary for the coordination of individual plans can only be gained at costs which are "unbearably" high. That is why the allocative mechanism fails in this situations. And therefore "entrepreneurial" behaviour which explores these opportunities becomes necessary: It's essential character lies in this "alertness to opportunities" which is more than a calculating procedure.[6]

Yet at the same time norms and institutions are developed which work towards a reduction of this constitutive ignorance so that processes towards a state close to equilibrium are started - "there is a tendency towards a system-endogenous reduction of informational and other transaction costs" (KUNZ, p. 153). Through this process equilibria emerge in which the allocative function of the market takes effect; these equilibria represent more or less stable temporary stages within the evolutionary process. But before the creative stage with the personal element of entrepreneurial activity is required which serves as the coordinator of non-homogenized knowledge.

It is this "sequential nature of real economies" (HAHN, 1973, p. 16) (which KUNZ converts into an "evolutionary nature") which lies at the basis of our interpretation: The specific stages represent states of more or less distance to equilibrium. The temporal one-after-the-other corresponds to a spatial side-by-side. We may also express our interpretation in the way that the states of distance to and close to equilibrium require specific spatial conditions: The state of distance to equilibrium with a high degree of constitutive ignorance requires conditions which aid to reduce this ignorance - e.g. the information-inducing effects of urbanization economies, the advantages of swift communication, the opportunities to gain knowledge by chance, the command over a wide set of data. Yet this adavantages of agglomerations can only persist if they are constantly renewed as externalities and not internalized by the market.[7]

The advantages of the "adaptive" regions consist of their low factor costs; as soon as the background becomes clearer the externalities of "urbanization economies" are no precondition for coordination, prices are sufficient parameters for decision making. They convey informations which can be handled by the allocative mechanism of the market. In these regions the allocative function of the market is dominating, here scarce ressources are put into efficient use. It is not the pioneering entrepreneur, but the imitator who populates this more stable market and pays attention to the parameters of price and quantity; these imitators do fulfill an essential role and produce a performance of adjustment which warrants survival.[8]

Old industrial areas give evidence that the temporary stability of the allocative equilibrium is framed by a two-sided distance to equilibrium. We called the behaviour within these regions as neither adaptable nor adaptive, as neither creative nor allocative. It is a state of distance to equilibrium marked - similar to the creative stage - by ignorance yet lacking the personal element of entrepreneurial activity and thus without a coordinative device to reduce ignorance.

6. Some Conclusions

(i) Economic activity does not only happen in real time, but also in real space. The sequential structure of economic processes also implies a sequence of preferred locations: There is not only a temporal, but also a spatial sequence. As logical time is not able to comprehend the irreversibility of economic processes, its being preconditioned through the past, in the same way an interpretation of space considering it merely as a cost factor is not able to come to terms with the phenomenon of regional inequality. The spatial dimension is of further influence for the decision of economic agents, it conditions the form of behaviour. The tradition of institutional economics has always emphasized that human action is neither independent of time nor of space: "For the institutional economist behavior is cultural behavior. Economic behavior takes place within a cultural milieu and is, in fact, a part of that cultural milieu. In the work of the institutionalist this is of extremely great importance; for he not only explains individual behavior in terms of culture, but he is concerned with economic behavior in general as an aspect of culture. In other words, economic behavior patterns are those aspects of cultural behavior that are concerned with earning a living" (HAMILTON, 1953, p. 81). In our interpretation the regional types stand for different environments which shape the behavior, they also stand for a "cultural milieu". In this sense there is a "culture of old industrial areas", a "culture of innovating centres" (GELDNER, 1987).

Because economic behaviour happens in real time and in real space under constitutive ignorance and uncertainty, it cannot be modeled in a precise, deterministic way. The spatial "cultural milieu" influences behaviour, it does not determine it. The elaboration of characteristics which indicate certain forms of behaviour (which we only hinted at in Chapter 4) is an attempt to confine the decisive influence in a probabilistic way: It is not sure that the centres have a higher rate of innovation, but it is more likely that innovations are going to happen there, it is not sure that industrial areas get "old" but under certain circumstances getting old becomes more probable. Regions differ with regard to their "selectional advantage", there is a regionally uneven distribution of the conditions for survival as emphasized by evolutionary theory. Innovations may happen everywhere; but the probabilty for their survival is not everywhere the same. The attempt to confine the unpredictability of behaviour by means of criteria which indicate these probabilities is an odysee-like wayout between the scylla of deterministic certainty (which is wrong) and the charybdis of theory obstructing ignorance (where a "nothing" and a "anything goes" is equally valid).

(ii) Regional inequality is based on different performances of adjustment. Not only the fittest survive. The evolutionary perspective rather postulates a non-survival of the non-fit. Different degrees of fitness may exist side by side, different performances of adjustment enable economic survival. Nevertheless they are the result of conscious effort, not the result of "blind" chance. In this respect economic interpretations of evolutionary processes differ from biological models: In these models there is dominance of a rigid form of adjustment, it is a natural selection, where acquired skills and characteristics cannot be passed on. Adjustment in these biological models only takes the form of a different intergenerational outcome of reproduction of phenotypes. In the socioeconomic realm variations and mutations caused by innovations are not completely chance events: Firms may choose consciously from a variety of routine actions; imitation is a conscious action; skill and experience can be passed on, anticipation of future developments is possible. The stream of variations does not have to stay constant: Satisfying behaviour does not undertake alterations if there is no need to do so (if e.g. profits stay the same). The selection process is also different from the biological one: Firms may die; but they cannot reproduce themselves. Their "multiplication" happens through expansion and imitation. The latter is rather artifical than natural as far as it requires a subjective, personal element of recognition and not only the objective element of different rates of survival. The performances of adjustment which are produced side by side in different types of regions represent a simultaneous survival of efficient and less efficient firms; not only the best, not only the fittest do survive.

The "survivors" do not have to maximize, they do not necessarily employ the best technique. A non-optimizing behaviour does not result in being eliminated by the market, the selection process does not eliminate all those who do not use the most efficient technique or the most efficient decision rule. A broad spectrum of behaviour can be sufficient for survival. This also means that what can be observed as a successful behaviour cannot only be explained by optimizing behaviour.

In assigning this evolutionary perspective to regions which are distinguished by specific forms of behaviour we further weakened the "biological" analogy: Regions do not die. Yet it can be shown that there are manyfold dimensions of stagnation in the type of region which is dominated by neither adaptable nor adaptive behaviour: decreasing employment, declining production, shrinking returns.

(iii) If the optimizing calculus is no precondition for economic activity, if not only maximizing behaviour leads to survival then the concept of equilibrium looses its contents. This has as a consequence that there is no stable final state, no equilibrium where the economic process ocurring in space moves to. This means that there is no onedimensional direction of development towards the better, no linear progress towards higher evaluated states. As a consequence regional development does not mean constant economic improvement, but stands for the possibility of a rise and a decline.

If competition, if the different performances of adjustment are considered as a process and not as a state, if economic activity happens in real time and in real space, if it is changing in the course of time through experience and if it is conditioned by the specific spatial environment, then there is no exact coordination of these activities, then the state towards which these activities move is no longer predictable. But then there is no optimal distribution of economic activities in space, then there is no objective standard as to the normative degree of equality or inequality of regions. At least this standard cannot be deduced form the calculus of efficiency.[9]

(iv) If regional inequality is regarded as a phenomenon which results from economic processes in disequilibrium, as a phenomenon the specific formation of which is dependent of chance events, then there is room for economic policy which is not restricted to cases of allocative market failure. Innovations are neither left to "blind" chance nor is the aging of regions until the state of stagnation inevitable. Because we see both of them as influenced by certain conditions economic policy becomes possible by creating or preventing those conditions. Within limits it is possible to control the probabilities of those processes. Or to use a metaphor: Chance events bringing about regional inequality are not stochastic in the sense of the roulette table where numbers and colours scatter according to the law of big numbers, but rather in the sense of a game of poker where it is possible to get quite different results with the same set of cards by consciously influencing action. Controlled probabilities lead to created space: The cultural milieu of a region may through conscious influence become a "self-fulfilling-prophecy", it is possible to increase the probability of growth centres.[10]

(v) Yet the kind of policy depends of the character of regional inequality. According to our interpretation they are a phenomenon of disequilibrium where regional types with adaptable behaviour, but also those with a neither adaptable nor adaptive behaviour incorporate states of distance to equilibrium. At best the adaptive type of region may be judged according to rules of equilibrium behaviour. In each case different functions of the market dominate each regional type. Thus there are for each case different rules of the market for the economic behaviour. For economic policy making this is a decisive difference. An undifferentiated call for "more market" which is not aware of its different functions may end in inprecise if not wrong policy instruments. The demand for a stronger function of the market with the emphasis on flexible wages, less influence of institutional regulations (in most cases exerted by unions), stronger "entrepreneurial climate" (where in most cases diminished workers protection, fewer environmental rules, reduced taxes are meant) are only oriented towards the allocative function of the market which should be freed of its blockages. Our interpretation of regional inequality underlines in contrast that these rules of market behaviour only apply to a limited realm of economic behaviour which is relevant to only a part of the regions. Rules concentrating on the allocative formation of the market do not apply to states of distance to equilibrium and their form of economic behaviour. It is incorrect to judge these situations according to causes for allocative market failures.

FOOTNOTES

[1] For a recent attempt to see spatial ordering not as pre-ordained but as process-dependent see Arthur (1986, 1987).

[2] This is also emphasized by Röpke (1980, p. 147): "Problems of allocative efficiency are - compared with problems of stability and evolution of the market economy - relativity minor: The market economy can tolerate a high degree of a waste of ressources (allocative inefficiency). But if it cannot stabilize itself and evolve then difficulties will appear compared to which problems of allocation are a "quantité negligeable".

[3] As a conceptual framework for cyclical models the product cycle hypothesis has increasingly been used also in regional sciences. The nucleus of this life cycle theory of products states that every product is subject to a process of aging and in the course of the cycle passes through several stages with different economic conditions. We use the product cycle as a heuristic device for our interpretation and take into account critical remarks and improvements especially by Suarez-Villa (1984), Markusen (1985) and Storper (1986).

[4] "This is an explicit change in behaviour from the locationally indecisive to the locationally decisive ... the locationally decisive behaviour becomes part of the industry as the time span from the time of invention increases and as the average and maximum concentration of firms in the industry increases" (Ellinger 1967, p. 297).

[5] Overspecialisation may lead to immobility because the adjustment was fit only for the needs of a very special niche.

[6] In this sense Kirzner (1979, p. 7) confronts the allocative mechanism of Robbinsian calculating with the creative entrepreneurial elements of Misesian action: "(Men) are not only calculating agents but are also alert to opportunities. Robbinsian theory only applies after a person is confronted with opportunities; for it does not explain how that person learns about opportunities in the first place. Misesian theory of human action conceives of the individual as having his eyes and ears open to opportunities that are "just around the corner". He is alert, waiting, continually receptive to something that may turn up. And when the prevailing price does not clear the market, market participants realize they should revise their estimates of prices bid or asked in order to avoid repeated disappointment. This alertness is the entrepreneurial element in

human action, a concept lacking in analysis carried out in exclusively Robbinsian terms."

[7] Meyer (1977) points to the fact that the nucleus of adavantages of agglomeration does not exist of economies of scale, but of externalities. As soon as they are mediated through the market and go down in lower prices they are a part of the individual decisions of the participants in the market.

[8] "As a market economist I cannot do without imitators. The real entrepreneurial economy has to comprise both the pioneers and the imitators! And this quite in the interest of a best possible use of the existing scarce ressource and the material supply of the population." (Helmstädter 1980, p. 73).

[9] Evolutionary theories imply "open" developments. For those one cannot find criteria of optimality: "In the context of the open market system which induces order there is no room for the optimizing calculus. The realm of optimizing is the closed system. To be able to evaluate the "optimality" of the results of evolutionary processes in open systems you would need as a standard a second evolutionary process which has the character of "optimality". Such a standard does not exist" (Kunz, 1985, p.3).

[10] In this sense Storper (1986, p. 67) speaks of "social structures of accumulation": "Above and beyond enabling factors ... the region is the essential milieu in which the social, political and institutional ressources can be marshaled so as to provide an environment taylored to the growth and development of these industries.

LITERATURE

Arthur, B., 1986: Industry Location Patterns and the Importance of History, Center for Economic Policy Research, Paper No. 84, Stanford University.

Arthur, B., 1987: Self-Reinforcing Mechanism in Economics, Mimeo, Stanford University.

Casson, M., 1982: The Entrepreneur. An Economic Theory, Robertson, Oxford.

Denison, E., 1967: Why Growth Rates Differ: Postwar Experience in Nine Western Countries, The Brookings Institution, Washington.

Ellinger, R., 1977: Industrial Location Behavior and Spatial Evolution, Journal of Industrial Economics 25, 295-312.

Funck, R., 1976: Entscheidungshilfen für die Regionalpolitik, in: Funck, R. ed, Heidenheimer Schriften. August Lösch in memoriam, 3, Stadt Heidenheim an der Brenz/Kulturamt, 8–29.

Funck, R., 1978: Multidimensionale Entscheidungsfunktionen bei der Planung der regionalen Infrastruktur, in: Buhr, W., Friedrich, P., eds, Konkurrenz zwischen kleinen Regionen, Nomos, Baden-Baden, 289–304.

Funck, R., 1986: August Lösch and the Concept of Region, in: Funck, R., Kuklinski, A., eds, 55–66.

Funck, R., Kowalski, J., 1986: Foreword, in: Funck, R., Kuklinski, A., eds, 1–4.

Funck, R., Kuklinski, A., eds, 1986: Space-Structure-Economy. A Tribute to August Lösch, von Loeper Verlag, Karlsruhe.

Geldner, N., 1987: Aspekte einer evolutionären Theorie wirtschaftlicher Entwicklung, in: Walter-Buchebner-Gesellschaft, Eds., Regionale Identität 1, Böhlau, Wien.

Hahn, F., 1973: On the Notion of Equilibrium in Economics. An Inaugural Lecture, Cambridge University Press, London.

Hamilton, D., 1953: Newtonian Classicism and Darwin Institutionalism – A Study of Change in Economic Theory, University of New Mexico Press, Albuquerque.

Helmstädter, E. 1980: Unternehmerwirtschaft oder unternehmerische Marktwirtschaft, (Koreferat), in: Streißler, E., Watrin, C., eds., 70–74.

Kaldor, N., 1972: The Irrelevance of Equilibrium Economics, The Economic Journal 82.

Kirzner, I., 1979: Perception, Opportunity, and Profit. Studies in the Theory of Entrepreneurship, University of Chicago Press, Chicago.

Kunz, H., 1985: Marktsystem und Information. "Konstitutionelle Unwissenheit" als Quelle von "Ordnung", Mohr/Siebeck, Tübingen.

Maddison, A., 1972: Explaining Economic Growth, Banca Nazionale del Lavoro, Quarterly Review 25, 211–62.

Markusen, A., 1985: Profit Cycles, Oligopoly, and Regional Development, MIT-Press, Cambridge, Mass.

Meyer, D., 1977: Agglomeration Economies and Urban Industrial Growth: A Clarification and Review of Concepts, Regional Science Perspectives 7, 80–92.

Nelson, R., Winter, S., 1982: An Evolutionary Theory of Economic Change, Harvard University Press, Cambridge.

Prisching, M., 1986: Krisen. Eine soziologische Untersuchung, Böhlau, Wien.

Röpke, J., 1980: Zur Stabilität und Evolution marktwirtschaftlicher Systeme aus klassischer Sicht, in: Streißler, E., Watrin, C., eds., 124–154.

Steiner, M., 1981: Zur Aussagekraft von Normalstrukturmodellen – Eine Note. Empirica, 111–27.

Steiner, M., 1989: Regionale Ungleichheit. Graz (forthcoming).

Storper, M., 1986: Technology and New Regional Growth Complexes: The Economics of Discontinous Spatial Development, in: Nijkamp, P., Ed., Technological Change, Employment and Spatial Dynamics, Springer, Berlin, 46–75.

Streißler, E., 1980: Kritik des neoklassischen Gleichgewichtsansatzes als Rechtfertigung marktwirtschaftlicher Ordnungen, in: Streißler, E., Watrin, C., eds, 36–69.

Streißler, E., Watrin, C., eds., 1980: Zur Theorie marktwirtschaftlicher Ordnungen, Mohr/Siebeck, Tübingen.

Suarez-Villa, L., 1984: Industrial Export Enclaves and Manufacturing Change, Papers of the Regional Science Association 54, 89–112.

Witt, U., 1986: Individualistische Grundlagen der evolutorischen Ökonomik, Mohr/Siebeck, Tübingen (quoted according to manuskript).

German quotations are translated by the author.

How not to Lie with Statistics in Regional Analysis

von

Wolfgang Eichhorn

Let $p_i=x_i/y_i$ be the ratio of two economic indicators observed in a region i of a country that is divided into n regions (i=1,...,n). In what follows, we think of p_i as the productivity of labour in region i, that is, x_i is the value added and y_i the labour applied in region i.

It is well known that the arithmetic mean of the n productivities, namely

$$\frac{p_1 + \ldots + p_n}{n} \tag{1}$$

is not necessarily the productivity of labour, p say, in this country.

One can only show that p is a mean, i.e.,

$$\min\{p_1,\ldots,p_n\} \le p \le \max\{p_1,\ldots,p_n\}. \tag{2}$$

A proof of (2) follows from the fact that p is a convex combination of the p_i's:

$$p = \frac{x_1 + \ldots + x_n}{y_1 + \ldots + y_n} = \frac{y_1}{y_1 + \ldots + y_n}\, p_1 + \ldots + \frac{y_n}{y_1 + \ldots + y_n}\, p_n. \tag{3}$$

From (3) it follows that p cannot be determined from $p_1,\ldots,p_n$ if the figures y_i are not at hand.

Nevertheless, a regional analyst would not be lying if he said that p is some value of the p_i's. Surprisingly enough, he would be mistaken if he extended this assertion to growth rates of (labour) productivities. We show here, by an example, that the growth rate of a country's productivity is not necessarily a mean of the productivity growth rates in its regions. It may even happen that the country's productivity growth rate is negative, although the productivity growth rates of its regions are positive, see table 1.

Table 1: Labour productivities and their growth rates

Regions / Years	A	B	Economy = A∪B
1988	$\frac{800}{20} = 40$	$\frac{600}{20} = 30$	$\frac{1400}{40} = 35$
1989	$\frac{764}{19} = 40{,}211$	$\frac{634}{21} = 30{,}190$	$\frac{1398}{40} = 34{,}96$
Growth rates 1989/1988	$\frac{40{,}211}{40} =$ 1,00526, i.e., .526 %	$\frac{30{,}190}{30} =$ 1,00635, i.e. .635 %	$\frac{34{,}95}{35} =$,998571, i.e. minus .143 %

This 'paradox', which can easily be extended to the case of n regions, has its root in the strange looking rule for the 'addition' O of productivities, namely

$$\frac{x_1}{y_1} \circ \frac{x_2}{y_2} = \frac{x_1 + x_2}{y_1 + y_2} \tag{4}$$

(Don't present this formula to children.)

A consequence of this result is another 'paradox': A (labour) productivity-oriented wage policy in each of the regions of a country can mean wage increases in every region, whereas the same policy applied to the country as a whole might require wage decrease. This example shows that there exist situations in which a (labour) productivity-oriented wage policy is by no means a good recommendation. (Another argument against such a policy is: Labour productivity takes only labour and value added into account, i.e., ignors the fact that the value added also depends on the capital.)

We have learnt that in a situation as described above a statistician would be lying if he asserted that the (total) growth rate connected with two (partial) growth rates always lies somewhere between them. However, there are also situations, where especially the arithmetic mean of n growth rates gives sense. For example, let $x_1, x_2, \ldots, x_T, x_{T+1}$ be a time series. Then

$$r_t = \frac{x_{t+1} - x_t}{x_t} \quad (t = 1, \ldots, T) \tag{5}$$

are the T growth rates from term x_t to term x_{t+1} of the series, and

$$r = \frac{x_{T+1} - x_1}{x_1} \tag{6}$$

is the growth rate from the first to the last term.

Obviously,

$$1+r = \frac{x_{T+1}}{x_1} = \frac{x_{T+1}}{x_T}\frac{x_T}{x_{T-1}} \cdots \frac{x_3}{x_2}\frac{x_2}{x_1} \tag{7}$$

$$= (1+r_T)\ (1+r_{T-1})\ \cdots\ (1+r_2)\ (1+r_1)$$

$$= 1+r_1+r_2+\cdots+r_{T-1}+r_T+f(r_1,r_2,\cdots,r_{T-1},r_T)$$

$$= 1+T\,\frac{r_1+\cdots+r_T}{T} + f(r_1,r_2,\cdots,r_{T-1},\ r_T).$$

Here, $f(r_1, r_2,\ldots,r_{T-1}, r_T)$ is a sum of products of two or more of the r_t's, i.e., f is small compared with $r_1+\ldots+r_T$, whenever the growth rates r_t are sufficiently small. In this case the growth rate from the first to the last term is approximately the sum of the growth rates r_t, i.e., one has

$$r \approx T\,\bar{r},\ \text{where}\ \bar{r} = \frac{r_1+\cdots+r_T}{T} \tag{8}$$

is the arithmetic mean of the growth rates r_t.

Notice that from (7) and (8) we obtain

$$x_{T+1} = x_1\ (1+r_1)\ (1+r_2)\ \cdots\ (1+r_{T-1})\ (1+r_T) \approx x_1\ (1+\bar{r})^T. \tag{9}$$

Notice further that for T = 6, say, growth rates r_t between 1,8 % and 3,4 % are not sufficiently small for satisfactorily applying formula (8), whereas formula (9) is alright; see table 2.

Table 2: GNP = Gross National Product at prices of 1980, Federal Republic of Germany, in billions of DM

Year	GNP	Growth rates in percent	Particular growth rates and functions of growth rates, in percent
1982	1471.0		sum of the growth rates:
1983	1497.8	$r_1 = 1.82$	$r_1+r_2+r_3+r_4+r_5+r_6$ = 14.63
1984	1548.1	$r_2 = 3.36$	Growth rate from 1982 to 1988: 15.53
1985	1578.1	$r_3 = 1.94$	Arithm.mean of the growth rates: 2.44
1986	1614.7	$r_4 = 2.32$	Average yearly growth rate
1987	1643.7	$r_5 = 1.80$	yielding 1699.4 in 1988 from
1988	1699.4	$r_6 = 3.39$	1471.0 in 1982: 2.44

A New Approach in Decision Theory

by

Harry Böttcher and Johannes Siemons

1. Introduction

Classical decision models select optimal solutions among alternativ options which are realizable under the prevailing conditions. The selection process itself is based upon preference scalings given by utility functions incorporating the norms and the value judgements with regard to public and individual objectives. In this way a classical decision-making model is based upon the seperation of constraints and utilities (BÖTTCHER and SIEMONS, 1989).

In this paper we consider normative aspects of constraints: the extent to which constraints are exhausted we suggest should in itself have an immediate impact on the decision process. This can be achieved through what we call acceptability functions. In this paper we introduce these functions and begin to investigate the resulting utility-constraint interdependence in decision models.

2. Constraints and Acceptability

Options in a decision model are represented as vectors in IR^n. Once the contraints of the model are formulated we assume that the decision maker is able to express his willingness to accept an option in view of the criteria constraints as a function $C:\mathbb{R}^n \to [0,1]$. Thus, $C(x) \approx 1$ if the option **x** is fully acceptable in view of the constraints and $C(x) \approx 0$ if **x** is not acceptable. Such a function shall be called an *acceptability function.*

Usually several constraint criteria are to be considered. In this case the decision maker is supposed to formulate acceptability functions $C_1, \ldots, C_i, \ldots$ for each i in a corresponding index set. For such a family of acceptability functions, representing all constraints of the decision problem, a global acceptability function is defined as the product

$$C(x) = C_1(x) \cdot \ldots \cdot C_i(x) \cdot \ldots$$

of the individual acceptability functions. (To motivate this definition note that for strict 0-1-variables the logical "and"-conjunction is just the product of the variables.)

Usually constraints are represented as criteria functions $c_i: \mathbb{R}^n \to \mathbb{R}$ and restrictions $c_i(x) \le \bar{c}_i$ for each i in the index set. In section 3 we shall discuss a particular form of acceptability function related to such criteria restrictions.

We assume that the decision maker is able to formulate the utility of an alternative as a scalar function $U: \mathbb{R}^n \to \mathbb{R}^+$, the set of positive real numbers. The expression $U(x) \cdot C(x)$ is $\approx U(x)$, the utility of **x**, if this option is fully acceptable. On the other hand $U(x) \cdot C(x) \approx 0$ if the option is not acceptable in view of the constraints. Thus a solution to the decision program should be selected as a solution of

P^*: **Max** $U(x) \cdot C(x)$, $x \in \mathbb{R}^n$, unconstrained.

In a classical decision model with criteria restrictions $c_i(x) \le \bar{c}_i$ and scalar utility $U(x)$ as above solutions are obtained as

P: **Max** $U(x)$, with $c_i(x) \le \bar{c}_i$, for each i (COHON, 1978; DINKELBACH, 1982).

If $F = \{ x \mid c_i(x) \le \bar{c}_i \text{ for all } i\}$ is the set of feasible option, define the *characteristic function* $C_F : \mathbb{R}^n \to [0,1]$ of **F** by $C_F(x) = 1$ for **x** in **F** and

$C_F(\mathbf{x}) = 0$ otherwise. Note that this is a particular example of an acceptability function and that in this case **P** is a particular form of **P***. For most of the paper we shall however assume that acceptability functions are continuous and differentiable.

Ab initio acceptability functions are defined independent of utility. In the formulation of the program **P*** an amalgamation of utility and acceptability takes place. This and in paticular the behavior at an optimum will be analysed in section 4.

3. Acceptability Functions

We suppose that the constraints of a decision problem are represented as criteria restrictions $c_i(\mathbf{x}) \leq \bar{c}_i$ for each i in some index set. A particular form of acceptability functions for such a constraint is defined as

$$C_i(\mathbf{x}) = (1 + \exp(\alpha_i \cdot (c_i(\mathbf{x}) - \bar{c}_i)))^{-1}. \tag{1}$$

This function is the composition of the *logit function* $(1 + \exp(y))^{-1}$ with the "constraint" $0 \geq y = \alpha_i \cdot (c_i(\mathbf{x}) - \bar{c}_i)$ in which $\alpha_i \geq 0$ is a weight parameter. It determines the rate of change of $C_i(\mathbf{x})$ in terms of $c_i(\mathbf{x}) - \bar{c}_i$. Note for instance that in the limit $\alpha_i \to \infty$ the function $C_i(\mathbf{x})$ approaches the characteristic function of the set $F_i = \{\mathbf{x} \mid c_i(\mathbf{x}) \leq \bar{c}_i\}$. Note also that $C_i(\mathbf{x})$ is a differentiable function of $\mathbf{x}$ if $c_i(\mathbf{x})$ has this property.

As indicated in section 2 a family of criteria restrictions $c_i(\mathbf{x}) \leq \bar{c}_i$ then gives rise to a global acceptability function

$$C(\mathbf{x}) = C_1(\mathbf{x}) \cdot \ldots \cdot C_i(\mathbf{x}) \cdot \ldots \tag{2}$$

which depends on the multi-parameter $\boldsymbol{\alpha} = (\alpha_1, \ldots, \alpha_i, \ldots)$. To indicate the dependency on $\boldsymbol{\alpha}$ we denote this acceptability function by $C_{\boldsymbol{\alpha}}$.

Example: The functional form of (2) becomes quite clear in the following simple example. We take $n = 2$, restrictions $-x_1 =: c_1(\mathbf{x}) \leq 0$, $-x_2 =: c_2(\mathbf{x}) \leq 0$ and $2/3\, x_1^2 + 1/3\, x_2^2 =: c_3(\mathbf{x}) \leq 1$.
Figure 1 plots $C_\alpha(\mathbf{x})$ with $\alpha = (75, 75, 75)$, *figure 2* plots $C_\beta(\mathbf{x})$ with $\beta = (50, 50, 5)$.

Figure 1:

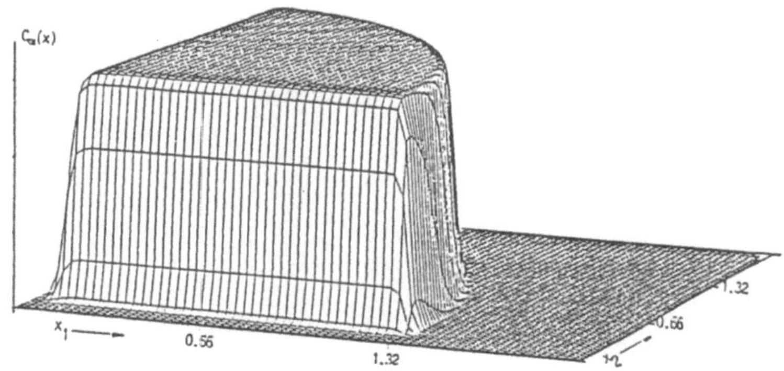

Figure 2:

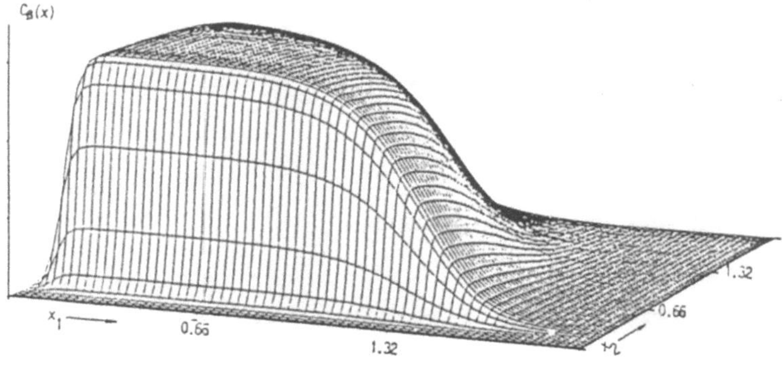

Assuming that utility is given by $U(\mathbf{x}) = (x_1 + x_2)^{1/2}$, in *figures 3* and 4 $U(\mathbf{x}) \cdot C_\alpha(\mathbf{x})$ and $U(\mathbf{x}) \cdot C_\beta(\mathbf{x})$ is shown.

Figure 3:

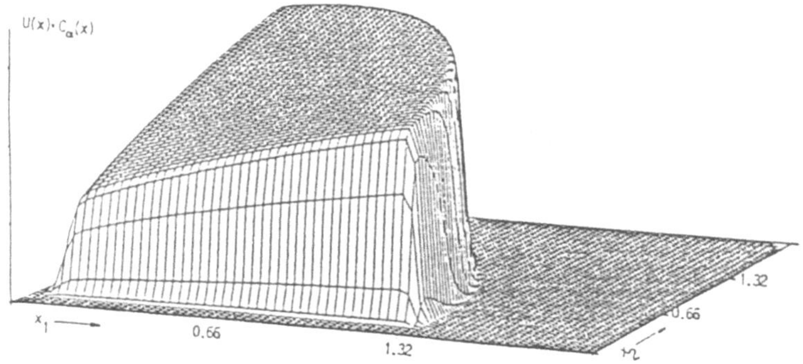

Figure 4:

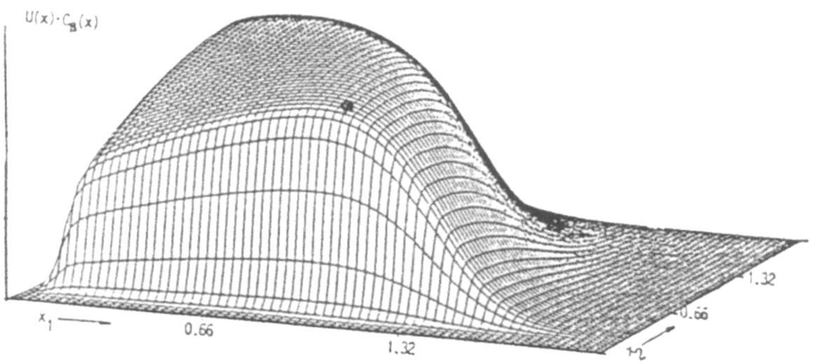

4. The Mathematical Analysis at an Optimum

Acceptability functions can take a variety of different forms. The class defined in section 3 is quite special but does exhibit general principles which we now investigate. Throughout we assume that C_α $(\mathbf{x})$ is defined as in (1) and (2) of section 3. We put G $(\mathbf{x},\alpha)$ = U$(\mathbf{x})$ · C_α $(\mathbf{x})$ where U$(\mathbf{x})$ is some utility function.

(i) As before let $\mathbf{F} = \{\mathbf{x} \mid c_i(\mathbf{x}) \leq \bar{c}_i$ for all i$\}$ be the set of feasible options for a family of criteria restrictions $c_i(\mathbf{x}) \leq \bar{c}_i$. We assume that **F** is bounded. If all $c_i(\mathbf{x})$ are continuous functions then **F** is closed so that it is a compact set. Let

$$C_\infty(\mathbf{x}) = \lim_{\alpha\to\infty} C_\alpha(\mathbf{x}).$$

It is easy to show that $C_\infty(x) = 0$ if $\mathbf{x}$ is not in $\mathbf{F}$, $C_\infty(x) = 1$ if $\mathbf{x}$ is an interior of $\mathbf{F}$ and $C_\infty(x) = 2^{-r}$ for $\mathbf{x}$ on the boundary $\partial\mathbf{F}$ of $\mathbf{F}$ when r is the number of constraints with $c_i(x) = \bar{c}_i$. Thus $C_\infty(x)$ agrees with the characteristic function C_F, defined in section 2, at all points in $\mathbb{R}^n \setminus \partial\mathbf{F}$.

(ii) Now suppose that U(x) is a continuous function defined on some open bounded set O containing $\mathbf{F}$. As $G(\mathbf{x},\alpha)\to 0$ with $\alpha \to \infty$ for points in $\mathbf{O}\setminus\mathbf{F}$, the function $G(\mathbf{x},\alpha)$ has a global optimum $\mathbf{x}^*$ on O for all sufficiently large values of the α_i. It follows from (1) that α than can be chosen large enough to guarantee that global màxima belong to $\mathbf{F}$.

(iii) Now suppose that U(x) and all $c_i(x)$ are differentiable and that some α has been fixed. Then a necessary first order condition at an optimum of $G(\mathbf{x},\alpha)$ is $\partial_j G_\alpha(\mathbf{x}^*) = 0$ for all $j = 1, \ldots, n$ (here ∂_j stands for $\frac{\partial}{\partial x_j}$). This gives

$$\left.\frac{\partial_j U}{U}\right|_{\mathbf{x}^*} = - \sum_i \left.\frac{\partial_j C_i}{C_i}\right|_{\mathbf{x}^*} \tag{3}$$

for each j. This can be viewed as an equilibrium equation between *utility elasticity* and *acceptability elasticities* of the constraints.

Formally equation (3) replaces the Lagrangian equation

$$\left.\partial_j U\right|_{\mathbf{x}'} = - \sum_i \lambda_i \cdot \left.\partial_j C_i\right|_{\mathbf{x}'} \tag{4}$$

which holds at an optimum of the constrained program $\mathbf{P}$. Equation (4), in contrast, is the equilibrium condition between *marginal utility* and *shadow prices.* We note in particular that in (3) the distinction between binding and non-binding constraints is absent.

(iv) For the particular form of the acceptability functions $C_i(\mathbf{x})$, i.e. (1) in section 3, we note that

$$\frac{\partial_j C_i}{C_i} = (1 - C_i)\cdot\alpha_i\cdot\partial_j c_i \qquad (5)$$

$$= \bar{C}_i\cdot\alpha_i\cdot\partial_j c_i \qquad (6)$$

where $\bar{C}_i(\mathbf{x}) = (1 - C_i(\mathbf{x}))$ is the acceptability function of the "mirrored" constraint $c_i(\mathbf{x}) \geqq \bar{c}_i$. This shows that at an optimum the utility elasticity in (3) is expressed in terms of "mirrored" acceptability functions and shadow prices, weighted by the parameters α_i.

5. Concluding Remarks

Vector maximization models (VMM) in the literature are transformed into the classical decision model **P** of section 2 in several ways: either through a process of utility amalgamation by selecting a solution of minimal distance from individual utility optima. Methodically both VMM and **P** are based on the separation between utility on the one hand and on feasibility on the other (HWANG and MASUD, 1979; GAL, 1983; HAFKAMP and NIJKAMP, 1981; BÖTTCHER, 1989a).

Under the rather general assumption "more is better" an optimal solution of the classical decision model automatically lies on the boundary of the set of feasible alternatives (this is **F** of section 4). This shows that the form of the feasible set largely predetermines solutions and so feasibility turns into a normative concept (BÖTTCHER, 1989 b). In the case of VMM the determination of the feasible set already may lead to conflicting aims.

We now turn to the optimization model $\mathbf{P}^*$ based on acceptability functions. There $U(\mathbf{x})$ should be viewed as a measure of intrinsic utility of options independent of constraints and costs. The acceptability functions in $\mathbf{P}^*$ represent an assessment of willingness to carry the burden and costs necessary to realize alternatives. We will assume that this is done inde-pently of expected utility. The function $G(\mathbf{x},\alpha) = U(\mathbf{x})\cdot C(\mathbf{x})$ then becomes a measure of utility in relation to costs.

The behavior of $G(x,\alpha)$ at an optimum has been discussed in section 4. Equation (3) and (6) show that optimal points are characterized as an equilibrium of utility elasticity and cost accaptability.

LITERATURE

Böttcher, H.; Siemons, J., 1989: Akzeptanzfunktionen - eine Antwort auf lösungsraumgebundene Konzepte der klassischen Entscheidungsmethodologie, Jahrbuch für Regionalwissenschaft, 9. Jg.

Böttcher, H., 1989 a: Decision and Appraisal Methods of Public Investment Projects - Some Methodological Reflections. Zaklad Badan' Statystyczno-Ekonomicznych, Polskiej Akademii Nauk, Zeszyt 32, Warszawa.

Böttcher, H., 1989 b: Der Antagonismus in der Diskussion multikriterieller Entscheidungsverfahren - Einige grundsätzliche Überlegungen zur Einordnung und Anwendung "Interaktiver" und "Nutzwertanalytischer" Verfahren in der Regionalwissenschaft - In: FUNCK, R. (Hrsg.): Karlsruher Beiträge zur Wirtschaftspolitik und Wirtschaftsforschung, H. 9 und 10, Karlsruhe.

Cohon, J.L., 1978: Multiobjective Programming and Planning, Vol. 140, Mathematics in Science and Engineering, New York/San Francisco/London.

Dinkelbach, W., 1982: Entscheidungsmodelle, Berlin/New York.

Gal, T., 1983: On efficient sets in vector maximum problems - a brief survey, in: Hansen, P. (Hrsg.), Essay and Surveys on Multiple Criteria Decision Making, Berlin/HeidelbergNew York.

Hafkamp, W.; Nijkamp, P., 1981: Multiobjective modelling for economic - environmental policies, Environment and Planning A, volume 13, number 1, London.

Hwang, C.-L.; Masud, A.S.M., 1979: Multiple Objective Decision Making-Methods and Applications. A State-of-the-Art Survey, Lecture Notes in Economics and Mathematical Systems, Band 164, Berlin/Heidelberg/New York.

The Silver Sheen on the Netherlands
A New Perspective for Regional Policy

by

Leo H. Klaassen and Jacob A. van der Vlist

1. The Present

Dutch economic science seems suddenly to have rediscovered how essential demography is to the economic development in our country and the western countries in general. The renewed interest is expressed among other things in publications, congresses and discussions concerning the significance of an aging population and its gloomy consequences for the active population, likely to go on diminishing steadily in future. For, as is pointed out, the latter group finds itself confronted with a steadily expanding group of senior citizens or senior non-workers, who tend to live longer than in the past, thus spelling serious trouble to the younger generations. Such worried statements are seldom heard from people who know how an economy functions, how for instance the system of saving up for pensions fits into it, and why at the time a system of apportionment for old-age pensions was preferred to one of capital reserve; probably that is precisely why they are heard so frequently.

The report of the Reconsideration Workgroup on demographic developments and the collective sector (Report of the RECONSIDERATION WORKGROUP, 1986) sketches the (separate) effects of the expected demographic developments on the expenses on social security, education, health care and social welfare, justice and housing, on the assumption that the corresponding policis remain unaltered. The general conclusion is that collective expenditure in 1985 will increase by demographic influences alone (medium variant of the Central Planning Bureau) from 117.000 million Dutch guilders in 1985 to 118.000 million in 2000 and to 142.000 million in 2030. The increase up to 2030 is mostly due to the growth of social security by some 24.000 million and of health care by 6.000 million. The cost of education will drop by 7.700 million (NEI, Huishoudens, van nu tot 2000, October 1982).

To consider demographic influences in isolation means of course to leave economic growth out of account. However, the report also gives some attention to the mutations in "demographic expenditure" in a situation of economic growth. At the medium estimate of the CPB for 1985-2000, collective expenditure appears to decline, in percentages of national income, from 33.25 in 1985 to 27 in the year 2000. The decline apparently occurs in all categories of collective expenditure. For social security, the largest expense item, a decline from 15.75 per cent to 13.5 per cent of national income ist estimated.

Other reports, too, give attention to the influence of demographic changes on the economy (NEI, 1983; NCW, 1986; DEMOS, Various publications, PRB, 1986). Moreover, the "Vereniging voor Staathuishoudkunde" devotes this year's annual meeting to the same subject. Indeed, the subject is getting more and more attention, fortunately also more expert attention.

Nevertheless, we believe that some favourable aspects of aging may be usefully pointed out. The question may even be posed whether to attract senior citizens would not be an effective tool with which municipal authorities can promote prosperity and welfare in their municipalities. Without pretending to an exhaustive treatment of the problems, we will try to highlight some elements of major importance for their appreciation, and draw a general conclusion.

1.1 Income and Expenditure of Citizens over 50

Some figures borrowed from the 1984 Budget Survey of the Central Statistical Bureau produce, after some processing, an interesting picture of the income position by member of family according to the age category of the heads of family. Table 1 presents the figures.

Table 1: Age of heads of family, annual family income and annual income by member of family, in 1984, in Dutch guilders

Age of family head	Family income	Income by family member
20-29	32,300	15,380
30-39	39,200	11,529
40-49	44,700	11,461
50-64	39,400	16,416
65 and over	28,500	16,764
Average	36,600	13,555

Although the family income of senior citizens, especially those over 65, is lower than that of the other groups, their income by member of the family proves to be the highest of all categories (40 per cent higher than that in the 30-49 category). Naturally, the effects can be observed in a number of spending categories. Compared to people in their thirties and fourties, senior citizens spend 80 per cent more on holidays, 40 per cent more on food, 16 per cent more on education and transport, 75 per cent more on the home and 50 per cent more on hygiene and medical care. The expenditure pattern of the over-65s is very similar to that of the 50-59 age group. The difference in income per head between the two categories is only a fraction (2 per cent higher for those over 65 than for those in their fifties). Persons over 65 do seem to save a bit more. According to the journal "Geldkoers" of the Postal Bank (September 1987), between 7 and 15 per cent of the income of the over-65s is either saved or given away (GELDKOERS, 1987).

So, in terms of income, the senior group is very attractive, especially in comparison with those between 30 and 49 years old. Their spendings can make a considerable contribution to local or regional income and employment. That is the more interesting as senior citizens do not themselves claim a job and therefore need neither unemployment benefit nor services from the local employment agencies. Areas like the "Utrechtse Heuvelrug", Bloemendaal and surroundings, and Wassenaar have always attracted affluent senior citizens, many of whom went to live there before or after their retirement. The regional economy has profited greatly, for these people brought their income and capital with them, which (if spent in the area) has generated income and employment. The turnover of personal services as well as local government revenues (rates, land sales, etc.) increase, and such social institutions as churches also benefit financially and otherwise, for senior citizens tend to take on much voluntary and administrative work.

1.2 Social Services

Other effects besides the spending impulses which the settling of senior citizens gives to the regional economy, deserve our attention. There are, for instance, the effects on the costs of various social services, such as transport, justice, police. Some have already been investigated by the official workgroup referred to before; in the area of the law, only Child Welfare has been considered so far. We will now present some data which convey an idea of the appeal which the different age categories make to such services.

Transport
Table 2 gives some figures about the number of disqualifications from driving by age category; the figures refer to the year 1984.

Table 2: Number of disqualifications from driving in the Netherlands, 1984

Age category	Holders of drivers licenses (thousands)	Disqualification	(3)/(2)	Index (NL = 100)
(1)	(2)	(3)	(4)	(5)
18-24	566	2,511	4.44	147
25-29	520	2,780	5.35	177
30-39	990	3,735	3.77	125
40-49	762	2,128	2.80	93
50-59	584	826	1.41	47
60 and over	638	291	0.46	16
NL total	4,060	12,271	3.02	100

Sources: CBS, Criminaliteit en strafrechtpleging, 1985, and Statistisch Zakboek, 1986.

Apparently, drivers under 24 score 47 per cent above the Dutch average, and the category between 25 and 29 years old even 77 per cent above average. The group between 50 and 59 scores 53 per cent and the group of 60 and over 85 per cent below the national average. A shift in the age distribution in favour of the older age groups will therefore undeniably affect the volume (and no doubt also the nature) of trafic accidents. The figures refer to males only, which seems justified because males account for 95 per cent of the total number of disqualifications from driving.

The victims of traffic accidents - in many cases also the ones responsible for the accidents - were in 1985 distributed among the age categories as shown in table 3. In particular young people between 15 and 24 tend to get involved in traffic accidents. Ceteris paribus, the relative growth of the number of persons of 55 and older will have a favourable influence on the number of traffic casualties. Apparently, senior citizens tend to behave better in traffic than, for instance, the category of twens.

Table 3: Traffic casualties in 1985, by age

Age category	Number of wounded	Size of age category (thousands)	(23)/(3)	Index (NL = 100)
(1)	(2)	(3)	(4)	(5)
0-9	2,475	1,762.8	1.40	42
10-14	3,054	1,087.4	2.80	84
15-24	20,121	2,502.1	8.04	242
25-34	7,081	2,347.6	3.02	91
35-54	8,379	3,636.6	2.30	69
55-64	3,092	1,387.4	2.23	67
65 and over	3,824	1,729.7	2.21	67
NL total	48,026	14,453.6	3.32	100

Source: CBS, Statistisch Zakboek, 1986.

In other respects, too, the participation of senior citizens in traffic is less costly than that of younger (working) people. For one thing, their use of infrastructure as well as public transport will mostly fall outside peak hours, with lower marginal costs than, for instance, home-to-work traffic of the active population.

Incidentally, a similar argument applies to other government services, such as libraries, to which senior citizens will appeal less frequently than others during weekends. The same holds for municipal and post-office services.

Justice

Another effect of the rising number of senior citizens is (ceteris paribus) a relative decline of criminalism. Table 4 represents the number of, wholly or partly, unsuspended prison sentences, by age, in 1985. Because the number of prison sentences imposed on women is only 4.4 per cent of the total, the table refers to male persons only. The age group under 24 is understood to refer to those between 15 and 24 years old. The table shows clearly that there are hardly any criminals among the over-sixties. Even among those between 50 and 59 years old the number of unsuspended sentences amounts to only 8 per cent of that in the age group between 25 and 29. Dutch senior citizens to not seem to go in for crime much.

Table 4: Prison sentences in 1985, by age category*

Age category	Non-suspended prison	Size of age category	(2)/(3)	Index (Total=100)
(1)	(2)	(3)	(4)	(5)
15-24	5,299	1.277	4.15	152
25-29	4,224	611	6.91	257
30-39	3,913	1.194	3.28	122
40-49	1,418	875	1.62	60
50-59	393	719	0.55	20
60 and over	62	1.015	0.06	2
Total	15,309	5.691	2.69	100

* Exclusive of persons of unknown age.

Source: CBS, Criminaliteit en strafrechtspleging, 1985.

The expectation being that the number of persons in the age group between 15 and 39 will decrease by some 11.5 per cent in the 1985-2000 period, ceteris paribus, the number of unsuspended prison sentences will drop by some 1,530, or 10 per cent of the present number.

These are only a few examples of the influences which demographic changes, in particular the declining number of young people and the rising number of senior citizens, will exert on some social phenomena. There are, of course, other adverse factors which render the category of young people progressively expensive to society, in spite of their declining number. In 1986, the American journal The Public Interest (UHLENBERG and EGGERBEEN, 1986) published some figures concerning the death causes of young whites between 15 and 19 in the United States. They showed that the total death probability in this group, despite a 24-percent decline due to the decreased chances of natural death, showed a net rise of 16 per cent due to a 42-percent increase in the number of motoring accidents, a 232-percent rise in murder cases, and a 140-percent rise in suicide cases. Society bears the greater part of the costs of these death causes. In the foregoing, no account is taken of such developments through time, only the isolated influence of demographic changes being considered.

1.3 The Costs of the Very Old

Whenever the disadvantages of an aging population come under discussion, the costs involved in housing and caring for those advanced in years (often conveniently identified with those over 65) tend to be emphasized. To be fair, we will set the advantages of an aging population enumerated in the previous sections against the disadvantage of higher costs due to aging of the population. As far as these higher costs are borne by the aged themselves, they do not count societally, for in that case they simply concern services paid for instantly, the same as services in other areas. What matters are the services which cannot be paid for in full by those involved and therefore have to be supplied by society as a whole.

While most discussions of the problems of aged persons refer to persons of 65 and over, there are those who are already worrying about people of 55 and over. Actually, most senior citizens are perfectly able to fend for themselves. From an investigation which the Netherlands Economic Institute carried out in Zeeland a few years ago (STUURGROEP, 1984, and VAN DER VLIST, 1984), more than half of the total group of 65-year-olds appeared to need no help at all, and more than a quarter just needed some simple help (cleaning etc.). Only about one in five needed daily help. Naturally, the need for help increases with the years. Of the senior citizens under 75, 12 per cent need daily help, against more than 80 per cent of the over-nineties.

The most expensive services for the aged - the homes for the elderly and the nursing homes - are mostly used by the very old (75 and over), the percentages being 90 for the homes for the elderly, and 74 per cent for the nursing homes. The latter are also occupied for persons under 65 (nearly 10 per cent).

Table 5 reviews the specific costs of people over the age of 65 in the area of social services and health care; the figures have been taken from a recent investigation of the Netherlands Economic Institute (NEI, 1987).

Table 5: Estimated annual costs of the use made of certain services by senior citizens (1985)

Service	Costs in millions of Dutch guilders
Hospitals	4,240
Homes for the Aged	4,142
Nursing homes	3,197
Family help	935
Psychiatric hospitals	640
District nursing service	326
Specialist care	323
Family doctor	212
Co-ordinated aid to senior citizens	96
RIAGG (Regional Institutes for Mental Health Care)	40
Flanking Services	39
General social services	14
Total	14,204

The total amount of some 14,200 million Dutch guilders involved in homes and nursing homes includes the expenses made for permanent housing, food, etc., estimated at 1,800 million Dutch guilders. The remaining net costs amount to 12,200 million Dutch guilders a year, which equals some 7,100 a year for each aged person.

Health services rendered by hospitals, specialists, family doctors, etc., are almost entirely paid for by the users, who have paid normal premiums for national health and insurance. Those premiums have been calculated on the basis of the cost of medical care throughout the insured person's life. A large portion of the costs of homes for the elderly, district nursing systems and family care are paid from public funds. The expenses for other services are almost entirely for the account of these sources.

From percentages of public and social-security contributions referred to in the "Preliminary Inventory" mentioned in note 7, we can calculate that in 1985 a total of 7,100 million Dutch guilders had to be contributed by society. That means a gross "burden" of almost 4,100 a year per citizen of over 65. Divided among persons under 65, it would mean an annual gross sum of about 560 Dutch guilders, or ten guilders a week. "Gross" refers to the fact that the senior citizens themselves pay taxes like everyone else, and therefore contribute to the financing of these costs. They are at the expense of society as a whole, inclusive of the aged population.

We have already referred to the fact that persons over 65 have an average income of approximately 17,000 Dutch guilders a year. On the assumption that about 10 per cent is saved or given away, total average expenditures amount to more than 15,000 Dutch guilders a year. In addition, society spends about 4,1000 guilders on facilities for the aged. The total annual spendings on aged persons thus come to 19,500 guilders.

The attraction of one senior citizen to a region generates a spending impulse of nearly 20,000 Dutch guilders a year, which in turn encourages employment. From NEI estimates, the positive employment effect can be calculated at 20 to 30 employment years for every 100 senior citizens.

1.4 Preliminary inventory

Obviously, from the previous sections, the packet of services for aged Dutchmen differs from that of their younger compatriots. The fact is not new, but the authors have chosen to give it a bit more relief than is usual. Often, the attention is on one aspect of aging only. Reality is more complex and also more favourable than is often thought. The elderly have a relatively high income per head, consume more than others, pay a considerable amount of taxes, save, give money to their children, create employment without needing a job themselves, claim no unemployment payment, hardly make appeal to police and justice (except as victims), are safe traffic participants, use traffic infrastructure outside peak hours, use other public and private services outside peak days, lengthen the tourist season, make no claims on the government for costly education and sports facilities, are active in clubs and associations, and so on, and so forth, and eventually leave a legacy.

On the other side of the account, only a modest amount of some 4,000 Dutch guilders a year per person over 65 has to be supplied by society towards the care of our senior citizens. And remember that they themselves make no mean contribution to its financing through their payment of income and capital taxes, VAT and death duties.

A more thorough, more general, and scientifically more justified analysis of the societal costs and benefits of the various age categories in our country would, of course, be meaningful. Our impression is that the result of such a deeper analysis could be much more positive than is mostly suggested now. If that were true, we should be grateful to those under 65 for raising the future number of over-65s. That future, however, deserves closer examination.

2. The Future

Recently, Professor J. BERTING of Erasmus University contributed an essay to the series "Denkers over dilemma's; Veranderende ethiek en politieke verantwoordelijkheid" (BERTING, 1987). While the entire article is fascinating to read, we will quote some important observations that are particularly relevant to the present article (our translation):

> "The social and economic developments of the last few decades have created a new category of senior citizens who differ in certain respects from the majority of the present senior population. The latter have past through the economic depression of the 1930s and the war period and were influenced by the authoritative conditions of the time. Time and again they figure in research reports as a category of people who are on the whole satisfied with their situation in life, which to the outsider may seem far from enviable. The new type of senior citizens, on the contrary, might well be the vanguard of a more exacting and better organised generation. This new generation will probably remain 'sprightly' to a riper age and be able to profilate their wishes more clearly than their predecessors. Perhaps something like a 'culture of

> seniors' may develop, the participants in which will be able to express more and more clearly that they are not a rest category in the margins of the performance society, but people with objectives, aspirations, desires and relation patterns who share *equal* rights of existence with others. In that sense, the senior citizens as a category could be compared to the youth culture or cultures which developed in the first decades after World War II."

So far our quotation from Professor BERTINGs article. He has added a new dimension to the discussions of future senior citizenship. While so far these discussions were concerned almost exclusively with the growing numbers of senior citizens and the amount of care they would need, Professor BERTING puts into relief their future position in society, their growing self-confidence and independence. Following his direction, the present contribution sets out to complete the vision of future senior citizenship.

2.1 The Future Number of the Over-65

However, we will not ignore entirely the quantitative aspect, and present a couple of figures taken from the population forecasts for the years 2000 and 2010 which the Central Bureau of Statistics published in 1987. According to these forecasts, in the period from 1987 to 2010 the number of people over 65 in the Netherlands will grow from 12.4 to 14.7 per cent of total population. Certainly no figures to cause loss of sleep, the less so as the percentage share of the over-75s in the number of over-65s will rise only very slightly, namely, from 41.8 in 1987 to 42.9 in 2010, an increase of one per cent.

As pointed out before, this paper will certainly not go into the details of proportional demographic changes expected in the Netherlands. Nevertheless, to point at the relative triviality of these changes as far as the over-65s are concerned, seemed right to us. Indeed, no spectacular changes on that score are imminent in the next 20 years.

2.2 Mobility

We have already pointed out that the senior citizen of the future will be different from the present one. We can meaningfully substantiate that statement, for instance by comparing the present mobility pattern of the over-65s with that of the total population, and trying to trace the future changes in that pattern.

From the mobility investigation carried out by the CBS for 1984 (CBS, 1985), the first conclusion to be drawn is that senior citizens travel an average 15.60 km a day (men 18.9, women 12.3), as compared to an average 29.32 km for the entire population. The difference of 11.7 km can be explained largely by the fact that the majority of senior citizens no longer need to travel from home to work and back (this accounts for 6.7 km or 57 per cent of the difference); not going to school saves 14 per cent; travelling less for leisure and sports purposes accounts for 15.6 per cent of the difference, and making fewer other moves for 8.1 per cent. The distribution among the most important other motives is as follows.

Table 6: Distance travelled by sex, age class and three motives

	Average	Over 65	45 - 65
Calling/Staying	7.33	7.55	8.38
Shopping	2.77	2.03	3.17
Touring/Walking	1.40	1.70	1.49
Total	11.50	11.28	13.04

For these age-neutral motives the distances covered appear to be nearly equal for all age categories.

For the overall Dutch population, over 72 per cent of the total number of trips were made by motorcar (as drivers or passengers). For the over-65s the corresponding percentage was 68. The percentages of public transport were 11.5 and 17.8, respectively, for the two categories, which reflects a (moderate) preference for public transport among the over-65s. That is one variable likely to change in the future, since in 1984, 61 per cent of the 45-65-year-olds had a driving license (1,822,000 driving licences for a population of 2,967,000). Among the total population of 1,729,700 over-65s, 462,000 possessed a driving license, that is 26.7 per cent. Most probably, more and more senior citizens will be holders of driving licenses and hence increasingly automobile, and therefore cover more distance, relative to the overall Dutch average, for the age-neutral motives mentioned in table 1. The overall distance covered by all Dutch citizens, in particular for the three purposes specified, is also likely to increase as drivership expands.

2.3 Income Developments

Naturally, the future mobility of the over-65s is also dependent on the way their incomes develop. The article quoted in the introduction already pointed out that, if the households in the Netherlands are arranged according to the ages of the heads of families, people over 65 turn out to have the highest income per member of family (16,764 guilders in 1984, against an overall Dutch average of households of 13,555 Dutch guilders per family member, and an average of 11,641 Dutch guilders per family member of the age group between 40 and 50). Against these figures the objection could be raised that especially in the younger families there are many very young children whose consumption volume cannot be compared with that of adult family members in the households of over-65s. To deal with that objection, we have checked how the figures change if corrected for the age factor, that is to say, if so-called consumption units are counted instead of family members. Even then, the net income of the over-65s (by consumption unit) appears to be the highest of all categories, namely 20,357 Dutch guilders a year, as compared to an average income of 13,969 per consumption unit in the 40-50 age category, and 18,672 in the 50-60 age group.

An other interesting observation is that - naturally as a consequence of their generous incomes - the over-65s also pay relatively handsome sums in wage and income tax: 2,765 Dutch guilders per family member in 1984, as compared with an average of 2,536 for the Netherlands as a whole. Only those between the ages of 50 and 60 pay more tax: 3,208 Dutch guilders. The income tax per member of the family paid by over-65s surpasses that of the 40-50 age group by 42 per cent, the 30-40 age group by 62 per cent, and the 20-30 group by 54 per cent.

To make conclusive statements about the future income position of senior citizens is not a simple matter, but a few main lines can be traced on the strength of reasonable expectations.

A first expectation is that for pensions to go up, from two causes. For one thing, at present, pension rights are not always transferable in case of changing to another job; as a result, the total pension sum may be lower than could have been reached by staying in the same occupation and hence with the same pension funds all of one's working life. (Actually, that phenomenon will decrease in importance, for the tendency is for all pension rights to be made transferable.) For another, pensions have been or are being uprated everywhere, both nominally and because they have been made inflation-proof or even welfare-indexed. Thanks to both factors, an increasing portion of the senior citizen's income is pension money. Moreover, the pensions stand to rise regularly with the rising incomes in the Netherlands, for they are always in proportion to the income earned in a given last work period.

Another expectation is that as their incomes rise, people will tend to save more, and to possess more securities and houses; from their 65th year onwards they will enjoy income from their savings and stocks, while needing less for their own accommodation. From a publication in GELDKOERS (1987), the over-65s reserve between 7 and 15 per cent of their income, probably for their children and grandchildren. The expectation is that people will always have a natural desire to make donations to their children and grandchildren, but on the other hand the need for such gifts and leaving them a legacy will decrease as the children's income position improves and their number diminishes. Besides, in future senior households, more than in the present ones, husband and wife will probably both have held jobs and both receive pensions.

All in all, the future income position of the Dutch over-65s does not look at all bad.

However, there is one point left to consider. In 1992 the Common Market will be a fact, and all over Europe people are preparing themselves for the consequences. The need for tax harmonisation is a major element of the unification of Europe. Tax tariffs vary widely from one country to another. But what many people fail to realise is that to consider taxes in isolation is not right; indeed, equally important is what happens with the tax revenues, what they are spent on. In our country a generous proportion of the taxes, as far as they are higher than elsewhere, is spent on social provisions and social benefits, parts of which accrue, naturally, to the seniors. Voices are already heard here and there to say that in a united Europe, the Netherlands cannot afford to remain a social paradise, and that, therefore, these benefits must be cut down.

Such statements are remarkable for their narrowness. Perhaps it is no concidence that other European countries with a comparable social system, like Austria, Switzerland, Sweden and Norway, are no members of the EC. If they were, and sooner or later they undoubtedly will be, the Netherlands would no longer be a unique social paradise in Europe: there would indeed be quite a few small social paradises, which naturally will never renounce their principles just because a few other countries have more conservative, that is, more ruthless ideas about social matters. The possibility must not even be excluded that other EC member states change their social policy to match ours.

For the time being, there does not seem to be much reason for concern. The Netherlands will not immediately give way when the care for our less privileged compatriots - and the senior citizens among them - is at stake.

Living conditions

From the previous considerations the conclusion can be drawn that the material circumstances of the future over-65s need not cause concern. Health is of equal, or even greater significance. We can easily ascertain that the health situation has steadily improved in the past and that the Netherlands ranks among the best in the world in that respect, together with such countries as New Zealand, Norway and Japan. The trend is likely to continue, also because the growing number of very old persons will draw increasing attention from medical science.

Besides, the future senior category will have enjoyed better medical care and followed healthier eating habits than the present one, and as a result will be livelier and healthier. Moreover, technological progress will see to it that people can live independently and function normally to a higher age. Things that come to mind are the development of automatic wheelchairs, stairlifts, microwave ovens, adapted cars, etc.

3. The Elderly in the Region

What with technical progress, greater vitality, higher mobility and higher incomes, the question is warranted whether future over-65s will not also be inclined to choose a residential environment better suited to their new situation. To young families, the location of their workplace(s) and of their children's schools dominate residential choice. Once a family no longer counts professionally active persons among its members, family, nor any schoolgoing children, the choice will be made from other motives, the residence of the children perhaps being one. In the United States, Florida is a clear example of an area which owes its original flourishing to the settlement of old-age pensioners from the north-east parts of the US. They preferred the quiet and sun of Florida to the unrest and cold of their area of origin. Admittedly, our country offers little choice in terms of climate, although the province of Zeeland and the Wadden Islands do enjoy considerably more sun than the rest of the country. But there are still quiet places to be found, and the more scenic and quieter parts of our country may profit from senior citizens preference.

Initiatives in that direction will also be taken by the regions. In the province of Zeeland in particular, voices have been raised lately to plead the attraction of senior citizens as a welcome instrument of regional policy. From a recent publication issued on the occasion of the 40th Anniversary of the Stichting Zeeland (STICHTING ZEELAND, 1988) we quote the following statement about citizens over 55 who are no longer practising an occupation (our translation):

> "Their financial capacity is reasonable to strong thanks to their incomes from early-retirement or regular pensions, old-age state pensions, and/or other sources. They form in fact a new generation of senior citizens: they are better educated, more responsible and more exacting than the previous generation; on the whole they also have a greater need for choice concerning housing, relaxation, culture, etc. Presumably they will prefer to live in or near the larger service centres and in a natural environment: beach, sand downs, woods. If that profile of the pension migrant is correct - and why shouldn't it be ? - we need to know whether Zeeland can satisfy his wants, wishes and needs, and if so on what points; and if possible uniquely, in other words, how Zeeland scores on things that elsewhere are absent or not accessible.
> Zeeland possesses some strong points which can be emphasized without exaggeration or provincial chauvinism: space, quiet, beach, water and a relatively clean environment, little delinquency, historic towns, possi-

> bilities for daytrips and longer sojourns, a favourable situation with respect to Randstad Holland and Belgium".

And further:

> "...that the Zeeland economy stands to profit from the settlement of a relatively affluent category of people is obvious: retail trade, commercial services, craft industries and the catering trade will get new impulses. The tourist industry can profit from friends and relations of the newly settled coming to visit for several days."

Such publications are refreshing to read. Especially because they spring from a corner from which somewhat more conservative ideas would perhaps be expected, namely, from provincial civil servants. Very likely, more such sounds will be heard in the near future. They will perhaps wake up private enterprise to the fact that they would do well to give up at least part of their American-inspired nervous, so-called dynamic advertising oriented exclusively to the shrinking group of young people, and address themselves to the more well-to-do and fast-growing category of Dutch senior citizens. Initiatives to that end could be taken by marketing teachers at Universities, starting perhaps at Rotterdam University. In their lectures on marketing they can teach their students how to cash in, for the benefit of the company in which they will find employ, on the commercial prospects opened by the ageing of the Netherlands population.

Perhaps some more thought will then also be given to the productive contribution which experienced seniors could make to societal development in general, for instance in terms of mutual aid. That, too, can be important to private industry as well as local authorities.

It may be some time before the so-called grey wave will be generally looked upon as a major potential and growing market, also by the medical profession. And, now that Europe 1992 gives rise to some economic rethinking, there is reason to hope that private enterprise, like the Zeeland government, will recognize and exploit its own interests on this market, which is also a European market. Innovative thinking need not be confined to high technology.

LITERATURE

Berting, J., 1987: Ouder worden in: Denkers over Dilemma's. Veranderende ethiek en politieke verantwoordelijkheid, Erasmus University, Rotterdam.

Centraal Bureau voor de Statistiek, 1985: De mobiliteit van de Nederlandse Bevolking in 1984, Staatsuitgeverij, The Hague.

Demos, various articles, The Hague.

Geldkoers, 1987: Zo besteden wij ons geld, Postbank, The Hague, no. 3.

Heroverwegingsgroep demografische ontwikkelingen en collectieve sector, 1986: Demografische veranderingen en de ontwikkelingen van de collectieve sector, Part I, Synthetic Final Report, The Hague.

Nederlands Christelijk Werkgeversverbond, 1986: Kijken naar de toekomst; een verkenning naar de gevolgen van demografische ontwikkelingen en individualisering, The Hague.

Nederlands Economisch Instituut, 1982: Huishoudens van nu tot 2000, Rotterdam.

Nederlands Economisch Instituut, 1983: Een verkenning van de sociaal-economische gevolgen van veranderingen in de leeftijdsstructuur, Rotterdam, January.

Nederlands Economisch Instituut, 1987: Inventariserend vooronderzoek naar dienstverlening voor ouderen, Rottterdam.

Population Reference Bureau, 1986: World Population Data Sheet, Washington.

Stichting Zeeland, 1988: Kan het wel minder? Middelburg.

Stuurgroep Onderzoek Bejaardenvoorzieningen in Zeeland, 1984: Oud en zelfstandig, een onderzoek naar de voorzieningen voor ouderen in Zeeland, Province of Zeeland, Middelburg.

Uhlenberg, P.; Eggebeen, D., 1986: The Declining Well-Being of American Adolescents, The Public Interes.

Vlist, J.A. van der, 1984: Oud en zelfstandig, Economisch Statistische Berichten.

New Jobs in Connection with Technology Centres and Research Parks

- A Case Study from Baden-Württemberg/West Germany -

by

Jürgen Morlok

1. Economic Facts and Figures of the State of Baden-Württemberg/West Germany

The federal state of Baden-Württemberg was formed as late as 1952 when the states of Baden, Württemberg-Baden and Württemberg-Hohenzollern were united. It is thus a rather young state.

The geographical situation of this Federal State which borders Austria, Switzerland and France offers many advantages to producers as well as consumers. However, the majority of Europe's industrial centres are situated within a radius of 1000 kilometres.

Baden-Württemberg lies in the geographical heart of Europe, so to speak, and from the state's capital - Stuttgart - all of the Federal Republic, France, most of Great Britain, Italy, the Benelux, Switzerland, Austria and Denmark as well as several of the most important states in Eastern Europe can be reached within a day. Once the unified European market comes into existence in 1992, Baden-Württemberg will be the right place from which 324 million people can be served.

Baden-Württemberg's industry is highly developed and diversified. On the one hand, there are large enterprises with a worldwide reputation. In this context, I would like to mention the automobile manufacturers Daimler-Benz, Porsche and Audi who produce 25 % of the total automobile output of the Federal Republic of Germany. I would also like to refer to such large international known corporations as Bosch, Zeiss, Voith as well as IBM, Hewlett Packard, SEL/Alcatel, and Sony. The headquarters of all of these companies are situated in Baden-Württemberg.

In the wake of such big companies there are numerous subcontractors in innovative and future-oriented fields of industry and service. This is reflected by the fact that 50 % of the State's GNP comes from small and medium-sized enterprises. Baden-Württemberg is characterized by a large number of prosperous middle-sized companies. Every second work-place and most of the apprenticeships are found in companies of that size.

Baden-Württemberg companies are autonomous, internationally recognized and highly competitive. They are therefore the ideal international partner.

This State is highly export-oriented, with a total trade volume of 154 billion Deutschmarks in 1987 and an export volume of 96 billion Deutschmarks.

The State's research and technology policy is to improve the scientific and technological infrastructure. The state of Baden-Württemberg contributes 22 % of the entire West Germany industrial research, even though its share of the total population is just 15 %. Eight out of every 1000 inhabitants in Baden-Württemberg are working full-time in R & D. This figure is higher than in any other state of the Federal Republic of Germany.

However, even the most sophisticated innovation is useless if it cannot be translated into a high standard of production. For this end, qualified and skilled labour are necessary. The high education and training of the Baden-Württemberg labourforce contribute greatly towards our stability. The so-called "dual system" forms the backbone of the vocational training.

This "dual system" stands for a cooperative partnership between business and schools. Young people entering business life spend some days of the week on the job and the other days at the vocational school. In this way they have the opportunity of acquiring practical training and theoretical background at the same time. After the apprenticeship they are skilled workers as well as educated technicians.

2. New Technologies – A Challenge for Industry and Government

After the technical evolutions at the end of the 19th century and in the first half of this century, the industrial countries are faced by the new challenges caused by technological progress and the structural changes. Above all, the "new or key technologies" have accelerated this process in recent years.

The term "new technologies" or "key technologies" refers to those technologies which either have a market by themselves, such as computers, or represent the key to a certain market, such as the control for a machine tool.

The most important new technologies are the following:
- Microelectronics
- Information and communications technology
- Manufacturing technology
- Biotechnology
- New materials

It is interesting to note that almost all countries in the world (developing countries, threshold countries or industrial countries) attach great importance to these technologies. The industry has realized that the strategic answer to technological challenge can only be a more rapid innovation and implementation of new technological findings in terms of products and processes. Only the export of technologically highquality goods and the marketing of technological knowledge can ensure a leading position in the world market where technology has become one of the most important standard for performance.

Structural change does not stop at the State level. The government must react to the technological challenge with new strategies and instruments: The aim must be to set up the conditions necessary to assist companies in overcoming structural problems.

3. The Basic Principles of the Baden–Württemberg Economic and Technology Policies

In general, every government has three instruments at its disposal to provide economic assistance:

(i) It can influence general conditions

(ii) It can improve the infrastructure

(iii) It can grant subsidies

The Baden–Württemberg government realized from the outset that to improve the competitiveness of small and medium–sized companies sophisticated R & D facilities are the most important infrastructural measure.

So what does the research and development scene in Baden–Württemberg look like? Baden–Württemberg has got 9 universities, 23 technical colleges, 3 large research establishments, 14 Max–Planck–Institutes, 14 institutes of the Fraunhofer–Gesellschaft, 10 institutes for joint industrial research and 78 transfer centres of the Steinbeis–Stiftung, amounting to well over 100 research and development centres. In this context, research institutes owned and controlled by industry must be considered as well.

There are, for example, the Institut für Mikroelektronik (Institute for Microelectronics) in Stuttgart, the Forschungszentrum Informatik (Computer Science Research Centre) in Karlsruhe and the Laserinstitut (Laser Institute) in Ulm. The State of Baden–Württemberg has set aside roughly 200 million DM for just these three newly–founded establishments alone.

The really novel move in Baden–Württemberg's economic and technology policy was the establishment of operational units for tasks which require state involvement and cannot be solved under public law. These units are (i) the Government Commissioner for Technology Transfer and (ii) the Stiftung Außenwirtschaft (Trade and Investment Office).

The Government Commissioner is in charge of the technical and economic assessments of individual technology–oriented projects and sponsors applications brought before the State. In personal union the Government Commissioner is also chairman of the management board of the private–law Stein–

beis-Stiftung carrying out projects for the Baden-Württemberg economy.

The Stiftung Außenwirtschaft (Trade and Investment Office) with its two divisions is responsible for the State's foreign economic activities (including export promotion and support of economic delegations) as well as for the industrial settlement and mediation of partnerships between medium-sized companies.

As a further measure to encourage the foundation of businesses in future-oriented fields, Baden-Württemberg has sponsored the establishment of technology centres and research parks.

4. Technology Centres: Instruments of Economic Policy

Of course, technology centres are not invented by Baden-Württemberg. Following the spirit of the age, the Baden-Württemberg Government was inspired by Silicon Valley, Cambridge Science Park and Sofia-Antipolis, yet recognized very quickly that it had to find its own specific way. To the Baden-Württemberg Government - and this is how its concept has to be understood - technology centres and research parks are effective only if they are in harmony with the other transfer instruments.

Technology centres and research parks by themselves cannot be the only incentives for a florishing economy. However, before explaining the structure of Baden-Württemberg technology parks it is necessary to define the word because the meaning for these joint locations ranges from trades centres, foundation centres, technology factory, technology centre to research park, and everyone understands something different by each term.

Trades centres
This term denotes a joint location of predominantly small and medium-sized companies in a building complex which is built and administered by a sponsor company. There is no specific occupation policy and so there's a heterogeneous industry structure. Often, the companies have strong ties to the location and fulfill the function of stopping gaps where services were lacking

before.

Foundation centre

This is a joint location of newly-founded companies, usually engaged in the sectors of processing trade and production-oriented services. Here, an attempt is made to increase the chances for growth and survival of these companies in the starting phase by providing variable accomodation space, joint office and administration facilities as well as management, consulting and support services.

Technology centre / Technology factory / Technology park

The word stands for a joint location for mostly young businesses and companies which develop and market new technological products and processes which have a high market risk, require high capital investment and for whose development both service and consultation facilities and a regionally significant network of research, consultation and financing institutions are of advantage.

Research park

This is a park-like industrial estate in which research-oriented young companies or research and development departments of large companies find attractive working conditions either as a result of the proximity to a technical college or research establishment or solely thanks to the quality of the premises. Central joint establishments are not provided.

Companies or research establishments settling in research parks are permitted production on a small scale only, that is it should not exceed the stage of pilot series production and prototypes, in addition to applied research.

What does the structure in Baden-Württemberg look like?

The majority of all joint locations in Baden-Württemberg are mixed forms of the foundation centre/technology centre /research park types. Since the establishment of the Karlsruhe Technology Factory in 1984, which will be dealt with in more detail later on, several technology centres have been created in Baden-Württemberg. We therefore have a sufficient number of such establishments for the time being. Figure 1 shows that the principle of regionalization has been applied when setting up these technology centres.

Moreover they have been built outside the densely populated regions - for example at St. Georgen in the Black Forest or at Aalen.

Figure 1: Technology Centers in the Federal Republic of Germany

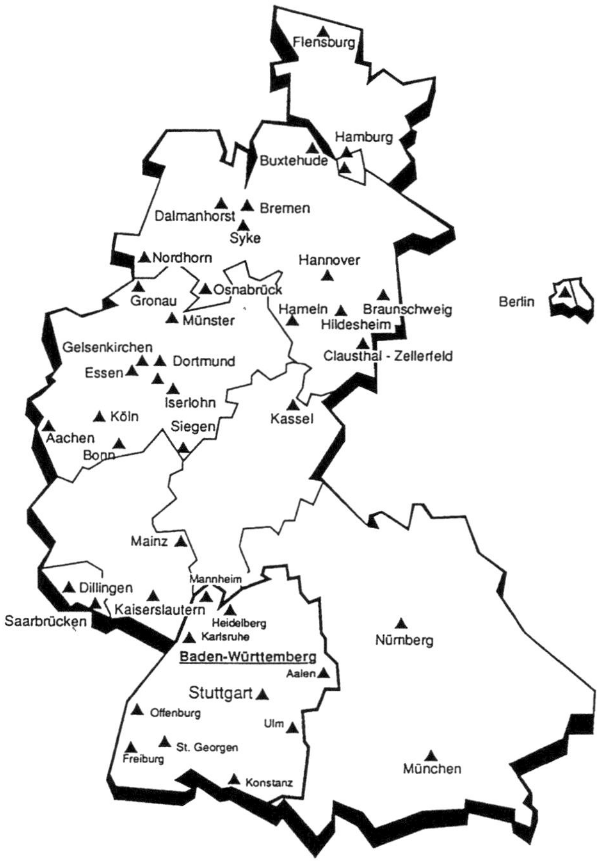

When discussing Baden-Württemberg sponsorship policy, one must mention that the objective was to support and promote on-the-spot activities in order to permit a large variety of operating or sponsoring agents namely:

- The Land (Regional State)
- Communes
- Chambers of Industry and Commerce
- Chambers of Trade and Crafts
- Financial Institutions
- Private companies
- Trade associations.

To give an impression of how technology parks work, I would like to refer to the Heidelberg and St. Georgen Technology Parks at this point.

Heidelberg, with its large variety of scientific institutes in the fields of medicine, organic chemistry, immunology and serology, is an exceptional base for founding new businesses in the future technologies such as biotechnology and genetic technology. When founding this technology park the aim was to develop and promote interesting research projects up to the stage of production maturity and to then introduce them on the market. All these endeavours are carried out in close cooperation with Heidelberg university.

Almost all of the founders of Heidelberg technology park are young scientists. Up to now, only very few offshoots from existing companies in Heidelberg have moved to the premises there. However, more and more renowned companies participate in spin-off foundations from the university, and this is a guarantee for success in most cases.

Actual production takes place in separate production parks which the town set up on abandoned industrial sites.

The structure of the technology centre in St. Georgen, on the other hand, is quite different. This small town in the middle of the Black Forest some 20 km away from the nearest technical college in Furtwangen, is known for its clock and precision mechanics industries. In the early eighties when the previously world-famous company Dual went bankrupt, this town suffered a considerable loss of jobs and image.

Faced by this situation the city of St. Georgen, the owner of the Dual real property, local industry and banks joined efforts and founded the "Technology company".

That company's goal is to sponsor and attract technology-orientated enterprises in order to create new jobs for those lost and to re-establish the image of a prospering industry because the other companies in the region had been affected by the Dual bankruptcy, too.

In early 1984, the first enterprise moved into the previous Dual factory site. Today, there are 11 companies with almost 120 employees in the technology centre. The spectrum of founder companies covers drive and control engineering, development and production of electronic circuits, software technology and environmental engineering.

What's so exciting about 120 new jobs? The question is justified. However, one should not just see the 120 new jobs but realize that, above all, the established local companies are profiting form these activities. A number of cooperative partnerships between founder companies have been set up. However, after getting into contact with the modern new enterprises in the technology centre, many old established firms became aware of what the new technologies have to offer to their own companies, and they have taken up the challenge.

In additon, a cooperation agreement was signed between the technology centre and the Furtwangen technical college. The aim is to translate scientific findings into innovative products more rapidly, and the old established companies are profiting from this also. In additon, the Steinbeis-Stiftung has set up a transfer centre in the technology centre.

The overall picture shows that all of these efforts have paid off for the town. While some years ago St. Georgen had a very high unemployment rate of some 10 %, it now holds one of the best places in unemployment statistics with just 5 % unemployed.

This favourable figure cannot be attributed to the creation of the technology centre only, but without the impulse of the technology centre the figures would certainly be less impressive.
After having made some statements about the quality of technology centres, I would now like to turn to quantitative considerations. Table 1 gives a survey of all the technology centres in Baden-Württemberg. The figures on the number of companies and the industrial floor space are based on surveys conducted in January 1988. At this point, I would like to add a few remarks about the subject of "location". When discussing preconditions related to location, it is necessary to distinguish between technology centres and research parks, because different location requirements are necessary for the success of the respective concepts.

Table 1: Technology and Research Centers in Baden Württemberg

Technology centers / Research parks	Companies	Employees	Area
Mannheim	6	72	2.350
Heidelberg	14	280	6.000
Karlsruhe	31	600	20.000
Offenburg	8	87	1.727
Freiburg	7	40	1.300
Konstanz	9	35	2.200
Ulm	10	15	2.150
Aalen	7	11	1.600
Stuttgart	18	60	4.000
St. Georgen	11	110	3.700
Mosbach	3	30	2.000
Total	124	1.340	47.027

In view of their philosophy, technology centres can be established only in regions where there are technical colleges, universities or research institutes. Experience has been that - as a rule - a close cooperation between a university, a research institute and the companies in the technology centre will work only if there have been established contacts between industry and university already before.

Another location precondition is an industrial environment with a large catchment area in conjunction with a developed infrastructure. A further prerequisite is the availability of qualified workers. It is thus clear that technology centres are not suited for poorly developed regions. This is even more true for research parks where preconditions are even more exacting.

5. Case History: Karlsruhe Technology Factory

I would now like to demonstrate how a technology centre stimulates both the labour market and the image of a region. The Karlsruhe Technology Factory, opened in Juli 1984, was the 3rd technology centre in the Federal Republic of Germany and the first in Baden-Württemberg. The Karlsruhe area, with its high concentration of state, semi-state and private research institutes, was the ideal place for a technology centre.

The Karlsruhe region features
- the oldest German technical university,
- the old state technical school which is today the largest technical college in Baden-Württemberg with
- the largest branch of the Steinbeis-Stiftung,
- the Nuclear Research Centre,
- three institutes of the Fraunhofer-Gesellschaft,
- the Federal Research Institute of Nutrition and Water-Supply Engineering,
- the Regional Institute for Environmental Protection,
- the European Research Centre for Air Pollution Control,
- the Technical Information Centre for Energy, Physics, Mathematics,
- the Computer Science Research Center.

This concentration is reflected in labour statistics: Out of 1000 employees, 36 are involved in research - the state average being 24.4. While this exceptional research and technological potential was known, it was not used for attracting industries. In addition, the founders of the technology centre benefited from circumstances they could not influence.

The premises of the Singer sewing machine factory, located centrally and bordering the university campus, were vacant. The Landeskreditbank (the state owned bank) developed these premises and the initiators decided that the future technology factory should be run by an operator company, namely

IHK-Unternehmens- und Technologieberatung Karlsruhe GmbH (UTB). After these two important measures were taken, the first period of erection was completed in 1984 and companies started to move in. Since then, the Karlsruhe Technology Factory has been expanding, covering an area of around 20.000 sqm today. It has become the larges West German technology centre. Its policy is to promote new ventures in technology-orientated fields promising commercial success: Potential tenants are required to meet the following conditions:

- the company's product or service must be a technical novelty
- its technical feasibility must be confirmed by an expertise
- a study must have verified its marketability
- the company must present a strategy for the first 3 years
- financing must be guaranteed during the phase of development and market introduction
- the company must leave the technology factory once its product attains market maturity to make space for other interested parties. It should not stay for more than 5 years.

There are no further prerequisits and there is no obligation to relocate in Karlsruhe or in the Karlsruhe region later on.

The spectrum of companies in the Karlsruhe technology factory reflects the scientific potential of the region. The leading position of Karlsruhe University in computer sciences is reflected by the fact that 21 companies work in computer science and microelectronics. 10 of these companies are mainly hardware-oriented, 11 concentrate on software. Six software companies develop instruments for the CAD/CAM fields.

Four companies work in sensories and physical process engineering while two others are involved in assembly and robotics.

Today, the Karlsruhe Technology Factory comprises 31 companies with a total of 600 employees and an overal turnover of 80 million DM. Four companies with 115 employees have already moved out and have settled in the Karlsruhe region. These figures do not cover the numerous suppliers and service companies working for these 31 companies. As to the employment statistics in the Karlsruhe region, one figure is outstanding. Since 1983, 17 % of the newly created jobs are created in the Karlsruhe Technology Factory, and all of them are highly-qualified and thus permanent positions.

The effects on quality in general are striking as well. Compared with the seventies, Karlsruhe is considered a high-tech location today. A large number of companies in the Karlsruhe region were founded as a result of the economic impetus.

In summary: Technology centres and research parks are certainly not the remedy to solve unemployment problems caused by structural change. However, they do provide stimulation so that a region or economy can cope with structural change more effectively. Technology centres permit people to translate their ideas into marketable products more rapidly and they do speed up technology transfer from universities to companies.

Poland – the Challenging Laboratory of Space Economy

by

Antoni Kukliński

1. Introduction

The summer 1989 is a unique experience in Polish and maybe not only Polish history. With some exaggeration we can speak about the Second Miracle on the Vistula.[1] Anyway this is a summer of qualitative change of the Polish political system – the end of the III. Republic and the birth of the IV. Republic. The situation is very complicated, dangerous and full of hope.

The most challenging danger is the cohabitation of the paradise of political change and the hell of a very deep economic structural crisis. If the new government is able to sustain the existence of the political paradise and change the economic hell into a purgatory – then the Polish summer 1989 will open a new successfull period in the Polish history. In these conditions Poland of the nineties will be a most challenging laboratory of social sciences in general and of space economy in particular.

In this paper I would like to outline 12 topics of comparative studies on the transformation of Polish space economy in the nineties:

(i) The transformation of the political system

(ii) The creation of a new system of a real local self-government

(iii) The creation of a new institutional system of space economy integrating the spontaneous change, generated by the market forces, and the guided change promoted by the central government and by the local selfgovernment

(iv) The creation of a new system of physical, architectural and cultural planning which will be liberal and imperative at the same time. This system should stop the negative processes of the destruction of Polish landscape and create a set of proper inducements for maintenance and investment activities in order to pen a new period in the history of Polish landscape

(v) The creation of a new integrated system of markets

(vi) The creation of a new geography of the Polish enterprise and of a new industrial space

(vii) The creation of a new comparative geography of national income

(viii) The creation of a new geography of national wealth

(ix) The transformation of the Upper Silesia

(x) The transformation of Greater Poland / The Poznań region

(xi) The transformation of Warsaw into a semi-global city

(xii) The creation of the Third Paradigm of the Polish School of Regional Studies

2. The Transformation of the Political System in Spatial Perspective

The transformation of the state is naturally a predominating macro-political process. But this process is spatially differential. Not all regions are participating in this process with the same intensity.

If we look at the geography of positive promotion of political change in Poland and at the geography of conservative barriers - then we see how deeply differentiated this process is.

The result of the June election is a strong inducement to develop in Poland studies in the field of electoral geography.

So the political framework of Polish space economy is a phenomenon also differentiated in regional and local dimensions. It is worth while to study the special role of the region of Gdansk in the political transformation of Poland.

3. The Creation of a New System of a Real Local Self-Government

An effective and honest local government is an important factor in the development of rational space economy. There are very few countries which - like Switzerland, Netherlands, Sweden, and Western Germany - have a really good local self-government. Poland is now trying very hard to build up a new system of real local government. The interlocal and interregional differentation of Poland is very deep. It is very easy to anticipate that, in the nineties, a very complicated geography of success and failure will emerge in the field of the development of local self-government in Poland.

4. The New Institutional System of Space Economy

The desillusion about perfect planning has created illusions about perfect market in Poland. These illusions are particularly dangerous in the field of the organisation of space where the experience of Western Europe has provided many solutions in the integration of the mechanism of spontaneous and guided change. In the nineties, Poland will build up a new institutional system of space economy promoting the forces of spontaneous change but keeping, at the same time, those forces under efficient supervision of the proper instruments of spatial policy and planning. This system of supervision and guidance must involve coordinated activities of the central and regional government and of the local self-government. Let us hope that the hard reality will induce the creative Polish thinking in the field of new institutional solutions related to an efficient organisation of space.

5. The Transformation of the Polish Landscape

In the Polish experiences of the last forty years we find many paradoxes, including that of the landscape. The stiff system of centralised policy and planning created a sad picture of destruction and anarchy in the Polish landscape. In the new political and moral atmosphere it is possible now to convince our public opinion that the attitude of the Polish society, economy and State in relation to the landscape must be changed very deeply. We must also build up a new system of policy and planning in this field integrating the approaches of physical planning, architecture, cultural anthropology, sociology, economics, geography, and history.

6. The Creation of a New Integrated System of Markets

The renaissance of the market economy on the Polish territory is a process differentiated in both time and space. The nineties will be a turning point in this field.

This is a challenging topic for regional and local studies in Poland: the emergence of different types of markets in different areal scales.

Table: Polish markets in the nineties

type of market	areal scale			
	local	regional	national	global
commodity market	x	x	x	x
labour market	x	x	x	x
capital market	?	?	x	x
information market	?	?	x	x

In the studies of this process the theory of the diffusion of innovation may be very useful. We are, for example, advanced somehow in the creation of commodity markets but we have not even started to develop the processes which lead to the emergence of an effective capital market. So we can envisage the development of market economy in Poland as growing systems of islands of different types of markets in different areal dimensions.

Naturally the dream about perfect markets is equally unrealistic as the old dream about perfect planning. The imperfection of the markets is multiplied by the conditions of the economy of scarcity which will dominate the Polish economic scene for a few years.

7. The Creation of a New Geography of Enterprises in Poland and the Emergence of the New Industrial Space

The industrial space created in Poland after the World War II is a sui generis Stalinistic caricature of the Fordist space[2] and the Swedish welfare state. The enterprise was also a bureucratic shadow of a real enterprise where economic functions were subordinated to social and political functions. In the nineties, we shall observe in Poland a challenging process of the transformation of Polish enterprises into real economic units able to survive and expand in competitive market conditions. The spatial behaviour of the enterprises in Poland in the emerging post-Fordist space is an open question for interdisciplinary studies.

8. The Creation of a New Comparative Geography of National Income in Poland and in Spain

In the sixties and early seventies, the Polish school of Regional Studies built up a tradition in the field of computation and analysis of the regional differentiation of national income in Poland. That tradition was interrupted in the early eighties. In the framework of the Research Programme: Regional Development - Local Development - Territorial Self-government the continuation of that tradition was restablished again. The team of L. Zienkowski has prepared computations and studies to compare the regional differentiation of national income in Poland and in Spain in the years 1976-1986. This experience in comparative regional studies will be expanded in the nineties.

9. The Creation of a New Geography of National Wealth

Unfortunately, the tradition of computation and studies of national wealth in Poland is totally neglected. Any attempt to discuss the problems of national wealth in Poland was destroyed by the Marxian theory of value. I think we can now cross this ideological barrier and start an immense effort to find out what are the volume structure and distribution of national wealth in Poland. Only in this context the decapitalization processes in Poland, which are one of the most depressing features of the eighties, can be studied in a responsible way.

10. The Transformation of the Upper Silesia

The Upper Silesia, the biggest concentration of mining and heavy industry in Poland, is probably one of the most tragic problem areas in Europe. The technological, economic, social, and ecological reconstruction of this region must be started in the nineties. The role of applied regional studies in this process should be very great. Especially important are the experiences of Western Europe and the United States in this field.

11. The Transformation of Greater Poland - Pozsnań Region

Greater Poland is the most promising region in Poland from the point of view of an efficient transformation of the space economy and of the development of a good territorial self-government.

12. The Transformation of Warsaw into a Semi-Global City[3]

In the nineties, a well-guided process must be started to transform Warsaw into a city directly participating in the processes of global development. It is impossible to build a modern technical infrastructure for the whole country in a few years - but it is possible and necessary to do it for one city - Warsaw. The structure of the Warsaw economy must be changed very deeply. Inter alia, the Warsaw Steel Works must totally be dismantled in order to use the Warsaw district as a place for a high-technology industrial estate. Warsaw should use the experience of the most eminent global cities which are real nodal regions of world economy.

13. The Creation of the III. Paradigm of the Polish School of Regional Studies[4]

If we look at the history of the Polish School of Regional Studies after 1918 we can indicate three periods of especially intense paradigmatic change.

The first paradigm emerged in the twenties as a response to the challenges of the new political and economic reality of Polonia Restituta. The second paradigm is a product of the fifties and a response to the demand created by the particular shape of the III. Republic. It is interesting to note that the first paradigm was annihilated in the late forties and early fifties by the Polish version of the Stalinistic ideology. The second paradigm emerged in the glorious process of the Polish October in the middle and late fifties.

Strangely enough the Sturm and Drang Periode of the eighties was not able to create the III. paradigm. The resistance of the conservative elements of our academic establishment was too strong and the preassure of the progressive elements too weak. So the task to create the III. paradigm of the Polish School of Regional Studies will be implemented in the conditions of the early stages of development of the IV. Republic. I think there will be some important similarities in Poland between the twenties and the nineties. The mechanism of Polonia Restituta is very strong also in the field of Regional Studies.

14. Conclusion

This is a panoramic view of some anticipated transformations of space economy in Poland and of the response of the Polish School of Regional Studies to deal with those transformations in the Sturm and Drang Periode of the nineties. There is no doubt Poland will be one of more interesting laboratories in our field in the nineties. We should try to look at the transformations of space economy in Poland with the classical eyes of August LÖSCH.[5] We will try to organise in Warsaw an international conference on "The ideas of August Lösch and the transformations of Polish space economy".

FOOTNOTES

[1] The first Miracle on the Vistula has been presented by N. Davies (1981), pp. 393-401.

[2] See Roobeek (1987).

[3] Compare Lambooy (1987).

[4] Compare Kukliński (1986), Johnston and Claval (1984), Geoforum (1989).

[5] Compare Funck, Kukliński (1986).

LITERATURE

Davies, N., 1981: God's Playground - A History of Poland, Vol. II, Clarendon Press, Oxford.

Funck, R.; Kukliński, A., Eds., 1986: Space-Structure-Economy. A Tribute to August Lösch, von Loeper Verlag, Karlsruhe.

Geoforum, 1989: Creating Geographies - Society and Space in Britain and Poland, Vol. 20, No. 2, Special Issue.

Johnston, R.J.; Claval, P., Eds., 1984: Geography since the Second World War. Croom Helm, London.

Kukliński, A., Ed., 1986: Regional Studies in Poland, Warsaw.

Lambooy, J.G., 1987: Global Cities and the World Economic System. Research Memorandum No. 8803, Amsterdam.

Roobeek, A.J.M., 1987: The Crisis of Fordism and the Rise of a New Technological Paradigm. Futures.

Economic Reforms in Centrally Planned Economies and their Consequences for Regional Development

by

Jan S. Kowalski

1. Introduction

One important issue which has been largely ignored in the literature on regional development and policy in Centrally Planned Economies (CPEs) concerns the possible impacts, if any, of the attempted economic reforms of the CPEs on future regional development in these countries (one notable exception from this neglect is GORZELAK, 1989).

In all the CPEs the declared goal of regional policy and planning has been and remains the decrease in the interregional inequality in development levels, coupled with an overall increase of the GNP (comp. DZIEWONSKI, MALISZ, 1978 and DEMKO, 1984). These goals of spatial policy have not been achieved. In the first 10 to 15 years after the World War II some equalization of regional development levels was attained but for the last 20-25 years evidence is available that a deepening of the differences in the regional production, investment and living standards took place (KOWALSKI, 1986a).

In this contribution we shall reflect on the question if reform efforts undertaken at present in CPEs will help to stop and reverse this tendency for polarization of space in CPEs between strong, industrial regions and weak agriculturally dominated ones.

What were the reasons for polarization processes in Centrally Planned Economies, where in principle the central planning authorities should be able to influence spatial development of the country?

We postulated elsewhere (KOWALSKI, 1986a, 1989) that CPEs are in reality unplanned economic systems. The reasons for spatial polarization are found in the ways these systems function, and this in turn is influenced mainly by "lawless", voluntaristic decision-making in the economy on the macro level coupled with soft budget constraints of enterprises at the micro level. This way of decision-making at the macro-level is possible because of the more or less harshly executed totalitarian one-party systems, permitting to disregard the legal rules of the game theoretically prescribed by the CPEs. Soft budget constraints lead in turn to expansion drive of enterprises, which are not limited in their growth wishes by demand considerations, and to widespread shortages of inputs. Enterprises facing soft budget constraints exert an almost insatiable demand for inputs in order to secure plan implementation. This phenomenon results in conflicts in informal bargaining for scarce resources.

In the spatial perspective this conglomerate of factors leads to the advantageous position of these regions and sectors which are governed by politically more powerfull lobbies, and these as a rule happen to be the strong industrial regions, dominated by traditional heavy industries. Regardless of plans calling for equalization of regional development levels, centrally controlled investment expenditures continue to flow to these strong regions. Weak, peripheral areas are as a rule also the political underdogs and hence they lose the competition for resources, competition which has very little to do with economic efficiency, performance or the declared goals of regional policy.

2. Main Directions of Economic Reform Efforts

Since Michail Gorbatschew proclaimed "perestrojka" and "glasnosc" in the Soviet Union it seems ideed possible that, for the first time after the Soviet Stalinist model of the CPE became the dominant mode of production, real changes in the ways of functioning of society and economic systems in countries of Eastern Europe can be implemented.

It is often forgotten in the present discussions on this subject, that there have been numerous attempts to reform the economic system since Stalin's death in 1953. Hungarian economy broke with the classical Stalinist model in 1968, and has moved since then some steps towards a more market oriented economy. Polish economists elaborated various reform schemes since 1956, some of which have been realized in the 70's and even more intensively in the 80's. In the 60's several changes in the direction of decentralization of the economic decision-making and increased reliance on economic parameters, like prices, wage rate, rate of interest etc., instead of commands have been introduced in the Soviet Union, the GDR and Bulgaria. (A useful overview of the past attempts to reform the CPEs can be found in CASSEL, 1983 as well as KOSTA, 1984, 1989; DOBIAS, 1986; LASKI, 1985).

With respect to the reasons for very limited succes or even failure of these earlier reform efforts in CPEs most of the authors point to the strong resistance against reforms of "nomenklatura", i.e. those members of the ruling communist parties who occupy, not by virtue but by party nomination, important posts in all spheres of the economy. Also in these earlier reform efforts the main thrust of the changes was directed at improving the existing central planning system, at making it more efficient and able to cope with the growing complexities of present day economic life. Only very recently the awarness spread that central planning as a way of regulating and coordinating economic systems must be decisively reduced in scope or even eliminated altogether (LAVOI, 1985).

It is beyond the scope of this contribution to analyze in depth the present reform efforts in the CPEs, which in any case have not yet changed the ways of functioning of these economies to such extent as to justify calling them something else than "CPEs". We shall also to a large degree refrain from speculations about the chances that these reforms will be really implemented in the future. We shall limit ourselves to a brief synopsis of the directions of reforms and focus on their possible impacts on regional development.

First of all institutional reforms should be mentioned. In all countries experimenting with reforms the weight and size of the Central Planning Commission is reduced, sometimes it gets renamed, as a rule the political rank of the head of this body significantly decreases. These measures are coupled with a considerable decrease in the number of ministries responsible for various sectors of the economy. For example in Hungary the number of these ministries sunk from about 20 in 1968 to one "super" ministry for manufacturing at present, accompanied by the finance ministry responsible for fiscal and monetary policies. In Poland the respective numbers decreased from about 30 to 5. In the Soviet Union, where in the past over 120 ministries were responsible at the central level for various branches of the economy, only about 30 of them remain and certainly some of them will disappear before this study gets published. It should be however noticed that not all of the officials from the dissolved ministries are fired. Actually nobody is fired. About 60% of all the employees are taken over by the remaining ministries, including ministers, who as a rule become vice-ministers in the new larger ministries, the rest of the officials from the "old" ministries retire or change employment.

Coupled with these institutional changes a significant evolution in the content and scope of plans is taking place. In Hungary and Poland already at present, and in the Soviet Union in the near future, the central plan should be limited to the overall determination of the directions of development. No longer are the plans of the enterprises given them from above by the ministries and the central planners. The enterprises must still elaborate plans, but are more or less free to do it on their own. As a rule the managements are more free than before to hire and fire workers and are able to pay them other than centrally determined wages. The intermediate administrative level of associations which functioned earlier as an "umbrella" between the central planning level and enterprises has been weakened or even abolished. Managements are also to some limited extent free to choose suppliers of their inputs and buyers of their output.

Directors of enterprises which are not strategically important (the rules determining which are and which are not strategic remain rather hazy) should be chosen in an open competition by committees comprising representatives of the employees and the ministries responsible for a given enterprise. This rule applied as a rule up till now only to smaller enterprises, but the establishement of a legal precedent is important.

Another important legal change concerns bankruptcy laws which at least in theory enable firms to go bust in some of the socialist countries. As already mentioned above these laws are not applied in real life at all or applied very selectively, so that subsidies to unprofitable enterprises still consume 30 to 40% of the state budgets. Actually it would be surprising if the bankruptcy laws were implemented seriously, as long as some of the most important parameters influencing profitability of firms are still determined by the authorities. The most important of those concern prices of inputs and output. Within the framework of reforms the share of prices determined by the market processes rose in most the CPEs, but even in Hungary about 60% of prices are directly or indirectly controlled by the government.

Another substantial change in the institutional environment of the CPEs concerns the reform of the banking system. In the past banks were considered to be a sort of clearing houses for enterprises wishing to get credits. They could not in practice refuse the money. In Poland, Hungary and to lesser extent in other socialist countries, at least according to the letter of law, banks should in future be guided by efficiency criteria in credit activities, and they should themselves watch that the credits get repayed, otherwise the banks themselves can go bust. The role of banks will also be increased by some efforts to implement at least a limited convertibility of currencies which have been conducted in Poland and Hungary, and are under discussion in the other Eastern European countries.

In all the CPEs the reform efforts change also the way foreign trade and foreign investment are conducted. The state monopolies in this area are abolished, joint ventures, i.e. projects involving domestic private or state capital as well as investors from abroad are allowed and fostered through favourable tax treatment and increasingly attractive conditions for reimbursement of profits in hard currencies. Private activities in exporting and importing of goods and services by foreign and own nationals are increasingly tolerated or even encouraged. The main problem at present, which renders most efforts to intensify economic relations with "hard currency" countries is the high indebtedness of some CPEs, especially those which lead on the road to market economy orientation.

Expansion of private sector, if not strongly supported (like in Hungary and Poland), is at least no longer considered an anathema even in the Soviet Union, Bulgaria or the German Democratic Republic. Apart from the expansion of the private sector in this sense that new firms are established (ROSTOWSKI, 1989), various schemes for privatisation of state enterprises are discussed, as a precondition for the spread of market mechanisms (WINIECKI, 1988). As a further step in this direction establishement of stock exchanges is considered to facilitate flow of assets between various actors and various sectors in the economy.

3. The Short Term Results of the Reform Efforts

In the last decade the awareness grew that from the economic point of view softness of budget constraints of the enterprises, meaning that they could spend more money than they earn, was one of the most important, or even the decisive factor leading to weaknesses of CPEs (for a review of this approach to CPEs see among others: KORNEI, 1980, 1982, 1986; KOWALSKI, 1983, 1986b, 1988; LASKI, 1985). Softness of budget constraints of the enterprises is explained by the dominant political system - one party, nomenklatura based system with the state ownership of most fixed assets. In such a system it is natural that central planning authorities "help" financially these enterprises which are unable to survive. Soft budget constraints lead in turn to insatiable demand for inputs and final goods, hoarding of reserve inputs by many enterprises and final goods by households, coexisting with widespread shortages on input and final goods markets, lack of responsiveness to consumer needs, irrelevance of money and monetary variables as factors explaining behavior of enterprises, and many other "plagues" characteristic for the mature CPEs.

Thus, it should not surprise that if we look at the reforms of functioning of the economic systems in Eastern Europe outlined in the previous section one common denominator becomes visible. They all aim at hardening of budget constraints of the economic agents at all levels of the organizational hierarchy, they should make money play a more active role in the economy, they try to force economic units to consider the financial effects of their activities, i.e., costs, profits and so forth. In other words the reforms attempt to introduce market economy mechanisms to centrally planned economic systems. The main level at which the hard budget constraints should change the behavior of economic actors is the level of enterprises, but the policy of "scarce" money should discipline the behavior patterns also of ministries and other central institutions.

At least at the present stage the effect of the reforms is a partial abolition of the institutional-legal framework of the CPEs and of the directive-rationing model of control connected with it. As has been shown in the previous section this elimination of some central planning attributes was not accompanied by a sufficient substitution in their place of the market economy mechanisms, and of clear "rules of the game" garanteed by law and independent courts, in a scope which would permit them to influence the functioning of the economy. The resulting system (in Hungary and Poland) is characterized by the widening of the scope of the "free for all" political bargaining for access to resources and "hard" money (i.e. convertible currencies).

With the *partial* introduction of economic reforms, i.e. with harder but not *very hard* budget constraints and without the profound changes in the political decision-making i.e. without the situation when everybody must observe existing laws and plans, bargaining and competition between ministries, and enterprises gets hotter, more pardonless then before.

As the recent experiences in Poland, and Hungary show (and to some extent the same becomes increasingly visible in the Soviet Union), economic reforms introduced there force enterprises to achieve maximal financial effects, without ensuring them the real freedom to take most of the decisions i.e. to choose their suppliers and buyers, to calculate and select input and output compositions and levels.

On the one side profit has indeed become the main measuring yardstick for the performance of the enterprises; salaries and wages depend to a higher degree than before on the financial results of the enterprise (on the rules for wage formation in the Polish enterprises since 1982 see DZIEMBOWSKI, 1983, on the principles valid in the Hungarian system comp. CASSEL, 1983, the new rules in the Soviet Union are reviewed in KOSTA, 1984).

On the other side, in the daily practice, the freedom of choice of the enterprises is severly limited by lawful and unlawful means. The lawful, but not market-conform measures include forcing the enterprises to participate in the government "market-supply" programs, obliging them to produce certain goods in prescribed quantities. Another legal form of reducing the freedom of decision-taking by managements of enterprises concerns production for defence requirements. It is well known that almost everything can be related to the defence sector in the CPEs. It must be also mentioned, that very often the managers of the involved enterprises are not unhappy about being told to participate in such programs. It releases them of the responsibility to take hard decisions, lessens the risk connected with the market economy, ensures better access to the still numerous centrally allocated inputs.

Simultaneously, the managements of the state, and to lesser extent even private enterprises, are confronted daily with numerous attempts to influence their decisions, pursued outside the legal channels, in an informal manner, by highly placed party officials controlling rationing channels for inputs, hard currencies and disposing of all sorts of patronage (KOWALSKI, 1983; WINIECKI, 1988). In this respect the changes that took place in the CPEs as result of reforms are still very superficial.

As a consequence, enterprises which must achieve profits, but are not free in the usual economic sense to pursue their goals, try harder then before to benefit from the disposable network of contacts, informal links, possibilities of access to political leaders etc. To the traditional motive for behavior of managements of the enterprises, i.e. to power hunger coupled with the wish to lead comfortable life, an additional element of economic coercion has been added. The crux of the matter is however that as long as the political decision-making remains as it is, and the budget constraints are not very hard but only hardened, the degree of their hardness being to a large extent politically and not economically determined, this strengthening of the economic coercion, forcing enterprises to report profits will in all probability lead to exacerbation of the mechanisms of informal decision making and influence peddling.

In the same time it will result in very strong inflationary tendencies. Rising prices of ones products is the easiest way to achieve short-term profits in a shortage economy. It has been mentioned above that a considerable part of prices is state-controlled even after the partial introduction of economic reforms. This could lead to the conclusion that these price increases can be held in check. It should however be born in mind, that it has always been possible for the enterprises to justify their applications to the state price commission to increase their prices by the fact that prices of inputs were rising. Also, the above mentioned political intermingling of the nomenklatura groups makes it as a rule very difficult for the price authorities not to grant the desired price changes to the applicants.

Discarding for a moment our reserve with respect to predictions for the near future developments in the Soviet Union and other CPEs, it seems that in order to eliminate the political, voluntaristic decision-making in economic matters, such profound and far-reaching changes in the political superstructure of these countries are needed, that in the short and medium time perspective we can safely express scepticism with regard to the prospects of success in this area.

4. Consequences of Economic Reforms for Regional Development

In the first approximation it would seem that by diminishing the insatiable demand for inputs and investment expenditures, i.e. by curbing the expansion drive regardless of costs involved, the economic reforms should contribute to the lessening of the polarization tendencies between regions. In reality, as long as the economic reforms in the CPEs do not succeed in an almost complete hardening of the budget constraints of the firms (and they do not) and as long as they are not accompanied by changes in the ways important decisions are taken at the central level and with respect to the individual enterprises,so that they would not transgress written laws and would not be taken on an ad hoc voluntaristic basis, the effects of these reforms are apt to even strengthen sectoral and spatial inequalities in the CPEs.

From the spatial perspective this means that the partial introduction of economic reforms will:

(i) Strengthen manufacturing sectors versus regions. Top officials from ministries responsible for manufacturing, as well as the top managers will be able to justify their disregard for regional interests and interests of the spatial organization of the country by quoting economic necessities forced upon them by economic reforms (GORZELAK, 1989). This means that these regions which already record higher levels of manufacturing activities, especially these activities which traditionally dominate in the economic structure of the CPEs, like heavy machines, mining, defence production can expect to see their advantageous position strengthened against the weak agriculturally dominated regions.

(ii) It will also strengthen strong versus weak regions, through all the mechanisms leading to spatial polarization like the distribution of investment expenditures for infrastructure projects favouring strong regions, better provision of inputs for current production to these strong regions etc.

Thus, in the short and medium term perspective our forcast is rather pessimistic with respect to the possibilities of bringing about a reversal in the polarization tendencies in CPEs. In the long run the success or failure to implement the privatization efforts in the Eastern European economies will determine future spatial developments. But we should always bear in mind the famous reflection by JOHN MAYNARD KEYNES about the long run, which seems especially apt when applied to the reform prospects in CPEs.

Post scriptum: Recent events in Eastern Europe show that History is moving faster than most of us expected. It seems that there is indeed a chance that constitutional states, where written laws are observed, can be established in Poland, Hungary, German Democratic Republic and elsewhere. It remains to be seen if political changes and the transformation of economic systems will proceed at the same pace. If, as seems probable, the economic reforms will lag behind, polarization in space will continue.

LITERATURE

Cassel, G. (ed), 1984: Wirtschaftspolitik im Systemvergleich, Verlag Vahlen, München.

Dobias, P., 1986: Die Wirtschaftssysteme Osteuropas, Darmstadt.

Demko, G., 1984: Regional Development - Problems and Policies in Eastern and Western Europe, Croom Helm, Beckenham, Kent.

Dziembowski, Z., 1983: Przedsiebiorstwo a reforma gospodarcza, (Enterprises and economic reforms), Ksiazka i Wiedza, Warszawa.

Dziewonski, K.; Malisz, B., 1978: Przeksztalcenia przetrzennogospodarczej struktury kraju, (Evolutions in the spatial-economic structure of the country), Studia Kpzkpan, vol. 62.

Gorzelak, G., 1989: Rozwoj regionalny Polski w warunkach kryzysu i reformy, (Regional development in Poland in the conditions of crisis and reforms), Uniwersytet Warszawski, Instytut Gospodarski Przestrzennej, Warszawa.

Kornai, J., 1980: Economics of Shortages, North Holland-Elsevier, Amsterdam.

Kornai, J., 1982: Growth, Shortage and Efficiency, University of California Press, Berkeley, Cal.

Kornai, J., 1986: The Hungarian reform process: visions, hopes and reality, Journal of Econ. Literature, XXIV, Dec. 1986, pp. 1687-1737.

Kosta, J., 1984: Wirtschaftssysteme des realen Sozialismus: Probleme und Alternativen, Köln.

Kosta, J., 1989: The Soviet reform concept and its impact on economic reforms in Eastern Europe, Vierteljahresberichte des Forschungsinstituts der Friedrich-Ebert Stiftung, No. 115, pp. 45-57.

Kowalski, J., 1983: On the relevance of the concept of the Centrally Planned Economies, Jahrbuch für Sozialwissenschaft, 2/83, pp. 255-266.

Kowalski, J., 1986a: Regional conflicts in Poland: spatial polarisation in a centrally planned economy, Environment and Planning A, 5/1986, pp. 599-619.

Kowalski, J., 1986b: Rational expectations of economic agents in Centrally Planned Economies, Jahrbuch für Sozialwissenschaft, Heft 1, pp. 84-104.

Kowalski, J., 1988: Expectations, shortages and behaviour, Jahrbuch für Sozialwissenschaft, Heft 3, 1988, pp. 376-384.

Kowalski, J., 1989: Spatial polarization processes in Soviet-type economies with voluntaristic decision-making: the Polish case, Manuscript, University of Karlsruhe.

Laski, K., 1985: Wirtschaftsreformen in Ost-Europa, Vierteljahreshefte zur Wirtschaftsforschung, DIW, Berlin, Heft 2.

Lavoi, D., 1985: Rivalry and Central Planning, Cambridge Univ. Press, Cambridge/Mass.

Rostowski, J., 1989: The decay of socialism and the growth of private enterprise in Poland, Soviet Studies, vol. XLI, No. 2, pp. 194-214.

Winiecki, J., 1988: The Distorted World of Soviet-type Economies, Routledge and Kegan Paul, London, New York.

A Theory Explaining Lösch–Cycles

by

Bernhard Felderer

1. Introduction

ROLF FUNCK was always very clear when he was talking about AUGUST LÖSCH. He considered him as one of the most important German economists who was not only pioneer in the field of regional science but who made also important contributions in several other areas of economics. With the book "Bevölkerungswellen und Wechsellagen" (LÖSCH, 1936) he presented pathbreaking research in demographical economics. As the book was never translated into English the knowledge of his empirical research and the theoretical explanations for his findings on the relation of population changes and business–cycles remained restricted to German students.

Method and content of his research is similar to the work of Simon KUZNETS. A number of new ideas for which KUZNETS has been credited later can be found in LÖSCH's book (FELDERER, 1986).

LÖSCH's work was first of all empirical. He investigated the data and interpreted his results usually by advancing some plausible theory explaining his findings. But as the relations between population and economic cycles and growth are very complex and interdependent, many of his explanations can be reinterpreted from the point of view of modern economic theory. In the following paper I will first present the empirical evidence and interpretations of LÖSCH and then present a consistent model of an economy where population and economic growth vary endogenously. So LÖSCH's "Bevölkerungswelle" (population cycle) needs not necessarily be a consequence of exogenous disturbances. The difference between the agricultural and industrial economy emphasized by LÖSCH can be shown to be structurally not completely different but a consequence of preference changes within the same model.

2. LÖSCH-Cycles

In his book LÖSCH presents data refering especially to Germany. He shows the existence of birth cycles since the wars against Napoleon. He points out that the cyclical decline in the birth rate is partly due to previous events like wars, famine and plagues.

This reduction of birth leads to another decline in the absolute number of birth about 33-34 years later. This echoeffect weakens after each generation and is superposed by new events influencing the birth rate. Till the end of the 19th century LÖSCH's data display a very good correlation between crops, births and marriages for Sweden and Germany. This holds for the absolute quantities of crops but also for the yield per hectar and the respective growth rates. In other words the data support the hypothesis of a positive correlation between technical progress and population growth. Another interesting finding is that the probability of marriage increases when the labour force grows. As the probability of marriage and the birth rate are closely related LÖSCH finds for the time up to the end of the 19th century, a period which he labels "agricultural economy", empirical support for the following causal relationship: When the economic situation improves, marriages and births increase. Per-capita income does not, as neoclassical theory would suggest, decrease but increase as efficiency increases. This again accelerates population growth. As a consequence the speed of technical progress cannot follow the increasing rate of population growth: Production reaches an upper bound given by ressource restrictions. This introduces a downward movement of marriages, births and income.

This modified Malthusian cycle, which I will call LÖSCH-cycle, is according to LÖSCH characteristic for the agricultural economy. In an industrial economy, where the share of the industrial production dominates the share of agriculture, the causal relationships of the LÖSCH-cycle continues to exist. But LÖSCH points out that industrial production cannot have the same close relation to marriages and birth rates. An increase in the number of workers will not automatically augment production in the short run, because an entrepreneur must decide to hire workers. Consequently business cycles may exist independently from changes in the labour force.

LÖSCH has great merits in emphasizing the positive correlation between population growth and technical progress. Though he was not the first who advanced this idea, the degree of precision and the empirical evidence presented make his contribution outstanding. Classical economists had generally assumed that per-capita-income will always oscillate around the subsistence level. Population growth must necessarily depress per-capita-income. This position was justified by the law of diminishing returns in production and was made most popular by MALTHUS. Neoclassical theory of economic growth in the nineteenfifties and -sixties was still a Malthusian theory in this sense. KUZNETS (1958, 1965) seems to be the first economist after LÖSCH who found empirical evidence for the dependence of technical progress on the population variable. Later HAYAMI and RUTTAN (1971); BEN-ZION and RUTTAN (1978); BOSERUP (1981); FELDERER and SAUGA (1988) added new results and ideas.

However, technical progress is a long run phenomenon and should be used for the explanation of cycles, even if they are 30 - 35 years long, only with great caution. Obviously the existence of a LÖSCH-cycle is not necessarily based on the assumption of technical progress as long as accelerating population growth is assumed.

On the following pages I will develop a model which can explain the cyclical variations of fertility without assuming technical progress. But, of course, progress can be added to make the model more realistic. The variations of fertility can be shown to be the result of rational individual decisions on the basis of given preferences. But the model can do better: It is possible to show that the difference between the agricultural and the industrial economy discerned by LÖSCH is a consequence of changing preferences which adopted to changing social conditions.

3. The Model

I have recently proposed a model which can produce endogenous cycles very similar to those discovered and described by LÖSCH (FELDERER, 1989). It is based on microeconomic decisions of a typical household and a representative firm. The assumption of rational behaviour of the household and the firm gives rise to the behavioural functions of both representative agents. These functions serve as the basis for the macro-model.

I then show the existence of a steady state solution. A graphical exposition demonstrates the most important comparative static properties. Afterwards (Part IV.) I will explain how cyclical fluctuations will be generated.

The utility function of the representative individual is defined as

$$U = U\ (c_t^1,\ c_{t+1}^2,\ e_t) \qquad (1)$$

where

$$e_t = \frac{N_{t+1}}{N_t}$$

N_t = size of labour force in time period t

c_t^1 = consumption in the active period of life in time period t

c_{t+1}^2 = consumption in the retirement period in time period t+1

U (...) is assumed to be strictly monotonically increasing and strictly quasi-concave; c_t^1, c_{t+1}^2 and e_t are normal goods.

The concept of the model is a simple life-cycle in discrete time. We do not take account of any government activities. The representative individual or household lives over two time periods: One in which he raises children and works receiving labour income and another in which he is retired and lives on his savings and possibly on subsidies from his children. Altruism in the sense of BARRO (1974) or BECKER and BARRO (1988); BARRO and BECKER (1989) is not assumed but could be taken care of by including the utility of children in the utility function of the parents.

The utility maximizing household has to consider two restrictions:

$$c_t^1 + s_t + q_t \cdot e_t = w_t - b_t \qquad (2)$$

and

$$c_t^2+1 = (1+r_t) \cdot s_t + b_{t+1} \cdot e_t \qquad (3)$$

s_t = savings of the representative household; we assume $s_t \geq 0$;

q_t = "quality" of investment cost for one child; q_t is exogenous in this version of the model

w_t = labour income in period t

b_t = subsidies payed to parents

r_t = rate of interest

b_{t+1} = subsidies received from children during retirement

$\frac{b_{t+1}}{q_t} - 1$ = rate of return on investment in children;

we assume

$$\frac{b_{t+1}}{q_t} - 1 \geq 0.$$

c_t^1 and c_{t+1}^2 are the representative or average consumption of all families or households. e_t is the average number of children in period t: N_{t+1} is the absolute number of children in period t equal to the population in period t+1. b_t and b_{t+1} are exogenous and not related to each other. We simply assume the possibility of transfer payments between generations.

Children are considered to be a consumption good, because they are an argument in the utility function, but they are also an investment good, because they can provide income during retirement.

Maximizing utility function (1) subject to restrictions (2) and (3) yields the following demand functions:

$$N_{t+1} = e_t \underset{(-)}{(q_t}, \underset{(+)}{b_{t+1}}, \underset{(\pm)}{r_t}, \underset{(+)}{w_t)} \cdot N_t \qquad (4)$$

The signs of the first derivatives have been calculated by comparative statics. In the case of the interest rate income and substitution effect have opposite signs.

Population in period t+1 or the supply of labour in the next period is determined by the present size of population multiplied by e_t. If e_t is equal to one and constant the population is stationary. e_t is depending on different parameters, which are given to the household. However, w_t and r_t are endogenous variables in the macroeconomic model demonstrated below.

Maximizing function (1) subject to (2) and (3) yields another function:

$$S_t = \underset{(\pm)}{s_t (q_t}, \underset{(-)}{b_{t+1}}, \underset{(\pm)}{r_t}, \underset{(+)}{w_t}) \cdot N_t \tag{5}$$

As s_t stands for savings per household/family, $N_t \cdot s_t$ is total savings S_t.

To obtain an interior solution the coefficients in this model cannot assume any arbitrary value. In particular the assumptions of the model imply an interesting relation. The first order conditions of maximizing function (1) subject to (2) and (3) gives:

$$\frac{\partial U}{\partial e_t} = \frac{\partial U}{\partial c_t^1} \left(q_t - \frac{b_{t+1}}{1+r_t}\right) \tag{6}$$

As both partial derivative in (6) are positive by assumption, we obtain:

$$q_t - \frac{b_{t+1}}{1+r_t} > 0$$

or

$$r_t > \frac{b_{t+1}}{q_t} - 1 \tag{7}$$

We can easily see that if

$$r > \frac{b_{t+1}}{q_t} - 1$$

savings will be greater than zero. But if

$$r \leq \frac{b_{t+1}}{q_t} - 1 \tag{8}$$

savings will be equal to zero. If equation (8) holds retirement income will be provided only by children.

To close the model we need assumptions about the representative firm. We state:

$$Y_{t+1} = F(N_{t+1}, K_t) \tag{9}$$

A change in the size of the labour force in period t+1 will affect production in the same period. But the firm has to decide on its capital input one period earlier: A new capital stock can only be effective for production in the next period.

The decision on N_{t+1} by the household is also taken in period t. Thus the decision on both factors of production are made one period before they are used as inputs in the production process.

The production function is assumed to be linear homogeneous in both factors and has positive first partial derivatives and negative second partial derivatives with respect to both factors of production. The firm is assumed to maximize profits π:

$$\pi_{t+1} = F\ (N_{t+1}, K_t) - w_{t+1}\ N_{t+1} - r_t\ K_t \tag{10}$$

The usual necessary conditions for a maximum give demand functions of the firm for both factors of production.

The behavioural functions of the representative agents can now be used as the basis for the macroeconomic model. From equation (9) we obtain:

$$\frac{Y_t}{N_t} = Y_t = f\ (k_t) \qquad \text{where } k_t = \frac{K_{t-1}}{N_t} \tag{11}$$

If

$$\Delta\, k_t = k_{t+1} - k_t = \frac{K_t}{N_{t+1}} - \frac{K_{t-1}}{N_t} \tag{12}$$

or

$$K_t = (\Delta\, k_t + k_t) \cdot N_{t+1} \tag{13}$$

and assuming equilibrium on the capital market ($s_t = k_t$) we can substitute the demand functions (4) and (5) into equation (13). We thus have:

$$\Delta\, k_t = \frac{s_t}{e_t} - k_t \tag{14}$$

If in steady state $\Delta\, k_t$ is equal to zero, equation (11) becomes:

$$s_t\ (r_t(k_t^*),\ w_t(k_t^*)) = e_t\ (r_t(k_t^*),\ w_t(k_t^*)) \cdot k_t^* \tag{15}$$

where k_t* ist the steady state value of k_t.
The relation of s_t and e_t with r_t and w_t is explained above, those of r_t and w_t with k_t are derived by maximizing equation (7).

The existence of a steady state solution of equation

$$s_t(k_t^*) = e_t\ (k_t^*) \cdot k_t^* \tag{16}$$

can be shown as follows:
e_t (k_t) y k_t is a monotonically increasing continuous function in k:

$$\frac{d\ (e(k) \cdot k)}{dk} = e'(k) \cdot \underset{(+)}{k} + e > 0 \tag{17}$$

We assume e'(k) to be positive because in equation (15) $\partial w/\partial k > 0$ and $\partial e/\partial w > 0$. Furthermore $\partial r/\partial k < 0$, and we assume $\partial e/\partial r < 0$. This assumption can be justified as follows: An increase in k will decrease the rate of interest and increase investment in children. But the income effect has an opposite sign: A decline in the interest rate has a negative effect on fertility. However, as the decline of the rate of interest is a consequence of a high capital-per-capita, it is save to assume that this secondary effect will be dominated by the original income effect.

Furthermore we assume that for some k

$$s_t > e_t k_t \tag{18}$$

This is a necessary condition for existence of an economy because if s is never greater than ek there is no capital-per-capita in the long run.

With increasing k the interest rate will decline. When the rate of interest comes down to $b_{t+1}/q_t - 1$ savings will vanish as we have shown above. We can thus conclude that the positive slope of ek and the final decline of s to zero guarantee the existence of a steady state.

Equation (16) can be depicted as follows:

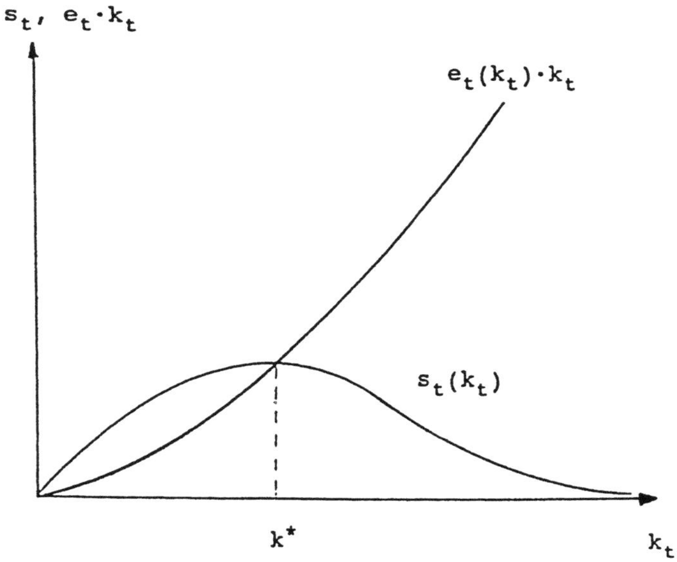

If e_t is assumed to be constant, such that

$$e_t = e = 1 + n, \qquad \text{where } n = \frac{N_{t+1} - N_t}{N_t} \qquad (19)$$

equation (16) is identical to equation (13) of a model recently published (FELDERER, 1988). In this model the savings decision was defined to be endogenous while the rate of population growth remained exogenous.

4. Explaining LÖSCH-Cycles

We are now back to LÖSCH's hypothesis explaining his empirical findings. He can show that changes in the number of births are caused by changes in marriages (this may not always be true in an industrial economy). When eliminating the effects of age-structure, that is also the effects of previous catastrophic events, changes in marriages (births) are a consequence of a changing economic situation. Increments of per-capita-income lead to accelerating population growth which will finally be stopped by some sort of upper bound given by the limits of production and by the incapability of technical progress to keep up with population growth.

We have now to examine the question whether the model presented above can explain the cycles discovered by LÖSCH and whether LÖSCH's explanation is still valid.

If in the model above we increase exogenously capital-per-capita and consequently per-capita-income fertility will be up. But at the same time also savings and capital-formation will increase.

Assuming again equilibrium on the capital market we have from equation (11):

$$k_t = \frac{K_{t-1}}{N_t} = \frac{S_{t-1}}{N_t} = \frac{S_{t-1}/N_{t-1}}{N_t\ /N_{t-1}} = \frac{s_{t-1}}{e_{t-1}} \tag{20}$$

Capital intensity k in period t is equal to the ratio of per-capita savings to the average number of children in period t-1. We know that

$$y_t = f(k_t) = f\left(\frac{s_{t-1}\ (w_{t-1}(y_{t-1}), r_{t-1}(y_{t-1})\ )}{e_{t-1}\ (w_{t-1}(y_{r-1}), r_{t-1}(y_{t-1})\ )}\right) \tag{21}$$

$$= g\ (y_{t-1})$$

We thus have a causal relationship in time between all y_t or k_t for $t = 1, 2, \ldots, n$:

A given capital intensity k_{t-1} in period t-1 leads to a certain y_{t-1}. Both s_{t-1} and e_{t-1} react positively to changes in w_t or y_{t-1} but probably with different intensity. The elasticities of s_{t-1} and e_{t-1} depend on the specific shape of the utility function.

There are three possibilities:

$$\frac{d\ k_t}{d\ y_{t-1}} = 0 \tag{22}$$

In this special case both savings and reproduction react with the same elasticity. Thus k_t remains unchanged in time.

or

$$\frac{d\ k_t}{s\ y_{t-1}} \geq 0 \qquad (23)$$

This is the case if savings react stronger than reproduction.

$$\frac{d\ k_t}{d\ y_{t-1}} \leq 0 \qquad (24)$$

Here the elasticity of reproduction with respect to income is greater than the elasticity of savings to income.

The answer to the question whether LÖSCH's theory is still valid is now easy. If capital intensity of period t does not react to changes in income of period t-1 (equation 22) there cannot be any fertility waves because capital formation and procreation react with the same elasticity with respect to income changes. If inequality (23) holds, fertility cannot endogenously decline. If in this case capital intensity and consequently per-capita-income is cut down exogenously, fertility will decline only in the first period because of this intervention. Afterwards there will be no further decline in fertility. So, if the utility function of the representative household can adequately be described by inequality (23), there will be no endogenous fertility waves. Only in the case of inequality (24) endogenous fertility waves are possible. As a decrease in capital intensity and income in period t-1 will lead to an increase of capital intensity in period t, fertility will go up and down with each generation. The birth rate will be low every second generation.

In LÖSCH's explanation of the cycle the accelerated increase of fertility is stopped by the limits of production. This sight corresponds in the model to inequality (24): In the model reproduction reacts stronger to income changes than capital formation does. This situation corresponded typically to reality till the end of the 19th century and represents therefore an economy which LÖSCH labeled as "agricultural". From LÖSCH's empirical findings emerged his conclusion that in "agricultural" economies fertility waves do exist and they are closely related to marriages and income. The reason why in "agricultural" economies the elasticity of fertility was higher than the elasticity of savings is simple: When income increased people also wanted to increase their income and security when they were old. As for most people the capital market was not accessible, the only investment for the old age were children.

LÖSCH also found out that in the "industrial" economy of the 20th century fertility waves are not so closely related to income and marriage. The LÖSCH-cycle became weaker and disappeared and the long-term decline of fertility dominated the statistical picture. This description corresponds to inequality (24) which is typical for a modern economy because the elasticity of savings with respect to income is greater than the elasticity of reproduction. This change in the utility function of the modern representative household can easily be explained: An increase in income does increase savings and capital formation to have a higher income when retired. But as children have lost their investment good property, increases in fertility will be relatively small.

We can consequently conclude that LÖSCH's theoretical explanations of his empirical findings are still a valid description of fertility waves. Modern economic theory with its logical power supports his hypothesis.

LITERATURE

Barro, R.J., 1974: Are Government Bonds Net Wealth, Journal of Political Economy, LXXXII.

Barro, R.J.; G.S. Becker, 1989: Fertility Choice in a Model of Economic Growth, in: Econometrica, Vol. 57, No. 2, March, 1989.

Becker, G.S.; R.J. Barro, 1988: A Reformulation of the Economic Theory of Fertility, The Quarterly Journal of Economics, CIII.

Ben-Zion, U.; V.W. Ruttan, 1978: Aggregate Demand and the Rate of Technical Change, in: Binswanger, Hans P. and Vernom W. Ruttan (Eds.), Induced Innovation, Technology, Institutions and Development, Baltimore and London.

Boserup, E., 1981: Population and Technology, Oxford

Felderer, B., 1986: August LÖSCH on Population Waves, in: R. Funck and Kuklinski (Eds.), A Tribute to August LÖSCH, Karlsruhe.

Felderer, B.; M. Sauga, 1988: Bevölkerung und Wirtschaftsentwicklung, Frankfurt.

Felderer, B., 1988: The Existence of a Neoclassical Steady State when Population Growth is Negative, Journal of Economics - Zeitschrift für Nationalökonomie, Vol. 48.

Felderer, B., 1989: Neoclassical Growth with Micro-foundations, discussion paper, University of Cologne.

Hayami, Y.; V.W. Ruttan, 1971: Agricultural Development, Baltimore and London.

Kuznets, S., 1958: Long Swings in the Growth of Population and in Related Economic Variables, in: Proceedings of the Americal Philosophical Society, Vol. 102.

Kuznets, S., 1965: Population and Aggregate Output, in: Kuznets, S., Economic Growth a

Lösch, A., 1936: Bevölkerungswellen und Wechsellagen, Jena.

August Lösch as a Population Analyst

by

Rainer Mackensen*

1. Introduction

Regional science and, particularly, location theory has been heavily influenced by AUGUST LÖSCH, and his name has become well known to all students dealing with regional analysis and theory. They are also well aware in general of his studies on interactions between population development and economic cycles, but may be less interested in this topic. The interests in space and in time must not coincide in every scholar. At least, while the literature on LÖSCH's theorems in regional economics is abundant, and quotations on them can be met almost everywhere in regional science contexts, the reaction on LÖSCH's writings by population analysts must be regarded less than meagre. MACKENROTH (1953, 506) mentions LÖSCH only twice (109,

* I have been long interested to critically evaluate LÖSCH's population writings and introduce them into the population literature where he is practically absent. This volume in honour of ROLF FUNCK is a wellcome opportunity for me to finally do so at least to some extent, as ROLF FUNCK must be considered as one of the outstanding revivalists of LÖSCH's regional science writings, and I had a chance to accompany him in many opportunities in this context. It was also in Heidenheim, the locus of most of LÖSCH's studies, and of FUNCK's frequent organization of "August Lösch-Tage", where I have had the opportunity to meet Mrs. M. KNKELE-LÖSCH, LÖSCH's widow, and the chance to lend from her some of LÖSCH's writings and private notes which are more difficult or even impossible to get at in libraries. I want to thank her at this opportunity for supporting my interests.

123), and only as a source for factual information (as in III, 1955, once, and only in the bibliography, concerning "special peaces of research, of particular importance": p.36); and neither W.G. HOFFMANN - who knew him like MACKENROTH from the Kiel Institute - nor HOHORST mention him at all, even while pursuing topics which LÖSCH must have, more or less, introduced.

It is true that LÖSCH's studies on population problems were his first, published mainly at the age of 26 to 32, and his regional studies are characteristic more for his later period, published from the age of 32 onwards, the main d'oeuvre at 34. Thus, one explanation for such different reactions by the scientific community might be hypothesized to be his relative youth when dealing with population, and his more advanced abilities when dealing with regional analysis. But then, he did not have the chance to continue, as he died with 39 already, in 1945. And the age difference is too small to attribute the obvious difference in reaction to his studies only to his scientific development during these few years. The main d'oeuvre of LÖSCH was wellcomed by W. EUCKEN (his Freiburg professor: see LÖSCH 24, 68; ZOTTMANN 1949, STOLPER, 1959 FUNCK-PARR, 1978; v. BÖVENTER, 1987) immediately as "an eminent achievement" (LÖSCH 24, 101) and the theses of him have been taken up by a long series of eminent scholars. He cannot have been much less intelligent and creative two years earlier, particularly as LÖSCH applies the same economic know-how and analytical approach to both realms of inquiry.

The evaluation might also have changed over time. ZOTTMANN, who contributed a memorial article on LÖSCH to the journal of the Institut für Weltwirtschaft, where LÖSCH had been employed from 1940 onwards, in 1949 (only: it was the first volume after the war: ZOTTMANN, 1971, 23), regarded LÖSCH there primarily as a population scientist, and introduced the paragraph on his regional theory with the somewhat diminishing notion, "But LÖSCH has improved modern theory also in a second field ...". STOLPER, in his 1953 introduction to the "Räumliche Ordnung" (see 1971, 61) devoted quite some space to the explanation of LÖSCH's population studies, which he explicitly praised. And in his biographical article (1959), STOLPER devoted almost the same space to LÖSCH's population studies as to the regional theory. Later, in his 1971 sketch, p. 30, also ZOTTMANN evaluated the "Räumliche Ordnung der Wirtschaft" as his "undoubtedly most important contribution to economic science". But K. SCHILLER (in RIEGGER 1971, 13-15); FUNCK-PARR (1978); v.

BÖVENTER (1987), in their biographical articels, hardly mention LÖSCH's population studies at all.

The following discussion pleades for a different explanation. "Die räumliche Ordnung der Wirtschaft" of 1940 is a much larger manuscript, resting on more than four years of statistical and theoretical investigations, in comparison to his relatively spontaneous early study on the "Geburtenrückgang" (1932, a HELFFERICH prize study - see PEARL, "notes" - the reward of which he used to get the study being printed - see ZOTTMANN, 1971, 29; the second volume, obviously, served as a Bonn doctoral dissertation, as the official list of dissertations in 1932 of the University of Bonn suggests), and "Bevölkerungswellen und Wechsellagen" of 1936, his venia legendi theses, which he had only about two years to prepare. So there is not only the difference in age and advancement, but also in intensity devoted to a particular topic. But, additionally, it seems me to be not only the different scientific weight or value of LÖSCH's contributions to two topical fields which is due for the diverging reactions, but even more so the different states of the respective scientific communities interested in the two topics.

To discuss this hypotheses, I will briefly review the main population studies of LÖSCH, the reactions in the academic literature to them, and the further contributions of LÖSCH to the population discussion of his time. I will then resume his main theses and findings, and consider their actuality. Finally I will consider changes in population theory and analysis from his time to our's, to demonstrate why there has been so little interest in LÖSCH's writings on population during his lifetime, as is equally in the present, in contrast to the quality, and creativity, of his studies, and the continuing substantial value of his contributions.

2. The "Geburtenrückgang" Analysis

The study on the case of birth reduction was published in 1932, but the first - textual - part finished in May 1930 (LÖSCH 24, 68). The main text consists of about 50 pages, and the second volume of almost a hundred pages of notes, appendices, and commentaries (his doctoral dissertation in 1932).

Therefore, it is not a straightforward peace of scientific reasoning, but much more a contribution to a substantial discussion. MACKENROTH called it (1937) "difficult to get at, and little noticed, probably also because of its somewhat unsystematic external form"; with the notion of "external", MACKENROTH presumedly wanted to characterize the organization of the text, in contrast to the organization of thought. The table of contents is not paralleled by the organization of the text itself.

The study is rather complex. It deals with all possible economic and political consequences of the birth reduction in Germany, and also with its causes. The argumentation is chiefly economical, but also political: By demonstrating that the economic impact of population stagnation is rather positive in respect to private and public income, as compared with the economic impact of population growth, he tries to show that also the political situation of Germany will be rather more improved, internally as well as externally, by reduction of births, in contrast to population growth.

To build up his argument, LÖSCH compares the economic and political implications of a static situation of population with those of population growth. After explaining in general the impact of population growth on the emergence of capitalism and on the development of modern technology, he describes the respective consequences of a lack of population growth. Later, he discusses his argument in more detail, demonstrating demographic consequences on foreign trade, on size, distribution, and utilization of the national product. He also enumerates causes of birth reduction, and concludes that these causes operate against the economic consequences, modifying them to a certain degree. – The appendices in the second volume elaborate the proofs for his argument. He shows that in considering economic dynamics he can systematically draw on more details, and more dependably so, than by considering just the possible upper limits of population (like MALTHUS) or the population optimum (like WICKSELL). In discussing different old age security systems, he concludes that the economy is better off with descreasing than with increasing younger generations. In general, he demonstrates, the economy will save by reduction of births, and be induced to more expenditures with growing life expectancy. Therefore, "population decrease means increase of wealth, in any case, generally speaking" (LÖSCH (1) I 60, 69). Consequently, LÖSCH does not consider unemployment as the mayor obstacle or problem. In a special chapter LÖSCH deals with the military power of Germany in respect to decreasing population growth, and argues that it is not the numbers that constitute military power, but economic wealth (l.c. I 56), and thus a state with less population might be in advantage also in war, at least for a certain period (l.c. II 84). This consideration leads him to an European population "review", where he compares population density and growth of nations as indicators for their respective situation and power in a geopolitical perspective. One of his conclusions, with respect to Italy, is that Faschist pronatalist population policy at least risks war (l.c. II 78). As he quotes ROSS: The true enemy of the peace dove is not the eagle of pride or the vulture of greed, but the stork (l.c. II 83). In this context, he also considers international migration, as a consequence of population pressure (l.c. II 77). Discussing the forecast of the national statistical office, he concludes that the prevalent situation of low fertility will in itself have consequences with the effect of regained population growth in the long run (l.c. I 69).

The architecture of the study is one of increasing specification. There is still much more detailed analysis behind the study than LÖSCH ever exposes. LÖSCH gives his conclusions first, and in short, and explaines his reasons later in more detail, mostly in the annexes of the second volume. This does give the impression that the study is primarily motivated politically, and economics is the tool of his argument more than his goal.

Nonetheless, the author gradually generates conclusions which he did not expect, and to some extent not even likes. He says himself (l.c., note 1, II 5) that it was not his intention to write an apology of birth reduction, which he also tended to refuse beforehand; but he had been forced by rational argumentation to accept the contradicting position. Thus, the study had become an argument in opposition to the "Zeitgeist" (l.c. I 9), defending the reduction of births.

LÖSCH's argument is mainly economic reasoning, and it cannot be my task to evaluate or criticise it. I can also not evaluate or criticise his final conclusions. But it is not economic reasoning alone in the study. His political position, which comes to a fore not only in his considerations of the international situation of Germany at that time, is confessedly the one of F. NAUMANN, that is national-liberal. Explicitly he argues in favour of liberal democracy, as well as the idea of the Reich, and of liberal means to reduce social tensions (l.c. II 64f), and these means, he resumes, are all being strengthened by reduced birth rates. QUANTE (1934), however, as an exampel, did not feel absolutely convinced by LÖSCH's argument of economic advantages of the reduced birth rate, although he joins him in criticizing BURGDÖRFER's tendency to condemn the reduction of births alltogether. There is more political thought in the back of rational economic argumentation also than LÖSCH himself would have been ready to admit. Part of it is certainly due to the school of economic thought which he belongs to, characterized by his teachers EUCKEN, SPIETHOFF, and SCHUMPETER, and also by scholars like SCHELER and NAUMANN, whom he tends to quote rather frequently.

Similarly, his interpretation of demographic development is influenced by concepts of M. WEBER; W. SOMBART; J. WOLF: The reduction of the birth rate is due to rational reproductive behaviour (l.c. I 22, passim), allthough it has been mostly a reaction to reduced mortality in the 19th century (l.c., note 33, II 60). LÖSCH's social differentiation is also paralleled to political categories (l.c., notes 4, 32, 40 in II 6, 11, 13). Via SCHELER, there are elements of "Gesinnungsethik" in his sociological considerations (l.c. I 25, 44 ff, passim), which allow him to follow H. HARMSEN in his interpretation of the French demographic development (l.c., note 38, II 13; note 45, II 76). The prominant French discussion of the time (LANDRY, SAUVY) is not explicitly being considered.

The method of his argument is model construction (l.c. I 9,26), or – as a sociologist would tend to call it – an ideal type experiment in thinking. His model is one of the isolated state; there is almost no consideration of foreign trade (except very roughly in the first paragraph of this part, l.c. I 26 f) and of international migration (except in annex 8, as a political argument, l.c. II 73 ff). Therefore, he does not consider the reduction of the birth rate in respect to possible changes in age compositon by migration (except l.c. II 86), but generally regards it as synonymous with reduced population growth, and this, precisely, is the fundamental demographic phenomenon for his economic argumentation.

This does not mean that LÖSCH is not familiar with demographic analysis. He uses mainly publications of the national statistical office (i.e., of BURGDÖRFER), and he does so with competence. On the other hand, his own demographic calculations are relatively simple – in this study, more linear trend extrapolations than differentiating model analyses, and there is little awareness in him of the shortcomings of rough measures.

In this respect, it may be understandable that this study did not influence the general discussion of the problem, allthough it did have the potential to do so by his economic reasoning, had this been taken up. But, as he argues in favour of population decrease, or stagnation at least, it was not in accord to the general oppinion of the time, also among demographers. It would be worthwhile to reconsider his line of argument even more so under present conditions.

3. The "Wechsellagen" Hypothesis

Four years after his first study, LÖSCH presented another one, concentrating on long and short processes, i.e. "population waves" and business cycles. The study has been prepared mainly in Bonn, where it served as a theses for his "habilitation", but probably finished in Cambridge/Mass., where he stayed twice after 1934, and where SCHUMPETER taught since 1932. Although the material had been prepared by "Bonn con-students" (LÖSCH (10) 1), the text might also have been influenced by his American impressions (LÖSCH 24, 86 ff).

Whereas the first study appears quite as a political essay, this is a thorough academic exercise. The logic and sequence of argumentation is clear. The text of 67 pages (including many analytical tables and graphs) is supplemented by another 57 pages of fundamental tables and explanations of calculating procedures. Every step is well documented and can be easily duplicated by the reader.

There are statistical parallels between "population waves" and business cycles (LÖSCH uses the terms of SPIETHOFF, where "Wechsellagen" stands for Juglar cycles, including periods of prosperity and of depression, l.c. 19, 39, note 1; he also consideres "Wechselspannen", i.e. Kondratieffs waves), during the 19th century, demostrated particularly for Sweden and Germany, but also present in other European nations (l.c. 22 ff). The parallels become visible especially after levelling out short term irregularities in the rough birth rate as well as in the annual increase of the economically active population ("Erwerbswillige", i.e. the "Erwerbspersonen" of public German statistics, including the unemployed). Then it became possible also to distinguish two or even three (Sweden; l.c. 4) sequences of overlapping population waves. The German waves start around 1800 (Napoleon wars) and around 1815 (Liberation wars), and continue through all of the 19th century, levelling off after three cycles of about generation length (33 years), each cycle reinforced by political or economic events.

The relation between population and economic waves can be interpreted psychologically: economic activity becomes animated by synchrone activities of many people strengthening their selfconfidence, and the task of integrating unusually many people into economy stimulating higher performance (l.c. 19, 33). There have been proposed even cultural influences of population waves (by PINDER, l.c.), which have some evidence as well in Lsch's own tests (l.c. 32). But this causality can work better in agricultural than in industrial production, as workers have more influence on economic activity in agriculture than in factories (l.c. 20, 36 ff). There is a triangular relationship between marriage and birth frequency, economically active population, and harvest returns (l.c. 29). In agriculture, the correlations between population and harvest returns become even more evident when additional factors are being considered: weather does have direct consequences for harvest returns, but also indirect effects on fertility, mortality, and migration (l.c. 25). And the "Lebensraum" (we would now tend to speak rather of "Tragfähigkeit", carrying capacity) can be enlarged by additionally used land, by increased productivity, and by emigration (26 f), as was the case in certain historical periods in Germany.

There are other correlations as well with both time series, besides climate, such as political events, and it has to be considered how they may act together. Whatever the particular causalities: they all seem to work in the same direction, stimulating or hindering economic activities, as the case may be.

In industrial times (after 1880), population waves may be a condition for economic changes (l.c. 40), but they may have more a function as a relaese then as a causal factor; LSCH shows this by testing industrial wages, production of consumption and production goods, and building activities (l.c. 49

ff). The correlations do not fit well enough each, but by taking all of the economic activities together, he finds a "absolute parallelity" (l.c. 58), with certain cases of time lag, up to 1932. Particularly, the correlation of economic situations with marriage frequency he feels to be convincing (l.c. 62 f), while the correlation with fertility has been loosened by the reduction of the birth rate, i.e. the increasing arbitrariness in the decision between competing opportunities (l.c. 59, 64), and mortality shows no significant correlation.

As a result, LÖSCH states that population changes, particularly in population size and size of the working force, as well as in the number of existing marriages, do effect economic cycles, and economic cycles have an effect on marriages, if also a different one in agriculture and in industry. The effects may not be causal, but more an occasion which can be taken up or not, but for economic upward swings, population growth seems to be the only necessary condition (l.c. 62). LÖSCH felt that WOLF and ÅKERMAN particularly in their quantitative studies come close to his results (l.c. 120).

4. Reactions to LÖSCH's Population Writings

This study did provoke a series of reactions in the literature in its time, and more than the first. Regarding the first study, there have been also a couple of prominant reactions, as by SCHUMPETER (1931: "A husar's attack – he had not seen something as keen on that topic", see LÖSCH (24) 72, and again in an statement from Harvard 1932, see "notes") and PEARL who, interestingly enough, stressed that the "presentation is free from political bias" (1936) which he regarded as "singularly", and pronounced that the conclusions "are not as pessimistic as so many treatises on this subject". MOMBERT stated that "LÖSCH partially argues substantially more fundamental than BURGDÖRFER" (all from "notes"). QUANTE (1934) reviewed the study thoroughly, and more critical. Asserting that he considers BURGDÖRFER's argumentation regarding reduction of births as not convincing, and even questionable, he also states that he feels LÖSCH to tend to overestimate the costs of population growth, because he attributes the formation of capital necessary for the outfit of additional labour solely to the preceding generation while it can be also be produced by the same generation. Nevertheless he agrees with all the arguments stressing the difficulties arising from a period of particularly fast population growth. His main achievement QUANTE regards to be the evaluation of savings of different form stemming from a decline in the birth rate. If not agreeing with LÖSCH in general, QUANTE considers the study to contribute constructively to a debate where more emotions than arguments are being invested, and thus stimulating more analytical work in this field, either pro or contra.

The scientific reaction, as far as I can see, regarding the first study has been accepting in general, particularly because of the multitude of aspects considered, and the originality of the line of thought pursued, in contrast to the main stream of contributions to the topic at the time (LÖSCH's study has become indexed in 1935, see LÖSCH (24) 94). STOLPER (1954, see 1971, 60) called it "probably the best existing essay on the problem, in the style of classical, i.e. pre-Keynesian, theory". – The reactions to the second study seem to be even more sincere and widespread. MOST, NEUMARK, and VORWERCK considered the second study as important, particularly because of the bulk of newly elaborated statistical materials used in the argumentation, J. MÜLLER (in the official review of statistics) appreciated the inventing phantasy seeing connections even where they cannot been proved for the time being (all from "notes"), and also some more practically oriented papers printed consensually referring reviews.

A more detailed review has been published by MACKENROTH in the journal of the Kiel Institute in 1937, i.e. before LÖSCH's employment there. He is, although also critical, enthusiastic in general and agrees with analysis and results. He stresses, particularly, the use of annual increments of growth in the labour force as an indicator, for the psychological mechanisms proposed. The growth in the productive labour force, and in the consuming population, in contrast, can also prevail in periods of stagnating harvest returns, and effects production positively only over longer periods. MACKENROTH is less convinced of the psychological mentalities attributed to subsequent generations by LÖSCH, as the time distance between changes seems too short for such changes in mentality proposed as an explanation. He also counterargues against LÖSCH's presentation of demographic consequences of economic changes (MACKENROTH 1937, 59). Even if births have become less a necessary consequence of marriages their frequency is still influenced by economic states, as D.S. THOMAS (London, 1925; l.c.) had proven for England, and R. HEBERLE (in the "Archiv für Bevölkerungswissenschaft" 1937, p.23 ff; l.c.) for Germany: The correlation of births with economic indicators, in contrast to that of marriages, does increase with industrial development. Therefore, the statement of LÖSCH, that the reduction in the birth rate has mainly other than economic reasons (l.c. 65), must be relativated, more in respect to population cycles than to birth reduction in general, which are of still more importance even under industrial conditions than LÖSCH would be aware of, although his own statistical material proves to the contrary. A differentiation of industrial and rural population would have been helpfull for the analysis, as only industrial population can show the correlation with industrial production correctly. Thus, the critical comments by MACKENROTH are substantial, and continue the argument and analysis as invented by LÖSCH.

More relevant for LÖSCH and, possibly, for further discussions of his topic, had probably been the review by ÅKERMAN (1937 I), published in Schmollers Jahrbuch. LÖSCH (17) got the chance to react to ÅKERMAN's critique, and ÅKERMAN (1937 II) responded "final" to this reaction in the same issue of the journal. ÅKERMAN consisted in his judgement, to see population waves as one of the driving factors in Kondratieff's waves, but not in conjunctioral cycles. He agreed with LÖSCH in the argument that these, produced increasingly by technological developments and capital investment, and decreasingly by changes in population, do have a continuous impact on weddings and births. Both authors presented new materials, and additional arguments from current literature, in their reactions; but ÅKERMAN - whom LÖSCH had mentioned in the introduction to his study - insisted in his point of view and suggested that further research on the topic, rightously taken up by LÖSCH's study, would be helpful, including more arguments from current conjunctions theory. - This exchange of arguments, as well as the invention by QUANTE, do give the impression that LÖSCH's writings this far, although considered intelligent, original and well elaborated, have not totally convinced the economic scholarship of his time. He is being accused of overweighting his empirical and analytical arguments, in the view of other relevant literature available, and kind of too enthusiastic about his positions. Consequently, the subsequent articles on this theme (20; 21) must be considered under this viewpoint of critical reactions against previous statements.

5. Further Contributions of LÖSCH to the Population Debates of his Time

LÖSCH has, beyond the two studies discussed, participated in the current economic population debate, by book reviews and articles. There are, in his bibliographies (ZOTTMANN 1949, and by MERKER in: RIEGGER 1971: the most complete one) five book reviews, and three brief articles between the time of publication of both the studies, and seven articles concerning population after the second study, including the response to ÅKERMAN, all - with one exception - devoted to the topic of the second study, and seven book review articles.

The early brief articles, 1933 and 1934, have been published in the journal "Deutscher Lebensraum - Blätter für deutsche Raum- und Bevölkerungspolitik", which he reviewed in 1933 for Schmollers Jahrbuch (7). There he criticises the tendency of the journal, representing the situation before the first war, in regretting the lack of space, whereas - in LÖSCH's oppinion - the lack of people is going to be the real future problem; demanding new settlements, while the lack of capital is the real reason for unemployment; overestimating possibilities for futher colonialization; criticising capitalism instead of wishing it to function better. "The interconnections of liberal economy have become too subtle and complicated to us, we do mistrust them, and that is why they ceased to function. We long back to more originative, more primitive, less efficient, but more dependable, more easy to survey conditions." These sentences seem to characterize, not only that specific journal, but much of the feelings of that time, and, at the same time, LÖSCH's position against them, and may be able to explain a lot about himself - why he did publish in this joural anyhow, continuously opposing its mainstream thoughts (including BURGDÖRFER's, who used to write quite frequently in that journal); and why he took up the topics he dealt with. Furthermore, both - the general thinking of the time, and the topics necessary to deal with - do have so much similarity with the present situation.

He also pleaded for more simply formulated, but more scientifically founded contributions. So his own small articles must be regarded as this kind of public information. In the first one (2) he reformulated the primary findings of his study on birth reduction: that continuous population growth would have led to even more unemployment, and more need for capital, which is lacking anyhow, that reduced population growth leads to economic relief, and that future political power will not so much depend on numbers of soldiers but on technical equipment, that is capital again.

The next article (3) establishes that a radical change of population development did already take place, and does not have to be invented by politics, which indeed is unable to influence population development very much. The reduction of birth frequency is a symptom of increased individual freedom, under the unfavourable conditions of the past decades; and with improved economic circumstances, the same individual freedom might react in a contrary direction. "Population politics does have many possibilities, ... but the efficiency of such measures is little. The state only acts after the relativity of having children has been widespread. Than political action is not just without effect but meaningless alltogether." Instead of quantitative, qualitative population politics becomes necessary, that is, more concern about education and employment, instead for numbers of offspring.

A minor contribution to the Journal Deutscher Volkswirt in 1933 (7a) reports again on the main findings of his Geburtenrückgang study. He stresses that the growth of population during the 19th century is due to the reduced death rate – the transition hypothesis – and that increases in the birth rate have their cause also mainly in it, by letting more mothers live. And he explaines his theses that more longevity costs more to the national economy than more births. He discusses the possibility to cope wirh these excess costs before and after the war. For further development, he pleades for technological progress as the only source, after the reduction in the birth rate.

In his book reviews, his own position contrasts explicitly with the mainstream argumentation. His first review (4) is on MOMBERT (1932), also in the "Lebensraum" journal. LÖSCH finds that MOMBERT has changed his position, as compared with earlier writings, and now achieved the "clear insight" that "reduction of births = enforced formation of capital = increased purchasing power" (LÖSCH (4) 96). Within his text, though, many remnants of his previous oppinions may still be found, disturbing the clearness of his argument, especially also regarding future developments.

He has reviewed three of BURGDÖRFER's books, one in 1933 (5), and others in 1937 (19) and in 1943 (23). These reviews might be particularly usefull to understand LÖSCH's position and his development during the Nazi regime. The review of "Volk ohne Jugend" (5) has been written before the first study of LÖSCH had been published, but after completion of the manuscript. So he attacks BURGDÖRFER with the results of his own investigation, namely, that ageing costs less, and birth reduction gains more for national economy than BURGDÖRFER has assumed. Increasing costs for old age maintenance must not be accounted to reduced offspring, but to inflation, and to growing life expectancy. The dependency ratio becomes more realistic by calculating economically active against inactive parts of the population, instead just taking the rough age brackets. And education becomes more costly by the time, contrary to BURGDÖRFER's assumption. From these propositions he also questions BURGDÖRFER's forecast. LÖSCH believes that rising conjunction will also improve the birth account.

This is also one of his arguments (in 6) against BALÀS (1932) position. LÖSCH answers BALÀS proposition that population stagnation improves culture whereas growth strengthens power with his hypothesis that modern war is more a capital than a number game, and capital will be consumed by increase. Particularly, LÖSCH questions BALÀS idea of an "international population cartel" mainly on the basis that no population optimum can be set scientifically.

During the later period, after the 1936 publication of the second book, LÖSCH reviewed more writings also of other prominent population scientists. He asserts this in calling W. WINKLER, "the BURGDÖRFER for Austria" (14). The "disastrous reduction of births" reported by WINKLER for Austria, is being surmounted only by France. But WINKLER, "cautiously", separates the cases where economic necessity, in contrast to comfortable life, must be regarded as a cause. With "a clarity rare to population pessimists", WINKLER explanes that "for renewed population growth an enlargement of carrying capacity will be necessary, and can be achieved by simplification of standard of living, by equalization of family burdens, and by organized emigration, for the time when enlargement of production will not yet be possible". LÖSCH wishes for WINKLER's writings in Austria the same distribution as for BURGDÖRFER's in Germany.

Particularly interesting is the review (18) of HAUFE's study (1936) on regional differences in population development in Europe since 1815. HAUFE (who died during the war) distinguished between urban and rural developments in two periods, 1815-1875-1925. Regarding this periodization, LÖSCH notes that the first - the pre-transition period - was more homogeneous than the second - the transitional period, which could have been decomposited into a period of growing increase before 1900, and one of diminishing increase, thereafter. Despite of this sacrifice of still more differentiation, and of several unclear points in calculations, LÖSCH appreciates the results. HAUFE had proven that, during the first period, only a quarter of the increase - in population growth regions, comprizing only one seventh of all of Europes population - went into industry, the rest remained in agricultural zones, where (in contrast also to frequent statements even now) in "Ostelbien" and "Pommerellen", as a consequence of the "Bauernbefreiung", and of additional enlargement of agrarable soil, and plotting, population carriage had been more than doubled. During the second period, four quarters of European population became incorporated into the industrial zones propper, and into the reserve zones around them. But even in this period there had been "enormous rural increase", particularly in Russia. HAUFE also demonstrated that same laws of succession, both "real division" and "total farmsteads succession", operated differently in different regions. The rich material, as appreciated by LÖSCH, should still serve as basic for further investigations into the regional developments in Europe during the 19th century.

Symptomatic for the discussion of the time is also the review (21) on UNGERN-STERNBERG's study (1937) on population developments in Northern Europe. LÖSCH reports that UNGERN-STERNBERG (who obviously had been interested in physical anthropology) had found less people of Nordic Race in these countries than expected, and among them especially the lowest birth rate all over the world - a result which must have been depressing to all the Nazi populationists of the time. Sweden, particularly, he considers to have had much additional carrying capacity in reserve, and indeed immigration surplus over several years. LÖSCH only regrets that UNGERN-STERNBERG, in considering the present population situation, had not been aware of the fact that northern countries, having been neutral over the war, lack the special population wave of the war participating countries.

One of his later review articles (22) covers several publications, and reports agreeably on their empirical contents. It may be interesting for present investigations on population reduction to compare results and considerations on this topic by SIEMERING (1937, a "reference book") or BARTH (1936, an "introduction") where population decrease is being regarded a main source for unemployment. And also to observe the general interest, shared by LÖSCH, in qualitative demographic questions, including biological aspects. But a major point for LÖSCH is obviously the parallel evaluation of the Report of the Committee on Population Problems to the National Ressources Committee, including contributions by LORIMER, THOMPSON, THOMAS, OGBURN a.o., on population decrease and ageing in the United States to be "on the whole beneficial to the life of the nation", as contrasted to the evaluation by German authors. He criticises these for accounting unemployment to birth reduction, and counters with his repeated argument that annual increases, and not changes in live stock, are relevant, and prove with different results, the effects of population changes on economic performance.

In the second BURGDÖRFER review (19: 1937), LÖSCH praises him: "His writings belong among the most impressive and best in our population literature, and need for long no recommendation any longer." BURGDÖRFER, in this study (of 1936) consideres the demographic effects of military service, and the changes in its regulations after 1933, and of the first world war. LÖSCH corrects BURGDÖRFER's calculations on the resulting birth deficit by considering the population waves, and thus reduces the birth deficit by one to three and a half million. He mentions one of BURGDÖRFER's results as particularly noteworthy: In a comparison of men able for military service between Europes nations, he had calculated the German position to be most favourable in 1940.

LÖSCH's latest book review on population matters (23: 1941) is again concerned with a BURGDÖRFER publication (of 1940), dealing with the population losses during both wars in main European countries - a "continuation and confirmation of BURGDÖRFER's former study" of 1936 (comp. LÖSCH 19). BURGDÖRFER, in this study, considers with concern the population loss of the second war against the background of general tendencies towards decrease because of birth reduction in Germany, Britain, and France, in contrast to the effects of the first war in the situation of population growth in spite of the reduction in births beginning in 1877 already. The only ciriticism by LÖSCH is that BURGDÖRFER consideres the losses only, and does not see the continuing effects from population waves.

Some of the articles (LÖSCH 11, 12a, 13, 15, 20) are more or less either short versions of his argument in the study on population waves, or contributions to a discussion stemming from it. The first of these (also from 1936) containes not much new substance, in comparison to the book, but can be regarded as a more compact, and possibly more exact, formulation of his main lines of argumentation (in 10). - The second title in the journal "Geistige Arbeit" stresses his interpolations to reconstruct the time series of the labour force in the 19th century. He reports on his result that "population waves are possibly one of the causes, but not results of conjunctional cycles". - The third of these small articles is just a brief contribution to the formerly FRIEDRICH NAUMANN journal. It is interesting in that it presents (in October 1936) a graph which has not been a part of his book: LÖSCH seems to have been rather fond of not only statistical, but also graphical presentations; many of his graphs are rich in content, and easy to read. They must have been part of his analytical and, equally, theoretical investigations. The graph hes been taken up in his next, American, presentation. LÖSCH, therefore, must have continued to work on the subject, not only using his finished studies for eventual publications. It is a curve showing the annual increase in German population from 1820 to 1930. He has used the index before, applied to the labour force; now he demonstrates that, not the population, but her relative annual increase, represents best his argument that not economy influences reproductive behaviour (his formulation is rather strict in this article) but the other way round, whereas population is more subject to political events. These, as he has argued before, have constituted the offspring of the population waves starting shortly after 1800, and ruling over most of the 19th century, and now have started an new series of waves with the First World War. - The third paper (read at a Chicago meeting in December, 1936) can also be considered a report on his German population analysis in the book to the American auditorium. Noteworthy here though seems the fact that he devotes more attention to PEARL, and also to CORRADO GINI. Both have not even been worth a footnote to him in his original studies. Now he explicitly attacs them both (15, 651 ff) with purely demographic arguments against their both biological perspective and secular views. Although PEARL's logistic curve seems to fit the German data between 1820 and 1920 pretty well, LÖSCH shows with the graph presented in the previous article that this curve is too broad to take account of the underlying swings. - LÖSCH's contribution to the 1937 Paris international congress of population (the one organized as a replica to that one of 1936 in Berlin, presided by H.HARMSEN, which indeed had been in accord to the resolution of the council of the International Union for the Scientific Study of Population Problems under the presidency of R. PEARL - the than IUSSPP - but opposed by a mainly American group which argued that scientific discussions on population were no longer possible in Germany) is again a short version of his main argument from the second study, but concentrated now on the point that populatin stagnation is not the cause of unemployment - the argument explicitly stated as anti-Keynesian. The most profound effect, he says, of population on economy is not through a change in numbers of consumers but on production via capital formation. The paper is too short to sufficiently cover his findings on population waves, birth reduction, and ageing, altogether, which he tries to present briefly at least.

Two deviant demographic topics have been taken up by LÖSCH, in comparison with the two mayor studies, one (1934) on the demographic consequences of the first world war, and one (1936) on the ageing of the German population.

The article (LÖSCH 12) on ageing is particularly interesting from the demographic point of view. In the first part, it presents a strict life table analysis - also proving that LÖSCH had become rather familiar with this instrument. He argues that the ageing of the German population (between 1870 and 1933) has been due mainly to the increase in life expectancy, not (as generally devoted) to the drop in the frequency of births, which was certainly true in this time. The difference to ours can be demonstrated by the fact that in the national statistical office forecast used by LÖSCH, the proportion of people of age 65 and over for 1975 was calculated to be 13,3 p.c. - but was actually 15,7 p.c. in 1980 (for both parts of Germany), and will still increase till about 2020, and not drop after 1980, as expected in that time. Thus, the further reduction in the birth rate, from 1965 onwards, did have a more profound impact on ageing than LÖSCH could have observed (not accounting for the role of World War II and subsequent migration history).

Also of current interest is LÖSCH's consideration of the impossibility to counteract increasing life expectancy by pronatalist politics, as was popular in that time, to compensate for increasing old age burden. He did see - like few at that time - that this idea was rather ridiculous, demographically (as it still is in the present). - Another of his arguments may seem to be more debatable in view also of QUANTE's critique: He expects an increase of national income by way of ageing. One of his arguments may be rather familiar even now: It depends on the economic activity of people beyond 65, which he expects to increase - contrary to factual developments and political majority intention in the present. The validity of the other argument, where he takes up his argument from the first study (II 26 ff) depends on the evaluation of QUANTE's opposing oppinion: He feels that ageing has an positive effect on the accumulation of capital. He also pleads for a program of wages to prevent aged unemployment.

Up till 1938, LÖSCH had published preferredly in Schmollers Jahrbuch, the journal edited by his second Bonn professor SPIETHOFF till 1937 (although he does not seem to have been on too good terms with him, in the beginning at least, because SPIETHOFF wanted to "educate" him, and was also too adjusted politically, in LÖSCH's mind; see LÖSCH 24, 73f); there he introduced his understanding of the reduction of births, in review articles, and also announced his second book with a substantial article (11). From now on, he used the Weltwirtschaftliches Archiv for articles and reviews, starting with an article on the Wechsellagen book after it had been reviewed a couple of times. At this time he was aware of the criticism he had provoked, but had affirmed his own position in the discussion, and with new arguments and materials. He had become more selfconfident; the article (11) is not just a new stilistic formulation, but also a more sovereign general statement of population-economy relations. Population does stimulate "inner" and "external expansions of the Lebensraum"; this concept may be helpfull for later scholars, who tend to understand the "Lebensraum" concept only geopolitically, concerning territorial expansion with constant internal carrying capacity (a Malthusian view), whereas LÖSCH - and others in that time - regard it equivalent to variabel "Tragfähigkeit" and territorial basis. But the stimulation of size or speed of territorial expansion by population change is only additional to other causes, particularly rationalization, and effective evenly on urban and rural economy. "Only", he states, "because accidentally population growth during the last century coincided with the opening up of fertile soil overseas, population growth effected internally mainly urban professions" (l.c. 455). Whereas in agrarian societies more people need more land, in industrial they need more machinery, in any case: "Population growth requires capital" (l.c.). And this holds true for growth through increased life expectancy as well as over population waves, and vice versa also for the reduction of births. His argument is explicitly anti-Keynesian: It is not the change in consuming capacity by population change that does have the main impact on economy, and it is not the number of consumers that count in the long run but their purchasing power which effects production capacity. The population effect on capital formation is much more massive in comparison to the effect on the production of consumer goods, and it may be caused by growth waves produced from age structure as well as from fertility change. The argument is backed by new figures, calculations, and literature contributions, and defended against ÅKERMAN's, and MACKENROTH's questions.

6. LÖSCH's Position in Population Analysis, the Present Population Question, and the Continuity of Science

It is true, LÖSCH sticked to a few topics only during the few years concerned regarding population. The question how birth reduction effects economic development was the topic of his first study, considering the question with model arguments, and pointing to the then present situation. It has been taken up in a broader context in the second. There, he discussed it on an empirical historical basis, discovering in his analysis the population waves. And his findings were also concentrated on a few arguments concerning the ways of impact of population on economy, and vice versa under different circumstances. But these were differentiated enough, and analysed and defended with a bulk of materials built up in own calculations and recalculated again and again.

In handling these topics, though, LÖSCH included all prominent aspects of population analysis – the frequency of births, and of marriages, social differentiation, the prolongation of the life span, effects of ageing, and migration. This, indeed, he tackled in international contexts only; internal migration was to become an argument in his later studies on location. His interest was not directly into population analysis, but into the effects of population changes on economy. While he discussed this in detail, he had nevertheless to touch on all aspects of population change.

In doing so, it is obvious that LÖSCH in the beginning just took advantage of published materials and only the minimum necessary recalculations arguing also with rough ratios. But, while he intruded the topic more and more thoroughly, he also became familiar with more elaborate demographic measuring, and calculations. He introduced specifications, not usual in demography, such as relative annual increase in births, and in economic active population, to be able to more specifically demonstrate his causes. And he tended, well educated in economic analysis as he was, to select and calculate those enumerators which seemed to be most relevant for the interrelations he wanted to argue with. Thus, without being a demographer in the narrow sense, he still became a population analyst of high capacity.

He dealt with phenomena of actuality in his time, but no less relevant to ours, even if under changed conditions. Birth reduction, and ageing, financial consequences of both, effects of international migration and causes of unemployment are really the main topics for population science also now. And he did contribute observations, and explanations to all these questions. They must be considered relevant also nowadays, and analysed with scrutiny, which is not always done.

It cannot been overlooked that LÖSCH did insist on positions which have not been accepted by the mayority of scholars dealing with similar questions in his time. But it is also true that the empirical proofs for his positions have been accepted to a certain degree, and no better proofs for adversial positions offered either. The core of the controversial discussions LÖSCH entered was not so much a scientific one, although he tried to make it such, but a political point of view. His constant energetic trials to search for the rational argument behind his beliefs was, and is always, one of the prominent sources for scientific inquiry, in the social sciences at least. And the debate has not been finished, the rational argument, and empirical proof not been investigated and discussed to a point where all scientists concerned could have agreed on the proofs, as were established, and on the points not to be proved analytically.

The reason is obvious. It was the time of the establishment of the Nazi regime. And more or less all of those scientists involved in demographic analysis have been also involved in Nazi ideology to a higher or lesser degree. It is fascinating to observe, in following the ways of LÖSCH, how the differentiation emerged over time. In the beginning, say: around 1933, there might have been little difference among the social scientists dealing with questions of national concern. But in establishing their positions gradually, the differentiation became more and more clear and obvious, scientifically and politically. So LÖSCH did not have his chance in his own time, and he died immediately after the war, so he did not get his chance even thereafter.

As a population analyst LÖSCH became forgotten. Not because of the irrelevance of his writings, but because of reading cartels: he was not in. Those population scientists who pursued the topic after the war had been the same who would have refused him before, and on political grounds mainly, not on analytical ones. And as nobody took him up afterwards, and he was not able to continue to defend his arguments, he stayed forgotten.

He has been recovered by regional scientists, because he contributed to this field also in the States, and some of his friends continued to deal with his theses overseas. But not in population - he did not succeed, and definitely did not want to, to make friends to the than prominant population scientists even there. He did not have a chance to get into contact with those of them he might have appreciated (LÖSCH 22), but it was too late for him (1939) when he discovered them, after his visits to the States, and at the time of the war beginning.

May be there is another reason involved also. Economics have changed since. And questions such as LÖSCH had posed, which had been prominent in economics at his time, and for long periods before, are hardly pursued any longer. Are they settled, really? Others will have to decide about this question.

Population analysis did change as well, with the war and after. It is mainly the American tradition which has overcome, and the research questions from this tradition overwhealm. German tradition has vanished - with LÖSCH, and with MACKENROTH. It has not had power enough to intrude international discussion. One of the reasons, certainly, has been the involvement of German population analysis in Nazi ideology, and the personal continuation after the war. LÖSCH had been a singular exception. It is too simple really to ask only whether, or to what degree, German population scientists have participated in Nazi crimes or in active support of the policy concerned. Reading cartels are a more subtle differentiation of political positions, and they have overcome.

As a result, we can conclude that the early death of LÖSCH has been not only a loss to regional science, but to population analysis as well. Being five years younger than MACKENROTH, and with the kind of training and interests he exhibited, he might have become a teacher to German demography even after the also early death of MACKENROTH, and during a period where population analysis in Germany was mainly observed by individuals more or less strongly involved in Nazi thinking. LÖSCH's own political position might have become clear enough through many of his statements; it is unquestionable in some sentences - and impressive in the documentation of his development till 1945 - in his diary, these dated April 1933: "A black Friday. Imperium fuit. I rage from pain and fury: The Reich is gone, and everyone now dances as the one fool whistles. In *this* Germany only creatures will be tolerated. But I hope, and will produce, for *that* Germany that will come *thereafter*, if God wants it!" (LÖSCH 24, 78, and passim similarly; his own accentuations). The demographic discipline might have taken a different development with him.

And, additionally, we tried to show that it would be worthwile to reconsider his analyses on the reciprocal effects between population and economy. They are testimonies of observation and evaluation before and at the beginning of the Nazi regime, from a time in which we have in general only onesided reports, and he was in opposition to them. Furthermore, he discussed topics of high actuality also to our time, including population decrease and ageing, earlier than others, and under different circumstances, but in a way that could contribute a couple of unconventional arguments to present discussions even now.

LITERATURE

(Only titels are mentioned I used for this review; other authors mentioned in the text have been either quoted from the literatur used, and may be verified there, or are generally well known.)

Åkerman, Johan, 1936: (I) review of Lösch, August: Bevlökerungswellen und Wechsellagen, Jena 1936, in: Schmollers Jahrbuch 61 (1937) 91-98 and (II) "Abschließende Äußerung" 460-467.

Böventer, Edwin von, 1987: Lösch, August - Artikel, Neue Deutsche Biographie 15 (59f.

Funck, Rolf; J.B.Parr, 1978: Introduction, to: The Analysis of Regional Structure - Essays in Honour of August Lösch, Karlsruhe: Pion (Karlsruhe Papers in Regional Science 2), 1-5.

Hohorst, Gerd, 1975: Wirtschaftswachstum und Bevölkerungsentwicklung in Preußen 1816 bis 1914, Diss. Münster.

Lösch, August (1), 1932: Was ist vom Geburtenrückgang zu halten? Heidenheim: Lösch, I: 73p, II: 95p.

Lösch, August (2),1933: Bevölkerungskrise? in: Deutscher Lebensraum 1, 53f.

Lösch, August (3), 1933: Umbruch der Bevölkerungsentwicklung! in: Deutscher Lebensraum 1, 82f.

Lösch, August (4), 1933: review of Mombert, Karl: Bevölkerungsentwicklung und Wirtschaftsgestaltung, in: Deutscher Lebensraum 1, 96f.

Lösch, August (5), 1933: review of Burgdröfer, Friedrich: Volk ohne Jugend, Heidelberg: Vowinkel 1932, 450 p, in: Schmollers Jahrbuch 57, 278-281.

Lösch, August (6), 1933: review of Balás, Karl von: Das neue Bevölkerungsproblem, Budapest 1932, in: Schmollers Jahrbuch 57 , 633-635.

Lösch, August (7), 1933: review of Deutscher Lebensraum - Blätter für Deutsche Raum- und Bevlkerungspolitik, in: Schmollers Jahrbuch 57, 157f.

Lösch, August (8), 1934: Die Bevölkerungsbilanz des Krieges, in: Deutscher Lebensraum 2, 47f.

Lösch, August (9), 1934: review of Huber, Michel: La population de la France pendant la guerre, Paris 1931, in: Schmollers Jahrbuch 58, 89-92.

Lösch, August (10), 1936: Bevölkerungswellen und Wechsellagen, Jena (Beiträge zur Erforschung der wirtschaftlichen Wechsellagen Aufschwung, Krise, Stockung 13), X + 124 p.

Lösch, August (11), 1936: Wirtschaftsschwankungen als Folge von Bevölkerungswellen, in: Schmollers Jahrbuch 60, 551-564.

Lösch, August (12), 1936: Die Vergreisung - wirtschaftlich gesehen, in: Schmollers Jahrbuch 60, 577-685.

Lösch, August (13), 1936: Bevölkerungswellen, in: Die Hilfe - Zeitschrift für Politik, Wirtschaft und geistige Bewegung, Begründer: Friedrich Naumann, 42, 468f.

Lösch, August (14), 1936: review of Winkler, Wilhelm: Der Geburtenrückgang in Österreich, in: Schmollers Jahrbuch 60.

Lösch, August (15),1936/37: Population cycles as a cause of business cycles, in: Quart.Journ.Econ. 51, 649–662.

Lösch, August (16), 1937: Die Bedeutung der Bevölkerungsbewegung für das Wirtschaftsleben, in: Congrès international de la population, Paris: Hermann, 78–84.

Lösch, August (17), 1937: Noch einmal: Bevölkerungswellen und Wechsellagen – Erwiderung an Professor Johan Åkerman, in: Schmollers Jahrbuch 61, 455–460.

Lösch, August (18), 1937: review of Haufe, Helmut: Die Bevölkerung Europas – Stadt und Land im 19. und 20. Jahrhundert, in: Schmollers Jahrbuch 61, 102 ff.

Lösch, August (19), 1937: review of Burgdörfer, Friedrich: Volks- und Wehrkraft, Krieg und Rasse, Berlin 1936, in: Schmollers Jahrbuch 61.

Lösch, August (20), 1938: Das Problem der Wechselwirkung zwischen Bevölkerungs- und Wirtschaftsentwicklung, (the reprint with the titel: Bevölkerung und Wirtschaft – ein Wechselspiel) in: Weltwirtschaftliches Archiv 48, 454–469.

Lösch, August (21), 1939: review of Ungern-Sternberg, Roderich von: Bevölkerungsverhältnisse in Schweden, in: Schmollers Jahrbuch 63, 115f.

Lösch, August (22), 1939: Neuerscheinungen über Bevölkerungsfragen (review article) in: Weltwirtschaftliches Archiv 50, 107*–113*.

Lösch, August (23), 1943: review of Burgdörfer, Friedrich: Krieg und Bevölkerungsentwicklung, München und Berlin 1940, in: Weltwirtschaftliches Archiv 57, 20*–22*.

Lösch, August (24), 1971: Aus Tagebüchern und Briefen 1925–1945, zusammengestellt von Marga Künkele-Lösch, mit Anmerkungen von Volker Riegger, in: Riegger, 67–127.

Mackenroth, Gerhard, 1937: Bevölkerungsprobleme des In- und Auslandes: Lösch, Haufe, Carr-Sounders, Kuczinski, Sakar (review article) in: Weltwirtschaftliches Archiv 46, 19*–26*.

Mackenroth, Gerhard, 1953: Bevölkerungslehre, Berlin: Springer.

notes: private excerpts from review articles concerning publications of August Lösch, by himself, six sheets of typoscript in the possession of Mrs. Künkele-Lösch.

Quante, Peter, 1934: review of Lösch, August: Was ist vom Geburtenrückgang zu halten? Heidenheim: Lösch 1932, in: Schmollers Jahrbuch 58, 213–215.

Riegger, Roland ed., 1971: August Lösch in memoriam, Heidenheim: Meuer 1, 137 p.

Stolper, Wolfgang F., 1971: August Lösch in memoriam, a preface to the American edition of "The Economics of Location", 1954; again in: Riegger, 59-65.

Stolper, Wolfgang F., 1959: August Lösch, in: Handwörterbuch der Sozialwissenschaften VI (1959), 643 f.

Stolper, Wolfgang F., 1971: Begegnung mit August Lösch, in Riegger, 53-57.

Vorwerck, 1936: review of Lösch, August: Bevölkerungswellen und Wechsellagen, Jena 1936, in: Reichsarbeitsblatt 24, 368.

Zottmann, Anton, 1949: Dr.habil August Lösch, gestorben am 30. Mai 1945, Nachruf in: Weltwirtschaftliches Archiv 62, 28-31.

Zottmann, Anton, 1971: Erinnerungen an August Lösch, in: Riegger, 17-32.

All translations of quotations from German writings of Lösch are mine, and deliberately very loose.

It has not been possible for me to inspect the following of Lösch's publications, also concerning population matters:

--- Geburtenrückgang und Kapitalmarkt, in: Deutscher Volkswirt 1933
--- Die Bevölkerungsbilanz des Krieges, in: Deutscher Lebensraum 2 (1934), 47f
--- Bevölkerungswellen - Konjunkturursache, in: Geistige Arbeit 1936
--- Ahnenrad für neun und zwölf Generationen, Heidenheim o.J.
--- Wie liest man Ahnentafeln?, in: Württembergische Schulwarte 1937

Development of Economy and Traffic

by

Wilhelm Leutzbach

Traffic is defined as a change of place of persons and goods. The nature and the volume of freight transport are directly influenced by the nature and the volume of the production of these goods. Passenger traffic as a realisation of potential wishes of mobility certainly depends on the supply of means of transport and, among other factors, on the standard of living. Thus, there is a connection between the development of economy and that of traffic. The analysis of some selected examples will show, if and how such connections can be illustrated.

Figure 1[1] shows the trend (and partly also the assumed development for the future) of the motorization in the USA, the Federal Republic of Germany and the German Democratic Republic, expressed by inhabitants per car (car ownership rate). If for the USA the time of the world depression in the 1930's and the time of World War II. is neglected, there are three very similar curves. Between the curve for the USA and the curve for the FRG there is an interval of about 40 years and between the FRG and the GDR there is one of about 15 years.

When searching for the economic reasons of this similarity of development, one will obviously draw first upon the development of the Gross Domestic Product per capita, or, depending on the data situation, the Gross National Product. Figure 2 shows the historical development of the Gross Domestic Product per capita in national currency in three countries under considera-

[1] All figures are given at the end of the paper

tion. With comparisons of this kind there are two problems:

- The Gross National Product has to be shown in its real value; i.e. without the influence of inflationary factors.
- When comparing different countries the Gross National Product has to be represented in a common currency.

Particularly the second problem is known to be extraordinarily difficult. Since an attempt to draw such a comparison on the basis of the US-Dollar failed due to the problems connected with the exchange rate, the Special Drawing Right (SDR) of the international monetary fund has been used as a basis of comparison (at this point the GDR could not be considered).

Figure 3 shows the similarity of the Gross Domestic Products per capita for the USA and the FRG. Figure 4 confirms the assumption that in both countries there exists a close connection between the motorization in cars per inhabitant and the now comparable Gross Domestic Product per capita.

A study by the World Bank (Figure 5) shows that there also exists a remarkable linear relation if - instead of the motorization index - the annual average vehicle mileage per head is inserted above the Gross National Product per capita (in US-Dollars).

If one looks for similarities in the area of freight transport, one finds that the same study of the World Bank has proven a similar seemingly linear relation between the total volume of the freight transport and the total Gross National Product (again in US-Dollars), (Figure 6).
It has to be considered, however, that the arising linearity is a result of the double logarithmic scale in which the quantities are represented. In a regular cartesic system of co-ordinates the connection is less pronounced; the tendency, however, is still preserved.

In India a relation has been discovered between the capacity of the passenger traffic of the road (CPT R) and the Gross National Product (GNP) in terms of

$$\ln \text{CPT R} = -21{,}6330 + 2{,}5912 \cdot \ln \text{GNP}$$

and between the capacity of the passenger traffic of the railway (CPT RW) and the Gross National Product (GNP) in terms of

$$\ln CPT\ RW = 7{,}8799 + 1{,}2127 \cdot \ln GNP$$

(BRANDT, 1989). This means, that at a rise of 1.0 % of the Gross National Product the road traffic has increased by about 1.2 %.

The Domestic Product per capita thus influences the motorization index which in turn influences the modal split between private transportation and public passenger transport, as VOIGT (1988) has shown (Figure 7).

The degree of the road extension (percentage of the fixed roads) seems, however, also to be influenced by the rate of the specific Gross Domestic Product (SOMBAI 1988), (Figure 8).

Certainly the Gross National Product is an essential but also a rather summary indicator of the economic power of a country. In order to conduct more thorough analysis, additional information on the economic structure of the respective country is required. Here the term "structure" here refers to the various productive activities of the country's population; i.e. the distribution of its potential labour supply over the various characteristic sectors of production. A classification into such sectors is discretionary, in the FRG one most commonly distinguishes (according to FOURASTIE (1954) and others) between a primary, a secondary and a tertiary sector (Figure 9). In their study of the economic development of countries HERMAN and MONTROLL (1972) have used up to six such characteristic sectors.

The existing economic structure of a given country at a given moment can thus be expressed graphically in a plane, with rays emanating from a centre point. These n rays form an angle of about $2\pi/n$ each. Then every ray is attributed to a fixed parameter (sector of production), with the length of the ray j being aquivalent to the portion f_j of this parameter of the entire productive power (= 100 %). Linking the final points of these rays by a line a geometric figure (polygon) will result in, which was called 'snowflake' by HERMAN and MONTROLL. Figure 10 shows very clearly how the shape of this 'snowflake' changes according to the modification of the proportion of the individual parameters. It is thus demonstrated whether the economic power of a country is still mainly determined by agriculture (for example Yugoslavia) or whether it is already heavily industrialized. (In this figure the rays represent agriculture (1), traffic (2), trade (3), industry (4), services (5), and construction trade (6).)

Such 'snowflakes' are also a vivid medium to illustrate the development of the economic structure of a country as time moves on. In this occasion HERMAN and MONTROLL have characterised the 'snowflake' of each country in a given year by the centre of gravity of its plane. Figure 11 shows a typical movement of those centres through the system of co-ordinates during the development from an agricultural to an industrial society. The figure also shows, however, the high degree of industrialization of Great Britain already in 1841 - just at the beginning of statistical records.

As mentioned before, FOURASTIE describes the development of the economic structure of a country by the temporal development of the proportion of the labour force employed in the three sectors of production. According to HERMAN and MONTROLL this development can be described by 'snowflakes' too: this time they have a shape of a three-cornered polygon. It is also possible, however, to represent the proportions of the three sectors for each year by one point in a three-cornered system of co-ordinates (Figure 12).

Such a representation makes it possible to classify the economic structure of countries immediately. Ten countries with different structures are represented in this way (Figure 13). More clearly one can distinguish three different categories, each with a similar stage of development: on the far right side there is the category of countries with an high proportion of the primary sector and on the far left side there is the category of countries with only a small proportion of the primary sector, but instead with an high proportion of the tertiary sector. Connecting the points of successive sectors results in a line of development. Figure 14 shows as an example (in such a system of co-ordinates) the line of development of the German Reich respectively the FRG and Brazil. The example of Brasil clearly shows that countries with a later development do not necessarily have to show an high proportion of the secundary sector, but that the transition from the primary to the tertiary sector can happen more directly.

Through the classification in three sectors of production it is possible to not only consider the Gross National Product but also to analyse more closely the relation between the structure of the economy and traffic. As an example Figure 15 indicates the modal split between road and rail of freight transport over the rate of employes in the primary sector. With a decreasing share of people employed in the primary sector the ratio of freight carried on the road is increasing.

This of course reflects only the affinity of special economic goods for special means of transport which is typical for the development of any national economy. Figure 15 also shows, however, deviations from this trend caused by political and geographical differences.

It would certainly be interesting to continue the empirical analysis of the connection between the development of the economy and traffic, which here was only been briefly outlined.

LITERATURE

Brandt, J., 1989: Probleme des Straßenverkehrs und ihre Lösung - Gegenstand eines Straßenkongresses in Indien. In: Die Straße, Heft 4.

Fourastie, J., 1954: Die große Hoffnung des zwanzigsten Jahrhunderts. Bund-Verlag, Köln-Deutz.

Herman, R.; Montroll, E.W., 1972: A Manner of Characterizing the Development of Countries. Proceedings of the National Academy of Sciences of the USA; Vol. 69, Nr. 10, pp. 3019-3023.

Herman, R. et. al., 1988: The Dynamic Characterization of Cities. In: Cities and their vital systems. Infrastructure past, present and future. Editors: Herman, R. and Ausobel, J.H., National Academy Press, Washington D.C.

Sombai, A., 1989: Drei Vorschläge zur Weiterentwicklung der Forschungsarbeit des Straßenverkehrs. Györ, Ungarn, Manuskript.

Voigt, W., 1988: Grundsätze der Aufgabenteilung im städtischen Personenverkehr der DDR. In: 19. Budapester Internationale Wissenschaftliche Beratung für Verkehrsplanung und Verkehrstechnik, Budapest, April.

Weltbank: China; Long Term Issues and Options. Rep. No. 5206 - CHA.

Figure 1: The trend of the motorization in the USA, FRG and GDR

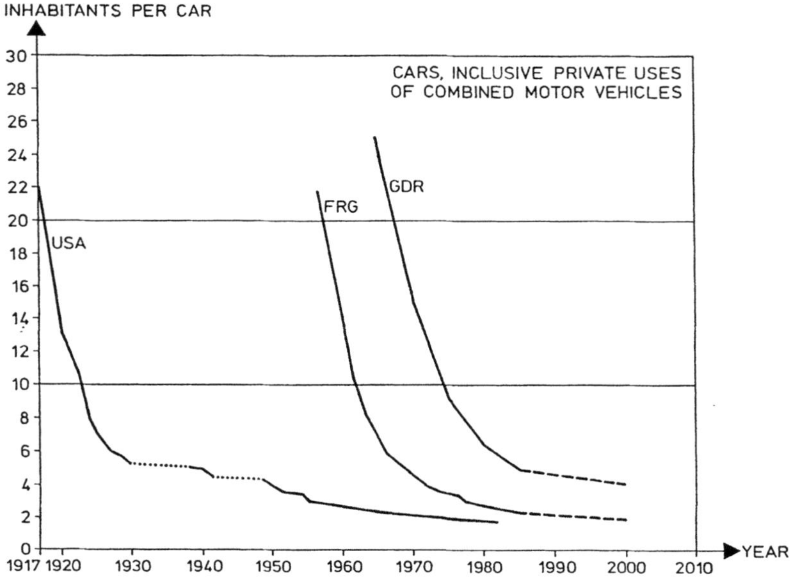

Figure 2: Development of the Gross Domestic Product (GDP) in the USA, FRG and GDR

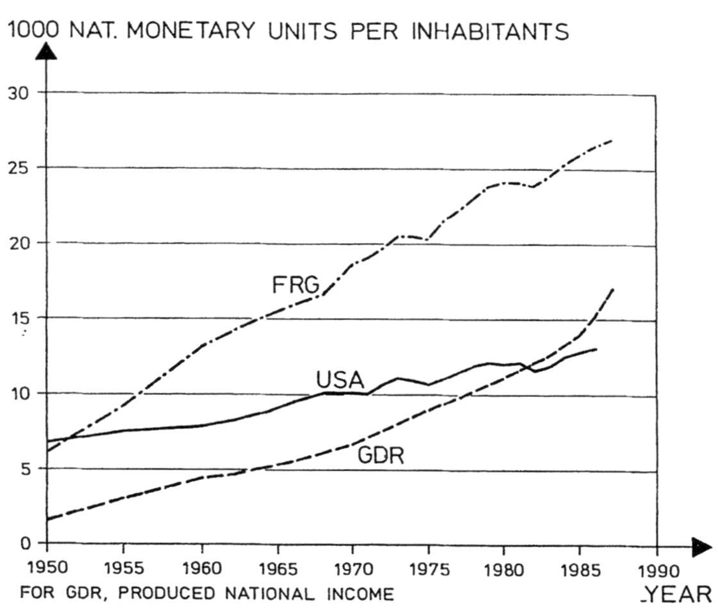

Figure 3: Development of the Gross Domestic Product (GDP) in the USA and the FRG expressed by the Special Drawing Right (SDR)

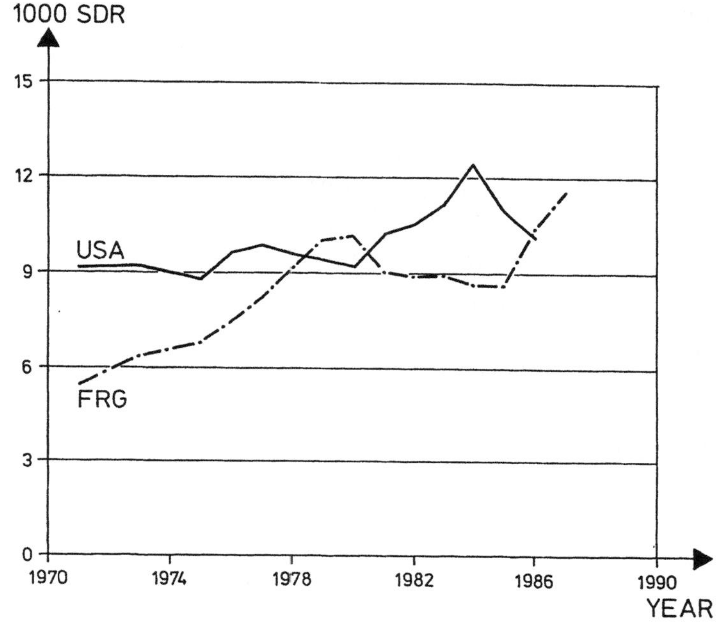

Figure 4: Motorization as a function of the GDP – expressed in Spezial Drawing Right (SDR) for the USA and the FRG

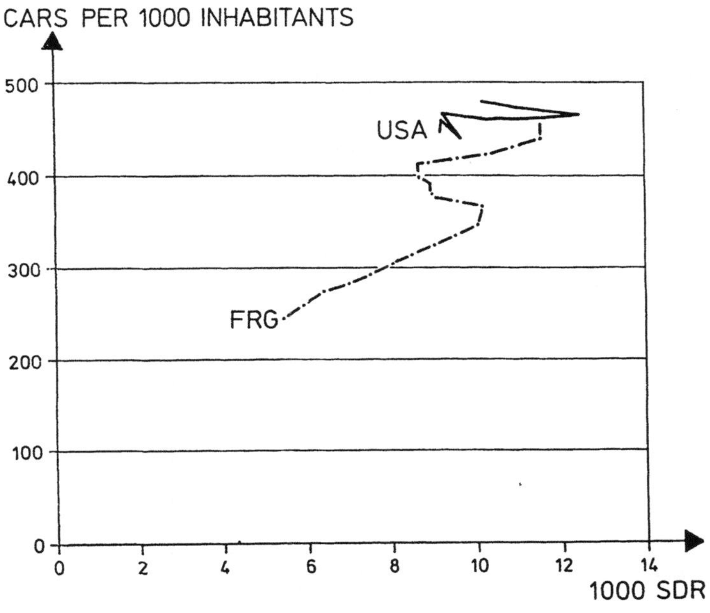

Figure 5: **Passenger traffic and Gross National Product (GNP) per capita Source: World Bank**

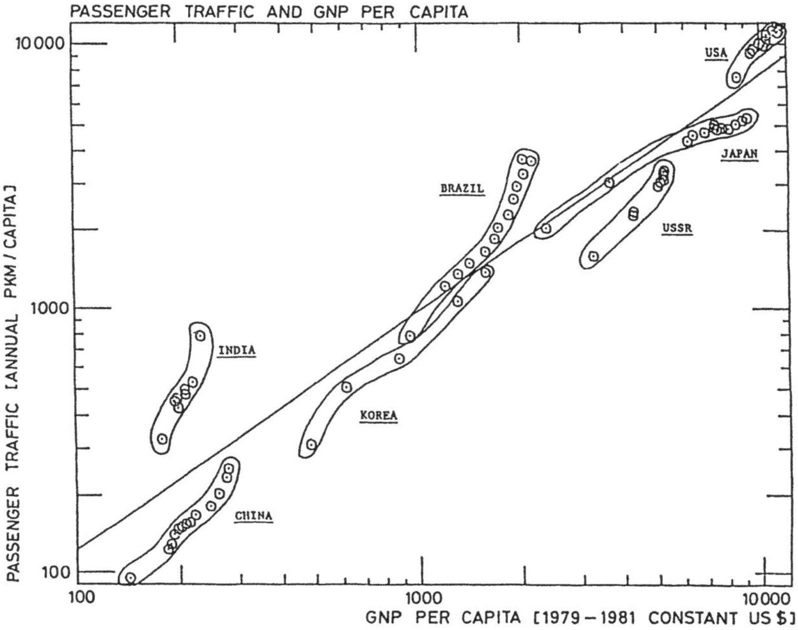

Figure 6: **Total freight and Gross National Product (GNP) for selected countries, 1960–1981 – Source: World Bank**

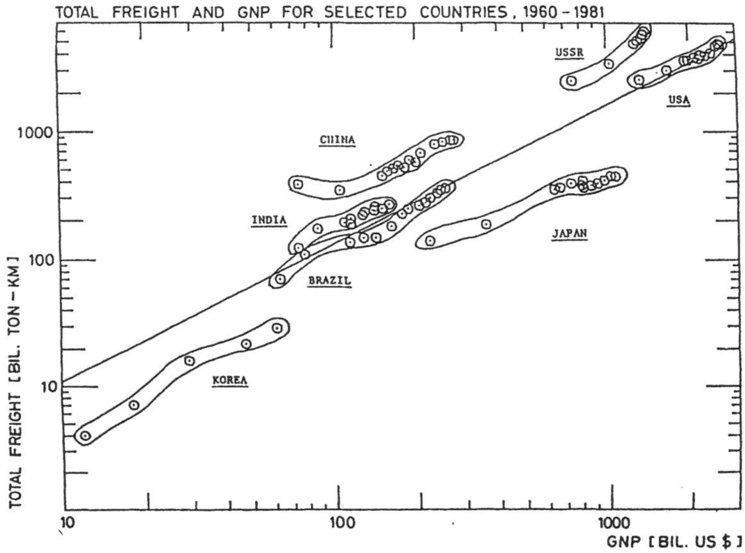

Figure 7: **Modal split as a function of rate of motorization**
Source: Voigt, 1988

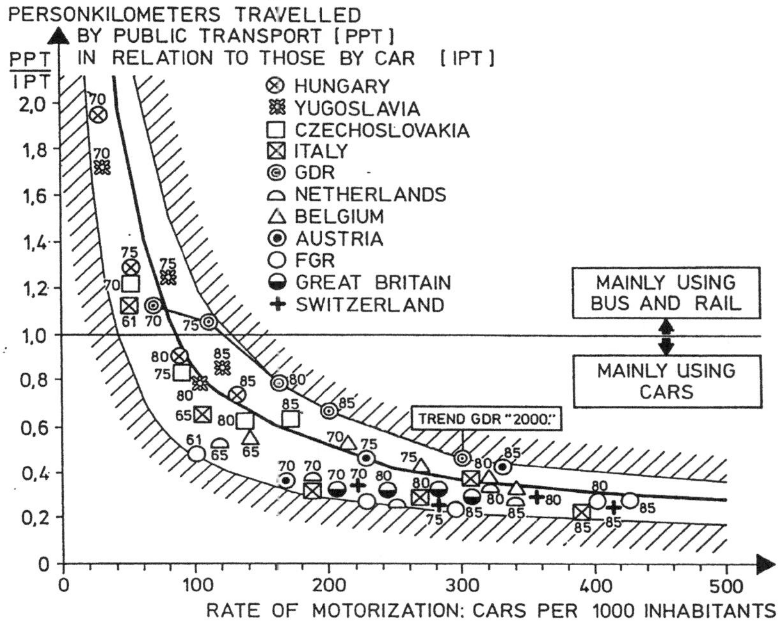

Figure 8: **Percentage of the paved roads influenced by the rate of the specific Gross Domestic Product (GDP) per capita**
Source: Sombai, 1989

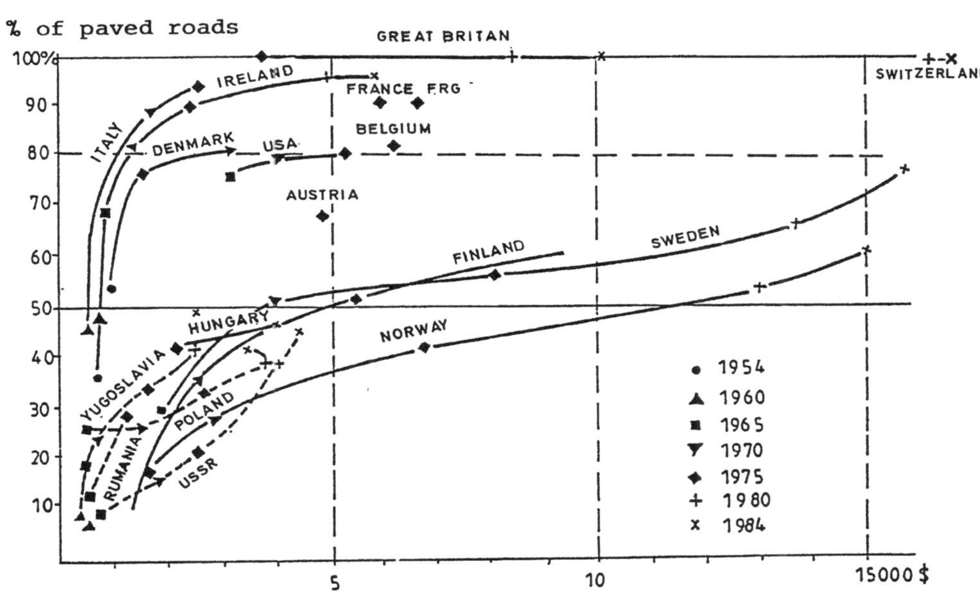

Figure 9: The development of the economical structure of nations, shown by the FRG - Source: Fourastie, 1954

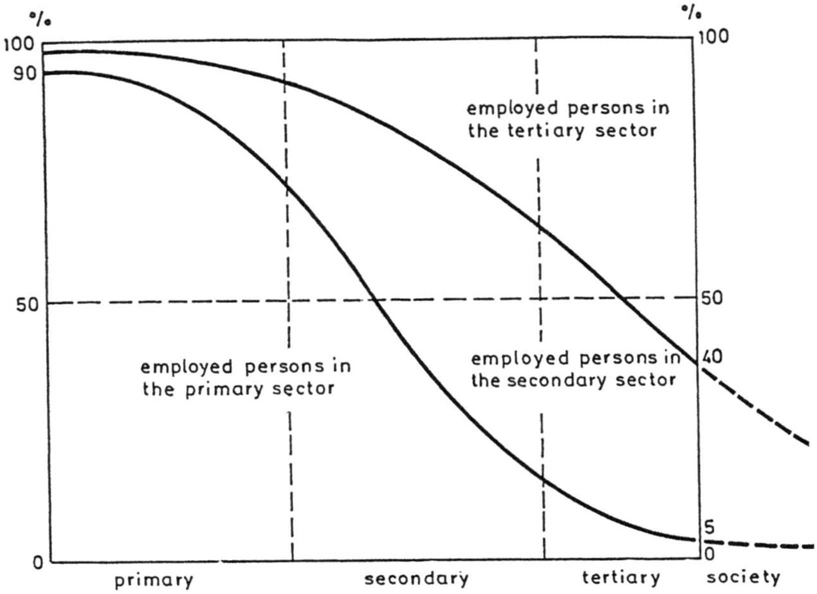

Figure 10: "Snowflakes" for different countries
Source: Herman and Montroll, 1972

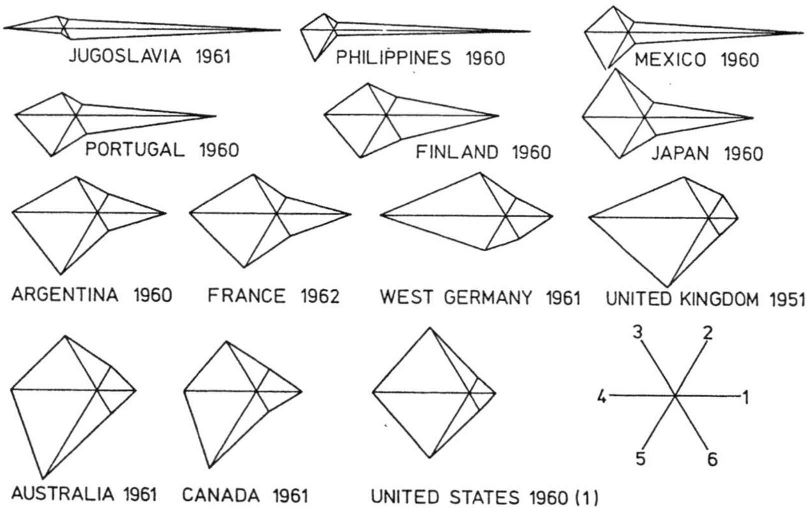

Figure 11: Development of the economical structure for some countries
Source: Herman and Montroll, 1972

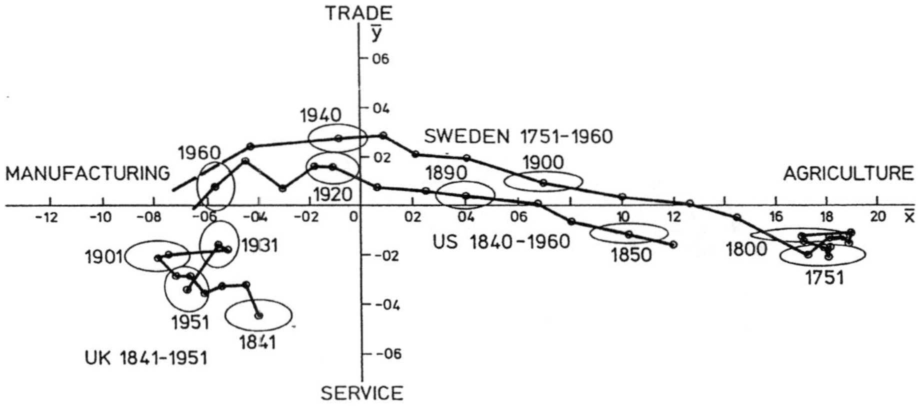

Figure 12: The shape of the economical structure represented as a point in a three-cornered co-ordinates system

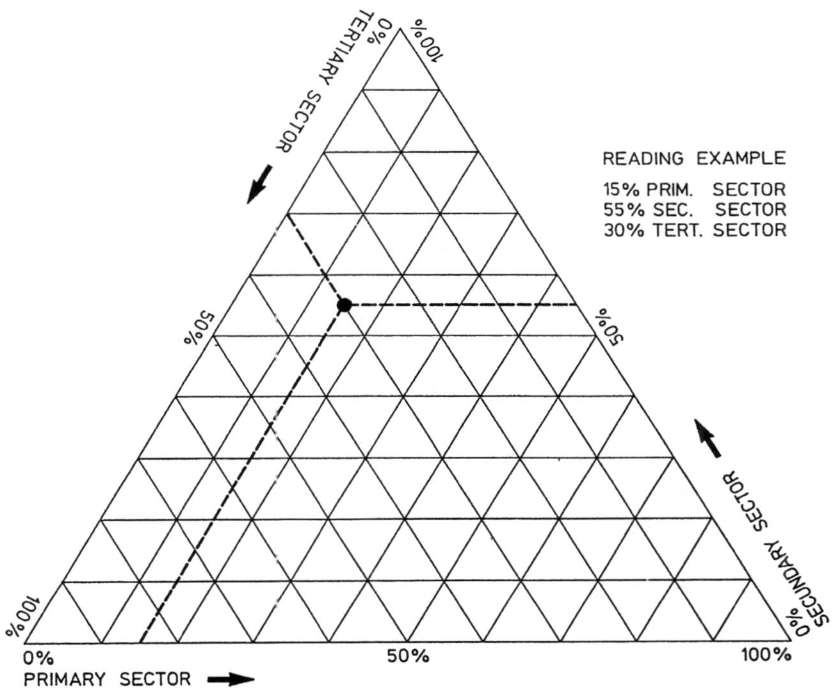

Figure 13: Economical structure of ten countries

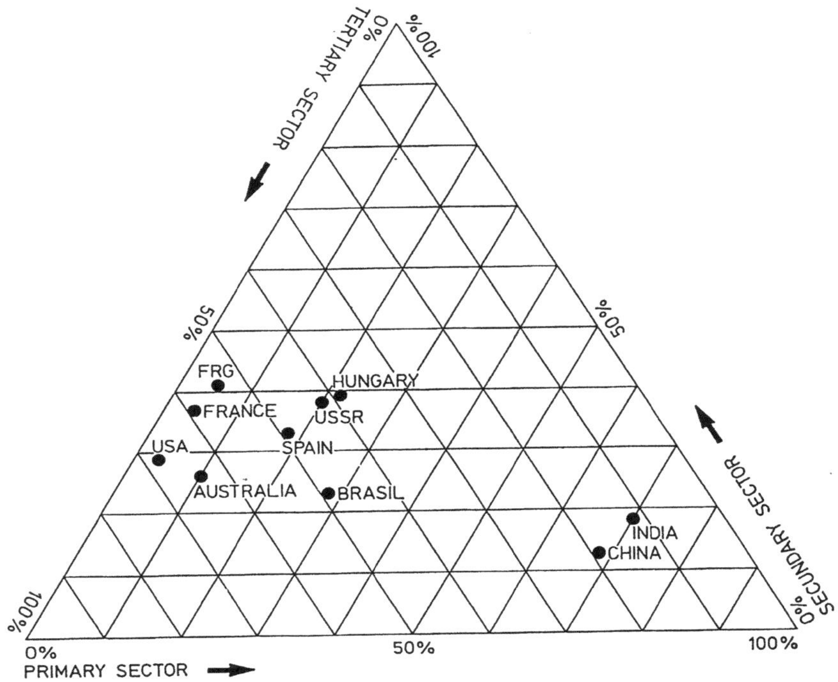

Figure 14: Development of the economical structure for FRG and Brazil

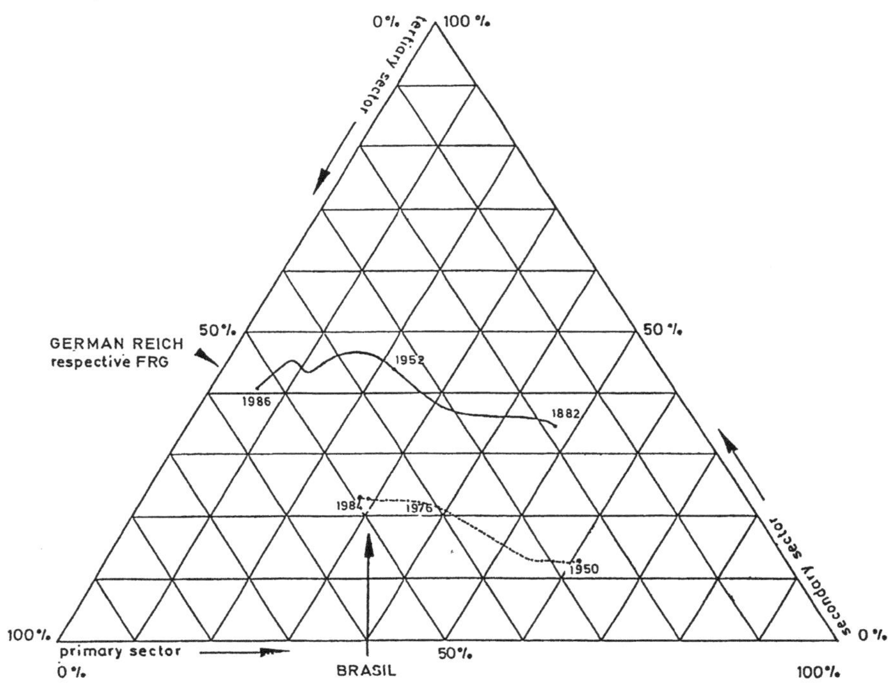

Figure 15: The modal split between road and rail of freight transport over the rate of employes in the primary sector

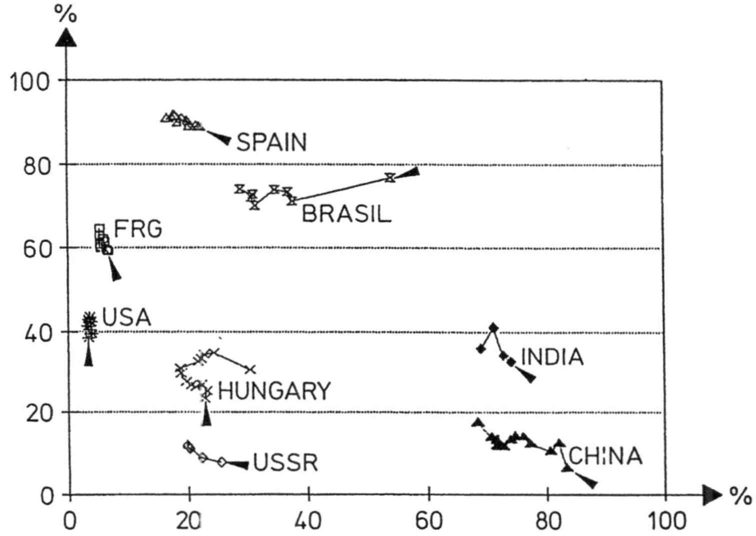

Are Car Purchases Savings?

An Analysis of German Households*

by

Ulrich C. H. Blum and Marc J.I. Gaudry

1. Introduction: the Problem

Savings, defined as increase in net wealth, are characterized in most studies as a portfolio of different financial assets or long term durable goods such as houses. Cars are normally not included, although they have typical properties of assets; for instance, they yield returns, change in value or are subject to risk just like bonds or shares. Germans, in particular, clearly take into account the resale values of cars when they make a car purchase, a 'stylized fact' that points to the usefulness of an analytical inquiry of this question: we want to explicitly test the hypothesis that car purchase expenditures are considered by households as part of their long-term investments, using German household data for 1983. Naturally, a savings function based on this extended definition should not look very different from a "classical" savings function.

*The research for this paper was supported in part by the F.C.A.R. program of Québec and by the S.S.H.R.C and the N.S.E.R.C. of Canada. A part of Gaudry's and Blum's contributions were made during their tenure respectively as Alexander von Humboldt Research Scholar in 1989 at the University of Bamberg and as Alexander von Humboldt Research Scholar in 1989 at the Université de Montréal where Catherine Laplante provided computing assistance.

There is a strong economic policy interest in the question: if cars are considered by households as consumption items, one would expect changes in their prices to influence strongly the demand for other consumption goods whereas, if they are more akin to savings, one would expect strong linkages with the structure of portfolios. In the latter case, financial market policies would have an impact on car investment behavior; car prices would in return affect capital markets.

In our comparison of the "classical" and the "enriched" specifications of the savings functions, we base our comparison on their different characteristics:

(i) the functional form,

(ii) the value of the partial derivatives of the dependent with respect to the independent variables,

(iii) the distributional properties of the error term, namely heteroskedasticity and autocorrelation.

Whereas the first and the third points are clear, the second needs some explanation. Assume that there exists a household investment portfolio including typical assets like bonds, shares, houses, precious metals, etc.. Certain types of investments can be given by linear combinations of other investments which are at the frontier of the portfolio. If car purchases are an investment that may be characterized by a combination of existing assets, then a "savings" function which includes these investments must be identical to a function which excludes them. By identical, we especially stress the partial derivatives because, at the margin, an additional DM earned could either be spent on a car or on a combination of assets with the same properties as the car investment.

If the two functions are different, we cannot classify and evaluate the function including car investments as it could either be the result of a new efficient type of investment or show that car purchase is consumption.

Equal partial derivatives of the two functions are therefore used as a sufficient but not necessary condition for the maintained hypothesis that car expenditures are a form of savings. Partial derivatives are a better indicator than elasticities, which are weighted by the dependent variable and necessarily smaller in absolute terms once car investments are included in the portfolio. Elasticities may, of course, give further results of interest, but are not

$$u_t = \exp[\beta_0 + \beta * Z_t^{(\lambda_Z)}]^{0.5} * v_t .\tag{4}$$

Autocorrelation of multiple orders may result from information in the endogenous variable not explained by the deterministic part of the savings function. In time series, a convenient approach is to identify the proper structures with Box-Jenkins techniques to determine the number of distinct orders required to obtain white noise residuals w_t.

$$v_t = \sum_{q=1}^{Q} \rho_q * v_{t-q} + w_t .\tag{5}$$

By defining a "residue impact criterion" stressing the fact that, in addition to the usual meaning of autocorrelation in time series, it is possible to give an economic interpretation to autocorrelation in cross sections, the correlation of residues between households of different (here: higher) income can be tested. Specifically, we defined a directly ordered impact criterion (DORIC) by relating each household to a richer group:

[Assume that each household looks up to a higher income household a given economic distance away] (6)

In our stratified sample, in which the number of households in each group was given, this involved repeating observations according to their frequency. The highest number of repetitions was found to be 12, thus looking up 13 observations never involved relating to the same group. Using 13th order autocorrelation implied that households looked up 3% higher in the distribution of income. A detailed description of the methodology can be found in GAUDRY and BLUM (1988) and BLUM and GAUDRY (1990).

3. Results

Table 1 contains 6 columns: the first three pertain to the explanation of savings defined without including capital expenditures on cars and the last three to similar experiments performed with the "enriched" definition of the ken from the same sample of households with the same distribution, the missing variable in one data set was replaced by the average value of the variable in the other set. In doing so, it is assumed that the resulting error is smaller than that obtained by keeping distinct data subsets and specifying models without either the age of the household head or the number of persons per household; in a linear model, our merging procedure will yield unbiased estimators. Various test with similar data for the year 1978 (BLUM and GAUDRY, 1990) were consistent with the appropriateness of the procedure.

2.2 Econometric Specification

In the absence of strong a priori views on the functional form of (1), it is appropriate to use the Box-Cox specification and correct simultaneously for heteroskedasticity and autocorrelation according to the strategy defined in GAUDRY and DAGENAIS (1979):

$$y_t^{(\lambda_y)} = a_0 + \sum_{j=1}^{m} a_j * X_{jt}^{(\lambda_j)} + u_t \tag{2}$$

where the a_j are coefficients of the variables and the transformation is defined as

$$X_j^{(\lambda_j)} = \begin{cases} \dfrac{X_j^{\lambda_j} - 1}{\lambda_j} & \text{if } \lambda_j \neq 0, \\ \ln X_j & \text{if } \lambda_j = 0. \end{cases} \tag{3}$$

The linear case is obtained by setting all λ_j to 1. Setting the same values to zero yields the log model. But it is clearly important to correct for autocorrelation and heteroskedasticity in order to obtain a final residual error of zero mean and constant variance.

Correcting for heteroskedasticity is of special importance when flexible forms are used on the variables of (1) and modify the error variance. We use the general error term specification:

$$u_t = \exp[\beta_0 + \beta * Z_t^{(\lambda_Z)}]^{0.5} * v_t . \quad (4)$$

Autocorrelation of multiple orders may result from information in the endogenous variable not explained by the deterministic part of the savings function. In time series, a convenient approach is to identify the proper structures with Box-Jenkins techniques to determine the number of distinct orders required to obtain white noise residuals w_t.

$$v_t = \sum_{q=1}^{Q} \rho_q * v_{t-q} + w_t . \quad (5)$$

By defining a "residue impact criterion" stressing the fact that, in addition to the usual meaning of autocorrelation in time series, it is possible to give an economic interpretation to autocorrelation in cross sections, the correlation of residues between households of different (here: higher) income can be tested. Specifically, we defined a directly ordered impact criterion (DORIC) by relating each household to a richer group:

[Assume that each household looks up to a higher income household a given economic distance away] (6)

In our stratified sample, in which the number of households in each group was given, this involved repeating observations according to their frequency. The highest number of repetitions was found to be 12, thus looking up 13 observations never involved relating to the same group. Using 13th order autocorrelation implied that households looked up 3% higher in the distribution of income. A detailed description of the methodology can be found in GAUDRY and BLUM (1988) and BLUM and GAUDRY (1990).

3. Results

Table 1 contains 6 columns: the first three pertain to the explanation of savings defined without including capital expenditures on cars and the last three to similar experiments performed with the "enriched" definition of the dependent variable. Columns 2 and 5 present the reference model of each function.

3.1 Reference Model for Both Savings Functions

In order to establish a reference model for both definitions of savings, a set of experiments was carried through:

(i) A linear regression was performed on all variables (fixed lambda = 1); results are given in columns 1 and 4. A log-linear model was also tried. Resulting log likelihood values (not shown) were very low: -2776.789 and -3006.091, respectively.

(ii) A model with a single lambda common to the endogenous and the exogenous variables was estimated; the result is given in columns 2 and 5. Attempts to use more lambdas, for instance one on the endogenous and the other one on all exogenous variables, were made but did not yield sufficient statistical gains to justify the additional parameter. The results shown in column 2 and column 5 were therefore retained as reference models. But before commenting on them, a description of the contents of the table is useful.

3.2 Reported Statistics and Reporting Format

The first part of the table contains partial derivatives of the expected value of the dependent variable and elasticities of the expected value of the dependent variable, both evaluated at the mean value of the variables, as well as t-statistics of the underlying regression coefficients computed conditionally upon the estimated value of the Box-Cox transformations used. An identifying group number for these transformations is also inscribed.

The second part contains functional form and residual error specification parameters and statistics. The third part contains general statistics associated with each experiment.

Table 1: Models of Household Savings, Germany 1983

PARTIAL DERIV. E(y) (EP) ELASTICITIES E(y) (EP) (COND. T-STATISTICS)	COLUMN DEP.VAR. =	1 TOT1SAV83	2 TOT1SAV83	3 TOT1SAV83	4 TOTSAVCAR83	5 TOTSAVCAR83	6 TOTSAVCAR83
HOUSEHOLD STRUCTURE							
age of head of household	TOTAGE83	-.174383e+01	.180637e+01	.171484e+01	-.202121e+01	.623733e-01	.116319e+00
		-.224	.273	.256	-.192	.007	.012
		(-3.14)	(4.84)	(4.47)	(-3.72)	(.15)	(.27)
		LAM 1	LAM 1	LAM 1	LAM 1	LAM 1	LAM 1
number of persons in household	TOTPERS83	.435697e+02	-.232798e+02	-.233087e+02	.322654e+02	-.207076e+02	-.209412e+02
		.240	-.151	-.149	.131	-.093	-.094
		(6.36)	(-3.73)	(-3.58)	(4.82)	(-3.33)	(-3.36)
		LAM 1	LAM 1	LAM 1	LAM 1	LAM 1	LAM 1
NET INCOME							
Total income before taxes.	TOTGINC83	.134794e+00	.174870e+00	.176867e+00	.180642e+00	.226416e+00	.225764e+00
		1.354	2.063	2.062	1.340	1.857	1.856
		(13.04)	(25.15)	(24.90)	(17.88)	(32.66)	(32.77)
		LAM 1	LAM 1	LAM 1	LAM 1	LAM 1	LAM 1
social security contributions	TOTSSC83	-.133028e+01	-.525563e+00	-.542335e+00	-.110521e+01	-.526531e+00	-.518242e+00
		-1.032	-.479	-.488	-.633	-.333	-.329
		(-26.29)	(-13.13)	(-12.63)	(-22.34)	(-16.85)	(-16.60)
		LAM 1	LAM 1	LAM 1	LAM 1	LAM 1	LAM 1
income taxes	TOTTAX83	.697139e+00	.200837e+00	.203754e+00	.625322e+00	.197420e+00	.196564e+00
		.857	.290	.291	.568	.198	.198
		(16.50)	(5.30)	(5.21)	(15.14)	(6.38)	(6.42)
		LAM 1	LAM 1	LAM 1	LAM 1	LAM 1	LAM 1
REGRESSION CONSTANT	CONSTANT	-.398535e+02	-.803878e+03	-.806790e+03	-.902331e+02	-.602839e+03	-.606632e+03
		-.097	-2.300	-2.281	-.162	-1.199	-1.210
		(-1.37)	(-21.26)	(-20.41)	(-3.18)	(-19.44)	(-19.42)

Table 1b: Models of Household Savings, Germany 1983

RTIAL DERIV. E(y) (EP) ASTICITIES E(y) (EP) COND. T-STATISTICS)	COLUMN DEP.VAR. =	1 TOT1SAV83	2 TOT1SAV83	3 TOT1SAV83	4 TOTSAVCAR83	5 TOTSAVCAR83	6 TOTSAVCAR83
USEHOLD STRUCTURE							
age of head of household	TOTAGE83	-.174383e+01	.180637e+01	.171484e+01	-.202121e+01	.623733e-01	.116319e+00
		-.224	.273	.256	-.192	.007	.012
		(-3.14)	(4.84)	(4.47)	(-3.72)	(.15)	(.27)
		LAM 1	LAM 1	LAM 1	LAM 1	LAM 1	LAM 1
number of persons in household	TOTPERS83	.435697e+02	-.232798e+02	-.233087e+02	.322654e+02	-.207076e+02	-.209412e+02
		.240	-.151	-.149	.131	-.093	-.094
		(6.36)	(-3.73)	(-3.58)	(4.82)	(-3.33)	(-3.36)
		LAM 1	LAM 1	LAM 1	LAM 1	LAM 1	LAM 1
T INCOME							
Total income before taxes.	TOTGINC83	.134794e+00	.174870e+00	.176867e+00	.180642e+00	.226416e+00	.225764e+00
		1.354	2.063	2.062	1.340	1.857	1.856
		(13.04)	(25.15)	(24.90)	(17.88)	(32.66)	(32.77)
		LAM 1	LAM 1	LAM 1	LAM 1	LAM 1	LAM 1
social security contributions	TOTSSC83	-.133028e+01	-.525563e+00	-.542335e+00	-.110521e+01	-.526531e+00	-.518242e+00
		-1.032	-.479	-.488	-.633	-.333	-.329
		(-26.29)	(-13.13)	(-12.63)	(-22.34)	(-16.85)	(-16.60)
		LAM 1	LAM 1	LAM 1	LAM 1	LAM 1	LAM 1
income taxes	TOTTAX83	.697139e+00	.200837e+00	.203754e+00	.625322e+00	.197420e+00	.196564e+00
		.857	.290	.291	.568	.198	.198
		(16.50)	(5.30)	(5.21)	(15.14)	(6.38)	(6.42)
		LAM 1	LAM 1	LAM 1	LAM 1	LAM 1	LAM 1
REGRESSION CONSTANT	CONSTANT	-.398535e+02	-.803878e+03	-.806790e+03	-.902331e+02	-.602839e+03	-.606632e+03
		-.097	-2.300	-2.281	-.162	-1.199	-1.210
		(-1.37)	(-21.26)	(-20.41)	(-3.18)	(-19.44)	(-19.42)

3.3 General Comments on the Reference Models

Some general comments on the results seem necessary at this point:

(i) Partial derivatives for all pairs of experiments are nearly identical and make very good economic sense. Only AGE, which is a socioeconomic and not a strictly economic variable, behaves differently. In both cases, the optimal functional form is the square root of all variables. All results are statistically highly significant with the exception of age in the function which includes car purchases. We conclude that car purchase expenditures are true investments within the households' portfolios.

(ii) Portfolio investments are a superior good; households spend about 20 Pfennigs of each additional DM of income on savings.

(iii) Households reduce savings because they make contributions to the Social Security System, as a portfolio approach and the extended life cycle theory leads us to expect: each additional DM spent on Social Security reduces savings by half of that value.

(iv) The more people live in a household, the more portfolio investments fall: each additional person reduces savings by slightly more than DM 20.

(v) Savings increase with age if cars are excluded from the portfolio, in apparent contradiction to the life cycle theory. However, once cars are included, age has no effect on savings. We suspect that older people reduce car purchases with age and, consequently, compensate the first influence.

(vi) Rising taxes increase savings: this unexpected result could easily be explained by a detailed analysis of household categories, such as that performed by the present authors with 1978 data. It appears that in a sample of the whole population, results that hold for low income people (whose savings are tax exempt) dominate the sample.

(vii) To develop this last point further, results obtained here for the population as a whole very closely resemble those found for low income white collar workers (as contrasted to self employed and to high income white collar, high and low income blue collar workers) with the 1978 sample. This is not surprising as they can be seen as an "average" category.

3.4 Heteroskedasticity and Autocorrelation Structures

The reference model assumes a spherical distribution of the residuals. However, it is by no means obvious that error variance is constant in the population or that autocorrelation is absent.

We found that, once we tested for the existence of heteroskedasticity, using either income or taxes as variable according to (4), the X'X matrix could not be inverted. We therefore had no choice but maintain our hypothesis that our residuals are homoskedastic.

We found no autocorrelation, in contrast to our earlier study where autocorrelation was either positive (high income white collar workers exhibiting a 'keeping up with the Joneses' behavior) or negative for other workers categories. As shown in columns 3 and 6, autocorrelation somehow averages out to zero. Of course, in a heterogeneous sample the notion of next richer household looses much of its economic relevance.

4. Conclusion

The evidence presented here is consistent with the view that car purchases are true investments in the households' portfolios. The only behavioral difference between the two specifications is associated with the age of the household head variable and consistent with a loss in relative importance of car purchases once people reach a certain age.

LITERATURE

Blum, Ulrich C.H., 1985: The Influence of the German Social Pension Insurance System on Private Savings and on Spatial Income Distribution, Geld, Banken und Versicherungen, H. Göppl and R. Henn ed., Verlag Versicherungswirtschaft, Karlsruhe.

Blum, Ulrich C.H.; Marc J.I. Gaudry, 1990: The Impact of Social Security Contributions on Savings, Jahrbuch für Sozialwissenschaft, forthcoming.

Gaudry, Marc J.I.; Marcel Dagenais, 1975: Heteroscedasticity and the Use of Box-Cox Transformations, Economics Letters 2/3, pp. 225 - 229.

Gaudry, Marc J.I.; Ulrich C.H. Blum, 1987: An Example of Correlation among Residuals in Directly Ordered Data, Economics Letters 26, pp. 335 - 340.

Statistisches Bundesamt (Federal Statistical Office), 1986: Einkommens- und Verbrauchsstichprobe 1983 (Sample of Income and Expenditures 1983), No. 4, Stuttgart - Mainz.

List of Authors

Nicholas W. Balabkins
Economics Department
Lehigh University
Bethlehem, PA 18015
USA

Martin J. Beckmann
Institut für Statistik und Unternehmensforschung
Technische Universität München
Arcisstraße 21
8000 München 2
Federal Republic of Germany

Ulrich Blum
Volkswirtschaftslehre, insbes. Wirtschaftspolitik
Universität Bamberg
Feldkirchenstraße 21
8600 Bamberg
Federal Republic of Germany

Dieter Bökemann
Institut für Stadt- und Regionalforschung
Technische Universität Wien
Karlsplatz 13
1040 Wien
Austria

Harry Böttcher
Institut für Wirtschaftspolitik und Wirtschaftsforschung
Universität Karlsruhe (TH)
Kollegium am Schloß, Bau IV
7500 Karlsruhe 1
Federal Republic of Germany

David E. Boyce
Urban Transportation Center (M/C 357)
University of Illinois at Chicago
1033 West Van Buren Street
Chicago, Illinois 60607-9940
USA

Frank Bruinsma
Faculty of Economics
Free University
P.O.B. 7161
1007 MC Amsterdam
The Netherlands

Manas Chatterji
School of Management
State University of New York at Binghamton
Binghamton, New York 13901
USA

John H.Ll. Dewhurst
Department of Economics
Dundee University
Dundee, DD1 4HN
United Kingdom

Ryszard Domański
Institute of Space Economy
Academy of Economics in Poznań
ul. Marchlewskiego 146/150
60-967 Poznań
Poland

Wolfgang Eichhorn
Institut für Wirtschaftstheorie
und Operations Research der
Universität (TH) Karlsruhe
Kollegium am Schloß, Bau IV
7500 Karlsruhe
Federal Republic of Germany

Bernhard Felderer
Staatswissenschaftliches Seminar
Universität zu Köln
Albertus-Magnus-Platz
5000 Köln 41
Federal Republic of Germany

Marc J.I. Gaudry
Centre de recherche sur les transports
Université de Montréal
C.P. 6128, Succursale A
Montréal,
Canada H3C 3J7

Maria Giaoutzi
Department of Geography and Regional Planning
National Technical University
Campus Zographou
Athens 157 10
Greece

John Greenhut
Arizona State University
Tempe, Arizona 85287
USA

Melvin L. Greenhut
Department of Economics
College of Liberal Arts
Texas A&M University
College Station
Texas 77843-4228
USA

Geoffrey J.D. Hewings
Department of Geography
University of Illinois at Urbana-Champaign
220 Davenport Hall
607 S. Mathews Avenue
Urbana, IL 61801
USA

Walter Isard
Department of Economics
College of Arts and Sciences
Cornell University
Uris Hall
Ithaca, New York 14853-7601
USA

Rodney C. Jensen
Department of Economics
University of Queensland
St. Lucia, Queensland 4067
Australia

Leo H. Klaassen
Netherlands Economic Institute
Burgemeester Oudlaan 50
3062 PA Rotterdam
The Netherlands

Jan S. Kowalski
Institut für Wirtschaftspolitik und Wirtschaftsforschung
Universität Karlsruhe (TH)
Kollegium am Schloß, Bau IV
7500 Karlsruhe 1
Federal Republic of Germany

J.H. Kuiper
Faculty of Economics
Erasmus University Rotterdam
P.O.B. 1738
3000 DR Rotterdam
The Netherlands

Antoni Kukliński
Institute of Space Economy
University of Warsaw
Krakowskie Przedmieście 30
00-927 Warszawa
Poland

Jan G. Lambooy
Institute of Economic Geography
University of Amsterdam
Jodenbreestraat 23
1011 NH Amsterdam
The Netherlands

Wilhelm Leutzbach
Institut für Verkehrswesen
Universität (TH) Karlsruhe
Kaiserstraße 12
7500 Karlsruhe 1
Federal Republic of Germany

Rainer Mackensen
Institut für Soziologie
Technische Universität Berlin
Dovestraße 1
1000 Berlin 10
Federal Republic of Germany

Jürgen Morlok
Stiftung Außenwirtschaft Baden-Württemberg
Schloß-Straße 25
7000 Stuttgart 1
Federal Republic of Germany

Peter Nijkamp
Faculty of Economics
Free University
P.O.B. 7161
1007 MC Amsterdam
The Netherlands

Hiroshi Ohta
Aoyama Gakuin University
School of International Politics, Economics and Business
Aoyama Campus
Shibuya, Tokyo 150
Japan

Jean H.P. Paelinck
Faculty of Economics
Erasmus University Rotterdam
P.O.B. 1738
3000 DR Rotterdam
The Netherlands

Karin Peschel
Institut für Regionalforschung der
Universität Kiel
Olshausenstraße 40-60
2300 Kiel 1
Federal Republic of Germany

Aisling J. Reynolds
Department of Economics
University College Dublin
Ireland

Piet Rietveld
Department of Economics
Free University
P.O.B. 7161
NL 1007 Amsterdam
The Netherlands

Kenneth E. Rosing
Faculty of Economics
Erasmus University Rotterdam
P.O.B. 1738
3000 DR Rotterdam
The Netherlands

Werner Rothengatter
Institut für Wirtschaftspolitik und Wirtschaftsforschung
Universität (TH) Karlsruhe
Kollegium am Schloß, Bau IV
7500 Karlsruhe 1
Federal Republic of Germany

Noburu Sakashita
Institute of Socio-Economic Planning
University of Tsukuba
Tsukuba, Ibaraki 305
Japan

Ashish Sen
Urban Transportation Center
University of Illinois at Chicago
USA

Johannes Siemons
School of Mathematics
University of East Anglia
Norwich
United Kingdom

Michael Steiner
Institut für Volkswirtschaft und Volkswirtschaftspolitik
Karl-Franzens-Universität Graz
Schubertstraße 6A
8010 G r a z
Austria

Guy R. West
Department of Economics
University of Queensland
St. Lucia, Queensland 4067
Australia

Jacob A. van der Vlist
Netherlands Economic Institute
Burgemeester Oudlaan 50
3062 PA Rotterdam
The Netherlands

Guy R. West
Department of Economics
University of Queensland
St. Lucia, Queensland 4067
Australia

Jacob A. van der Vlist
Netherlands Economic Institute
Burgemeester Oudlaan 50
3062 PA Rotterdam
The Netherlands

MIX
Papier aus verantwortungsvollen Quellen
Paper from responsible sources
FSC® C105338

If you have any concerns about our products,
you can contact us on
ProductSafety@springernature.com

In case Publisher is established outside the EU,
the EU authorized representative is:
Springer Nature Customer Service Center GmbH
Europaplatz 3, 69115 Heidelberg, Germany

Printed by Libri Plureos GmbH
in Hamburg, Germany